AEROELASTICITY

AEROELASTICITY

Raymond L. Bisplinghoff
Holt Ashley
Robert L. Halfman

DOVER PUBLICATIONS
Garden City, New York

Bibliographical Note

This Dover edition, first published in 1996, is a corrected republication of the work first published by Addison-Wesley Publishing Company, Cambridge, Mass., 1955. The authors have provided a number of corrections for the Dover edition.

Library of Congress Cataloging-in-Publication Data

Bisplinghoff, Raymond L.
  Aeroelasticity / Raymond L. Bisplinghoff, Holt Ashley, Robert L. Halfman.
    p.   cm.
  Corrected republication of the work originally published: Cambridge, Mass. : Addison-Wesley, 1955.
  Includes bibliographical references and index.
    ISBN-13: 978-0-486-69189-3 (pbk.)
    ISBN-10: 0-486-691896 (pbk.)

  1. Aeroelasticity. I. Ashley, Holt. II. Halfman, Robert L. III. Title.
TL574.A37B5397 1996
629.132'362—dc20   96-5412
                                                                      CIP

Manufactured in the United States of America
69189610   2023
www.doverpublications.com

PREFACE

The objective of the authors in writing *Aeroelasticity* has been to provide both a textbook for advanced engineering students and a reference book for practicing engineers. In selecting material for the book it was the authors' conviction that not only the practical aspects of aeroelasticity should be treated, but also the aerodynamic and structural tools upon which these rest. Accordingly, the book divides roughly into two halves; the first deals with the tools and the second with applications of the tools to aeroelastic phenomena. The authors' convictions concerning a need for further treatment of the tools do not stem from a feeling that they are inadequately treated elsewhere but rather from the realization that they are not treated from the point of view of the aeroelastician.

The first chapter emphasizes the role of aeroelasticity among the aeronautical sciences and its influence on modern design. Chapters 2, 3, and 4 are concerned with the deformation behavior of airplane structures under static and dynamic loads. These three chapters comprise the total treatment of the structural tools. The aerodynamic tools are treated in Chapters 5, 6, and 7. The reader will observe that although steady-state aerodynamics is discussed briefly, the primary emphasis is on unsteady phenomena. Chapter 8 brings together for the first time the aerodynamic and structural tools and treats the subject of static aeroelasticity. Problems of static aeroelasticity are characterized by the absence of the independent variable time, and they are introduced first because of their simplicity. Chapter 9 is concerned with flutter and Chapter 10 with dynamic response phenomena. Whereas the former entails essentially a harmonic dependence of the motion on time, the latter includes a class of problems in which the motion of the system may vary in a transient manner with time. Chapters 11 and 12 treat, respectively, the important subjects of aeroelastic model theory and model design and construction; the final chapter is concerned with experimental techniques for studying aeroelastic phenomena. Although the space devoted to experimental methods is relatively small, it is not the authors' intention to imply that experimental tools and techniques in aeroelasticity are of minor importance in the solution of practical problems. Indeed, aeroelastic phenomena encountered at the forefront of modern design often do not yield to analytical methods, and if solutions are to be obtained within a reasonable length of time the employment of experimental methods is absolutely necessary.

The authors have endeavored to write each chapter by progressing from the easy to the hard. Thus the engineering instructor who seeks to use this book as an elementary text in aeroelasticity will find that his purpose is served by merely using the first parts of selected chapters. For example, the book may be used as an introductory text in aeroelasticity for senior or graduate students in aeronautical engineering by using Chapter 1 and the

v

first parts of Chapters 2, 3, 5, 8, and 9. The mathematical prerequisites for an understanding of these portions of the book are the mathematics courses included in the usual engineering curriculum, through differential equations. The latter course should have at least an introduction to the notions of partial derivatives and partial differential equations. An introductory laboratory course in experimental aeroelasticity can be based upon Chapters 11, 12, and 13. Advanced courses in aeroelasticity may be based upon the latter parts of Chapters 2, 3, 5, 8, and 9, as well as Chapters 4, 6, 7, and 10. In general, the mathematical prerequisites for an understanding of the complete book include a course in advanced calculus in addition to the courses mentioned above.

The practicing engineer who uses the book for reference purposes will find that the authors have attempted to present applications of fundamentals instead of the compendium of standard tabular methods which may be in current favor. Although many numerical examples are included, it is unlikely that the practicing engineer will often find the particular problem that he is concerned with at the moment. However, it is hoped that the illustrative examples will always be of some value to the reader in perceiving how the fundamental tools may be applied to his case.

The authors arranged the material content of the book and the outlines of each chapter in close cooperation. Then each author worked on certain chapters independently, with R. L. Bisplinghoff concentrating on the structural tools, H. Ashley on the aerodynamic tools, and R. L. Halfman on the experimental aspects. Finally, the applications to aeroelastic phenomena were prepared jointly and the entire manuscript was worked over by the three authors to ensure continuity.

Acknowledgement is due a great many people who aided in bringing the book to completion. Professor Eric Reissner's counsel is gratefully acknowledged. A number of M.I.T. staff members and former students read portions of the manuscript and offered valuable advice and criticism. These include Professors Shatswell Ober, James Mar, Theodore Pian, Morton Finston, and Leon Trilling of the M.I.T. Department of Aeronautical Engineering; Garabed Zartarian, Hua Lin, John McCarthy, Kenneth Foss, and Robert Staley of the Aeroelastic and Structures Research Laboratory at M.I.T.; Mr. M. J. Turner of the Boeing Airplane Co., Mr. H. C. Johnson of the Glenn L. Martin Company, Professor H. C. Martin of the University of Washington, and Professor K. Washizu of the University of Tokyo. The numerical examples were worked out by Mr. John Martuccelli and Mr. Yechiel Shulman of the Aeroelastic and Structures Research Laboratory. The authors express their sincere appreciation to all of these people, and to Miss Nancy Ladd for so ably performing the seemingly endless chore of typing and preparing the manuscript.

R.L.B., H.A., R.L.H.

Cambridge, Mass.
February, 1955.

CONTENTS

CHAPTER 1

INTRODUCTION TO AEROELASTICITY

1-1 Definitions. The term *aeroelasticity* has been applied by aeronautical engineers to an important class of problems in airplane design. It is often defined as a science which studies the mutual interaction between aerodynamic forces and elastic forces, and the influence of this interaction on airplane design. Aeroelastic problems would not exist if airplane structures were perfectly rigid. Modern airplane structures are very flexible, and this flexibility is fundamentally responsible for the various types of aeroelastic phenomena. Structural flexibility itself may not be objectionable; however, aeroelastic phenomena arise when structural deformations induce additional aerodynamic forces. These additional aerodynamic forces may produce additional structural deformations which will induce still greater aerodynamic forces. Such interactions may tend to become smaller and smaller until a condition of stable equilibrium is reached, or they may tend to diverge and destroy the structure.

The term aeroelasticity, however, is not completely descriptive, since many important aeroelastic phenomena involve inertial forces as well as aerodynamic and elastic forces. We shall apply a definition in which the term aeroelasticity includes phenomena involving interactions among inertial, aerodynamic, and elastic forces, and other phenomena involving interactions between aerodynamic and elastic forces. The former will be referred to as *dynamic* and the latter as *static* aeroelastic phenomena.

Collar (Ref. 1-1) has ingeniously classified problems in aeroelasticity by means of a triangle of forces. Referring to Fig. 1-1, the three types of forces, aerodynamic, elastic, and inertial, represented by the symbols $A, E,$ and I, respectively, are placed at the vertices of a triangle. Each aeroelastic phenomenon can be located on the diagram according to its relation to the three vertices. For example, dynamic aeroelastic phenomena such as flutter, F, lie within the triangle, since they involve all three types of forces and must be bonded to all three vertices. Static aeroelastic phenomena such as wing divergence, D, lie outside the triangle on the upper left side, since they involve only aerodynamic and elastic forces. Although it is difficult to define precise limits on the field of aeroelasticity, the classes of problems connected by solid lines to the vertices in Fig. 1-1 are usually accepted as the principal ones. Of course, other borderline fields can be placed on the diagram. For example, the fields of mechanical vibrations, V, and rigid-body aerodynamic stability, DS, are connected to the vertices by dotted lines. It is very likely that in certain cases the dynamic stability

1

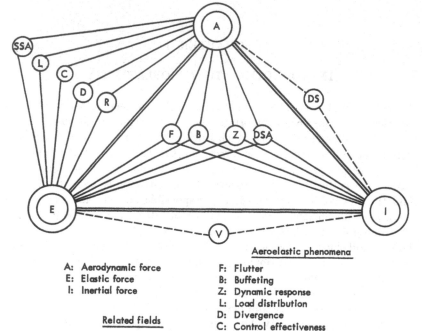

A: Aerodynamic force	Aeroelastic phenomena
E: Elastic force	F: Flutter
I: Inertial force	B: Buffeting
	Z: Dynamic response
	L: Load distribution
Related fields	D: Divergence
	C: Control effectiveness
V: Mechanical vibrations	R: Control system reversal
DS: Dynamic stability	DSA: Aeroelastic effects on dynamic stability
	SSA: Aeroelastic effects on static stability

Fig. 1–1.   The aeroelastic triangle of forces.

problem is influenced by airplane flexibility and it would therefore be moved within the triangle to correspond with *DSA*, where it would be regarded as a dynamic aeroelastic problem.

It will be convenient to state concise definitions of each aeroelastic phenomenon which appears on the diagram in Fig. 1–1.

Flutter, F.   A dynamic instability occurring in an aircraft in flight, at a speed called the flutter speed, where the elasticity of the structure plays an essential part in the instability.

Buffeting, B.   Transient vibrations of aircraft structural components due to aerodynamic impulses produced by the wake behind wings, nacelles, fuselage pods, or other components of the airplane.

Dynamic response, Z.   Transient response of aircraft structural components produced by rapidly applied loads due to gusts, landing, gun reactions, abrupt control motions, moving shock waves, or other dynamic loads.

Aeroelastic effects on stability, SA.   Influence of elastic deformations of the structure on dynamic and static airplane stability.

Load distribution, L.   Influence of elastic deformations of the structure on the distribution of aerodynamic pressures over the structure.

Divergence, D. A static instability of a lifting surface of an aircraft in flight, at a speed called the divergence speed, where the elasticity of the lifting surface plays an essential role in the instability.

Control effectiveness, C. Influence of elastic deformations of the structure on the controllability of an airplane.

Control system reversal, R. A condition occurring in flight, at a speed called the control reversal speed, at which the intended effects of displacing a given component of the control system are completely nullified by elastic deformations of the structure.

1–2 Historical background. Problems in aeroelasticity did not attain the prominent role that they now play until the early stages of World War II. Prior to that time, airplane speeds were relatively low and the load requirements placed on aircraft structures by design criteria specifications produced a structure sufficiently rigid to preclude most aeroelastic phenomena. As speeds increased, however, with little or no increase in load requirements, and in the absence of rational stiffness criteria for design, aircraft designers encountered a wide variety of problems which we now classify as aeroelastic problems.

Although aeroelastic problems have occupied their current prominent position for a relatively short period, they have had some influence on airplane design since the beginning of powered flight. Perhaps the first designer to be affected was Professor Samuel P. Langley of the Smithsonian Institution. In the light of modern knowledge, it seems likely that the unfortunate wing failure which wrecked Langley's machine on the Potomac River houseboat in 1903 could be described as wing torsional divergence. In going over the arguments put forward at the time, the best explanation of what happened (Ref. 1–2) was given by Griffith Brewer (Ref. 1–3), one time president of the Royal Aeronautical Society. Brewer described the phenomenon which wrecked the Langley monoplane in the same way as we describe wing torsional divergence today. Langley's misfortune occurred shortly before the Wright Brothers made the first sustained heavier-than-air flight.

Perhaps the success of the Wright biplane and the failure of the Langley monoplane was the original reason for the strong predilection for biplanes in the early days of airplane design. The technical arguments of biplane versus monoplane, which were prevalent for so many years, were undoubtedly influenced by the lack of a rational torsional stiffness criterion for monoplane wings. Although a number of externally braced monoplanes were constructed by the French and Germans prior to World War I, the monoplane as a military machine ceased to exist in 1917, and it was not until the mid-thirties that designers ventured to build high-performance monoplane military aircraft.

Fig. 1–2.   Handley Page 0/400 bomber.

The most widespread early aeroelastic problem in the days when military aircraft were almost exclusively biplanes was the tail flutter problem. One of the first documented cases of flutter occurred on the horizontal tail of the twin-engined Handley Page 0/400 bomber, shown by Fig. 1–2, at the beginning of World War I. Lanchester and Bairstow were asked to investigate the cause of violent oscillations of the fuselage and tail surfaces. They discovered (Refs. 1–4 and 1–5) that the fuselage and tail had two principal low-frequency modes of vibration. In one mode, the left and right elevators oscillated about their hinges 180° out of phase. This was possible because the elevators were not attached to the same torque tube, but were connected by a relatively weak spring provided by the long control cables through which each individual elevator was connected to the stick. In the second mode, the fuselage oscillated in torsion. The possibility of a self-excited oscillation involving coupling between the modes was diagnosed as the cause of the vibrations. One of the proposed remedial measures was that of connecting both elevators to the same torque tube. A second epidemic of tail flutter due to the same cause was experienced by the DH-9 airplane in 1917, and a number of lives were lost before it was cured. The cure was identical to that applied to the Handley Page airplane, and a torsionally stiff connection between elevators has been a design feature in airplanes ever since.

Aeroelastic wing problems appeared when designers abandoned biplane construction with its interplane bracing and relatively high torsional rigidity, in favor of monoplane types. The latter often had insufficient torsional rigidity, and flutter, loss of aileron effectiveness, and deformation effects on load distribution resulted. An early example of this kind arose during World War I in the development of the Fokker D-8 airplane shown in Fig. 1–3. In the initial design of this airplane, which was a high-wing cantilever monoplane, the torsional stiffness was determined by a criterion which had been applied to biplanes. The D-8 was put into production because of its superior performance, and was not in combat more than a few days before wing failures repeatedly occurred in high-speed dives. Since the best pilots and squadrons were receiving them first, it appeared possible that the flower of the German Air Corps would be wiped out. After a period in which the Army engineers and the Fokker Company each tried to place the responsibility on the other, the Army conducted static *strength* tests on half a dozen wings and found them sufficiently strong to support the required ultimate factor of 6. This produced a serious dilemma, and it was clearly up to Anthony Fokker to discover the cause or cease production on the D-8. Static tests were undertaken by the Fokker Company, and this time, *deflections* were carefully measured from tip to tip. In Fokker's words (Ref. 1–6), the following conclusions were drawn: "I discovered that with increasing load, the angle of incidence at the wing tips increased perceptibly. It suddenly dawned on me that this increasing

Fig. 1-3.   Fokker D-8 airplane.

angle of incidence was the cause of the wing's collapse, as logically the load resulting from the air pressure in a steep dive would increase faster at the wing tips than at the middle. The resulting torsion caused the wings to collapse under the strain of combat maneuvers." This seems to be the first documented case where static aeroelastic effects at a fairly high speed produced a redistribution in airload such that failure resulted.

In later experience with the D-8, subsequent to the war, U. S. Army Air Corps engineers at McCook Field, Dayton, Ohio, observed a violent but nondestructive case of wing bending-aileron flutter. This was cured by statically balancing the ailerons about the hinge line, a technique which seems to have been pointed out first by Baumhauer and Koning (Ref. 1-7) in 1922. Several of the monoplane racers of the 1920's and 1930's experienced forms of wing-aileron flutter; and mass balancing was a commonly applied cure and preventive measure.

The period of development of the cantilever monoplane seems to have been the period in which serious research in aeroelasticity commenced. In the earliest days of monoplane design, aeroelastic problems were overcome by cut-and-try methods. A theory of wing-load distribution and wing divergence was first presented in 1926 by Hans Reissner (Ref. 1-8). A theory of loss of lateral control and aileron reversal was published six years later by Roxbee Cox and Pugsley (Ref. 1-9) in 1932. The mechanism of potential flow flutter was understood sufficiently well for design use by 1935, largely through the early efforts of Glauert (Ref. 1-10), Frazer and Duncan (Ref. 1-11), Küssner (Ref. 1-12), and Theodorsen (Ref. 1-13). However, few designers were able to comprehend the theories in the early papers and the majority were reluctant to trust mathematicians to compute sizes of structural members to preclude aeroelastic effects.

1-3 Influence of aeroelastic phenomena on design. Aeroelastic phenomena in modern high-speed aircraft have profound effects upon the design of structural members and somewhat lesser but nonetheless important effects upon mass distribution, lifting surface planforms, and control system design.

Flutter. Flutter has perhaps the most far-reaching effects of all aeroelastic phenomena on the design of high-speed aircraft. Modern aircraft are subject to many kinds of flutter phenomena. The classical type of flutter is associated with potential flow and usually, but not necessarily, involves the coupling of two or more degrees of freedom. The nonclassical type of flutter, which has so far been difficult to analyze on a purely theoretical basis, may involve separated flow, periodic breakaway and reattachment of the flow, stalling conditions, and various time-lag effects between the aerodynamic forces and the motion. Preventive measures and cures usually involve either increased stiffness or decreased coupling by adjustments in mass distribution, or a combination of both. The most important stiffness parameter affected by flutter considerations is wing torsional stiffness. It is not uncommon for the flutter condition to control the selection of wing skin thickness. Of course, wing structural design is controlled by either a strength or a stiffness criterion. For example, if the torsion carrying structure of a wing is designed by a stiffness requirement, the wing would probably consist of a structure which carries its normal stresses in the wing skin with a minimum of stringers and flanges. This type of wing structure would require several spanwise webs in order to stabilize the heavily loaded cover skin. For a wing designed initially by strength considerations to carry a given load factor, it is obvious that a higher torsional stiffness and hence a higher flutter speed will result if the ratio of stiffener area to skin area is reduced to a minimum. In addition, the use of higher strength alloys, which have no corresponding increase in modulus of elasticity, tends to make flutter more critical for wings designed for strength only.

Heavy mass items in the wing are often located by considerations of optimum conditions for flutter prevention. For example, a given mass distribution may require higher wing stiffness and hence higher wing structural weight to prevent flutter than some other mass distribution. For this reason, analytical and model studies are often made in the design stages in order to determine the optimum mass distribution for flutter prevention.

Wing planform and aspect ratio also have significant effects on flutter characteristics. Decreases in wing aspect ratio and increases in sweep tend to raise flutter speeds, whereas increases in aspect ratio and decreases in sweep, including sweep forward, reduce flutter speeds.

Flutter considerations may affect control surface design in the determination of aerodynamic and mass balance, hinge location, and the degree of irreversibility required in the actuating system.

Buffeting. A serious buffeting phenomenon confronting designers is encountered by fighter aircraft during pull-ups to $C_{L\max}$ at high speed. This often results in rugged transient vibrations in the tail due to aerodynamic impulses from the wing wake. The principal problems are those of reducing the severity of these vibrations, and the provision of adequate strength. Designing for strength is very difficult. The problem of predicting dynamic stresses due to a given buffeting condition is still unsolved analytically. The principal obstacle has been a lack of knowledge of the properties of the wake behind a stalled wing. Designers have alleviated their buffeting problems up to the present time largely by proper positioning of the tail assembly and by clean aerodynamic design.

Dynamic loads problems. This class of aeroelastic problem has its primary influence on structural design. In the prediction of design loads on an airplane structure in an accelerated condition, it is usually assumed that the airplane is perfectly rigid. Structural components designed by loads computed on this basis may fail due to a dynamic overstress. External loads that are rapidly applied not only cause translation and rotation of the airplane as a whole, but tend to excite vibrations of the structure. The additional inertial forces associated with these vibrations produce the dynamic overstress. Dynamic stresses are usually manifested in the form of increased bending and torsional stresses in the wing and fuselage beams. The design of these beams must take account of dynamic stresses by increasing the normal and shear carrying areas. Perhaps the two most important dynamic response problems have been the gust and landing problems. The gust condition is usually the controlling strength condition in large aircraft. Aeroelastic effects may have an important influence on gust design conditions. For example, the designer can expect that the dynamic response of a straight wing in the gust condition may produce wing bending moments at the root 15% to 20% greater than those calculated on the assumption of a rigid wing. Response of a slender flexible swept wing when a gust is encountered is a matter of considerable practical interest, particularly in the case of large high-speed aircraft. Elastic deformations and vibratory response of the wing have important and complicating effects upon the load distribution. Load distributions predicted on the assumption of a rigid wing are often too much in error to be useful, and swept-wing designers are compelled to consider the principal aeroelastic effects. As an illustration of the profound influence of elasticity, consider the result given by Fig. 1–4. This figure illustrates a calculated result showing the difference between fuselage and wing-tip acceleration when a typical swept wing flying at 460 mph equivalent airspeed at 11,000 ft.

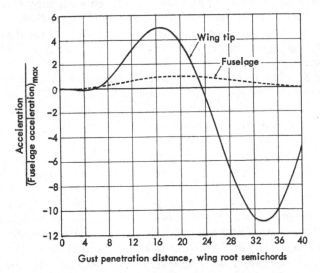

Fig. 1–4. Comparison of wing-tip and fuselage accelerations of swept-wing airplane during passage through a gust.

altitude strikes a gust. The dotted curve indicates the history of fuselage acceleration plotted in such a way that the peak value is one. The solid curve gives the ratio of the history of wing-tip acceleration to peak fuselage acceleration. The marked difference in accelerations shown in Fig. 1–4 indicates the important role of wing elasticity in determining correct wing-load distributions on swept wings.

Load distribution. During high-speed flight, deflections of the structure tend to redistribute the airloads, and may cause their distribution to be significantly different from that computed on the assumption of complete rigidity. It is, of course, necessary that structural members be designed for the airload distribution corresponding to the deformed structure. Let us consider briefly the case of a simple straight wing with the center of twist behind the aerodynamic center. The torsional moment about the center of twist due to the lift at the aerodynamic center tends to twist the wing, which increases the angle of attack and increases the lift further. The increased lift in turn causes another increment in twist. For speeds below a critical value, called the divergence speed, the increments of twist and lift become smaller until a condition of stable equilibrium is reached. The problem of finding the load distribution on a wing is one of computing the twist distribution along the wing corresponding to this condition of stable equilibrium. The same type of phenomenon occurs in every load distribution problem, whether it involves straight wings, swept wings, or other structural components. Figure 1–5 illustrates, for example, the relative airload distribution on an elastic and a rigid swept wing. In the case of a swept wing, it can be seen that the deformation effects are favorable in that they shift the center of pressure inboard rather than outboard, as in the case of a straight wing. This is a consequence of the decrease in streamwise angle of attack which accompanies the bending deflections of a swept wing.

Divergence. The most common divergence problem is wing torsional divergence of a straight wing. Let us consider the case of a simple straight wing with the center of twist behind the aerodynamic center. When this wing reaches its torsional divergence speed, the increment in aerodynamic torsional moment due to an arbitrary increment in twist angle is exactly equal to the increment in elastic restoring torque. When the speed exceeds the torsional divergence speed, the increment in aerodynamic torsional moment exceeds the increment in elastic restoring torque, and the wing becomes statically unstable. Design parameters affecting divergence of straight wings are primarily wing torsional stiffness and offset distance between center of twist and aerodynamic center. Raising the divergence speed of a given wing by increasing wing torsional stiffness is a costly process at the expense of considerable weight. An approach more frequently employed by designers is to proportion the wing structurally so as to move the center of twist forward and thus reduce the offset between

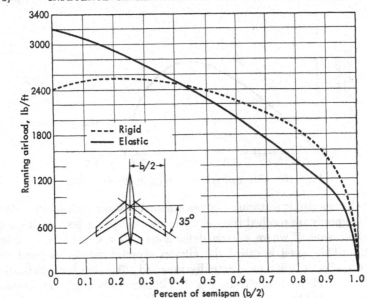

Fig. 1–5. Airload distribution on swept rigid and elastic wing during a 2g maneuvering condition.

center of twist and aerodynamic center. For example, a straight wing which carries its torsional load by a D box, has a forward center of twist location, and consequently a high divergence speed. Divergence speeds of swept-back wings are not a matter of vital concern to the designer. However, divergence speeds of swept-forward wings are so low that for this reason alone sweep forward is practically ruled out as a design feature.

When the aileron plays an important role in the process, the phenomenon is called wing-aileron divergence. The horizontal and vertical tail surfaces each have divergence speeds in which the elevator and rudder may play important roles.

Control effectiveness and reversal. Aircraft with conventional planforms may suffer serious loss of aileron, elevator, and rudder control effectiveness due to elastic deformations of the structure. In order to describe these phenomena qualitatively, let us consider the case of aileron effectiveness at subsonic speeds. Normally the lift on the wing with down aileron is increased and the lift on the wing with up aileron is decreased. Thus a rolling moment is produced. The down aileron produces a twisting moment on the wing which tends to twist the wing nose down and reduce the angle of attack, whereas the up aileron produces a torque which increases the angle of attack. It is apparent, then, that the wing twist tends to reduce the rolling moment to a value less than it would be for a rigid wing. Wing twisting moments due to deflected ailerons increase as the square of the

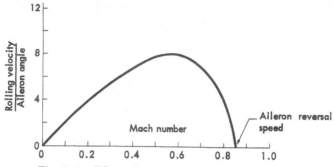

Fig. 1-6.   Effect of speed on aileron effectiveness.

speed, whereas elastic restoring torques remain constant with speed. Thus, as higher speeds are reached the rolling moments become less, until finally a speed is. reached where aileron deflections will not produce a rolling moment. This speed is called the aileron reversal speed. Beyond this speed, the effect of ailerons is actually reversed. To give some indication of the magnitude of these effects, Fig. 1-6 illustrates how aileron effectiveness, as measured by the ratio of rolling velocity to aileron angle, is affected by forward speed for a World War II fighter-type airplane at sea level.

Avoidance of aileron reversal in a straight wing with conventional ailerons is almost entirely a matter of providing sufficient wing torsional stiffness. In the case of swept-back wings, where aileron reversal is a very serious problem, bending stiffness must also be increased in order to raise the aileron reversal speed. Increases in bending and torsional stiffnesses are often accompanied by prohibitively large increases in weight, and hence other means must be sought. Alternative methods of producing rolling moments, such as spoilers and all-moving wing tips, have proved beneficial.

The phenomena of elevator and rudder control effectiveness and reversal are somewhat less critical than those of aileron effectiveness and reversal. They are complicated by the relatively large number of elastic elements which contribute to the total deflection of the tailplanes. For example, deformations of the fuselage and attachments between the tail and fuselage are often as important as deformations of the tailplane itself. Figure 1-7 illustrates the reduction in tail lift per degree of elevator deflection due to elastic deformations that can be expected in a fighter-type airplane at sea level.

It should be observed that flutter and divergence correspond to conditions of aeroelastic instability, and that speeds beyond critical flutter and divergence speeds will result in an eventual structural failure. However, control reversal is not a condition of instability and speeds beyond the control reversal speed will result only in a reversal of the action of the control system and not necessarily a failure of the structure.

Since the phenomenon of control effectiveness has a serious effect upon

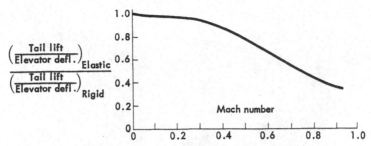

Fig. 1-7.   Effect of speed on elevator control effectiveness.

maneuverability, it is important that it be thoroughly understood by the control system designer. A severe loss in control effectiveness cannot be tolerated within the speed range of any airplane. It should be remarked that as airplanes approach sonic and supersonic speeds, other purely aerodynamic phenomena influence control effectiveness, in addition to the aeroelastic effects that have been considered here.

1-4 Comparison of wing critical speeds.   We have seen in the previous sections that a conventional wing has three critical speeds, each of which is important to the designer. They are the flutter speed, the divergence speed, and the aileron reversal speed. A comparison of their relative values is a necessary process in the design of a wing. In the case of straight unswept wings of conventional construction, wing torsional divergence usually occurs at a speed higher than aileron reversal speed, which is in turn higher than the bending-torsion flutter speed. In the case of swept-forward wings, the divergence speed can be expected to be lower than the flutter speed, which is in turn lower than the aileron reversal speed. For swept-back wings, the aileron reversal speed is lower than the flutter speed, which is in turn lower than the divergence speed. Figure 1-8 shows qualitatively the relation between critical speeds for a typical wing with varying amounts of rearward and forward sweep.

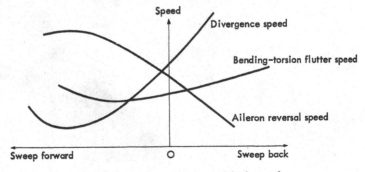

Fig. 1-8.   Comparison of wing critical speeds.

Fig. 2-1.   B-47 airplane.

CHAPTER 2

DEFORMATIONS OF AIRPLANE STRUCTURES UNDER STATIC LOADS

2-1 Introduction. Modern airplanes vary widely in geometric shape. On the one hand, we find high-aspect ratio wings which resemble slender beams, and on the other, low-aspect ratio delta wings which resemble plates. The Boeing B-47 airplane, illustrated in Fig. 2-1, is an example of the former, and the Convair XF-92A in Fig. 2-3 provides an excellent example of the latter. The fuselage may be long and slender, or it may be nonexistent, as in the case of a flying wing. The arrangements and types of load-carrying members also differ widely. As aircraft are put to wider uses, it is natural to expect larger differences among their geometric and structural configurations.

The object of the present chapter is to discuss methods of analysis of deformations of airplane structures under static loads. For example, consider an elastic airplane, as illustrated schematically by Fig. 2-2. An orthogonal set of axes is fixed at an arbitrarily selected point O on the center line such that the x- and z-axes lie in the plane of symmetry. For purposes of discussion of its elastic behavior, we assume that the structure is restrained against linear and angular deflections at the point O so that the elastic curves or surfaces generated by the deformed wing are tangent to the xy-plane. It is also assumed, without loss of generality, that the xz-plane is a plane of elastic symmetry as well as of geometric symmetry.

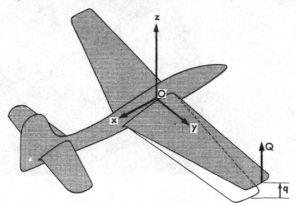

Fig. 2-2.   Discrete force applied to elastic structure.

15

Fig. 2-3. XF-92A airplane.

2-2 Elastic properties of structures. In this book it will be assumed that the aircraft structures under consideration are perfectly elastic. That is, when external forces are removed the structure resumes its initial form. Experiments on aircraft structures have indicated that within certain limits an applied force, Q, and its resulting deflection, q, as illustrated by Fig. 2-2, are related by

$$q = CQ, \tag{2-1}$$

where C is a constant of proportionality. Thus force and deflection are linearly related and the loaded structure may be referred to as a linear system.

In thin-skin aircraft structures elastic buckling may produce a discontinuity in the force-deflection diagram even though the materials which make up the structure are stressed at a relatively low level. We shall assume that the elastic behavior of our structures is defined in the range below the point of elastic buckling. Although this assumption is not restrictive for structures with thick skins, as in the case of military aircraft, its validity may require examination when applied to aeroelastic problems involving thin-skin structures.

A consequence of Eq. (2-1) is that the work done by the external force during application is transformed completely into strain energy in the structure. That is,

$$U = \tfrac{1}{2}Qq, \tag{2-2}$$

where U denotes the strain energy.

2-3 Deformation due to several forces. Influence coefficients. Consider a more general case, such as illustrated by Fig. 2-4, where several forces and moments act on the structure. We can express the total linear or angular deflection of any point as the sum of the deflections at that point

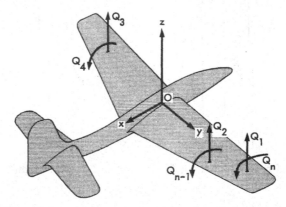

Fig. 2-4. Several discrete forces and moments applied to elastic structure.

produced by the individual forces and moments.   This is a statement of the *principle of superposition*, which is fundamental to the analysis of linear systems.   In Fig. 2–4, the symbol Q is assigned to an arbitrary force or moment called a "generalized force."   Similarly, the symbol q is assigned to the linear or angular displacement of the point of application of each generalized force and is called a "generalized coordinate."   Generalized coordinates are quantities which represent possible independent displacements of the system.   This means that they must not violate the geometric constraints imposed upon the system.

Applying the principle of superposition, the displacement of the point of application of the ith generalized force due to n generalized forces is given by

$$q_i = \sum_{j=1}^{n} C_{ij}Q_j, \qquad (i = 1, 2, \cdots, n), \qquad (2\text{--}3)$$

where the constants C_{ij} are called flexibility influence coefficients.

Conversely, the forces can be expressed as linear functions of the displacements by

$$Q_i = \sum_{j=1}^{n} k_{ij}q_j, \qquad (i = 1, 2, \cdots, n), \qquad (2\text{--}4)$$

where the constants k_{ij} are called stiffness influence coefficients.   In the case of a single force applied to the structure, the constant k_{ij} is the familiar spring constant.

Equations (2–3) and (2–4) can be represented in matrix notation as follows:

$$\begin{bmatrix} q_1 \\ q_2 \\ \cdot \\ \cdot \\ \cdot \\ q_n \end{bmatrix} = \begin{bmatrix} C_{11} & C_{12} & \cdots & C_{1n} \\ C_{21} & C_{22} & \cdots & C_{2n} \\ \cdot & \cdot & & \cdot \\ \cdot & \cdot & & \cdot \\ \cdot & \cdot & & \cdot \\ C_{n1} & C_{n2} & \cdots & C_{nn} \end{bmatrix} \begin{bmatrix} Q_1 \\ Q_2 \\ \cdot \\ \cdot \\ \cdot \\ Q_n \end{bmatrix}, \qquad (2\text{--}5)$$

$$\begin{bmatrix} Q_1 \\ Q_2 \\ \cdot \\ \cdot \\ \cdot \\ Q_n \end{bmatrix} = \begin{bmatrix} k_{11} & k_{12} & \cdots & k_{1n} \\ k_{21} & k_{22} & \cdots & k_{2n} \\ \cdot & \cdot & & \cdot \\ \cdot & \cdot & & \cdot \\ \cdot & \cdot & & \cdot \\ k_{n1} & k_{n2} & \cdots & k_{nn} \end{bmatrix} \begin{bmatrix} q_1 \\ q_2 \\ \cdot \\ \cdot \\ \cdot \\ q_n \end{bmatrix}. \qquad (2\text{--}6)$$

In shorthand matrix notation, Eqs. (2–5) and (2–6) become

$$\{q\} = [C]\{Q\}, \qquad (2\text{--}7)$$

$$\{Q\} = [k]\{q\}. \qquad (2\text{--}8)$$

The representation of sets of linear equations in matrix notation is frequently useful in manipulating the equations and in indicating orderly processes for carrying out numerical calculations. Matrix methods are widely used in aeroelasticity, but a knowledge of only a few of the elementary rules of matrix algebra is sufficient for most purposes. Appendix A develops these rules, and Ref. 2–1 is an excellent source of more complete information.

The square array of numbers, $[C]$, called the matrix of flexibility influence coefficients, is related mathematically to the matrix of stiffness influence coefficients, $[k]$. If the set of linear equations, Eq. (2–7), is solved for the Q's as linear functions of the q's, the set of linear equations given by Eq. (2–8) is obtained. This process is known as matrix inversion (Appendix A) and is represented symbolically by

$$[k] = [C]^{-1}, \qquad (2\text{--}9)$$

where the matrix $[k]$ is said to be the reciprocal of the matrix $[C]$.

For purposes of illustration, consider the problem of finding by direct calculation the influence coefficient matrices $[C]$ and $[k]$ of the uniform cantilever beam illustrated by Fig. 2–5. Figure 2–6 illustrates diagrammatically the problem to be solved in computing the elements C_{11} and C_{21}. A unit force is applied in place of Q_1, and the resulting deflections q_1 and q_2 are C_{11} and C_{21}, respectively.

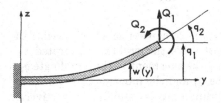 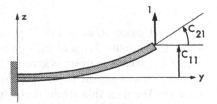

Fig. 2–5. Force and moment applied at tip of uniform cantilever beam.   Fig. 2–6. Unit force applied at tip of uniform cantilever beam.

We assume that the bending deformation of the beam is governed by the Bernoulli-Euler formula (Ref. 2–2).

$$M(y) = EIw''(y), \qquad (2\text{--}10)$$

where $M(y)$ is the bending moment (assumed positive for bottom fibers in tension), EI is the flexural rigidity, and $w(y)$ is the displacement of the neutral axis of the beam. Primes indicate differentiation with respect to y.

For a unit load at the tip, Eq. (2–10) becomes

$$l - y = EIw''(y). \qquad (2\text{--}11)$$

Integrating and introducing the boundary conditions $w(0) = w'(0) = 0$, we obtain

$$C_{21} = w'(l) = \frac{l^2}{2EI}, \quad (2\text{--}12)$$

$$C_{11} = w(l) = \frac{l^3}{3EI}. \quad (2\text{--}13)$$

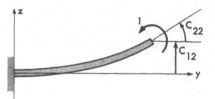

Fig. 2-7. Unit moment applied at tip of uniform cantilever beam.

In a similar way, Fig. 2–7 illustrates the problem to be solved in computing the elements C_{22} and C_{12}. For a unit moment at the tip, Eq. (2–10) becomes

$$1 = EIw''(y) \quad (2\text{--}14)$$

and integration yields

$$C_{22} = w'(l) = \frac{l}{EI}, \quad (2\text{--}15)$$

$$C_{12} = w(l) = \frac{l^2}{2EI}. \quad (2\text{--}16)$$

From Eqs. (2–12), (2–13), (2–15), and (2–16), the matrix of flexibility influence coefficients is

$$[C_{ij}] = \begin{bmatrix} l^3/3EI & l^2/2EI \\ l^2/2EI & l/EI \end{bmatrix}. \quad (2\text{--}17)$$

Direct calculation of the $[k]$ matrix is somewhat more involved. In order to compute k_{11} and k_{21}, we must solve the problem illustrated by Fig. 2–8. Here the beam is given a unit displacement in the coordinate q_1, and zero displacement in q_2. The force Q_1 and the moment Q_2 required to hold the beam in this strained configuration are k_{11} and k_{21}, respectively.

Introducing $M(y) = k_{11}(l - y) + k_{21}$ into Eq. (2–10), we have

$$EIw''(y) = k_{11}(l - y) + k_{21}. \quad (2\text{--}18)$$

Integrating and applying the boundary conditions, $w(0) = w'(0) = w'(l) = 0$ and $w(l) = 1$, we obtain the simultaneous equations

$$k_{11} + \frac{2}{l} k_{21} = 0, \qquad k_{11} + \frac{3}{2l} k_{21} = \frac{3EI}{l^3}, \quad (2\text{--}19)$$

which yield, upon solution,

$$k_{11} = \frac{12EI}{l^3}, \quad (2\text{--}20)$$

$$k_{21} = -\frac{6EI}{l^2}. \quad (2\text{--}21)$$

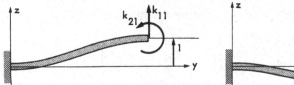

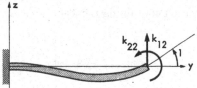

Fig. 2-8.   Uniform cantilever beam
with unit linear displacement at tip.

Fig. 2-9.   Uniform cantilever beam
with unit angular displacement at tip.

The problem to be solved in computing the elements k_{22} and k_{12} is illustrated by Fig. 2-9. Equation (2-10) becomes, in this case,

$$k_{12}(l - y) + k_{22} = EIw''(y). \tag{2-22}$$

Integrating and introducing the boundary conditions $w(0) = w'(0) = w(l) = 0$ and $w'(l) = 1$, we obtain the simultaneous equations

$$k_{22} + \frac{l}{2} k_{12} = \frac{EI}{l}, \qquad k_{22} + \frac{2l}{3} k_{12} = 0, \tag{2-23}$$

which yield, upon solution,

$$k_{12} = -\frac{6EI}{l^2}, \tag{2-24}$$

$$k_{22} = \frac{4EI}{l}. \tag{2-25}$$

The matrix of stiffness influence coefficients is therefore given by

$$[k] = \begin{bmatrix} 12EI/l^3 & -6EI/l^2 \\ -6EI/l^2 & 4EI/l \end{bmatrix}. \tag{2-26}$$

It is readily verified that the $[C]$ and $[k]$ matrices given by Eqs. (2-17) and (2-26) are reciprocal.

2-4 Properties of influence coefficients. Influence coefficients and their matrices have the important property of symmetry. This property is expressed by

$$C_{ij} = C_{ji}, \qquad k_{ij} = k_{ji}. \tag{2-27}$$

In addition, each matrix is equal to its transpose (Appendix A) as follows:

$$[C] = [C]', \qquad [k] = [k]'. \tag{2-28}$$

The symmetry properties are in accordance with the reciprocal theorem of Betti (Ref. 2-3).

Let us consider in some detail the properties of the $[C]$ matrix which apply to Fig. 2-4 when $n = 4$. It can be partitioned into four matrices, each containing different types of influence coefficients, as follows:

$$[C] = \begin{bmatrix} C_{11}{}^{\delta\delta} & C_{12}{}^{\delta\delta} & C_{13}{}^{\delta\alpha} & C_{14}{}^{\delta\alpha} \\ C_{21}{}^{\delta\delta} & C_{22}{}^{\delta\delta} & C_{23}{}^{\delta\alpha} & C_{24}{}^{\delta\alpha} \\ \hline C_{31}{}^{\alpha\delta} & C_{32}{}^{\alpha\delta} & C_{33}{}^{\alpha\alpha} & C_{34}{}^{\alpha\alpha} \\ C_{41}{}^{\alpha\delta} & C_{42}{}^{\alpha\delta} & C_{43}{}^{\alpha\alpha} & C_{44}{}^{\alpha\alpha} \end{bmatrix}. \tag{2-29}$$

The four different types of elements are

$$C_{ij}{}^{\delta\delta} = \text{linear deflection at } i \text{ due to unit force at } j,$$
$$C_{ij}{}^{\alpha\alpha} = \text{angular deflection at } i \text{ due to unit moment at } j,$$
$$C_{ij}{}^{\delta\alpha} = \text{linear deflection at } i \text{ due to unit moment at } j,$$
$$C_{ij}{}^{\alpha\delta} = \text{angular deflection at } i \text{ due to unit force at } j.$$

In order for the $[C]$ matrix to be symmetrical, the following reciprocal relations must hold:

$$C_{ij}{}^{\delta\delta} = C_{ji}{}^{\delta\delta}, \tag{2-30}$$

$$C_{ij}{}^{\alpha\alpha} = C_{ji}{}^{\alpha\alpha}, \tag{2-31}$$

$$C_{ij}{}^{\delta\alpha} = C_{ji}{}^{\alpha\delta}. \tag{2-32}$$

These relations can be shown to follow by considering an elastic body loaded at two points i and j by a force F_i and a moment M_j, as illustrated in Fig. 2–10. The strain energy in the body can be computed by two different sequences: apply the force F_i first and then the moment M_j, or reverse the order of application. We have for the two orders of load application:

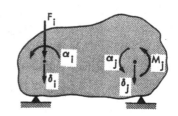

Fig. 2–10.   Elastic body loaded at two points.

$$U = \tfrac{1}{2}(C_{ii}{}^{\delta\delta}F_i)F_i + \tfrac{1}{2}(C_{jj}{}^{\alpha\alpha}M_j)M_j + (C_{ij}{}^{\delta\alpha}M_j)F_i, \tag{2-33}$$

$$U = \tfrac{1}{2}(C_{jj}{}^{\alpha\alpha}M_j)M_j + \tfrac{1}{2}(C_{ii}{}^{\delta\delta}F_i)F_i + (C_{ji}{}^{\alpha\delta}F_i)M_j. \tag{2-34}$$

Since the order of load application is immaterial,* Eqs. (2–33) and (2–34) can be equated to obtain the reciprocal relation given by Eq. (2–32),

$$C_{ij}{}^{\delta\alpha} = C_{ji}{}^{\alpha\delta}.$$

Other reciprocal relations can be proved in the same manner.

---

* This can be regarded as an application of the principle of conservation of energy to a conservative system.

2-5 Strain energy in terms of influence coefficients. In the application of energy theorems to aeroelastic systems, formulas for strain energy in terms of influence coefficients are useful. Referring to Fig. 2–4, we have the following expression for strain energy:

$$U = \tfrac{1}{2} \sum_{i=1}^{n} Q_i q_i. \qquad (2\text{–}35)$$

The introduction of Eq. (2–3) into Eq. (2–35) results in an expression for the strain energy in terms of the flexibility influence coefficients and the external loads,

$$U = \tfrac{1}{2} \sum_{i=1}^{n} \sum_{j=1}^{n} C_{ij} Q_i Q_j. \qquad (2\text{–}36)$$

Similarly, by introducing Eq. (2–4) into Eq. (2–35), we obtain an expression for strain energy in terms of stiffness influence coefficients and displacements,

$$U = \tfrac{1}{2} \sum_{i=1}^{n} \sum_{j=1}^{n} k_{ij} q_i q_j. \qquad (2\text{–}37)$$

In matrix notation, Eqs. (2–36) and (2–37) can be written in the forms

$$U = \tfrac{1}{2} \lfloor Q_1 \quad Q_2 \quad \cdots \quad Q_n \rfloor \begin{bmatrix} C_{11} & C_{12} & \cdots & C_{1n} \\ C_{21} & C_{22} & \cdots & C_{2n} \\ \vdots & \vdots & & \vdots \\ C_{n1} & C_{n2} & \cdots & C_{nn} \end{bmatrix} \begin{bmatrix} Q_1 \\ Q_2 \\ \vdots \\ Q_n \end{bmatrix}, \qquad (2\text{–}38)$$

$$U = \tfrac{1}{2} \lfloor q_1 \quad q_2 \quad \cdots \quad q_n \rfloor \begin{bmatrix} k_{11} & k_{12} & \cdots & k_{1n} \\ k_{21} & k_{22} & \cdots & k_{2n} \\ \vdots & \vdots & & \vdots \\ k_{n1} & k_{n2} & \cdots & k_{nn} \end{bmatrix} \begin{bmatrix} q_1 \\ q_2 \\ \vdots \\ q_n \end{bmatrix}. \qquad (2\text{–}39)$$

Equations (2–38) and (2–39) can be expressed in shorthand matrix notation by

$$U = \tfrac{1}{2} \lfloor Q \rfloor [C] \{Q\}, \qquad (2\text{–}40)$$

$$U = \tfrac{1}{2} \lfloor q \rfloor [k] \{q\}. \qquad (2\text{–}41)$$

2-6 Deformations under distributed forces. Influence functions. Preceding sections were concerned with discrete forces and moments applied to the structure. Consider next the results which are obtained if the struc-

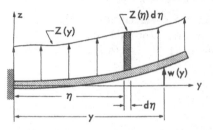

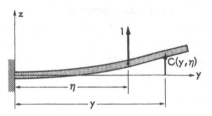

Fig. 2–11.   Uniform cantilever   Fig. 2–12.   Uniform cantilever
beam loaded by distributed side load.   beam loaded by unit load.

ture is loaded by a continuously distributed loading. As a simple illustration, we consider the cantilever beam in Fig. 2–11 loaded by a distributed side load, $Z(y)$. The infinitesimal deflections $dw(y)$ at a point y, due to an infinitesimal element of the load $Z(\eta)d\eta$ at a point η, can be expressed by

$$dw(y) = C(y,\eta)Z(\eta)d\eta, \qquad (2\text{–}42)$$

where $C(y,\eta)$ is a function giving the deflection at the point y due to a unit load at a point η. The deflection at y due to the entire side load is obtained by integrating Eq. (2–42) over the length of the beam:

$$w(y) = \int_0^l C(y,\eta)Z(\eta)d\eta. \qquad (2\text{–}43)$$

We call the function $C(y,\eta)$ a flexibility influence function or Green's function* in one dimension. Since it represents the same physical quantity as an influence coefficient, it is a symmetrical function. Thus,

$$C(y,\eta) = C(\eta,y). \qquad (2\text{–}44)$$

If shear deformation is neglected, the influence function for the bending of a cantilever beam can be computed by applying Eq. (2–10). Referring to Fig. 2–12, the bending moment on a cantilever beam due to a unit load applied at $y = \eta$ is given by

$$M(y,\eta) = (\eta - y), \qquad (y \leq \eta), \qquad (2\text{–}45)$$

$$M(y,\eta) = 0, \qquad (y \geq \eta). \qquad (2\text{–}46)$$

Introducing Eq. (2–45) into Eq. (2–10) gives

$$EIC''(y,\eta) = (\eta - y), \qquad (y \leq \eta). \qquad (2\text{–}47)$$

Integrating Eq. (2–47) and introducing the boundary conditions

$$C(0,\eta) = C'(0,\eta) = 0, \qquad (2\text{–}48)$$

---

* Named for the English physicist George Green (1793–1841), who applied it first to electrostatic problems.

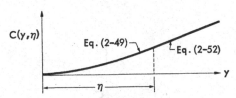

Fig. 2–13. Plot of the influence function, $C(y,\eta)$.

we obtain the influence function of a uniform beam for the range $y \leq \eta$:

$$C(y,\eta) = \frac{y^2}{6EI} (3\eta - y), \qquad (y \leq \eta). \tag{2-49}$$

Introducing Eq. (2–46) into (2–10), we obtain

$$EIC''(y,\eta) = 0, \qquad (y \geq \eta). \tag{2-50}$$

Integrating Eq. (2–50) and evaluating the constants of integration by putting $y = \eta$ in Eq. (2–49), with the resultant requirements that

$$C(\eta,\eta) = \frac{\eta^3}{3EI}, \qquad C'(\eta,\eta) = \frac{\eta^2}{2EI}, \tag{2-51}$$

we obtain

$$C(y,\eta) = \frac{\eta^2}{6EI} (3y - \eta), \qquad (y \geq \eta). \tag{2-52}$$

If y and η are interchanged in either Eq. (2–49) or Eq. (2–52), the other equation is obtained. The latter result merely verifies the reciprocal relation (Eq. 2–44). The first two derivatives of $C(y,\eta)$ are continuous at $y = \eta$, but the third derivative is discontinuous. A graph of Eqs. (2–49) and (2–52) is shown in Fig. 2–13 for a particular value of η.

Introduction of (2–49) and (2–52) into (2–43) gives

$$w(y) = \int_0^y \frac{\eta^2}{6EI} (3y - \eta)Z(\eta)d\eta + \int_y^l \frac{y^2}{6EI} (3\eta - y)Z(\eta)d\eta. \tag{2-53}$$

If, for example, $Z(\eta)$ is constant, Eq. (2–53) yields the result that

$$w(y) = \frac{Zy^2}{24EI} (y^2 - 4yl + 6l^2). \tag{2-54}$$

The notion of an influence function is useful also in two-dimensional systems such as plate members loaded normal to their surface. Assume, for example, that the wing of the airplane illustrated in Fig. 2–14 is loaded normal to its surface by a distributed load $Z(x,y)$. The deflection $w(x,y)$

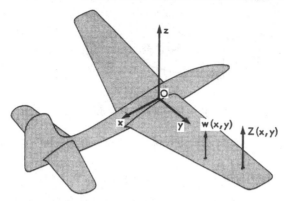

Fig. 2–14.   Airplane wing under distributed side load, $Z(x,y)$.

produced by $Z(x,y)$ is expressible as

$$w(x,y) = \iint_S C(x,y;\xi,\eta)Z(\xi,\eta)d\xi d\eta, \qquad (2\text{–}55)$$

where $C(x,y;\xi,\eta)$ is the two-dimensional influence function describing the deflection at x,y due to a unit force at ξ,η, and the integration is taken over the surface of the wing.

2–7 Properties of influence functions.   The properties of influence functions are similar to those of influence coefficients.   The property of symmetry already stated for the one-dimensional system (Eq. 2–44) applies generally and, for example, in the case of the two-dimensional system is

$$C(x,y;\xi,\eta) = C(\xi,\eta;x,y). \qquad (2\text{–}56)$$

Other properties will be stated in terms of the one-dimensional system, and these results may be easily generalized to two- or three-dimensional systems.   Equation (2–43) expressed the deflection of a beam in terms of an influence function as follows:

$$w(y) = \int_0^l C(y,\eta)Z(\eta)d\eta. \qquad (2\text{–}43)$$

If we regard Eq. (2–43) as a linear integral equation of the first kind (Ref. 2–4) in which $w(y)$ is known and $Z(\eta)$ is unknown, it follows in principle from the theory of integral equations that

$$Z(y) = \int_0^l k(y,\eta)w(\eta)d\eta, \qquad (2\text{–}57)$$

where $k(y,\eta)$ can be regarded as a hypothetical stiffness influence function. In general, the formal process of inverting Eq. (2–43) to obtain Eq. (2–57) is formidable. But in a practical problem this process does not have to be carried out in functional form, and our principal need here is a recognition of the dual relation between Eqs. (2–43) and (2–57).

The strain energy of a continuously loaded one-dimensional system can be expressed in the form

$$U = \tfrac{1}{2} \int_0^l Z(y)w(y)dy. \tag{2-58}$$

Substituting Eq. (2–43), we obtain an expression for the strain energy in terms of the flexibility influence function and the distributed load,

$$U = \tfrac{1}{2} \int_0^l Z(y) \int_0^l C(y,\eta)Z(\eta)d\eta dy. \tag{2-59}$$

Similarly, if Eq. (2–57) is substituted in Eq. (2–58), the strain energy can also be expressed by

$$U = \tfrac{1}{2} \int_0^l w(y) \int_0^l k(y,\eta)w(\eta)d\eta dy. \tag{2-60}$$

2-8 The simplified elastic airplane. Because of the complexity of aircraft structures, it is usually necessary to introduce simplifying assumptions in order to compute their elastic properties. In many cases, the wings, the fuselage, and the horizontal and vertical tails can be regarded as beams rigid in cross sections perpendicular to their lengthwise direction. In such cases, the airplane becomes a collection of beams which can be represented by the intersecting elastic lines shown in Fig. 2–15. The deformations of each line are described by two linear displacements and one angular displacement. For example, the displacement of a rigid cross section of the wing at a point y in Fig. 2–15 is described by the linear displacements $w(y)$ and $u(y)$ and by an angular displacement $\theta(y)$.

In cases which involve thin wings of very low aspect ratio, it may not be possible to assume that the wing is rigid in cross section, and therefore

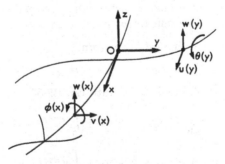

Fig. 2–15. Simplified elastic airplane.

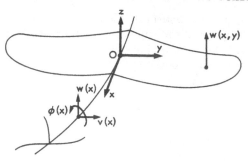

Fig. 2-16. Simplified elastic airplane.

allowance must be made for a chordwise bending. In such cases, the wing can often be regarded as a thin flat plate, and the simplified airplane can be represented by an elastic surface and an elastic line, as illustrated by Fig. 2-16. The wing deformation in Fig. 2-16 is described by the function $w(x,y)$, which represents deflections of the elastic surface with respect to the xy-plane.

2-9 Deformations of airplane wings. Let us consider in some detail the deformation of an elastic wing under the influence of a distributed normal load $Z(x,y)$. The deformation can be expressed by a single function $w(x,y)$, as defined by Eq. (2-55):

$$w(x,y) = \iint_S C(x,y;\xi,\eta)Z(\xi,\eta)d\xi d\eta. \tag{2-55}$$

In those cases where the wing is sufficiently slender so that chordwise segments of the wing parallel to the x-axis can be assumed rigid, the influence function can be written as follows:

$$C(x,y;\xi,\eta) = C^{zz}(y,\eta) - xC^{\theta z}(y,\eta) + \xi x C^{\theta\theta}(y,\eta) - \xi C^{z\theta}(y,\eta), \tag{2-61}$$

where the influence functions are defined by

$$C^{pq}(y,\eta) = \text{linear or angular deflection in the } p\text{-direction at } y$$
$$\text{due to a unit force or torque in the } q\text{-direction at } \eta.$$

The deflection can be expressed in the form

$$w(x,y) = w(y) - x\theta(y). \tag{2-62}$$

Substituting Eqs. (2-61) and (2-62) into Eq. (2-55), we obtain

$$w(y) = \int_0^l C^{zz}(y,\eta)Z(\eta)d\eta + \int_0^l C^{z\theta}(y,\eta)t(\eta)d\eta, \tag{2-63}$$

$$\theta(y) = \int_0^l C^{\theta z}(y,\eta)Z(\eta)d\eta + \int_0^l C^{\theta\theta}(y,\eta)t(\eta)d\eta, \tag{2-64}$$

where
$$Z(\eta) = \int_{\text{chord}} Z(\xi,\eta)d\xi, \qquad (2\text{-}65)$$

$$t(\eta) = -\int_{\text{chord}} \xi Z(\xi,\eta)d\xi. \qquad (2\text{-}66)$$

In deriving Eqs. (2–63) and (2–64), the effects of chordwise drag loads on the wing have been neglected. Equations (2–63) and (2–64) apply in general to the case of unswept or swept wings with structural discontinuities. The principal elastic effects of sweep and structural discontinuities are to couple the bending and torsional actions. It should be observed that the assumption of rigidity along segments parallel to the x-axis becomes less valid with increasing angles of sweep. The error involved seems to be small for slender wings with angles of sweep up to about 45°.

In Eqs. (2–63) and (2–64) the running streamwise torque $t(\eta)$ is referred to the y-axis, and the linear deflection $w(y)$ is computed along the y-axis. In dealing with swept wings, it is often convenient to conceive of a reference axis $\bar{y}$, as shown by Fig. 2–17. The running torque $t(\eta)$ can be referred to the $\bar{y}$-axis, and the linear deflections of the wing can be computed along the $\bar{y}$-axis. Thus, when the $\bar{y}$ reference axis is employed, the influence functions in Eqs. (2–63) and (2–64) are defined as those giving linear and streamwise angular deflections along the $\bar{y}$-axis due to lateral forces and to streamwise torques referred to the $\bar{y}$-axis. In this case, however, the running streamwise torque is defined by the formula

$$t(\eta) = -\int_{\text{chord}} (\xi - \eta \tan \Lambda)Z(\xi,\eta)d\xi. \qquad (2\text{-}67)$$

A further simplification in Eqs. (2–63) and (2–64) can be made for unswept wings which are uniform or which have a gradual change in section

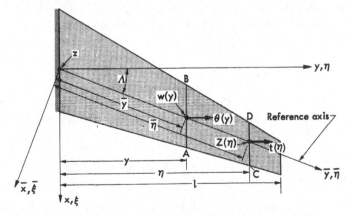

Fig. 2–17. Axis system for swept wing.

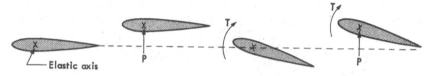

Fig. 2–18.   Force and torque applied to airplane wing.

along the spanwise axis. Such wings can be analyzed by the simple engineering theories which permit the bending and torsion to be separated into two separate actions with no coupling between them. As a consequence of the possibility of decoupling of bending and torsion, the concept of an elastic axis can be introduced. Forces applied to the wing on this axis produce translations of all sections of the beam without rotations, and torques produce pure twisting of all sections about this axis. Figure 2–18 illustrates the action which results when forces and torques are applied to the elastic axis. It is convenient to resolve the forces and moments on the beam into forces along and torques about the elastic axis. In this case, Eqs. (2–63) and (2–64) reduce to

$$w(y) = \int_0^l C^{zz}(y,\eta)Z(\eta)d\eta, \qquad (2\text{–}68)$$

$$\theta(y) = \int_0^l C^{\theta\theta}(y,\eta)t(\eta)d\eta. \qquad (2\text{–}69)$$

2–10 Integration by weighting matrices. The elastic properties of the majority of beams in aircraft structures are complicated functions of their lengthwise coordinates and cannot be represented conveniently in functional form. For this reason, it is usually not possible to derive explicit analytical expressions for influence functions. Numerical values of influence coefficients at specific points can, however, be computed or determined by test. It is often necessary, therefore, to employ methods whereby the deformation can be expressed in terms of influence coefficients even though the wing is loaded by continuously distributed loads.

For example, consider the bending displacement of the elastic axis as expressed by Eq. (2–68). We can compute the deflections at a finite number of points, say n, as shown by Fig. 2–19. From Eq. (2–68), the deflection at the ith station is given by

$$w_i = \int_0^l C^{zz}(y_i,\eta)Z(\eta)d\eta. \qquad (2\text{–}70)$$

An approximate numerical evaluation of the definite integral can be carried out which leads to a linear combination of the ordinates of the loading diagram, as follows:

$$w_i = \sum_{j=0}^{n} C_{ij}{}^{zz}\overline{W}_j Z_j,$$

$$(i = 0, 1, \cdots, n), \quad (2\text{--}71)$$

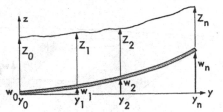

where $C_{ij}{}^{zz} = C^{zz}(y_i, \eta_j)$ are the influence coefficients associated with the n points and the $\overline{W}_j$ are weighting numbers (Ref. 2–5)

Fig. 2-19. Deflections of cantilever beam at n discrete points.

which depend upon the method of numerical integration that is employed. Equation (2–71) becomes, in matrix form,

$$\{w\} = [C^{zz}][\overline{W}]\{Z\}, \qquad (2\text{--}72)$$

where

$$\{w\} = \begin{bmatrix} w_0 \\ w_1 \\ w_2 \\ \cdot \\ \cdot \\ \cdot \\ w_n \end{bmatrix}, \quad [C] = \begin{bmatrix} 0 & 0 & 0 & \cdots & 0 \\ 0 & C_{11} & C_{12} & \cdots & C_{1n} \\ 0 & C_{21} & C_{22} & \cdots & C_{2n} \\ \cdot & \cdot & \cdot & & \cdot \\ \cdot & \cdot & \cdot & & \cdot \\ \cdot & \cdot & \cdot & & \cdot \\ 0 & C_{n1} & C_{n2} & \cdots & C_{nn} \end{bmatrix},$$

$$[\overline{W}] = \begin{bmatrix} \overline{W}_0 & 0 & 0 & 0 & \cdots & 0 \\ 0 & \overline{W}_1 & 0 & \cdot & \cdots & \cdot \\ 0 & 0 & \overline{W}_2 & \cdot & \cdots & \cdot \\ \cdot & & & & & \\ \cdot & & & & & \\ 0 & 0 & 0 & \cdots & & \overline{W}_n \end{bmatrix}, \quad \{Z\} = \begin{bmatrix} Z_0 \\ Z_1 \\ Z_2 \\ \cdot \\ \cdot \\ Z_n \end{bmatrix}.$$

Weighting number matrices for various methods of numerical integration are summarized in Appendix B.

The evaluation of torsional deflections (Eq. 2–69) can be carried out in exactly the same way to give

$$\{\theta\} = [C^{\theta\theta}][\overline{W}]\{t\}. \qquad (2\text{--}73)$$

Equations (2–63) and (2–64) can also be put in matrix form for approximate solutions by dividing the y-axis into a number of intervals. The linear and angular deflections of a streamwise segment of the wing at $y = y_i$ are

$$w_i = \sum_{j=1}^{n} C_{ij}{}^{zz}\overline{W}_j Z_j + \sum_{j=1}^{n} C_{ij}{}^{z\theta}\overline{W}_j t_j, \qquad (2\text{--}74)$$

$$\theta_i = \sum_{j=1}^{n} C_{ij}{}^{\theta z}\overline{W}_j Z_j + \sum_{j=1}^{n} C_{ij}{}^{\theta\theta}\overline{W}_j t_j. \qquad (2\text{--}75)$$

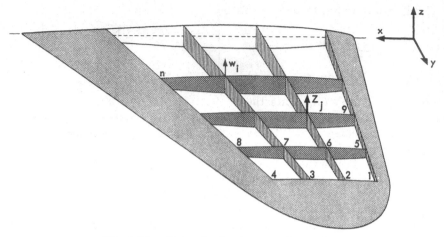

Fig. 2-20.   Network of points on wing surface.

In matrix notation, Eqs. (2-74) and (2-75) become

$$
\begin{bmatrix} \{w\} \\ \hline \{\theta\} \end{bmatrix} = \begin{bmatrix} [C^{zz}] & \vdots & [C^{z\theta}] \\ \hline [C^{\theta z}] & \vdots & [C^{\theta\theta}] \end{bmatrix} \begin{bmatrix} [\overline{W}] & \vdots & [0] \\ \hline [0] & \vdots & [\overline{W}] \end{bmatrix} \begin{bmatrix} \{Z\} \\ \hline \{t\} \end{bmatrix}. \tag{2-76}
$$

The weighting matrices in Eq. (2-76) are the same as those in Eqs. (2-72) and (2-73). As a consequence of the reciprocal theorem (Section 2-4), we observe that

$$
[C^{\theta z}] = [C^{z\theta}]'. \tag{2-77}
$$

The $[C^{z\theta}]$ and $[C^{\theta z}]$ matrices couple bending and torsion, and if they are put equal to zero, Eq. (2-76) reduces to Eqs. (2-72) and 2-73), which represent the case of a straight slender wing without coupling between bending and torsion.

The principles of numerical integration by weighting matrices apply also to low-aspect ratio wings which have curvature in the chordwise as well as the spanwise direction. If a network of n points is located on the surface of the wing, as illustrated in Fig. 2-20, the deflection at the ith point is given by Eq. (2-55) as

$$
w_i = \iint_S C(x_i, y_i; \xi, \eta) Z(\xi, \eta) d\xi d\eta. \tag{2-78}
$$

Numerical integration yields the matrix relation,

$$
\{w\} = [C][\overline{W}]\{Z\}, \tag{2-79}
$$

where $[C]$ is the matrix of flexibility influence coefficients associated with the n stations, $\{Z\}$ are the ordinates to the loading function at the n stations, and $\lceil \overline{W} \rceil$ is a matrix of weighting numbers. The form of the $\lceil \overline{W} \rceil$ matrix depends, of course, on the scheme of numerical integration that is employed.

2-11 Energy methods in deflection calculations. Energy methods find wide use in aeroelastic problems in the determination of deformation shapes under static and dynamic loads and in the calculation of influence functions and coefficients.

(a) *Deflections by the principle of minimum potential energy.* The principle of minimum potential energy, a useful tool in computing displacements of conservative elastic systems, is based on the principle of virtual work (Ref. 2–3). The *principle of virtual work*, applied to deformable bodies, can be stated as follows:

> If a body is in equilibrium under the action of prescribed external forces, the work (virtual work) done by these forces in a small additional displacement compatible with the geometric constraints (virtual displacement) is equal to the change in strain energy.

The principle of virtual work can be stated in the mathematical form

$$\delta W_e = \delta U, \tag{2–80}$$

where δW_e is the virtual work done by the external forces and δU is the change in strain energy resulting from a small virtual displacement of the body.

The principle of minimum potential energy, applicable only to conservative systems, can be derived by transposing Eq. (2–80):

$$\delta U - \delta W_e = \delta(U - W_e) = 0. \tag{2–81}$$

The *principle of minimum potential energy*, expressed mathematically by Eq. (2–81), can be stated as follows:

> Among all possible deformation configurations compatible with the geometric constraints, the configuration which satisfies the equations of equilibrium is the one which minimizes the potential energy, $U - W_e$.

The principle of minimum potential energy finds two principal uses in deflection analyses. The first application is the derivation of exact differential, integral, or algebraic equations of equilibrium, and the second, approximate solutions to deflection problems of continuous systems which are difficult to solve by exact methods.

(b) *The principle of minimum potential energy applied to continuous systems; Rayleigh-Ritz method.* We shall illustrate the application of the

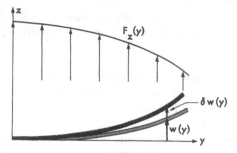

Fig. 2–21.   Virtual displacement of cantilever beam.

principle of minimum potential energy by referring to a cantilever wing beam under a side load $F_z(y)$, as illustrated by Fig. 2–21.

Let us assume that the lateral deflection of the beam can be written as a sum of independent deflection functions $\gamma_i(y)$, each of which satisfies the geometrical boundary conditions:

$$w(y) = \sum_{i=1}^{n} \gamma_i(y)q_i, \tag{2-82}$$

where the q_i are generalized coordinates whose values are to be determined. For a cantilever beam, the conditions of geometrical constraint on the functions $\gamma_i(y)$ are

$$\gamma_i(0) = \gamma_i{}'(0) = 0. \tag{2-83}$$

The coordinates q_i are seen to fall within the definition of generalized coordinates, since they act as independent coordinates which represent possible displacements and which do not violate the conditions of geometric constraint.

Next let us assume that the beam is given an arbitrary lateral virtual displacement, shown by Fig. 2–21, which has a magnitude

$$\delta w(y) = \sum_{i=1}^{n} \gamma_i(y)\delta q_i. \tag{2-84}$$

The corresponding work done by the external forces is given by

$$\delta W_e = \int_0^l F_z(y)\delta w(y)dy = \sum_{i=1}^{n} \left\{ \int_0^l F_z(y)\gamma_i(y)dy \right\} \delta q_i. \tag{2-85}$$

The strain energy due to bending in a slender beam can be expressed in terms of the lateral displacement by the following formula (Ref. 2–2):

$$U = \tfrac{1}{2} \int_0^l EI \left(\frac{d^2 w}{dy^2} \right)^2 dy. \tag{2-86}$$

When Eq. (2-82) is introduced, this formula becomes

$$U = \frac{1}{2} \sum_{i=1}^{n} \sum_{j=1}^{n} \left\{ \int_{0}^{l} EI \left(\frac{d^2\gamma_i}{dy^2} \right) \left(\frac{d^2\gamma_j}{dy^2} \right) dy \right\} q_i q_j. \tag{2-87}$$

From this follows for the change in strain energy due to coordinate changes δq_i,

$$\delta U = \sum_{i=1}^{n} \sum_{j=1}^{n} \left\{ \int_{0}^{l} EI \left(\frac{d^2\gamma_i}{dy^2} \right) \left(\frac{d^2\gamma_j}{dy^2} \right) dy \right\} q_j \delta q_i. \tag{2-88}$$

Substituting Eqs. (2-85) and (2-88) into Eq. (2-81) gives

$$\sum_{i=1}^{n} \left[\sum_{j=1}^{n} \left\{ \int_{0}^{l} EI \left(\frac{d^2\gamma_i}{dy^2} \right) \left(\frac{d^2\gamma_j}{dy^2} \right) dy \right\} q_j - \int_{0}^{l} F_z(y) \gamma_i(y) dy \right] \delta q_i = 0. \tag{2-89}$$

Since the δq_i are independent arbitrary quantities, Eq. (2-89) can be satisfied only if

$$\sum_{j=1}^{n} \left\{ \int_{0}^{l} EI \left(\frac{d^2\gamma_i}{dy^2} \right) \left(\frac{d^2\gamma_j}{dy^2} \right) dy \right\} q_j - \int_{0}^{l} F_z(y) \gamma_i(y) dy = 0,$$
$$(i = 1, 2, \cdots, n). \tag{2-90}$$

Equations (2-90) constitute a set of n simultaneous linear algebraic equations in the unknown generalized coordinates, $q_1, \cdots, q_n$. The final solution for the lateral deflection is obtained by substituting values of $q_1, \cdots, q_n$, obtained by solving Eqs. (2-90), into Eq. (2-82). This solution is approximate if n is a finite quantity. The process of finding an approximate solution in this manner is known as the Rayleigh-Ritz method (Ref. 2-8).

Let us consider next the restrained three-dimensional elastic body shown in Fig. 2-22. The x,y,z-axis system is fixed in space and the displacements in the directions of these axes are $u(x,y,z)$, $v(x,y,z)$, and $w(x,y,z)$. Let $F_x(x,y,z)$, $F_y(x,y,z)$, and $F_z(x,y,z)$ be components of surface force in-

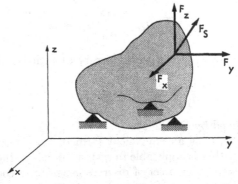

Fig. 2-22. Surface pressures applied to three-dimensional elastic body.

tensity. The work done by the surface forces during an arbitrary virtual displacement is then

$$\delta W_e = \int_S (F_x \delta u + F_y \delta v + F_z \delta w) dS, \tag{2-91}$$

where the integration is over the surface. Next let us assume that any displacement of the body can be expressed as a function of n discrete generalized coordinates q_i:

$$u = u(x,y,z,q_1,q_2,\cdots,q_n),$$
$$v = v(x,y,z,q_1,q_2,\cdots,q_n), \tag{2-92}$$
$$w = w(x,y,z,q_1,q_2,\cdots,q_n).$$

In the beam application discussed above, it can be observed that Eq. (2-82) is an example of Eqs. (2-92) in a specific case.

Introducing Eqs. (2-92) into Eq. (2-91), the virtual work due to the external forces reduces to the form

$$\delta W_e = \sum_{i=1}^{n} \left\{ \int_S \left(F_x \frac{\partial u}{\partial q_i} + F_y \frac{\partial v}{\partial q_i} + F_z \frac{\partial w}{\partial q_i} \right) dS \right\} \delta q_i = \sum_{i=1}^{n} Q_i \delta q_i, \tag{2-93}$$

where

$$Q_i = \int_S \left(F_x \frac{\partial u}{\partial q_i} + F_y \frac{\partial v}{\partial q_i} + F_z \frac{\partial w}{\partial q_i} \right) dS \tag{2-94}$$

is the generalized force which corresponds to the generalized coordinate q_i.

The strain energy can also be expressed as a function of the generalized coordinates $q_1, \cdots, q_n$, and the change in strain energy due to arbitrary virtual displacement of the generalized coordinates is

$$\delta U = \sum_{i=1}^{n} \frac{\partial U}{\partial q_i} \delta q_i. \tag{2-95}$$

Introducing Eqs. (2-94) and (2-95) into Eq. (2-81) yields

$$\sum_{i=1}^{n} \left\{ Q_i - \frac{\partial U}{\partial q_i} \right\} \delta q_i = 0. \tag{2-96}$$

Since the δq_i are independent and arbitrary quantities, this reduces to

$$\frac{\partial U}{\partial q_i} = Q_i, \qquad (i = 1, 2, \cdots, n), \tag{2-97}$$

where Q_i is defined by Eq. (2-94).

Equation (2-97) is an equivalent form of the principle of minimum potential energy that is applicable to systems in which the space configuration can be described by a set of discrete generalized coordinates. When n is finite, the process of applying Eq. (2-97) to a continuous system, called the Rayleigh-Ritz method, leads to approximate solutions. If n is increased without limit, it is possible for the equilibrium equations given by Eq. (2–97) to yield an exact solution.

(c) *Deflections by Castigliano's theorem.* Castigliano's theorem provides a method for computing deflections from the strain energy when the latter is expressed as a function of the applied loads. When Eq. (2–36) is differentiated with respect to Q_i, we obtain

$$\frac{\partial U}{\partial Q_i} = \frac{\partial}{\partial Q_i}\left(\tfrac{1}{2}\sum_{i=1}^{n}\sum_{j=1}^{n}C_{ij}Q_iQ_j\right) = \sum_{j=1}^{n}C_{ij}Q_j. \qquad (2\text{–}98)$$

Combining Eqs. (2–98) with Eq. (2–3), we obtain the mathematical statement of Castigliano's theorem:

$$q_i = \frac{\partial U}{\partial Q_i}. \qquad (2\text{–}99)$$

Castigliano's theorem states that the partial derivative with respect to one of the loads of the strain energy, expressed as a function of the applied loads, yields the deflection of the structure at the point of application and in the direction of that load.

To demonstrate Castigliano's theorem as a means of computing beam deflections, let us consider the problem of computing the deflection of the point a due to an arbitrary side load on a cantilever beam, as shown by Fig. 2–23. We apply an additional dummy load P at the point a at which we seek the deflection Δ. If the bending moment distribution due to a 1-lb load at a is given by m, then, by the principle of superposition, the bending moment distribution due to a load P at a is given by mP. The strain energy due to normal strains in a slender beam can.be expressed in terms of the applied bending moment distribution $\overline{M}$ by the following formula (Ref. 2–2):

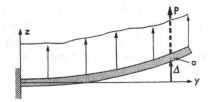

Fig. 2–23. Cantilever beam under arbitrary side load.

$$U = \int_0^l \frac{\overline{M}^2}{2EI}\,dy. \qquad (2\text{–}100)$$

The strain energy due to the given side load plus the dummy load P is obtained by substituting

$$\overline{M} = M(y) + mP \qquad (2\text{–}101)$$

into Eq. (2–100):

$$U = \int_0^l \frac{(M + mP)^2}{2EI}\,dy. \qquad (2\text{–}102)$$

The deflection at a resulting from the given side load plus the dummy load P is obtained by applying Castigliano's theorem:

$$\Delta = \frac{\partial U}{\partial P} = \int_0^l \frac{(M + mP)m}{EI} \, dy. \tag{2–103}$$

The deflection at a due to the given side load alone is obtained from Eq. (2–103) by putting $P = 0$:

$$\Delta = \int_0^l \frac{Mm}{EI} \, dy. \tag{2–104}$$

Equation (2–104) may, for example, be used to compute influence coefficients if written in the form

$$C_{ij} = \int_0^l \frac{m_i m_j}{EI} \, dy, \tag{2–105}$$

where

m_i = bending moment distribution due to 1-lb load at i,
m_j = bending moment distribution due to 1-lb load at j.

2–12 Deformations of slender unswept wings.   Let us consider a type of beam in which the skin and shear webs are assumed thin, that is, they have no bending rigidity, and the cross-sectional shape is preserved by closely spaced ribs which have infinite rigidity in their own plane but are completely free to warp out of their plane.   The beam may taper in the spanwise direction, and the properties of the cross section may vary with y. The reference axes are illustrated by Fig. 2–24.   It is convenient in the present discussion to assume that the y-axis is placed such that it pierces each cross section at the centroid of the effective normal stress-carrying area and that the x- and z-axes are oriented so that the xy- and yz-planes pass through the principal bending axes.   A point on the cross section is located by the spanwise coordinate y and the tangential coordinate s.   The tangential coordinate is measured positive in the counterclockwise direc-

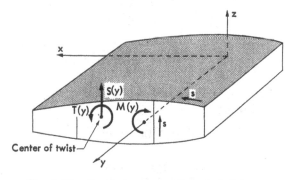

Fig. 2–24.   Loaded cross section of shell beam.

tion for the peripheral skin and is positive in the positive direction of the z-axis for the interior webs.

A loaded cross section at a distance y from the origin is acted on by a moment $M(y)$, a shear $S(y)$, and a torque $T(y)$, with positive directions as shown in Fig. 2–24. The point of application of $S(y)$ and $T(y)$ is at the center of twist or shear center of the section. Normal stresses, assumed positive in the y-direction, are denoted by σ, and shear flows, assumed positive in the s-direction, are denoted by q.

In discussing slender beams, we make two simplifying assumptions. First, the beam is permitted complete freedom to warp when torque loads are applied. This leads to the St. Venant solution of the torsion problem and permits the simplification that bending and torsion are separate uncoupled actions. Second, we assume that plane sections remain plane during bending. This allows application of the well-known engineering bending theory.

(a) *Bending and shearing deformations.* The wing is composed of skin elements in equilibrium under normal stresses σ and shear flows q, as illustrated by Fig. 2–25. Since the ribs are assumed rigid in their own planes, and we may neglect normal stresses in the chordwise direction, the strain energy in the wing in terms of stresses is

$$U = \frac{1}{2E} \int_0^l \int_{\text{c.s.}} \bar{\sigma}^2 t\, ds\, d\lambda + \frac{1}{2G} \int_0^l \int_{\text{c.s.}} \bar{q}^2 \frac{ds}{t}\, d\lambda, \qquad (2\text{–}106)$$

where t is the skin thickness, the integration on s is over the wing cross section, and the integration on λ is over the length of the wing beam. Let us compute first the lateral deflection $w(y)$ of the wing resulting from a bending moment distribution, $M(\lambda)$, and a shear distribution, $S(\lambda)$. In order to apply Castigliano's theorem, we must apply an additional external load P at the spanwise location $\lambda = y$. Then we denote the total normal stress due to the given bending moment $M(\lambda)$, plus P, by

$$\bar{\sigma}(\lambda,s) = \sigma(\lambda,s) + P \frac{\partial \bar{\sigma}}{\partial P}, \qquad (2\text{–}107a)$$

where $\sigma(\lambda,s)$ is the normal stress due to the given bending moment distribu-

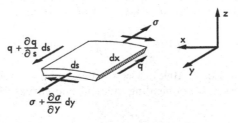

Fig. 2–25.   Element of wing skin.

tion $M(\lambda)$ and $\partial\bar{\sigma}/\partial P$ is the normal stress distribution due to $P = 1$ located at $\lambda = y$. Similarly, the total shear flow is denoted by

$$\bar{q}(\lambda,s) = q(\lambda,s) + P\frac{\partial\bar{q}}{\partial P}, \qquad (2\text{-}107\text{b})$$

where $q(\lambda,s)$ is the shear flow due to the given shear distribution $S(\lambda)$, and $\partial\bar{q}/\partial P$ is the shear flow due to $P = 1$ located at $\lambda = y$.

Substituting Eqs. (2–107a) and (2–107b) into Eq. (2–106), the total strain energy due to the given bending moment and shear distributions plus the dummy load P is

$$U = \frac{1}{2E}\int_0^l\int_{\text{c.s.}}\left[\sigma(\lambda,s) + P\frac{\partial\bar{\sigma}}{\partial P}\right]^2 t\,ds\,d\lambda$$

$$+ \frac{1}{2G}\int_0^l\int_{\text{c.s.}}\left[q(\lambda,s) + P\frac{\partial\bar{q}}{\partial P}\right]^2\frac{ds}{t}\,d\lambda. \qquad (2\text{-}107\text{c})$$

Differentiating with respect to P and setting $P = 0$, the lateral deflection $w(y)$ becomes

$$w(y) = \frac{1}{E}\int_0^l\int_{\text{c.s.}}\sigma(\lambda,s)\frac{\partial\bar{\sigma}}{\partial P}t\,ds\,d\lambda + \frac{1}{G}\int_0^l\int_{\text{c.s.}}q(\lambda,s)\frac{\partial\bar{q}}{\partial P}\frac{ds}{t}\,d\lambda. \quad (2\text{-}108)$$

The first term in Eq. (2–108) represents the bending deflection, which we denote by $\alpha(y)$:

$$\alpha(y) = \frac{1}{E}\int_0^l\int_{\text{c.s.}}\sigma(\lambda,s)\frac{\partial\bar{\sigma}}{\partial P}t\,ds\,d\lambda. \qquad (2\text{-}109\text{a})$$

If we assume, for example, that the engineering theory of beam bending can be applied, the normal stresses are represented by

$$\sigma(\lambda,s) = -\frac{M(\lambda)z}{I},$$

$$\frac{\partial\bar{\sigma}}{\partial P} = \begin{cases} -\dfrac{y-\lambda}{I}z, & (y > \lambda), \\[2mm] 0, & (y < \lambda), \end{cases} \qquad (2\text{-}109\text{b})$$

where I is the area moment of inertia of the cross section about the x-axis. Substituting Eqs. (2–109b) into Eq. (2–109a) results in

$$\alpha(y) = \int_0^y\frac{M(\lambda)(y-\lambda)}{EI}\,d\lambda. \qquad (2\text{-}109\text{c})$$

Differentiation of Eq. (2–109c) twice with respect to y produces the following relation between beam bending moment and curvature:

$$\alpha'' = \frac{M}{EI}. \qquad (2\text{-}110)$$

The second term in Eq. (2–108) represents the shearing deflection, which we denote by $\beta(y)$:

$$\beta(y) = \frac{1}{G} \int_0^l \int_{\text{c.s.}} q(\lambda,s) \frac{\partial \bar{q}}{\partial P} \frac{ds}{t} \, d\lambda. \qquad (2\text{–}111a)$$

If we assume that the shear flow distribution due to $S = 1$ is denoted by $u(\lambda,s)$, then

$$q(\lambda,s) = Su(\lambda,s),$$

$$\frac{\partial \bar{q}}{\partial P} = \begin{cases} u(\lambda,s), & (y > \lambda), \\ 0, & (y < \lambda). \end{cases} \qquad (2\text{–}111b)$$

Substituting Eqs. (2–111b) into Eq. (2–111a) gives

$$\beta(y) = \int_0^y \frac{S d\lambda}{GK}, \qquad (2\text{–}111c)$$

where

$$K = \frac{1}{\displaystyle\int_{\text{c.s.}} u^2 \frac{ds}{t}}$$

is a shearing constant of the beam cross section defined in terms of the shear flow distribution due to a unit shear force. The quantity GK in Eq. (2–111c) is termed the shearing rigidity of the beam. Differentiating Eq. (2–111c) with respect to y gives a relation between the shear and the first derivative of the shearing deflection, as follows:

$$\beta' = \frac{S}{GK}. \qquad (2\text{–}112)$$

(b) *Influence functions and coefficients.* Influence functions taking both bending and shearing deformations into account can be computed from Eq. (2–108). For example, to compute $C^{zz}(y,\eta)$ by applying the engineering theory of beam bending, we put

$$\sigma(\lambda,s) = -\frac{\eta - \lambda}{I} z, \qquad (0 < \lambda < \eta),$$

$$\frac{\partial \bar{\sigma}}{\partial P} = -\frac{y - \lambda}{I} z, \qquad (0 < \lambda < y), \qquad (2\text{–}113a)$$

and

$$q(\lambda,s) = u(\lambda,s), \qquad (0 < \lambda < \eta),$$

$$\frac{\partial \bar{q}}{\partial P} = u(\lambda,s), \qquad (0 < \lambda < y), \qquad (2\text{–}113b)$$

into Eq. (2–108). For the case of $\eta \geq y$, this yields

$$C^{zz}(y,\eta) = \int_0^y \frac{(\eta - \lambda)(y - \lambda)}{EI}\, d\lambda + \int_0^y \frac{d\lambda}{GK}, \qquad (\eta \geq y), \quad (2\text{–}114)$$

and for the case of $y \geq \eta$,

$$C^{zz}(y,\eta) = \int_0^\eta \frac{(\eta - \lambda)(y - \lambda)}{EI}\, d\lambda + \int_0^\eta \frac{d\lambda}{GK}, \qquad (y \geq \eta). \quad (2\text{–}115)$$

Influence coefficients can be obtained from Eqs. (2–114) and (2–115) by introducing explicit numerical values for y and η.

(c) *Torsional deformation and influence function.* If, during the application of torsional moments, the beam is free to warp, the strain energy is due entirely to shear stresses, as follows:

$$U = \frac{1}{2G} \int_0^l \int_{\text{c.s.}} \bar{q}^2\, \frac{ds}{t}\, d\lambda. \qquad (2\text{–}116)$$

By applying Castigliano's theorem in the same manner as for the bending problem, the angle of twist of the beam due to a given distribution of applied torque, $T(\lambda)$, can be expressed as

$$\theta(y) = \frac{1}{G} \int_0^l \int_{\text{c.s.}} q(\lambda,s)\, \frac{\partial \bar{q}}{\partial T}\, \frac{ds}{t}\, d\lambda, \qquad (2\text{–}117a)$$

where $q(\lambda,s)$ is the shear flow distribution due to the applied torques and $\partial \bar{q}/\partial T$ is the shear flow distribution due to a unit torque $T = 1$ applied at the spanwise station $\lambda = y$.

If we assume that the shear flow distribution due to $T = 1$ is denoted by $v(\lambda,s)$, then

$$q(\lambda,s) = T(\lambda)v(\lambda,s),$$

$$\frac{\partial \bar{q}}{\partial T} = \begin{cases} v(\lambda,s), & (y > \lambda), \\ 0, & (y < \lambda). \end{cases} \qquad (2\text{–}117b)$$

Substituting Eqs. (2–117b) into Eq. (2–117a) gives

$$\theta(y) = \int_0^y \frac{T(\lambda)d\lambda}{GJ}, \qquad (2\text{–}117c)$$

where

$$J = \frac{1}{\displaystyle\int_{\text{c.s.}} v^2\, \frac{ds}{t}}$$

is the torsion constant of the beam cross section. The quantity GJ is termed the torsional stiffness of the beam. Differentiating Eq. (2–117c)

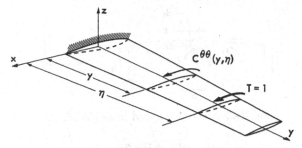

Fig. 2-26.   Cantilever wing under unit torque load.

with respect to y gives a relation between the applied torsional moment and the rate of twist, as follows:

$$\theta' = \frac{T}{GJ}. \tag{2-118}$$

The deflection problem to be solved in computing the torsional influence function is illustrated by Fig. 2-26. A unit torque about the elastic axis is applied at a distance η from the origin, and the resulting angular displacement at y is denoted by $C^{\theta\theta}(y,\eta)$. The influence function $C^{\theta\theta}(y,\eta)$ is obtained from Eq. (2-117a) by putting

$$q(\lambda,s) = v(\lambda,s), \qquad (0 < \lambda < \eta),$$

$$\frac{\partial \bar{q}}{\partial T} = v(\lambda,s), \qquad (0 < \lambda < y). \tag{2-119}$$

This gives, for $\eta \geq y$,

$$C^{\theta\theta}(y,\eta) = \int_0^y \frac{d\lambda}{GJ}, \qquad (\eta \geq y), \tag{2-120}$$

and for $y \geq \eta$,

$$C^{\theta\theta}(y,\eta) = \int_0^\eta \frac{d\lambda}{GJ}, \qquad (y \geq \eta). \tag{2-121}$$

The torsion constant J, defined in Eq. (2-120), can be easily evaluated at each section of the beam, providing the shear flow distribution due to a unit torque, $v(s,\lambda)$, is known. As a simple example of this, consider the case of a single-cell beam. Suppose that a torque T is applied to a single-cell box, as illustrated by Fig. 2-27. Since we assume complete freedom to warp, normal stresses cannot be induced. Referring to Fig. 2-25, which illustrated the free-body diagram of a skin element, we observe that when $\sigma = 0$, we have

$$\frac{\partial q}{\partial s} = 0. \tag{2-122}$$

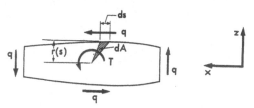

Fig. 2–27. Torque applied to single-cell box.

Equation (2–122) tells us that the shear flow q around a single-cell box is constant.

Referring to Fig. 2–27 and equating the applied torque to the moments of the applied shear flows about an arbitrary point, we obtain

$$T = \oint qr\,ds. \qquad (2\text{–}123)$$

Since q is constant around the section, Eq. (2–123) reduces to the result known as Bredt's formula,

$$T = 2Aq, \qquad (2\text{–}124)$$

where A is the cross-sectional area of the cell.

Putting this result into Eq. (2–120), we obtain a formula for the torsion constant of a single-cell section:

$$J = \frac{4A^2}{\oint \dfrac{ds}{t}}. \qquad (2\text{–}125)$$

In computing the torsional rigidity of a multicell box, we must compute the shear flow distribution in a statically indeterminate system due to the unit torque. Methods for accomplishing this are well known and may be found, for example, in Ref. 2–10.

(d) *Elastic axis.* We have derived formulas in Parts (a) and (b) above for computing bending and torsional flexibility influence coefficients of beams with straight elastic axes. In order to use these results, it is necessary to know the location of the elastic axis. The elastic axis is located by drawing a spanwise line through the shear centers of the cross sections of the beam. The shear center of each cross section is computed by establishing the point in the plane of the section at which a shear force can be applied to the section without producing a rate of twist of the section. Methods for computing the shear center location of multicell beams may be found in Refs. 2–2 and 2–10. Tests of full-scale wings and of models are frequently employed as a means of determining elastic axis locations for aeroelastic studies.

EXAMPLE 2–1. To calculate the influence coefficients of the wing of a jet transport. The planform of the wing is illustrated by Fig. 2–28. The elastic axis is straight and perpendicular to the root at 35% of the chord. The curves of bending rigidity EI, torsional rigidity GJ, and shearing rigidity GK have been computed and are plotted in Fig. 2–29. It is desired to calculate the five by five matrices of bending and torsional influence coefficients at wing stations 90, 186, 268, 368, and 458.

Solution.

(a) *Matrix of bending influence coefficients.* The bending influence coefficients are obtained by a numerical evaluation of Eqs. (2–114) and (2–115). Expanding

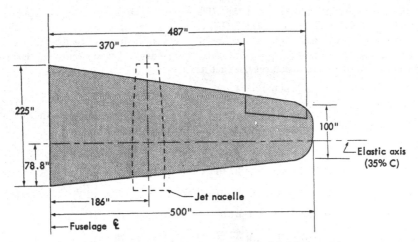

Fig. 2–28. Wing planform of typical jet transport.

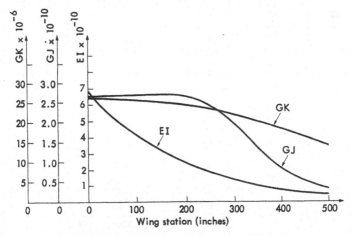

Fig. 2–29. Bending, torsional, and shear stiffness curves.

Eq. (2–115) and putting $y = y_i$ and $\eta = y_j$, we obtain an expression for the bending influence coefficient C_{ij}, as follows:

$$C_{ij} = y_iy_j\int_0^{y_j} \frac{d\lambda}{EI} - (y_i + y_j)\int_0^{y_j} \frac{\lambda d\lambda}{EI} + \int_0^{y_j} \frac{\lambda^2 d\lambda}{EI} + \int_0^{y_j} \frac{d\lambda}{GK}, \qquad (y_i \geq y_j). \quad \text{(a)}$$

Putting $y_j = 90$, we can, for example, compute the first column of the bending influence coefficient matrix from the following formula:

$$C_{i1} = 90y_i\int_0^{90} \frac{d\lambda}{EI} - (y_i + 90)\int_0^{90} \frac{\lambda d\lambda}{EI} + \int_0^{90} \frac{\lambda^2 d\lambda}{EI} + \int_0^{90} \frac{d\lambda}{GK}, \qquad \text{(b)}$$

where $y_i = 90, 186, 268, 368$, and 458. The definite integrals can be evaluated by plotting $1/EI$, λ/EI, λ^2/EI, and $1/GK$, as shown by Fig. 2–30, and taking the areas under the curves to the left of Sta. 90. A planimeter may be used, but the trapezoidal rule is easy to apply and usually provides sufficient accuracy. Evaluating the areas under the curves, we get for C_{i1},

$$C_{i1} = (.44126y_i - 4.369) \times 10^{-7} + 37.07 \times 10^{-7}, \qquad \text{(c)}$$

where the first term is due to the normal strains and the second term due to the shear strains. Putting y_i equal to 90, 186, 268, 368, and 458, we get

$$C_{11} = 35.3445 \times 10^{-7} + 37.07 \times 10^{-7} = 72.410 \times 10^{-7} \text{ in/lb,}$$

$$C_{21} = 77.7054 \times 10^{-7} + 37.07 \times 10^{-7} = 114.771 \times 10^{-7},$$

$$C_{31} = 113.889 \times 10^{-7} + 37.07 \times 10^{-7} = 150.955 \times 10^{-7}, \qquad \text{(d)}$$

$$C_{41} = 158.015 \times 10^{-7} + 37.07 \times 10^{-7} = 195.081 \times 10^{-7},$$

$$C_{51} = 197.728 \times 10^{-7} + 37.07 \times 10^{-7} = 234.794 \times 10^{-7}.$$

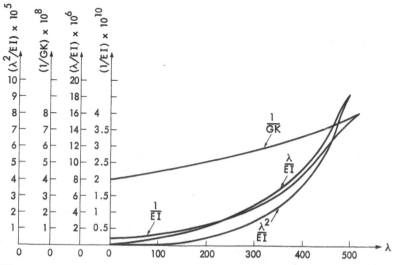

Fig. 2–30. Plots of integrands involved in influence coefficient calculations.

The shear deformations are a significant fraction of the total at the inboard stations, whereas they are relatively unimportant at the outboard stations. The results computed above give not only the first column of the matrix but also the first row, because of the reciprocal theorem. Other rows and columns can be computed in the same way. The final matrix of bending influence coefficients becomes

$$[C^{zz}] =$$
$$\begin{bmatrix} 0 & 0 & 0 & 0 & 0 & 0 \\ 0 & 72.410 & 114.771 & 150.995 & 195.081 & 234.794 \\ 0 & 114.771 & 253.306 & 461.299 & 714.949 & 943.234 \\ 0 & 150.995 & 461.299 & 1247.53 & 1911.22 & 2508.54 \\ 0 & 195.081 & 714.949 & 1911.22 & 2649.96 & 5237.42 \\ 0 & 234.794 & 943.234 & 2508.54 & 5237.42 & 8434.02 \end{bmatrix} \times 10^{-7} \text{ in/lb.} \quad (e)$$

(b) *Matrix of torsional influence coefficients.* The torsional influence coefficients are obtained from Eqs. (2–120) and (2–121) by calculating areas under the $1/GJ$ curve, as follows:

$$C_{11} = C_{12} = C_{13} = C_{14} = C_{15} = \int_0^{90} \frac{d\lambda}{GJ} = 36 \times 10^{-10} \text{ rad/in·lb,}$$

$$C_{22} = C_{23} = C_{24} = C_{25} = \int_0^{186} \frac{d\lambda}{GJ} = 75.36 \times 10^{-10},$$

$$C_{33} = C_{34} = C_{35} = \int_0^{268} \frac{d\lambda}{GJ} = 113.26 \times 10^{-10}, \quad (f)$$

$$C_{44} = C_{45} = \int_0^{368} \frac{d\lambda}{GJ} = 204.66 \times 10^{-10},$$

$$C_{55} = \int_0^{458} \frac{d\lambda}{GJ} = 409.49 \times 10^{-10}.$$

Constructing the matrix of torsional influence coefficients from the above, we obtain

$$[C^{\theta\theta}] = \begin{bmatrix} 0 & 0 & 0 & 0 & 0 & 0 \\ 0 & 36 & 36 & 36 & 36 & 36 \\ 0 & 36 & 75.36 & 75.36 & 75.36 & 75.36 \\ 0 & 36 & 75.36 & 113.26 & 113.26 & 113.26 \\ 0 & 36 & 75.36 & 113.26 & 204.66 & 204.66 \\ 0 & 36 & 75.36 & 113.26 & 204.66 & 409.49 \end{bmatrix} \times 10^{-10} \text{ rad/in·lb.} \quad (g)$$

2–13 Influence functions and coefficients of slender swept wings. In Section 2–9 the deformation of a swept wing was described by Eqs. (2–63) and (2–64) in terms of linear and angular displacements of streamwise segments due to running loads and torques applied along a reference axis $\bar{y}$. In the case of slender swept beams of large aspect ratio, an effective root can be assumed, as shown by Fig. 2–31. Thus, an elastic axis can be postulated, and the $\bar{y}$ reference axis can be located along the elastic axis.

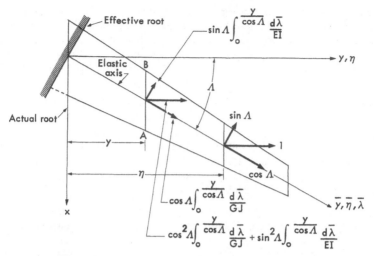

Fig. 2–31.   Slender swept beam with effective root.

If it is assumed that the $\bar{y}$-axis is a true elastic axis, the influence functions in Eqs. (2–63) and (2–64) can be computed by slender beam theory.

The influence function $C^{zz}(y,\eta)$ which describes the linear deflection of the elastic axis at y due to a unit load applied to the elastic axis at η follows directly from Eqs. (2–114) and (2–115):

$$C^{zz}(y,\eta) = \int_0^{y/\cos \Lambda} \frac{\left(\dfrac{\eta}{\cos \Lambda} - \bar{\lambda}\right)\left(\dfrac{y}{\cos \Lambda} - \bar{\lambda}\right)}{EI} \, d\bar{\lambda}$$

$$+ \int_0^{y/\cos \Lambda} \frac{d\bar{\lambda}}{GK}, \qquad (\eta \geq y), \qquad (2\text{–}126)$$

$$C^{zz}(y,\eta) = \int_0^{\eta/\cos \Lambda} \frac{\left(\dfrac{\eta}{\cos \Lambda} - \bar{\lambda}\right)\left(\dfrac{y}{\cos \Lambda} - \bar{\lambda}\right)}{EI} \, d\bar{\lambda}$$

$$+ \int_0^{\eta/\cos \Lambda} \frac{d\bar{\lambda}}{GK}, \qquad (y \geq \eta), \qquad (2\text{–}127)$$

where EI and GK are stiffness properties of the beam on sections normal to the elastic axis.

The function $C^{\theta\theta}(y,\eta)$ represents the angular deflection of the streamwise strip A–B about the y-axis due to a unit streamwise torque applied to the reference axis at η.   Figure 2–31 shows the unit streamwise torque at η divided into components parallel and perpendicular to the elastic axis and the angular deflections which these components produce at y when $y \leq \eta$.

The function $C^{\theta\theta}(y,\eta)$ is the resultant component of twist about the y-axis:

$$C^{\theta\theta}(y,\eta) = \int_0^{y/\cos \Lambda} \left(\frac{\cos^2 \Lambda}{GJ} + \frac{\sin^2 \Lambda}{EI} \right) d\bar{\lambda}, \qquad (\eta \geq y), \quad (2\text{--}128)$$

where EI and GJ are stiffness properties of the beam on sections normal to the elastic axis. It can also be seen that

$$C^{\theta\theta}(y,\eta) = \int_0^{\eta/\cos \Lambda} \left(\frac{\cos^2 \Lambda}{GJ} + \frac{\sin^2 \Lambda}{EI} \right) d\bar{\lambda}, \qquad (y \geq \eta). \quad (2\text{--}129)$$

Finally, $C^{\theta z}(y,\eta)$ is the angular deflection of the streamwise strip A–B about the y-axis due to a unit load applied to the reference axis at η:

$$C^{\theta z}(y,\eta) = -\sin \Lambda \int_0^{y/\cos \Lambda} \frac{\left(\dfrac{\eta}{\cos \Lambda} - \bar{\lambda} \right)}{EI} d\bar{\lambda}, \qquad (\eta \geq y), \quad (2\text{--}130)$$

$$C^{\theta z}(y,\eta) = -\sin \Lambda \int_0^{\eta/\cos \Lambda} \frac{\left(\dfrac{\eta}{\cos \Lambda} - \bar{\lambda} \right)}{EI} d\bar{\lambda}, \qquad (y \geq \eta). \quad (2\text{--}131)$$

Using the reciprocal relation given by Eq. (2–32), the influence function $C^{z\theta}(y,\eta)$ is obtained from $C^{\theta z}(y,\eta)$ by interchanging y and η:

$$C^{z\theta}(y,\eta) = C^{\theta z}(\eta,y). \qquad (2\text{--}132)$$

2–14 Deformations and influence coefficients of low aspect-ratio wings. In dealing with explicit methods of predicting structural deformations, we have so far confined our remarks to slender wings which can be treated by the engineering beam theory. Modern configuration trends in high-speed aircraft design, however, have been away from slender wings and towards low aspect-ratio surfaces of the type exemplified by the wing of the XF–92A airplane, illustrated in Fig. 2–3. The problem of predicting the elastic properties of low aspect-ratio surfaces for aeroelastic analyses presents exceptional difficulties and requires the expenditure of considerable time and effort in order to arrive at useful and accurate results.

Analyses of low aspect-ratio surfaces are significantly different from analyses of long slender wings in at least two important respects. First, the root restraint plays a much more important role in the deformations and, hence, must be accounted for more rigorously. This stipulation applies primarily to root restraint against warping, which has a profound effect upon the twisting action of the wing (Ref. 2–11). Second, camber bending of the surface may become a significant component in the deformation. As a result, the wing may behave in a manner similar to a plate with elastic curvature of appreciable magnitude in two directions. In recent years, considerable effort has been expended in deriving methods of analysis of deformations of both built-up and solid low aspect-ratio wings.

The intensive demand for low drag, low aspect-ratio, supersonic lifting surfaces has resulted in a continuous decrease in thickness ratio. Structural design trends accompanying this thickness ratio reduction have been towards thicker cover skins and multiple webs. The ultimate limit of this process is, of course, the completely solid wing. In an effort to reduce drag to an absolute minimum, supersonic missile designers have frequently employed solid lifting and control surfaces.

In general, it can be said that deformation analyses of low aspect-ratio wings can be divided into three fundamentally different methods of approach. The first method, which applies essentially to built-up wings, was described by Levy in 1947 (Ref. 2–12). This approach views the structure as an assemblage of a finite number of elastic components. Each component is placed in equilibrium and the forces acting on the various components are regarded as the unknown quantities. If the structure is statically determinate, the internal forces can be computed from the equations of equilibrium and the deflections can be obtained by applying Castigliano's theorem. If the structure is statically indeterminate, the correct internal force system must be obtained by application of the principle of minimum strain energy. Application of the latter is equivalent to imposing the requirement that the deformation of each member be compatible with the deformation of every other member. Subsequent papers by Lang and Bisplinghoff (Ref. 2–13), Langefors (Ref. 2–14), and Wehle and Lansing (Ref. 2–15) have made further contributions to Levy's method of approach. These contributions have consisted largely of methods for casting the problem in matrix form and improved techniques for solution.

The second method, which also applies essentially to built-up wings, is similar to the first in that it views the structure as an assemblage of interconnected elastic components. But in this method the deformations of the various discrete components of the structure are regarded as the unknown quantities, instead of the forces on the components. The deformation of each elastic component is required to be compatible with the deformation of every other component. The correct deformation pattern is obtained by applying the principle of minimum potential energy. Such an application is equivalent to the requirement that every component of the structure should be in static equilibrium. This method is sometimes called the method of direct stiffness calculation. Contributors to this method of approach have included Schuerch (Ref. 2–16), Levy (Ref. 2–17), and Turner, Clough, Martin, and Topp (Ref. 2–18).

A third method of approach applies to both built-up and solid wings. It consists of applying the principle of minimum potential energy in the form of either the Rayleigh-Ritz or the generalized Rayleigh-Ritz method. In these methods the space configuration of the deformed elastic surface is taken as a superposition of assumed deformation modes. The contribution of each assumed mode is obtained by application of the principle of minimum potential energy. The basic contributions to this method of approach have been made primarily by Reissner and Stein (Ref. 2–19) and Fung (Ref. 2–20).

Although the present chapter is devoted primarily to analysis, it should be remarked that structural models can be employed as a means of obtaining influence coefficients in many instances where analysis is not deemed sufficiently reliable. As a result of the widely different structural arrangements which are currently being developed, structural models will undoubtedly be used more extensively in the future. Faithful model reproductions are expensive, however, and take considerable time to construct. Moreover, they tend to become obsolete due to design changes if they are constructed too early in the airplane program. These factors, as well as others which relate to structural models, are discussed at greater length in Chapters 12 and 13.

2–15 Influence coefficients of complex built-up wings by the principle of minimum strain energy. An airplane structure, no matter what shape it may assume, is fundamentally a group of elastic elements joined together to form a system that is often statically indeterminate. The structure can usually be idealized so that it consists of structural elements such as axially loaded flange elements, shear panels, or other types of elementary members whose strain energy can be computed in terms of internal stresses. This idealization is essentially a process of replacing a continuous system by an equivalent lumped parameter system. The success of the method is largely due to the ability of the analyst to replace the actual structure by a simple idealized structure which retains the essential features of the actual structure. For example, it is common practice in the analysis of slender monocoque structures with thin cover sheets to consider the axial-load and shear-carrying capacities of the cover sheet separately. Thus, the axial-load carrying contribution of the cover sheet is handled by lumping the various portions of the cross-sectional area of the cover sheets with that of the neighboring spar caps and other axial reinforcements. This practice is based on the assumption that the stress in the cover sheet in the direction perpendicular to the spar caps and axial reinforcements is negligible. For high-aspect ratio wings with relatively little taper or sweep and with bulkheads and ribs perpendicular to the beam axis, this implicit assumption is nearly satisfied. In a multispar delta wing or other wing of low aspect ratio, the selection of a direction of negligible stress in the cover sheet is no longer possible (Ref. 2–17). Thus, in a low aspect-ratio wing, it is desirable to consider the load-carrying ability of the cover sheet both in the direction parallel to the spars and in a transverse direction.

Suppose, for example, that we wish to compute the influence coefficients associated with m points on the structure illustrated in Fig. 2–32. We can

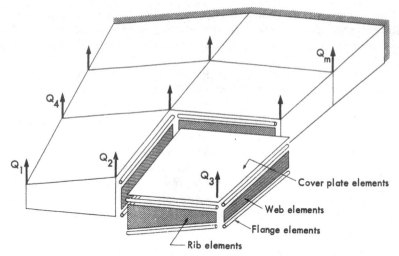

Fig. 2-32.   Wing structure partitioned into structural elements.

compute the deflection at the ith point due to a unit load at the jth point by
the principle of virtual work.   When the entire matrix of influence coefficients is desired, however, there is a somewhat more direct approach which
yields the matrix in one operation (Refs. 2-12 and 2-13).   Suppose that the
structure is partitioned into n separate structural elements (*cf.* Fig. 2-32).
If the loads in these elements can be represented by the n values $p_1, \cdots, p_n$,
the total strain energy in the structure is given by

$$U = \tfrac{1}{2}\lfloor p_1, \cdots, p_n \rfloor [F]\{p_1, \cdots, p_n\}, \tag{2-133}$$

where the $[F]$ matrix is a symmetrical matrix such that

$$[F] = [F]'. \tag{2-134}$$

Under the applied loads $Q_1, \cdots, Q_m$ on the structure, a stress analysis
yields the internal loads as a linear function of the applied loads, as follows:

$$\{p_1, \cdots, p_n\} = [G]\{Q_1, \cdots, Q_m\}. \tag{2-135}$$

Substituting Eq. (2-135) into (2-133), we obtain

$$U = \tfrac{1}{2}\lfloor Q_1, \cdots, Q_m \rfloor [G]'[F][G]\{Q_1, \cdots, Q_m\}. \tag{2-136}$$

Comparing Eq. (2-136) with Eq. (2-40), we see that the matrix of influence coefficients associated with the m points of load application is an
m by m matrix given by

$$[C] = [G]'[F][G]. \tag{2-137}$$

In most cases involving practical wings, the process described above is
complicated by the fact that the wing structure is statically indeterminate,

and the matrix $[G]$ cannot be computed from statics alone. Suppose that from the application of the principles of static equilibrium to each structural element we can construct only k independent equations of equilibrium as follows:

$$[E]\{p_1, \cdots, p_n, Q_1, \cdots, Q_m\} = 0. \tag{2-138}$$

The degree of redundancy of the structure is therefore

$$r = n - k. \tag{2-139}$$

If r redundant members are selected from among the n members, Eq. (2-138) can be solved for the n unknown internal loads in terms of the applied loads and the r redundant loads as follows:

$$\{p_1, \cdots, p_n\} = [G]\{Q_1, \cdots, Q_m, p_r\}, \tag{2-140}$$

where p_r indicates the loads in the redundant members. If the system is statically determinate and $p_r = 0$, Eq. (2-140) would, of course, constitute a complete solution of the internal stress distribution.

The redundant stresses can be computed by applying the principle of minimum strain energy. Substituting Eq. (2-140) into Eq. (2-133) gives

$$U = \tfrac{1}{2} \lfloor Q_1, \cdots, Q_m, p_r \rfloor [H] \{Q_1, \cdots, Q_m, p_r\}, \tag{2-141}$$

where

$$[H] = [G]'[F][G].$$

Applying the principle of minimum strain energy (Ref. 1-3) expressed by

$$\frac{\partial U}{\partial p_r} = 0, \tag{2-142}$$

there is obtained the column matrix of redundant loads

$$\{p_r\} = -[H_{pp}]^{-1}[H_{pQ}]\{Q_1, \cdots, Q_m\}. \tag{2-143}$$

The matrices $[H_{pp}]$ and $[H_{pQ}]$ are formed by partitioning:

$$[H] = \begin{bmatrix} H_{QQ} & \vdots & H_{Qp} \\ \cdots\cdots & \vdots & \cdots\cdots \\ H_{pQ} & \vdots & H_{pp} \end{bmatrix}. \tag{2-144}$$

From Eq. (2-143), we observe that

$$\{Q_1, \cdots, Q_m, p_r\} = \begin{bmatrix} [I] \\ \cdots\cdots\cdots\cdots \\ -[H_{pp}]^{-1}[H_{pQ}] \end{bmatrix} \{Q_1, \cdots, Q_m\}. \tag{2-145}$$

Introducing Eq. (2-145) into Eq. (2-141) and reducing, we obtain

$$U = \tfrac{1}{2} \lfloor Q_1, \cdots, Q_m \rfloor [[H_{QQ}] - [H_{Qp}][H_{pp}]^{-1}[H_{pQ}]]\{Q_1, \cdots, Q_m\}. \tag{2-146}$$

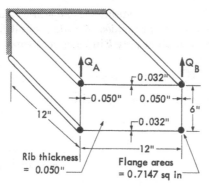

Fig. 2–33.   Simplified sheet-metal box.

Comparing Eq. (2–146) with Eq. (2–40), we see that the matrix of influence coefficients is expressible as

$$[C] = [[H_{QQ}] - [H_{Qp}][H_{pp}]^{-1}[H_{pQ}]].$$ (2–147)

Since the $[H]$ matrix can be easily formed and partitioned, Eq. (2–147) provides a convenient means of computing the matrix of flexibility influence coefficients of a statically indeterminate structure.   It should be observed, however, that evaluation of Eq. (2–147) requires inversion of the matrix $[H_{pp}]$.   Thus, if there are r redundant members, it is necessary to invert a matrix of r rows and columns.   It is apparent, therefore, that the computational labor required by the above method may become prohibitively great if the structure is highly redundant.

EXAMPLE 2–2.   To compute the matrix of flexibility influence coefficients associated with points A and B of the simple aluminum alloy sheet metal box shown by Fig. 2–33.   In carrying out the analysis, the following assumptions are introduced:

(a) Sheets carry only shear flows.
(b) Flanges carry only normal stresses.
(c) Normal stresses vary linearly along the flanges.

Solution.   The first step in the solution is to break the structure into its various component parts and draw a free-body diagram of each part, as shown by Fig. 2–34.   The $[G]$ matrix indicated by Eq. (2–140) is formed from the equations of equilibrium.   The independent equations of equilibrium are (*cf.* Fig. 2–34)

$$Q_A = p_1 + p_2, \qquad Q_B = -p_2 + p_3.$$ (a)

Since there are three unknown internal forces, p_1, p_2, and p_3, and only two equations of equilibrium, we see that one redundant quantity exists.   We shall select p_2 as the redundant quantity in this analysis.   The equilibrium equation can be used to form the following matrix equation:

$$\begin{bmatrix} p_1 \\ p_2 \\ p_3 \end{bmatrix} = \begin{bmatrix} 1 & 0 & -1 \\ 0 & 0 & 1 \\ 0 & 1 & 1 \end{bmatrix} \begin{bmatrix} Q_A \\ Q_B \\ p_2 \end{bmatrix}.$$ (b)

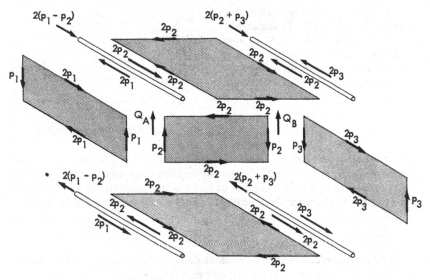

Fig. 2–34.   Simplified box exploded into structural elements.

The $[G]$ matrix is therefore

$$[G] = \begin{bmatrix} 1 & 0 & -1 \\ 0 & 0 & 1 \\ 0 & 1 & 1 \end{bmatrix}. \tag{c}$$

The $[F]$ matrix indicated in Eq. (2–133) is formed by computing the strain energy in terms of the internal forces p_1, p_2, and p_3. In order to do this, we need formulas for the strain energy in an axially loaded member in which the stress varies linearly along its length, and the strain energy in a sheet in pure shear.

Each axially loaded member in the structure is assumed to be in equilibrium, as shown in Fig. 2–35. The strain energy in an axially loaded bar can be expressed (Ref. 2–2) in the form

$$U = \int_0^l \frac{p^2(y)\,dy}{2AE}.$$

If the axial load varies linearly from u at one end to v at the other, as shown by Fig. 2–35, we have

$$p(y) = u + (v - u)\frac{y}{l}. \tag{d}$$

Inserting $p(y)$ in the integral given directly above and integrating, we obtain an expression for strain energy in terms of the applied end loads u and v:

$$U = \frac{l}{6AE}(u^2 + uv + v^2). \tag{e}$$

Fig. 2–35.   Axially loaded bar.

Each shear panel is in equilibrium, as shown by Fig. 2–36. In the case of the shear panels, it is shown in Ref. (2–2) that the strain energy in a thin rectangular sheet of width a, length b, and thickness t is given by

$$U = \frac{\tau^2 abt}{2G} \qquad \text{(f)}$$

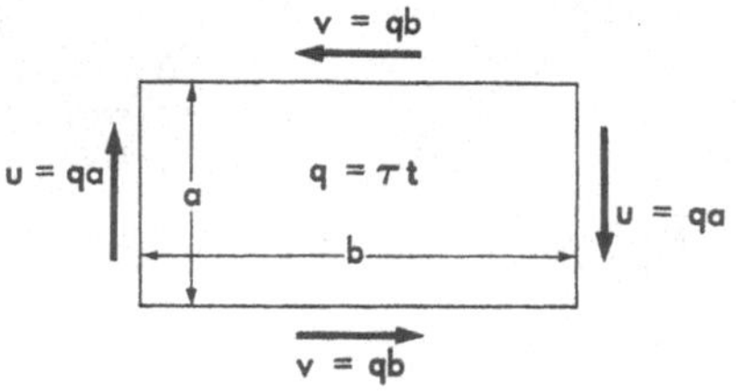

Fig. 2–36. Shear panel.

when the sheet is in a state of constant shear stress τ. If $q = \tau t$ is the constant shear flow in the sheet, then the externally applied edge shear loads required to keep the sheet in equilibrium are $u = qa$ and $v = qb$. Inserting this notation, we can write the strain energy in terms of the applied edge shear loads, as follows:

$$U = \frac{uv}{2Gt}. \qquad \text{(g)}$$

Applying the strain energy expressions derived above and given by Eqs. (e) and (g) to each element in Fig. 2–34, and summing the results, we obtain

$$U = \tfrac{1}{2}\lfloor p_1 p_2 p_3 \rfloor [F] \{ p_1 p_2 p_3 \}, \qquad \text{(h)}$$

where

$$[F] = \frac{1}{E} \begin{bmatrix} 144.8 & -44.8 & 0 \\ -44.8 & 815.2 & 44.8 \\ 0 & 44.8 & 144.8 \end{bmatrix}. \qquad \text{(i)}$$

The $[G]$ and $[F]$ matrices must, of course, be constructed by the engineer. The following operations leading to the influence coefficients are purely mechanical and can be carried out by trained computers.

The $[H]$ matrix is formed from Eq. (2–141) by introducing the matrices given by (c) and (i) above, as follows,

$$[H] = [G]'[F][G] = \frac{1}{E} \begin{bmatrix} 1 & 0 & 0 \\ 0 & 0 & 1 \\ -1 & 1 & 1 \end{bmatrix} \begin{bmatrix} 144.8 & -44.8 & 0 \\ -44.8 & 815.2 & 44.8 \\ 0 & 44.8 & 144.8 \end{bmatrix} \begin{bmatrix} 1 & 0 & -1 \\ 0 & 0 & 1 \\ 0 & 1 & 1 \end{bmatrix},$$

$$[H] = \frac{1}{E} \left[\begin{array}{cc|c} 144.8 & 0 & -189.6 \\ 0 & 144.8 & 189.6 \\ \hline -189.6 & 189.6 & 1284 \end{array} \right] = \left[\begin{array}{c|c} [H_{QQ}] & [H_{Qp}] \\ \hline [H_{pQ}] & [H_{pp}] \end{array} \right].$$

The final $[C]$ matrix is obtained from Eq. (2–147) by combining partitioned matrices from the $[H]$ matrix, as follows:

$$[C] = [H_{QQ}] - [H_{Qp}][H_{pp}]^{-1}[H_{pQ}],$$

$$[C] = \frac{1}{E} \begin{bmatrix} 144.8 & 0 \\ 0 & 144.8 \end{bmatrix} - \frac{1}{E} \begin{bmatrix} -189.6 \\ 189.6 \end{bmatrix} \times \frac{1}{1284} \times [-189.6 \quad 189.6],$$

$$[C] = \frac{1}{E} \begin{bmatrix} 119 & 25.8 \\ 25.8 & 119 \end{bmatrix} \text{in/lb.}$$

2–16 Influence coefficients of complex built-up wings by principle of minimum potential energy. The method of approach described here is sometimes referred to as the method of direct stiffness calculation (Ref. 2–18). Its principal feature is that the wing deformation is described in terms of discrete deformations of the individual elastic elements into which the wing has been exploded. Like the first method of approach, its usefulness depends upon the ability of the analyst to replace the actual structure by an idealized structure. The latter should be simple enough to avoid large volumes of computation yet elaborate enough to retain the essential elastic behavior characteristics of the actual structure.

Consider the problem of computing the influence coefficients of a two-spar wing with connecting torque tubes, as illustrated by Fig. 2–37. This type of construction is commonly applied to flutter models. The complete deformation configuration of the wing can be described by ten linear deflections and ten slopes of the spars, denoted respectively by $q_1, \cdots, q_{10}$ and $\alpha_1, \cdots, \alpha_{10}$ in Fig. 2–37. It will be assumed that the wing deforms by bending and twisting of the spars and twisting of the torque tubes. The torque tubes are assumed rigid in bending.

Equilibrium equations can be formed at each juncture of spar and torque tube by a direct summation of applied forces and moments, or they can be formed by application of the principle of minimum potential energy. The latter approach is somewhat more systematic and will be employed here.

The strain energy in each segment of spar and in each torque tube can be expressed in terms of the symbols denoting bending and twisting deformations.

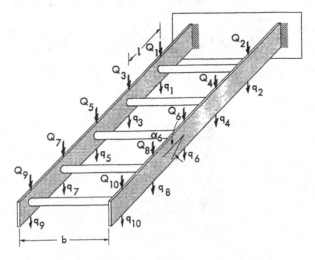

Fig. 2–37. Two-spar wing with connecting torque tubes.

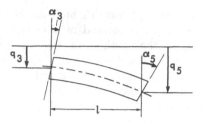

Fig. 2–38.   Bending deformation of spar segment.

(a) *Bending of spars.* Referring to Fig. 2–38, it can be seen by Eqs. (2–26) and (2–41) that the strain energy in the bending of a spar segment between, for example, stations 3 and 5 is expressible by

$$U = \tfrac{1}{2}\lfloor q_5 - q_3 - l\alpha_3,\ \alpha_5 - \alpha_3 \rfloor \begin{bmatrix} \dfrac{12EI}{l^3} & -\dfrac{6EI}{l^2} \\[2ex] -\dfrac{6EI}{l^2} & \dfrac{4EI}{l} \end{bmatrix} \{q_5 - q_3 - l\alpha_3,\ \alpha_5 - \alpha_3\}.$$

$$(2\text{--}148)$$

(b) *Twisting of spars.* The strain energy in torsion, according to the St. Venant torsion theory, can be expressed (Ref. 2–2) by

$$U = \int_0^l \frac{GJ}{2}\left(\frac{d\theta}{dy}\right)^2 dy. \tag{2-149}$$

Referring to Fig. 2–39, and assuming a uniform rate of twist, the strain energy in the twisting of spar segments 3–5 can be derived from Eq. (2–149) as follows:

$$U = \frac{GJ_s}{2l}(\Delta\theta)^2 = \frac{GJ_s}{2l}\left(\frac{q_5 - q_6}{b} - \frac{q_3 - q_4}{b}\right)^2, \tag{2-150}$$

where GJ_s is the torsional stiffness of the spar segment.

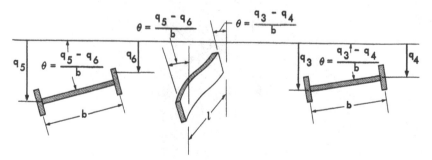

Fig. 2–39.   Twisting deformation of spar segment.

(c) *Twisting of torque tubes.* The strain energy due to the twisting of torque tube 3–4 is given, for example, by

$$U = \frac{GJ_t}{2l}(\Delta\alpha)^2 = \frac{GJ_t}{2b}(\alpha_3 - \alpha_4)^2, \qquad (2\text{--}151)$$

where GJ_t is the torsional stiffness of the torque tube.

(d) *Total strain energy.* When the expressions similar to those derived above are applied to each spar segment and torque tube, the results can be summed algebraically to give the total strain energy in terms of the twenty generalized coordinates:

$$U = \tfrac{1}{2}\lfloor q_1, \cdots, q_{10}, \alpha_1, \cdots, \alpha_{10} \rfloor [k]\{q_1, \cdots, q_{10}, \alpha_1, \cdots, \alpha_{10}\}, \qquad (2\text{--}152)$$

where $[k]$ is a matrix of stiffness influence coefficients.

(e) *Equations of equilibrium.* The equations of equilibrium are formed by substituting Eq. (2–152) into Eq. (2–97), to yield

$$[k]\{q_1, \cdots, q_{10}, \alpha_1, \cdots, \alpha_{10}\} = \{Q_1, \cdots, Q_{10}, 00000\ 00000\}. \qquad (2\text{--}153)$$

(f) *Influence coefficients.* Stiffness influence coefficients are of little direct practical value, and an inversion process must therefore be applied in order to derive the more useful flexibility influence coefficients. If the $[k]$ matrix is inverted, we obtain

$$\{q_1, \cdots, q_{10}, \alpha_1, \cdots, \alpha_{10}\} = [C]\{Q_1, \cdots, Q_{10}, 00000\ 00000\}, \qquad (2\text{--}154)$$

where $$[C] = [k]^{-1}.$$

The matrix $[C]$ can be partitioned as follows:

$$[C] = \begin{bmatrix} [C_{qq}] & [C_{q\alpha}] \\ \hline [C_{\alpha q}] & [C_{\alpha\alpha}] \end{bmatrix}, \qquad (2\text{--}155)$$

where $$[C_{\alpha q}] = [C_{q\alpha}].$$

The partitioned matrix $[C_{qq}]$ is the portion of the $[C]$ matrix of most practical interest. It can be computed from the partitioned $[k]$ matrix by the inversion formula

$$[C_{qq}] = [[k_{qq}] - [k_{q\alpha}][k_{\alpha\alpha}]^{-1}[k_{\alpha q}]]^{-1}. \qquad (2\text{--}156)$$

The elements of Eq. (2–156) are formed by partitioning the $[k]$ matrix as follows:

$$[k] = \begin{bmatrix} [k_{qq}]_{10\times10} & [k_{q\alpha}]_{10\times10} \\ \hline [k_{\alpha q}]_{10\times10} & [k_{\alpha\alpha}]_{10\times10} \end{bmatrix}. \qquad (2\text{--}157)$$

After the matrix $[k]$ has been constructed, the matrix operations indicated by Eq. (2–156) can be carried out as a routine computing operation. The most time-consuming steps are, of course, the inversions of the 10 by 10 matrices.

2–17 Calculation of deformations of solid wings of variable thickness and complex built-up wings by the Rayleigh-Ritz method. The classical Rayleigh-Ritz approach to problems in elasticity and, in particular, to plate problems has been applied extensively by engineers since its introduction in 1908 (Ref. 2–8), and it is a standard feature of textbooks on structures and elasticity (Refs. 2–2 and 2–8).

In applying the Rayleigh-Ritz method to the low aspect-ratio wing, two classes of displacement functions are often assumed. The first corresponds to the conventional Rayleigh-Ritz method, and we put

$$w(x,y) = \sum_{j=1}^{n} \gamma_j(x,y) C_j, \qquad (2\text{–}158)$$

where $\gamma_j(x,y)$ are assumed deflection shapes that satisfy the geometric boundary conditions on the cantilever plate.* These conditions are

$$\gamma_j(x,0) = \frac{\partial \gamma_j(x,0)}{\partial y} = 0. \qquad (2\text{–}159)$$

The constants C_j are obtained by solution of a set of simultaneous linear algebraic equations which result when the principle of minimum potential energy is applied in conjunction with the assumed solution given by Eq. (2–158).

A second and more useful class of displacement function corresponds to the generalized Rayleigh-Ritz method, in which we express the displacement by

$$w(x,y) = \sum_{j=1}^{n} \gamma_j(x) q_j(y), \qquad (2\text{–}160)$$

where the $\gamma_j(x)$ are assumed displacement functions in the x-direction and the $q_j(y)$ are functions of y to be computed. The latter are obtained by solution of a set of simultaneous linear differential equations which result when the principle of minimum potential energy is applied in conjunction with Eq. (2–160). Selection of the functions to be used depends upon the wing under consideration. The most widely exploited functions applied to problems of this kind are a power series in x such that Eq. (2–160) becomes

$$w(x,y) = q_0(y) + q_1(y)x + q_2(y)x^2 + \cdots + q_n(y)x^n, \qquad (2\text{–}161)$$

The class of solutions produced by Eq. (2–161) can be employed to calculate deflections to any desired degree of accuracy for a given low-aspect ratio wing, depending upon the number of terms taken. The case for

---

* Although the principle of minimum potential energy requires that only the geometric boundary conditions need be satisfied, a better approximation can be obtained by choosing functions that satisfy all the boundary conditions.

$n = 1$ assumes linear chordwise deflections, and the case for $n = 2$ takes into account parabolic chordwise curvature. The case of $n = 1$ is adequate for many practical problems, and will be employed in the illustration of the method which follows.

(a) *Solid wings of variable thickness.* The generalized Rayleigh-Ritz method has found wide application in the approximate solution of deflections of thin solid wings of variable thickness. These wings are often of very low aspect ratio and, as a result, analyses cannot be based upon beam theory, since the structural deformations more closely approach those of a plate. Let us consider a thin elastic isotropic solid wing of arbitrary planform and slowly varying thickness, as illustrated in Fig. 2–40. Suppose that the wing is loaded by an arbitrarily distributed side load $Z(x,y)$, and we wish to compute the resulting deformation $w(x,y)$. If our interest is restricted to small deformations normal to the plane of the plate, the strain energy in the plate (Ref. 2–8) can be shown to be

$$U = \tfrac{1}{2} \int_0^l \int_{c_1(y)}^{c_2(y)} D(x,y) \left\{ \left(\frac{\partial^2 w}{\partial x^2} + \frac{\partial^2 w}{\partial y^2} \right)^2 \right.$$

$$\left. + 2(1 - \nu) \left[\left(\frac{\partial^2 w}{\partial x \partial y} \right)^2 - \frac{\partial^2 w}{\partial x^2} \frac{\partial^2 w}{\partial y^2} \right] \right\} dx \, dy, \qquad (2\text{--}162)$$

where

$$D(x,y), \text{ the bending stiffness of the plate, } = \frac{Eh^3}{12(1 - \nu^2)}$$

and ν is Poisson's ratio.

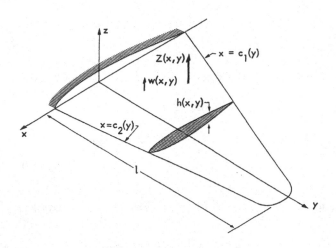

Fig. 2–40.   Thin solid wing.

The virtual work due to the distributed side load is given by

$$\delta W_e = \int_0^l \int_{c_1(y)}^{c_2(y)} Z(x,y)\delta w(x,y)dxdy. \qquad (2\text{-}163)$$

If Eqs. (2-162) and (2-163) are substituted into the expression for the principle of minimum potential energy given by Eq. (2-81), the exact differential equations of the plate are obtained. Solutions to these equations in closed form are possible only in very simple cases. Numerical solutions by means of finite difference and relaxation processes can, however, be employed in practical problems, as illustrated, for example, in Ref. 2-21.

We shall seek an approximate solution by taking the first two terms in Eq. (2-161) as an assumed deformation shape. To be consistent with previous wing notation, it will be convenient to express the first two terms as

$$w(x,y) = w(y) - x\theta(y), \qquad (2\text{-}164)$$

where the function $w(y)$ corresponds to the bending displacement of the wing along the y-axis, and the function $\theta(y)$ corresponds to the twist distribution along the y-axis. Introducing Eq. (2-164) into Eq. (2-162), we obtain the following expressions for strain energy in terms of the unknown functions $w(y)$ and $\theta(y)$:

$$U = \frac{1}{2}\int_0^l \int_{c_1(y)}^{c_2(y)} D(x,y)[(w''(y) - x\theta''(y))^2 + 2(1-\nu)(\theta'(y))^2]dxdy. \quad (2\text{-}165)$$

The virtual work expression given by Eq. (2-163), after substitution of Eq. (2-164), becomes

$$\delta W_e = \int_0^l [P_1(y)\delta w(y) + P_2(y)\delta\theta(y)]dy, \qquad (2\text{-}166)$$

where

$$P_1(y) = \int_{c_1(y)}^{c_2(y)} Z(x,y)dx, \qquad P_2(y) = -\int_{c_1(y)}^{c_2(y)} Z(x,y)xdx.$$

Introducing Eqs. (2-165) and (2-166) into Eq. (2-81):

$$\int_0^l \{a_1 w''\delta w'' + a_2 w''\delta\theta'' + a_2\theta''\delta w'' + a_3\theta''\delta\theta''$$
$$+ 2(1-\nu)a_1\theta'\delta\theta' - P_1\delta w - P_2\delta\theta\}dy = 0, \quad (2\text{-}167)$$

where

$$a_1(y) = \int_{c_1(y)}^{c_2(y)} D(x,y)dx, \qquad a_2(y) = -\int_{c_1(y)}^{c_2(y)} D(x,y)xdx,$$

$$a_3(y) = \int_{c_1(y)}^{c_2(y)} D(x,y)x^2dx.$$

Equation (2–167) can be integrated by parts to yield

$$\int_0^l \{[(a_1w'')'' + (a_2\theta'')'' - P_1]\delta w$$

$$+ [(a_3\theta'')'' + (a_2w'')'' - 2(1 - \nu)(a_1\theta')' - P_2]\delta\theta\}dy$$

$$- \{[(a_1w'')' + (a_2\theta'')']\delta w + [(a_3\theta'')' + (a_2w'')' - 2(1 - \nu)a_1\theta']\delta\theta$$

$$- [a_1w'' + a_2\theta'']\delta w' - [a_3\theta'' + a_2w'']\delta\theta'\}_0^l = 0. \qquad (2\text{–}168)$$

At $y = 0$, the geometric boundary conditions on w and θ are

$$w(0) = \theta(0) = w'(0) = \theta'(0) = 0. \qquad (2\text{–}169)$$

The variations in these quantities at the boundary are also zero:

$$\delta w(0) = \delta\theta(0) = \delta w'(0) = \delta\theta'(0) = 0. \qquad (2\text{–}170)$$

Everywhere in the plate except at the boundary $y = 0$, the conditions on w and θ are arbitrary. Therefore, $\delta w(y)$ and $\delta\theta(y)$ are independent and arbitrary quantities everywhere except at $y = 0$. In these circumstances, we may say that Eq. (2–168) and the geometric boundary conditions can be satisfied only if the following equations are satisfied:

$$(a_1w'')'' + (a_2\theta'')'' = P_1, \qquad (2\text{–}171)$$

$$(a_3\theta'')'' + (a_2w'')'' - 2(1 - \nu)(a_1\theta')' = P_2, \qquad (2\text{–}172)$$

$$[(a_1w'')' + (a_2\theta'')']_{y=l} = 0, \qquad (2\text{–}173)$$

$$[(a_3\theta'')' + (a_2w'')' - 2(1 - \nu)a_1\theta']_{y=l} = 0, \qquad (2\text{–}174)$$

$$[a_1w'' + a_2\theta'']_{y=l} = 0, \qquad (2\text{–}175)$$

$$[a_3\theta'' + a_2w'']_{y=l} = 0, \qquad (2\text{–}176)$$

$$w(0) = \theta(0) = w'(0) = \theta'(0) = 0. \qquad (2\text{–}177)$$

Equations (2–171) and (2–172) are fourth-order differential equations in $w(y)$ and $\theta(y)$, and Eqs. (2–173) through (2–177) give the eight boundary conditions necessary to obtain a complete solution to the differential equations. It is convenient to combine Eqs. (2–171) and (2–172) into a single differential equation in the dependent variable $\theta(y)$. Integrating Eq. (2–171) twice and applying boundary conditions (2–173) and (2–175), we obtain

$$w'' = -\frac{a_2}{a_1}\theta'' + \frac{1}{a_1}\int_y^l \int_y^l P_1 dy dy. \qquad (2\text{–}178)$$

Integrating Eq. (2–172) once and introducing boundary condition (2–174) gives

$$(a_3\theta'')' + (a_2w'')' - 2(1 - \nu)(a_1\theta')' = -\int_y^l P_2 dy. \qquad (2\text{–}179)$$

The dependent variable $w(y)$ can be eliminated from Eq. (2–179) by substituting Eq. (2–178). Thus,

$$-\left[a_3 - \frac{a_2^2}{a_1}\theta''\right]' + 2(1 - \nu)a_1\theta'$$
$$= \int_y^l P_2 dy + \left(\frac{a_2}{a_1}\int_y^l \int_y^l P_1 dy dy\right)'. \quad (2\text{–}180)$$

The form of Eq. (2–180) can be simplified somewhat by recognizing that

$$\int_y^l \int_y^l P_1 dy dy = M(y), \quad (2\text{–}181)$$

where $M(y)$ is the bending moment, and that

$$\int_y^l P_2 dy = T(y), \quad (2\text{–}182)$$

where $T(y)$ is the torsional moment. Thus, the differential equation which defines the twist distribution $\theta(y)$ is

$$-\left[\left(a_3 - \frac{a_2^2}{a_1}\right)\theta''\right]' + 2(1 - \nu)a_1\theta' = T(y) + \left(\frac{a_2}{a_1}M(y)\right)'. \quad (2\text{–}183)$$

Boundary conditions (2–173), (2–174), and (2–175) have already been employed in deriving Eq. (2–183). Therefore, the boundary conditions appropriate to a further solution of Eq. (2–183) can be obtained from Eqs. (2–176) and (2–177). Eliminating $w''(l)$ from Eq. (2–176) by means of Eq. (2–178), we find that

$$\left[\left(a_3 - \frac{a_2^2}{a_1}\right)\theta''\right]_{y=l} = 0. \quad (2\text{–}184)$$

The other two boundary conditions applicable to Eq. (2–183) come from Eq. (2–177):

$$\theta(0) = \theta'(0) = 0. \quad (2\text{–}185)$$

Thus, the twist distribution $\theta(y)$ is obtained by a solution to the third-order differential equation given by Eq. (2–183) subject to the boundary conditions (2–184) and (2–185). When an explicit result for the twist distribution $\theta(y)$ has been obtained, the spanwise deflection distribution can be computed by direct integration of Eq. (2–178). In order to integrate Eq. (2–178), the remaining boundary conditions, $w(0) = w'(0) = 0$, must be applied.

In general, the coefficients of Eq. (2–183) are complicated functions of y, and numerical solutions are often required in practical applications. A convenient method of obtaining numerical solutions to Eq. (2–183) consists of replacing it by its finite difference form and solving the resulting system of equations by matrix techniques.

EXAMPLE 2-3. To compute the spanwise twist distribution along a uniform
cantilever plate subjected to a concentrated torque at the free end, as illustrated by Fig. 2-41.

Solution. The plate stiffness D is
a constant given by

$$D = \frac{Eh^3}{12(1 - \nu^2)}, \qquad \text{(a)}$$

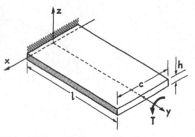

Fig. 2-41. Uniform cantilever plate.

where h is the uniform thickness of the plate. The quantities a_1, a_2, and a_3, which
appear in the equations and boundary conditions, are also constants, with the
following values:

$$a_1 = D \int_{-c/2}^{+c/2} dx \quad = Dc, \qquad \text{(b)}$$

$$a_2 = -D \int_{-c/2}^{+c/2} x\,dx = 0, \qquad \text{(c)}$$

$$a_3 = D \int_{-c/2}^{+c/2} x^2 dx \quad = \frac{Dc^3}{12}. \qquad \text{(d)}$$

Introducing these values for a_1, a_2, and a_3 into Eqs. (2-183), (2-184), and (2-
185), and putting $T(y) = T$ and $M(y) = 0$, we obtain the differential equation
and boundary conditions defining the twist $\theta(y)$:

$$\frac{Dc^3}{12} \theta'''(y) - 2(1 - \nu)Dc\theta'(y) = -T, \qquad \text{(e)}$$

$$\theta(0) = \theta'(0) = \theta''(l) = 0. \qquad \text{(f)}$$

A solution to the above differential equation satisfying the boundary conditions is

$$\theta(y) = \frac{Tl}{2(1 - \nu)Dc} \left[\frac{y}{l} - \frac{\sinh \dfrac{4\lambda y}{l}}{4\lambda} - \frac{\tanh 4\lambda}{4\lambda} \left(1 - \cosh \frac{4\lambda y}{l} \right) \right], \qquad \text{(g)}$$

where $\lambda = \dfrac{l}{c}\sqrt{\dfrac{3}{2}(1 - \nu)}$ is an aspect ratio parameter.

The corresponding twist distribution in terms of the elementary torsion theory
is obtained by taking only the first term in the result given above. Thus,

$$\theta(y)_{(\text{elementary})} = \frac{Ty}{2(1 - \nu)Dc}. \qquad \text{(h)}$$

It can be seen that the elementary result approaches the actual result for larger
values of the aspect-ratio parameter λ. This can be demonstrated in a simple way

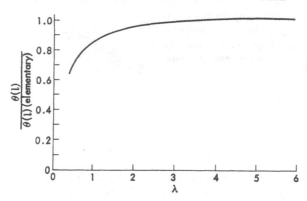

Fig. 2–42. Tip deflection of uniform cantilever plate versus λ.

by computing the ratio of the angles of twist at the tip, as follows:

$$\frac{\theta(l)}{\theta(l)_{\text{(elementary)}}} = 1 - \frac{\tanh 4\lambda}{4\lambda}. \tag{i}$$

A plot of this result is given in Fig. 2–42, where it can be seen that for values of the aspect-ratio parameter λ less than about 3, appreciable errors exist in the elementary result. These differences arise because the effects of warping restraint at the root are implicitly accounted for in the theory developed above and are neglected in the elementary theory.

(b) *Complex built-up wings.* The generalized Rayleigh-Ritz method can also be applied to complex built-up wings of low aspect ratio. Such applications have so far been confined to the first two terms of the series of Eq. (2–161). In principle, however, any number of terms can be taken. Application of the first two terms to the built-up wing problem has been referred to by Schuerch in Ref. 2–22 as a "wide beam theory."

CHAPTER 3

DEFORMATIONS OF AIRPLANE STRUCTURES UNDER DYNAMIC LOADS

3-1 Introduction. We have seen in Chapter 1 that many important problems in aeroelasticity involve external loads that vary with time. Dynamic loads on airplane structures not only produce translation and rotation of the airplane, but they tend also to excite vibrations of the structure. The process of predicting deformations of the structure is thus somewhat more complicated than in the case of static loads.

3-2 Differential equations of motion of a beam. We consider the differential equations of motion of a beam under the action of a dynamic transverse load of intensity $F_z(y,t)$, as shown in Fig. 3-1.

If the total lateral deflection $w(y,t)$ of the beam results from both bending strains and shearing strains, we may put

$$w(y,t) = \alpha(y,t) + \beta(y,t), \qquad (3-1)$$

where $\alpha(y,t)$ and $\beta(y,t)$ are the lateral deflections due to bending strains and shearing strains respectively.

Figure 3-2 illustrates a free-body diagram of a segment of the beam of length dy. In accordance with D'Alembert's principle, a lateral inertial load of intensity $m(y)\ddot{w}(y,t)$ and a rotary inertial moment $\mu(y)\ddot{\alpha}'(y,t)dy$ are applied to the beam segment in the directions shown by Fig. 3-2. The quantities $m(y)$ and $\mu(y)$ represent the mass per unit length and the mass moment of inertia per unit length, respectively. A dot over a quantity indicates a differentiation with respect to time, and a prime indicates

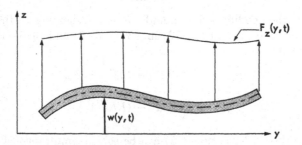

Fig. 3-1. Beam subjected to dynamic transverse load.

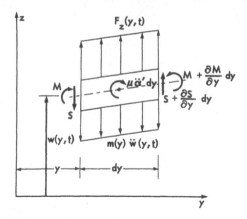

Fig. 3–2.   Free-body diagram of beam segment.

differentiation with respect to y.   Putting the sum of the vertical forces on the segment in Fig. 3–2 equal to zero yields

$$m\ddot{w} - \frac{\partial S}{\partial y} = F_z. \tag{3-2}$$

Similarly, when the sum of the moments on the segment about the center of gravity is put equal to zero, we obtain

$$\frac{\partial M}{\partial y} + S = \mu\ddot{\alpha}'. \tag{3-3}$$

The relations between bending moment and bending deflection and shear and shear deflection of a beam, respectively, are

$$\alpha'' = \frac{M}{EI}, \tag{2-110}$$

$$\beta' = \frac{S}{GK}, \tag{2-112}$$

where EI is the bending stiffness and GK is the shearing stiffness.   Substituting Eqs. (2–110) and (2–112) into Eqs. (3–2) and (3–3), the following differential equations can be derived:

$$m\ddot{w} + (EI\alpha'')'' - (\mu\ddot{\alpha}')' = F_z, \tag{3-4}$$

$$\beta' = \frac{1}{GK}\left[\mu\ddot{\alpha}' - (EI\alpha'')'\right]. \tag{3-5}$$

In principle, Eqs. (3–4) and (3–5) can be solved simultaneously together with Eq. (3–1) for arbitrary variations in elastic and inertial properties of the beam and for all practical boundary conditions.   In practice, closed solutions are difficult to obtain, except under the simplest assumptions. For example, the differential equation obtained by combining Eqs. (3–1), (3–4), and (3–5) for the simple case of free vibrations of a uniform beam has the complicated form

$$m\ddot{w} - \left(\frac{EIm}{GK} + \mu\right)\ddot{w}'' + \frac{\mu m}{GK}\ddddot{w} + EIw^{\text{IV}} = 0. \qquad (3\text{–}6)$$

(a) *Differential equation of free vibration of a slender beam.* Let us consider free vibrations of a slender beam in which the cross-sectional dimensions are small in comparison with the length, and for which rotary inertia effects and transverse shear deformations may be neglected. The partial differential equation of lateral vibration for the simple case is derivable from Eqs. (3–1), (3–4), and (3–5) by putting $G = \infty$ and $\mu = F_z(y,t) = 0$, which gives

$$(EIw'')'' + m\ddot{w} = 0. \qquad (3\text{–}7)$$

Equation (3–7) is a separable partial differential equation; that is, we can obtain solutions of the form

$$w(y,t) = W(y)T(t). \qquad (3\text{–}8)$$

Substitution of Eq. (3–8) into Eq. (3–7) yields

$$-\frac{\ddot{T}}{T} = \frac{(EIW'')''}{mW}. \qquad (3\text{–}9)$$

Since y and t are independent variables, both quotients in Eq. (3–9) are independent both of y and of t. They are equated to a separation constant, ω^2, and we get two independent ordinary differential equations, as follows:

$$\ddot{T} + \omega^2 T = 0, \qquad (3\text{–}10)$$

$$(EIW'')'' - m\omega^2 W = 0. \qquad (3\text{–}11)$$

Solutions to these equations provide the unknown functions $T(t)$ and $W(y)$, and the separation constant ω^2. Before solutions can be obtained, however, we must supply additional statements concerning the initial conditions and boundary conditions. Two initial conditions are required for a solution of Eq. (3–10), and they involve specification of the initial displacement and velocity of the beam when $t = 0$. They may have the general form

$$w(y,0) = f_1(y), \qquad \dot{w}(y,0) = f_2(y), \qquad (3\text{–}12)$$

where $f_1(y)$ and $f_2(y)$ are arbitrary functions of y. Four boundary conditions are required for solution of Eq. (3–11), and they involve specification of the support conditions on the two ends of the beam. The quantities ω^2 and $W(y)$ which satisfy Eq. (3–11) and the boundary conditions are called eigenvalues and eigenfunctions, respectively. There are, in fact, an infinite number of pairs of eigenvalues and eigenfunctions. These quantities have an important physical meaning. Every beam can vibrate in an infinite number of modes of vibration and each mode has a certain natural frequency. Each eigenfunction $W(y)$ represents the shape of a natural vibration mode, and the corresponding eigenvalue, ω^2, represents the square of the natural frequency of that mode. The exact nature of the quantities $W(y)$ and ω^2 obviously depends upon the specific boundary conditions on the beam as well as its stiffness and mass distribution.

The eigenfunctions or mode shapes have an interesting and useful property; they are orthogonal with respect to a weighting function $m(y)$, where $m(y)$ is the mass distribution of the beam. This property is demonstrated in the development which follows. Let us assume that an infinite set of pairs of eigenvalues and eigenfunctions are known which satisfy Eq. (3–11) and the boundary conditions. They are represented by

$$\omega_1 \qquad W_1(y)$$
$$\omega_2 \qquad W_2(y)$$
$$\vdots \qquad \vdots$$
$$\omega_\infty \qquad W_\infty(y)$$

If we substitute two different pairs of these quantities, distinguished by subscripts m and n, into Eq. (3–11), we obtain

$$(EIW_m'')'' - m\omega_m{}^2 W_m = 0, \tag{3–13}$$

$$(EIW_n'')'' - m\omega_n{}^2 W_n = 0. \tag{3–14}$$

Multiplying Eqs. (3–13) and (3–14) by W_n and W_m, respectively, and integrating from 0 to l yields

$$\int_0^l (EIW_m'')'' W_n \, dy = \omega_m{}^2 \int_0^l W_m W_n m \, dy, \tag{3–15}$$

$$\int_0^l (EIW_n'')'' W_m \, dy = \omega_n{}^2 \int_0^l W_m W_n m \, dy. \tag{3–16}$$

Subtracting Eq. (3–16) from (3–15) gives

$$(\omega_m{}^2 - \omega_n{}^2) \int_0^l W_m W_n m \, dy = \int_0^l [(EIW_m'')'' W_n - (EIW_n'')'' W_m] \, dy. \tag{3–17}$$

Integrating the right side by parts results in

$$(\omega_m{}^2 - \omega_n{}^2) \int_0^l W_m W_n m \, dy = \{[W_n(EIW_m'')' - W_m(EIW_n'')']$$
$$- EI(W_n' W_m'' - W_m' W_n'')\}_0^l. \tag{3–18}$$

The right side of Eq. (3–18) vanishes if at each end of the beam there is prescribed at least one of the following pairs of boundary conditions:

$$W = 0 \quad \text{and} \quad W' = 0,$$

$$W = 0 \quad \text{and} \quad EIW'' = 0,$$

$$W' = 0 \quad \text{and} \quad (EIW'')' = 0, \tag{3–19}$$

$$EIW'' = 0 \quad \text{and} \quad (EIW'')' = 0.$$

Assuming that at least one of the pairs of boundary conditions given by Eqs. (3–19) is applied at each end of the beam, Eq. (3–18) reduces to

$$\int_0^l W_m W_n m \, dy = 0, \quad \text{for} \quad m \neq n. \tag{3–20}$$

Equation (3–20) is known as the orthogonality condition for the natural mode shapes of the beam, and it is said that the functions $W_n(y)$ and $W_m(y)$ are orthogonal to each other with respect to a weighting function $m(y)$. It is important to observe that satisfaction of the orthogonality condition depends upon the existence of the boundary condition pairs prescribed by Eqs. (3–19).

Since Eq. (3–11) is homogeneous, if $W_n(y)$ is an eigenfunction, so is $\alpha W_n(y)$, where α is an arbitrary constant. Therefore, each mode shape function $W_n(y)$ can describe only relative displacements of the various parts of the beam, and cannot describe absolute values. The process of assigning absolute values to the function $W_n(y)$ is known as normalizing the mode, and it is one of multiplying it by a suitable constant. The selection of this constant is arbitrary. We may, for example, assign an arbitrary value, perhaps unity, to $W_n(a)$ at the particular point $y = a$. In this case, the normalized mode shape of the nth mode is given by

$$\phi_n(y) = A_n W_n(y), \tag{3–21}$$

where $A_n = 1/W_n(a)$. Another method of normalization which finds application in dynamic aeroelastic problems is defined by

$$\phi_n(y) = B_n W_n(y), \tag{3–22}$$

where $B_n^2 = M/\int_0^l W_n^2(y) m(y) \, dy$, and M is an arbitrarily assigned number, perhaps the mass of the beam. The orthogonality condition, Eq. (3–20), can, of course, be expressed also in terms of normalized mode shapes.

To illustrate an explicit solution of Eq. (3–7), let us consider the particular case of a uniform beam. The partial differential equation is

$$EIw^{IV} + m\ddot{w} = 0, \tag{3–23}$$

where EI and m are constants. For convenience, we can write Eq. (3–23) in the form

$$\ddot{w} + a^2 w^{IV} = 0, \tag{3–24}$$

where

$$a^2 = \frac{EI}{m}.$$

In this case, Eq. (3–9) becomes

$$-\frac{\ddot{T}}{T} = a^2 \frac{W^{IV}}{W}, \tag{3–25}$$

and Eqs. (3–10) and (3–11) simplify to

$$\ddot{T} + \omega^2 T = 0, \tag{3–26}$$

$$W^{IV} - \frac{\omega^2}{a^2} W = 0. \tag{3–27}$$

Solutions to Eqs. (3–26) and (3–27), respectively, are given by

$$T = A \sin \omega t + B \cos \omega t \tag{3–28}$$

and

$$W = C \sinh \sqrt{\frac{\omega}{a}}\, y + D \cosh \sqrt{\frac{\omega}{a}}\, y + E \sin \sqrt{\frac{\omega}{a}}\, y + F \cos \sqrt{\frac{\omega}{a}}\, y. \tag{3–29}$$

The complete solution to Eq. (3–23) is given by the product of Eqs. (3–28) and (3–29), as follows:

$$w = WT = \left[A \sin \omega t + B \cos \omega t \right]$$
$$\times \left[C \sinh \sqrt{\frac{\omega}{a}}\, y + D \cosh \sqrt{\frac{\omega}{a}}\, y + E \sin \sqrt{\frac{\omega}{a}}\, y + F \cos \sqrt{\frac{\omega}{a}}\, y \right]. \tag{3–30}$$

The quantities A, B, C, D, E, F, and ω in Eq. (3–30) must be obtained from the boundary conditions and initial conditions of the problem.

Restrained beams. Consider, for example, the boundary conditions for a simply supported uniform beam. They can be expressed in terms of end conditions on the function $W(y)$, as follows:

$$W(0) = 0, \qquad W(l) = 0, \qquad W''(0) = 0, \qquad W''(l) = 0. \tag{3–31}$$

Substituting Eq. (3–29) into the boundary conditions yields the following four equations:

$$D + F = 0, \tag{3–32}$$

$$C \sinh \sqrt{\frac{\omega}{a}}\, l + D \cosh \sqrt{\frac{\omega}{a}}\, l + E \sin \sqrt{\frac{\omega}{a}}\, l + F \cos \sqrt{\frac{\omega}{a}}\, l = 0, \tag{3–33}$$

$$D - F = 0, \tag{3-34}$$

$$C \sinh \sqrt{\frac{\omega}{a}} l + D \cosh \sqrt{\frac{\omega}{a}} l - E \sin \sqrt{\frac{\omega}{a}} l - F \cos \sqrt{\frac{\omega}{a}} l = 0. \tag{3-35}$$

Equations (3-32) and (3-34), taken together, show that $D = F = 0$. Adding and subtracting Eqs. (3-33) and (3-35) gives

$$C \sinh \sqrt{\frac{\omega}{a}} l = 0, \tag{3-36}$$

$$E \sin \sqrt{\frac{\omega}{a}} l = 0. \tag{3-37}$$

Since $\sinh \sqrt{\omega/a}\, l$ cannot be zero for a finite value of $\sqrt{\omega/a}\, l$, we conclude that $C = 0$. Moreover, $\sin \sqrt{\omega/a}\, l$ can be zero for a finite argument only when

$$\sqrt{\frac{\omega}{a}} l = n\pi, \qquad (n \text{ is an integer}). \tag{3-38}$$

Equation (3-38) provides a set of results for the frequencies:

$$\omega_n = \frac{n^2\pi^2}{l^2} \sqrt{\frac{EI}{m}}. \tag{3-39}$$

The corresponding mode shapes are obtained by putting $C = D = F = 0$, together with the result from Eq. (3-39), into Eq. (3-29). This yields the simple result that

$$W_n = E \sin \frac{n\pi y}{l}. \tag{3-40}$$

Since boundary conditions (3-31) are included as a special case of boundary conditions (3-19), the eigenfunctions given by Eq. (3-40) are orthogonal. The proof of orthogonality,

$$\int_0^l \sin \frac{n\pi y}{l} \sin \frac{m\pi y}{l} dy = \frac{l}{2} \delta_{mn}, * \tag{3-41}$$

is evident in this simple case. Thus the complete solution for the case of a uniform simply supported beam reduces to

$$w = [A \sin \omega_n t + B \cos \omega_n t] \left[E \sin \frac{n\pi y}{l} \right], \tag{3-42}$$

---

* The symbol δ_{mn} simply denotes a value 1 when $m = n$ and a value 0 when $m \neq n$.

and when the constants A, B, and E are combined, we obtain

$$w = [A' \sin \omega_n t + B' \cos \omega_n t] \sin \frac{n\pi y}{l}. \tag{3-43}$$

The lowest frequency corresponding to $n = 1$ is called the fundamental frequency, and its mode shape is called the fundamental mode shape. Figure 3–3 illustrates the vibration mode shapes and the corresponding frequencies of a simply supported beam.

Since Eq. (3–43) represents a solution for any integer n, we may say that the following sum also represents a solution:

$$w = \sum_{n=1}^{\infty} [A_n' \sin \omega_n t + B_n' \cos \omega_n t] \sin \frac{n\pi y}{l}, \tag{3-44}$$

where the constants A_n' and B_n' can be determined to satisfy a given set of initial conditions. For example, assume that at $t = 0$ the beam is subjected to the initial conditions stated by Eqs. (3–12). Putting Eq. (3–44) into Eqs. (3–12) gives

$$f_1(y) = \sum_{n=1}^{\infty} B_n' \sin \frac{n\pi y}{l}, \tag{3-45}$$

$$f_2(y) = \sum_{n=1}^{\infty} A_n' \omega_n \sin \frac{n\pi y}{l}. \tag{3-46}$$

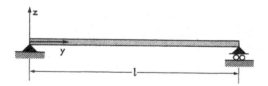

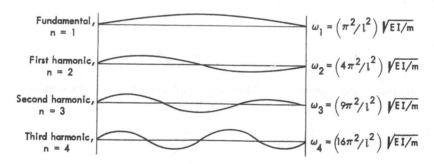

Fig. 3–3.   Mode shapes and frequencies of simply supported uniform beam.

The constants A_n' and B_n' can be found in the usual way common to Fourier series analysis (see, for example, Ref. 3–1) by multiplying both sides by $\sin(m\pi y/l)$ and integrating from zero to l:

$$\int_0^l f_1(y) \sin\frac{m\pi y}{l}\,dy = \sum_{n=1}^{\infty} B_n' \int_0^l \sin\frac{n\pi y}{l}\sin\frac{m\pi y}{l}\,dy, \qquad (3\text{–}47)$$

$$\int_0^l f_2(y) \sin\frac{m\pi y}{l}\,dy = \sum_{n=1}^{\infty} A_n'\omega_n \int_0^l \sin\frac{n\pi y}{l}\sin\frac{m\pi y}{l}\,dy. \qquad (3\text{–}48)$$

Applying the condition stated by Eq. (3–41), we obtain

$$A_n' = \frac{2}{\omega_n l}\int_0^l f_2(y)\sin\frac{n\pi y}{l}\,dy, \qquad (3\text{–}49)$$

$$B_n' = \frac{2}{l}\int_0^l f_1(y)\sin\frac{n\pi y}{l}\,dy. \qquad (3\text{–}50)$$

We see that the problem of free vibration of a simply supported beam is now completely defined in terms of the initial conditions and physical constants of the beam. The solution illustrated above for the simply supported beam typifies free vibration beam problems, and variations in the solution are the result of other boundary conditions and initial conditions.

EXAMPLE 3–1. To calculate the natural mode shapes and frequencies of a uniform cantilever beam of length l.

Solution. The boundary conditions on the natural mode shapes are

$$W(0) = 0, \qquad (a)$$

$$W'(0) = 0, \qquad (b)$$

$$W''(l) = 0, \qquad (c)$$

$$W'''(l) = 0. \qquad (d)$$

Substituting Eq. (3–29) into Eqs. (a) and (b) gives

$$D + F = 0, \qquad (e)$$

$$C + E = 0. \qquad (f)$$

Substituting Eq. (3–29) into Eqs. (c) and (d) results in

$$C\cosh\sqrt{\frac{\omega}{a}}l + D\sinh\sqrt{\frac{\omega}{a}}l - E\cos\sqrt{\frac{\omega}{a}}l + F\sin\sqrt{\frac{\omega}{a}}l = 0, \qquad (g)$$

$$C\sinh\sqrt{\frac{\omega}{a}}l + D\cosh\sqrt{\frac{\omega}{a}}l - E\sin\sqrt{\frac{\omega}{a}}l - F\cos\sqrt{\frac{\omega}{a}}l = 0. \qquad (h)$$

These four equations can be combined to yield two equations:

$$\left(\cosh\sqrt{\frac{\omega}{a}}l + \cos\sqrt{\frac{\omega}{a}}l\right)C + \left(\sinh\sqrt{\frac{\omega}{a}}l - \sin\sqrt{\frac{\omega}{a}}l\right)D = 0, \qquad (i)$$

$$\left(\sinh \sqrt{\frac{\omega}{a}}\, l + \sin \sqrt{\frac{\omega}{a}}\, l\right) C + \left(\cosh \sqrt{\frac{\omega}{a}}\, l + \cos \sqrt{\frac{\omega}{a}}\, l\right) D = 0. \qquad (j)$$

A nontrivial solution to this set of homogeneous equations in C and D is obtained by setting the determinant of their coefficients equal to zero. Expansion of the determinant and subsequent reduction gives a transcendental equation in the frequency ω:

$$\cos \sqrt{\frac{\omega}{a}}\, l = -\frac{1}{\cosh \sqrt{\frac{\omega}{a}}\, l}. \qquad (k)$$

Equation (k) is solved graphically for values of frequency ω which satisfy it in Fig. 3–4. These values are given by the abcissas of the points of intersection of the curves in Fig. 3–4, as follows:

$$\sqrt{\frac{\omega}{a}}\, l = 0.597\pi,\ 1.49\pi,\ \tfrac{5}{2}\pi,\ \tfrac{7}{2}\pi,\ \tfrac{9}{2}\pi \cdots \qquad (l)$$

Thus the frequencies of the various modes of vibration are

$$\omega_1 = (0.597)^2 \frac{\pi^2}{l^2} \sqrt{\frac{EI}{m}},$$

$$\omega_2 = (1.49)^2 \frac{\pi^2}{l^2} \sqrt{\frac{EI}{m}},$$

$$\cdot \qquad \cdot$$
$$\cdot \qquad \cdot \qquad\qquad\qquad (m)$$
$$\cdot \qquad \cdot$$

$$\omega_n = (n - \tfrac{1}{2})^2 \frac{\pi^2}{l^2} \sqrt{\frac{EI}{m}}, \qquad (n \text{ sufficiently large}).$$

$$\cdot \qquad \cdot$$
$$\cdot \qquad \cdot$$
$$\cdot \qquad \cdot$$

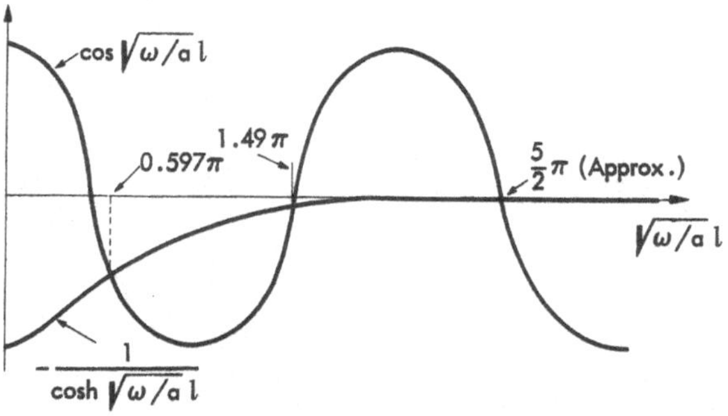

Fig. 3–4.   Graphical solution of transcendental equation of uniform cantilever beam.

Substituting each natural frequency in turn into Eq. (3–29) and using Eqs. (e) through (h) to express $\dot{C}$, E, and F in terms of D, we obtain the following expression for the natural mode shapes:

$$W_n(y) = D\left[\left(\frac{\sin\sqrt{\frac{\omega_n}{a}}\,l - \sinh\sqrt{\frac{\omega_n}{a}}\,l}{\cosh\sqrt{\frac{\omega_n}{a}}\,l + \cos\sqrt{\frac{\omega_n}{a}}\,l}\right)\left(\sinh\sqrt{\frac{\omega_n}{a}}\,y - \sin\sqrt{\frac{\omega_n}{a}}\,y\right)\right.$$

$$\left.+\left(\cosh\sqrt{\frac{\omega_n}{a}}\,y - \cos\sqrt{\frac{\omega_n}{a}}\,y\right)\right]. \tag{n}$$

The mode shape $W_n(y)$ corresponding to a particular natural frequency ω_n is obtained by substituting that natural frequency into Eq. (n). Since the boundary conditions are included as a special case of boundary conditions (3–19), the mode shapes given by Eq. (n) are orthogonal. Figure 3–5 illustrates a plot of the natural mode shapes of a uniform cantilever beam.

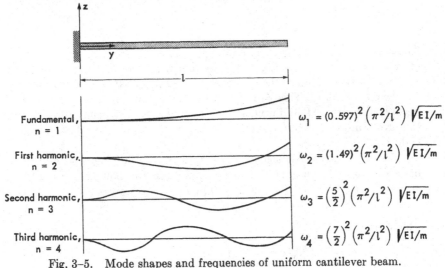

Fig. 3–5.   Mode shapes and frequencies of uniform cantilever beam.

Unrestrained beams. The principles illustrated above for restrained beams apply also to unrestrained beams. Let us consider the problem of computing the natural mode shapes and frequencies of a uniform beam with free ends. The boundary conditions applicable to this case are

$$W''(0) = 0, \tag{3–51a}$$

$$W'''(0) = 0, \tag{3-51b}$$

$$W''(l) = 0, \tag{3-51c}$$

$$W'''(l) = 0. \tag{3-51d}$$

Substituting Eq. (3-29) into boundary conditions (3-51a) and (3-51b), we obtain

$$C - E = 0, \qquad D - F = 0. \tag{3-52}$$

Substituting Eq. (3-29) into boundary conditions (3-51c) and (3-51d) and applying Eqs. (3-52) yields

$$\left(\sinh\sqrt{\frac{\omega}{a}}\,l - \sin\sqrt{\frac{\omega}{a}}\,l\right)C + \left(\cosh\sqrt{\frac{\omega}{a}}\,l - \cos\sqrt{\frac{\omega}{a}}\,l\right)D = 0,$$

$$\left(\cosh\sqrt{\frac{\omega}{a}}\,l - \cos\sqrt{\frac{\omega}{a}}\,l\right)C + \left(\sinh\sqrt{\frac{\omega}{a}}\,l + \sin\sqrt{\frac{\omega}{a}}\,l\right)D = 0. \tag{3-53}$$

Putting the determinant of the coefficients of C and D equal to zero gives, after simplification, the following transcendental equation in ω:

$$\cos\sqrt{\frac{\omega}{a}}\,l = \frac{1}{\cosh\sqrt{\frac{\omega}{a}}\,l}. \tag{3-54}$$

The roots of this equation are obtained graphically from the intersections of the curves plotted in Fig. 3-6. In this case, one of the intersections occurs at a zero value of the argument, indicating that one of the natural frequencies is zero. This situation arises as a result of the fact that the beam is unrestrained, and we may say that this zero frequency corresponds to a gross motion of the unrestrained beam, which we shall term a rigid-body mode of motion.

The rigid-body mode shape can be obtained by putting $\omega = 0$ into the basic differential equation (3-27) and applying the boundary conditions given by Eqs. (3-51). Putting $\omega = 0$ in Eq. (3-27) gives

$$W^{\mathrm{IV}} = 0, \tag{3-55}$$

which integrates to

$$W(y) = C_1 y^3 + C_2 y^2 + C_3 y + C_4. \tag{3-56}$$

Substituting boundary conditions (3-51), we obtain as the rigid-body mode shape

$$W_0(y) = C_3 y + C_4, \tag{3-57}$$

where C_3 and C_4 are arbitrary constants. Thus the rigid-body mode shape of an unrestrained beam can be described by a rigid translation and a rigid rotation.

The deformation mode shapes are obtained by applying Eqs. (3–52) and (3–53) to determine C, E, and F in terms of D and substituting the result in Eq. (3–29):

$$W_n(y) = D\left[\left(\frac{\cos\sqrt{\frac{\omega_n}{a}}\,l - \cosh\sqrt{\frac{\omega_n}{a}}\,l}{\sinh\sqrt{\frac{\omega_n}{a}}\,l - \sin\sqrt{\frac{\omega_n}{a}}\,l}\right)\left(\sinh\sqrt{\frac{\omega_n}{a}}\,y + \sin\sqrt{\frac{\omega_n}{a}}\,y\right)\right.$$

$$\left. + \left(\cosh\sqrt{\frac{\omega_n}{a}}\,y + \cos\sqrt{\frac{\omega_n}{a}}\,y\right)\right]. \qquad (3\text{–}58)$$

Deformation mode shapes $W_1(y)$, $W_2(y)$, $\cdots$, $W_n(y)$ can be computed by Eq. (3–58) by substituting corresponding frequency values ω_1, ω_2, $\cdots$, ω_n obtained from the points of intersection in Fig. 3–6. The various mode shapes of the unrestrained beam and the corresponding frequencies are illustrated in Fig. 3–7. It can be seen that the frequencies of an unrestrained beam approach those of a cantilever beam for high values of n.

It should be observed that for a freely vibrating, unrestrained beam in the absence of external forces, the summation of the inertial forces in the z-direction and the summation of the inertial moments about the x-axis should be zero:

$$\int_0^l \ddot{w}(y,t)dm = \ddot{T}(t)\int_0^l W_n(y)m\,dy = 0, \qquad (3\text{–}59)$$

$$\int_0^l \ddot{w}(y,t)y\,dm = \ddot{T}(t)\int_0^l W_n(y)my\,dy. \qquad (3\text{–}60)$$

Introducing Eq. (3–58) into Eqs. (3–59) and (3–60), integrating, and applying Eq. (3–54), we see that these conditions are indeed satisfied.

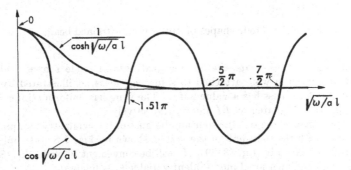

Fig. 3–6. Graphical solution of transcendental equation of uniform unrestrained beam.

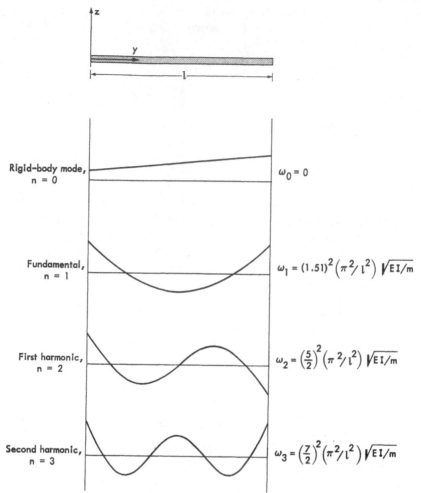

Fig. 3-7.   Mode shapes of uniform unrestrained beam.

EXAMPLE 3-2.   To calculate the symmetrical natural mode shapes and frequencies of a uniform wing attached to a fuselage mass, as illustrated by Fig. 3-8. The fuselage mass has a value $2M_F$. The wing has uniform stiffness and mass properties designated by EI and m, respectively.

Solution. In carrying out the solution, because of the symmetry assumption, we can work with the right half of the system shown by Fig. 3-8 and apply the general solution given by Eq. (3-29). It will be convenient to write Eq. (3-29) in terms of a nondimensional independent variable ξ, as follows:

$$W(\xi) = C \sinh kl\xi + D \cosh kl\xi + E \sin kl\xi + F \cos kl\xi, \tag{a}$$

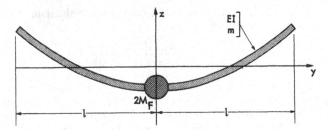

Fig. 3-8.   Uniform wing attached to fuselage mass.

where
$$k^4 = \frac{m\omega^2}{EI}, \qquad \xi = \frac{y}{l}.$$

Referring to Fig. 3-8, it can be seen that the four boundary conditions on the wing, when symmetrical motion is assumed, are

$$W'(0) = 0, \tag{b}$$

$$\frac{1}{l^3}\, W'''(0) = \frac{M_F}{m}\, k^4 W(0), \tag{c}$$

$$W''(1) = 0, \tag{d}$$

$$W'''(1) = 0. \tag{e}$$

Applying boundary condition (b) to the general solution (a), we obtain
$$C + E = 0,$$
so that
$$W(\xi) = F \cos kl\xi + E(\sin kl\xi - \sinh kl\xi) + D \cosh kl\xi. \tag{f}$$

Applying boundary conditions (c), (d), and (e) in turn to the above result, we obtain

$$F \cos kl + E(\sin kl + \sinh kl) - D \cosh kl = 0, \tag{g}$$

$$F\,\frac{M_F}{m}\,k + 2E + D\,\frac{M_F}{m}\,k = 0, \tag{h}$$

$$F \sin kl - E(\cos kl + \cosh kl) + D \sinh kl = 0. \tag{i}$$

A nontrivial solution to this set of homogeneous equations in the unknowns F, E, and D may be obtained by setting the determinant of the coefficients equal to zero.   This yields the determinantal equation

$$\begin{vmatrix} \cos \Omega & (\sin \Omega + \sinh \Omega) & -\cosh \Omega \\ R\Omega & 2 & R\Omega \\ \sin \Omega & -(\cos \Omega + \cosh \Omega) & \sinh \Omega \end{vmatrix} = 0,$$

where R and Ω are nondimensional parameters defined by

$$R = \frac{M_F}{ml}, \qquad \Omega = kl.$$

Upon expansion of the determinant and consequent reduction, the following frequency equation is obtained:

$$R = -\frac{\sinh \Omega \cos \Omega + \sin \Omega \cosh \Omega}{\Omega(1 + \cos \Omega \cosh \Omega)}.$$

If this transcendental equation is solved graphically, it is found to be satisfied by a zero value of the characteristic value Ω plus an infinite set of finite values of Ω. Each characteristic value determines a natural frequency ω_n, and a natural mode of vibration $W_n(\xi)$. Substituting each finite characteristic value in turn into Eq. (f), and using Eqs. (g), (h), and (i) to express E and D in terms of F, which is set equal to 1, we obtain the characteristic functions describing the natural deformation mode shapes in the form

$$W_n(\xi) = \cosh \Omega_n\xi + C_1{}^{(n)} \cos \Omega_n\xi + C_2{}^{(n)}(\sin \Omega_n\xi - \sinh \Omega_n\xi), \qquad (j)$$

where

$$C_1{}^{(n)} = -\frac{(2/R\Omega_n) \cosh \Omega_n + (\sin \Omega_n + \sinh \Omega_n)}{(\sin \Omega_n + \sinh \Omega_n)|+(-2/R\Omega_n) \cos \Omega_n},$$

$$C_2{}^{(n)} = \frac{\cosh \Omega_n + \cos \Omega_n}{(\sin \Omega_n + \sinh \Omega_n)+(-2/R\Omega_n) \cos \Omega_n}.$$

The form of Eq. (j) leads to small differences between large numbers. To avoid this difficulty, the following rearrangement is made:

$$W_n(\xi) = e^{-\Omega_n\xi} + C_1{}^{(n)} \cos \Omega_n\xi + C_2{}^{(n)} \sin \Omega_n\xi + (1 - C_2{}^{(n)}) \sinh \Omega_n\xi. \quad (k)$$

The quantity $(1 - C_2{}^{(n)})$ may be written as

$$1 - C_2{}^{(n)} = \frac{\sin \Omega_n - [(2/R\Omega_n) + 1] \cos \Omega_n - e^{-\Omega_n}}{\sin \Omega_n - (2/R\Omega_n) \cos \Omega_n + \sinh \Omega_n}.$$

Making the following substitutions:

$$K_1{}^{(n)} = C_1{}^{(n)}, \qquad K_2{}^{(n)} = 1 - C_2{}^{(n)},$$

Eq. (k) becomes

$$W_n(\xi) = e^{-\Omega_n\xi} + K_1{}^{(n)} \cos \Omega_n\xi + K_2{}^{(n)} \sinh \Omega_n\xi + (1 - K_2{}^{(n)}) \sin \Omega_n\xi. \quad (l)$$

We have mentioned that the process of assigning absolute values to the function $W_n(\xi)$ is known as normalization. Let us suppose that in the present example we determine a set of normalizing factors B_n, as defined by Eq. (3–22), in such a way that

$$B_n{}^2 = \frac{M}{\displaystyle\int_0^l W_n{}^2(y)m(y)dy} = \frac{ml(R + 1)}{ml\displaystyle\int_0^1 W_n{}^2(\xi)d\xi}, \qquad (m)$$

where M is half the mass of the total system shown in Fig. 3–8. Since the fuselage mass must be included in the integration, the factor B_n reduces to

$$B_n = \left\{ \frac{R + 1}{RW_n{}^2(0) + \displaystyle\int_0^1 W_n{}^2(\xi)d\xi} \right\}^{\frac{1}{2}}, \qquad (n)$$

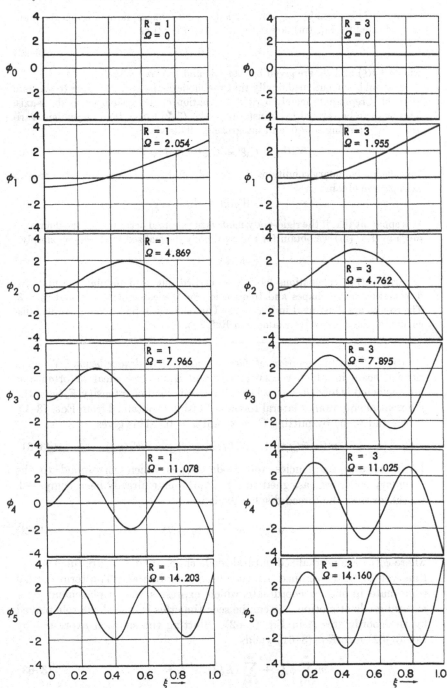

Fig. 3–9. Mode shapes and frequency parameters of uniform wing attached to fuselage mass.

where $W_n(\xi)$ is given by Eq. (l). Thus the mode shapes, normalized in the manner indicated by Eq. (m), are defined by Eq. (3–22) as

$$\phi_n(\xi) = B_n W_n(\xi), \tag{3-22}$$

where $W_n(\xi)$ and B_n are given by Eqs. (l) and (n) respectively.

It should be mentioned finally that the mode shape corresponding to the zero value of Ω represents merely a rigid translation of the system along the z-axis. This fact can be verified by transforming Eq. (3–27) from the independent variable y to ξ, putting $\omega = 0$, and integrating. This yields

$$W_0(\xi) = C_1\xi^3 + C_2\xi^2 + C_3\xi + C_4. \tag{o}$$

Introducing boundary conditions (b), (c), (d), and (e), appropriate to the present example, we obtain

$$W_0(\xi) = C_4. \tag{p}$$

It is apparent that if the rigid-body mode is normalized according to the definition given by Eq. (m) we obtain for the rigid-body normalized mode shape, simply,

$$\phi_0(y) = 1. \tag{q}$$

Arithmetical calculations have been carried out to determine the first five deformation mode shapes and frequencies for the cases of $R = 1$ and $R = 3$. The results are illustrated in Fig. 3–9. The reader will find a more detailed discussion of this illustrative example in Ref. 3–5.

(b) *Differential equation of forced motion of a slender beam.* When a slender beam in which rotary inertia and transverse shear deformations are neglected is loaded with a lateral dynamic load of intensity $F_z(y,t)$, the differential equation of lateral forced vibration is obtained from Eqs. (3–1), (3–4), and (3–5) by putting $G = \infty$ and $\mu = 0$. This gives

$$m(y)\ddot{w}(y,t) + [EI(y)w''(y,t)]'' = F_z(y,t). \tag{3-61}$$

If the natural frequencies and mode shapes which correspond to the boundary conditions assigned to Eq. (3–61) have already been computed, we can express the forced displacement by

$$w(y,t) = \sum_{i=1}^{\infty} \phi_i(y)\xi_i(t), \tag{3-62}$$

where $\phi_i(y)$ are normalized natural mode shapes and $\xi_i(t)$ are functions of time yet to be determined, called normal coordinates. The term normal coordinate implies a coordinate which expresses the displacement of a natural mode of motion. It can be seen that this is indeed the role played by the coordinates ξ_i in Eq. (3–62). Putting the solution expressed by Eq. (3–62) into Eq. (3–61) yields

$$m \sum_{i=1}^{\infty} \phi_i\ddot{\xi}_i + \sum_{i=1}^{\infty} (EI\phi_i'')''\xi_i = F_z(y,t). \tag{3-63}$$

The properties of the eigenfunctions can be applied to advantage if we multiply Eq. (3–63) through by $\phi_j(y)$ and integrate over the length:

$$\sum_{i=1}^{\infty} \ddot{\xi}_i \int_0^l \phi_i \phi_j m \, dy + \sum_{i=1}^{\infty} \xi_i \int_0^l (EI\phi_i'')'' \phi_j \, dy = \int_0^l F_z(y,t) \phi_j \, dy. \quad (3\text{–}64)$$

Applying Eq. (3–15) to the second term yields

$$\sum_{i=1}^{\infty} \ddot{\xi}_i \int_0^l \phi_i \phi_j m \, dy + \sum_{i=1}^{\infty} \xi_i \omega_i^2 \int_0^l \phi_i \phi_j m \, dy = \int_0^l F_z(y,t) \phi_j \, dy. \quad (3\text{–}65)$$

Using the orthogonality condition, (Eq. 3–20), and putting

$$\int_0^l \phi_i \phi_j m \, dy = M_j \delta_{ij}, \quad (3\text{–}66)$$

we obtain from Eq. (3–65)

$$M_j \ddot{\xi}_j + M_j \omega_j^2 \xi_j = \Xi_j, \quad (j = 1, \cdots, \infty), \quad (3\text{–}67)$$

where

$$M_j = \int_0^l \phi_j^2 m \, dy \text{ is called the generalized mass of the } j\text{th mode,}$$

and

$$\Xi_j = \int_0^l F_z(y,t) \phi_j \, dy \text{ is called the generalized force of the } j\text{th mode.}$$

Equation (3–67) defines the response of the jth mode of motion to a disturbing force $F_z(y,t)$. If $F_z(y,t)$ is independent of the motion of the beam, we see that the differential equations defining the response of the modes are uncoupled, and can be solved separately. If some component of $F_z(y,t)$ depends upon the motion of the beam (such as a damping force), the differential equations are coupled. The final solution is obtained by putting the solutions to Eqs. (3–67) into Eq. (3–62).

To illustrate the principles outlined above, consider a simply supported uniform beam under the influence of a lateral dynamic load given by

$$F_z(y,t) = F_z(y) \sin \Omega t. \quad (3\text{–}68)$$

The solution is given by Eq. (3–62), where $\phi_i(y) = \sin (i\pi y/l)$ and where the $\xi_i(t)$ are given by solutions to Eq. (3–67). The generalized mass is

$$M_j = m \int_0^l \sin^2 \frac{j\pi y}{l} \, dy = \frac{ml}{2}, \quad (3\text{–}69)$$

and the generalized force is

$$\Xi_j = \sin \Omega t \int_0^l F_z(y) \sin \frac{j\pi y}{l} \, dy. \quad (3\text{–}70)$$

Putting Eqs. (3–69) and (3–70) into Eq. (3–67), we obtain

$$\ddot{\xi}_j + \omega_j^2 \xi_j = \left(\frac{2}{ml} \int_0^l F_z(y) \sin \frac{j\pi y}{l} \, dy \right) \sin \Omega t, \quad (j = 1, 2, \cdots, \infty). \quad (3\text{-}71)$$

If it is assumed that the initial beam displacement is zero, $w(y,0) = 0$, then the normal coordinates must also be zero at time zero, i.e., $\xi_j(0) = 0$. This can be verified by introducing $t = 0$ into Eq. (3–62) and applying the initial conditions:

$$w(y,0) = 0 = \sum_{i=1}^{\infty} \phi_i(y)\xi_i(0). \quad (3\text{-}72)$$

The quantities $\xi_i(0)$ can be computed by multiplying through Eq. (3–72) by $m\phi_j(y)$ and integrating from 0 to l, as follows:

$$0 = \sum_{i=1}^{\infty} \xi_i(0) \int_0^l \phi_i(y)\phi_j(y)m(y)dy. \quad (3\text{-}73)$$

Applying the orthogonality condition, Eq. (3–20), to Eq. (3–73), we obtain

$$\xi_j(0) = 0. \quad (3\text{-}74)$$

It follows also by similar reasoning that if $\dot{w}(y,0) = 0$, then

$$\dot{\xi}_j = 0. \quad (3\text{-}75)$$

Solving Eq. (3–71) subject to the initial conditions $\xi_j(0) = \dot{\xi}_j(0) = 0$, results in

$$\xi_j = \frac{2}{ml} \left(\int_0^l F_z(y) \sin \frac{j\pi y}{l} \, dy \right) \frac{\sin \Omega t - (\Omega/\omega_j) \sin \omega_j t}{\omega_j^2 [1 - (\Omega^2/\omega_j^2)]}. \quad (3\text{-}76)$$

The response of the beam is obtained by substituting Eq. (3–76) together with $\phi_i(y) = \sin(i\pi y/l)$ into Eq. (3–62):

$$w(y,t) = \frac{2}{ml} \sum_{i=1}^{\infty} \sin \frac{i\pi y}{l} \left(\int_0^l F_z(y) \sin \frac{i\pi y}{l} \, dy \right) \frac{\sin \Omega t - (\Omega/\omega_i) \sin \omega_i t}{\omega_i^2 [1 - (\Omega^2/\omega_i^2)]}. \quad (3\text{-}77)$$

For the case of a uniformly distributed dynamic load where $F_z(y) = f =$ constant, the solution reduces to

$$w(y,t) = \frac{4f}{m\pi} \sum_{i=1,3,5,\cdots}^{\infty} \frac{\sin \dfrac{i\pi y}{l}}{i} \frac{[\sin \Omega t - (\Omega/\omega_i) \sin \omega_i t]}{\omega_i^2 [1 - (\Omega^2/\omega_i^2)]}, \quad (3\text{-}78)$$

where it can be observed that only symmetrical mode shapes (odd values of i) appear.

When the load is a concentrated dynamic load $F \sin \Omega t$ applied at a point $y = \eta$ along the beam, we represent the function $F_z(y)$ by

$$F_z(y) = F\delta(y - \eta), \quad (3\text{-}79)$$

where the function $\delta(y - \eta)$ is the Dirac delta-function (Ref. 3–6). This is a function which has the unique property of being zero everywhere except at $y = \eta$, where it is infinite. In addition, for any assumed function $g(y)$, the Dirac delta-function has the integral properties

$$\int \delta(y - \eta)dy = 1, \qquad \int g(y)\delta(y - \eta)dy = g(\eta) \qquad (3\text{-}80)$$

if η is included in the region of integration. If η is not included in the region of integration, the function has the properties that

$$\int \delta(y - \eta)dy = 0, \qquad \int g(y)\delta(y - \eta)dy = 0. \qquad (3\text{-}81)$$

Prior to substituting Eq. (3–79) into Eq. (3–77), we first evaluate

$$\int_0^l F\delta(y - \eta) \sin \frac{i\pi y}{l} \, dy = F \sin \frac{i\pi\eta}{l} \qquad (3\text{-}82)$$

by applying Eq. (3–80). Substituting Eq. (3–79) into (3–77) and applying Eq. (3–82), we obtain the response of the beam to the concentrated dynamic load $F \sin \Omega t$:

$$w(y,t) = \frac{2F}{ml} \sum_{i=1}^{\infty} \sin \frac{i\pi y}{l} \sin \frac{i\pi\eta}{l} \frac{\left[\sin \Omega t - (\Omega/\omega_i) \sin \omega_i t\right]}{\omega_i^2[1 - (\Omega^2/\omega_i^2)]}. \qquad (3\text{-}83)$$

Equations (3–78) and (3–83) express the familiar resonance condition in which the response approaches infinity as the exciting frequency Ω approaches one of the natural frequencies ω_i.

3-3 Integral equation of motion of a slender beam. When a restrained beam is loaded by a lateral load of intensity $Z(y)$ and by an applied running moment of intensity $\bar{m}(y)$, as illustrated by Fig. 3–10, the lateral deflection and slope can be expressed respectively by

$$w(y) = \int_0^l [C^{zz}(y,\eta)Z(\eta) + C^{z\alpha}(y,\eta)\bar{m}(\eta)]d\eta, \qquad (3\text{-}84a)$$

$$w'(y) = \int_0^l [C^{\alpha z}(y,\eta)Z(\eta) + C^{\alpha\alpha}(y,\eta)\bar{m}(\eta)]d\eta, \qquad (3\text{-}84b)$$

where $C^{zz}(y,\eta)$ = lateral deflection at y due to a unit load at η,
$\quad\quad C^{z\alpha}(y,\eta)$ = lateral deflection at y due to a unit moment at η,[*]
$\quad\quad C^{\alpha z}(y,\eta)$ = slope at y due to a unit load at η,[*]
$\quad\quad C^{\alpha\alpha}(y,\eta)$ = slope at y due to a unit moment at η.

---

[*] It can easily be seen that

$$C^{\alpha z}(y,\eta) = \partial C^{zz}(y,\eta)/\partial y$$

and by the reciprocal relations (cf. Section 2–4)

$$C^{z\alpha}(y,\eta) = C^{\alpha z}(\eta,y) = \partial C^{zz}(y,\eta)/\partial \eta.$$

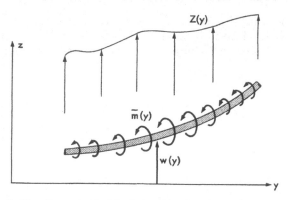

Fig. 3-10.   Beam loaded by lateral load and running moment.

The influence functions may, in general, include both bending and shearing deflections.
    Introducing

$$Z(\eta) = F_z(\eta,t) - m(\eta)\ddot{w}(\eta,t)$$

and
$$\bar{m}(\eta) = -\mu(\eta)\ddot{\alpha}'(\eta,t) \tag{3-85}$$

into Eqs. (3-84), we obtain the simultaneous integral equations of forced motion of a restrained beam in which rotary inertia effects are taken into account:

$$w(y,t) = \int_0^l \{C^{zz}(y,\eta)[F_z(\eta,t) - m(\eta)\ddot{w}(\eta,t)] - C^{z\alpha}(y,\eta)\mu(\eta)\ddot{\alpha}'(\eta,t)\}d\eta, \tag{3-86}$$

$$w'(y,t) = \int_0^l \{C^{\alpha z}(y,\eta)[F_z(\eta,t) - m(\eta)\ddot{w}(\eta,t)] - C^{\alpha\alpha}(y,\eta)\mu(\eta)\ddot{\alpha}'(\eta,t)\}d\eta. \tag{3-87}$$

Equations (3-86) and (3-87) are called integral equations, since the unknown quantities appear under the integral sign. If we assume $\alpha' = w'$, the problem is simplified to one involving the solution of a single integral equation derived from Eq. (3-86):

$$w(y,t) = \int_0^l \{C^{zz}(y,\eta)[F_z(\eta,t) - m(\eta)\ddot{w}(\eta,t)] - C^{z\alpha}(y,\eta)\mu(\eta)\ddot{w}'(\eta,t)\}d\eta. \tag{3-88}$$

    (a) *Integral equation of free vibration of a slender restrained beam.* The integral equation of free vibration of a slender restrained beam for which rotary inertia effects may be neglected derives from Eq. (3-88) by putting $\mu = F_z(y,t) = 0$.

$$w(y,t) = -\int_0^l C^{zz}(y,\eta)\ddot{w}(\eta,t)m(\eta)d\eta. \tag{3-89}$$

Like its differential equation counterpart, Eq. (3–89) is separable, and we can put as a solution

$$w(y,t) = W(y)T(t). \tag{3-90}$$

Substitution of Eq. (3–90) into Eq. (3–89) yields

$$-\frac{T}{\ddot{T}} = \frac{1}{W(y)}\int_0^l C(y,\eta)W(\eta)m(\eta)d\eta. \tag{3-91}$$

The superscripts on the influence function are dropped in Eq. (3–91) for brevity. Since y and t are independent variables, it follows that

$$\ddot{T} + \omega^2 T = 0, \tag{3-92}$$

$$W(y) = \omega^2\int_0^l C(y,\eta)W(\eta)m(\eta)d\eta. \tag{3-93}$$

The time-dependent equation, Eq. (3–92), is identical to Eq. (3–10). The space-dependent equation that defines the mode shapes, Eq. (3–93), is now a homogeneous linear integral equation instead of a differential equation. The boundary conditions are implicit in the influence function $C(y,\eta)$ and need not be specified in addition to Eq. (3–93).

According to the Hilbert-Schmidt theory of integral equations (Ref. 3–7), an infinite set of pairs of eigenfunctions $W_n(y)$ and eigenvalues ω_n satisfy Eq. (3–93), and they are the same set that satisfy Eq. (3–11) providing the assigned boundary conditions on Eq. (3–11) and $C(y,\eta)$ are identical. The orthogonality relation between eigenfunctions of the integral equation can be demonstrated by substituting $W_n(y)$ and $W_m(y)$ into Eq. (3–93).

$$W_n(y) = \omega_n{}^2\int_0^l C(y,\eta)W_n(\eta)m(\eta)d\eta, \tag{3-94}$$

$$W_m(y) = \omega_m{}^2\int_0^l C(y,\eta)W_m(\eta)m(\eta)d\eta. \tag{3-95}$$

Multiplying Eqs. (3–94) and (3–95) by $W_m(y)m(y)$ and $W_n(y)m(y)$ respectively, and integrating from 0 to l, yields

$$\frac{1}{\omega_n{}^2}\int_0^l W_m W_n m\,dy = \int_0^l W_m(y)m(y)\int_0^l C(y,\eta)W_n(\eta)m(\eta)d\eta dy, \tag{3-96}$$

$$\frac{1}{\omega_m{}^2}\int_0^l W_n W_m m\,dy = \int_0^l W_n(y)m(y)\int_0^l C(y,\eta)W_m(\eta)m(\eta)d\eta dy. \tag{3-97}$$

Since the influence function is symmetrical, and therefore $C(y,\eta) = C(\eta,y)$, the right sides of Eqs. (3-96) and (3-97) are identical, and

$$\left(\frac{1}{\omega_n{}^2} - \frac{1}{\omega_m{}^2}\right)\int_0^l W_m W_n m\, dy = 0, \qquad (m \neq n), \qquad (3\text{-}98)$$

which reduces to the orthogonality condition previously given by Eq. (3-20).

(b) *Integral equation of free vibration of a slender unrestrained beam.* In Section (a) above, we have applied integral equations to the solution of free vibration problems of slender restrained beams. When integral equation solutions are applied to a beam which is unrestrained, some further consideration is required. The physical significance of an influence function as applied to an unrestrained beam is not immediately obvious. The problem of computing an influence function for a restrained beam is a simple statics problem. When the beam is unrestrained, the application of a force will induce motion. We are then concerned with the problem of what the reactions are and how deflections are to be measured.

The dynamic problem can be reduced to an equivalent static problem in the conventional manner by the application of D'Alembert's principle. That is, we consider the steady condition of uniform acceleration due to an applied force, and put the resulting inertial forces into equilibrium with this force. The influence function is then calculated for this force system, with inertial forces taking the place of support reactions.

The natural modes of vibration of an unrestrained beam will include two rigid-body modes which describe the motion of its principal inertial axes. The deformation modes, on the other hand, are characterized by displacements relative to the principal inertial axes. Consequently, we can eliminate the rigid-body modes from the solution by measuring deflections relative to the principal inertial axes in the computation of the influence function.

Let us consider an unrestrained beam free to move in the plane of the paper, as illustrated by Fig. 3-11. The axis system is taken such that the z-axis is at midspan, and the beam is assumed symmetrical with respect to the z-axis. The origin is taken in coincidence with the center of gravity and the yz-axes in coincidence with the principal inertial axes

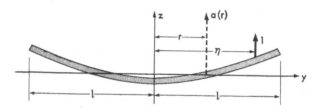

Fig. 3-11. Axis system of unrestrained beam.

of the vibrating beam. Since the beam is vibrating freely, the yz-axes can be considered fixed in space. The summations of inertial forces along the z-axis and inertial moments about the x-axis are zero. Thus,

$$\int_{-l}^{l} \ddot{w}(y,t)dm = \ddot{T}(t) \int_{-l}^{l} W_n(y)m\,dy = 0, \qquad (3\text{-}99\text{a})$$

$$\int_{-l}^{l} \ddot{w}(y,t)y\,dm = \ddot{T}(t) \int_{-l}^{l} W_n(y)my\,dy = 0. \qquad (3\text{-}99\text{b})$$

Equations (3–99a) and (3–99b) verify that the origin of the yz-axis system remains always at the center of gravity of the deformed beam and that the yz-axes are principal inertial axes of the deformed beam.

If we apply a unit load in the z-direction at point $y = \eta$, as shown by Fig. 3–11, the resulting acceleration of a rigid beam at point $y = r$ is given by

$$a(r) = \frac{1}{M} + \frac{\eta r}{I_x}, \qquad (3\text{-}100)$$

where M is the total mass of the beam and I_x is its moment of inertia about the x-axis; η and r are dummy variables in the coordinate y.

The inertial force due to an element of mass $m\,dr$ at the point r is

$$df(r) = -\left(\frac{1}{M} + \frac{\eta r}{I_x}\right)m\,dr. \qquad (3\text{-}101)$$

It can be seen, therefore, that when a unit force is applied to an unrestrained beam, the beam is in dynamic equilibrium under the three force systems illustrated in Fig. 3–12.

The displacement of the beam with respect to a straight line drawn tangent to the beam at the origin, due to the unit force and the distributed

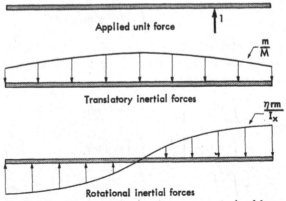

Applied unit force

Translatory inertial forces

Rotational inertial forces

Fig. 3–12. Force systems acting on unrestrained beam.

inertial forces, may be written as

$$H(y,\eta) = C(y,\eta) - \int_{-l}^{l} C(y,r) \left(\frac{1}{M} + \frac{\eta r}{I_x} \right) m(r)dr, \qquad (3\text{-}102)$$

where $C(y,\eta)$ is the conventional influence function representing the displacement in the z-direction of a point at y due to a unit force at η, when the origin of the y-axis is clamped.

If we measure the beam deflection with respect to the y-axis rather than with respect to an axis tangent to the beam at its center line, we obtain

$$G(y,\eta) = H(y,\eta) + W(0) + \frac{\partial W(0)}{\partial y} y, \qquad (3\text{-}103)$$

where $H(y,\eta)$ is defined by Eq. (3-102), and $W(0)$ and $\partial W(0)/\partial y$ are respectively the deflection and slope of the beam at its center line. Since all displacements of the beam must satisfy the conditions given by Eqs. (3-99a) and (3-99b), we are afforded conditions for computing $W(0)$ and $\partial W(0)/\partial y$. Substituting Eq. (3-103) into Eq. (3-99a) gives

$$\int_{-l}^{l} \left[H(r,\eta) + W(0) + \frac{\partial W(0)}{\partial y} r \right] m(r)dr = 0. \qquad (3\text{-}104)$$

Integrating Eq. (3-104), applying the condition that the origin is at the center of gravity, and solving for $W(0)$ yields

$$W(0) = - \frac{1}{M} \int_{-l}^{l} H(r,\eta)m(r)dr. \qquad (3\text{-}105)$$

Substituting Eq. (3-103) into Eq. (3-99b) gives

$$\int_{-l}^{l} \left[H(r,\eta) + W(0) + \frac{\partial W(0)}{\partial y} r \right] m(r)rdr = 0, \qquad (3\text{-}106)$$

which upon integration and application of the center of gravity condition yields

$$\frac{\partial W(0)}{\partial y} = - \frac{1}{I_x} \int_{-l}^{l} H(r,\eta)m(r)rdr. \qquad (3\text{-}107)$$

Substituting Eqs. (3-102), (3-105), and (3-107) into Eq. (3-103) yields an expression for the influence function of an unrestrained beam with respect to its principal inertial axes.

$$G(y,\eta) = C(y,\eta) - \int_{-l}^{l} C(y,r) \left[\frac{1}{M} + \frac{\eta r}{I_x} \right] m(r)dr$$

$$- \int_{-l}^{l} C(r,\eta) \left[\frac{1}{M} + \frac{yr}{I_x} \right] m(r)dr$$

$$- \int_{-l}^{l} \int_{-l}^{l} C(r,p) \left[\frac{1}{M^2} + \frac{\eta p + ry}{MI_x} + \frac{\eta pry}{I_x^2} \right] m(r)m(p)drdp.$$

$$(3\text{-}108)$$

The integral equation governing the free vibration of a slender unrestrained beam becomes

$$W(y) = \omega^2 \int_{-l}^{l} G(y,\eta)W(\eta)m(\eta)d\eta, \tag{3-109}$$

where $G(y,\eta)$ is the influence function given in Eq. (3-108).

It can be observed that in the second and fourth integrals in Eq. (3-108), η appears only to the zero or first power. Consequently, these terms will not contribute anything when substituted into Eq. (3-109), since by virtue of Eqs. (3-99a) and (3-99b) we have

$$\int_{-l}^{l} W(\eta)md\eta = \int_{-l}^{l} W(\eta)m\eta d\eta = 0. \tag{3-110}$$

As a result, the influence function $G(y,\eta)$ simplifies to the form

$$G(y,\eta) = C(y,\eta) - \int_{-l}^{l} C(r,\eta)\left[\frac{1}{M} + \frac{yr}{I_x}\right]m(r)dr. \tag{3-111}$$

Thus the integral equation (3-109), together with the influence function given by Eq. (3-111), can be employed to derive the natural deformation mode shapes and frequencies of an unrestrained beam.

(c) *Integral equation of forced motion of a slender restrained beam.* If the rotary inertia term in Eq. (3-88) is put equal to zero, we obtain the integral equation of forced motion of a slender restrained beam, as follows:

$$w(y,t) = \int_{0}^{l} C(y,\eta)[-m(\eta)\ddot{w}(\eta,t) + F_z(\eta,t)]d\eta. \tag{3-112}$$

Following the method of solution applied to the differential equation, we put as a solution the series given by Eq. (3-62):

$$w(y,t) = \sum_{i=1}^{\infty} \phi_i(y)\xi_i(t), \tag{3-62}$$

where the $\phi_i(y)$ are eigenfunctions of the homogeneous form of Eq. (3-112), that is, Eq. (3-93). Substituting Eq. (3-62) into Eq. (3-112) yields

$$\sum_{i=1}^{\infty} \phi_i\xi_i = -\sum_{i=1}^{\infty} \ddot{\xi}_i\int_{0}^{l} C(y,\eta)\phi_i(\eta)m(\eta)d\eta + \int_{0}^{l} C(y,\eta)F_z(\eta,t)d\eta. \tag{3-113}$$

When Eq. (3-93) is introduced, Eq. (3-113) reduces to

$$\sum_{i=1}^{\infty} \phi_i\left(\xi_i + \frac{\ddot{\xi}_i}{\omega_i^2}\right) = \int_{0}^{l} C(y,\eta)F_z(\eta,t)d\eta. \tag{3-114}$$

Multiplying through by $\phi_j(y)m(y)$ and integrating from 0 to l gives

$$\sum_{i=1}^{\infty}\left(\xi_i + \frac{\ddot{\xi}_i}{\omega_i^2}\right)\int_{0}^{l} \phi_i\phi_j m\,dy = \int_{0}^{l} \phi_j(y)m(y)\int_{0}^{l} C(y,\eta)F_z(\eta,t)d\eta dy. \tag{3-115}$$

Applying the orthogonality condition, Eq. (3–115) reduces to

$$M_j\ddot{\xi}_j + M_j\omega_j{}^2\xi_j = \omega_j{}^2 \int_0^l \phi_j(y)m(y) \int_0^l C(y,\eta)F_z(\eta,t)d\eta dy, \quad (3\text{–}116)$$

where

$$M_j = \int_0^l \phi_j{}^2 m \, dy.$$

Applying the reciprocal relation $C(y,\eta) = C(\eta,y)$, and the condition that $\phi_j(y)$ satisfies Eq. (3–93), there is finally obtained

$$M_j\ddot{\xi}_j + M_j\omega_j{}^2\xi_j = \Xi_j, \quad (3\text{–}117)$$

where

$$\Xi_j = \int_0^l F_z(y,t)\phi_j(y)dy.$$

We see, therefore, that the problem of forced motion from the point of view of an integral equation reduces to the same set of total differential equations already derived from the partial differential equation formulation of the problem. The only practical difference between the integral and differential equation approaches is that the eigenfunctions are defined by an integral equation on the one hand [Eq. (3–93)] and a differential equation on the other [Eq. (3–11)].

EXAMPLE 3–3. To deduce a relation between the influence function of a beam and its eigenfunctions and eigenvalues.

Solution. Suppose that a restrained beam, initially at rest with arbitrary end conditions, is subjected to a unit concentrated harmonic force at the point $y = \eta$, as follows:

$$F(\eta,t) = \delta(y - \eta) \sin \Omega t. \quad (a)$$

For this case, Eq. (3–117) becomes

$$M_j\ddot{\xi}_j + M_j\omega_j{}^2\xi_j = \sin \Omega t \int_0^l \delta(y - \eta)\phi_j(y)dy = \phi_j(\eta) \sin \Omega t. \quad (b)$$

If the beam is initially at rest, the solution to Eq. (b) is

$$\xi_j = \frac{\phi_j(\eta)}{M_j\omega_j{}^2[1 - (\Omega^2/\omega_j{}^2)]}\left(\sin \Omega t - \frac{\Omega}{\omega_j}\sin \omega_j t\right). \quad (c)$$

The lateral displacement of the beam is obtained by substituting Eq. (c) into Eq. (3–62):

$$w(y,t) = \sum_{i=1}^{\infty} \frac{\phi_i(y)\phi_i(\eta)}{M_i\omega_i{}^2[1 - (\Omega^2/\omega_i{}^2)]}\left(\sin \Omega t - \frac{\Omega}{\omega_i}\sin \omega_i t\right). \quad (d)$$

If we suppose that the forcing frequency Ω approaches zero, we have a condition approaching static loading, and Eq. (d) reduces to a result representing the deflection at y due to a unit static load at η:

$$C(y,\eta) = \sum_{i=1}^{\infty} \frac{\phi_i(y)\phi_i(\eta)}{M_i\omega_i^2} . \tag{e}$$

Equation (e) expresses an interesting equivalence between the influence function of the beam and its natural mode shapes and frequencies. For example, applying Eq. (e) to the case of a uniform simply supported beam, the influence function can be expressed by the infinite series

$$C(y,\eta) = \sum_{i=1}^{\infty} \frac{\sin (i\pi y/l) \sin (i\pi\eta/l)}{(ml/2)\omega_i^2} . \tag{f}$$

3–4 Dynamic equilibrium of slender rotating beams in bending. Previous sections of this chapter have been concerned with the formulation and solution of the equations of free and forced motion of slender non-rotating beams. An important class of problems arises in the case of rotary wing aircraft where it is necessary to consider rotating beams. Here the centrifugal forces on the rotating beam exert an important influence on the bending stiffness.

(a) *Differential equation of free vibration of slender rotating beams.* Figure 3–13 is a diagram of the lateral inertial forces and centrifugal forces on a slender beam rotating at constant angular velocity Ω about the z-axis. The y-axis is assumed to rotate with the beam, and the boundary conditions defining the attachment to the z-axis will be left undefined in the development of the equations of motion. From Fig. 3–13 it can be seen that the moment at any point y along the beam is given by

$$M(y) = \int_y^R \{Z(\eta,t)(\eta - y) - Y(\eta)[w(\eta,t) - w(y,t)]\}d\eta, \tag{3–118}$$

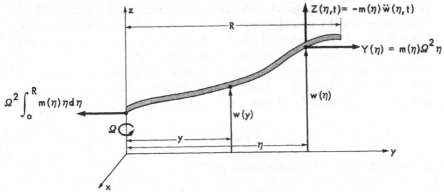

Fig. 3–13.   Rotating beam coordinate system.

where

$$Z(\eta,t) = -m(\eta)\ddot{w}(\eta,t), \qquad Y(\eta) = m(\eta)\Omega^2\eta. \qquad (3\text{-}119)$$

In writing Eq. (3-119), it is assumed that the flapping and bending displacements of the rotating beam are small quantities. Substituting Eqs. (3-119) into Eq. (3-118) and putting $M(y) = EIw''(y)$ we obtain an equation of motion, as follows:

$$EIw''(y,t) = \int_y^R \{-m(\eta)\ddot{w}(\eta,t)(\eta - y) - m(\eta)\Omega^2\eta[w(\eta,t) - w(y,t)]\}d\eta.$$
$$(3\text{-}120)$$

In order to bring Eq. (3-120) into a form comparable to Eq. (3-7), it is differentiated twice with respect to y:

$$m(y)\ddot{w}(y,t) + [EIw''(y,t)]''$$
$$+ \Omega^2\left\{m(y)yw'(y,t) - w''(y,t)\int_y^R m(\eta)\eta d\eta\right\} = 0. \quad (3\text{-}121)$$

Equation (3-121) is a separable partial differential equation, and we can put as a solution

$$w(y,t) = W(y)T(t). \qquad (3\text{-}122)$$

Substituting Eq. (3-122) into Eq. (3-121) yields

$$\ddot{T} + \omega^2 T = 0, \qquad (3\text{-}123)$$

$$(EIW'')'' - m\omega^2 W + \Omega^2\left\{m(y)yW' - W''\int_y^R m(\eta)\eta d\eta\right\} = 0, \quad (3\text{-}124)$$

where ω is the natural vibration frequency of the rotating beam.

Natural mode shapes are obtained by solutions to Eq. (3-124) together with four appropriate boundary conditions. The natural mode shapes of a rotating beam are orthogonal. This fact can be shown by the same method as employed for the case of the nonrotating beam. In the case where the beam is hinged at the root, and is free to flap in the yz-plane, we would suspect the existence of a rigid-body flapping mode. However, unlike other nonrotating free systems, this mode has a finite frequency equal to the rotational frequency. This fact can be verified by introducing a normalized expression for the rigid flapping mode,

$$W_0(y) = \frac{y}{R}, \qquad (3\text{-}125)$$

into Eq. (3-124). This substitution gives

$$-m\omega_0^2\left(\frac{y}{R}\right) + \Omega^2 my\left(\frac{1}{R}\right) = 0, \qquad (3\text{-}126)$$

which indicates that the differential equation is satisfied by the function given by Eq. (3-125) providing $\omega_0 = \Omega$.

(b) *Differential equation of forced motion of rotating beams.* When a slender rotating beam is subjected to distributed lateral forces $F_z(y,t)$, the differential equation of forced lateral motion becomes

$$m(y)\ddot{w}(y,t) + [EIw''(y,t)]''$$

$$+ \Omega^2 \left\{ m(y)yw'(y,t) - w''(y,t) \int_y^R m(\eta)\eta \, d\eta \right\} = F_z(y,t). \quad (3\text{-}127)$$

Following the method used for nonrotating beams, we express the displacement by

$$w(y,t) = \sum_{i=1}^{\infty} \phi_i(y)\xi_i(t), \quad (3\text{-}128)$$

where the $\phi_i(y)$ are normalized natural bending mode shapes of the rotating beam, and are eigenfunctions of Eq. (3-124) with prescribed boundary conditions. Substituting Eq. (3-128) into Eq. (3-127) gives

$$\sum_{i=1}^{\infty} \left\{ m\phi_i\ddot{\xi}_i + \left[(EI\phi_i'')'' + \Omega^2 \left(my\phi_i' - \phi_i'' \int_y^R m\eta \, d\eta \right) \right] \xi_i \right\} = F_z(y,t). \quad (3\text{-}129)$$

Multiplying through by $\phi_j(y)$ and integrating from 0 to R yields

$$\sum_{i=1}^{\infty} \left\{ \ddot{\xi}_i \int_0^R \phi_i\phi_j m \, dy + \xi_i \int_0^R \phi_j \left[(EI\phi_i'')'' \right. \right.$$

$$\left. \left. + \Omega^2 \left(my\phi_i' - \phi_i'' \int_y^R m\eta \, d\eta \right) \right] dy \right\} = \int_0^R \phi_j F_z(y,t) dy. \quad (3\text{-}130)$$

Upon application of Eq. (3-124), and if the boundary conditions are such that the orthogonality condition holds in the present case, Eq. (3-130) can be simplified to

$$M_j\ddot{\xi}_j + M_j\omega^2\xi_j = \Xi_j, \quad (j = 1, 2, \cdots, \infty), \quad (3\text{-}131)$$

where

$$M_j = \int_0^R \phi_j{}^2 m \, dy, \qquad \Xi_j = \int_0^R F_z(y,t)\phi_j \, dy.$$

Thus the forced motion of a rotating beam can be treated by the same equations of motion as hold for a nonrotating beam provided the solution is expressed in terms of natural mode shapes of the rotating beam. When Eqs. (3-131) are applied to the case of a rotating beam with a hinged root (flapping blade), it should be observed that one of the equations represents the rigid flapping motion. That is, we have a rigid-body mode shape and its associated frequency defined by $\phi_1(y) = y/R$ and $\omega_1 = \Omega$, respectively.

When these values are substituted into Eq. (3–131), we obtain

$$I_1\ddot{\beta} + I_1\Omega^2\beta = \int_0^R F_z(y,t)y\,dy,   (3\text{–}132)$$

where $\beta = \xi_1/R$ is the rigid flapping motion and I_1 is the moment of inertia of the beam about its flapping hinge. In addition to Eq. (3–132), we have an infinite number of equations of motion of the form of Eqs. (3–131), each equation corresponding to a natural deformation mode of the rotating beam.

3–5 Dynamic equilibrium of slender beams in torsion. The problem of the twisting vibration of slender beams can be treated in a manner analogous to the treatment of beams in bending in terms of differential or integral equations. The following remarks are confined to the differential equation approach. The steps necessary to express the problem in terms of an integral equation can easily be constructed by referring to the integral equation approach to beam bending.

(a) *Differential equation of free vibration.* The torsional deformation of a slender beam, according to the St. Venant torsion theory, is given by

$$T(y) = GJ(y)\theta'(y),   (3\text{–}133)$$

where $T(y)$, $GJ(y)$, and $\theta(y)$ are respectively the applied torque, the torsional rigidity, and the angle of twist. Let us consider a free-body diagram of a unit length of torsion member, as illustrated by Fig. 3–14. Here we see that for equilibrium of torsional moments about the center of twist

$$T'(y) + t(y,t) = 0,   (3\text{–}134)$$

where $t(y,t)$ is an applied external torsional moment per unit length, and is taken positive in the positive θ-direction, as shown by Fig. 3–14. When the beam is vibrating freely,

$$t(y,t) = -I_0\ddot{\theta}(y,t),   (3\text{–}135)$$

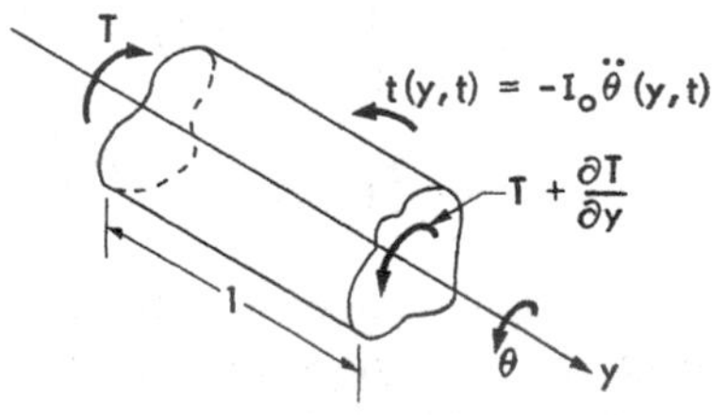

Fig. 3–14.   Free-body diagram of unit segment of torsion member.

where I_0 is the moment of inertia per unit length about the center of twist. Differentiating Eq. (3–133) with respect to y and introducing Eqs. (3–134) and (3–135), we obtain

$$[GJ\theta'(y,t)]' - I_0\ddot{\theta}(y,t) = 0,   (3\text{–}136)$$

the partial differential equation governing free vibration in torsion. Equation (3–136) is separable and can be solved by the same type of solution used to solve the bending differential equation. Putting

$$\theta(y,t) = \Theta(y)T(t),   (3\text{–}137)$$

we obtain, after separating variables,

$$\ddot{T} + \omega^2 T = 0, \tag{3-138}$$

$$(GJ\Theta')' + I_0\omega^2\Theta = 0, \tag{3-139}$$

where ω^2 is a separation constant. Equation (3-138) must be supplemented by two initial conditions and Eq. (3-139) by two boundary conditions. Equation (3-139), together with its boundary conditions, is satisfied by an infinite number of pairs of frequencies and orthogonal mode shapes.

EXAMPLE 3-4. To calculate the natural mode shapes and frequencies of a uniform torsion member with a fixed and a free end.

Solution. Equation (3-139) reduces, for the case of a uniform member, to

$$\Theta'' + \frac{\omega^2}{a^2}\Theta = 0, \tag{a}$$

where

$$a^2 = \frac{GJ}{I_0}.$$

The solution is

$$\Theta = C \sin\frac{\omega}{a}y + D \cos\frac{\omega}{a}y. \tag{b}$$

Boundary conditions on $\Theta(y)$ are

$$\Theta(0) = 0, \qquad \Theta'(l) = 0. \tag{c}$$

Introducing Eq. (b) into the boundary conditions gives

$$D = 0, \tag{d}$$

$$\cos\frac{\omega}{a}l = 0. \tag{e}$$

Equation (e) is satisfied when

$$\omega_n = \frac{n\pi a}{2l} = \frac{n\pi}{2l}\sqrt{\frac{GJ}{I_0}}, \qquad (n = 1, 3, 5, \cdots, \infty), \tag{f}$$

which corresponds to mode shapes

$$\Theta_n(y) = C \sin\frac{n\pi}{2l}y, \qquad (n = 1, 3, 5, \cdots, \infty). \tag{g}$$

These mode shapes are illustrated by Fig. 3-15.

EXAMPLE 3-5. To calculate the natural modes and frequencies of a uniform cantilever airplane wing with tip tank, as illustrated by Fig. 3-16. The wing is assumed to have uniform torsional rigidity GJ, and uniform moment of inertia I_0 per unit length about its elastic axis. The latter is assumed coincident with the y-axis. The rigid tip tank has a total pitching moment of inertia of $\bar{I}$ about the elastic axis.

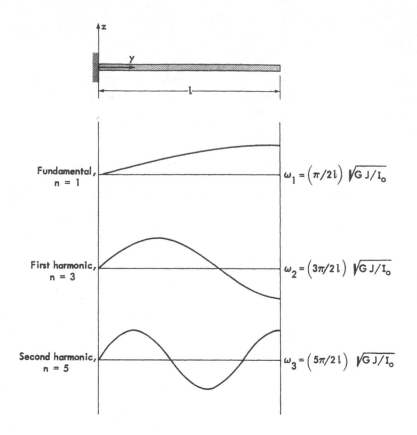

Fig. 3–15.   Mode shapes and frequencies of uniform cantilever torsion member.

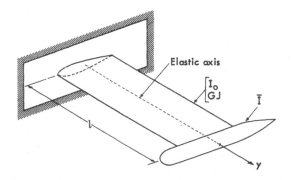

Fig. 3–16.   Uniform cantilever airplane wing with tip tank.

Solution. Boundary conditions on the wing are

$$\Theta(0) = 0, \tag{a}$$

$$GJ\Theta'(l) = \bar{I}\omega^2\Theta(l). \tag{b}$$

Equation (b), Example 3–4, which applies to the case of a uniform torsion member, when substituted into boundary conditions (a) and (b) results in

$$D = 0, \tag{c}$$

$$\frac{I_0 l}{\bar{I}} = \frac{\omega l}{a} \tan \frac{\omega l}{a}. \tag{d}$$

The transcendental frequency equation, Eq. (d), is plotted in Fig. 3–17. This plot illustrates graphically the dependence of the argument $\omega l/a$ upon the ratio $I_0 l/\bar{I}$. For example, in the case of a cantilever wing without a tip tank, we put $\bar{I} = 0$. In this case, the various branches in Fig. 3–17 approach the vertical dashed ordinates asymptotically, and we have

$$\frac{\omega l}{a} = \frac{n\pi}{2}, \qquad (n = 1, 3, 5, \cdots, \infty). \tag{e}$$

This is the case already considered in Example 3–4. If the tip tank is very large, and if $\bar{I}$ approaches infinity, we approach the case of a uniform wing fixed at both ends. This corresponds to intercepts of the branch curves on the horizontal axis, and

$$\frac{\omega l}{a} = n\pi, \qquad (n = 0, 1, 2, 3, \cdots, \infty). \tag{f}$$

The mode shapes corresponding to arbitrary finite values of $I_0 l/\bar{I}$ are truncated sine curves, as illustrated by Fig. 3–18. The mode shapes in the general case derive from Eq. (b), Example 3–4, by putting $D = 0$:

$$\Theta_n(y) = C \sin \frac{\omega_n}{a} y, \tag{g}$$

where the ω_n are picked off the graph of Fig. 3–18 for a specified value of $I_0 l/\bar{I}$.

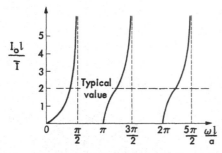

Fig. 3–17. Graphical solution of transcendental equation of uniform cantilever wing with tip tank.

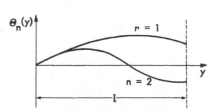

Fig. 3–18. Mode shapes of uniform cantilever wing with tip tank.

(b) *Differential equation of forced motion.* When a torsion member is subjected to an externally applied time-dependent torsional moment per unit length, $t(y,t)$, the partial differential equation of forced motion becomes

$$I_0(y)\ddot{\theta}(y,t) - [GJ(y)\theta'(y,t)]' = t(y,t). \tag{3-140}$$

Solutions to Eq. (3-140) are obtained in the same manner as for the bending differential equation by putting

$$\theta(y,t) = \sum_{i=1}^{\infty} \phi_i(y)\xi_i(t), \tag{3-141}$$

where the $\phi_i(y)$ are normalized mode shapes in torsion which satisfy the assigned boundary conditions, including a rigid-body mode if the beam is unrestrained, and the $\xi_i(t)$ are normal coordinates. The latter are determined by solutions to the ordinary differential equations

$$M_j\ddot{\xi}_j + M_j\omega_j{}^2\xi_j = \Xi_j, \tag{3-142}$$

where

$$M_j = \int_0^l \phi_j{}^2 I_0 \, dy$$

is the generalized mass and

$$\Xi_j = \int_0^l t(y,t)\phi_j \, dy$$

is the generalized force of the jth natural torsional vibration mode.

3-6 Dynamic equilibrium of restrained airplane wing. If we represent an airplane wing by an elastic surface, with thickness small compared with chord and span, as shown by Fig. 3-19, we have seen in Chapter 2 that its deformation under load can be expressed in terms of a two-dimensional

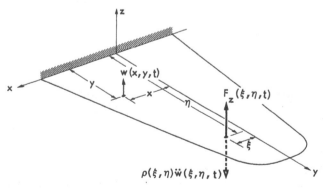

Fig. 3-19. Airplane wing represented by an elastic surface.

influence function. The lateral deformation under a distributed side load of intensity $Z(x,y)$ is given in Chapter 2, Eq. (2–55), as

$$w(x,y) = \iint_S C(x,y;\xi,\eta)Z(\xi,\eta)d\xi d\eta, \qquad (2\text{–}55)$$

where the integrations are taken over the surface of the wing. Under a dynamic side load $F_z(\xi,\eta,t)$, the total side load, including inertial forces,* is given by

$$Z(\xi,\eta,t) = -\rho(\xi,\eta)\ddot{w}(\xi,\eta,t) + F_z(\xi,\eta,t), \qquad (3\text{–}143)$$

where $\rho(\xi,\eta)$ is the mass per unit area. The integral equation of motion is obtained by substituting Eq. (3–143) into Eq. (2–55):

$$w(x,y,t) = \iint_S C(x,y;\xi,\eta)[-\rho(\xi,\eta)\ddot{w}(\xi,\eta,t) + F_z(\xi,\eta,t)]d\xi d\eta. \quad (3\text{–}144)$$

Equation (3–144) is solved in terms of its natural modes of vibration by putting

$$w(x,y,t) = \sum_{i=1}^{\infty} \phi_i(x,y)\xi_i(t), \qquad (3\text{–}145)$$

where the $\phi_i(x,y)$ are normalized natural modes of vibration of the wing, and are eigenfunctions of the homogeneous two-dimensional integral equation

$$\phi_i(x,y) = \omega_i^2 \iint_S C(x,y;\xi,\eta)\phi_i(\xi,\eta)\rho(\xi,\eta)d\xi d\eta. \qquad (3\text{–}146)$$

Following the methods developed for the one-dimensional systems, the time histories of the normal coordinates, $\xi_i(t)$, can easily be shown to be governed by solutions to the differential equations

$$M_j\ddot{\xi}_j + M_j\omega_j^2\xi_j = \Xi_j, \qquad (j = 1, 2, \cdots, \infty), \qquad (3\text{–}147a)$$

where

$$M_j = \iint_S \phi_j^2 \rho \, dx dy \qquad (3\text{–}147b)$$

is the generalized mass, and

$$\Xi_j = \iint_S F_z(x,y,t)\phi_j \, dx dy \qquad (3\text{–}147c)$$

is the generalized force of the jth natural vibration mode. This solution assumes that Eq. (3–146) can be solved for its eigenvalues and eigenfunctions. This is a formidable problem in cases of practical wings, and

---

* We assume a wing sufficiently thin so that rotary inertia effects can be neglected.

approximate numerical solutions are usually required. Such solutions are discussed in detail in Chapter 4.

If chordwise segments of the wing are assumed rigid, Eq. (3–144) can be reduced to two coupled one-dimensional integral equations. In this case, the deformation is described by a curvature in the yz-plane and a rotation about the y-axis, as discussed in Section 2–9.

Introducing Eqs. (2–61) and (2–62) into Eq. (3–144), we obtain, after some reduction, the following coupled integral equations of motion:

$$
w(y,t) = \int_0^l C^{zz}(y,\eta)[-m(\eta)\ddot{w}(\eta,t) + S_y(\eta)\ddot{\theta}(\eta,t) + F_z(\eta,t)]d\eta
$$
$$
+ \int_0^l C^{z\theta}(y,\eta)[-I_y(\eta)\ddot{\theta}(\eta,t) + S_y(\eta)\ddot{w}(\eta,t) + t(\eta,t)]d\eta,
\qquad (3\text{–}148)
$$

$$
\theta(y,t) = \int_0^l C^{\theta z}(y,\eta)[-m(\eta)\ddot{w}(\eta,t) + S_y(\eta)\ddot{\theta}(\eta,t) + F_z(\eta,t)]d\eta
$$
$$
+ \int_0^l C^{\theta\theta}(y,\eta)[-I_y(\eta)\ddot{\theta}(\eta,t) + S_y(\eta)\ddot{w}(\eta,t) + t(\eta,t)]d\eta,
\qquad (3\text{–}149)
$$

where

$$
m(\eta) = \int_{\text{chord}} \rho(\xi,\eta)d\xi \qquad = \text{mass per unit span,}
$$

$$
S_y(\eta) = \int_{\text{chord}} \rho(\xi,\eta)\xi d\xi \qquad = \text{static moment per unit span about } y\text{-axis,}
$$

$$
I_y(\eta) = \int_{\text{chord}} \rho(\xi,\eta)\xi^2 d\xi \qquad = \text{moment of inertia per unit span about } y\text{-axis,}
$$

$$
F_z(\eta,t) = \int_{\text{chord}} F_z(\xi,\eta,t)d\xi \qquad = \text{applied external force per unit span,}
$$

$$
t(\eta,t) = -\int_{\text{chord}} F_z(\xi,\eta,t)\xi d\xi = \text{applied external torque per unit span.}
$$

Equations (3–148) and (3–149) are coupled inertially and elastically. The elastic coupling is represented by the influence functions $C^{z\theta}(y,\eta)$ and $C^{\theta z}(y,\eta)$, and the inertial coupling by the static unbalance terms $S_y(\eta)$. These equations are applicable to wings that can be assumed rigid in the x-direction, including swept wings and wings with cut-outs. If the wing is unswept and without structural discontinuities, we can usually assume an elastic axis. In this case, the elastic coupling is eliminated, and the equations of motion become, simply,

$$
w(y,t) = \int_0^l C^{zz}(y,\eta)[-m(\eta)\ddot{w}(\eta,t) + S_y(\eta)\ddot{\theta}(\eta,t) + F_z(\eta,t)]d\eta, \quad (3\text{–}150)
$$

$$\theta(y,t) = \int_0^l C^{\theta\theta}(y,\eta)[-I_y(\eta)\ddot{\theta}(\eta,t) + S_y(\eta)\ddot{w}(\eta,t) + t(\eta,t)]d\eta. \qquad (3\text{-}151)$$

Equations (3–150) and (3–151) are coupled inertially, and this coupling can be eliminated only in those cases where the center of gravity of each chordwise section falls on the elastic axis.

The simplified form given by Eqs. (3–150) and (3–151) frequently arises in dynamic analyses of wings with straight elastic axes. Since no elastic coupling exists between bending and twisting, it is often more convenient to work with differential equations in terms of bending rigidity EI, and torsional rigidity GJ rather than integral equations. The differential equation forms can be derived most conveniently from elementary equilibrium considerations. Referring to Fig. 3–20, which illustrates the inertial forces and applied forces on a wing with straight elastic axis, we see that the total force per unit length, including inertial forces, is given by

$$Z(y,t) = F_z(y,t) + m(y)s(y)\ddot{\theta}(y,t) - m(y)\ddot{w}(y,t), \qquad (3\text{-}152)$$

and the rate of change of torque with respect to y about the elastic axis is

$$T'(y,t) = -t(y,t) + s(y)[m(y)s(y)\ddot{\theta}(y,t) - m(y)\ddot{w}(y,t)] + I_{c.g.}(y)\ddot{\theta}(y,t).$$
$$(3\text{-}153)$$

If rotary inertia and shear deformation are neglected, we may substitute these results into

$$[EIw''(y,t)]'' = Z(y,t)$$
$$[GJ\theta'(y,t)]' = T'(y,t) \qquad (3\text{-}154)$$

and

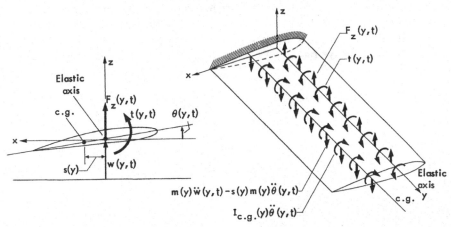

Fig. 3–20.   Airplane wing with rigid chordwise segments and straight elastic axis.

and obtain the partial differential equations of equilibrium:

$$m(y)\ddot{w}(y,t) - S_y(y)\ddot{\theta}(y,t) + [EIw''(y,t)]'' = F_z(y,t), \quad (3\text{--}155)$$

$$I_y(y)\ddot{\theta}(y,t) - S_y(y)\ddot{w}(y,t) - [GJ\theta'(y,t)]' = t(y,t), \quad (3\text{--}156)$$

where $S_y = ms$ is the static mass moment per unit length about the elastic axis and $I_y = I_{c.g.} + ms^2$ is the mass moment of inertia about the elastic axis. These differential equations are equivalent to the integral equations (3–150) and (3–151), and are in some cases more convenient for practical analyses of actual airplane wings.

3–7 Dynamic equilibrium of the unrestrained elastic airplane. Practical problems in aeroelasticity usually involve aircraft in flight. It is necessary, therefore, to use as a basis for such problems equations of motion of the unrestrained airplane. In the most general development of these equations it must, of course, be recognized that the airplane is a three-dimensional elastic object completely free to translate and rotate in space. In the present discussion, however, it will be assumed that the airplane is compressed into a flat elastic plate, as illustrated by Fig. 3–21, and that, in addition to its elastic degrees of freedom, it is permitted small displacements in vertical translation, pitch, and roll. The elastic plate is assumed rigid in its own plane. Although some loss of generality is incurred by these assumptions, they permit a concise development of the equations of motion. These equations are applicable to a majority of the practical aeroelastic problems involving aircraft in flight.

(a) *Integral equation of free vibration.* The deformation of the elastic airplane assumed here can be treated conveniently in terms of a two-dimensional influence function and its associated integral equation. Let us assume reference axes oriented with respect to the airplane, as illustrated by Fig. 3–21. The origin of the axis system is located such that it remains at the center of gravity of the deformed airplane. In addition, the xyz-axes always remain in coincidence with the principal inertial axes

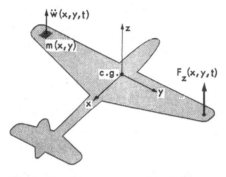

Fig. 3–21.   Unrestrained elastic airplane.

of the deformed airplane. Since the airplane is vibrating freely, the xyz-axes can be considered fixed in space. The deformation is described by a function $w(x,y,t)$ which represents the displacement of the weaving elastic surface with respect to the fixed xy-plane.

Four equations of motion are required. For equilibrium of total forces along the z-axis, in the absence of external forces,

$$\iint_S \ddot{w}(x,y,t)\rho(x,y)dxdy = 0, \qquad (3\text{--}157)$$

where the integration is over the surface area of the airplane. For moment equilibrium of free vibration about the y-axis,

$$\iint_S \ddot{w}(x,y,t)x\rho(x,y)dxdy = 0, \qquad (3\text{--}158)$$

and for moment equilibrium about the x-axis,

$$\iint_S \ddot{w}(x,y,t)y\rho(x,y)dxdy = 0. \qquad (3\text{--}159)$$

A fourth equation of equilibrium relating elastic forces and inertial forces is obtained by applying Eq. (2–55):

$$w(x,y,t) - w(0,0,t) - x\frac{\partial w(0,0,t)}{\partial x}$$
$$- y\frac{\partial w(0,0,t)}{\partial y} = -\iint_S C(x,y;\xi,\eta)\rho\ddot{w}d\xi d\eta, \quad (3\text{--}160)$$

where $w(0,0,t)$ is the deflection and $\partial w(0,0,t)/\partial x$ and $\partial w(0,0,t)/\partial y$ are the slopes of the elastic surface at the point pierced by the z-axis.

Introducing as a solution

$$w(x,y,t) = W(x,y)T(t), \qquad (3\text{--}161)$$

the equations of motion become

$$\iint_S W(x,y)\rho(x,y)dxdy = 0, \qquad (3\text{--}162)$$

$$\iint_S W(x,y)x\rho(x,y)dxdy = 0, \qquad (3\text{--}163)$$

$$\iint_S W(x,y)y\rho(x,y)dxdy = 0, \qquad (3\text{--}164)$$

$$T\left\{W(x,y) - W(0,0) - x\frac{\partial W(0,0)}{\partial x} - y\frac{\partial W(0,0)}{\partial y}\right\}$$
$$= -\ddot{T}\iint_S C(x,y;\xi,\eta)\rho(\xi,\eta)W(\xi,\eta)d\xi d\eta. \quad (3\text{--}165)$$

Equation (3-165) is separable and reduces to two separate equations, as follows:

$$\omega^2 \iint_S C(x,y;\xi,\eta) W(\xi,\eta) \rho(\xi,\eta) d\xi d\eta$$

$$= W(x,y) - W(0,0) - x \frac{\partial W(0,0)}{\partial x} - y \frac{\partial W(0,0)}{\partial y}, \qquad (3\text{-}166a)$$

$$\ddot{T} + \omega^2 T = 0, \qquad (3\text{-}166b)$$

where ω^2 is a separation constant which represents physically the frequencies of the natural vibration modes.

Equations (3-162), (3-163), (3-164), and (3-166a) can be combined into a single equation in which the amplitude $W(0,0)$, $\partial W(0,0)/\partial x$ and $\partial W(0,0)/\partial y$ are eliminated (Ref. 3-8). Multiplying both sides of Eq. (3-166a) by $\rho(x,y)$ and integrating over the surface of the airplane, we obtain

$$\omega^2 \iint_S \rho(x,y) \iint_S C(x,y;\xi,\eta) W(\xi,\eta) \rho(\xi,\eta) d\xi d\eta dx dy = \iint_S W(x,y) \rho(x,y) dx dy$$

$$- W(0,0) \iint_S \rho(x,y) dx dy - \frac{\partial W(0,0)}{\partial x} \iint_S x\rho(x,y) dx dy$$

$$- \frac{\partial W(0,0)}{\partial y} \iint_S y\rho(x,y) dx dy. \qquad (3\text{-}167)$$

We observe that the first integral on the right side is zero, by virtue of Eq. (3-162). The third and fourth integrals on the right side are zero, since the origin of coordinates is, by definition, at the center of gravity. The remaining terms may be solved for $W(0,0)$ as follows:

$$W(0,0) = - \frac{\omega^2}{M} \iint_S \rho(x,y) \iint_S C(x,y;\xi,\eta) W(\xi,\eta) \rho(\xi,\eta) d\xi d\eta dx dy. \qquad (3\text{-}168)$$

When Eq. (3-166a) is multiplied by $x\rho(x,y)$ and integrated over the surface, we obtain

$$\omega^2 \iint_S x\rho(x,y) \iint_S C(x,y;\xi,\eta) W(\xi,\eta) \rho(\xi,\eta) d\xi d\eta dx dy$$

$$= \iint_S W(x,y) x\rho(x,y) dx dy - W(0,0) \iint_S x\rho(x,y) dx dy - \frac{\partial W(0,0)}{\partial x}$$

$$\times \iint_S x^2 \rho(x,y) dx dy - \frac{\partial W(0,0)}{\partial y} \iint_S xy\rho(x,y) dx dy. \qquad (3\text{-}169)$$

The first integral on the right side of Eq. (3-169) is zero, according to Eq. (3-163). The second integral is zero, since the origin of coordinates is at the center of gravity. The fourth integral is zero since the xz-plane is a plane of symmetry. Thus, the remaining terms in Eq. (3–169) can be solved for

$$\frac{\partial W(0,0)}{\partial x} = -\frac{\omega^2}{I_y} \iint_S x\rho(x,y) \iint_S C(x,y;\xi,\eta)W(\xi,\eta)\rho(\xi,\eta)d\xi d\eta dxdy, \quad (3\text{–}170)$$

where

$$I_y = \iint_S x^2\rho(x,y)dxdy.$$

Finally, when Eq. (3–166a) is multiplied by $y\rho(x,y)$ and integrated over the surface, the result can be reduced in a similar manner to

$$\frac{\partial W(0,0)}{\partial y} = -\frac{\omega^2}{I_x} \iint_S y\rho(x,y) \iint_S C(x,y;\xi,\eta)W(\xi,\eta)\rho(\xi,\eta)d\xi d\eta dxdy, \quad (3\text{–}171)$$

where

$$I_x = \iint_S y^2\rho(x,y)dxdy.$$

When Eqs. (3–168), (3–170), and (3–171) are substituted into Eq. (3–166a), we obtain a homogeneous linear integral equation in $W(x,y)$ as follows:

$$\omega^2 \iint_S G(x,y;\xi,\eta)W(\xi,\eta)\rho(\xi,\eta)d\xi d\eta = W(x,y), \quad (3\text{–}172)$$

where

$$G(x,y;\xi,\eta) = C(x,y;\xi,\eta) - \iint_S C(r,s;\xi,\eta)\left[\frac{1}{M} + \frac{ys}{I_x} + \frac{xr}{I_y}\right]\rho(r,s)drds.$$
$$(3\text{–}173)$$

$G(x,y;\xi,\eta)$ is the influence function of the unrestrained airplane, and it can be computed from Eq. (3–173) if the elastic and inertial properties are known. Equation (3–172) is satisfied by an infinite number of pairs of deformation mode shapes $W_j(x,y)$ and frequencies ω_j. If Eqs. (3–172) and (3–173) are specialized to the case of a beam, the result is identical to that already given in Section 3–3(b).

Although the influence function $G(x,y;\xi,\eta)$ is not a symmetrical function, it can be shown that the natural mode shapes of the unrestrained airplane are orthogonal. For two different modes denoted by m and n, we can write

$$\omega_n{}^2 \iint_S G(x,y;\xi,\eta)W_n(\xi,\eta)\rho(\xi,\eta)d\xi d\eta = W_n(x,y) \quad (3\text{–}174)$$

and

$$\omega_m{}^2 \iint_S G(x,y;\xi,\eta)W_m(\xi,\eta)\rho(\xi,\eta)d\xi d\eta = W_m(x,y). \quad (3\text{–}175)$$

Multiplying Eqs. (3–174) and (3–175) by $W_m\rho(x,y)$ and $W_n\rho(x,y)$, respectively, and integrating over the area, we obtain

$$\frac{1}{\omega_n{}^2} \iint_S W_n W_m \rho\, dx dy = \iint_S W_m \rho \left[\iint_S GW_n \rho\, d\xi d\eta \right] dx dy, \quad (3\text{–}176)$$

$$\frac{1}{\omega_m{}^2} \iint_S W_m W_n \rho\, dx dy = \iint_S W_n \rho \left[\iint_S GW_m \rho\, d\xi d\eta \right] dx dy. \quad (3\text{–}177)$$

Interchanging the variables of integration on the right side of Eq. (3–177) results in

$$\frac{1}{\omega_m{}^2} \iint_S W_m W_n \rho\, dx dy$$

$$= \iint_S W_m(x,y)\rho(x,y) \left[\iint_S G(\xi,\eta;x,y) W_n(\xi,\eta)\rho(\xi,\eta)\, d\xi d\eta \right] dx dy. \quad (3\text{–}178)$$

The following expression can be constructed from Eq. (3–173):

$$G(\xi,\eta;x,y) = G(x,y;\xi,\eta) + \iint_S C(r,s;\xi,\eta) \left[\frac{1}{M} + \frac{ys}{I_x} + \frac{xr}{I_y} \right] \rho(r,s)\, dr ds$$

$$- \iint_S C(r,s;x,y) \left[\frac{1}{M} + \frac{\eta s}{I_x} + \frac{\xi r}{I_y} \right] \rho(r,s)\, dr ds. \quad (3\text{–}179)$$

Introducing Eq. (3–179) into Eq. (3–178) and applying Eqs. (3–162), (3–163), and (3–164) we obtain

$$\frac{1}{\omega_m{}^2} \iint_S W_m W_n \rho\, dx dy = \iint_S W_m \rho \left[\iint_S GW_n \rho\, d\xi d\eta \right] dx dy. \quad (3\text{–}180)$$

Since the right sides of Eqs. (3–176) and (3–180) are identical, we can conclude that

$$\left(\frac{1}{\omega_m{}^2} - \frac{1}{\omega_n{}^2} \right) \iint_S W_m W_n \rho\, dx dy = 0, \quad (3\text{–}181)$$

which reduces, for $m \neq n$, to the orthogonality condition

$$\iint_S W_m W_n \rho\, dx dy = 0. \quad (3\text{–}182)$$

It can be said that the process of computing the integral equation of the free unrestrained airplane given by Eq. (3–172) is one of "sweeping out" the rigid-body modes of motion. These rigid-body modes can be represented by

$$W_1(x,y) = C_1, \quad (3\text{–}183)$$

which defines a translation of the origin of coordinates along the z-axis:

$$W_2(x,y) = C_2 x, \qquad (3\text{-}184)$$

which defines a small rotation of the axis system about the y-axis, and

$$W_3(x,y) = C_3 y, \qquad (3\text{-}185)$$

which represents a small rotation of the axis system about the x-axis.

We may consider then, that an airplane, under the assumptions considered here, has three rigid-body modes of zero frequency given by expressions (3-183), (3-184), and (3-185), plus an infinite number of modes of finite frequency defined by solutions to Eq. (3-172). It is important to notice that the orthogonality condition stated by Eq. (3-182), which was derived from considerations of the deformation modes only, also encompasses the rigid-body modes. That is, the rigid-body modes are orthogonal to each other and orthogonal to the deformation modes. This fact, which can easily be verified by substitution, rests upon the assumption that the xyz-axis system is the principal centroidal axis system of the deformed airplane.

(b) *Forced motion in terms of normal modes.* A solution to the problem of the forced motion of an unrestrained airplane can be formulated in terms of normal modes of the free, unrestrained airplane. Let us consider at the outset, for simplicity, the forced motion of an elastic airplane free to plunge but restrained against pitching and rolling. If the applied dynamic force per unit area is denoted by $F_z(x,y,t)$, we have, for equilibrium of total forces along the z-axis,

$$\iint_S \ddot{w}(x,y,t)\rho(x,y)dxdy = \iint_S F_z(x,y,t)dxdy. \qquad (3\text{-}186)$$

Applying Eq. (2-55) and assuming that the airplane is free to plunge, the following equilibrium equation relates external, elastic, and inertial forces:

$$w(x,y,t) - w(0,0,t)$$
$$= -\iint_S C(x,y;\xi,\eta)[\rho(\xi,\eta)\ddot{w}(\xi,\eta,t) - F_z(\xi,\eta,t)]d\xi d\eta. \qquad (3\text{-}187)$$

In analyzing the forced motion of an unrestrained airplane, it is convenient to employ the concept of rigid-body modes. This means, in the present example, that the total displacement of the airplane is described in terms of a translation of the xy-plane in the z-direction, that is, a translation of the center of gravity plus elastic deformation with respect to the xy-plane. Thus, we represent the total displacement from any suitable horizontal reference plane by

$$w(x,y,t) = \sum_{i=1}^{\infty} \phi_i(x,y)\xi_i(t), \qquad (3\text{-}188)$$

where the translation of the xy-plane is represented by a rigid-body mode of zero frequency denoted by

$$\phi_1(x,y) = 1, \qquad \omega_1 = 0, \tag{3-189}$$

and the elastic deformation with respect to the xy-plane is represented by a superposition of normal modes of finite frequency denoted by

$$
\begin{array}{cc}
\phi_2(x,y) & \omega_2 \\
\cdot & \cdot \\
\cdot & \cdot \\
\cdot & \cdot \\
\phi_\infty(x,y) & \omega_\infty
\end{array}
\tag{3-190}
$$

It should be apparent that the meaning of the displacement function $w(x,y,t)$ is different in the case of forced motion than in the case of free vibration discussed in the previous section. In the case of free vibration, $w(x,y,t)$ represents a displacement with respect to the principal inertial axes which can be assumed fixed in space, whereas for forced motion, $w(x,y,t)$ represents the total displacement of the elastic surface which may, in general, include small vertical translation, pitching, and rolling displacements of the principal inertial axes.

It is apparent from the previous section that $\phi_2(x,y), \cdots, \phi_\infty(x,y)$, in the present example, are solutions of the homogeneous equations

$$\omega_i^2 \iint_S C(x,y;\xi,\eta)\phi_i(\xi,\eta)\rho(\xi,\eta)d\xi d\eta = \phi_i(x,y) - \phi_i(0,0),$$

$$(i = 2, 3, \cdots, \infty), \quad (3\text{-}191)$$

with the additional conditions

$$\iint_S \phi_i \rho(x,y)dxdy = 0. \tag{3-192}$$

Introducing Eq. (3-188) into Eq. (3-186) and applying Eqs. (3-189) and (3-192), we obtain

$$\ddot{\xi}_1 \iint_S \rho(x,y)dxdy = M\ddot{\xi}_1 = \iint_S F_z(x,y,t)dxdy, \tag{3-193}$$

where M is the mass of the airplane.

Substituting Eq. (3-188) into Eq. (3-187), and putting $\phi_1(x,y) = \phi_1(0,0)$ yields

$$\sum_{i=2}^{\infty} [\phi_i(x,y) - \phi_i(0,0)]\xi_i$$

$$= -\iint_S C(x,y;\xi,\eta)\left[\rho(\xi,\eta)\sum_{i=1}^{\infty}\phi_i(\xi,\eta)\ddot{\xi}_i - F_z(\xi,\eta,t)\right]d\xi d\eta. \tag{3-194}$$

By applying Eq. (3–191), we can reduce Eq. (3–194) to

$$\sum_{i=2}^{\infty} [\phi_i(x,y) - \phi_i(0,0)]\xi_i + \sum_{i=2}^{\infty} [\phi_i(x,y) - \phi_i(0,0)]\frac{\ddot{\xi}_i}{\omega_i^2}$$

$$= -\iint_S C(x,y;\xi,\eta)[\rho(\xi,\eta)\ddot{\xi}_1 - F_z(\xi,\eta,t)]d\xi d\eta. \quad (3\text{–}195)$$

Multiplying through by $\phi_j(x,y)\rho(x,y)$, integrating over the area S, and applying the orthogonality condition, we obtain

$$M_j\left(\frac{\ddot{\xi}_j}{\omega_j^2} + \xi_j\right) = \iint_S \phi_j(x,y)\rho(x,y)$$

$$\times \iint_S C(x,y;\xi,\eta)[F_z(\xi,\eta,t) - \rho(\xi,\eta)\ddot{\xi}_1]d\xi d\eta dx dy, \quad (3\text{–}196)$$

where

$$M_j = \iint_S \phi_j^2(x,y)\rho(x,y)dx dy.$$

Putting $C(x,y;\xi,\eta) = C(\xi,\eta;x,y)$ allows Eq. (3–196) to be reduced to

$$M_j\left(\frac{\ddot{\xi}_j}{\omega_j^2} + \xi_j\right) = \iint_S [F_z(x,y,t) - \rho(x,y)\ddot{\xi}_1]$$

$$\times \iint_S C(x,y;\xi,\eta)\phi_j(\xi,\eta)\rho(\xi,\eta)d\xi d\eta dx dy. \quad (3\text{–}197)$$

The right side of Eq. (3–197) can be further simplified by applying Eq. (3–191):

$$M_j\left(\frac{\ddot{\xi}_j}{\omega_j^2} + \xi_j\right) = \frac{1}{\omega_j^2} \iint_S [F_z(x,y,t) - \rho(x,y)\ddot{\xi}_1][\phi_j(x,y) - \phi_j(0,0)]dx dy.$$

$$(3\text{–}198)$$

Expanding the right side and making use of Eqs. (3–191) and (3–193) results finally in

$$M_j\ddot{\xi}_j + M_j\omega_j^2\xi_j = \iint_S F_z(x,y,t)\phi_j(x,y)dx dy, \quad (j=2, 3, \cdots, \infty). \quad (3\text{–}199)$$

The equations of motion of the elastic airplane with freedom to plunge consist therefore of a single equation describing the vertical translation of the center of gravity,

$$M_1\ddot{\xi}_1 = \Xi_1, \quad (3\text{–}200)$$

and an infinite set of equations describing displacements of the natural modes of motion,

$$M_j\ddot{\xi}_j + M_j\omega_j^2\xi_j = \Xi_j, \quad (j = 2, 3, \cdots, \infty). \quad (3\text{–}201)$$

In Eqs. (3–200) and (3–201), the generalized mass and generalized force are obtainable from the following formulas:

$$M_j = \iint_S \phi_j{}^2(x,y)\rho(x,y)dxdy,$$

$$\Xi_j = \iint_S F_z(x,y,t)\phi_j(x,y)dxdy. \tag{3-202}$$

The derivation given above can be extended to include small angular displacements in pitching and rolling by introducing the following additional rigid-body modes of zero frequency:

Pitching: $\phi_2 = -x, \quad \omega_2 = 0.$

Rolling: $\phi_3 = +y, \quad \omega_3 = 0.$ $\tag{3-203}$

A development similar to that given above leads, in the more general case, including plunging, pitching, and rolling, to the following differential equations:

$$M_1\ddot{\xi}_1 = \Xi_1, \qquad M_2\ddot{\xi}_2 = \Xi_2, \qquad M_3\ddot{\xi}_3 = \Xi_3,$$

$$M_j\ddot{\xi}_j + M_j\omega_j{}^2\xi_j = \Xi_j, \qquad (j = 4, 5, \cdots, \infty), \tag{3-204}$$

where M_j and Ξ_j are defined by Eqs. (3–202). It can be seen that the generalized mass M_1 in Eqs. (3–204) represents the mass of the airplane, and ξ_1 the plunging displacement. Similarly, M_2 and M_3 represent the pitching and rolling moments of inertia, and ξ_2 and ξ_3 the small angular displacements in pitch and roll, respectively.

3–8 Energy methods. The earlier parts of this chapter provide an illustration of one general approach to the computation of the response of a structure to dynamic loads. It was based on D'Alembert's principle, used in conjunction with the equations of equilibrium. A different and occasionally more useful approach is provided by energy methods. Such methods can be applied to the determination of the equations of equilibrium by formulating the problem in terms of a vanishing variation of certain work or energy expressions in the same manner as applied to static problems in Section 2–11. Thus the equations of equilibrium are obtained directly from the work and energy expressions, and it is not necessary to apply the equations of equilibrium explicitly.

(a) *The principle of virtual work applied to dynamic systems.* The fundamental basis for energy methods is the *principle of virtual work*, introduced in Section 2–11(a). In applying the principle of virtual work to dynamics problems, we proceed in exactly the same way as in the static application except that we now introduce D'Alembert's principle and include the inertial forces with the prescribed external forces. Thus, we have the following statement of the principle of virtual work for dynamic systems:

$$\delta W_e + \delta W_{in} = \delta U, \tag{3–205}$$

where δW_{in} is the virtual work due to the inertial forces during an arbitrary virtual displacement of the system.

(b) *Lagrange's equation.* An alternative but somewhat specialized form of the principle of virtual work can be derived, and is called Lagrange's equation.* It is applicable to systems in which the space configuration can be described by a set of discrete generalized coordinates. We have seen in Chapter 2, in connection with statics problems, that generalized coordinates are independent coordinates which represent possible displacements of the system. Thus, in applying the principle of virtual work, it is possible to satisfy the requirement that each generalized coordinate can be given a virtual displacement δq_i without changing the value of the other coordinates, and without violating the geometrical constraints on the system. In dynamic aeroelastic systems it is usually possible to select such a set of coordinates by inspection. Systems which can be described in terms of generalized coordinates are often called *holonomic* systems.†

In order to derive Lagrange's equation, let us consider an unrestrained three-dimensional continuous elastic body subjected to aerodynamic pressures over its outer surface. As a result of these aerodynamic pressures, the body is assumed to undergo small displacements $u(x,y,z,t)$, $v(x,y,z,t)$, and $w(x,y,z,t)$ with respect to an axis system fixed in space. If we represent the surface aerodynamic pressure by $F_s(x,y,z,t)$, with components $F_x(x,y,z,t)$, $F_y(x,y,z,t)$, and $F_z(x,y,z,t)$, the virtual work due to the external forces during arbitrary virtual displacements δu, δv, and δw is given by

$$\delta W_e = \int_S (F_x \delta u + F_y \delta v + F_z \delta w) dS, \tag{3–206}$$

where the integration is over the surface of the body. If the mass density is $\rho(x,y,z)$, the virtual work due to the inertial forces is given by

$$\delta W_{in} = - \int_V \rho(\ddot{u}\delta u + \ddot{v}\delta v + \ddot{w}\delta w) dV, \tag{3–207}$$

---

* The principle of virtual work also forms the basis for Hamilton's principle (Ref. 3–9), which applies to continuous systems, as well as those that are described by a number of discrete coordinates.

† When the constraints on the system are such that this is not possible, the system is called nonholonomic. Such systems are not considered here.

where the integration is over the volume. Representing the strain energy per unit volume (strain energy density) by the symbol $U_0(x,y,z)$, the change in internal strain energy can be expressed symbolically by

$$\delta U = \int_V \delta U_0 dV. \qquad (3\text{–}208)$$

We assume next that any displacement of the body can be expressed in terms of a set of n discrete generalized coordinates. Thus, we can, in principle, write the following transformation relations:

$$u(x,y,z,t) = u(x,y,z,q_1,q_2,\cdots,q_n),$$
$$v(x,y,z,t) = v(x,y,z,q_1,q_2,\cdots,q_n), \qquad (3\text{–}209)$$
$$w(x,y,z,t) = w(x,y,z,q_1,q_2,\cdots,q_n),$$

where the generalized coordinates are functions of time.

Introducing Eqs. (3–209) into Eq. (3–206), we obtain an expression for the virtual work due to the external forces, as follows:

$$\delta W_e = \sum_{i=1}^{n} Q_i \delta q_i, \qquad (3\text{–}210)$$

where

$$Q_i = \int_S \left(F_x \frac{\partial u}{\partial q_i} + F_y \frac{\partial v}{\partial q_i} + F_z \frac{\partial w}{\partial q_i} \right) dS \qquad (3\text{–}211)$$

is the generalized force corresponding to the generalized coordinate q_i.

Introducing Eqs. (3–209) into Eq. (3–207) gives

$$\delta W_{\text{in}} = -\sum_{i=1}^{n} \int_V \rho \left(\ddot{u} \frac{\partial u}{\partial q_i} + \ddot{v} \frac{\partial v}{\partial q_i} + \ddot{w} \frac{\partial w}{\partial q_i} \right) dV \delta q_i. \qquad (3\text{–}212)$$

Equation (3–212) can be rewritten in the form

$$\delta W_{\text{in}} = -\frac{d}{dt} \sum_{i=1}^{n} \int_V \rho \left(\dot{u} \frac{\partial u}{\partial q_i} + \dot{v} \frac{\partial v}{\partial q_i} + \dot{w} \frac{\partial w}{\partial q_i} \right) dV \delta q_i$$
$$+ \sum_{i=1}^{n} \int_V \rho \left(\dot{u} \frac{\partial \dot{u}}{\partial \dot{q}_i} + \dot{v} \frac{\partial \dot{v}}{\partial \dot{q}_i} + \dot{w} \frac{\partial \dot{w}}{\partial \dot{q}_i} \right) dV \delta q_i. \qquad (3\text{–}213)$$

Equation (3–213) can be simplified by introducing the kinetic energy of the body, defined by

$$T = \tfrac{1}{2} \int_V \rho (\dot{u}^2 + \dot{v}^2 + \dot{w}^2) dV. \qquad (3\text{–}214)$$

From Eq. (3–214), we observe that

$$\frac{\partial T}{\partial \dot{q}_i} = \int_V \rho \left(\dot{u} \frac{\partial \dot{u}}{\partial \dot{q}_i} + \dot{v} \frac{\partial \dot{v}}{\partial \dot{q}_i} + \dot{w} \frac{\partial \dot{w}}{\partial \dot{q}_i} \right) dV, \qquad (3\text{–}215)$$

and since, for example,

$$\frac{\partial \dot{u}}{\partial \dot{q}_i} = \frac{\partial}{\partial \dot{q}_i}\left(\sum_{j=1}^{n}\frac{\partial u}{\partial q_j}\dot{q}_j\right) = \frac{\partial u}{\partial q_i}, \tag{3-216}$$

Eq. (3-215) becomes

$$\frac{\partial T}{\partial \dot{q}_i} = \int_V \rho\left(\dot{u}\frac{\partial u}{\partial q_i} + \dot{v}\frac{\partial v}{\partial q_i} + \dot{w}\frac{\partial w}{\partial q_i}\right)dV. \tag{3-217}$$

In addition, it can be seen that

$$\frac{\partial T}{\partial q_i} = \int_V \rho\left(\dot{u}\frac{\partial \dot{u}}{\partial q_i} + \dot{v}\frac{\partial \dot{v}}{\partial q_i} + \dot{w}\frac{\partial \dot{w}}{\partial q_i}\right)dV. \tag{3-218}$$

Inserting Eqs. (3-217) and (3-218) into Eq. (3-213), we obtain the following virtual work expression for the inertial forces in terms of the kinetic energy:

$$\delta W_{\text{in}} = -\frac{d}{dt}\sum_{i=1}^{n}\frac{\partial T}{\partial \dot{q}_i}\delta q_i + \sum_{i=1}^{n}\frac{\partial T}{\partial q_i}\delta q_i. \tag{3-219}$$

Since the strain energy per unit volume is also expressible as a function of the generalized coordinates, Eq. (3-208) can be written in the form

$$\delta U = \int_V\left(\sum_{i=1}^{n}\frac{\partial U_0}{\partial q_i}\delta q_i\right)dV = \sum_{i=1}^{n}\frac{\partial U}{\partial q_i}\delta q_i, \tag{3-220}$$

where

$$\frac{\partial U}{\partial q_i} = \frac{\partial}{\partial q_i}\int_V U_0 dV.$$

Introducing Eqs. (3-210), (3-219), and (3-220) into the virtual work expression given by Eq. (3-205) yields

$$\sum_{i=1}^{n}\left\{Q_i - \frac{d}{dt}\left(\frac{\partial T}{\partial \dot{q}_i}\right) + \frac{\partial T}{\partial q_i} - \frac{\partial U}{\partial q_i}\right\}\delta q_i = 0, \tag{3-221}$$

which, since the δq_i are completely independent and arbitrary finite quantities, is equivalent to a set of equations of the form

$$\frac{d}{dt}\left(\frac{\partial T}{\partial \dot{q}_i}\right) - \frac{\partial T}{\partial q_i} + \frac{\partial U}{\partial q_i} = Q_i, \qquad (i = 1, \cdots, n). \tag{3-222}$$

This equation is called Lagrange's equation of motion. Lagrange's equation in this form can be used as a substitute for the more fundamental principle of virtual work as a means of writing the differential equations of equilibrium of a system whose configuration can be expressed in terms of discrete generalized coordinates.* In the above development, we assumed

---

* It should be apparent that Lagrange's equation plays the same role in dynamics problems that is played by Eq. (2-97) in statics problems, namely, $\partial U/\partial q_i = Q_i$. In fact, the latter can be seen to be a special case of Lagrange's equation when the kinetic energy is zero.

for simplicity that any displacement of the body can be expressed in terms of a set of n discrete generalized coordinates. This is generally not exactly true for a continuous system, and in order to obtain an exact solution a complete system of functions compatible with the boundary conditions has to be taken, and n must be increased without limit. If the latter conditions are fulfilled, Lagrange's equation forms the basis for deriving an exact solution in terms of a denumerably infinite set of generalized coordinates. However, if n remains finite, the procedure gives only an approximate solution, and its accuracy depends completely on the choice of generalized coordinates. For a lumped parameter system with n degrees of freedom, Lagrange's equation gives exact differential equations of equilibrium.

Lagrange's equation finds wide application in dynamic aeroelastic problems. The first principal application to be mentioned is in the derivation of the exact differential equations of equilibrium of a lumped parameter system. This is illustrated in Section 3–8(c), which follows. A second principal application is in deriving differential equations of equilibrium appropriate to continuous systems. These results may be exact if n is increased without limit. This feature of Lagrange's equation is illustrated in Section 3–8(d). Approximate results are obtained by taking a finite value of n, which forms the basis for the Rayleigh-Ritz method as applied to dynamics problems. This last, but perhaps most important, application of Lagrange's equation is illustrated in Section 3–10.

(c) *Lagrange's equation applied to a lumped parameter system.* In illustrating the application of Lagrange's equation to a lumped parameter system, let us consider, for example, the system consisting of a spring-supported rigid wing segment illustrated by Fig. 3–22. If the wing segment is permitted freedom to execute *small* vertical and angular displacements, the coordinates h and α shown by Fig. 3–22 are convenient generalized coordinates * for describing the displaced configuration of the system. Thus, the transformation equations, corresponding to Eqs. (3–209), between wing displacement and generalized coordinates are

$$u = 0, \qquad v = 0, \qquad w = -h - r\alpha, \qquad (3\text{–}223)$$

where r is a running coordinate along the chord measured positive in the aft direction from the elastic axis. The kinetic energy of the wing segment is obtained by applying Eq. (3–214):

$$T = \tfrac{1}{2}\int_{\text{chord}} \dot{w}^2 dm = \tfrac{1}{2}\int_{\text{chord}} (\dot{h} + r\dot{\alpha})^2 dm. \qquad (3\text{–}224)$$

---

* It is evident by inspection that these coordinates can each be varied independently without violating the geometrical constraints on the system. That is, they are not related by equations of constraint. It can be shown (Ref. 3–9) that this is the least possible number of coordinates that will completely describe the configuration of the system. This number is often called the number of degrees of freedom of the system.

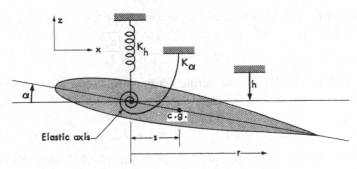

Fig. 3-22.   Spring-supported rigid-wing segment.

Integration of Eq. (3–224) gives a quadratic function of the velocities $\dot{h}$ and $\dot{\alpha}$, as follows:

$$T = \tfrac{1}{2}m\dot{h}^2 + \tfrac{1}{2}I_\alpha\dot{\alpha}^2 + S_\alpha\dot{\alpha}\dot{h}, \qquad (3\text{–}225)$$

where

m = mass of wing segment,

I_α = mass moment of inertia of wing segment about the elastic axis, and

$S_\alpha = ms$ = static mass moment of wing segment about the elastic axis.

If the stiffnesses of the bending and torsion springs are represented by K_h and K_α, respectively, the strain energy is given by

$$U = \tfrac{1}{2}K_h h^2 + \tfrac{1}{2}K_\alpha \alpha^2. \qquad (3\text{–}226)$$

Lagrange's equations applied to this problem are

$$\frac{d}{dt}\left(\frac{\partial T}{\partial \dot{h}}\right) + \frac{\partial U}{\partial h} = 0, \qquad \frac{d}{dt}\left(\frac{\partial T}{\partial \dot{\alpha}}\right) + \frac{\partial U}{\partial \alpha} = 0. \qquad (3\text{–}227)$$

It should be observed that the terms $\partial T/\partial q_i$ do not appear in the above equations, since T is a function of the velocity terms only in the present example. Inserting Eqs. (3–225) and (3–226) into Eqs. (3–227), the following equations of motion are obtained:

$$m\ddot{h} + S_\alpha\ddot{\alpha} + K_h h = 0, \qquad S_\alpha\ddot{h} + I_\alpha\ddot{\alpha} + K_\alpha \alpha = 0. \qquad (3\text{–}228)$$

It is sometimes convenient, particularly in flutter analyses, to write Eqs. (3–228) in terms of the frequencies of the uncoupled system. The system can be uncoupled by putting the inertial coupling terms, S_α, equal to zero. Thus,

$$m\ddot{h} + K_h h = 0, \qquad I_\alpha\ddot{\alpha} + K_\alpha \alpha = 0. \qquad (3\text{–}229)$$

Inserting as solutions

$$h = h_0 \sin \omega_h t, \qquad \alpha = \alpha_0 \sin \omega_\alpha t, \qquad (3\text{–}230)$$

we obtain the following natural frequencies of the uncoupled system:

$$\omega_h = \sqrt{\frac{K_h}{m}}, \qquad \omega_\alpha = \sqrt{\frac{K_\alpha}{I_\alpha}}. \qquad (3\text{-}231)$$

Solving Eqs. (3-231) for the spring constants in terms of the uncoupled natural frequencies and substituting the results into Eqs. (3-228), we obtain

$$m\ddot{h} + S_\alpha\ddot{\alpha} + m\omega_h^2 h = 0, \qquad S_\alpha\ddot{h} + I_\alpha\ddot{\alpha} + I_\alpha\omega_\alpha^2\alpha = 0. \qquad (3\text{-}232)$$

We have seen in the above illustration, Eq. (3-225), that the kinetic energy is a positive definite quadratic function of the velocities $\dot{\alpha}$ and $\dot{h}$. This is a property of systems with small displacements, and it can be shown that the kinetic energy is always expressible in this form, providing the displacements are small. If we consider, for simplicity, a lumped parameter system with the transformation equations

$$u = u(q_1,q_2,q_3), \qquad v = v(q_1,q_2,q_3), \qquad w = w(q_1,q_2,q_3), \qquad (3\text{-}233)$$

the kinetic energy is given by Eq. (3-214) as

$$T = \tfrac{1}{2}\int_V \rho \left\{ \left(\sum_{i=1}^{3} \frac{\partial u}{\partial q_i}\dot{q}_i \right)^2 + \left(\sum_{i=1}^{3} \frac{\partial v}{\partial q_i}\dot{q}_i \right)^2 + \left(\sum_{i=1}^{3} \frac{\partial w}{\partial q_i}\dot{q}_i \right)^2 \right\} dV. \qquad (3\text{-}234)$$

Expanding Eq. (3-234) gives

$$T = \tfrac{1}{2}\int_V \rho \left\{ \sum_{i=1}^{3}\sum_{j=1}^{3} \frac{\partial u}{\partial q_i}\frac{\partial u}{\partial q_j}\dot{q}_i\dot{q}_j + \sum_{i=1}^{3}\sum_{j=1}^{3} \frac{\partial v}{\partial q_i}\frac{\partial v}{\partial q_j}\dot{q}_i\dot{q}_j \right.$$
$$\left. + \sum_{i=1}^{3}\sum_{j=1}^{3} \frac{\partial w}{\partial q_i}\frac{\partial w}{\partial q_j}\dot{q}_i\dot{q}_j \right\} dV, \qquad (3\text{-}235)$$

which can be written as

$$T = \tfrac{1}{2}\sum_{i=1}^{3}\sum_{j=1}^{3} m_{ij}\dot{q}_i\dot{q}_j, \qquad (3\text{-}236)$$

where

$$m_{ij} = \int_V \rho \left(\frac{\partial u}{\partial q_i}\frac{\partial u}{\partial q_j} + \frac{\partial v}{\partial q_i}\frac{\partial v}{\partial q_j} + \frac{\partial w}{\partial q_i}\frac{\partial w}{\partial q_j} \right) dV. \qquad (3\text{-}237)$$

Since u, v, and w are, in general, functions of q_1, q_2, and q_3, each of the derivatives in Eq. (3-237) is a function of q_1, q_2, and q_3. However, if q_1, q_2, and q_3 are small and are measured from the position of equilibrium about which the system vibrates, we can with small error put $q_1 = q_2 = q_3 = 0$ in each of the derivatives. Equation (3-237) thus becomes a constant. It is also apparent from Eq. (3-237) that the inertial parameters, m_{ij}, enjoy the reciprocal property

$$m_{ij} = m_{ji}. \qquad (3\text{-}238)$$

(d) *Lagrange's equation applied to a continuous system.* Let us consider the problem of lateral deflections of slender beams under arbitrary side forces. We select as the functional relation between beam displacement and generalized coordinates

$$w(y,t) = \sum_{i=1}^{\infty} \phi_i(y)\xi_i(t), \qquad (3\text{–}239)$$

where the $\phi_i(y)$ are normalized natural mode shapes of the beam under the prescribed boundary conditions, and the $\xi_i(t)$ are normal coordinates. Introducing Eq. (3–239) into the expression for the kinetic energy of the beam, we obtain

$$T = \frac{1}{2}\int_0^l \dot{w}^2(y,t)m(y)dy = \frac{1}{2}\sum_{i=1}^{\infty}\sum_{j=1}^{\infty}\int_0^l \phi_i(y)\phi_j(y)m(y)dy\dot{\xi}_i\dot{\xi}_j. \quad (3\text{–}240)$$

Since the natural mode shapes are orthogonal, Eq. (3–240) reduces to

$$T = \frac{1}{2}\sum_{i=1}^{\infty} M_i\dot{\xi}_i^2, \qquad (3\text{–}241)$$

where

$$M_i = \int_0^l \phi_i^2(y)m(y)dy.$$

Thus the kinetic energy reduces to a function of squares of the generalized coordinates in the case where the generalized coordinates are normal coordinates.

The strain energy can be calculated in terms of the normal coordinates, as follows:

$$U = \frac{1}{2}\int_0^l EI(y)\left(\frac{\partial^2 w}{\partial y^2}\right)^2 dy = \frac{1}{2}\sum_{i=1}^{\infty}\sum_{j=1}^{\infty}\int_0^l EI(y)\phi_i''\phi_j''dy\xi_i\xi_j. \quad (3\text{–}242)$$

Equation (3–242) can be reduced by applying Eq. (3–15). The latter can be written in terms of normalized natural mode shapes:

$$\int_0^l (EI\phi_i'')''\phi_j dy = \omega_i^2\int_0^l \phi_i\phi_j m dy. \qquad (3\text{–}243)$$

Integrating the left side by parts gives

$$\left[\phi_j(EI\phi_i'')' - \phi_j'(EI\phi_i'')\right]_0^l + \int_0^l EI\phi_i''\phi_j''dy = \omega_i^2\int_0^l \phi_i\phi_j m dy. \quad (3\text{–}244)$$

For all practical end conditions, we see that the bracketed term is zero, and Eq. (3–244) reduces by the orthogonality condition to

$$\int_0^l EI\phi_i''\phi_j''dy = \begin{cases} M_i\omega_i^2, & \text{for } j = i, \\ 0, & \text{for } j \neq i. \end{cases} \qquad (3\text{–}245)$$

Applying Eq. (3–245) to Eq. (3–242), we obtain the following result for the strain energy:

$$U = \tfrac{1}{2} \sum_{i=1}^{\infty} M_i \omega_i^2 \xi_i^2. \tag{3–246}$$

Thus we see that the employment of normal coordinates reduces the kinetic and potential energy expressions to sums of squares. This is a general property of normal coordinates.

The form of Lagrange's equation applicable to the present problem is

$$\frac{d}{dt}\left(\frac{\partial T}{\partial \dot{\xi}_j}\right) + \frac{\partial U}{\partial \xi_j} = \Xi_j, \qquad (j = 1, \cdots, \infty). \tag{3–247}$$

Substituting the expressions for kinetic and potential energy given by Eqs. (3–241) and (3–246) into Lagrange's equation results in

$$M_j \ddot{\xi}_j + M_j \omega_j^2 \xi_j = \Xi_j, \qquad (j = 1, \cdots, \infty). \tag{3–248}$$

The generalized force Ξ_j is obtained by applying Eq. (3–206) in the following way:

$$\delta W_e = \int_0^l F_z(y,t)\delta w(y,t)dy = \int_0^l F_z(y,t) \sum_{i=1}^{\infty} \phi_i(y)\delta\xi_i dy = \sum_{i=1}^{\infty} \Xi_i \delta\xi_i, \tag{3–249}$$

which results, since the $\delta\xi_i$ are arbitrary and finite, in

$$\Xi_j = \int_0^l F_z(y,t)\phi_j(y)dy. \tag{3–250}$$

Equations (3–248) and (3–250) are identical to those previously derived from the partial differential and integral equations of forced vibrations of slender beams.

EXAMPLE 3–6. To derive the differential equations of forced motion of an unrestrained elastic airplane in terms of normal coordinates by means of Lagrange's equation.

Solution. We introduce the transformation

$$w(x,y,t) = \sum_{i=1}^{\infty} \phi_i(x,y)\xi_i(t), \tag{a}$$

where $w(x,y,t)$ is the small vertical displacement of the elastic surface representing the airplane measured from an arbitrarily positioned horizontal plane. In this transformation, displacement of the hypothetical airplane is represented by three rigid-body modes of zero frequency plus a superposition of normalized modes of the unrestrained airplane, as follows:

$$\phi_1 = 1, \qquad \omega_1 = 0,$$
$$\phi_2 = -x, \qquad \omega_2 = 0, \tag{b}$$
$$\phi_3 = +y, \qquad \omega_3 = 0.$$

The normalized deformation modes and frequencies of the unrestrained airplane are solutions to Eq. (3–172) and they are represented by

$$\phi_4 \qquad \omega_4$$
$$\vdots \qquad \vdots$$
$$\phi_\infty \qquad \omega_\infty$$

(c)

Lagrange's equation can be applied to this problem by computing the kinetic and strain energies in terms of normal coordinates. The kinetic energy can be reduced as follows:

$$T = \tfrac{1}{2} \iint_S \dot{w}^2 \rho \, dx \, dy = \tfrac{1}{2} \iint_S \rho \left[\sum_{i=1}^{\infty} \phi_i \dot{\xi}_i \right]^2 dx \, dy.$$

(d)

Expanding and applying Eqs. (3–162), (3–163), (3–164), and (3–182), the kinetic energy reduces to

$$T = \tfrac{1}{2} \sum_{i=1}^{\infty} M_i \dot{\xi}_i^2,$$

(e)

where

$$M_i = \iint_S \phi_i^2 \rho \, dx \, dy.$$

The strain energy can be expressed symbolically in terms of a stiffness influence function $k(x,y;\xi,\eta)$ by applying the two-dimensional form of Eq. (2–60), as follows:

$$U = \tfrac{1}{2} \iint_S \left[w(x,y) - w(0,0) - x \frac{\partial w(0,0)}{\partial x} - y \frac{\partial w(0,0)}{\partial y} \right]$$
$$\times \iint_S k(x,y;\xi,\eta) \left[w(\xi,\eta) - w(0,0) - \xi \frac{\partial w(0,0)}{\partial x} - \eta \frac{\partial w(0,0)}{\partial y} \right] d\xi \, d\eta \, dx \, dy.$$

(f)

Substituting Eq. (a) into Eq. (f) results in

$$U = \tfrac{1}{2} \iint_S \sum_{i=4}^{\infty} \left[\phi_i(x,y) - \phi_i(0,0) - x \frac{\partial \phi_i(0,0)}{\partial x} - y \frac{\partial \phi_i(0,0)}{\partial y} \right] \xi_i$$
$$\times \iint_S k(x,y;\xi,\eta) \sum_{j=4}^{\infty} \left[\phi_j(\xi,\eta) - \phi_j(0,0) - \xi \frac{\partial \phi_j(0,0)}{\partial x} \right.$$
$$\left. - \eta \frac{\partial \phi_j(0,0)}{\partial y} \right] \xi_j \, d\xi \, d\eta \, dx \, dy.$$

(g)

The lower limits on the summations are put at 4, since for $i < 4$ the sums within the square brackets are zero. Equation (g) can be reduced by applying Eq. (3–166a). Using the reciprocal relation between $C(x,y;\xi,\eta)$ and $k(x,y;\xi,\eta)$, we observe that Eq. (3–166a), expressed in terms of normalized mode shapes, can be inverted to give

$$\rho(x,y)\phi_i(x,y)\omega_i^2 = \iint_S k(x,y;\xi,\eta) \left[\phi_i(\xi,\eta) - \phi_i(0,0) - \xi \frac{\partial \phi_i(0,0)}{\partial x} \right.$$
$$\left. - \eta \frac{\partial \phi_i(0,0)}{\partial y} \right] d\xi \, d\eta.$$

(h)

Introducing Eq. (h) into Eq. (g) yields

$$U = \tfrac{1}{2} \sum_{i=4}^{\infty} \sum_{j=4}^{\infty} \left\{ \iint_S \left[\phi_i(x,y) - \phi_i(0,0) - x\frac{\partial\phi_i(0,0)}{\partial x} - y\frac{\partial\phi_i(0,0)}{\partial y} \right] \right. $$
$$\left. \times \phi_j(x,y)\rho(x,y)dxdy \right\} \omega_j^2 \xi_i \xi_j. \quad \text{(i)}$$

Applying the orthogonality condition to Eq. (i) results in

$$U = \tfrac{1}{2} \sum_{i=4}^{\infty} M_i \omega_i^2 \xi_i^2, \tag{j}$$

where

$$M_i = \iint_S \phi_i^2 \rho \, dx \, dy.$$

Substituting Eqs. (e) and (j) into Lagrange's equation, we obtain the following differential equations in terms of normal coordinates,

$$M_1\ddot{\xi}_1 = \Xi_1, \qquad M_2\ddot{\xi}_2 = \Xi_2, \qquad M_3\ddot{\xi}_3 = \Xi_3, \tag{k}$$

$$M_j\ddot{\xi}_j + M_j\omega_j^2\xi_j = \Xi_j, \qquad (j = 4, \cdots, \infty). \tag{l}$$

If the airplane is subjected to a distributed force $F_z(x,y,t)$, the generalized forces are computed from the following equation, derived from Eq. (3–206):

$$\delta W_e = \iint_S F_z(x,y,t)\delta w \, dx \, dy = \sum_{i=1}^{\infty} \Xi_i \delta\xi_i, \tag{m}$$

which results in

$$\Xi_j = \iint_S F_z(x,y,t)\phi_j(x,y)dxdy, \qquad (j = 1, 2, \cdots, \infty). \tag{n}$$

3–9 Approximate methods of solution to practical problems.

Previous sections in the present chapter have been concerned with the formulation of the equations of motion of a continuous system and their exact solution in terms of a denumerably infinite set of normal coordinates. Let us consider next some of the processes involved in obtaining solutions to practical problems. Such solutions, in order to be tractable, must, of course, involve some degree of approximation. Generally, the approximation consists of assuming that the space configuration of the deformed structure, which is actually an infinite degree of freedom system, can be approximated by some equivalent system with a finite number of degrees of freedom. The success of this initial step is dependent largely upon the ingenuity and judgment of the analyst. Once the step is taken, the behavior of the system is defined by a finite set of simultaneous total differential equations in the independent variable time.

The first method of approximation which suggests itself is one of simply taking a finite number of normal coordinates rather than an infinite number, and employing the solutions which have already been stated in previous sections of the present chapter. This is perhaps the most widely used approach to aeroelastic problems, and it is usually successful, since only a relatively small number of the lower natural modes of a structure are actually needed to describe its deformation in most cases. Because of the fundamental importance of this approach in the solution of aeroelastic problems, Chapter 4 has been devoted entirely to methods for computing natural mode shapes.

A second method of approximation, which has found wide application because of its simplicity, is that of approximating the space configuration of a structure by a superposition of a finite number of *assumed* mode shapes. We shall refer to this as the Rayleigh-Ritz method, and it corresponds, of course, to the similar procedure employed in Section 2-11 in the solution of static deformation problems. In this method, each assumed deformation shape or mode represents a degree of freedom, and the multiplier that determines the amount of its contribution to any general deformation represents the generalized coordinate corresponding to that mode. Application of the Rayleigh-Ritz method to dynamic problems is discussed in some detail in Section 3-10.

The first and second methods described above of reducing the continuous system to a finite degree of freedom system are based upon representation of the actual deformation shape as a superposition of explicitly defined continuous functions or modes. A third and somewhat different approach which is often employed is termed a lumped parameter method, in which the deformation of a continuous structure is approximated by a finite number of discrete generalized displacements of various parts of the structure. The lumped mass or rigid segment method in which the structure is divided into a number of rigid segments with interconnecting weightless springs is a prominent example. The lumped parameter method is discussed in Section 3-11.

The methods outlined above serve to reduce the problem from one involving partial differential or integral equations with an infinity of degrees of freedom, to one involving a finite number of simultaneous total differential equations. Methods for solving simultaneous total differential equations are considered in some detail in the particular applications where they arise in subsequent chapters.

3-10 Approximate solutions by the Rayleigh-Ritz method. The technique of expressing a solution in terms of a finite set of assumed modes, known as the Rayleigh-Ritz method, can be applied to dynamics problems in much the same manner as applied to statics problems in Section 2-11. The contribution of each assumed mode to the total motion is derived by applying the principle of virtual work or, more specifically, the form of this principle known as Lagrange's equation.

(a) *Restrained airplane wing.* Consider, for example, the problem of computing an approximate solution to the forced response of a restrained airplane wing fixed at the root under a dynamic side load. We take as the assumed deflection shape the finite series

$$w(x,y,t) = \sum_{i=1}^{n} \gamma_i(x,y) q_i(t), \qquad (3\text{-}251)$$

where $\gamma_i(x,y)$ are assumed displacement functions that satisfy at least the geometrical boundary conditions at the wing root.* For a thin cantilever wing, the latter are

$$\gamma_i(x,0) = 0, \qquad \gamma_i'(x,0) = 0. \qquad (3\text{-}252)$$

The quantities $q_i(t)$ are generalized coordinates, to be determined, which measure the contribution of each of the assumed displacement functions to the total displacement of the wing surface. Inserting Eq. (3–251) into the expression for the kinetic energy of the restrained wing, we obtain

$$T = \tfrac{1}{2}\iint_S \dot{w}^2 \rho\, dx\, dy = \tfrac{1}{2}\sum_{i=1}^{n}\sum_{j=1}^{n} m_{ij}\dot{q}_i\dot{q}_j, \qquad (3\text{-}253\text{a})$$

where

$$m_{ij} = \iint_S \gamma_i\gamma_j\rho\, dx\, dy \qquad (3\text{-}253\text{b})$$

are inertial coupling terms between generalized coordinates with the property $m_{ij} = m_{ji}$.

The strain energy can be derived by inserting Eq. (3–251) into the expression for the strain energy expressed in terms of a stiffness influence function, which gives

$$U = \tfrac{1}{2}\iint_S w(x,y,t)\iint_S k(x,y;\xi,\eta)w(\xi,\eta,t)\,d\xi\,d\eta\,dx\,dy = \tfrac{1}{2}\sum_{i=1}^{n}\sum_{j=1}^{n} k_{ij}q_iq_j, \quad (3\text{-}254)$$

where

$$k_{ij} = \iint_S \gamma_i(x,y)\iint_S k(x,y;\xi,\eta)\gamma_j(\xi,\eta)\,d\xi\,d\eta\,dx\,dy \qquad (3\text{-}255)$$

are elastic coupling terms between generalized coordinates with the property $k_{ij} = k_{ji}$. Substituting Eqs. (3–253a) and (3–254) into Lagrange's equation, we obtain n total differential equations of the form

$$\sum_{j=1}^{n} m_{ij}\ddot{q}_j + \sum_{j=1}^{n} k_{ij}q_j = Q_i, \qquad (i = 1, \cdots, n). \qquad (3\text{-}256)$$

---

* Although the principle of virtual work and Lagrange's equation require that only the geometrical boundary conditions need be satisfied, a better approximation can be obtained by choosing functions that satisfy the free-edge boundary conditions as well.

The generalized force Q_i is computed as follows:

$$\delta W_e = \sum_{i=1}^{n} \left\{ \iint_S F_z(x,y,t)\gamma_i(x,y)dxdy \right\} \delta q_i = \sum_{i=1}^{n} Q_i \delta q_i,$$
$$(i = 1, \cdots, n), \quad (3\text{–}257)$$

which results in

$$Q_i = \iint_S F_z(x,y,t)\gamma_i(x,y)dxdy, \qquad (i = 1, \cdots, n). \quad (3\text{–}258)$$

Solutions to Eqs. (3–256) for $q_1(t), \cdots, q_n(t)$ are substituted into Eq. (3–251) to obtain the final approximate solution. It is important to observe that differential equations (3–256) are coupled both inertially and elastically. We have seen that both these couplings disappear if the assumed modes are selected as natural vibration modes.

The degree of approximation to the exact solution provided by the Rayleigh-Ritz method depends entirely upon the choice and number of generalized coordinates. It can be stated, however, that when n is increased without limit, the difference between the approximate solution and the exact solution approaches zero.

Practical computations involving Eqs. (3–256) require a numerical evaluation of the coefficients m_{ij} and k_{ij}, as well as the generalized force Q_i. The computation of k_{ij} is the most tedious, and varies with different types of structures. The simplest case involves a slender wing with rigid chordwise cross sections. In this case, the strain energy has the form

$$U = \int_0^l \frac{EI}{2}\left(\frac{\partial^2 w}{\partial y^2}\right)^2 dy + \int_0^l \frac{GJ}{2}\left(\frac{\partial^2 w}{\partial y \partial x}\right)^2 dy, \quad (3\text{–}259)$$

where $\partial w/\partial x$ is an angle of twist of the wing assumed rigid in the chordwise direction. Introducing Eq. (3–251), we obtain

$$k_{ij} = \int_0^l EI(y)\left(\frac{\partial^2 \gamma_i}{\partial y^2}\right)\left(\frac{\partial^2 \gamma_j}{\partial y^2}\right)dy + \int_0^l GJ(y)\left(\frac{\partial^2 \gamma_i}{\partial y \partial x}\right)\left(\frac{\partial^2 \gamma_j}{\partial y \partial x}\right)dy. \quad (3\text{–}260)$$

A more complex application involves a low-aspect ratio wing acting as a deformable plate, where the flexibility influence coefficient matrix is defined at n discrete points. In this case, we calculate a numerical approximation to Eq. (3–255) by expressing it in the matrix form

$$k_{ij} = \lfloor \gamma_i \rfloor [\overline{W}][k][\overline{W}]\{\gamma_j\}, \quad (3\text{–}261)$$

which can be expressed in terms of flexibility influence coefficients as follows:

$$k_{ij} = \lfloor \gamma_i \rfloor [\overline{W}][C]^{-1}[\overline{W}]\{\gamma_j\}, \quad (3\text{–}262)$$

where

$\lfloor \gamma_i \rfloor$ = row matrix of amplitudes of ith mode shape,

$\{\gamma_j\}$ = column matrix of amplitudes of jth mode shape,

$[\overline{W}]$ = weighting matrix,

$[k]$ = $[C]^{-1}$ = matrix of stiffness influence coefficients.

(b) *Unrestrained elastic airplane.* Because of the labor involved in computing natural modes of the unrestrained airplane, it is often convenient to assume that the deformations of the structure are approximated by a superposition of assumed or artificial modes. We may, for example, assume that the motion is described by the small linear and angular displacements of a plane tangent to the elastic surface of the airplane (Fig. 3–21) at the point pierced by the z-axis, plus assumed elastic modes representing displacements with respect to the tangent plane. Thus

$$w(x,y,t) = \sum_{i=1}^{n} \gamma_i(x,y)q_i(t), \tag{3-263}$$

where $w(x,y,t)$ is the total vertical displacement of the elastic surface measured with respect to an arbitrarily positioned horizontal plane. In Eq. (3–263), the first three assumed modes represent the linear and angular displacements of the plane tangent to the deformed surface, and the remaining modes are assumed elastic displacements with respect to the tangent plane. Inserting Eq. (3–263) into the expression for the kinetic energy of the unrestrained airplane, we obtain

$$T = \tfrac{1}{2}\iint_S \dot{w}^2 \rho\, dx dy = \tfrac{1}{2}\sum_{i=1}^{n}\sum_{j=1}^{n} m_{ij}\dot{q}_i\dot{q}_j, \tag{3-264}$$

where

$$m_{ij} = \iint_S \gamma_i\gamma_j\rho\, dx dy.$$

The strain energy is expressed by

$$U = \tfrac{1}{2}\sum_{i=4}^{n}\sum_{j=4}^{n} k_{ij}q_i q_j, \tag{3-265}$$

where the k_{ij} can be defined in the same way as in Eq. (3–255). Substituting Eqs. (3–264) and (3–265) into Lagrange's equation, we obtain n ordinary differential equations:

$$\sum_{j=1}^{n} m_{ij}\ddot{q}_j = Q_i, \qquad (i = 1, 2, 3),$$

$$\sum_{j=1}^{n} m_{ij}\ddot{q}_j + \sum_{j=4}^{n} k_{ij}q_j = Q_i, \qquad (i = 4, \cdots, n), \tag{3-266}$$

where

$$Q_i = \iint_S F_z(x,y,t)\gamma_i\, dx dy. \tag{3-267}$$

Since the $\gamma_i(x,y)$ are arbitrarily assumed modes, the above differential equations are, in general, coupled both inertially and elastically.

3-11 Approximate solutions by the lumped parameter method. The simplest application of the lumped parameter method assumes that the deformation of the continuous structure can be approximated by linear deflections at a finite number of discrete points. That is, the function $w(x,y,t)$ describing, for example, a deformed wing surface, is approximated by a column matrix as follows:

$$w(x,y,t) = \begin{bmatrix} w_1(t) \\ w_2(t) \\ w_3(t) \\ \cdot \\ \cdot \\ \cdot \\ w_n(t) \end{bmatrix}. \tag{3-268}$$

To illustrate the principles involved in applying Eq. (3-268), consider the restrained low aspect-ratio wing shown by Fig. 3-23. Let us assume that the wing is subjected to a distributed time-dependent external force $F_z(x,y,t)$, and that the resulting deformation can be described by the deformations of n discrete points on the wing surface. The integral equation forms a convenient basis for forming the total differential equations in terms of $w_1, w_2, \cdots, w_n$. We have seen by Eq. (3-144) that the pertinent integral equation of equilibrium is

$$w(x,y,t) = \iint_S C(x,y;\xi,\eta)[-\rho(\xi,\eta)\ddot{w}(\xi,\eta,t) + F_z(\xi,\eta,t)]d\xi d\eta. \tag{3-144}$$

The wing deflection at each of the n discrete points can be obtained from Eq. (3-144) as

$$w_i = \iint_S C(x_i,y_i;\xi,\eta)[-\rho(\xi,\eta)\ddot{w}(\xi,\eta,t) + F_z(\xi,\eta,t)]d\xi d\eta,$$
$$(i = 1, \cdots, n). \tag{3-269}$$

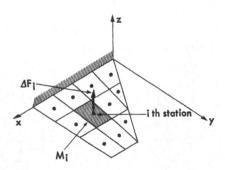

Fig. 3-23.   Low aspect-ratio wing divided by a surface gridwork.

We have seen by Eq. (2–79) that Eq. (3–269) can be integrated numerically by means of a weighting matrix:

$$\{w\} = -[C][\rho][\overline{W}]\{\ddot{w}\} + [C][\overline{W}]\{F_z\}, \qquad (3\text{--}270)$$

where $[\overline{W}]$ is a diagonal matrix of weighting numbers depending upon the scheme of numerical integration that is employed. The method employed above of satisfying an equation at a discrete number of points is sometimes referred to as a collocation method.

It can be observed that if the wing mass and external force within each grid boundary in Fig. 3–23 are assumed concentrated at a point within the boundary, Eq. (3–270) takes on a simpler form appropriate to a concentrated mass system:

$$\{w\} = -[C][M]\{\ddot{w}\} + [C]\{\Delta F_z\}, \qquad (3\text{--}271)$$

where $[M]$ is a diagonal matrix of point concentrated masses and $\{\Delta F_z\}$ is a column matrix of discrete applied forces.

In dealing with slender wings where the wing structure can be assumed rigid along chordwise sections, the deformation of the wing can be described by the linear and angular displacements of n spanwise stations. Thus, we can apply a collocation process to Eqs. (3–148) and (3–149). At the ith station, the linear and angular deflections of the wing, according to Eqs. (3–148) and (3–149), can be written in the following forms:

$$w_i = \int_0^l C^{zz}(y_i,\eta)[-m(\eta)\ddot{w}(\eta,t) + S_y(\eta)\ddot{\theta}(\eta,t) + F_z(\eta,t)]d\eta$$

$$+ \int_0^l C^{z\theta}(y_i,\eta)[-I_y(\eta)\ddot{\theta}(\eta,t) + S_y(\eta)\ddot{w}(\eta,t) + t(\eta,t)]d\eta, \qquad (3\text{--}272)$$

$$\theta_i = \int_0^l C^{\theta z}(y_i,\eta)[-m(\eta)\ddot{w}(\eta,t) + S_y(\eta)\ddot{\theta}(\eta,t) + F_z(\eta,t)]d\eta$$

$$+ \int_0^l C^{\theta\theta}(y_i,\eta)[-I_y(\eta)\ddot{\theta}(\eta,t) + S_y(\eta)\ddot{w}(\eta,t) + t(\eta,t)]d\eta. \qquad (3\text{--}273)$$

Integrating Eqs. (3–272) and (3–273) numerically by means of weighting matrices, we obtain the single matrix equation

$$\left[\begin{array}{c} \{w\} \\ \hline \{\theta\} \end{array}\right] = \left[\begin{array}{c|c} [C^{zz}] & [C^{z\theta}] \\ \hline [C^{\theta z}] & [C^{\theta\theta}] \end{array}\right] \left[\begin{array}{c|c} [\overline{W}] & [0] \\ \hline [0] & [\overline{W}] \end{array}\right]$$

$$\times \left\{ \left[\begin{array}{c|c} -[m] & [S_y] \\ \hline [S_y] & -[I_y] \end{array}\right] \left[\begin{array}{c} \{\ddot{w}\} \\ \hline \{\ddot{\theta}\} \end{array}\right] + \left[\begin{array}{c} \{F_z\} \\ \hline \{t\} \end{array}\right] \right\}, \qquad (3\text{--}274)$$

where $[\overline{W}]$ is a diagonal matrix of weighting numbers.

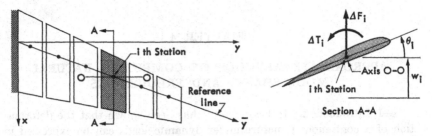

Fig. 3-24.   Large aspect-ratio wing divided into rigid segments.

Finally, let us conceive of the wings as divided into n rigid segments, as illustrated by Fig. 3-24. If the inertial properties of each wing segment are assumed concentrated at its wing station, Eq. (3-274) reduces to the form appropriate to a lumped mass system:

$$
\begin{bmatrix} \{w\} \\ \hline \{\theta\} \end{bmatrix} =
\begin{bmatrix} [C^{zz}] & [C^{z\theta}] \\ \hline [C^{\theta z}] & [C^{\theta\theta}] \end{bmatrix}
$$

$$
\times \left\{ \begin{bmatrix} -[M] & [S] \\ \hline [S] & -[I] \end{bmatrix} \begin{bmatrix} \{\ddot{w}\} \\ \hline \{\ddot{\theta}\} \end{bmatrix} + \begin{bmatrix} \{\Delta F_z\} \\ \hline \{\Delta t\} \end{bmatrix} \right\}, \qquad (3\text{-}275)
$$

where $[M]$, $[I]$, and $[S]$ are diagonal matrices of mass, moment of inertia, and static unbalance, respectively, of the rigid segments. The moment of inertia and static unbalance of each segment are measured with respect to an axis parallel to the y-axis and passing through the ith station. This axis is designated by the line $O-O$ in Fig. 3-24. $\{\Delta F_z\}$ and $\{\Delta t\}$ are column matrices of the total forces and moments, respectively, on each rigid segment.

Equation (3-275) contains both inertial and elastic coupling. The inertial coupling can be eliminated if the center of rotation of each wing segment, that is, the axis $O-O$ in Fig. 3-24, is taken at the center of gravity of the segment, thus eliminating the static unbalance. The elastic coupling between bending and twisting, $[C^{z\theta}]$ and $[C^{\theta z}]$, can be eliminated if the axis of rotation, $O-O$, is along the elastic axis.

CHAPTER 4

APPROXIMATE METHODS OF COMPUTING NATURAL MODE SHAPES AND FREQUENCIES

4–1 Introduction. In Chapter 3 it has been shown that the deformation of a continuous structure under dynamic loads can be expressed in terms of natural modes and frequencies of vibration. Except in very special cases, these cannot be determined exactly, and many approximate methods have been devised for their computation. The object of the present chapter is to present a few representative methods in a systematic manner. An attempt has been made to illustrate one or more methods from each important class in a manner such that the reader can perceive how to combine the features of several methods. Part of the present chapter has been based upon Ref. 4–1, which is a comprehensive survey and comparison of approximate methods.

We have seen that an actual airplane structure is a continuous system with an infinite number of degrees of freedom. There are, consequently,. an infinite number of natural modes and frequencies. Approximate methods of analysis serve to give approximations only to a relatively small number of the lower modes and their associated frequencies. These methods can be divided generally into three groups, according to the manner in which the problem is formulated: energy methods, integral equation methods, and differential equation methods. In a few cases, classification in this manner is difficult; however, this grouping forms a convenient basis for organizing the material of the present chapter. In Sections 4–2, 4–3, and 4–4, each method is described, for simplicity, by application to a restrained tapered beam. In these sections it is shown that most of the methods lead to a set of linear homogeneous simultaneous equations. Techniques for solving such equations are discussed in Section 4–5. Finally, Sections 4–6 and 4–7 are devoted to methods applicable to complex airplane structures and to rotating beams, respectively.

4–2 Natural modes and frequencies by energy methods. Energy methods of computing natural modes and frequencies derive from the principle of virtual work or, more specifically, from the form of this principle known as Lagrange's equation. Since we are dealing with free vibrations and small displacements, the applicable form of Lagrange's equation can be derived from Eq. (3–222) by putting $\partial T/\partial q_i = Q_i = 0$:

$$\frac{d}{dt}\left(\frac{\partial T}{\partial \dot{q}_i}\right) + \frac{\partial U}{\partial q_i} = 0, \qquad (i = 1, \cdots, n). \tag{4–1}$$

Lagrange's equation is applied by approximating the deformation of the structure in such a way that it can be described in terms of a finite number of generalized coordinates $q_1, \cdots, q_n$.

(a) *The Rayleigh-Ritz method.* Considering, for purposes of illustration, the case of a restrained cantilever beam with varying cross section, as shown by Fig. 4–1, we represent the deflection curve of the neutral axis by the expression

$$w(y,t) = \sum_{i=1}^{n} \gamma_i(y)q_i(t). \qquad (4\text{--}2)$$

The functions $\gamma_i(y)$ are assumed displacement functions that satisfy the geometrical boundary conditions $\gamma_i(0) = \gamma_i{}'(0) = 0$, and they are selected such that it is possible to obtain a good approximation to each of the required natural modes by superposition. The quantities $q_i(t)$ are generalized coordinates representing the contributions of each of the assumed functions.

Neglecting rotary inertia effects, the kinetic energy of a slender beam is given by

$$T = \tfrac{1}{2} \int_0^l m(y)\dot{w}^2(y,t)dy. \qquad (4\text{--}3)$$

Neglecting shear deformation, the strain energy of a slender beam can be expressed in terms of the bending stiffness, as follows:

$$U = \tfrac{1}{2} \int_0^l EI(y)(w'')^2 dy, \qquad (4\text{--}4)$$

where primes indicate differentiations with respect to y. Introducing Eq. (4–2) into Eqs. (4–3) and (4–4) and reducing, we obtain the following quadratic forms:

$$T = \tfrac{1}{2} \sum_{i=1}^{n} \sum_{j=1}^{n} m_{ij}\dot{q}_i\dot{q}_j, \qquad (4\text{--}5)$$

$$U = \tfrac{1}{2} \sum_{i=1}^{n} \sum_{j=1}^{n} k_{ij}q_i q_j, \qquad (4\text{--}6)$$

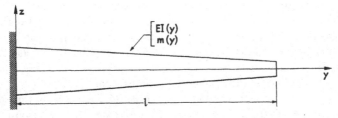

Fig. 4–1.   Cantilever beam with arbitrary spanwise properties.

where
$$m_{ij} = \int_0^l \gamma_i \gamma_j m\, dy, \qquad (4\text{–}7)$$

$$k_{ij} = \int_0^l EI\gamma_i'' \gamma_j'' dy. \qquad (4\text{–}8)$$

Substituting Eqs. (4–5) and (4–6) into Lagrange's equation (4–1), we obtain the total differential equations of free vibrations:

$$\sum_{j=1}^n m_{ij}\ddot{q}_j + \sum_{j=1}^n k_{ij}q_j = 0, \qquad (i = 1, \cdots, n). \qquad (4\text{–}9)$$

Putting as a solution

$$q_i = \bar{q}_i \sin(\omega t + \psi), \qquad (i = 1, \cdots, n), \qquad (4\text{–}10)$$

where $\bar{q}_i$ is the amplitude of the displacement, ω is the frequency, and ψ is a phase angle, we obtain

$$\sum_{j=1}^n (k_{ij} - m_{ij}\omega^2)\bar{q}_j = 0, \qquad (i = 1, \cdots, n). \qquad (4\text{–}11\text{a})$$

This set of characteristic equations can be solved for n discrete approximate values of ω^2. These approximate values $\omega_1^2, \omega_2^2, \cdots, \omega_n^2$ have the property

$$\lambda_1^2 \leq \omega_1^2, \qquad \lambda_2^2 \leq \omega_2^2, \qquad \cdots, \qquad \lambda_n^2 \leq \omega_n^2, \qquad (4\text{–}11\text{b})$$

where $\lambda_1^2, \lambda_2^2, \cdots, \lambda_n^2$ are the smallest n exact eigenvalues of the original problem (Ref. 4–2).

Associated with each eigenvalue is a set of values of $\bar{q}_1, \cdots, \bar{q}_n$. The $\bar{q}$'s are not determined absolutely; only their ratios can be obtained. The natural modes are given by

$$W_r(y) = \sum_{i=1}^n \gamma_i(y)\bar{q}_i^{(r)}, \qquad (r = 1, \cdots, n). \qquad (4\text{–}12)$$

These solutions represent only approximations to the true natural modes, since they are limited by the number and the nature of the assumed functions. The approximation is best for the lower modes and becomes progressively worse for the higher modes. It cannot be expected that all n modal solutions obtained in this way will give good approximations to the true modes, and it is usually necessary to discard some of the higher frequency modes. The success of the method depends upon the ingenuity of the analyst in selecting assumed functions. Since the selection of assumed functions is an extremely important part of this and other methods of analysis, it is discussed in some detail in Section 4–2(d).

Characteristic equations similar to Eqs. (4–11a) are obtained in a majority of the methods of analysis. Methods for their solution are considered in Section 4–5.

The characteristic equations given by Eqs. (4–11a) can be easily put into a form suitable for numerical calculation, using matrix notation. If we divide the beam into n spanwise stations, the mass and stiffness terms given by Eqs. (4–7) and (4–8), respectively, can be written in the following matrix forms:

$$[m_{ij}] = [\gamma][m][\overline{W}][\gamma]', \tag{4–13}$$

$$[k_{ij}] = [\gamma''][EI][\overline{W}][\gamma'']', \tag{4–14}$$

where

$$[\gamma] = \begin{bmatrix} \gamma_1(y_1) & \cdots & \gamma_1(y_n) \\ & & \\ \cdot & & \cdot \\ \cdot & & \cdot \\ \cdot & & \cdot \\ \gamma_n(y_1) & \cdots & \gamma_n(y_n) \end{bmatrix}; \quad [\gamma]' = \text{transpose of } [\gamma],$$

$$[\gamma''] = \begin{bmatrix} \gamma_1''(y_1) & \cdots & \gamma_1''(y_n) \\ & & \\ \cdot & & \cdot \\ \cdot & & \cdot \\ \cdot & & \cdot \\ \gamma_n''(y_1) & \cdots & \gamma_n''(y_n) \end{bmatrix},$$

$$[m] = \begin{bmatrix} m(y_1) & & & \\ & m(y_2) & & \\ & & \cdot & \\ & & & \cdot \\ & & & & m(y_n) \end{bmatrix};$$

$$[EI] = \begin{bmatrix} EI(y_1) & & & \\ \cdot & EI(y_2) & & \\ & & \cdot & \\ & & & \cdot \\ & & & & EI(y_n) \end{bmatrix}.$$

$[\overline{W}]$ is a matrix of weighting numbers corresponding to the n spanwise stations.

The matrix form of Eq. (4–11a) is

$$\omega^2[m_{ij}]\{\bar{q}\} = [k_{ij}]\{\bar{q}\}. \tag{4–15}$$

By introducing Eqs. (4–13) and (4–14), we obtain

$$\omega^2[\gamma][m][\overline{W}][\gamma]'\{\bar{q}\} = [\gamma''][EI][\overline{W}][\gamma'']'\{\bar{q}\}. \tag{4–16}$$

The form of Eq. (4–16) is particularly convenient for computation. A severe limitation on the method lies in the manner of expressing the strain energy, however. Since Eq. (4–16) involves derivatives of the assumed modes, the modes should be known as analytical functions if their derivatives are to be known accurately. An illustrative example of the Rayleigh-Ritz method is given by Example 4–1.

(b) *Rayleigh's method.* When a single deformation mode is assumed in the Rayleigh-Ritz method described in the previous section, the method reduces to a result commonly known as Rayleigh's method. Putting $i = j = 1$ in Eq. (4–11a), we obtain a formula for the frequency:

$$\omega_1 = \sqrt{\frac{k_{11}}{m_{11}}} = \sqrt{\frac{\int_0^l EI(\gamma_1'')^2 dy}{\int_0^l \gamma_1^2 m \, dy}}. \tag{4–17}$$

Equation (4–17) is useful in estimating the lowest natural bending frequency of a beam. In applying Rayleigh's method, the choosing of a single deflection mode is equivalent to introducing additional constraints which reduce the system to one having a single degree of freedom. These additional constraints can only increase the rigidity of the system, and hence the frequency of vibration. Thus, approximate values obtained by Rayleigh's method are higher than exact values.

(c) *The modified Rayleigh-Ritz method.* A useful variation of the Rayleigh-Ritz method expresses the strain energy in terms of the inertial loading. This method was formulated by E. Reissner using the concept of complementary energy (Ref. 4–3). Considering the case of the restrained cantilever beam, we may write the strain energy in terms of the applied bending moment:

$$U = \tfrac{1}{2} \int_0^l \frac{M^2}{EI(y)} \, dy, \tag{4–18}$$

where M is the bending moment in the beam. The bending moment resulting from the inertial loading due to free vibration can be obtained by integration:

$$M = -\int_y^l \int_\eta^l m(\eta) \ddot{w}(\eta, t) \, d\eta \, dy. \tag{4–19}$$

Putting, as before,

$$w(y, t) = \sum_{i=1}^n \gamma_i(y) q_i(t), \tag{4–2}$$

where the $\gamma_i(y)$ $(i = 1, 2, \cdots, n)$ are assumed functions satisfying the geometrical boundary conditions, and combining Eqs. (4–18) and (4–19), there is obtained an expression for strain energy in terms of the inertial loading due to free vibration, as follows:

$$U = \tfrac{1}{2} \sum_{i=1}^n \sum_{j=1}^n L_{ij} \ddot{q}_i \ddot{q}_j, \tag{4–20}$$

where

$$L_{ij} = \int_0^l \frac{1}{EI(y)} \left\{ \int_y^l \int_\eta^l m(\eta)\gamma_i(\eta)d\eta dy \right\} \left\{ \int_y^l \int_\eta^l m(\eta)\gamma_j(\eta)d\eta dy \right\} dy. \quad (4\text{-}21)$$

Since the q's vary in simple harmonic fashion, we can put $\ddot{q}_i = -\omega^2 q_i$, and

$$U = \tfrac{1}{2}\omega^4 \sum_{i=1}^n \sum_{j=1}^n L_{ij}q_iq_j. \quad (4\text{-}22)$$

The kinetic energy is the same as given previously by Eq. (4-5):

$$T = \tfrac{1}{2} \sum_{i=1}^n \sum_{j=1}^n m_{ij}\dot{q}_i\dot{q}_j. \quad (4\text{-}5)$$

Substituting Eqs. (4-22) and (4-5) into Lagrange's equation and introducing Eq. (4-10), we obtain

$$\sum_{j=1}^n (m_{ij} - \omega^2 L_{ij})\ddot{q}_j = 0, \qquad (i = 1, \cdots, n). \quad (4\text{-}23)$$

It is seen that this set of equations has the same form as Eqs. (4-11a) except that now the factor ω^2 multiplies the coefficients L_{ij} rather than the coefficients m_{ij}.

It is possible to express the coupling terms L_{ij} in terms of the influence function of the beam rather than the bending stiffness. The distribution of inertial forces for simple harmonic motion can be written in the form

$$Z(y,t) = \omega^2 \sum_{i=1}^n m(y)\gamma_i(y)q_i. \quad (4\text{-}24)$$

The beam displacement resulting from this inertial force distribution can be obtained from Eq. (2-43):

$$w(y,t) = \int_0^l C(y,\eta) \left\{ \omega^2 \sum_{i=1}^n m(\eta)\gamma_i(\eta)q_i \right\} d\eta. \quad (4\text{-}25)$$

Equations (4-24) and (4-25) can be used to compute the strain energy by applying Eq. (2-58), as follows:

$$U = \tfrac{1}{2} \int_0^l Z(y,t)w(y,t)dy. \quad (2\text{-}58)$$

Substituting Eqs. (4-24) and (4-25) into Eq. (2-58) yields

$$U = \tfrac{1}{2}\omega^4 \sum_{i=1}^n \sum_{j=1}^n L_{ij}q_iq_j, \quad (4\text{-}26a)$$

where

$$L_{ij} = \int_0^l \gamma_j(y)m(y)\int_0^l C(y,\eta)\gamma_i(\eta)m(\eta)d\eta dy. \quad (4\text{-}26b)$$

Thus Eq. (4–26b) may be used as an alternative expression in place of Eq. (4–21). If we evaluate the integrals numerically for m_{ij} (Eq. 4–13) and L_{ij} (Eq. 4–26b), we can write Eqs. (4–23) in matrix form:

$$[\gamma][m][\overline{W}][\gamma]'\{\bar{q}\} = \omega^2[\gamma][m][\overline{W}][C][m][\overline{W}][\gamma]'\{\bar{q}\}. \quad (4\text{–}27)$$

Equations (4–27) are the characteristic equations for the modified Rayleigh-Ritz method expressed in terms of influence coefficients.

The difference between the Rayleigh-Ritz method and the modified Rayleigh-Ritz method is due essentially to the manner in which the strain energy is expressed. In the former, the strain energy is obtained for a state of strain which is consistent with the assumed deformations. In the latter, the strain energy is consistent with the inertial loading experienced in a vibration in the assumed mode of deformation. In an exact analysis, where an infinite number of modes is assumed, the two methods are equivalent. In an approximate analysis, where a finite number of modes is taken, the states of stress and strain cannot be compatible simultaneously with the assumed deformation and the inertial loading. The nature of the approximation is thus seen to be different in the two methods. An important advantage of the modified Rayleigh-Ritz method over the classical Rayleigh-Ritz method lies in the manner of expressing the elastic properties of the structure. The expression of these properties in influence coefficient form gives the modified Rayleigh-Ritz method additional generality and flexibility.

(d) *Choice of assumed functions in energy methods.* We have seen that in the energy methods, the deflection of the structure is expressed as a superposition of assumed deformation modes. The choice of the functions representing the assumed modes is very important, and the success of the methods depends largely upon this choice.

Two general requirements for the assumed functions are necessary for best results. They should satisfy the boundary conditions, and they should be linearly independent of each other. Satisfaction of all the free-edge boundary conditions by each function individually is not absolutely necessary, since the tendency in the approximation process will be to superpose the functions in such a way as to satisfy these boundary conditions in the final result.

Linear independence requires that it be impossible to express any one function as a superposition of any or all of the remaining functions. The further the functions are from being linearly dependent, the better conditioned they will be for use in the analysis. Examples of functions which are linearly independent are those represented by the terms in a power series or a Fourier series.

Perhaps the simplest useful functional form is that of a polynomial expression. Standard formulas can be derived which will satisfy the boundary conditions in specific problems. For example, for a cantilever beam Duncan (Ref. 4–4) suggests the following polynomial expression:

$$\gamma_j(y) = \tfrac{1}{6}(j+2)(j+3)\left(\frac{y}{l}\right)^{j+1} - \tfrac{1}{3}j(j+3)\left(\frac{y}{l}\right)^{j+2}$$

$$+ \tfrac{1}{6}j(j+1)\left(\frac{y}{l}\right)^{j+3}, \qquad (j = 1, \cdots, n). \qquad (4\text{–}28)$$

It can easily be shown by substitution that Eq. (4–28) satisfies the boundary conditions of a cantilever beam. Functions obtained from Eq. (4–28) by putting $j = 1$ and 2 are illustrated graphically in Fig. 4–2.

Polynomial functions of a different type, as well as other functions, may be constructed by a method suggested by Williams (Ref. 4–5). An expression in series form is written for the highest derivative of the displacement function for which the boundary conditions are defined. This expression is integrated successively until the displacement function is obtained, and is made to satisfy the remaining boundary conditions at appropriate stages. For example, for a uniform cantilever beam, we can start with the following expression, which satisfies the condition that the shear is zero at the tip:

$$\gamma_j'''(y) = \left(1 - \frac{y}{l}\right)^j. \qquad (4\text{–}29)$$

Integrating and applying the boundary conditions $\gamma_j(0) = \gamma_j'(0) = \gamma_j''(l) = 0$ yields

$$\gamma_j(y) = \frac{l^3\left[1 - (j+3)\frac{y}{l} - \left(1 - \frac{y}{l}\right)^{j+3}\right]}{(j+1)(j+2)(j+3)}. \qquad (4\text{–}30)$$

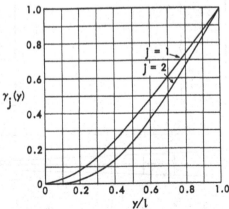

Fig. 4–2. Polynomial functions applicable to cantilever beam.

Since the magnitude of each function is arbitrary, we may write Eq. (4-30) more simply:

$$\gamma_j(y) = 1 - (j + 3)\frac{y}{l} - \left(1 - \frac{y}{l}\right)^{j+3} \tag{4-31}$$

When the natural modes are known for a structure which is similar in its characteristics, and has the same boundary conditions as the structure under consideration, these modes are well suited for use as assumed functions. For example, the natural modes of a uniform cantilever beam may be used very effectively in the analysis of a cantilever beam with variable elastic and inertial properties. When the beam carries concentrated masses, it may be beneficial to include assumed modes representing the static deflection modes due to concentrated loads applied at the masses.

EXAMPLE 4-1. To compute by the Rayleigh-Ritz method the first two natural modes and frequencies of the cantilever beam of unit width and linearly tapering depth illustrated by Fig. 4-3. Compare with the exact solution.

Solution. As described in Section 4-2(a), the Rayleigh-Ritz method involves the use of assumed modes. Let us select as the assumed modes the polynomial functions given by Eq. (4-28) for $j = 1$ and 2. These functions are illustrated graphically in Fig. 4-2. If the beam is divided into ten equal segments, as illustrated by Fig. 4-3, we have eleven spanwise stations, including the root, and the assumed lateral displacement may be represented by the matrix equation

$$\{W\} = \{\gamma_1\}\bar{q}_1 + \{\gamma_2\}\bar{q}_2, \tag{a}$$

where

$$\{\gamma_1\} = \begin{bmatrix} 0 \\ .0187 \\ .0699 \\ .1467 \\ .2423 \\ .3542 \\ .4752 \\ .6027 \\ .7339 \\ .8667 \\ 1.000 \end{bmatrix}, \quad \text{(b)} \qquad \{\gamma_2\} = \begin{bmatrix} 0 \\ .0030 \\ .0217 \\ .0654 \\ .1382 \\ .2396 \\ .3658 \\ .5111 \\ .6690 \\ .8335 \\ 1.000 \end{bmatrix}. \quad \text{(c)}$$

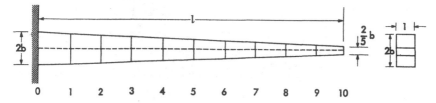

Fig. 4-3. Cantilever beam of unit width and linearly tapering depth.

The elements of the column matrices (b) and (c) are ordinates to the curves in Fig. 4–2.

The characteristic equations are given by Eq. (4–16):

$$\omega^2[\gamma][m][\overline{W}][\gamma]'\{\bar{q}\} = [\gamma''][EI][\overline{W}][\gamma'']'\{\bar{q}\} \tag{4–16}$$

The matrix $[\gamma]$ in Eq. (4–16) is formed from Eqs. (a) and (b), as follows:

$$[\gamma] =
\begin{bmatrix}
0 & .0187 & .0699 & .1467 & .2423 & .3542 & .4752 & .6027 & .7339 & .8667 & 1.000 \\
0 & .0030 & .0217 & .0654 & .1382 & .2396 & .3658 & .5111 & .6690 & .8335 & 1.000
\end{bmatrix}. \tag{d}$$

The matrix $[\gamma]'$ is the transpose of the matrix given above in Eq. (d). The matrix $[\gamma'']$ is obtained by differentiating Eq. (4–28) twice with respect to y and evaluating the result at the spanwise stations. This produces the result

$$[\gamma''] =
\frac{1}{l^2}
\begin{bmatrix}
4.00 & 3.24 & 2.56 & 1.96 & 1.44 & 1.00 & .640 & .360 & .160 & .040 & 0 \\
0 & 1.62 & 2.56 & 2.94 & 2.88 & 2.50 & 1.92 & 1.26 & .640 & .180 & 0
\end{bmatrix}. \tag{e}$$

The matrix $[\gamma'']'$ is the transpose of the matrix given in Eq. (e).

The mass matrix $[m]$ is a diagonal matrix representing the mass per unit length at the eleven spanwise stations:

$$[m] = 2b\rho
\begin{bmatrix}
1.0 \\
& .92 \\
& & .84 \\
& & & .76 \\
& & & & .68 \\
& & & & & .60 \\
& & & & & & .52 \\
& & & & & & & .44 \\
& & & & & & & & .36 \\
& & & & & & & & & .28 \\
& & & & & & & & & & .20
\end{bmatrix}. \tag{f}$$

In the matrix (f) the symbol ρ represents the mass per unit volume of the material of the beam. The weighting matrix $[\overline{W}]$ is taken from Lagrange's interpolation formula for $n = 10$, as given in Appendix B.

$$[\overline{W}] = 0.02683l$$

$$\times
\begin{bmatrix}
1 \\
& 6.616 \\
& & -3.020 \\
& & & 16.954 \\
& & & & -16.216 \\
& & & & & 26.599 \\
& & & & & & -16.216 \\
& & & & & & & 16.954 \\
& & & & & & & & -3.020 \\
& & & & & & & & & 6.616 \\
& & & & & & & & & & 1
\end{bmatrix}. \tag{g}$$

The rigidity matrix $[EI]$ is a diagonal matrix representing the product of the modulus of elasticity and the area moment of inertia of the beam about its neutral axis at the eleven spanwise stations:

$$[EI] = \frac{2b^3}{3} E$$

$$\times \begin{bmatrix} 1 & & & & & & & & & & \\ & .7787 & & & & & & & & & \\ & & .5927 & & & & & & & & \\ & & & .4389 & & & & & & & \\ & & & & .3144 & & & & & & \\ & & & & & .2160 & & & & & \\ & & & & & & .1406 & & & & \\ & & & & & & & .08518 & & & \\ & & & & & & & & .04665 & & \\ & & & & & & & & & .02195 & \\ & & & & & & & & & & .0080 \end{bmatrix} \cdot \quad (h)$$

Substituting the matrices given above into Eq. (4–16), we obtain the following matrix equation:

$$.05367b\rho l\omega^2 \begin{bmatrix} 3.4282 & 2.9401 \\ 2.9401 & 2.5799 \end{bmatrix} \begin{bmatrix} \bar{q}_1 \\ \bar{q}_2 \end{bmatrix}$$

$$= .01789 \frac{Eb^3}{l^3} \begin{bmatrix} 81.3634 & 49.2526 \\ 49.2526 & 53.5682 \end{bmatrix} \begin{bmatrix} \bar{q}_1 \\ \bar{q}_2 \end{bmatrix}. \quad (i)$$

Transposition gives the determinantal form

$$\begin{bmatrix} \left(0.18396b\rho l\omega^2 - 1.4555 \frac{b^3 E}{l^3}\right) & \left(0.1578b\rho l\omega^2 - 0.8811 \frac{b^3 E}{l^3}\right) \\ \left(0.1578b\rho l\omega^2 - 0.8811 \frac{b^3 E}{l^3}\right) & \left(0.1384b\rho l\omega^2 - 0.9583 \frac{b^3 E}{l^3}\right) \end{bmatrix} = 0. \quad (j)$$

Expansion of the determinant results in a second degree polynomial in ω^2, as follows:

$$0.0005766b^2\rho^2 l^2\omega^4 - 0.09979 \frac{\rho Eb^4}{l^2} \omega^2 + 0.6185 \frac{E^2 b^6}{l^6} = 0. \quad (k)$$

Factoring Eq. (k) by the quadratic formula gives the first two natural frequencies:

$$\omega_1 = 2.5372 \sqrt{\frac{Eb^2}{\rho l^4}}, \qquad \omega_2 = 12.9085 \sqrt{\frac{Eb^2}{\rho l^4}}.$$

Substituting ω_1 into Eq. (i) and solving for the ratio $\bar{q}_1/\bar{q}_2$, we obtain for the first mode,

$$\frac{\bar{q}_1}{\bar{q}_2} = 0.4968. \quad (l)$$

In a similar manner, by using ω_2 for the second mode,

$$\frac{\bar{q}_1}{\bar{q}_2} = -0.8702. \quad (m)$$

The column matrix which represents the first mode shape is obtained by putting $\bar{q}_2 = 1$ and $\bar{q}_1 = 0.4968$ in Eq. (a).

$$\{W^{(1)}\} = 0.4968\{\gamma_1\} + \{\gamma_2\} = 1.4968 \begin{bmatrix} 0 \\ .00822 \\ .03765 \\ .09240 \\ .17308 \\ .2776 \\ .4021 \\ .5415 \\ .6905 \\ .8445 \\ 1.0000 \end{bmatrix}. \tag{n}$$

The second mode shape is obtained by putting $\bar{q}_2 = 1$ and $\bar{q}_1 = -.8702$ in Eq. (a).

$$\{W^{(2)}\} = -0.8702\{\gamma_1\} + \{\gamma_2\} = 0.1298 \begin{bmatrix} 0 \\ -.1022 \\ -.3016 \\ -.4795 \\ -.5654 \\ -.5287 \\ -.3679 \\ -.1032 \\ .2342 \\ .6108 \\ 1.0000 \end{bmatrix}. \tag{o}$$

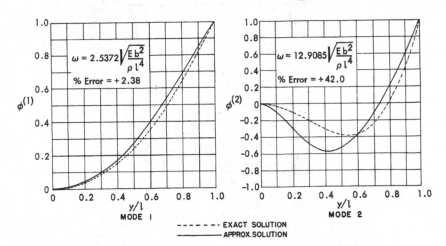

Fig. 4–4. Rayleigh-Ritz method (two polynomials).

The final modal columns in Eqs. (n) and (o) have been normalized such that the tip deflections have value unity. A comparison of these results with the exact solution as given by Ref. 4–1 is shown in Fig. 4–4. It is apparent that in this case only the first-mode solution can be considered satisfactory, and even here the errors are appreciable. Figure 4–5 illustrates the results obtained when three polynomial functions γ_j ($j = 1, 2, 3$) defined by Eq. (4–28) are applied. It is apparent that the first-mode solution is much better than in the two-mode approximation; however, the second- and third-mode results are generally unsatisfactory. Other assumed functions such as uniform cantilever beam modes may be expected to yield superior results to those given above; however, the functions describing the mode shapes may be more unwieldy to apply than the polynomial function.

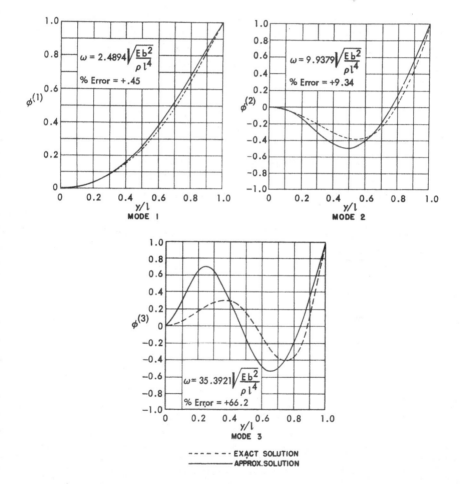

Fig. 4–5. Rayleigh-Ritz method (three polynomials).

EXAMPLE 4–2. To compute the first two natural modes and frequencies of the cantilever beam illustrated in Fig. 4–3 by the modified Rayleigh-Ritz method. Compare with the exact solution.

Solution. The modified Rayleigh-Ritz method can be applied to the tapered beam by constructing the simultaneous equations given by matrix Eq. (4–27).

$$[\gamma][m][\overline{W}][\gamma]'\{\bar{q}\} = \omega^2[\gamma][m][\overline{W}][C][m][\overline{W}][\gamma]'\{\bar{q}\}. \qquad (4\text{–}27)$$

Using the same polynomial functions as used in Example 4–1, the left side of matrix Eq. (4–27) can be formed from the data given in Example 4–1. The right side requires the matrix of influence coefficients of the beam. These are computed by applying Eqs. (2–114) and (2–115), neglecting shearing deformations, with the following result:

$$[C] = \frac{15}{128}\frac{l^3}{Eb^3}$$

$$\times \begin{bmatrix}
0 & 0 & 0 & 0 & 0 & 0 & 0 & 0 & 0 & 0 \\
0 & .00450 & .01146 & .01841 & .02538 & .03232 & .03929 & .04624 & .05320 & .06015 & .06711 \\
0 & .01146 & .03875 & .06923 & .09970 & .13017 & .16065 & .19113 & .22161 & .25209 & .28256 \\
0 & .01841 & .06923 & .14125 & .21704 & .29283 & .36861 & .44441 & .52020 & .59599 & .67177 \\
0 & .02538 & .09970 & .21704 & .36150 & .51209 & .66268 & .81326 & .96385 & 1.11444 & 1.26502 \\
0 & .03232 & .13017 & .29283 & .51209 & .77075 & 1.03742 & 1.30409 & 1.57075 & 1.83741 & 2.10409 \\
0 & .03929 & .16065 & .36861 & .66268 & 1.03742 & 1.46825 & 1.91133 & 2.35440 & 2.79747 & 3.24056 \\
0 & .04624 & .19113 & .44441 & .81326 & 1.30409 & 1.91133 & 2.60475 & 3.31748 & 4.03020 & 4.74294 \\
0 & .05320 & .22161 & .52020 & .96385 & 1.57075 & 2.35440 & 3.31748 & 4.42150 & 5.55928 & 6.69705 \\
0 & .06015 & .25209 & .59599 & 1.11444 & 1.83741 & 2.79747 & 4.03020 & 5.55928 & 7.34400 & 9.19542 \\
0 & .06711 & .28256 & .67177 & 1.26502 & 2.10409 & 3.24056 & 4.74294 & 6.69705 & 9.19542 & 12.23600
\end{bmatrix} \cdot \quad (a)$$

Combining the various matrices in Eq. (4–27) by applying data from Example 4–1 together with the $[C]$ matrix given above, we obtain the following characteristic equations:

$$.05367b\rho l \begin{bmatrix} 3.4282 & 2.9401 \\ 2.9401 & 2.5799 \end{bmatrix}\begin{bmatrix} \bar{q}_1 \\ \bar{q}_2 \end{bmatrix}$$

$$= .0003375\,\frac{\rho^2 l^5}{Eb}\,\omega^2 \begin{bmatrix} 87.3827 & 76.3653 \\ 76.3653 & 66.8419 \end{bmatrix}\begin{bmatrix} \bar{q}_1 \\ \bar{q}_2 \end{bmatrix}. \quad (b)$$

Solution of Eqs. (b) by expansion of the determinant yields the following frequencies and amplitude ratios:

$$\omega_1 = 2.4769\,\frac{b}{l^2}\sqrt{\frac{E}{\rho}}\,; \qquad \frac{\bar{q}_1}{\bar{q}_2} = 0.116. \qquad (c)$$

$$\omega_2 = 9.4832\,\frac{b}{l^2}\sqrt{\frac{E}{\rho}}\,; \qquad \frac{\bar{q}_1}{\bar{q}_2} = -0.875. \qquad (d)$$

The first two mode shapes are obtained by normalizing the column matrices given by

$$\{W^{(1)}\} = 0.116\{\gamma_1\} + \{\gamma_2\}\,, \qquad (e)$$

$$\{W^{(2)}\} = -0.875\{\gamma_1\} + \{\gamma_2\}\,, \qquad (f)$$

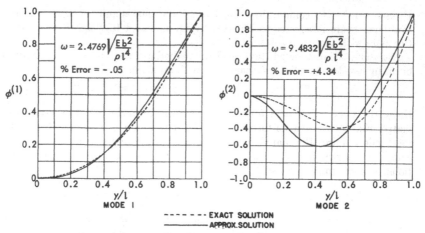

Fig. 4–6.   Modified Rayleigh-Ritz method (two polynomials).

where $\{\gamma_1\}$ and $\{\gamma_2\}$ are given in Example 4–1. These results are compared with the exact results in Fig. 4–6. The comparison shows a superiority of the modified Rayleigh-Ritz method over the Rayleigh-Ritz method. Frequencies are obtained more accurately; however, the mode shapes are not noticeably better.

4–3 Natural mode shapes and frequencies derived from the integral equation. We have seen in Chapter 3 that it is possible to express the natural mode shapes and frequencies of a structure in terms of solutions to homogeneous integral equations. For example, in the case of a beam with varying properties, the appropriate integral equation is given by

$$W(y) = \omega^2 \int_0^l C(y,\eta) W(\eta) m(\eta) d\eta. \qquad (4\text{–}32)$$

The integral equation lends itself conveniently to approximate numerical solutions, and is, in general, a better basis for such solutions than the corresponding differential equation. It can be said that this is due essentially to the fact that the basic operation performed on assumed solutions is an integration rather than a differentiation.

(a) *The integral equation applied to concentrated mass systems.* Perhaps the crudest approximate solution to Eq. (4–32) is obtained by arbitrarily replacing the continuous beam by a weightless beam supporting n concentrated masses. The integral in this case reduces to the form of a summation, and the equation is satisfied at each of the masses, yielding the following set of characteristic equations:

$$W_i = \omega^2 \sum_{j=1}^{n} C_{ij} M_j W_j, \qquad (i = 1, \cdots, n), \qquad (4\text{–}33)$$

where M_j represents the concentrated masses. Equations (4–33) can be written in the matrix form:

$$\{W\} = \omega^2[C][M]\{W\}. \tag{4-34}$$

Expression (4–34) is similar to the characteristic equations derived in the Rayleigh-Ritz method. However, the position of ω is different, and this fact is very important for numerical calculation by successive approximation. Numerical examples of the concentrated mass method are given by Examples 4–4, 4–5, 4–6, and 4–7 of the present chapter.

(b) *Collocation using numerical integration.* An approximate numerical solution to a homogeneous integral equation can be obtained by satisfying the equation at a finite number of selected points. This process is known as collocation. Let us suppose that we wish to satisfy the integral equation of free vibrations (Eq. 4–32) at n points along the beam. Thus we may write

$$W_i = \omega^2 \int_0^l C(y_i, \eta) m(\eta) W(\eta) d\eta, \qquad (i = 1, \cdots, n). \tag{4-35}$$

If we integrate Eq. (4–35) numerically by means of weighting numbers, we obtain

$$W_i = \omega^2 \sum_{j=1}^n C_{ij} m_j \overline{W}_j W_j, \tag{4-36}$$

where $W_j = W(y_j)$. In matrix notation,

$$\{W\} = \omega^2[C][m][\overline{W}]\{W\}. \tag{4-37}$$

This method of constructing the simultaneous equations represents a direct solution to the problem. The equations may be set up very easily and rapidly when the variation of inertial and elastic properties is smooth. When this is not the case, the method still remains useful, but a considerable amount of judgment and ingenuity may be required. Since there are as many equations as collocation stations, and since there must be sufficient collocation stations to provide ordinates for plotting the deflection curves, there will usually be a large number of simultaneous equations to solve. The solution of these equations then represents the major part of the work. A comparison between the results obtained by an eleven-station collocation solution and the exact solution is shown by Fig. 4–7. It can be seen that the first three mode shapes compare fairly well with the exact results, but only the first two frequencies are satisfactory. The sizeable error in frequency of the third mode is probably due to errors in the numerical integration process. It is reasonable to expect these errors to increase progressively in going to the higher modes.

It should be observed that an eleven-station collocation solution of a cantilever beam involves the solution of ten linear homogeneous equations. Solutions to such systems cannot conveniently be obtained by the

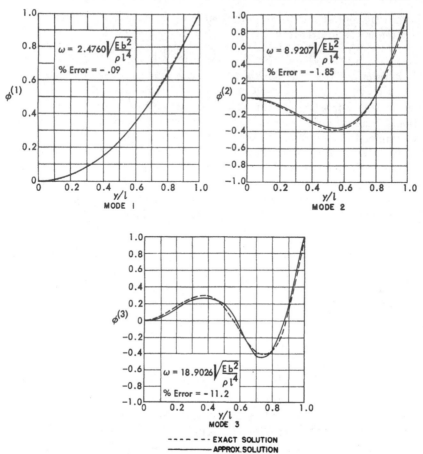

Fig. 4–7.   Collocation using numerical integration—11 stations.

simple methods applied in Examples 4–1 and 4–2.   Section 4–5 discusses other more powerful methods of solving these equations.

(c) *Collocation using assumed functions and station functions.*   Another variation of the collocation method involves representing the deflection as a superposition of assumed modes, as follows:

$$W(y) = \sum_{j=1}^{n} \gamma_j(y)\bar{q}_j, \tag{4–38}$$

where $\gamma_j(y)$ $(j = 1, 2, \cdots, n)$ are assumed functions that satisfy the boundary conditions.   This is substituted into the integral equation (Eq. 4–32) to give

$$\sum_{j=1}^{n} \gamma_j(y)\bar{q}_j = \omega^2 \sum_{j=1}^{n} \left\{ \int_0^l C(y,\eta)m(\eta)\gamma_j(\eta)d\eta \right\} \bar{q}_j. \tag{4–39}$$

We now require that Eq. (4–39) be satisfied at n stations, including the root station, each of which is identified by the subscript i, as follows:

$$\sum_{=1}^{n} \gamma_j(y_i)\bar{q}_j = \omega^2 \sum_{j=1}^{n} \left\{ \int_0^l C(y_i,\eta)m(\eta)\gamma_j(\eta)d\eta \right\} \bar{q}_j, \qquad (i = 0, 1, \cdots, n).$$

$$(4\text{–}40)$$

Since the root station is designated as the zeroth station in Fig. 4–3, the subscript i includes all stations from zero to n. The integral in Eq. (4–40) may be evaluated numerically by means of weighting numbers giving the characteristic equations

$$\sum_{j=1}^{n} \gamma_{ij}\bar{q}_j = \omega^2 \sum_{j=1}^{n} \sum_{k=0}^{n} C_{ik}m_k\overline{W}_k\gamma_{kj}\bar{q}_j, \qquad (i = 0, 1, \cdots, n), \quad (4\text{–}41)$$

where $\gamma_{ij} = \gamma_j(y_i)$ and $C_{ik} = C(y_i,y_k)$. In matrix notation,

$$[\gamma]\{\bar{q}\} = \omega^2[C][m][\overline{W}][\gamma]\{\bar{q}\}. \qquad (4\text{–}42)$$

The collocation process can be simplified by the use of interpolation functions, that is, functions which permit the expression of $W(y)$ in terms of a set of ordinates W_j at specified points:

$$W(y) = \sum_{j=0}^{n} \gamma_j(y)W_j. \qquad (4\text{–}43)$$

The summation in Eq. (4–43) commences with zero to include the root station. A form such as this is possible only if the functions $\gamma_j(y)$ satisfy the conditions

$$\gamma_j(y_i) = \begin{cases} 0, & \text{when } i \neq j, \\ 1, & \text{when } i = j. \end{cases} \qquad (4\text{–}44)$$

Introducing Eq. (4–43) into Eq. (4–32) and requiring that it be satisfied at n stations, we have

$$W_i = \omega^2 \sum_{j=0}^{n} \left\{ \int_0^l C(y_i,\eta)m(\eta)\gamma_j(\eta)d\eta \right\} W_j, \qquad (i = 0, 1, \cdots, n), \quad (4\text{–}45)$$

and, in matrix form, using numerical integration,

$$\{W\} = \omega^2[C][m][\overline{W}][\gamma]\{W\}. \qquad (4\text{–}46)$$

A useful form of interpolation function is called a station function. Rauscher (Ref. 4–6) conceived the idea of station functions as a refinement of the lumped mass type of analysis, taking the continuous mass distribution into account but retaining the simplicity of a finite set of ordinates as the generalized coordinates. In their simplest form, station functions are constructed from the deflection curves due to concentrated loads at selected stations along the beam. For the cantilever beam,

$$\gamma_i(y) = \sum_{j=1}^{n} C(y,\eta_j)P_{ij}, \qquad (i = 1, \cdots, n), \qquad (4\text{–}47)$$

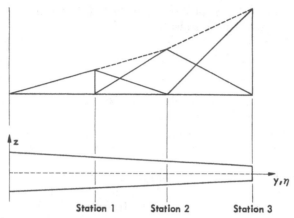

Fig. 4-8. Triangular loading curves for refined computation of station functions.

where P_{ij} is the applied concentrated load at the jth station. The condition that $\gamma_i(y)$ be an interpolation function is

$$\gamma_i(y) = \begin{cases} 1, & \text{at} \quad y = y_i, \\ 0, & \text{at} \quad y = y_k, \ (y_k \neq y_i). \end{cases} \qquad (4\text{-}48)$$

Applying this condition to Eq. (4-47), we have

$$\sum_{j=1}^{n} C_{kj}P_{ij} = \begin{cases} 1, & \text{for} \quad k = i, \\ 0, & \text{for} \quad k \neq i. \end{cases} \qquad (4\text{-}49)$$

Equations (4-49) comprise a set of simultaneous equations which may be solved for the P_{ij}'s which, inserted into Eq. (4-47), yield the interpolation or station functions $\gamma_i(y)$.

The station functions may be further refined by using, in place of a concentrated loading, triangular loads with the apex of each at one of the stations and the base extending to the two adjacent stations, as shown in Fig. 4-8. The broken line in the figure represents the sum of such loadings. This type of loading may be expected to yield a deflection curve closer to the shape of a vibration mode for a beam with distributed mass than would the concentrated loads. Still further refinement may be obtained by multiplying the triangular loadings by the distributed mass and using the resultant loadings in determining the station functions. If there are concentrated masses on the beam, the latter procedure will take account of these masses.

EXAMPLE 4-3. To compute the first two natural modes and frequencies of the tapered cantilever beam illustrated in Fig. 4-3 by the collocation method using station functions. Compare with the exact solution.

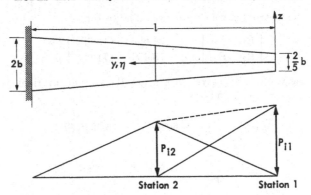

Fig. 4–9.   Triangular loading curves for two station functions.

Solution.   The characteristic equations for the station function method are derived from Eq. (4–46).

$$\{W\} = \omega^2 [C][m][\overline{W}][\gamma]\{W\}. \tag{4–46}$$

We shall consider two stations; one at the mid-span and one at the tip, as illustrated by Fig. 4–9. The station functions in the present example will be found by the use of triangular loadings multiplied by the distributed mass of the beam. Thus we must compute the beam deflections resulting from the loading curves illustrated in Fig. 4–9. The origin is taken at the tip for convenience. The two station functions are defined by

$$\gamma_1(\bar{y}) = P_{11}f_1(\bar{y}) + P_{12}f_2(\bar{y}), \tag{a}$$

$$\gamma_2(\bar{y}) = P_{21}f_1(\bar{y}) + P_{22}f_2(\bar{y}), \tag{b}$$

where

$$f_1(\bar{y}) = \int_0^{l/2} C(\bar{y},\bar{\eta}) \left(1 - \frac{2\bar{\eta}}{l}\right) m(\bar{\eta})d\bar{\eta}, \tag{c}$$

$$f_2(\bar{y}) = \int_0^{l/2} C(\bar{y},\bar{\eta}) \left(\frac{2\bar{\eta}}{l}\right) m(\bar{\eta})d\bar{\eta} + \int_{l/2}^{l} C(\bar{y},\bar{\eta}) \left(2 - \frac{2\bar{\eta}}{l}\right) m(\bar{\eta})d\bar{\eta}. \tag{d}$$

The P_{ij} terms are computed from the following condition, derived from Eq. (4–49):

$$\begin{bmatrix} P_{11} & P_{12} \\ P_{21} & P_{22} \end{bmatrix} \begin{bmatrix} f_1(0) & f_1\left(\dfrac{l}{2}\right) \\ f_2(0) & f_2\left(\dfrac{l}{2}\right) \end{bmatrix} = \begin{bmatrix} 1 & 0 \\ 0 & 1 \end{bmatrix}. \tag{e}$$

Evaluating Eqs. (c) and (d) numerically, we find that

$$\begin{bmatrix} f_1(0) & f_1\left(\dfrac{l}{2}\right) \\ f_2(0) & f_2\left(\dfrac{l}{2}\right) \end{bmatrix} = \begin{bmatrix} 0.1385 & 0.0307 \\ 0.1510 & 0.0473 \end{bmatrix} \frac{\rho l^4}{Eb^2}. \tag{f}$$

Inverting the matrix (f), we obtain, according to Eq. (e),

$$\begin{bmatrix} P_{11} & P_{12} \\ P_{21} & P_{22} \end{bmatrix} = \begin{bmatrix} 24.602 & -15.937 \\ -78.450 & 71.938 \end{bmatrix} \frac{Eb^2}{\rho l^4}. \tag{g}$$

Using the values of P_{ij} given in Eq. (g), the station functions are obtained from Eqs. (a) and (b), as follows:

$$\gamma_1(\bar{y}) = \frac{Eb^2}{\rho l^4} \{24.602 f_1(\bar{y}) - 15.937 f_2(\bar{y})\}, \tag{h}$$

$$\gamma_2(\bar{y}) = \frac{Eb^2}{\rho l^4} \{-78.450 f_1(\bar{y}) + 71.938 f_2(\bar{y})\}, \tag{i}$$

where $f_1(\bar{y})$ and $f_2(\bar{y})$ are defined by Eqs. (c) and (d), respectively. Evaluating Eqs. (h) and (i) numerically at eleven equally spaced spanwise stations, we obtain the following results for the station functions:

$$\{\gamma_1\} = \begin{bmatrix} 1 \\ .70122 \\ .43248 \\ .22252 \\ .08023 \\ 0 \\ -.03251 \\ -.03399 \\ -.02071 \\ -.00634 \\ 0 \end{bmatrix}, \quad (j) \qquad \{\gamma_2\} = \begin{bmatrix} 0 \\ .49084 \\ .88830 \\ 1.10964 \\ 1.13677 \\ 1 \\ .76038 \\ .48460 \\ .23455 \\ .06247 \\ 0 \end{bmatrix}. \quad (k)$$

The station functions $\gamma_1(\bar{y})$ and $\gamma_2(\bar{y})$ are plotted in Fig. 4–10.

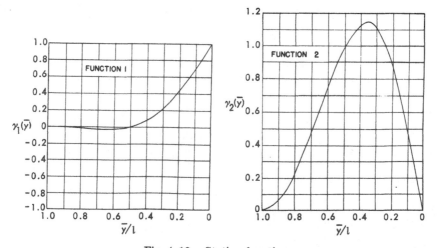

Fig. 4–10. Station functions.

From Eq. (4–43), the assumed deflection of the beam is represented by

$$\{W\} = \{\gamma_1\} W_1 + \{\gamma_2\} W_2,$$ (1)

where $\{\gamma_1\}$ and $\{\gamma_2\}$ are the column matrices of station functions given by Eqs. (j) and (k), respectively. The matrix $[\gamma]$ in the characteristic equation (4–46) thus becomes

$$[\gamma] = \begin{bmatrix} 1 & 0 \\ .70122 & .49084 \\ .43248 & .88830 \\ .22252 & 1.10964 \\ .08023 & 1.13677 \\ 0 & 1 \\ -.03251 & .76038 \\ -.03399 & .48460 \\ -.02071 & .23455 \\ -.00634 & .06247 \\ 0 & 0 \end{bmatrix}.$$ (m)

The matrix of influence coefficients in Eq. (4–46) is a 2×11 matrix obtained by taking rows from Eq. (a) in Example 4–2 corresponding to stations at the mid-span and the tip, as follows:

$$[C] = \frac{15}{128} \frac{l^3}{Eb^3}$$

$$\times \begin{bmatrix} 12.23600 & 9.19542 & 6.69705 & 4.74294 & 3.24056 & 2.10409 & 1.26502 & .67177 & .28256 & .06711 & 0 \\ 2.10409 & 1.83741 & 1.57075 & 1.30409 & 1.03742 & .77075 & .51209 & .29283 & .13017 & .03232 & 0 \end{bmatrix}.$$ (n)

The mass matrix in Eq. (4–46) is obtained from (f) in Example 4–1 by inverting the order of the stations so that station zero is at the tip. This gives

$$[m] = 2b\rho \begin{bmatrix} .20 & & & & & & & & & & \\ & .28 & & & & & & & & & \\ & & .36 & & & & & & & & \\ & & & .44 & & & & & & & \\ & & & & .52 & & & & & & \\ & & & & & .60 & & & & & \\ & & & & & & .68 & & & & \\ & & & & & & & .76 & & & \\ & & & & & & & & .84 & & \\ & & & & & & & & & .92 & \\ & & & & & & & & & & 1.0 \end{bmatrix}.$$ (o)

The weighting matrix in Eq. (4–46) is identical to matrix (g) in Example 4–1. Substituting matrices (m), (n), and (o), given directly above, together with matrix (g) from Example 4–1, into Eq. (4–46), and expanding, we obtain the following characteristic equations:

$$\begin{bmatrix} W_1 \\ W_2 \end{bmatrix} = \omega^2 \begin{bmatrix} 17.0952 & 37.1150 \\ 3.59281 & 10.7742 \end{bmatrix} \begin{bmatrix} W_1 \\ W_2 \end{bmatrix} \times 0.00628924 \frac{\rho l^4}{Eb^2}.$$ (p)

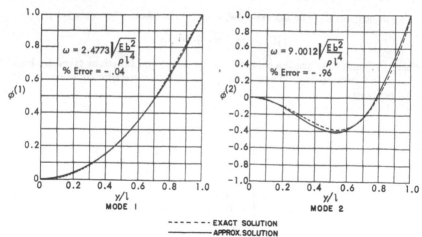

Fig. 4–11. Collocation using two station functions.

Solutions to Eqs. (p) are

$$\omega_1 = 2.4773 \frac{b}{l^2} \sqrt{\frac{E}{\rho}} \, ; \qquad \frac{W_1}{W_2} = 4.21178; \qquad (q)$$

$$\omega_2 = 9.0012 \frac{b}{l^2} \sqrt{\frac{E}{\rho}} \, ; \qquad \frac{W_1}{W_2} = -2.45261. \qquad (r)$$

Putting $W_2 = 1$ and $W_1 = 4.21178$ into Eq. (l), we obtain the first mode shape, and putting $W_2 = 1$ and $W_1 = -2.45261$ into Eq. (l) gives the second mode shape. Normalized values of these mode shapes are compared with the exact mode shapes in Fig. 4–11. The comparison shows that the collocation method yields a highly accurate result when two station functions are applied to the computation of the first two modes. In fact, it can be concluded in general that excellent results can be obtained by applying the same number of station functions as desired modes.

(d) *Direct iteration (method of Stodola and Vianello).* An iterative process can be applied directly to Eq. (4–32). In this method, a function $f_1(y)$ is assumed for the first mode. This is substituted into the right side of the integral equation (4–32), and the integral is evaluated, giving

$$f_2(y) = \int_0^l C(y,\eta) m(\eta) f_1(\eta) \, d\eta. \qquad (4-50)$$

Since the natural modes can be computed only to within a constant factor, we have dropped ω^2 in writing Eq. (4–50). The process is repeated using $f_2(y)$ as the assumed function in the integral, and so on until satisfactory convergence has been established. The process will converge to the natural mode with the lowest frequency.

It is common practice after each iteration to normalize the mode for unit displacement at a station where the deflection is large, for example, the free end in the case of a cantilever beam. Representing the normalized mode by placing a bar over the function, we have

$$\bar{f}_n(y) = \frac{f_n(y)}{f_n(l)}. \tag{4-51}$$

Thus, after $n - 1$ iterations, we have

$$f_n(y) = \int_0^l C(y,\eta)m(\eta)\bar{f}_{n-1}(\eta)d\eta. \tag{4-52}$$

However, if the approximation improves as the iteration proceeds, we have an approximate solution after $n - 1$ iterations, as follows:

$$\bar{f}_{n-1}(y) \cong \omega^2 \int_0^l C(y,\eta)m(\eta)\bar{f}_{n-1}(\eta)d\eta. \tag{4-53}$$

Comparing Eqs. (4-52) and (4-53), we see that

$$\omega^2 \cong \frac{\bar{f}_{n-1}(y)}{f_n(y)}. \tag{4-54a}$$

The value of the right side of the above expression is a function of y. Now, if we choose, as an approximation, the value at $y = l$, then

$$\omega^2 \cong \frac{1}{f_n(l)} \tag{4-54b}$$

and

$$\omega^2 = \lim_{n \to \infty} \frac{1}{f_n(l)}. \tag{4-54c}$$

The convergence will be most rapid if the first assumed displacement satisfies the boundary conditions, but this is not a necessary condition. A convenient function to use at the beginning is the influence function for a specific value of η, for example, $\eta = a$.

$$f_1(y) = C(y,a). \tag{4-55}$$

If a is chosen so that it is located in the vicinity of the free end of the beam, the influence function will have the same general shape as the first vibration mode.

The iteration will always converge to the first mode. This can be demonstrated by expanding the assumed function as a series in the natural modes $\phi_r(y)$ of the system, as follows:

$$f_1(y) = \sum_{r=1}^{\infty} \phi_r(y)\xi_r, \tag{4-56}$$

where the ξ_r's are coefficients determining the component of each natural mode in the assumed function. Substituting this expression into Eq. (4–50), we obtain

$$f_2(y) = \sum_{r=1}^{\infty} \left\{ \int_0^l C(y,\eta) m(\eta) \phi_r(\eta) d\eta \right\} \xi_r. \qquad (4\text{–}57)$$

Since $\phi_r(y)$ is a natural mode, Eq. (4–57) reduces to

$$f_2(y) = \sum_{r=1}^{\infty} \phi_r(y) \frac{\xi_r}{\omega_r^2}. \qquad (4\text{–}58)$$

After the nth iteration without normalization, we would have

$$f_n(y) = \phi_1(y) \frac{\xi_1}{(\omega_1^2)^n} + \phi_2(y) \frac{\xi_2}{(\omega_2^2)^n} + \cdots + \phi_\infty(y) \frac{\xi_\infty}{(\omega_\infty^2)^n}. \qquad (4\text{–}59)$$

Since ω_1 is the lowest frequency, and $\omega_1 < \omega_2 < \cdots < \omega_\infty$, it is seen that as n increases, terms beyond the first become smaller relative to the first. Thus,

$$f_n(y) \cong \phi_1(y) \frac{\xi_1}{(\omega_1^2)^n}. \qquad (4\text{–}60)$$

Since the method described above converges to the first mode, additional considerations must be introduced if we wish to compute the second mode by the same process. To obtain the second mode, it is necessary to remove all of the first-mode component from the assumed displacement. This may be done by taking any assumed function $h(y)$ and subtracting out the first-mode component in the following way:

$$f_1^{(2)}(y) = h(y) - \phi_1(y)\xi_1. \qquad (4\text{–}61)$$

In order to evaluate ξ_1 so that $f_1^{(2)}(y)$ will have no first-mode component, we make $f_1^{(2)}(y)$ orthogonal to the first mode with respect to a weighting function $m(y)$. Thus,

$$\int_0^l f_1^{(2)}(y)\phi_1(y)m(y)dy = \int_0^l h(y)\phi_1(y)m(y)dy - \xi_1 \int_0^l \phi_1^2(y)m(y)dy = 0,$$
$$(4\text{–}62)$$

so that

$$\xi_1 = \frac{\displaystyle\int_0^l h(y)\phi_1(y)m(y)dy}{\displaystyle\int_0^l \phi_1^2(y)m(y)dy}. \qquad (4\text{–}63)$$

The process of purifying the deformation function of its first-mode component must be carried out after each iteration. This is because the first mode may not have been determined with sufficient accuracy, and because the integral will probably be evaluated numerically, and as a consequence slight errors will be introduced. The presence of any first-mode component in the assumed function will be magnified by iteration. The third mode is obtained similarly by using an assumed function which is purified of all first- and second-mode components, and so on for higher modes. The process is sometimes called "sweeping" (Ref. 4–7). It should be observed that the convergence of the iteration process may become extremely poor if two or more modes have frequencies nearly the same.

When the integral equation is cast in matrix form, and the iterative process applied, it is referred to as matrix iteration. This technique, which is widely used as a practical tool, is discussed in this chapter in Section 4–5.

(e) *Galerkin's method.* Galerkin's method of solving differential and integral equations (Ref. 4–8), can be applied to the homogeneous integral equation of free vibration. We proceed by representing the deflection as a superposition of assumed functions in the same manner as employed by the collocation method using assumed functions.

$$W(y) = \sum_{j=1}^{n} \gamma_j(y)\bar{q}_j, \qquad (4\text{--}38)$$

where $\gamma_j(y)\,(j = 1, 2, \cdots, n)$ are assumed functions satisfying the boundary conditions. Substituting Eq. (4–38) into the integral equation (4–32) results in an error function $\mathcal{E}(y)$ as follows,

$$\mathcal{E}(y) = \sum_{j=1}^{n} \left\{ \gamma_j(y) - \omega^2 \int_0^l C(y,\eta)\gamma_j(\eta)m(\eta)d\eta \right\} \bar{q}_j. \qquad (4\text{--}64)$$

According to the Galerkin method, we require $\mathcal{E}(y)$ to be orthogonal to $\gamma_i(y)m(y)$ $(i = 1, 2, \cdots, n)$, that is, the average value with respect to the weighting function $\gamma_i(y)m(y)$ is zero.

$$\int_0^l \mathcal{E}(y)\gamma_i(y)m(y)dy = 0. \qquad (4\text{--}64a)$$

Substituting Eq. (4–64) into (4–64a) gives

$$\sum_{j=1}^{n} \bar{q}_j \int_0^l \gamma_i(y)\gamma_j(y)m(y)dy = \omega^2 \sum_{j=1}^{n} \bar{q}_j \int_0^l \gamma_i(y)m(y) \int_0^l C(y,\eta)\gamma_j(\eta)m(\eta)d\eta dy,$$
$$(i = 1, \cdots, n). \qquad (4\text{--}65)$$

Equations (4–65) constitute a set of characteristic equations of the form

$$\sum_{j=1}^{n} (m_{ij} - \omega^2 L_{ij})\bar{q}_j = 0, \qquad (i = 1, \cdots, n), \qquad (4\text{--}66)$$

where

$$m_{ij} = \int_0^l \gamma_i\gamma_j m\,dy, \qquad (4\text{--}67a)$$

$$L_{ij} = \int_0^l \gamma_i(y)m(y) \int_0^l C(y,\eta)\gamma_j(\eta)m(\eta)d\eta dy. \qquad (4\text{--}67\text{b})$$

It can be observed by comparison of Eqs. (4–23) and (4–66) that Galerkin's method applied to the integral equation produces a result identical to the modified Rayleigh-Ritz method when the latter is expressed in terms of influence functions.

(f) *Iterated modified Rayleigh-Ritz (Galerkin) method.* We have seen in Example 4–2 that the modified Rayleigh-Ritz method yields frequency values which are closer to the exact values than does the Rayleigh-Ritz method. However, the mode shape results are not noticeably better. It is interesting to observe that the mode shapes obtained by the modified Rayleigh-Ritz method can be vastly improved by a single iteration of the integral equation. Thus, a given mode shape obtained by the modified Rayleigh-Ritz or its equivalent Galerkin method,

$$W_r(y) = \sum_{i=1}^n \gamma_i(y)\bar{q}_i^{(r)}, \qquad (4\text{--}12)$$

is substituted into the right side of the integral equation (4–32) to yield

$$W_r(y)_{(\text{iterated})} = \sum_{i=1}^n \bar{q}_i^{(r)} \int_0^l C(y,\eta)\gamma_i(\eta)m(\eta)d\eta. \qquad (4\text{--}68)$$

In this manner an improved mode shape is obtained by a single cycle of the process of direct iteration. The matrix form of Eq. (4–68) is

$$\{W_r\}_{(\text{iterated})} = [C][m][\overline{W}][\gamma]'\{\bar{q}^{(r)}\}. \qquad (4\text{--}69)$$

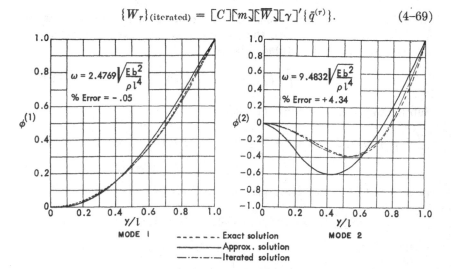

Fig. 4–12. Iterated modified Rayleigh-Ritz method (two polynomials).

As an illustration of the improvement in mode shapes that can be obtained by this operation, the mode shapes obtained in Example 4–2 have been iterated once numerically by means of Eq. (4–69). The results are illustrated by Fig. 4–12. It can be seen that the first two mode shapes obtained in this manner are in close agreement with the exact solution. Although the scheme of applying a single iteration as a means of improving a mode shape has been illustrated here by application to the modified Rayleigh-Ritz method, it can, of course, be applied to modes obtained by any method. The iteration is effective only for the lower frequency modes, and may, in fact, produce more deviation from the exact values if applied to the higher frequency modes.

4–4 Natural mode shapes and frequencies derived from the differential equation. The natural mode shapes and frequencies of slender beams in bending and torsion can be derived from differential equations. For example, we have seen in Chapter 3 that the mode shapes and frequencies of a slender beam in bending are governed by the following differential equation:

$$(EIW'')'' - m\omega^2 W = 0. \tag{3–11}$$

In general, the differential equation does not lend itself to approximate solutions of irregular structures nearly so well as the equivalent integral equation (4–32). For example, collocation methods can be applied to Eq. (3–11) in much the same manner as illustrated in Sections 4–3(b) and 4–3(c). However, if the beam has irregular properties, the results will be generally unsatisfactory unless many stations are taken, since no account is taken of the properties between selected stations. This is in contrast to the integral equation where integrations are performed which reflect the properties of the beam between stations.

(a) *Direct iteration (method of Stodola and Vianello).* A process of direct iteration can be applied to the differential equation in the same manner as the integral equation. The method is applied to the result obtained by successively integrating Eq. (3–11). For example, for a cantilever beam, successive integrations of Eq. (3–11) yield

$$W(y) = \omega^2 \int_0^y \int_0^y \frac{1}{EI(y)} \int_y^l \int_y^l m(y)W(y)dydydydy. \tag{4–70}$$

The iteration process proceeds in the same manner as for the integral equation, as discussed in Section 4–3(d). An initial function $f_1(y)$ is assumed and substituted into the right side of Eq. (4–70), yielding

$$f_2(y) = \int_0^y \int_0^y \frac{1}{EI(y)} \int_y^l \int_y^l m(y)f_1(y)dydydydy, \tag{4–71}$$

where $f_2(y)$ is a closer approximation to the first mode if $f_1(y)$ is properly chosen. The factor ω^2 has been dropped in writing Eq. (4–71), since the mode shape can be determined only to within a constant factor. In the case of irregular beams, Eq. (4–71) must be evaluated numerically or graphically.

The iteration process proceeds by substituting $f_2(y)$ into the right side, and is continued in this manner until convergence to the first mode has been established. The frequency is determined as described in Section 4–3(d) by normalization according to Eq. (4–54a). Higher modes are obtained by purifying the assumed modes of lower mode components after each iteration, as described in Section 4–3(d). Direct iteration of the differential equation can thus be seen to be equivalent to iteration of the integral equation with the exception that the elastic properties are expressed differently in the basic equations. In this respect, the integral equation provides a more general approach, since influence functions can include such effects as shear deformation, which are not included in Eq. (4–70).

(b) *Holzer-Myklestad method.* A useful and practical method of mode shape and frequency analysis, known as the Holzer method, has been used by engineers for many years. It was developed by Holzer (Ref. 4–9) primarily for application to torsional vibration problems. An extension by Myklestad of Holzer's method to the beam bending problem (Ref. 4–10) has proven useful in analyses of airplane wings. This method will be referred to as the Holzer-Myklestad method. Both Holzer's original method and Myklestad's extension are effectively step-by-step solutions of the differential equation of a lumped parameter system.

Let us consider the cantilever beam concentrated-mass system illustrated by Fig. 4–13. The deformation of a segment of the beam between stations i and $i + 1$ is illustrated by Fig. 4–14. We define the following quantities in connection with Fig. 4–14:

S_i = shear in beam immediately to the right of the ith mass,

M_i = bending moment in beam immediately to the right of the ith mass,

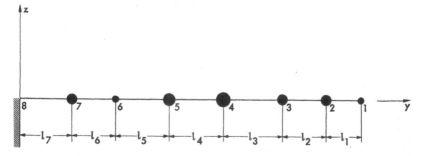

Fig. 4–13. Concentrated mass system employed by Holzer-Myklestad method.

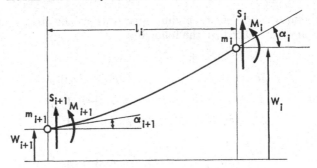

Fig. 4–14. Deformation of beam segment between concentrated masses.

α_i = slope of the beam at the ith mass,

W_i = deflection of the ith mass.

The positive directions of the various physical quantities are shown in Fig. 4–14. Assuming harmonic vibrations of frequency ω, the equations relating quantities at adjacent stations can be written

$$S_{i+1} = S_i + m_i\omega^2 W_i,\tag{4–72}$$

$$M_{i+1} = M_i + S_i l_i + m_i\omega^2 W_i l_i,\tag{4–73}$$

$$\alpha_{i+1} = \alpha_i - M_i C_i^{\alpha\alpha} - S_i C_i^{\alpha z} - m_i\omega^2 W_i C_i^{\alpha z},\tag{4–74}$$

$$W_{i+1} = W_i - M_i C_i^{z\alpha} - S_i C_i^{zz} - m_i\omega^2 W_i C_i^{zz} - \alpha_{i+1} l_i,\tag{4–75}$$

where

C_i^{zz} = linear deflection at i relative to $i + 1$ due to a unit force at i when station $i + 1$ is regarded as fixed,

$C_i^{z\alpha}$ = linear deflection at i relative to $i + 1$ due to a unit moment at i when station $i + 1$ is regarded as fixed,

$C_i^{\alpha z}$ = angular deflection at i relative to the slope at $i + 1$ due to a unit force at i,

$C_i^{\alpha\alpha}$ = angular deflection at i relative to the slope at $i + 1$ due to a unit moment at i.

Thus, knowing the shear, bending moment, slope, and deflection at one station, it is possible to compute the corresponding values at the next station. Assuming a value for the frequency, and starting with the shear, bending moment, slope, and deflection at one end of the beam, we can proceed from mass to mass until the corresponding quantities at the other end are obtained. If the frequency chosen is not the correct one, it will not be possible to satisfy the boundary conditions at both ends of the beam simultaneously. In general, the shear, bending moment, slope, and deflection at the left end of the beam in Fig. 4–13 may be expressed as linear functions of the corresponding quantities at the right end. These functions are represented by the following:

$$S_8 = C_1 S_1 + C_2 M_1 + C_3 \alpha_1 + C_4 W_1, \tag{4-76a}$$

$$M_8 = D_1 S_1 + D_2 M_1 + D_3 \alpha_1 + D_4 W_1, \tag{4-76b}$$

$$\alpha_8 = E_1 S_1 + E_2 M_1 + E_3 \alpha_1 + E_4 W_1, \tag{4-77a}$$

$$W_8 = F_1 S_1 + F_2 M_1 + F_3 \alpha_1 + F_4 W_1, \tag{4-77b}$$

where C, D, E, and F are functions of ω.

In the case of a cantilever beam, we may introduce the following boundary conditions into Eqs. (4–76) and (4–77):

$$\alpha_8 = W_8 = 0, \qquad M_1 = S_1 = 0, \tag{4-78}$$

which reduce to

$$S_8 = C_3 \alpha_1 + C_4 W_1, \tag{4-79a}$$

$$M_8 = D_3 \alpha_1 + D_4 W_1, \tag{4-79b}$$

$$0 = E_3 \alpha_1 + E_4 W_1, \tag{4-80a}$$

$$0 = F_3 \alpha_1 + F_4 W_1. \tag{4-80b}$$

Equations (4–80a) and (4–80b) provide the conditions from which the natural frequencies can be computed. That is, in order for the quantities α_1 and W_1 to be finite, the determinant of their coefficients must be zero:

$$\begin{vmatrix} E_3 & E_4 \\ F_3 & F_4 \end{vmatrix} = 0. \tag{4-81}$$

Since the elements of the determinant are functions of natural frequency, the value of the determinant can be plotted versus frequency, as illustrated by Fig. 4–15. The points of intersection of the curve with the zero axis define the natural frequencies of the beam. In a practical problem, these intersection points are computed by a trial-and-error process.

Having computed the natural frequencies of the system, the mode shape corresponding to each frequency can be obtained by a simple cal-

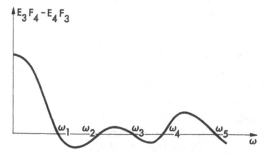

Fig. 4–15. Value of frequency determinant plotted versus frequency.

culation. We observe first from Eqs. (4–80) that the deflection W_1 and slope α_1 at the outer end of the beam, station 1, are related by

$$\alpha_1 = -\frac{E_4}{E_3} W_1 = -\frac{F_4}{F_3} W_1. \tag{4-82}$$

For a given value of frequency, we commence the mode shape calculation by assigning an arbitrary value, say one, to the quantity W_1, and computing the corresponding value of α_1 from Eq. (4–82). Thus, we obtain explicit numerical values for α_1 and W_1 at station 1. Substituting these values into Eqs. (4–74) and (4–75), we are able to compute α_2 and W_2. From station 2, we can progress to station 3 by means of Eqs. (4–72) through (4–75), and so on until values of α and W have been obtained at all stations.

Since the Holzer-Myklestad method involves replacement of the structure by a system of lumped masses, it suffers from the limitations of lumped mass systems. The principal advantage of the method is that each mode can be obtained independently. The necessity of using lower modes in the calculation of higher modes, a factor which causes difficulty in other methods, is eliminated here. The method has been used satisfactorily in conjunction with automatic computing machines of the punched-card type. It is particularly suited to use in this manner, since the calculations, while large in volume, are routine and repetitive. An illustration of the results that can be obtained by the Holzer-Myklestad method is shown by Fig. 4–16. This figure compares the first two exact modes of the tapered beam, illustrated by Fig. 4–3, with the first two modes computed by the Holzer-Myklestad method. In applying the Holzer-Myklestad

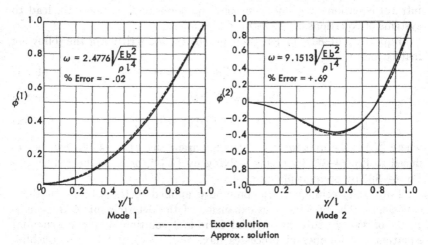

Fig. 4–16.   Holzer-Myklestad method—ten lumped masses.

method, the beam was divided into ten equal segments, and each segment
was replaced by a lumped mass at the mid-point of the segment.

4–5 Solution of characteristic equations. The most general form of
the characteristic equations given by the various methods discussed in
the previous sections involves both inertial and elastic coupling, as follows:

$$\omega^2 \sum_{j=1}^{n} m_{ij}\bar{q}_j = \sum_{j=1}^{n} k_{ij}\bar{q}_j, \qquad (i = 1, \cdots, n). \qquad (4\text{--}83)$$

Some of the methods yield equations directly in simpler forms in which
the elastic or inertial coupling terms do not appear. When elastic coupling
is eliminated, the equations are of the form

$$\omega^2 \sum_{j=1}^{n} D_{ij}\bar{q}_j = \bar{q}_i, \qquad (i = 1, \cdots, n), \qquad (4\text{--}84)$$

and when inertial coupling is eliminated, they appear as

$$\omega^2 \bar{q}_i = \sum_{j=1}^{n} E_{ij}\bar{q}_j, \qquad (i = 1, \cdots, n). \qquad (4\text{--}85)$$

For example, the Rayleigh-Ritz method normally gives equations of the
form of Eqs. (4–83). However, if the assumed functions are suitably
chosen, the coupling of the coordinates can be eliminated on one side
(Ref. 4–1). Solution of the integral equation by collocation methods using
assumed functions also leads to equations of the form of Eqs. (4–83).
We have seen that interpolation or station functions can be employed in
this case to eliminate the coupling on one side and reduce the results to
the form of Eq. (4–84). Concentrated mass systems and solutions of the
integral equation by collocation using numerical integration both lead to
systems of equations of the form of Eq. (4–84).

Equations (4–83), (4–84), and (4–85) can be written in the following
matrix forms:

$$([k] - \omega^2[m])\{\bar{q}\} = 0, \qquad (4\text{--}86)$$

$$([I] - \omega^2[D])\{\bar{q}\} = 0, \qquad (4\text{--}87)$$

$$([E] - \omega^2[I])\{\bar{q}\} = 0, \qquad (4\text{--}88)$$

where $[I]$ is the unit matrix. It is apparent that Eq. (4–86) can be reduced to Eq. (4–87) by premultiplying by $[k]^{-1}$, or it can be reduced to
Eq. (4–88) by premultiplying by $[m]^{-1}$.

Two basic techniques are commonly applied to the solution of these
equations. The first involves expansion of the determinant of the coefficients of the $\bar{q}$'s into polynomial form and solution of the polynomial
equation. A number of processes have been devised for accomplishing
this task. Wayland (Ref. 4–11) gives an excellent summary of these

processes. For example, in simple problems such as Examples 4-1, 4-2, and 4-3 involving two degrees of freedom, direct expansion of the determinant and solution by the quadratic formula provides a convenient method. This technique is, in fact, satisfactory for systems up to three or four degrees of freedom. For higher order systems, more elaborate methods must be applied. Of these, Danielewsky's method (Ref. 4-11) and Samuelson's method (Ref. 4-11) are mentioned as efficient methods for higher order systems.

The second basic technique, known as matrix iteration, provides a useful and practical engineering tool. It is directly applicable only to the forms of Eqs. (4-87) and (4-88). Let us consider its application to a system of the form of Eq. (4-87), as follows:

$$\frac{1}{\omega^2}\{W\} = [D]\{W\}, \tag{4-89}$$

where the column matrix $\{W\}$ represents ordinates to the mode shape curve. This system may be the direct result of, for example, a lumped mass formulation of the problem, or it may be the result of multiplying Eq. (4-86) by $[k]^{-1}$.

Let us suppose that a set of ordinates to the first mode is arbitrarily selected. These are designated by W_{1j}, where $j = 1, \cdots, n$. One of the W_{1j}'s, say W_{1n}, is set equal to 1. Substituting this assumed modal column matrix into the right side of Eq. (4-89) gives

$$[D]\{W_{1j}\} = \{N_{1j}\}. \tag{4-90}$$

The column matrix thus obtained is normalized by dividing by N_{1n}, so that the last element is again equal to 1:

$$\{N_{1j}\} = N_{1n}\left\{\frac{N_{1j}}{N_{1n}}\right\} = N_{1n}\{W_{2j}\}. \tag{4-91}$$

The process is repeated, this time using $\{W_{2j}\}$ as the approximation to the mode shape. If this iterative process is continued, successive values of the $\{W\}$ matrix will converge to the lowest natural mode, and the quantity $1/N_{mn}$ will converge to ω_1^2, when m becomes large. ω_1 is the lowest frequency of the system. Iteration of Eq. (4-88) in a similar fashion will converge on the natural mode with the highest frequency.

Convergence may be demonstrated in the same manner as demonstrated in the case of direct iteration of the integral equation in Section 4-3(d). In fact, matrix iteration can be regarded as exactly the same process except that now the integral equation has been replaced by its equivalent approximate matrix form. Expanding the displacements in the first assumed mode as a series in the normalized natural modes, we have

$$\{W_{1j}\} = \sum_{r=1}^{n} \{\phi^{(r)}\}\xi_r, \tag{4-92}$$

where the ξ_r's are coefficients determining the contributions of the individual natural modes, and the column matrices $\{\phi^{(r)}\}$ are the normalized natural mode shapes. Premultiplying both sides of Eq. (4–92) by $[D]$, we obtain

$$[D]\{W_{1j}\} = \sum_{r=1}^{n} [D]\{\phi^{(r)}\}\xi_r. \tag{4–93}$$

Since $\{\phi^{(r)}\}$ is one of the natural mode shapes, it satisfies

$$\frac{1}{\omega_r^2} \{\phi^{(r)}\} = [D]\{\phi^{(r)}\}. \tag{4–94}$$

Introducing Eq. (4–94) into the right side of Eq. (4–93), we have

$$[D]\{W_{1j}\} = \sum_{r=1}^{n} \{\phi^{(r)}\} \frac{\xi_r}{\omega_r^2}. \tag{4–95}$$

Equation (4–95) represents one cycle in the iteration process. Further cycles involve further premultiplications by the $[D]$ matrix, so that after the mth iteration, we have

$$[D]^m\{W_{1j}\} = \left(\frac{1}{\omega_1^2}\right)^m \{\phi^{(1)}\}\xi_1 + \cdots + \left(\frac{1}{\omega_n^2}\right)^m \{\phi^{(n)}\}\xi_n. \tag{4–96}$$

Inspection of Eq. (4–96) shows that since ω_r increases with r, all terms involving modes higher than the first tend to become small relative to the first. The mode obtained after each iteration has not been normalized in Eq. (4–96), but this is unimportant in this demonstration since it does not affect the convergence properties.

It can be seen from Eq. (4–96) that the convergence of the iteration process can be markedly affected in a case where two modes have nearly the same frequency. This can sometimes result in a complete breakdown of the method because of poor or erratic convergence. Methods are available for accelerating the convergence of an iteration process. Isakson, for example, in Ref. 4–12, uses the values obtained in three successive iterations to predict the final solution by means of an exponential extrapolation.

In the discussion above we have seen how to use an iteration process to obtain the mode with the lowest frequency. In practical problems, we are usually interested in the lowest two or three modes. It is therefore necessary to extend the method to include modes above the lowest. This is accomplished by application of the orthogonality condition as a means of purifying assumed modes of lower mode components. The most general form of the orthogonality condition between sets of ordinates $W_i^{(r)}$ and $W_j^{(s)}$ can be written in the form

$$\sum_{i=1}^{n} \sum_{j=1}^{n} A_{ij} W_i^{(r)} W_j^{(s)} = 0. \tag{4–97}$$

If the $W_i^{(r)}$'s are known explicitly, the condition that the $W_j^{(s)}$'s must satisfy in order for the two modes to be orthogonal is

$$\sum_{j=1}^{n} B_j^{(r)} W_j^{(s)} = 0, \qquad (4\text{-}98)$$

where
$$B_j^{(r)} = \sum_{i=1}^{n} A_{ij} W_i^{(r)}. \qquad (4\text{-}99)$$

It is apparent then that a mode shape $W_j^{(s)}$ can be assumed which is orthogonal to a given natural mode shape $W_i^{(r)}$ by assuming all of the ordinates of $W_j^{(s)}$ except one, and determining that ordinate by means of Eq. (4–98). Thus, if we select arbitrarily a set of ordinates $W_2^{(s)}$, $W_3^{(s)}, \cdots, W_n^{(s)}$, and if $W_1^{(s)}$ is obtained by applying Eq. (4–99), as follows:

$$W_1^{(s)} = -\frac{B_2^{(r)}}{B_1^{(r)}} W_2^{(s)} - \frac{B_3^{(r)}}{B_1^{(r)}} W_3^{(s)} - \cdots - \frac{B_n^{(r)}}{B_1^{(r)}} W_n^{(s)}, \quad (4\text{-}100)$$

then the complete set of $W_i^{(s)}$'s can be said to be orthogonal to the set of $W_i^{(r)}$'s. We may express in matrix form a condition that an arbitrarily selected set of $W_i^{(2)}$'s which are intended to represent a second mode be orthogonal to a given set of $W_i^{(1)}$'s, representing a first mode, by the following:

$$\begin{bmatrix} W_1 \\ W_2 \\ W_3 \\ \cdot \\ \cdot \\ \cdot \\ W_n \end{bmatrix} = \begin{bmatrix} 0 & -K_{12}^{(1)} & -K_{13}^{(1)} & \cdots & -K_{1n}^{(1)} \\ 0 & 1 & 0 & \cdots & 0 \\ 0 & 0 & 1 & \cdots & 0 \\ \cdot & \cdot & \cdot & \cdots & \cdot \\ \cdot & \cdot & \cdot & \cdots & \cdot \\ \cdot & \cdot & \cdot & \cdots & \cdot \\ 0 & 0 & 0 & \cdots & 1 \end{bmatrix} \begin{bmatrix} W_1 \\ W_2 \\ W_3 \\ \cdot \\ \cdot \\ \cdot \\ W_n \end{bmatrix}, \quad (4\text{-}101)$$

where
$$K_{1i}^{(1)} = \frac{B_i^{(1)}}{B_1^{(1)}}. \qquad (4\text{-}102)$$

The square matrix in Eq. (4–101) can be called a "sweeping" matrix, and it is designated by $[S^{(1)}]$. It is apparent that if any arbitrary set of numbers $W_1, \cdots, W_n$ is substituted into the right side of Eq. (4–101), the matrix product yields a set of numbers $W_1, \cdots, W_n$ which is orthogonal to the first mode. If this mode is used in place of the original arbitrarily assumed mode in the iteration process, convergence to the second mode will be obtained. It is, of course, necessary that this "sweeping" operation be performed once during each cycle of iteration. The process is handled most efficiently by multiplying the original matrix $[D]$ by the sweeping matrix $[S^{(1)}]$:

$$[D][S^{(1)}]\{W\} = [D^{(2)}]\{W\}. \qquad (4\text{-}103)$$

The matrix $[D^{(2)}]$ converges to the second mode in the same manner as the matrix $[D]$ converges to the first. It is evident that the first column of the $[D^{(2)}]$ matrix is composed of all zeros, and the order of the matrix is thus reduced by one.

In proceeding to the third mode, we make use of the orthogonality relations between the assumed mode and the first natural mode, and the assumed mode and the second natural mode. In this way, a sweeping matrix can be established which eliminates both the first- and second-mode components. It will be of the form

$$[S^{(2)}] = \begin{bmatrix} 0 & 0 & K_{13}^{(2)} & K_{14}^{(2)} & \cdots & K_{1n}^{(2)} \\ 0 & 0 & K_{23}^{(2)} & K_{24}^{(2)} & \cdots & K_{2n}^{(2)} \\ 0 & 0 & 1 & 0 & \cdots & 0 \\ 0 & 0 & 0 & 1 & \cdots & 0 \\ \cdot & \cdot & \cdot & \cdot & \cdots & \cdot \\ \cdot & \cdot & \cdot & \cdot & \cdots & \cdot \\ \cdot & \cdot & \cdot & \cdot & \cdots & \cdot \\ 0 & 0 & 0 & 0 & \cdots & 1 \end{bmatrix}, \qquad (4\text{-}104)$$

where the first two columns contain only zeros, so that the order of the matrix,

$$[D^{(3)}] = [D][S^{(2)}], \qquad (4\text{-}105)$$

is reduced by two from that of the original $[D]$ matrix. In general, for each successive mode, a new row of K's is added to the sweeping matrix and, at the same time, the number of K's in each row is reduced by one. As the number of modes increases, the computation of the K's becomes increasingly laborious. A systematic way of computing the K's for the $(r + 1)$th mode from the K's already determined for the rth mode, together with the orthogonality relation between the assumed mode and the rth mode, is given in Ref. 4–13.

An alternative method of computing the higher modes by matrix iteration has been communicated to the authors by Mr. M. J. Turner of the Boeing Airplane Company. If the simultaneous equations of free vibration are of the form

$$[C][m]\{W\} = \frac{1}{\omega^2}\{W\}, \qquad (4\text{-}106a)$$

and if the first mode is represented by the column of elements $\{\phi^{(1)}\}$, then it is easily seen that the second mode may be obtained by iteration from the matrix equation

$$([C][m] - \frac{1}{\omega_1^2}\{\phi^{(1)}\}\lfloor\phi^{(1)}\rfloor[m])\{W\} = \frac{1}{\omega^2}\{W\}. \qquad (4\text{-}106b)$$

Thus, in the determination of the higher modes it suffices to modify the dynamical matrix at the beginning of each iteration by subtraction of a simple triple matrix product. Modes are normalized so that $\lfloor \phi^{(i)} \rfloor [m] \{\phi^{(i)}\} = 1$. This procedure is considered particularly suitable for automatic computation.

EXAMPLE 4-4. To compute the first two natural modes and frequencies of the tapered cantilever beam illustrated in Fig. 4–3 by applying the concentrated mass method [Section 4–3(a)]. Solve the characteristic equations by matrix iteration.

Solution. In applying the concentrated mass method, we arbitrarily divide the beam into five equal segments, and concentrate the mass of each segment at the center of the segment, as shown by Fig. 4–17. Thus we replace the continuous beam by a weightless beam supporting five concentrated masses. The characteristic equations are given by Eq. (4–34):

$$\{W\} = \omega^2 [C][M]\{W\}. \tag{4-34}$$

The mass matrix, the diagonal terms of which represent the lumped masses in Fig. 4–17, is given by

$$[M] = b\rho l \begin{bmatrix} .368 & 0 & 0 & 0 & 0 \\ 0 & .304 & 0 & 0 & 0 \\ 0 & 0 & .240 & 0 & 0 \\ 0 & 0 & 0 & .176 & 0 \\ 0 & 0 & 0 & 0 & .112 \end{bmatrix}. \tag{a}$$

The influence coefficient matrix representing the influence coefficients associated with the five stations in Fig. 4–17 is obtained by taking alternate rows and columns from Eq. (a) in Example 4–2.

$$[C] = \frac{15}{128} \frac{l^3}{Eb^3} \begin{bmatrix} .00450 & .01841 & .03232 & .04624 & .06015 \\ .01841 & .14125 & .29283 & .44441 & .59599 \\ .03232 & .29283 & .77075 & 1.30409 & 1.83741 \\ .04624 & .44441 & 1.30409 & 2.60475 & 4.03020 \\ .06015 & .59599 & 1.83741 & 4.03020 & 7.34400 \end{bmatrix}. \tag{b}$$

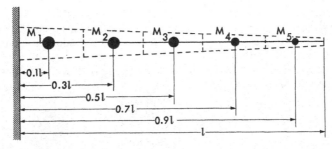

Fig. 4–17. Replacement of tapered beam by concentrated mass system.

Putting Eqs. (a) and (b) into Eq. (4–34) results in the following characteristic equations:

$$\frac{128}{15}\frac{Eb^2}{\rho l^4}\frac{1}{\omega^2}\{W\} = [D]\{W\},\tag{c}$$

where

$$[D] = \begin{bmatrix} .001656 & .0055966 & .0077568 & .0081382 & .0067368 \\ .0067748 & .042940 & .0702792 & .0782162 & .0667509 \\ .0118938 & .0890203 & .184980 & .229520 & .205790 \\ .0170163 & .135101 & .312982 & .458436 & .451382 \\ .0221352 & .181181 & .440978 & .709315 & .822528 \end{bmatrix}.\tag{d}$$

In commencing the iteration process, let us insert as a first trial a column matrix of ones into the right side of Eq. (c). We obtain, after multiplication and normalization, the following column matrix:

$$\begin{bmatrix} 1 \\ 1 \\ 1 \\ 1 \\ 1 \end{bmatrix} \xrightarrow[\text{Trial}]{\text{First}} \begin{bmatrix} .029884 \\ .264961 \\ .721204 \\ 1.374917 \\ 2.176137 \end{bmatrix} = 2.176137 \begin{bmatrix} .0137328 \\ .1217576 \\ .3314148 \\ .631815 \\ 1 \end{bmatrix}.$$

Using the normalized column matrix thus obtained as a second trial, and continuing until the process converges, we obtain the following results:

$$\xrightarrow[\text{Trial}]{\text{Second}} \begin{bmatrix} .0151360 \\ .1447820 \\ .4231116 \\ .8614391 \\ 1.4391950 \end{bmatrix} = 1.4391950 \begin{bmatrix} .0105292 \\ .1005993 \\ .2939918 \\ .5985562 \\ 1 \end{bmatrix} \cdots$$

$$\xrightarrow[\text{Trial}]{\text{Sixth}} \begin{bmatrix} .0144178 \\ .1381565 \\ .4053817 \\ .8292775 \\ 1.3918115 \end{bmatrix} = 1.3918115 \begin{bmatrix} .0103590 \\ .0992638 \\ .2912619 \\ .5958260 \\ 1 \end{bmatrix}$$

$$\xrightarrow[\text{Trial}]{\text{Seventh}} \begin{bmatrix} .0144177 \\ .1381564 \\ .4053813 \\ .8292767 \\ 1.3918104 \end{bmatrix} = 1.3918104 \begin{bmatrix} .0103590 \\ .0992638 \\ .2912619 \\ .5958259 \\ 1 \end{bmatrix}.$$

We observe that the normalized column matrix introduced at the beginning of the seventh trial coincides almost identically with the normalized column matrix resulting from the seventh trial.

The fundamental mode shape is given by the normalized matrix resulting from the seventh trial.

$$\{\phi^{(1)}\} = \begin{bmatrix} .010359 \\ .0992638 \\ .2912619 \\ .5958259 \\ 1 \end{bmatrix}.$$

The frequency corresponding to this column matrix is obtained from the identity

$$\frac{128}{15}\frac{Eb^2}{\rho l^4}\frac{1}{\omega_1{}^2} = 1.3918104,$$

which yields as the first mode frequency

$$\omega_1 = 2.47612\sqrt{\frac{Eb^2}{\rho l^4}}.$$

A matrix which converges on the second mode can be obtained by constructing a sweeping matrix according to Eq. (4–101):

$$[S^{(1)}] = \begin{bmatrix} 0 & -\dfrac{M_2\phi_2{}^{(1)}}{M_1\phi_1{}^{(1)}} & -\dfrac{M_3\phi_3{}^{(1)}}{M_1\phi_1{}^{(1)}} & -\dfrac{M_4\phi_4{}^{(1)}}{M_1\phi_1{}^{(1)}} & -\dfrac{M_5\phi_5{}^{(1)}}{M_1\phi_1{}^{(1)}} \\ 0 & 1 & 0 & 0 & 0 \\ 0 & 0 & 1 & 0 & 0 \\ 0 & 0 & 0 & 1 & 0 \\ 0 & 0 & 0 & 0 & 1 \end{bmatrix}$$

$$= \begin{bmatrix} 0 & -7.91588 & -18.3371 & -27.5085 & -29.3801 \\ 0 & 1.0 & 0 & 0 & 0 \\ 0 & 0 & 1.0 & 0 & 0 \\ 0 & 0 & 0 & 1.0 & 0 \\ 0 & 0 & 0 & 0 & 1.0 \end{bmatrix}. \quad (e)$$

The matrix equation which converges to the second mode is obtained by inserting this sweeping matrix into Eq. (c), as follows:

$$\frac{128}{15}\frac{Eb^2}{\rho l^4}\frac{1}{\omega^2}\{W\} = [D][S^{(1)}]\{W\} = [D^{(2)}]\{W\}, \quad (f)$$

which yields for $[D^{(2)}]$:

$$[D^{(2)}] = [D][S^{(1)}] =$$

$$\begin{bmatrix} 0 & -.0075121 & -.022609 & -.037418 & -.041917 \\ 0 & -.010689 & -.053951 & -.108148 & -.132293 \\ 0 & -.0051296 & -.033118 & -.097661 & -.143651 \\ 0 & .0004020 & .0009524 & -.0096569 & -.048559 \\ 0 & .0059614 & .035083 & .100409 & .172194 \end{bmatrix}. \quad (g)$$

Starting, as before, with a column matrix of ones as the first trial, we obtain after a total of eight iterations the following normalized mode shape and frequency of the second mode:

$$\{\phi^{(2)}\} = \begin{bmatrix} -.051472 \\ -.393380 \\ -.7390130 \\ -.4603381 \\ 1 \end{bmatrix}; \quad \omega_2 = 9.34574\sqrt{\frac{Eb^2}{\rho l^4}}.$$

A comparison of these results with the exact solution is given in Fig. 4–18. It can be seen that a good result is obtained for the first mode and only a fair result for the second mode. If higher modes were required, it would be necessary to take a larger number of masses in order to obtain acceptable results.

In the computation of two modes, as illustrated above, matrix iteration can be applied satisfactorily. The process becomes increasingly difficult and compli-

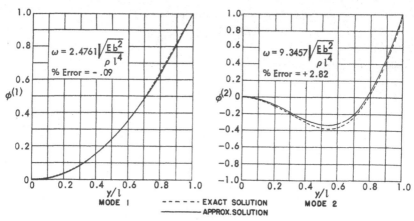

Fig. 4–18.   Lumped mass method—five masses.

cated, however, as we go to higher modes, because of the necessity of sweeping out the lower modes, and because the convergence is poorer.   The iteration process should be started with a relatively large number of significant figures, since accuracy is lost in proceeding to the higher modes.   It has been found that there is a loss in accuracy of roughly one significant figure in proceeding from one mode to the next.

4–6 Natural modes and frequencies of complex airplane structures. The methods of analysis presented in the earlier parts of this chapter have been described by application to a cantilever beam in bending.   Most of these same methods can be extended to more complex structures.   The main difficulty in computing modes and frequencies of complex structures is in choosing a reasonably small number of generalized coordinates which will yield a satisfactory solution for the desired number of modes.   In most cases, it is necessary to replace the actual structure by a simpler model which still retains the significant features of the original structure.

Although an airplane is basically a three-dimensional object, the assumption introduced in Chapter 3 of compressing the airplane structure into an elastic surface in the xy-plane is valid in a great majority of cases. We have seen that a convenient approach to the problem of computing the natural modes of a two-dimensional elastic surface is through the integral equation.   The integral equation defining the unrestrained free motion of an elastic surface is given in Chapter 3 as

$$W(x,y) = \omega^2 \iint_S G(x,y;\xi,\eta) W(\xi,\eta) \rho(\xi,\eta) d\xi d\eta, \qquad (3\text{--}172)$$

where

$$G(x,y;\xi,\eta) = C(x,y;\xi,\eta) - \iint_S C(r,s;\xi,\eta) \left[\frac{1}{M} + \frac{ys}{I_x} + \frac{xr}{I_y} \right] \rho(r,s) dr ds.$$

$$(3\text{--}173)$$

The methods of approximate solution to homogeneous integral equations described in Section 4–3 can be extended to the solution of Eq. (3–172). The detailed techniques required to extend these methods depend largely upon the nature of the structure under consideration. For example, if collocation using numerical integration is applied, Eq. (3–172) reduces to the matrix form

$$\{W\} = \omega^2 [G][\rho][\overline{W}]\{W\}. \tag{4–107}$$

The technique of numerical integration to be applied here depends upon the nature of the mass distribution. If the distribution is smooth, weighting numbers derived, for example, by interpolation functions can be applied. If the distribution is discontinuous, as is often the case in airplane structures carrying fuel tanks and other concentrated masses, the form of the weighting matrix will depend upon the particular conditions encountered, and no general rules can be given.

Collocation using assumed functions can also be applied in the same manner as it is applied to the beam in Section 4–3(c). If the assumed functions are station functions, we obtain

$$\{W\} = \omega^2 [G][\rho][\overline{W}][\gamma]\{W\}. \tag{4–108}$$

In this case, the station functions could be determined from the deflection modes for loadings in the form of a pyramid with its apex at one station and the corners of its base at the four diagonally adjacent stations, as suggested in Ref. 4–6.

The influence function of the unrestrained airplane given by Eq. (3–173) can be written in terms of influence coefficients and weighting numbers as follows:

$$G_{ij} = C_{ij} - \frac{1}{M}\sum_{k=1}^{n} C_{kj}\rho_k\overline{W}_k - \frac{y_i}{I_x}\sum_{k=1}^{n} C_{kj}\rho_k\overline{W}_k y_k - \frac{x_i}{I_y}\sum_{k=1}^{n} C_{kj}\rho_k\overline{W}_k x_k. \tag{4–109}$$

Values computed from Eq. (4–109) are used to form the $[G]$ matrix which appears in Eqs. (4–107) and (4–108).

A concentrated mass approximate solution to Eq. (3–172), although comparatively crude compared with other methods, offers a quick and convenient method of approach when only a few of the lower natural modes are required. The concentrated mass approximation is applied by replacing the structure by a set of concentrated masses in the xy-plane. With this approximation, Eq. (3–172) reduces to the matrix form

$$\{W\} = \omega^2 [G][M]\{W\}, \tag{4–110}$$

where $[M]$ is a diagonal matrix of concentrated masses and the elements of the influence coefficient matrix, $[G]$, are given by

$$G_{ij} = C_{ij} - \frac{1}{M}\sum_{k=1}^{n} C_{kj}M_k - \frac{y_i}{I_x}\sum_{k=1}^{n} C_{kj}M_k y_k - \frac{x_i}{I_y}\sum_{k=1}^{n} C_{kj}M_k x_k. \tag{4–111}$$

Replacement of the continuous airplane structure by concentrated masses is a somewhat arbitrary process, and the success achieved depends upon the judgment of the analyst. The process is generally one of dividing the structure into rigid segments and concentrating the total mass of each segment at its center of gravity. In carrying out the process, advantage should be taken of the peculiarities of the particular structure being analyzed. Concentrated masses should be located where masses such as engines and fuel tanks appear on the structure. Division between segments should be located at sections of discontinuity in the structure. Experience has shown that it is usually necessary to have at least twice as many masses as required modes. (See, for example, Example 4-4.)

Levy (Ref. 4-14) has proposed an analytical method for selecting the location and magnitude of concentrated masses. This method is based upon the requirement that the dynamic loads due to the concentrated masses produce nearly the same deflection at distant parts of the structure as do the distributed masses. Let us consider the airplane wing illustrated by Fig. 4-19 vibrating in a natural mode. It is assumed that the mode shape throughout the crosshatched region S can be approximated by the polynomial

$$W(\xi,\eta) = a_1 + b_1\eta + c_1\xi + d_1\eta^2 + e_1\xi^2 + \cdots. \qquad (4\text{-}112)$$

The influence function between a point (x,y), distant from region S, and a point (ξ,η) within the region S, is approximated by the polynomial

$$C(x,y;\xi,\eta) = a_2 + b_2\eta + c_2\xi + d_2\eta^2 + e_2\xi^2 + \cdots, \qquad (4\text{-}113)$$

where a_2, b_2, c_2, d_2, and e_2 are functions of x and y. The deflection at (x,y) due to the distributed inertial loads over the region S, while the wing is vibrating in a natural mode, is given by

$$W(x,y) = \omega^2 \iint_S C(x,y;\xi,\eta)W(\xi,\eta)\rho(\xi,\eta)d\xi d\eta. \qquad (4\text{-}114)$$

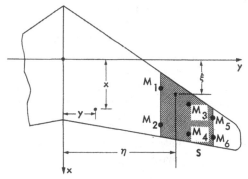

Fig. 4-19. Replacement of airplane wing by concentrated mass system.

Substituting Eqs. (4–112) and (4–113) into Eq. (4–114), we obtain

$$W(x,y) = \omega^2 a_1 a_2 \iint_S \rho\, d\xi d\eta + \omega^2 (a_1 b_2 + a_2 b_1) \iint_S \rho\eta\, d\xi d\eta$$

$$+ \omega^2 (a_1 c_2 + a_2 c_1) \iint_S \rho\xi\, d\xi d\eta + \omega^2 (a_1 e_2 + a_2 e_1 + c_1 c_2) \iint_S \rho\xi^2\, d\xi d\eta + \cdots .$$

$$(4\text{–}115)$$

We next assume that the distributed mass over the region S has been replaced by concentrated masses M_1, M_2, $M_3 \cdots$ with coordinates (ξ_1,η_1), (ξ_2,η_2), $(\xi_3,\eta_3) \cdots$. The deflection at (x,y) due to these concentrated masses is given by

$$W(x,y) = \omega^2 [C(x,y;\xi_1,\eta_1)M_1 W_1(\xi_1,\eta_1) + C(x,y;\xi_2,\eta_2)M_2 W_2(\xi_2,\eta_2) + \cdots].$$

$$(4\text{–}116)$$

Substituting Eqs. (4–112) and (4–113) into Eq. (4–116), and simplifying, yields

$$W(x,y) = \omega^2 [a_1 a_2 (M_1 + M_2 + \cdots) + (a_1 b_2 + a_2 b_1)(M_1 \eta_1 + M_2 \eta_2 + \cdots)$$
$$+ (a_1 c_2 + a_2 c_1)(M_1 \xi_1 + M_2 \xi_2 + \cdots)$$
$$+ (a_1 e_2 + a_2 e_1 + c_1 c_2)(M_1 \xi_1^2 + M_2 \xi_2^2 + \cdots) + \cdots]. \quad (4\text{–}117)$$

Equating Eqs. (4–115) and (4–117), we obtain the conditions that the deflection at (x,y) due to the distributed masses be equal to the deflection at (x,y) due to the concentrated masses. If, for example, six mass points are selected within the area S, as illustrated by Fig. 4–19, the following six conditions must be satisfied:

$$M_1 + M_2 + \cdots + M_6 = \iint_S \rho\, d\xi d\eta,$$

$$M_1 \eta_1 + M_2 \eta_2 + \cdots + M_6 \eta_6 = \iint_S \rho\eta\, d\xi d\eta,$$

$$M_1 \xi_1 + M_2 \xi_2 + \cdots + M_6 \xi_6 = \iint_S \rho\xi\, d\xi d\eta,$$

$$(4\text{–}118)$$

$$M_1 \eta_1^2 + M_2 \eta_2^2 + \cdots + M_6 \eta_6^2 = \iint_S \rho\eta^2\, d\xi d\eta,$$

$$M_1 \xi_1^2 + M_2 \xi_2^2 + \cdots + M_6 \xi_6^2 = \iint_S \rho\xi^2\, d\xi d\eta,$$

$$M_1 \xi_1 \eta_1 + M_2 \xi_2 \eta_2 + \cdots + M_6 \xi_6 \eta_6 = \iint_S \rho\xi\eta\, d\xi d\eta.$$

The first three equations above require the total mass and the center of gravity of the concentrated masses to be the same as those of the distributed masses. The last three equations require the moments and products of inertia of the concentrated masses to be the same as the distributed masses. The six concentrated masses are obtained by a simultaneous solution of Eqs. (4–118).

EXAMPLE 4–5. To compute the following pure bending modes (uncoupled bending modes) of the wing of the jet transport introduced in Example 2–1:

(a) First cantilever bending mode.
(b) First unrestrained symmetrical bending mode.
(c) Second unrestrained symmetrical bending mode.
(d) First antisymmetrical bending mode.

Solution. Pure bending modes, sometimes referred to as uncoupled bending modes, are obtained by assuming that the masses of all sections of the wing are concentrated along the elastic axis. With this assumption, the wing is, of course, capable of executing only pure bending vibrations. The mass distribution of the wing of the jet transport is assumed to be replaced by the concentrated mass system illustrated by Fig. 4–20. Since it is possible to obtain the answer we seek by working with only half of the wing, the mass on the airplane center line in Fig. 4–20 represents one-half of the fuselage mass. The mass matrix of the system in Fig. 4–20 is

$$[M] = \frac{1}{386} \begin{bmatrix} 17,400 & & & & & \\ & 6,039 & & & & \\ & & 10,200 & & & \\ & & & 4,200 & & \\ & & & & 3,400 & \\ & & & & & 680 \end{bmatrix} \frac{\text{lb·sec}^2}{\text{in.}} . \qquad (a)$$

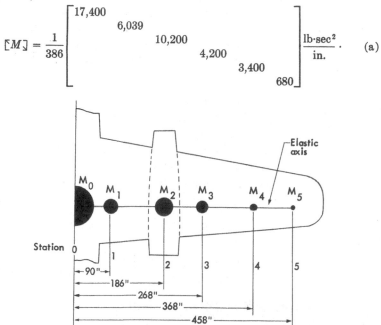

Fig. 4–20. Concentrated mass system for computation of bending modes—jet transport.

The bending influence coefficient matrix $[C^{zz}]$ associated with the mass points in Fig. 4–20 has been computed in Example 2–1.

(a) *First cantilever bending mode.* The characteristic equations for cantilever bending modes are (cf. Eq. 4–34)

$$\{W\} = \omega^2[C^{zz}][M]\{W\} = \omega^2[D]\{W\}, \tag{b}$$

where the influence coefficient matrix $[C^{zz}]$ is given by Eq. (e) of Example 2–1, and the mass matrix $[M]$ is given by Eq. (a). Combining these matrices, Eq. (b) reduces to

$$\frac{1}{\omega^2}\begin{bmatrix} W_1 \\ W_2 \\ W_3 \\ W_4 \\ W_5 \end{bmatrix} = \frac{10^{-3}}{386}\begin{bmatrix} 43.279 & 117.07 & 63.401 & 66.327 & 15.966 \\ 69.310 & 258.37 & 193.74 & 243.08 & 64.139 \\ 91.162 & 470.52 & 523.96 & 649.81 & 170.58 \\ 117.81 & 729.25 & 802.71 & 1240.98 & 356.14 \\ 141.79 & 962.09 & 1053.58 & 1780.72 & 573.51 \end{bmatrix}\begin{bmatrix} W_1 \\ W_2 \\ W_3 \\ W_4 \\ W_5 \end{bmatrix}. \tag{c}$$

Applying matrix iteration to Eq. (c), we obtain the following mode shape and frequency for the first cantilever bending mode:

$$\{\phi^{(1)}\} = \begin{bmatrix} .04466 \\ .1417 \\ .3792 \\ .6935 \\ 1 \end{bmatrix}; \qquad \omega_1 = 12.799 \text{ rad/sec.}$$

The mode shape is plotted in Fig. 4–21 (a).

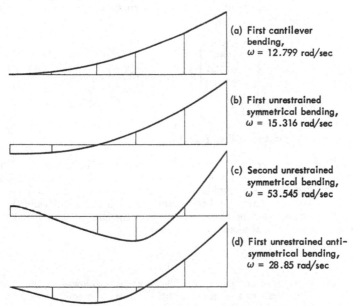

(a) First cantilever bending, $\omega = 12.799$ rad/sec

(b) First unrestrained symmetrical bending, $\omega = 15.316$ rad/sec

(c) Second unrestrained symmetrical bending, $\omega = 53.545$ rad/sec

(d) First unrestrained anti-symmetrical bending, $\omega = 28.85$ rad/sec

Fig. 4–21. Pure bending modes of jet transport wing.

(b) *First unrestrained symmetrical bending mode.* The characteristic equations for unrestrained symmetrical bending modes are (cf. Eq. 4–110)

$$\{W\} = \omega^2[G][M]\{W\} = \omega^2[D]\{W\}, \tag{d}$$

where the elements of the $[G]$ matrix are obtained from Eq. (4–111) as follows:

$$G_{ij} = C_{ij}{}^{zz} - \frac{1}{M} \sum_{k=0}^{5} C_{kj}{}^{zz} M_k. \tag{e}$$

In deriving Eq. (e) from Eq. (4–111), only the plunging degree of rigid-body freedom is included. Since we are working with half of the airplane, the quantity M in Eq. (e) represents half of the total mass of the airplane. Constructing the elements of the $[G]$ matrix from the mass and influence coefficient matrices previously given, we obtain

$$[G] =$$

$$\begin{bmatrix}
0 & -7.3119 & -19.768 & -45.470 & -69.346 & -107.63 \\
0 & -.070436 & -8.2908 & -30.374 & -57.948 & -84.149 \\
0 & 4.1656 & 5.5627 & .66020 & -5.9616 & -13.306 \\
0 & 7.7840 & 26.362 & 79.283 & 113.66 & 143.22 \\
0 & 12.196 & 51.727 & 145.65 & 287.54 & 416.11 \\
0 & 16.168 & 74.555 & 205.38 & 446.28 & 735.77
\end{bmatrix} \times 10^{-6}. \tag{f}$$

Inserting $[M]$ and $[G]$, given numerically by matrices (a) and (f), respectively, into Eq. (d), and iterating, we obtain for the first unrestrained symmetrical mode shape and frequency *

$$\{\phi^{(1)}\} = \begin{bmatrix}
-.1683 \\
-.1286 \\
-.01530 \\
.2513 \\
.6277 \\
1.000
\end{bmatrix}; \qquad \omega_1 = 15.316 \text{ rad/sec.}$$

The mode shape is plotted in Fig. 4–21(b).

---

* An alternative procedure sometimes followed in unrestrained airplane calculations is that of omitting the first row of the $[D] = [G][M]$ matrix when the iteration process is applied. Thus the iteration yields the modal deflections $\phi_1{}^{(1)}, \phi_2{}^{(1)}, \cdots, \phi_n{}^{(1)}$, and the quantity $\phi_{,}{}^{(1)}$ is left undetermined. The latter is obtained by applying the condition that the sum of the inertial forces in the unrestrained normal mode must be zero:

$$\phi_0{}^{(1)} = -\frac{1}{M_0} \sum_{i=1}^{5} M_i \phi_i{}^{(1)}.$$

Although the alternative procedure and the procedure followed in the example should, in principle, lead to the same result, the former is sometimes preferred when it is desired to ensure with a high order of accuracy that the sum of the inertial forces is zero.

(c) *Second unrestrained symmetrical bending mode.* The matrix equation which converges when iterated to the second mode is given by (cf. Eq. 4–103)

$$\{W\} = \omega^2[G][M][S^{(1)}]\{W\} = \omega^2[D^{(2)}]\{W\}, \tag{g}$$

where $[S^{(1)}]$ is a sweeping matrix as defined by Eq. (4–101). The first row of the sweeping matrix is merely a representation of the orthogonality relation between the first and second modes. The orthogonality relation between first and second modes of a concentrated mass system is given by

$$\sum_{i=0}^{5} M_i W_i^{(1)} W_i^{(2)} = 0. \tag{h}$$

Since the system is unrestrained, we have the additional requirement that the inertial forces due to free vibration in the second mode be in equilibrium. This can be expressed by

$$\sum_{i=0}^{5} M_i W_i^{(2)} = 0. \tag{i}$$

Expression (i) also states that the second mode is orthogonal with the rigid-body mode. Combining Eqs. (h) and (i), an orthogonality condition can be constructed in which the quantity $W_0^{(2)}$ has been eliminated.

$$\sum_{i=1}^{5} M_i(W_i^{(1)} - W_0^{(1)})W_i^{(2)} = 0. \tag{j}$$

Equation (j) requires that the first and second modes be orthogonal, and in addition requires that the inertial forces due to free vibrations in the second mode be in equilibrium. Equation (j) corresponds to Eq. (4–98) in the general development of the sweeping process, and we see therefore that the $K_{1i}^{(1)}$ terms in the sweeping matrix (Eq. 4–101) are given by

$$K_{1i}^{(1)} = \frac{M_i(W_i^{(1)} - W_0^{(1)})}{M_1(W_1^{(1)} - W_0^{(1)})} = \frac{M_i(\phi_i^{(1)} - \phi_0^{(1)})}{M_1(\phi_1^{(1)} - \phi_0^{(1)})}. \tag{k}$$

Introducing numerical values from previous results, the sweeping matrix becomes

$$[S^{(1)}] = \begin{bmatrix} 0 & -6.4515 & -7.2709 & -11.1595 & -3.2751 \\ 0 & 1 & 0 & 0 & 0 \\ 0 & 0 & 1 & 0 & 0 \\ 0 & 0 & 0 & 1 & 0 \\ 0 & 0 & 0 & 0 & 1 \end{bmatrix}. \tag{l}$$

Substituting this result, together with the values of $[G]$ and $[M]$ given above, into Eq. (g), and iterating, there is obtained for the second unrestrained symmetrical mode shape and frequency:

$$\{\phi^{(2)}\} = \begin{bmatrix} .1667 \\ -.04153 \\ -.1742 \\ -.4578 \\ .1092 \\ 1.000 \end{bmatrix}; \quad \omega_2 = 53.545 \text{ rad/sec.}$$

The mode shape is plotted in Fig. 4–21(c).

(d) *First antisymmetrical bending mode.* The characteristic equations for antisymmetrical bending modes are (*cf.* Eq. 4–110)

$$\{W\} = \omega^2[G][M]\{W\} = \omega^2[D]\{W\}, \tag{m}$$

where the elements of the $[G]$ matrix, obtained from Eq. (4–111), are

$$G_{ij} = C_{ij}{}^{zz} - \frac{y_i}{I_x}\sum_{k=0}^{5} C_{kj}M_k y_k . \tag{n}$$

In deriving Eq. (n) from Eq. (4–111) only the rolling degree of rigid-body freedom is included. The quantity I_x in Eq. (n) represents one-half of the rolling moment of inertia about the principal x-axis of the entire airplane. The rolling moment of inertia of the fuselage about the x-axis should be included in this quantity. In this particular example, $I_x = 135.02 \times 10^7$ lb $\cdot$ in². Applying the mass and influence coefficient data given previously, the $[G]$ matrix appropriate to pure antisymmetrical bending becomes

$$[G] =$$

$$\begin{bmatrix} 0 & 0 & 0 & 0 & 0 \\ 0 & .22802 & -.35243 & -2.1793 & -4.5895 & -6.9312 \\ 0 & .12247 & -.56723 & -3.0107 & -6.3671 & -9.7445 \\ 0 & .032313 & .14591 & 1.4906 & -.36330 & -2.5458 \\ 0 & -.077635 & 1.01558 & 4.0288 & 9.7571 & 14.433 \\ 0 & -.17659 & 1.7982 & 6.3131 & 19.0914 & 37.119 \end{bmatrix} \times 10^{-5}. \tag{f}$$

Substituting for $[M]$ and $[G]$ the matrices given by (a) and (f), respectively, into Eq. (m) and iterating, we obtain the first antisymmetrical bending mode shape and frequency, as follows:

$$\{\phi^{(1)}\}^* = \begin{bmatrix} 0 \\ -.2175 \\ -.2932 \\ -.06840 \\ .4330 \\ 1.000 \end{bmatrix} ; \qquad \omega_1 = 28.85 \text{ rad/sec.}$$

The mode shape is plotted in Fig. 4–21(d).

EXAMPLE 4–6. To compute the first pure torsion mode of the wing of the jet transport.

Solution. Pure torsion modes (uncoupled torsion modes) are obtained by assuming that the centers of gravity of all chordwise sections of the wing lie

---

* It should be observed that the rolling angular displacement of the fuselage can be computed from the mode shape data given by the column matrix $\{\phi^{(1)}\}$ by applying the condition that the moment of the inertial forces about the x-axis in an unrestrained antisymmetrical mode must be zero.

$$\frac{\partial\phi^{(1)}}{\partial y} = -\frac{2}{I_f}\sum_{i=1}^{5} M_i y_i \phi_i^{(1)},$$

where I_f is the rolling moment of inertia of the fuselage about the x-axis.

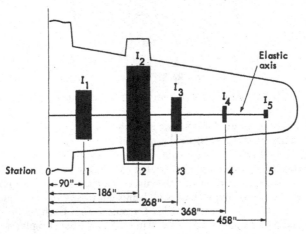

Fig. 4-22. Lumped mass system for computation of torsion modes—jet transport.

along the elastic axis, so that there can be no coupling between bending and torsional vibrations. The actual spanwise mass moment of inertia distribution along the span of the jet transport can be replaced by the lumped mass moments of inertia illustrated by Fig. 4-22. The fuselage is assumed to have infinite pitching moment of inertia. Thus the analysis can be conducted as though the elastic axis is fixed at the airplane center line. The mass moment of inertia matrix of the system in Fig. 4-22 is

$$[I] = \frac{1}{386} \begin{bmatrix} \infty & & & & \\ & 8.723 & & & \\ & & 93.16 & & \\ & & & 3.725 & \\ & & & & 2.775 \\ & & & & & 0.40 \end{bmatrix} \text{lb·in·sec}^2. \qquad (a)$$

The characteristic equations for pure torsional vibration of a lumped mass system are

$$\{\theta\} = \omega^2 [C^{\theta\theta}][I]\{\theta\}, \qquad (b)$$

where the influence coefficient matrix $[C^{\theta\theta}]$ is given by Eq. (g) of Example 2–1, and the moment of inertia matrix $[I]$ is given by Eq. (a) above. Forming the matrix product in Eq. (b) and iterating, we obtain the following mode shape and frequency of the first torsion mode:

$$\{\phi^{(1)}\} = \begin{bmatrix} 0 \\ .4489 \\ .91932 \\ .95216 \\ .98939 \\ 1.000 \end{bmatrix}; \qquad \omega_1 = 22.357 \text{ rad/sec.}$$

The mode shape is plotted in Fig. 4-23.

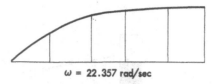

$\omega = 22.357$ rad/sec

Fig. 4–23.   First pure torsion mode of jet transport wing.

EXAMPLE 4–7.   To compute the first two symmetrical unrestrained normal modes (coupled bending-torsion modes) of the wing of the jet transport.

Solution.   It will be assumed in this computation that the airplane is free to translate vertically but is restrained against pitching and rolling.   The wing is replaced by the system of concentrated masses illustrated by Fig. 4–24.   Each wing segment is replaced by a "dumbbell" unit consisting of two equal concentrated masses attached at the two ends of a rigid arm.   Each "dumbbell" unit has a total mass equal to the mass of the wing segment that it replaces.   In addition, it is placed in a chordwise location such that its moment of inertia and static unbalance about the elastic axis are the same as those of the wing segment.

In order to construct the system illustrated by Fig. 4–24, we need, in addition to data given in Examples 4–5 and 4–6, the static unbalance S_α of each wing segment.   The following table is used to calculate the properties of the "dumbbell" units:

Station	M	I_α	S_α	$e = S_\alpha/M$	m	$d = 2\sqrt{\dfrac{I_\alpha}{M} - e^2}$
(in)	(lb)	(lb·in²)	(lb·in)	(in)	(lb·)	(in)
0	17,400	∞	—	—	17,400	—
90	6,039	8.72×10^6	$-42,273$	-7	3,019.5	74.7
186	10,200	93.16×10^6	$-140,083$	-13.734	5,100	189.15
268	4,200	3.72×10^6	8,400	2.0	2,100	59.39
368	3,400	2.77×10^6	6,800	2.0	1,700	56.95
458	680	0.40×10^6	2,720	4.0	340	47.84

In the above table, S_α is the static unbalance of each "dumbbell" unit about the elastic axis, and e is the distance from the elastic axis to the c.g. of each unit.

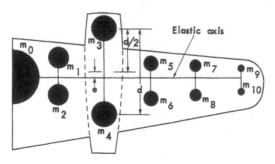

Fig. 4–24.   Dynamic model of jet transport wing.

Both of these quantities are positive when the c.g. is aft of the elastic axis. m is the mass of each of the concentrated masses of the "dumbbell" units and is one-half the total mass of the wing segment at all wing stations except the center-line station, where it is equal to the total mass. d is the length of the rigid arm of the "dumbbell" unit.

The characteristic equations for the unrestrained symmetrical normal modes are (cf. Eq. 4-110)

$$\{W\} = \omega^2[G][m]\{W\} = \omega^2[D]\{W\}, \tag{a}$$

where the elements of the $[G]$ matrix are obtained from Eq. (4-111) as follows:

$$G_{ij} = C_{ij}{}^{zz} - \frac{1}{M} \sum_{k=0}^{10} C_{kj}{}^{zz} M_k. \tag{b}$$

Since the system consists of eleven concentrated masses, the $[G]$ and $[m]$ matrices consist of eleven rows and columns. The quantity M in Eq. (b) represents half the total airplane mass.

The influence coefficients associated with the concentrated masses in Fig. 4-24 can be computed by application of the $[C^{zz}]$ and $[C^{\theta\theta}]$ matrices given respectively by matrices (e) and (g) of Example 2-1. This is accomplished simply by a process of placing a unit load on a single mass point and computing the deflection at all mass points by use of the $[C^{zz}]$ and $[C^{\theta\theta}]$ matrices. The process is repeated until the unit load has been successively placed on all mass points.

When the iteration process is applied to Eq. (a), it converges on the fundamental unrestrained normal mode. The normalized mode shape and frequency are as follows:

$$\{\phi^{(1)}\} = \begin{bmatrix} \phi_0{}^{(1)} \\ \phi_1{}^{(1)} \\ \cdot \\ \cdot \\ \cdot \\ \phi_{10}{}^{(1)} \end{bmatrix} = \begin{bmatrix} -.167241 \\ -.130234 \\ -.125428 \\ -.0279016 \\ -.00513142 \\ .246037 \\ .254583 \\ .619040 \\ .629625 \\ .989446 \\ 1.0000 \end{bmatrix} ; \quad \omega_1 = 15.310 \text{ rad/sec.}$$

Each numerical value in the above column matrix represents the modal displacement of a concentrated mass. The node line of the first unrestrained normal mode is plotted on the wing planform in Fig. 4-25(a).* The first normal mode is very nearly a pure bending mode and its frequency is but slightly different from the pure bending mode frequency computed in Example 4-5(b). This similarity is typical of reasonably slender wings with small static unbalance. It should be observed that in this particular example, even though a large engine mass is carried by the wing, the static unbalance is small.

---

* The node line is plotted by a simple interpolation of the deflection curves along spanwise lines between "dumbbell" units m_3, m_4, and m_5, m_6.

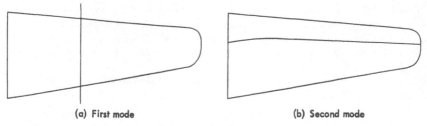

| (a) First mode | (b) Second mode |

Fig. 4–25. Node lines of unrestrained symmetrical normal modes of jet transport wing.

The second unrestrained symmetrical normal mode is obtained by introducing a sweeping matrix in Eq. (a),

$$\{W\} = \omega^2[G][m][S^{(1)}]\{W\} = \omega^2[D^{(2)}]\{W\}, \tag{c}$$

where $[S^{(1)}]$ is a sweeping matrix defined by Eq. (4–101). Iteration of Eq. (c) converges on the second unrestrained symmetrical normal mode, as follows:

$$\{\phi^{(2)}\} = \begin{bmatrix} -\ .0531214 \\ +\ .165048 \\ -\ .179550 \\ +1.0000 \\ -\ .789320 \\ +\ .265021 \\ -\ .316863 \\ +\ .289834 \\ -\ .289804 \\ +\ .249535 \\ -\ .242188 \end{bmatrix} ; \qquad \omega_2 = 22.410 \text{ rad/sec.}$$

The node line of the second unrestrained symmetrical normal mode is plotted on the wing planform in Fig. 4–25(b). This mode is nearly a pure torsion mode and its mode shape and frequency are very close to the pure torsion mode results computed in Example 4–6. This again is a typical result for reasonably slender wings with small static unbalance.

4–7 Natural modes and frequencies of rotating beams. Many of the methods of analysis discussed in the preceding sections can be applied also to rotating beams. Let us consider, for example, application of Galerkin's method (Ref. 4–8) to the solution of the homogeneous differential equation of motion of a rotating beam. Considering a uniform rotating beam, for simplicity, the differential equation (3–124) can be reduced to

$$EIW^{IV} - \frac{m\Omega^2}{2}(R^2 - y^2)W'' + m\Omega^2 yW' - m\omega^2 W = 0. \tag{4–119}$$

We nondimensionalize Eq. (4–119) by putting $\psi = W/R$ and $\eta = y/R$.

The differential equation which results is

$$\frac{d^4\psi}{d\eta^4} - \frac{K}{2}(1 - \eta^2)\frac{d^2\psi}{d\eta^2} + K\eta\frac{d\psi}{d\eta} - \left(\frac{\omega}{\Omega}\right)^2 K\psi = 0, \qquad (4\text{--}120)$$

where K is a dimensionless constant defined by

$$K = \frac{m\Omega^2 R^4}{EI}.$$

Assume that the deformed rotating beam can be represented by the series

$$\psi(\eta) = \sum_{j=1}^{n} \gamma_j(\eta)\bar{q}_j, \qquad (4\text{--}121)$$

where the $\gamma_j(\eta)$ are assumed functions satisfying the boundary conditions. Introducing Eq. (4–121) into Eq. (4–120) yields

$$\sum_{j=1}^{n}\left[\gamma_j{}^{IV}(\eta) - \frac{K}{2}(1 - \eta^2)\gamma_j{}''(\eta) + K\eta\gamma_j{}'(\eta) - \left(\frac{\omega}{\Omega}\right)^2 K\gamma_j(\eta)\right]\bar{q}_j = 0,$$
$$(4\text{--}122)$$

where primes now denote differentiations with respect to η. Multiplying through by $\gamma_i(\eta)$ and integrating from 0 to 1,

$$\sum_{j=1}^{n}\left[\int_0^1 \gamma_j{}^{IV}\gamma_i d\eta - \frac{K}{2}\int_0^1 (1 - \eta^2)\gamma_j{}''\gamma_i d\eta \right.$$
$$\left. + K\int_0^1 \eta\gamma_j{}'\gamma_i d\eta - \left(\frac{\omega}{\Omega}\right)^2 K\int_0^1 \gamma_i\gamma_j d\eta\right]\bar{q}_j = 0. \quad (4\text{--}123)$$

Integrating the first integral by parts,

$$\int_0^1 \gamma_j{}^{IV}\gamma_i d\eta = [\gamma_j{}'''\gamma_i - \gamma_j{}''\gamma_i{}']_0^1 + \int_0^1 \gamma_i{}''\gamma_j{}'' d\eta. \qquad (4\text{--}124)$$

If we assume that the boundary conditions are such that the bracketed term in Eq. (4–124) vanishes, Eq. (4–123) reduces to

$$\sum_{j=1}^{n}\left[k_{ij} - \left(\frac{\omega}{\Omega}\right)^2 m_{ij}\right]\bar{q}_j = 0, \qquad (i = 1, \cdots, n), \qquad (4\text{--}125)$$

where

$$m_{ij} = \int_0^1 \gamma_i\gamma_j d\eta,$$

$$k_{ij} = \frac{1}{K}\int_0^1 \gamma_i{}''\gamma_j{}'' d\eta - \frac{1}{2}\int_0^1 (1 - \eta^2)\gamma_j{}''\gamma_i d\eta + \int_0^1 \eta\gamma_j{}'\gamma_i d\eta.$$

The characteristic equations given by (4–125) are similar in form to Eqs. (4–11a) and the various methods of solution described in Section 4–5 are applicable. It is often convenient to put Eq. (4–125) in the matrix form

$$\left(\frac{\omega}{\Omega}\right)^2 [m_{ij}]\{\bar{q}\} = [k_{ij}]\{\bar{q}\}, \tag{4–126}$$

where

$$[m_{ij}] = [\gamma][\overline{W}][\gamma]',$$

$$[k_{ij}] = \frac{1}{K}[\gamma''][\overline{W}][\gamma'']' - \tfrac{1}{2}[\gamma''][\overline{W}][(1 - \eta^2)][\gamma]' + [\gamma'][\overline{W}][\eta][\gamma]'.$$

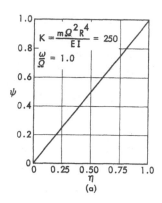

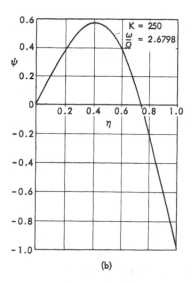

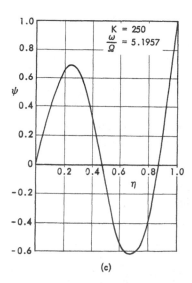

Fig. 4–26.   (a) Rigid-flapping mode of uniform rotating beam.   (b) First bending mode of uniform rotating beam.   (c) Second bending mode of uniform rotating beam.

The form of the matrices $[\gamma]$, $[\gamma']$, and $[\gamma'']$ is the same as illustrated previously in connection with Eqs. (4–13) and (4–14).

Displacement functions $\gamma_j(\eta)$ are selected such that the boundary conditions are satisfied as nearly as possible. For example, a common application is that of a hinged-free rotating beam. We may apply Duncan's polynomial expression (Ref. 4–4) by putting

$$\gamma_j = \eta, \qquad (j = 1), \tag{4–127}$$

$$\gamma_j = \tfrac{1}{6}(j+2)(j+3)\eta^{j+1} - \tfrac{1}{3}j(j+3)\eta^{j+2} + \tfrac{1}{6}j(j+1)\eta^{j+3}, \qquad (j \geq 2). \tag{4–128}$$

Figure 4–26 illustrates results for the first two bending modes obtained by applying four of the displacement functions ($j = 1, 2, 3, 4$) given by Eqs. (4–127) and (4–128). In computing these results, the quantity K is assumed equal to 250. The weighting matrices are constructed by dividing the hinged-free rotating beam into ten equal spanwise segments and applying Simpson's rule. Since the dimensionless constant K is the only physical quantity that is involved, Fig. 4–26 applies, of course, to all uniform rotating beams which have a value of $K = 250$.

CHAPTER 5

AERODYNAMIC TOOLS: TWO- AND THREE-DIMENSIONAL INCOMPRESSIBLE FLOW

5-1 Fundamentals: the concept of small disturbances. The foregoing chapters have described the physical properties and motion of those structural solids with which the aeroelastician has to work. We shall now consider the dynamics of fluids, so as to complete all necessary introductory material for the subject of interactions between elastic deformations and aerodynamic loads imposed by fluid motion. Since so many excellent references and textbooks have been written on fluid dynamics, a word of justification seems in order for once more reviewing some of the fundamentals. One important reason is the nature of the aerodynamic tools employed in practical analysis of aeroelastic problems. Except in certain special cases like stall flutter, where nonlinearity is of the essence, these tools must perforce be mathematically linear to ensure the possibility of finding solutions with reasonable time and effort. It is therefore necessary to outline the severe linearization procedures that underlie most of the results to be used. Moreover, from the aeroelastician's standpoint, a defect of nearly all books on fluid dynamics is their preoccupation with steady-flow phenomena. Even such a highly regarded work as Ferri's *Elements of Aerodynamics of Supersonic Flows* commences the first section of the first chapter with the words, "Consider a perfect compressible flow in steady motion," and its author abides by this limitation throughout. In discussing unsteady flow, certain other writers stress one-dimensional disturbance propagation, whereas the aeroelastician usually needs information on two- and three-dimensional external flows, especially those over bodies which produce lift-type forces.

Except when the assumption of incompressibility is made for temporary mathematical simplicity, the fluid considered here is a frictionless perfect gas. The mechanical properties of a perfect gas are summarized by stating that the force acting across any interior or bounding surface is normal to that surface and compressive. All thermodynamic phenomena are deducible from the equation of state relating ambient pressure p, density ρ, and absolute temperature T:

$$p = R\rho T, \tag{5-1}$$

in combination with the assumption of constant isobaric specific heat c_p and the observation that, with a few specific exceptions, all processes are reversible.

Since the reader is presumed to have some familiarity with general aerodynamic theory, he is referred to books like Refs. 5-1 and 5-2 for derivations of certain of the basic equations. Flow of a perfect gas can be completely described by specifying, as functions of time t and position (x,y,z) in rectangular Cartesian coordinates, the ambient quantities p, ρ, and T, and the Cartesian velocity components (u,v,w) of the particles. These six dependent variables are always related by the condition of mass conservation, the continuity equation,

$$\frac{\partial \rho}{\partial t} + \frac{\partial (\rho u)}{\partial x} + \frac{\partial (\rho v)}{\partial y} + \frac{\partial (\rho w)}{\partial z} = 0, \tag{5-2}$$

and by the three components of the vector equation of motion in inertial coordinates:*

$$\frac{Du}{Dt} = \frac{\partial u}{\partial t} + u \frac{\partial u}{\partial x} + v \frac{\partial u}{\partial y} + w \frac{\partial u}{\partial z} = -\frac{1}{\rho} \frac{\partial p}{\partial x}, \tag{5-3a}$$

$$\frac{Dv}{Dt} = -\frac{1}{\rho} \frac{\partial p}{\partial y}, \tag{5-3b}$$

$$\frac{Dw}{Dt} = -\frac{1}{\rho} \frac{\partial p}{\partial z}. \tag{5-3c}$$

In Eqs. (5-3) gravity and other external forces have been omitted, as is customary for heavier-than-air craft.

To produce a mathematically determinate system of equations for the flow field, Eqs. (5-1) through (5-3) must be supplemented by the first two experimental laws of equilibrium thermodynamics. As shown, for example, in Ref. 5-3, these are equivalent to the definition of three new state variables known as the specific energy, the specific enthalpy, and the specific entropy of a pure substance.

It must be pointed out that in most aerodynamic theory of interest to aeroelasticians the thermodynamic properties of the gas enter only through the specific heat ratio γ, and the state is represented by a disturbance pressure, or difference between local ambient pressure and some reference value. In this sort of theory an essential part is played by certain functions which reduce the number of dependent variables, notably the velocity potential $\phi(x,y,z,t)$ and the acceleration potential or pressure function $\psi(x,y,z,t)$. The existence of the former function depends on the condition of irrotationality of the flow, which means physically that all fluid particles have zero angular momentum about their own center-of-gravity axes. This condition is expressed mathematically by the dis-

---

* The substantial derivative $D(\ )/Dt$ denotes the time rate of change of a quantity for an individual fluid particle.

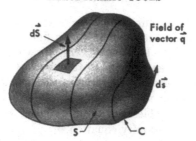

Fig. 5–1. Surface S, bounded by curve C, in a vector field. The areal element vector $d\vec{S}$ is normal to the area dS with which it corresponds. Line element vector $d\vec{s}$ is positive in the sense of right-handed rotation about $d\vec{S}$.

appearance of the curl* of vector q, or in component form,

$$\frac{\partial v}{\partial x} - \frac{\partial u}{\partial y} = 0, \qquad \frac{\partial w}{\partial y} - \frac{\partial v}{\partial z} = 0, \qquad \frac{\partial u}{\partial z} - \frac{\partial w}{\partial x} = 0. \qquad (5\text{–}4)$$

The uniform vanishing of curl q throughout most flows of practical interest is by no means intuitively obvious, but it can be proved, for unsteady as well as steady motions, through simultaneous application of theorems named after Stokes and Kelvin. Stokes' theorem states a purely geometrical property of any vector field, namely, that the integral of the normal component of curl q over any surface S in the field (Fig. 5–1) equals the line integral of q around the closed curve C which bounds S. As proved in any text on advanced calculus or vector analysis,

$$\iint_S \operatorname{curl} \mathbf{q} \cdot d\mathbf{S} = \oint_C \mathbf{q} \cdot d\mathbf{s} = \Gamma, \qquad (5\text{–}5)$$

where the vector symbols are defined in Fig. 5–1. The line integral on the right is known in fluid mechanics as the circulation Γ. The usefulness of Stokes' theorem is that it establishes the truth of Eqs. (5–4) once one is able to show that Γ vanishes for all paths wholly within a simply connected flow region, since otherwise any fluid particle with finite curl q can immediately be shown to imply a nonzero Γ by proper choice of the surface S.

Kelvin's theorem (Ref. 5–2), a consequence of the laws of dynamics, concerns the rate of change of Γ with time for a closed path consisting always of the same aggregate of fluid elements. In its most general form for a frictionless fluid, it reads

$$\frac{D\Gamma}{Dt} = -\oint_C \frac{dp}{\rho}. \qquad (5\text{–}6)$$

---

* The vector notation $\mathbf{q} = u\mathbf{i} + v\mathbf{j} + w\mathbf{k}$ will be used here to simplify the writing of a few equations. Vector operators divergence, gradient, and curl are given their customary definitions (Phillips, Ref. 5–4).

Clearly, this reduces to

$$\frac{D\Gamma}{Dt} = 0 \qquad (5\text{-}7)$$

either for incompressible fluid or when the flow is barotropic—possessing a unique pressure-density relation—throughout.   The barotropic situation prevails in a large majority of all cases practically important to the aeroelastician.   In particular, when the fluid originates in a reservoir under uniform stagnation conditions (continuous or intermittent wind tunnel) or when there are straight, parallel streamlines with constant ambient properties upstream (rectilinear flight through the atmosphere), it can be deduced that a reversible flow will be isentropic with the relation $p = \text{const} \cdot \rho^{\gamma}$ valid everywhere.   Only strong shocks or intense heating can spoil this uniformity.   When Eq. (5-7) does hold for all time and all paths, Γ remains constant for any aggregate of particles.   Since Γ is initially zero around any such path in flows starting either from a reservoir or from parallel streamlines, the continued vanishing of circulation, together with Stokes' theorem, assures a wholly irrotational flow.

In view of the above reasoning, we normally assume that Eqs. (5-4) are true throughout the interior of a flow.*   Vanishing of the curl in a vector field is necessary and sufficient to assure that the vector is the gradient of some scalar function (cf. Ref. 5-4, pages 100-101).   This function in the present case is the velocity potential ϕ.   The scalar components of

$$\mathbf{q} = \text{grad } \phi \qquad (5\text{-}8)$$

are the familiar relations

$$u = \frac{\partial \phi}{\partial x}, \qquad v = \frac{\partial \phi}{\partial y}, \qquad w = \frac{\partial \phi}{\partial z}, \qquad (5\text{-}9)$$

which reduce the number of dependent variables in aerodynamic problems by two.   Equations (5-9) are also seen to imply Eqs. (5-4) through the equality of three pairs of cross partial derivatives of ϕ.

The velocity potential is the principal unknown in a multitude of theoretical investigations.   It is related to the pressure, a quantity of more direct physical significance, through the equations of motion (5-3) and a first integral of them known as Kelvin's equation or the unsteady Bernoulli's equation.   Using vector shorthand, we can readily find this integral by observing that Eqs. (5-3) read

$$\frac{\partial \mathbf{q}}{\partial t} + (\mathbf{q} \cdot \text{grad}) \, \mathbf{q} = -\frac{1}{\rho} \, \text{grad } p, \qquad (5\text{-}10)$$

---

* The principal exception to this assumption occurs when surfaces of tangential-velocity discontinuity are shed from the trailing edges of lifting wings, a subject that will be discussed below.

which is equivalent in irrotational, barotropic flow to

$$\frac{\partial}{\partial t}\operatorname{grad} \phi + \operatorname{grad}\left(\frac{q^2}{2}\right) = -\frac{1}{\rho}\operatorname{grad} p \qquad (5\text{--}11)$$

or

$$\operatorname{grad}\left[\frac{\partial \phi}{\partial t} + \frac{q^2}{2} + \int \frac{dp}{\rho}\right] = 0. \qquad (5\text{--}12)$$

Integration of Eq. (5–12) shows that the sum of these three quantities is a constant throughout the flow at any particular instant of time, so this sum can at most equal some function of time:

$$\frac{\partial \phi}{\partial t} + \frac{q^2}{2} + \int \frac{dp}{\rho} = F(t). \qquad (5\text{--}13)$$

The function $F(t)$ may be eliminated from the right side of Kelvin's equation by a redefinition of the velocity potential. Thus ϕ may be replaced by $[\phi - \int F(t)dt]$ without altering the velocity field in any respect. This substitution is unnecessary in most practical cases, however, because $F(t)$ vanishes when a barotropic flow is connected to a large reservoir. Moreover,

$$F(t) = \tfrac{1}{2}U^2 \qquad (5\text{--}14)$$

whenever the remote fluid motion consists of parallel streamlines with velocity U. Under isentropic conditions, a simple evaluation of the pressure integral is possible.

A partial differential equation for the velocity potential can be derived by explicitly eliminating the state variables between Eq. (5–13) and the continuity condition (5–2), which we put in the vectorial form

$$\frac{1}{\rho}\frac{\partial \rho}{\partial t} + \frac{\mathbf{q}}{\rho}\cdot \operatorname{grad} \rho + \operatorname{div} \mathbf{q} = 0. \qquad (5\text{--}15)$$

If ϕ exists, we take advantage of the fact that the divergence of the gradient is equal to Laplace's operator $\nabla^2(\ \ )$:

$$\operatorname{div}\,(\operatorname{grad}\,(\ \ )) = \nabla^2(\ \ ) = \frac{\partial^2(\ \ )}{\partial x^2} + \frac{\partial^2(\ \ )}{\partial y^2} + \frac{\partial^2(\ \ )}{\partial z^2}. \qquad (5\text{--}16)$$

Equation (5–15) then becomes

$$\frac{1}{\rho}\frac{\partial \rho}{\partial t} + \frac{\mathbf{q}}{\rho}\cdot \operatorname{grad} \rho + \nabla^2\phi = 0. \qquad (5\text{--}17)$$

Before proceeding, it is noteworthy to observe that when the fluid is effectively incompressible this simplifies to Laplace's equation

$$\nabla^2\phi = 0. \qquad (5\text{--}18)$$

Not only do familiar and thoroughly developed mathematical tools (e.g., Refs. 5–2 and 5–5) then become available, especially complex-variable techniques in the two-dimensional case, but the problem is actually separated into two distinct parts. Equation (5–18) is the purely geometrical outcome of the concurrent continuity and irrotationality conditions; the laws of mechanics enter only when we calculate pressures from the velocity field by means of the algebraic, incompressible-flow counterpart of Kelvin's equation

$$p + \frac{\rho q^2}{2} + \rho \frac{\partial \phi}{\partial t} = \rho F(t). \tag{5–19}$$

Returning to the general equation governing ϕ, we remove the two density terms, using Eq. (5–13), after disposing of $F(t)$ in one of the ways described above. The partial derivative with respect to time of Eq. (5–13) is

$$-\frac{\partial}{\partial t}\left(\frac{\partial \phi}{\partial t} + \frac{q^2}{2}\right) = \frac{\partial}{\partial t}\int \frac{dp}{\rho} = \frac{1}{\rho}\frac{\partial p}{\partial t} = \frac{1}{\rho}\frac{dp}{d\rho}\frac{\partial \rho}{\partial t} = \frac{a^2}{\rho}\frac{\partial \rho}{\partial t}. \tag{5–20}$$

Here

$$a = \sqrt{\frac{dp}{d\rho}} \tag{5–21}$$

is the speed of sound or local speed of propagation of small disturbances, which may be regarded as a convenient substitute for the state variables. Application of the gradient operator to Kelvin's equation leads to

$$-\mathrm{grad}\left(\frac{\partial \phi}{\partial t} + \frac{q^2}{2}\right) = \frac{1}{\rho}\,\mathrm{grad}\,p = \frac{a^2}{\rho}\,\mathrm{grad}\,\rho, \tag{5–22}$$

whence

$$\frac{\mathbf{q}}{\rho} \cdot \mathrm{grad}\,\rho = -\frac{1}{a^2}\left[\mathbf{q}\cdot\frac{\partial}{\partial t}\,\mathrm{grad}\,\phi + \mathbf{q}\cdot\mathrm{grad}\left(\frac{q^2}{2}\right)\right]$$

$$= -\frac{1}{a^2}\left[\frac{\partial}{\partial t}\left(\frac{q^2}{2}\right) + \mathbf{q}\cdot\mathrm{grad}\left(\frac{q^2}{2}\right)\right]. \tag{5–23}$$

If Eqs. (5–20) and (5–23) are inserted into Eq. (5–17), we get

$$\nabla^2\phi - \frac{1}{a^2}\left[\frac{\partial^2\phi}{\partial t^2} + \frac{\partial q^2}{\partial t} + \mathbf{q}\cdot\mathrm{grad}\left(\frac{q^2}{2}\right)\right] = 0. \tag{5–24}$$

Since $\mathbf{q} = \mathrm{grad}\,\phi$, this is the exact, nonlinear, unsteady-flow equation to be satisfied by the velocity potential. The presence of a^2 creates a question as to whether ϕ is the only dependent variable involved in Eq. (5–24). For steady flow a^2 is easily eliminated using a familiar consequence of the first law of thermodynamics,

$$a^2 + \tfrac{1}{2}(\gamma - 1)q^2 = \mathrm{const.} \tag{5–25}$$

It can be demonstrated, however, that this kind of relation cannot be applied when time variations occur. This complication has little practical import in aeroelasticity, because the nonlinear equations cannot be solved anyway. We can replace a^2 by its constant, uniform-stream value in linearized versions of Eq. (5–24), thus removing the difficulty.

We next turn to the second part of the definition of a complete mathematical description of flow problems, namely, the boundary conditions. When the fluid mass is unbounded, these ordinarily comprise two types: conditions at infinity, and conditions arising from the presence of one or more submerged bodies which set the fluid in motion. The former depend on the nature of the partial differential equation governing ϕ. If it is Laplace's equation, as in incompressible flow, they require only that the fluid be at rest or in some specified uniform motion at remote points. If the differential equation resembles the wave equation, as it does when the motion is unsteady and compressibility is accounted for, they must include in addition the statement that wavelike disturbances are propagating outward away from their sources. These matters are treated more fully in subsequent sections.

The condition at the boundary of a typical body states simply that, over its surface, the perpendicular component of fluid velocity $\partial\phi/\partial n$ is fixed by the body's motion. Particles in contact with the surface must share its normal motion, for no holes can form in the flow except due to liquid cavitation, which we rule out here. Lamb shows (Ref. 5–2, Chapter 1) that if the equation of the surface of a body moving in a time-dependent fashion is

$$F(x,y,z,t) = 0, \qquad (5\text{--}26)$$

the boundary condition reads

$$\frac{DF}{Dt} = \frac{\partial F}{\partial t} + u\,\frac{\partial F}{\partial x} + v\,\frac{\partial F}{\partial y} + w\,\frac{\partial F}{\partial z} = 0 \qquad (5\text{--}27)$$

over the area of F. In other words, the rate of change of the numerical value of the function F is zero when we follow the motion of a particular fluid element (substantial derivative), so that the element continually touches the surface $F = 0$. For steady flow, F is independent of time t, and Eq. (5–27) reduces to the expression

$$\mathbf{q} \cdot \operatorname{grad} F = 0. \qquad (5\text{--}28)$$

In physical terms, the component of velocity normal to F vanishes.

It is just the form of boundary conditions (5–27) and (5–28) for bodies of importance in aeronautical engineering that gave rise to the small-disturbance concept underlying most techniques for linearizing the aerodynamic problem. We here concentrate on wings, which are of most interest to aeroelasticians, but the remarks we make are readily extended to cover slender fuselages, etc. We think of the wing or airfoil as submerged in an infinite mass of fluid, which translates steadily with velocity U in the positive x-direction of Cartesian coordinates. The wing is fixed to the Cartesian system in such a way that it lies close to the xy-plane. We then find that the variable z can be explicitly separated from Eq. (5–26), and this formula can be written in two parts, one describing the upper surface of the wing, the other the lower surface:

$$F_U = z - z_U(x,y,t) = 0, \qquad (5\text{–}29a)$$

$$F_L = z - z_L(x,y,t) = 0. \qquad (5\text{–}29b)$$

Equations (5–29) hold between the leading edge, whose trace on the xy-plane is $x = x_l(y)$, and the trailing edge, whose trace is $x = x_t(y)$. Since $\partial F/\partial z = 1$, we are able to solve Eq. (5–27) for the values of vertical velocity w over the wing surface:

$$w = \frac{\partial z_U}{\partial t} + u\,\frac{\partial z_U}{\partial x} + v\,\frac{\partial z_U}{\partial y}\,; \qquad \text{for } z = z_U,\ (x,y) \text{ in } R_a, \quad (5\text{–}30a)$$

$$w = \frac{\partial z_L}{\partial t} + u\,\frac{\partial z_L}{\partial x} + v\,\frac{\partial z_L}{\partial y}\,; \qquad \text{for } z = z_L,\ (x,y) \text{ in } R_a, \quad (5\text{–}30b)$$

where R_a is the portion of the xy-plane covered by the projection of the planform.

So far Eqs. (5–30) are exact, but we approximate them by observing that over almost the entire area of nearly all wings (1) the slopes $\partial z_U/\partial x$, $\partial z_U/\partial y$, etc. are very small compared with unity, and (2) the resultant fluid velocity vector $\mathbf{q}$ differs only slightly in direction and magnitude from the free-stream velocity $U\mathbf{i}$. We adopt observation (2) as the basic assumption of the theory of small disturbances by defining a disturbance velocity potential ϕ', which is obtained from the total velocity potential ϕ by separating out the contribution of the uniform flow,

$$\phi = \phi' + Ux. \qquad (5\text{–}31)$$

The disturbance velocity components

$$u - U = u' = \frac{\partial \phi'}{\partial x}, \qquad v = \frac{\partial \phi'}{\partial y}, \qquad w = \frac{\partial \phi'}{\partial z} \qquad (5\text{–}32)$$

are assumed to satisfy the order-of-magnitude requirement

$$u',\, v,\, w \ll U. \qquad (5\text{–}33)$$

Condition (5–33) is, of course, closely related to the existence of small slopes over the airfoil surface. If conditions (1) and (2) above are actually fulfilled, the terms $u'(\partial z_U/\partial x)$ and $v(\partial z_U/\partial y)$ in Eq. (5–30a) can be neglected by comparison with the much larger term $U(\partial z_U/\partial x)$, and similarly in Eq. (5–30b). This leads to

$$w = \frac{\partial z_U}{\partial t} + U \frac{\partial z_U}{\partial x}, \qquad \text{for } z = z_U, \ (x,y) \text{ in } R_a, \qquad (5\text{–}34\text{a})$$

$$w = \frac{\partial z_L}{\partial t} + U \frac{\partial z_L}{\partial x}; \qquad \text{for } z = z_L, \ (x,y) \text{ in } R_a. \qquad (5\text{–}34\text{b})$$

However, condition (1) is equivalent to saying that z_U and z_L are very small compared with the wing chord, which fact enables us to go one step further. It turns out, for purposes of solving the boundary-value problem, that we are able to replace the actual wing with an infinitesimally thick surface of discontinuities in u, v, w, and pressure p. Since this is mathematically possible, we locate this discontinuity surface on the xy-plane. We expand w (Eqs. 5–34) in Maclaurin series about its values just above and below the xy-plane:

$$w(x,y,z_U,t)$$
$$= w(x,y,0^+,t) + z_U \frac{\partial w(x,y,0^+,t)}{\partial z} + \frac{z_U^2}{2!} \frac{\partial^2 w(x,y,0^+,t)}{\partial z^2} + \cdots, \qquad (5\text{–}35\text{a})$$

$$w(x,y,z_L,t)$$
$$= w(x,y,0^-,t) + z_L \frac{\partial w(x,y,0^-,t)}{\partial z} + \frac{z_L^2}{2!} \frac{\partial^2 w(x,y,0^-,t)}{\partial z^2} + \cdots. \qquad (5\text{–}35\text{b})$$

If the derivatives $\partial w/\partial z$, etc. are sufficiently well behaved that their products with small quantities $z_U(\partial w/\partial z)$, etc. may be neglected by comparison with w itself, all but the first terms on the right of Eqs. (5–35) can be omitted. This also we assume, and boundary conditions (5–30) finally take the linear, homogeneous forms

$$w = \frac{\partial z_U}{\partial t} + U \frac{\partial z_U}{\partial x}; \qquad \text{for } z = 0^+, \ (x,y) \text{ in } R_a, \qquad (5\text{–}36\text{a})$$

$$w = \frac{\partial z_L}{\partial t} + U \frac{\partial z_L}{\partial x}; \qquad \text{for } z = 0^-, \ (x,y) \text{ in } R_a. \qquad (5\text{–}36\text{b})$$

It is important to recognize that the usefulness of Eqs. (5–36) relies entirely on the possibility of approximating the actual wing with a mathematical plane surface across which appropriate discontinuities can be constructed.

As far as the simplification of boundary conditions goes, assumption (5–33) need hold only in the neighborhood of the surface. It is natural to expect, however, that if it is fulfilled there it is also valid throughout the entire flow. In fact, Lamb (Ref. 5–2) proves for an *incompressible* fluid that the disturbance must be a maximum on the boundaries. Extending this hypothesis to cover the whole field of compressible flow, we are led to a procedure for linearizing the differential equation (5–24) which is consistent with Eqs. (5–36). The terms in Laplace's operator are already linear, and there is no reason to expect any one to be much larger or smaller than the others. Into the three terms in brackets, we substitute

$$\mathbf{q} = (U + u')\mathbf{i} + v\mathbf{j} + w\mathbf{k}$$
$$= U\mathbf{i} + \text{grad } \phi' \tag{5-37}$$

and omit all portions which are manifestly much smaller than those retained:

$$\frac{\partial^2 \phi}{\partial t^2} + \frac{\partial q^2}{\partial t} + \mathbf{q} \cdot \text{grad} \left(\frac{q^2}{2} \right)$$

$$= \frac{\partial^2 \phi'}{\partial t^2} + 2[U\mathbf{i} + \text{grad } \phi'] \cdot \frac{\partial}{\partial t} [U\mathbf{i} + \text{grad } \phi']$$

$$+ [U\mathbf{i} + \text{grad } \phi'] \cdot \text{grad} \left[\frac{U^2}{2} + U\mathbf{i} \cdot \text{grad } \phi' + \tfrac{1}{2} |\text{grad } \phi'|^2 \right]$$

$$= \frac{\partial^2 \phi'}{\partial t^2} + 2[U\mathbf{i} + \text{grad } \phi'] \cdot \left[\frac{\partial^2 \phi'}{\partial x \partial t} \mathbf{i} + \frac{\partial^2 \phi'}{\partial y \partial t} \mathbf{j} + \frac{\partial^2 \phi'}{\partial z \partial t} \mathbf{k} \right]$$

$$+ [U\mathbf{i} + \text{grad } \phi'] \cdot \left[U \frac{\partial u'}{\partial x} \mathbf{i} + \tfrac{1}{2} \left(\frac{\partial u'^2}{\partial x} \mathbf{i} + \frac{\partial v^2}{\partial x} \mathbf{i} + \cdots \right) \right]$$

$$\cong \frac{\partial^2 \phi'}{\partial t^2} + 2U \frac{\partial^2 \phi'}{\partial x \partial t} + U^2 \frac{\partial^2 \phi'}{\partial x^2}. \tag{5-38}$$

Obviously, we have made, in addition to Eq. (5-33), a series of assumptions about the smallness of first derivatives of the velocity components with respect to space and time coordinates. For example, the third term in the last bracket of Eq. (5-38) is neglected because

$$U \frac{\partial u'}{\partial x} \gg v \frac{\partial v}{\partial x}, \tag{5-39}$$

etc. All these may be thought of as following from Eq. (5-33), provided we assume that the small quantities change sufficiently gradually in all directions, and no time variations are too rapid.

Linearization of Eq. (5-24) is completed by disposing of the factor $1/a^2$. For steady flow, this is done through Eq. (5-25), noting that any variations in a^2 from its free-stream value a_∞^2 produce factors on the bracket in Eq. (5-24) that generate terms like the ones neglected in Eq. (5-38).* When the flow is unsteady, we select an appropriate form of the thermodynamic energy equation and substitute Eq. (5-20) into it:

$$\frac{D}{Dt} \left[\frac{a^2}{\gamma - 1} + \frac{q^2}{2} \right] = \frac{1}{\rho} \frac{\partial p}{\partial t} = - \frac{\partial^2 \phi}{\partial t^2} - \frac{\partial}{\partial t} \left(\frac{q^2}{2} \right). \tag{5-40}$$

---

* This is true only so long as U is not too large compared with a_∞, so we must rule out hypersonic flows.

Similar approximations to those which led to form (5–24) of the boundary conditions permit the substantial derivative to be replaced as follows:

$$\frac{D}{Dt}[\cdots] \cong \frac{\partial}{\partial t}[\cdots] + U\frac{\partial}{\partial x}[\cdots]. \tag{5–41}$$

The smallness of various derivatives of the disturbance velocity components shows that

$$\frac{\partial}{\partial t}\left(\frac{q^2}{2}\right) = \frac{\partial}{\partial t}\left[\frac{(U+u')^2}{2} + \frac{v^2}{2} + \frac{w^2}{2}\right] \cong U\frac{\partial u'}{\partial t}. \tag{5–42}$$

Substituting Eqs. (5–41) and (5–42) into Eq. (5–40), and replacing ϕ by ϕ', we get

$$\frac{\partial}{\partial t}\left(\frac{a^2}{\gamma-1}\right) + U\frac{\partial}{\partial x}\left(\frac{a^2}{\gamma-1}\right) = -\frac{\partial^2\phi'}{\partial t^2} - 2U\frac{\partial u'}{\partial t} - U^2\frac{\partial u'}{\partial x}. \tag{5–43a}$$

If we define a disturbance sound velocity a' by $a = a_\infty + a'$ and multiply Eq. (5–43a) by $(\gamma - 1)/2a_\infty^2$, keeping only the largest terms,

$$\frac{\partial}{\partial t}\left(\frac{a'}{a_\infty}\right) + U\frac{\partial}{\partial x}\left(\frac{a'}{a_\infty}\right)$$

$$= -\frac{\gamma-1}{2}\left[\frac{1}{a_\infty^2}\frac{\partial^2\phi'}{\partial t^2} + \frac{2U}{a_\infty^2}\frac{\partial u'}{\partial t} + \frac{U^2}{a_\infty^2}\frac{\partial u'}{\partial x}\right]. \tag{5–43b}$$

When Eq. (5–43b) is integrated over the interval of time it takes for a fluid particle to traverse the disturbed region, which is of the order of 5 or 10 times the product of U by the wing chord, we are led to the conclusion that a'/a_∞ always remains very small compared with unity. We insert Eq. (5–38) into Eq. (5–24) and observe that the smallness of a' relative to a_∞ means that a in (5–24) can be replaced by a_∞. All terms neglected in this way are at least one order higher than those retained. Hence we come finally to

$$\nabla^2\phi' - \frac{1}{a_\infty^2}\left[\frac{\partial^2\phi'}{\partial t^2} + 2U\frac{\partial^2\phi'}{\partial x\partial t} + U^2\frac{\partial^2\phi'}{\partial x^2}\right] = 0. \tag{5–44}$$

This is the linearized partial differential equation for unsteady, compressible flow. Its applications for calculating the important aerodynamic tools of aeroelasticity will be the main topic of this and the following two chapters.

We conclude the present section by mentioning that a second function, of more limited usefulness than ϕ, is Prandtl's acceleration potential. Its existence is assured for barotropic flow by the vector equation of motion

$$\frac{D\mathbf{q}}{Dt} = -\frac{1}{\rho}\operatorname{grad} p = -\operatorname{grad}\int\frac{dp}{\rho}. \tag{5–3}$$

This shows that the acceleration vector is the gradient of a scalar function, which we designate $\psi(x,y,z,t)$:

$$\frac{D\mathbf{q}}{Dt} = \text{grad } \psi, \tag{5-45}$$

with components

$$\frac{Du}{Dt} = \frac{\partial \psi}{\partial x}, \quad \text{etc.} \tag{5-46}$$

From Eqs. (5-3) and (5-45),

$$\text{grad } \psi + \text{grad} \int \frac{dp}{\rho} = 0, \tag{5-47}$$

so that these quantities differ at most by a function of time

$$\psi = -\int \frac{dp}{\rho} + G(t). \tag{5-48}$$

$G(t)$ has the same properties as $F(t)$ in Eq. (5-13) and therefore reduces to a constant in most practical circumstances. For example, when the fluid is incompressible and ψ is assigned the value zero in the remote flow where $p = p_\infty$, Eq. (5-48) reads

$$\psi = \frac{p_\infty - p}{\rho}. \tag{5-49}$$

Here, and also in certain compressible flows such as those with only small disturbances, ψ differs only by a constant factor from the disturbance pressure $(p - p_\infty)$. Hence it is often referred to as the pressure function; in fact, the entire theory of the acceleration potential as used in practice could have been derived simply by introducing disturbance pressure as the unknown to be solved for. The relation between ψ, ϕ, and the velocity is found by inserting Eq. (5-47) into Eq. (5-12). When the flow is uniform, with velocity U at infinity, the resulting differential equation integrates to give

$$\psi = \frac{\partial \phi}{\partial t} + \tfrac{1}{2}(q^2 - U^2). \tag{5-50a}$$

If the flow obeys the small-perturbation assumption (5-33), Eq. (5-50a) can be linearized to

$$\psi = \frac{\partial \phi'}{\partial t} + Uu' = \frac{\partial \phi'}{\partial t} + U \frac{\partial \phi'}{\partial x}. \tag{5-50b}$$

In view of the interchangeability of the operator $[\partial/\partial t + U(\partial/\partial x)]$ with the other linear operators in Eq. (5-44), this also proves that ψ satisfies the same partial differential equation as ϕ'.

5–2 Properties of incompressible flow with and without circulation.
Mathematical solution of incompressible potential flow problems is a
straightforward matter in principle. Laplace's equation (5–18) is to be
solved subject to a determinate set of boundary conditions, which might
resemble Eqs. (5–27), (5–28), or (5–36) over the surface of one or more
submerged bodies, while specifying that at remote points the liquid is
either at rest (ϕ = const.) or in uniform translation. These matters are
treated more fully by the first seven chapters of Lamb (Ref. 5–2). Certain
results are worth restating, however, so as to emphasize their significance
in aeroelastic applications.

There exist two general categories, to one of which any incompressible
potential flow can be assigned; these are flows without circulation, and
circulatory flows with their attendant lines or sheets of vortices. The
former are vastly easier to deal with because of their readily demonstrated
property that the entire field of particle velocities depends only on the instantaneous motion of the submerged bodies, quite independently of the
past history of that motion. This behavior stems from the fact that the
speed of sound is effectively infinite, so that any change of boundary conditions is propagated instantaneously to all the particles. For example,
consider a smooth, rigid solid which translates in a fixed direction or ro-
tates about a fixed axis within an otherwise undisturbed infinite liquid
mass.* The linearity of the differential equation and the linear relationship between the boundary conditions and the translational or angular
velocity of the solid establish that the instantaneous velocity of the liquid
particle in any particular position relative to the solid will be directly
proportional to the magnitude of the latter's motion. Consequently,
the total kinetic energy imparted to the liquid will vary directly with the
square of the linear velocity of the solid during translation, and with the
square of the angular velocity during rotational motion. Since there are
no dissipative agencies in the system, any change in the velocity of the
solid would require that it do an amount of work on the liquid equal to
the change of this kinetic energy. If the acceleration of the entire system
were produced by an externally applied force or moment, some constant
fraction thereof would go into altering the energy of the liquid, an effect
which could be explained physically by imagining the mass or moment of
inertia of the solid to be augmented by a fixed amount. These incremental
masses and moments of inertia are known as virtual or apparent masses;
they are directly proportional to the density of the liquid but vary, in
general, with the orientation of the axis of translation or rotation relative
to the solid. Care must be exercised when applying the virtual-mass concept for combinations of translations, or for simultaneous rotations about

---

* More general combined motions of this type are discussed quantitatively
in Chap. VI of Ref. 5–2 and by Munk (Ref. 5–6), whose work is concentrated
on ellipsoidal solids.

different axes, because cross-coupling terms then appear in the kinetic energy. This difficulty can be resolved, however, as shown by Lamb (Ref. 5–2, Chap. VI), and certain coupling terms disappear when the solid has axes or planes of symmetry.

The virtual masses can be calculated analytically for many two- and three-dimensional bodies of practical aeronautical interest. They are very useful in estimating the forces and moments on streamlined aircraft components which do not develop circulation, such as fuselages, jet pods, and external tanks. It is also convenient to identify noncirculatory portions of the aerodynamic loads on lifting surfaces. These effects become increasingly important for larger values of acceleration; for example, in very high-frequency oscillations they are proportional to the second power of the frequency and dominate the motion. Since one can compute the noncirculatory loads so easily, it is unfortunate that the compressibility of air renders the concept increasingly inaccurate at larger frequencies. The problem merges into one of acoustic wave propagation, which will be discussed more fully in subsequent chapters.

It would seem more straightforward to introduce the concept of virtual mass by reference to the momentum, rather than the kinetic energy, imparted to the infinite liquid mass by bodies moving through it. However, as shown by Lamb (Ref. 5–2), Theodorsen (Ref. 5–7), and others, this is not the case. In fact, a change in the motion of any body can give rise to a pressure distribution over an imaginary bounding surface at infinity that alters the total momentum by an infinite amount (since this bounding surface may be regarded as fixed, no work is done there, and the body work can still be equated to the kinetic energy of the liquid). There is, however, an easy way to calculate the *impulse* applied in any direction by an accelerated body to the liquid. To do this, we ask what system of impulsive pressures would produce the entire state of motion instantaneously from rest. We are able to find these by integrating the vector equation (5–3), with $\rho = $ const., over the infinitesimal duration of such an impulse:

$$\rho \int \frac{D\mathbf{q}}{Dt} \, dt = -\int [\mathrm{grad}\ p] dt \qquad (5\text{–}51a)$$

or

$$\rho\mathbf{q} = -\mathrm{grad} \int p dt. \qquad (5\text{–}51b)$$

It is also true, of course, that

$$\rho\mathbf{q} = \mathrm{grad}\ (\rho\phi). \qquad (5\text{–}8)$$

Hence, if we specify that the velocity potential ϕ must vanish when the liquid is everywhere at rest, we obtain

$$\int p dt = -\rho\phi. \qquad (5\text{–}52)$$

This states that the impulse associated with any change in the motion is directly proportional to the change in the value of ϕ. The size of this impulse is independent of whether the motion is produced suddenly, as assumed in the above derivation, or brought about more gradually, since ϕ depends only on the end states. The total impulse applied during starting by any body to the liquid in any given direction is found by integrating the appropriate component of the impulsive forces $-\rho\phi d\mathbf{S}$ over the body's surface S. For example,

$$I_x = \int F_x dt = -\rho \iint_S \phi \mathbf{i} \cdot d\mathbf{S}, \tag{5-53}$$

which leads, after differentiation, to

$$F_x = -\rho \frac{d}{dt} \iint_S \phi \mathbf{i} \cdot d\mathbf{S}. \tag{5-54}$$

The negative of F_x would represent the resistance of the liquid to the acceleration of the body in the x-direction.

As an illustration of the application of these ideas, consider the rectilinear motion of a slender ellipsoid of revolution (Fig. 5-2), which might represent an externally mounted fuel tank or pod. Lamb (Ref. 5-2, pp. 152-155) shows that the impulsive forces associated with velocities U_x and U_y parallel, respectively, to the x-axis of symmetry and to any lateral y-axis are as follows:

$$\begin{aligned} I_x &= k_1 U_x \rho \cdot \text{(volume)}, \\ I_y &= k_2 U_y \rho \cdot \text{(volume)}, \end{aligned} \tag{5-55}$$

where k_1 and k_2 are inertia coefficients, dependent only on the fineness ratio. When the ellipsoid becomes very long and slender, k_1 vanishes, while k_2 approaches unity (Table 5-1). Many interesting conclusions can be drawn from Eqs. (5-55). For example, a longitudinal acceleration dU_x/dt will be resisted by a drag type of force

$$D_x = -\frac{dI_x}{dt} = -k_1 \rho \cdot \text{(volume)} \cdot \frac{dU_x}{dt}. \tag{5-56}$$

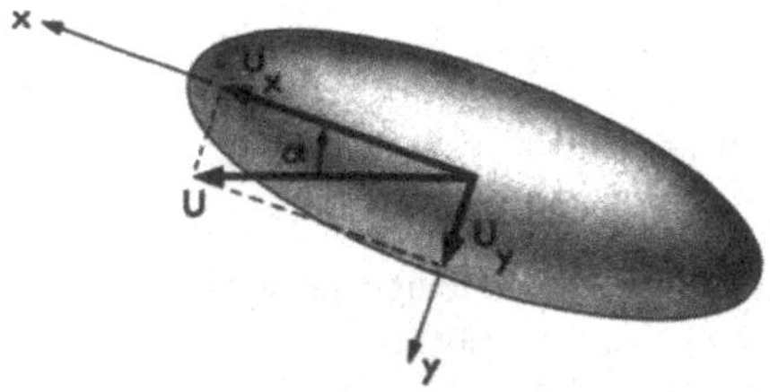

Fig. 5-2. Streamlined ellipsoid of revolution (shown moving in translation through an infinite mass of incompressible fluid).

A lateral oscillation $U_y = \bar{U}_y \sin \omega t$, such as might be produced by bending vibration of the wing, would give rise to a side force

$$F_y = -\frac{dI_y}{dt} = -k_2 \rho \omega \bar{U}_y \cos \omega t \cdot (\text{volume}). \qquad (5\text{--}57)$$

Equations (5–56) and (5–57) reveal that the virtual masses for accelerations in the x- and y-directions are $k_1 \rho \cdot (\text{volume})$ and $k_2 \rho \cdot (\text{volume})$, the latter being very nearly equal to the mass of liquid that would occupy the volume of the body for streamlined ellipsoids of practical interest.

If the ellipsoid is translating steadily with resultant airspeed U and angle of attack α (Fig. 5–2), the components of impulse parallel and perpendicular to the flight direction are

$$[k_1 U \cos^2 \alpha + k_2 U \sin^2 \alpha]\rho \cdot (\text{volume})$$

and

$$[k_2 U \sin \alpha \cos \alpha - k_1 U \cos \alpha \sin \alpha]\rho \cdot (\text{volume}),$$

TABLE 5–1

INERTIA COEFFICIENTS FOR AN ELLIPSOID OF
REVOLUTION IN INCOMPRESSIBLE FLOW

Fineness ratio (length/diameter)	k_1	k_2
1	0.5	0.5
1.5	0.3037	0.6221
2.0	0.2100	0.7042
2.5	0.1563	0.7619
3.0	0.1220	0.8039
3.5	0.0985	0.8354
4.0	0.0816	0.8598
4.5	0.0689	0.8789
5.0	0.0591	0.8943
5.5	0.0514	0.9068
6.0	0.0452	0.9171
6.5	0.0401	0.9258
7.0	0.0358	0.9331
7.5	0.0323	0.9393
8.0	0.0292	0.9447
8.5	0.0266	0.9494
9.0	0.0244	0.9535
9.5	0.0224	0.9571
10	0.0207	0.9602
15	0.0109	0.9787
20	0.00679	0.9866
∞	0	1

respectively. Although the latter component remains constant in magnitude, its line of action is translating steadily normal to itself with velocity U, in such a way as continually to decrease the angular momentum of the liquid about a z-axis fixed in space. To impart the additional negative angular momentum to the liquid particles, the ellipsoid must apply a couple in the negative z sense equal to the rate of change of this momentum. The liquid reacts on the ellipsoid with a positive couple

$$M_z = U[k_2 U \sin \alpha \cos \alpha - k_1 U \cos \alpha \sin \alpha]\rho \cdot \text{(volume)}$$

$$= \tfrac{1}{2}\rho U^2[k_2 - k_1] \sin 2\alpha \cdot \text{(volume)}. \tag{5-58}$$

This is the familiar destabilizing moment experienced by elongated bodies of revolution like airplane fuselages. Its actual measured value in incompressible flow falls somewhat below that given by Eq. (5–58), so that the expression on the right is multiplied by an efficiency factor between 0.8 and 0.9 for practical applications. The discrepancy is due to separation of the boundary layer from the upper rear portion of the body. Separation also generates a small steady-state lift force (cf. Ref. 5–8), although the theory predicts none.

5–3 Vortex flow. When we turn to the topic of incompressible fluid motion with vortices and circulation, we no longer find the attractive unique relationship between flow pattern and instantaneous boundary conditions which is described above. An elementary situation where the uniqueness disappears occurs when liquid is flowing about a two-dimensional body (any cylindrical object extending to infinity in both directions or embedded at its ends in parallel, plane walls which bound the flow). Stokes' theorem (Eq. 5–5) and Kelvin's theorem (Eq. 5–7) do not apply to a closed curve surrounding the body, because a surface bounded by such a curve must pass through the body and hence out of the velocity field. Accordingly, the circulation Γ around the curve may be assigned any magnitude whatever. We must realize, of course, that Stokes' theorem requires Γ to be instantaneously the same for all these circumscribed curves, but there is sometimes nothing in the boundary conditions of a physical problem to fix it uniquely. When two or more two-dimensional bodies are present, as in a cascade of turbine blades, circulation must be specified for each of them.

The best known aeronautical examples where physical considerations dictate nonzero circulation are those of airfoils or similar bodies having cusped or pointed trailing edges. Kutta's hypothesis that the circulation should be chosen to make the fluid velocity finite at the sharp edge, thus producing a smooth joining between upper and lower portions of the flow, has now been well verified for subsonic flight. Compatibility with Stokes' theorem may be demonstrated by imagining the airfoil replaced by a mass of liquid at rest, and the boundary between it and the external flow by a narrow shear region approximating the actual viscous boundary layer. The net strength of vortices in this layer then equals the circulation outside, and a mechanism for generating them is provided by the viscous forces due to large tangential velocity gradients in the not quite frictionless gas or liquid. Naturally, these forces were not accounted for in Kelvin's theorem.

It is readily shown by studying a series of large circles surrounding a body with circulation that the velocity must drop off in inverse proportion to the distance. That is, the remote flow pattern approximates one due to a line vortex, whose velocity potential in cylindrical coordinates is

$$\phi_V = \frac{\Gamma\theta}{2\pi}. \tag{5-59}$$

Now, the fact is that this flow possesses infinite angular momentum and energy. One is led to the conclusion that no system of finite forces, acting for a finite time, can generate an isolated line vortex, a result which is verified if one attempts to do so experimentally. On the other hand, a pair of equal and opposite parallel vortices, separated a finite distance d (Fig. 5–3) has finite kinetic energy per unit axial length. The impulse per unit axial length is also known to have a magnitude (Ref. 5–9, pp. 102–104 and 325–326)

$$I_N = \rho\Gamma d \tag{5-60}$$

and to be directed normal to the vortex lines and to the perpendicular between them. Such a vortex pair therefore presents no physical obstacles to its generation. We are led to the assumption that, whenever circulation develops around a two-dimensional body or changes its magnitude, an equal countervortex must appear somewhere in the flow. When the body is an airfoil, the natural place to look for the appearance of a countervortex is just behind the trailing edge, where a wake of vortices from upper and

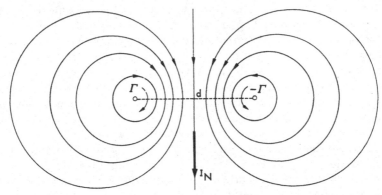

Fig. 5–3. Vortex pair separated a distance d, showing the direction of impulse $I_N = \rho\Gamma d$.

lower boundary layers is continually being shed. This wake should exhibit locally a net negative circulation whenever the net positive strength of vortices over the airfoil is augmented. Since a free vortex must always be composed of the same fluid particles and must move with the surrounding medium, any countervortex will pass on downstream as a part of the wake. We usually assume that it moves very nearly with the velocity and direction of the general flow past the airfoil.

If the motion is steady and the circulation constant, the existence of countervortices presents no difficulty. We assume that at some remote instant in the past the airfoil began its flight at a distant point. The correct Kutta circulation Γ then rapidly developed, leaving a short trail of countervortices of total strength $-\Gamma$. Once the flow became steady, these were too far downstream to affect conditions in the neighborhood of the airfoil. They furnish a convenient explanation of the existence of a lift force, however, since the separation d between the "bound" vortex on the airfoil and the "starting" countervortices is increasing at a steady rate U equal to the airspeed of the airfoil. Hence the vortex system must apply a running force to the liquid in the direction of impulse I_N (Fig. 5–3). This running force has the value

$$\frac{dI_N}{dt} = \rho\Gamma \frac{d(d)}{dt} = \rho\Gamma U. \tag{5–61}$$

Since the airfoil is the only solid body present, it must exert this force; the reaction to it is the lift per unit span $\rho\Gamma U$.

A lifting airfoil in unsteady motion will experience continual changes of the circulation around it. Only when these alterations are very gradual is it possible to assume that the associated countervortices pass downstream to infinity so fast that their influence on the pattern of liquid velocities around the airfoil is negligible. Otherwise, their presence endows the flow with a "memory," in that its present characteristics are affected by the past history of the bound circulation. The phenomenon is similar to the action of the history of deformation under load of a plastic solid in determining the present stress distribution, or like the effect of the details of previously impressed magnetic fields on the instantaneous degree of polarization in an electromagnet subject to hysteresis. Such behavior always leads to a mathematical representation involving an integral equation with time as the independent variable, and this will be seen later to occur whenever the wake countervortices play a significant part in a flow.

Although the foregoing considerations were limited to two-dimensional liquid motion, it is also true that finite lifting wings give rise to sheets of vortices, which have their origin in boundary-layer viscous forces. The creation of three-dimensional vortex wakes can also be explained by studying the system of impulses applied by the wing onto the liquid, as is done

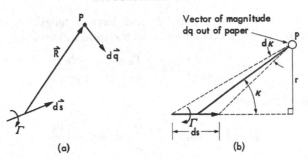

Fig. 5–4. Two views of the velocity dq induced by an element ds of a vortex line with circulation Γ. (a) Vector quantities involved. (b) Plan view to illustrate lengths r, R, and angle κ.

particularly well by von Kármán and Burgers in Vol. II of *Aerodynamic Theory* (Ref. 5–9). Free vortices in incompressible fluid obey certain well-known laws (Ref. 5–2, Chap. VII): they remain constant in circulation and are always composed of the same liquid particles; they cannot end within the flow, but must either form closed loops, extend to infinity, or abut against a solid boundary. These rules are listed here without proof. We shall occasionally find them useful in dealing with steady and unsteady motion of wings. We also give without derivation an important result of potential theory, called the Biot-Savart law.* As utilized in fluid mechanics, this law states that an elementary length ds of a vortex line of circulation Γ induces a velocity

$$dq = \frac{\Gamma ds \times R}{4\pi R^3} \qquad (5\text{–}62\text{a})$$

at a point P located a vector distance $\mathbf{R}$ from ds. The direction of ds must be taken such that the circulation is positive around it, in accordance with the right-hand rule, as illustrated in Fig. 5–4(a). The scalar form of this law reads

$$dq = \frac{r}{4\pi R^3}\,\Gamma ds = \frac{\sin\kappa}{4\pi R^2}\,\Gamma ds. \qquad (5\text{–}62\text{b})$$

The perpendicular length r from P to the axis of the vortex segment, and the angle κ between R and that axis are shown in Fig. 5–4(b).

As a simple application of Eq. (5–62b), consider a straight, two-dimensional vortex line of infinite extent and constant circulation Γ_0. At point P, a distance r from the vortex, all velocities induced thereby will point in the tangential or θ-direction of a set of cylindrical coordinates whose polar axis coincides with the vortex. The element ds at an angle κ will lie at a

---

* One of the best derivations of this law appears in Sect. 4.4, pp. 230–232 of Stratton's *Electromagnetic Theory* (Ref. 5–10).

distance $R = r/\sin \kappa$ from point P and have a length $ds = Rd\kappa/\sin \kappa$ $= rd\kappa/\sin^2 \kappa$. Hence its contribution to the total induced velocity q_θ is

$$dq_\theta = \frac{\sin \kappa}{4\pi R^2} \Gamma ds = \frac{\Gamma_0}{4\pi} \frac{\sin^3 \kappa}{r^2} \frac{rd\kappa}{\sin^2 \kappa}.\tag{5-63}$$

Simple integration yields

$$q_\theta = \int_{\kappa=0}^{\kappa=\pi} dq_\theta = \frac{\Gamma_0}{4\pi r} \int_0^\pi \sin \kappa d\kappa = \frac{\Gamma_0}{2\pi r},\tag{5-64}$$

a well-known formula which might also have been found by differentiating Eq. (5–59). The results expressed by Eqs. (5–64) and (5–62) will be applied in subsequent sections for calculation of the aerodynamic loading on two- and three-dimensional wings in steady and unsteady motion.

5–4 Thin airfoils in steady motion. From the standpoint of the aero-elastician, practically all important characteristics of very large aspect-ratio wings are predicted by the so-called thin airfoil theory. This method has its foundation in the assumption of small disturbances, which is fully discussed and applied to the differential equation and boundary conditions in Section 5–1. There exists a disturbance velocity potential ϕ'; for incompressible flow, the speed of sound becomes relatively large, and Eq. (5–44) reduces to Laplace's equation

$$\nabla^2 \phi' = 0.\tag{5-65}$$

This could have been deduced directly from Eq. (5–18), thus showing that the small-disturbance idea actually does not affect the differential equation.

For steady flow, the exact specification of tangency of the flow to the airfoil surfaces

$$z = z_U(x) \qquad \text{and} \qquad z = z_L(x)\tag{5-66}$$

reads

$$\frac{w}{u} = \frac{dz_U}{dx}; \qquad \text{for } z = z_U, \ -b \le x \le b,\tag{5-67}$$

$$\frac{w}{u} = \frac{dz_L}{dx}; \qquad \text{for } z = z_L, \ -b \le x \le b,\tag{5-68}$$

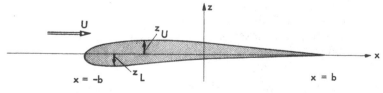

Fig. 5–5.　Cross section of a thin airfoil or wing lying in the xy-plane.

where $2b$ is the chord. A typical cross section is shown in Fig. 5–5. Since we are dealing with a two-dimensional airfoil, z_U and z_L are functions of x only and this picture is the same at all stations along the span. As demonstrated in Eqs. (5–36) the boundary conditions can be simplified to

$$w = U\frac{dz_U}{dx}; \quad \text{for } z = 0^+, -b \leq x \leq b, \tag{5–69}$$

$$w = U\frac{dz_L}{dx}; \quad \text{for } z = 0^-, -b \leq x \leq b. \tag{5–70}$$

In the present problem the assumption of small disturbances faces the immediate objection that it is invalid around stagnation points where the liquid is brought to rest, such as occur near the leading edge and at the (finite-angle) trailing edge of any real airfoil. Experience has shown, however, that the neighborhoods of such points in which the particle speeds differ substantially from U are quite small. The theory can be used with remarkable success for predicting lift, pitching moment, and even the distribution of pressure everywhere except within a few percent chord lengths of the forbidden points.

For incompressible flow, Laplace's equation impresses the geometrical conditions of continuity and irrotationality on the motion. The laws of dynamics are introduced when Kelvin's equation (5–19) is employed to compute pressure from a known velocity and potential pattern. Its form may be simplified by observing that the uniform motion at infinity makes

$$\rho F(t) = \tfrac{1}{2}\rho U^2 + p_\infty. \tag{5–71}$$

The consistent application of Eq. (5–33) calls for the substitution

$$q^2 = (U + u')^2 + v^2 + w^2 \cong U^2 + 2Uu'. \tag{5–72}$$

With these insertions, Eq. (5–19) furnishes a simple formula for the dimensionless coefficient of pressure C_p:

$$C_p = \frac{p - p_\infty}{\tfrac{1}{2}\rho U^2} = -\frac{2u'}{U} - \frac{2}{U^2}\frac{\partial\phi'}{\partial t}. \tag{5–73}$$

Since the motion of the airfoil is here taken to be steady, Eq. (5–73) reduces to

$$C_p = -\frac{2u'}{U} = -\frac{2}{U}\frac{\partial\phi'}{\partial x}. \tag{5–74}$$

Equations (5–73) and (5–74) are linear relations between the dependent variables. They and the boundary conditions represent the important linearization accomplished by our assumptions. It should be emphasized that additional requirements on the velocity derivatives, like Eq. (5–39), are not needed here as they are for compressible fluids, when the exact differential equation is nonlinear. The more stringent simplifications of the true physical picture in the latter case will be taken up in Chapter 6.

We next examine more closely the boundary-value problem posed by Eqs. (5–65), (5–69), and (5–70). We split z_U and z_L into an even part z_a and an odd part z_t, as follows:

$$z_U = z_a + z_t, \qquad z_L = z_a - z_t, \qquad (5\text{–}75)$$

where z_t describes a shape symmetrical about the xy-plane and gives the chordwise distribution of thickness, whereas z_a contains the angle of attack and camber. Since Eqs. (5–69) and (5–70) are linear and satisfied along a fixed line, the boundary-value problems associated with z_t and z_a can be solved separately. In physical terms, the airfoil properties may be regarded as the superposition of those of a symmetrical airfoil at zero incidence and a cambered, inclined mean line of zero thickness, as pictured in Fig. 5–6. Before proceeding to solve various special cases of the problem, we point out that Eqs. (5–69), (5–70), and (5–75) apply equally well for a finite wing in steady motion. The functions z_U, z_L, z_a, and z_t then depend on both x and y, so that the derivatives are partials with respect to x. The small-disturbance assumption also implies that all slopes of the wing in the spanwise direction are numerically small. To this we add the statement that almost every wing of aeronautical interest is nearly a plane, so that there remains no objection to specifying the boundary conditions on $z = 0$.

The symmetrical or thickness part of the problem is of little interest to the aeroelastician, so we dispose of it quickly for airfoils and finite wings alike. Looking at the most general case, we must solve the differential equation (5–65) subject to the boundary conditions

$$w = \frac{\partial \phi'}{\partial z} = U \frac{\partial z_t}{\partial x}; \qquad \text{for } z = 0^+, \ (x,y) \text{ in } R_a, \qquad (5\text{–}76)$$

$$w = \frac{\partial \phi'}{\partial z} = -U \frac{\partial z_t}{\partial x}; \qquad \text{for } z = 0^-, \ (x,y) \text{ in } R_a, \qquad (5\text{–}77)$$

where R_a is a region consisting of the projection of the planform on the xy-plane. Because of the wing's symmetry, we expect a flow pattern

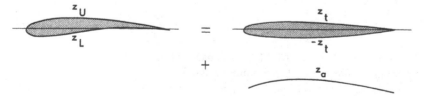

Fig. 5–6. Thin airfoil or wing shown as the superposition of a symmetrical body at zero incidence and a cambered, inclined mean line.

which is completely symmetrical with respect to the xy-plane, having no discontinuities in pressure or velocity components across that plane except for the jump $2U(\partial z_t/\partial x)$ in w which is specified over R_a. Familiarity with the elementary tools of fluid mechanics suggests that all these requirements are met by distributing over R_a a sheet of those solutions of Laplace's equation known as point sources. A single, concentrated source centered at $x = \xi$, $y = \eta$, $z = \zeta$ has the velocity potential

$$\phi_s = \frac{-H}{4\pi\sqrt{(x - \xi)^2 + (y - \eta)^2 + (z - \zeta)^2}}. \tag{5–78}$$

Equation (5–78) describes motion with spherical symmetry around its center, from which the source is readily shown to be emanating H ft^3/sec (in engineering units) of liquid into the surrounding space. H is called the strength of the source. A sheet of these spread continuously over the surface R_a, having strength $H(\xi,\eta)$ per unit area in the neighborhood of the point $(\xi,\eta,0)$, possesses the disturbance potential

$$\phi'(x,y,z) = -\frac{1}{4\pi} \iint_{R_a} \frac{H(\xi,\eta)d\xi d\eta}{\sqrt{(x - \xi)^2 + (y - \eta)^2 + z^2}}. \tag{5–79}$$

To see how this can be made to satisfy condition (5–76), we calculate

$$w(x,y,0^+) = \frac{\partial \phi'(x,y,0^+)}{\partial z}$$

$$= -\frac{1}{4\pi} \lim_{z \to 0^+} \frac{\partial}{\partial z} \iint_{R_a} \frac{H(\xi,\eta)d\xi d\eta}{\sqrt{(x - \xi)^2 + (y - \eta)^2 + z^2}}$$

$$= \frac{1}{4\pi} \lim_{z \to 0^+} z \iint_{R_a} \frac{H(\xi,\eta)d\xi d\eta}{[(x - \xi)^2 + (y - \eta)^2 + z^2]^{\frac{3}{2}}}. \tag{5–80}$$

Here the differentiation may be taken inside the integral sign, since the limits of R_a do not depend on z. As z assumes smaller positive values, the integral is caused to vanish by its multiplying factor, except in the vicinity of point $\xi = x$, $\eta = y$, where the integrand tends to infinity. We isolate this region with a small square of side 2ϵ, obtaining

$$w(x,y,0^+) = \frac{1}{4\pi} \lim_{z \to 0^+} z \int_{y-\epsilon}^{y+\epsilon} \int_{x-\epsilon}^{x+\epsilon} \frac{H(\xi,\eta)d\xi d\eta}{[(x - \xi)^2 + (y - \eta)^2 + z^2]^{\frac{3}{2}}}. \tag{5–81}$$

Being a continuous function, $H(\xi,\eta)$ over the entire square differs from its central value $H(x,y)$ by an amount of order ϵ. Hence, if we neglect these tiny variations and introduce temporary integration variables $\xi' = (x - \xi)$, $\eta' = (y - \eta)$, we find

$$
\begin{aligned}
w(x,y,0^+) &= \frac{H(x,y)}{4\pi} \lim_{z \to 0^+} z \int_{-\epsilon}^{\epsilon} \int_{-\epsilon}^{\epsilon} \frac{d\xi' d\eta'}{[\xi'^2 + \eta'^2 + z^2]^{\frac{3}{2}}} \\
&= \frac{H(x,y)}{4\pi} \lim_{z \to 0^+} \left[2 \tan^{-1} \left(\frac{\epsilon^2}{z \sqrt{2\epsilon^2 + z^2}} \right) - 2 \tan^{-1} \left(\frac{-\epsilon^2}{z \sqrt{2\epsilon^2 + z^2}} \right) \right].
\end{aligned}
$$

$$(5\text{-}82)$$

A sophisticated study of the limiting process involved here shows that we must let z and ϵ approach zero in such a way that the ratio ϵ/z becomes indefinitely large. We have to make ϵ vanish to ensure that terms omitted in going from Eq. (5–81) to Eq. (5–82) are truly negligible, but, unless ϵ remains large compared with z, we get nonzero contributions from the region R_a outside the small square. Limiting in the correct way causes the inverse tangents to approach $+\pi/2$ and $-\pi/2$, so that

$$w(x,y,0^+) = \tfrac{1}{2} H(x,y).$$

$$(5\text{-}83a)$$

Coming up to the source sheet from the lower side reverses the algebraic signs of the arguments of the inverse tangents, leading to

$$w(x,y,0^-) = -\tfrac{1}{2} H(x,y).$$

$$(5\text{-}83b)$$

Therefore, we see that the sources do produce the symmetrical discontinuity specified by Eqs. (5–76) and (5–77), which can both be satisfied by setting

$$H(x,y) = 2U \frac{\partial z_t(x,y)}{\partial x}.$$

$$(5\text{-}84)$$

The same conclusion might also have been reached in a qualitative physical way by arguing that the sources discharge H ft^3 of liquid per unit area per unit time. This liquid can escape only by moving away from the sheet, half of it going upward with normal velocity $H/2$, the other half downward with velocity $-H/2$. The phenomenon is sometimes described by saying that any source in such a sheet is a point function with respect to the velocity component normal to the sheet.

Solution of the problem is completed by substituting Eq. (5–84) into Eq. (5–79):

$$
\phi'(x,y,z) = - \frac{U}{2\pi} \iint_{R_a} \frac{\dfrac{\partial z_t(\xi,\eta)}{\partial x} \, d\xi d\eta}{\sqrt{(x - \xi)^2 + (y - \eta)^2 + z^2}}.
$$

$$(5\text{-}85)$$

All properties of the flow are deducible from Eq. (5–85). For example, the distribution of pressure over the wing comes from differentiating with respect to x, setting $z = 0$, and inserting into Eq. (5–74). When it is a case of two-dimensional flow past an airfoil, we start from a two-dimensional source sheet along the chord and find, in a manner exactly analogous to the above,

$$\phi'(x,z) = \frac{U}{2\pi} \int_{-b}^{b} \frac{dz_t(\xi)}{dx} \ln \left[(x - \xi)^2 + z^2 \right] d\xi. \qquad (5\text{–}86)$$

The situations described by Eqs. (5–85) and (5–86) are good examples of noncirculatory flows in the sense of Section 5–2. Accordingly, the same formulas are perfectly valid when the boundary conditions are time dependent, except that then Eq. (5–84) must be modified to account for vertical velocities due to the local variation of z_t (*cf.* Section 5–1):

$$H(x,y,t) = 2 \frac{\partial z_t(x,y,t)}{\partial t} + 2U \frac{\partial z_t(x,y,t)}{\partial x}, \qquad (5\text{–}87)$$

with corresponding changes in Eqs. (5–85) and (5–86). The solutions thus obtained relate to the unlikely event of pulsations in the wing's thickness. Subsonic panel flutter and time-dependent skin buckling are the only aeroelastic problems to which they might have practical application, but these have proved unimportant in ordinary flight circumstances.

Because of symmetry the thickness contributes nothing to the lift, pitching moment, and rolling moment experienced by a thin airfoil or wing. We therefore expect to gain much more useful information from the antisymmetrical part of the problem, associated with the function z_a. Here we no longer know the values of w over the entire xy-plane (they were zero off R_a in the symmetrical case), and we must solve a much more difficult boundary-value situation of mixed type. It is fortunate that the way is relatively clear in the two-dimensional steady-flow case, which we deal with first. ϕ' must satisfy Laplace's equation in two independent variables, subject to the condition

$$w = \frac{\partial \phi'}{\partial z} = U \frac{dz_a}{dx}; \qquad \text{for } z = 0, \; -b \leq x \leq b. \qquad (5\text{–}88)$$

To supplement Eq. (5–88), it is also worth emphasizing that no discontinuity of pressure, hence no jump in u', can exist across the xy-plane except along the airfoil chord. We examine the symmetries of this disturbance flow pattern by observing that Eq. (5–88) can be handled with a w-distribution which is a continuous, even function of z. Recalling that integration reverses the evenness or oddness of a function of any particular variable, this means that ϕ' and $u' = \partial \phi'/\partial x$ would be odd functions of z. The disturbed motion would therefore be antisymmetrical with respect to the xy-plane. Such a flow can be generated and an agency for sustaining lift simultaneously introduced, by putting a sheet of two-dimensional vortices along the x-axis. Equation (5–59) in Cartesian coordinates, with the sign reversed so as to give positive circulation around the positive y-direction, gives for a concentrated vortex along the line $x = \xi$, $z = \zeta$:

$$\phi_V = -\frac{\Gamma}{2\pi} \tan^{-1}\left(\frac{z-\zeta}{x-\xi}\right). \tag{5-89}$$

A distribution along the x-axis ($\zeta = 0$) with circulation $\gamma_a(\xi)$ per unit distance has the disturbance potential

$$\phi'(x,z) = -\frac{1}{2\pi} \int \gamma_a(\xi) \tan^{-1}\left(\frac{z}{x-\xi}\right) d\xi. \tag{5-90}$$

To decide what should be the extent of the vortex sheet, we make use of the close relationship between γ_a and the local disturbance velocity u'. This could be done by differentiating Eq. (5–90) with respect to x and letting z approach zero. Greater physical clarity is achieved, however, by noting that the existence of circulation is equivalent to a discontinuity in u' across the sheet. If the value is u_U' for $z = 0^+$, then by antisymmetry it must be $-u_U'$ for $z = 0^-$. This state of affairs is illustrated in Fig. 5–7, which shows an element of the sheet with differential length dx. Calculating the circulation around the rectangular path outlined in Fig. 5–7, we obtain

$$(U + u_U')dx - wdz - (U - u_U')dx + wdz = 2u_U'dx. \tag{5-91}$$

By the definition of γ_a, this circulation must also equal $\gamma_a dx$, so that

$$u_U' = \tfrac{1}{2}\gamma_a. \tag{5-92}$$

We may say that any vortex in such a sheet is a point function with respect to the tangential velocity component. Comparison with Eq. (5–74) demonstrates that there must also be a pressure discontinuity across the vortex sheet, given by

$$\frac{p_U - p_L}{\tfrac{1}{2}\rho U^2} = -\frac{2u_U'}{U} + \frac{2(-u_U')}{U} = -\frac{2\gamma_a}{U}. \tag{5-93}$$

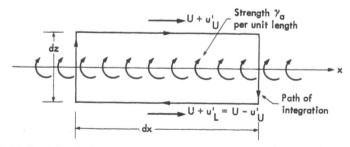

Fig. 5–7. Length dx of a vortex sheet along the x-axis simulating the disturbance produced by the camber and angle of attack of a thin airfoil. Path of integration for relating u_U' and γ_a is shown.

Since in steady-flow circulation, pressure and tangential velocity jumps go hand in hand, we can have vortices only over that portion of the x-axis occupied by the airfoil chord. Hence the integration limits in Eq. (5–90) are $\xi = -b$ and $\xi = b$. We also see that γ_a is the principal unknown to be determined, for on it depend the pressure distribution, lift, and moment of the airfoil.

A relatively simple integral equation for γ_a is derived by calculating w on the airfoil surface and applying Eq. (5–88). In general,

$$w(x,z) = \frac{\partial \phi'}{\partial z} = -\frac{1}{2\pi} \int_{-b}^{b} \frac{[x - \xi]\gamma_a(\xi)d\xi}{(x - \xi)^2 + z^2}. \qquad (5\text{–}94)$$

When z is set equal to zero, the singularity of the integrand at $\xi = x$ makes this integral appear to be a mathematically undefined quantity. However, the physical requirement that w must remain a continuous function of z throughout the limiting process leads to the conclusion that the uniquely correct w is found by taking the Cauchy principal value of the integral:

$$w(x,0) = U\frac{dz_a}{dx} = -\frac{1}{2\pi} \oint_{-b}^{b} \frac{\gamma_a(\xi)}{x - \xi} d\xi. \qquad (5\text{–}95)$$

Cauchy's principal value is calculated by isolating the singular point $\xi = x$ with a small interval symmetrical about it (i.e., $x - \epsilon \le \xi \le x + \epsilon$), evaluating the pieces of the integral, and letting ϵ approach zero. An illuminating discussion of this point, and of the evaluation of improper integrals with even stronger singularities by the physical requirement of continuity, is given by Mangler in Ref. 5–11.

Before solving Eq. (5–95), we mention that it might also have been derived starting from the concept (Section 5–3) of an airfoil as a mass of liquid at rest, separated top and bottom from the external flow by narrow shear regions. These boundary layers resemble very closely surfaces of tangential-velocity discontinuity, so the equivalence which led to Eq. (5–92) suggests that we replace them with two-dimensional vortex sheets. Suppose the upper sheet to have circulation γ_U and the lower one γ_L per unit horizontal distance [Fig. 5–8(a)]. γ_U will be a large positive quantity, while for an airfoil developing lift, γ_L will be a slightly smaller

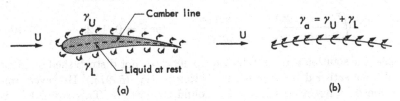

(a)   (b)

Fig. 5–8. Two stages in the replacement of a thin airfoil by vortex sheets.

negative one. Assuming we can handle the effects of thickness by a separate investigation, we reduce the vertical dimensions of the airfoil to zero and cause the two vortex sheets to merge into a single sheet, along the camber line, with net strength $\gamma_a = \gamma_U + \gamma_L$. This second stage of approximation is shown in Fig. 5–8(b). If we write the integral equation which requires that the vertical velocity w induced along the camber line by this slightly curved vortex sheet must everywhere equal $U dz_a/dx$, and if we approximate all cosines of the small angular inclinations of lines connecting various points along the sheet by unity, we find ourselves back at Eq. (5–95).

For mathematical convenience, the independent variables in Eq. (5–95) are replaced by dimensionless equivalents

$$x^* = \frac{x}{b}, \qquad \xi^* = \frac{\xi}{b}, \tag{5–96}$$

leading to

$$w(x^*,0) = -\frac{1}{2\pi} \oint_{-1}^{1} \frac{\gamma_a(\xi^*)}{x^* - \xi^*} \, d\xi^*. \tag{5–97}$$

As mentioned in Section 5–3, the solution of Eq. (5–97) is indeterminate to the extent that an arbitrary amount of circulation can be placed around the airfoil without violating the boundary conditions. This is seen analytically by observing that the function

$$\gamma_{ac}(\xi^*) = \frac{\Gamma_c}{\pi b \sqrt{1 - \xi^{*2}}}, \tag{5–98}$$

when substituted into the right side of Eq. (5–97), yields zero. For, if we make the change of variable $\xi^* = \cos\theta$,

$$\oint_{-1}^{1} \frac{\gamma_{ac}(\xi^*)}{x^* - \xi^*} \, d\xi^* = \frac{\Gamma_c}{\pi b} \oint_{-1}^{1} \frac{d\xi^*}{(x^* - \xi^*)\sqrt{1 - \xi^{*2}}}$$

$$= \frac{\Gamma_c}{\pi b} \oint_{0}^{\pi} \frac{d\theta}{x^* - \cos\theta} = 0, \tag{5–99}$$

this being a special case for $r = 0$ of the formula of Glauert (Ref. 5–1, Chapter IX), which is so familiar to aeronautical scientists:

$$\oint_{0}^{\pi} \frac{\cos r\theta \, d\theta}{\cos\theta - \cos\phi} = \frac{\pi \sin r\phi}{\sin\phi}; \qquad \text{for } r = 0, 1, 2, \cdots. \tag{5–100}$$

Hence, the solution expressed by Eq. (5–98) does not disturb the boundary conditions satisfied by any other solution of Eq. (5–97). However, this solution does put a circulation Γ_c around the airfoil. This circulation is found by summing the contributions of all bound vortices along the chord,

$$\Gamma = \int_{-b}^{b} \gamma_a d\xi = b \int_{-1}^{1} \gamma_{ac} d\xi^*$$

$$= \frac{\Gamma_c}{\pi} \int_{-1}^{1} \frac{d\xi^*}{\sqrt{1 - \xi^{*2}}} = \frac{\Gamma_c}{\pi} \int_{0}^{\pi} d\theta = \Gamma_c. \qquad (5\text{-}101)$$

To determine the actual magnitude of Γ_c, we have recourse to Kutta's hypothesis of smooth flow off the sharp trailing edge. Continuous motion of the liquid can occur in that region only if no pressure discontinuity is met when passing rearward off the airfoil. Therefore, Eq. (5–93) leads to

$$\gamma_a = 0; \quad \text{for } x = b \quad (x^* = 1). \qquad (5\text{-}102)$$

With the final condition (5–102), Eq. (5–97) can be solved by appropriate Fourier-series substitutions for the known and unknown quantities (e.g., Allen, Ref. 5-12). However, Söhngen (Ref. 5-13) has given a very convenient inversion formula derived from basic potential theory. Since his method is useful in unsteady-flow problems, it will be illustrated here. Incidentally, his results can also be deduced from the Fourier-series solution, but with less generality in the types of airfoil camber lines to which they rigorously apply. Söhngen proves, for any two functions f and g of engineering interest,† that the unique solution to the integral equation

$$g(x^*) = \frac{1}{2\pi} \oint_{-1}^{1} \frac{f(\xi^*)}{x^* - \xi^*} d\xi^* \qquad (5\text{-}103)$$

for which $f(1)$ is finite or zero, is

$$f(x^*) = -\frac{2}{\pi} \sqrt{\frac{1 - x^*}{1 + x^*}} \oint_{-1}^{1} \sqrt{\frac{1 + \xi^*}{1 - \xi^*}} \frac{g(\xi^*)}{(x^* - \xi^*)} d\xi^*. \qquad (5\text{-}104)$$

Since f can be identified with γ_a and g with $-w$, the inverted form of Eq. (5–97) is

$$\gamma_a(x^*) = \frac{2}{\pi} \sqrt{\frac{1 - x^*}{1 + x^*}} \oint_{-1}^{1} \sqrt{\frac{1 + \xi^*}{1 - \xi^*}} \frac{w(\xi^*,0)}{(x^* - \xi^*)} d\xi^*$$

$$= \frac{2U}{\pi} \sqrt{\frac{1 - x^*}{1 + x^*}} \oint_{-1}^{1} \sqrt{\frac{1 + \xi^*}{1 - \xi^*}} \frac{\dfrac{dz_a(\xi^*)}{dx}}{(x^* - \xi^*)} d\xi^*. \qquad (5\text{-}105)$$

Once we know the camber line and angle of attack of the airfoil, we integrate Eq. (5–105) and insert the answer into Eq. (5–93) to find the

---

† By this we mean that f and g must be continuous, except possibly for a finite number of finite discontinuities and/or a finite number of integrable singularities. An integrable singularity at $x^* = x_0^*$ is one which, in the neighborhood of this point, can be shown to behave like $1/(x^* - x_0^*)^\beta$, where β is a positive number less than unity.

chordwise pressure distribution. If the cosines of small angles are approximated by unity, as is justified by the assumptions of the theory, the lift and pitching moment (about an axis at $x = ba$) per unit span are given by

$$L = -\int_{-b}^{b} [p_U - p_L]dx = \rho U \int_{-b}^{b} \gamma_a dx = \rho U \Gamma, \qquad (5\text{--}106)$$

$$M_y = \int_{-b}^{b} [p_U - p_L][x - ba]dx. \qquad (5\text{--}107)$$

Here Γ denotes the actual bound circulation on the airfoil. Substituting Eq. (5–105) into Eq. (5–106), inverting order of integration, and evaluating the known integral over x^*, we obtain Munk's integral formula

$$L = -2\rho U^2 b \int_{-1}^{1} \sqrt{\frac{1 + \xi^*}{1 - \xi^*}} \frac{dz_a(\xi^*)}{dx} d\xi^*. \qquad (5\text{--}108)$$

The foregoing results supply especially valuable information when applied to a flat plate at angle of attack α, for which

$$z_a = -\alpha x = -\alpha b x^*. \qquad (5\text{--}109)$$

Inserted into Eq. (5–105), this yields the so-called flat-plate chordwise loading

$$\gamma_a = 2U\alpha \sqrt{\frac{1 - x^*}{1 + x^*}} = 2U\alpha \sqrt{\frac{b - x}{b + x}}. \qquad (5\text{--}110)$$

Equation (5–110) exhibits the well-known singularity at the leading edge $x = -b$, which is the penalty we must pay for the simplifications behind our theory. Nevertheless, it compares very favorably with the measured chordwise distribution of pressure difference over a slightly inclined, thin, uncambered airfoil developing the same lift. The only significant discrepancies come from within a few percent chord lengths of the singularity.

Equations (5–106) and (5–107) give, per unit span,

$$L = 2\pi\rho U^2 b \alpha, \qquad (5\text{--}111)$$

$$M_y = -L[ba + \tfrac{1}{2}b]. \qquad (5\text{--}112)$$

In the form of sectional lift and moment coefficients, defined according to standard aeronautical practice, these read

$$c_l = \frac{L}{\frac{1}{2}\rho U^2 (2b)} = 2\pi\alpha, \qquad (5\text{--}113)$$

$$c_m = \frac{M_y}{\frac{1}{2}\rho U^2 (2b)^2} = -\pi\alpha[a + \tfrac{1}{2}]. \qquad (5\text{--}114)$$

Thus the sectional lift-curve slope works out to be

$$a_0 = \frac{dc_l}{d\alpha} = 2\pi. \tag{5–115}$$

The aerodynamic center, or chordwise axis position for which the pitching moment is independent of angle of attack, lies exactly at the quarter-chord point $a = -\frac{1}{2}$.

Equation (5–115) and the aerodynamic center location are the same for all thin airfoils, regardless of camber or thickness distribution. This is because the flat plate results give the effect on pressure, lift, and moment of a *change* $\Delta\alpha$ in the angle of attack of any airfoil. These results can be superimposed directly because of the linearity of the differential equation and boundary conditions of the problem. Equations (5–113) and (5–114) describe what the National Advisory Committee for Aeronautics has named the "additional" lift and moment. The "basic" lift and moment of an airfoil or wing do depend on its camber. For example, an airfoil at zero geometrical angle of attack with a parabolic camber line would have

$$z_a = z_{a0}\left[1 - \frac{x^2}{b^2}\right], \tag{5–116}$$

where $z_{a0}/(2b)$ is sometimes regarded as the quantitative measure of camber or maximum camber. Substituted into Eqs. (5–108) and (5–107), this leads to

$$c_l = 2\pi \frac{z_{a0}}{b}, \tag{5–117}$$

$$c_m = \pi a \frac{z_{a0}}{b}. \tag{5–118}$$

In particular, for the aerodynamic center $a = -\frac{1}{2}$ the moment invariant under angle of attack changes is

$$c_{mAC} = \frac{-\pi z_{a0}}{2b}. \tag{5–119}$$

Negative values of c_{mAC} are characteristic of airfoils with normal camber lines. Lift and moment at other angles of incidence are found by adding Eq. (5–113) to Eq. (5–117), or Eq. (5–114) to Eq. (5–118), respectively. However, in the following, α will usually denote angle of attack measured from the attitude in which the airfoil develops zero lift.

We should be misleading the reader if we left him with the impression that the foregoing represents the only way whereby airfoil properties in incompressible flow can be theoretically predicted. On the contrary, as most effectively summarized in Ref. 5–14, conformal-transformation methods yield appreciably more accurate estimates of pressure distribution, aerodynamic center location, and most other important quantities. Furthermore, no theory is fully satisfactory for describing phenomena like drag and stall. The point to be made here is that we have presumed the reader to possess some general knowledge of wings, and we are speaking entirely from the aeroelastician's standpoint. The exact theory of arbitrary airfoils, and even such ingenious simplifications as Allen's scheme for modifying thin-airfoil results to account for thickness (Ref. 5–12), are non-linear in the sense that their handling of the boundary conditions does not permit superposition of the effects of angle of attack, camber changes, etc. on lift and other integrals of the pressure distribution. The aeroelastician needs complete linearity in both his structural and aerodynamic relations to facilitate his analysis of complex interaction problems. He can best account for weaknesses in linear theory by empirically modifying the constants in his equations, as will be illustrated by numerous examples in subsequent chapters. Drag forces are of minor importance to him, and he has made little progress in dealing theoretically with situations such as high-angle flutter where stalling is involved.

We close this section on airfoils by summarizing a few facts that will be useful in later applications. Formulas for straight airfoils can be adapted to swept wings of sufficiently large span as follows: if the wing is swept through an angle Λ, and if cross sections are taken perpendicular to the span with all airfoil properties measured in those cross sections, then U is everywhere replaced by the normal component $U \cos \Lambda$ of the airspeed. When the airplane flies faster than about 300 mph, the influence of compressibility must be accounted for. This usually appears through the introduction of the flight Mach number M into equations for lift and moment. Detailed results for steady and unsteady flow are given in Chapter 6.

For any wing and Mach number, there is a limited range of angles of attack (some 15°–30° between the negative and positive stall conditions) within which lift varies linearly with α, while drag varies in a roughly parabolic fashion. Also within this α-range, the aerodynamic center remains quite clearly defined, although it tends to move about rapidly with increasing Mach number in the transonic flight range. Lift, moment, and drag, for a wing of given shape, are observed to be proportional to wing plan area S; this is why S (or chord $2b$ in the case of airfoils of unit span) is used as the reference area in defining dimensionless coefficients associated with these quantities.

At subsonic airspeeds measured lift-curve slopes run about 90–95% of values predicted by thin-airfoil theory. Predictions of c_{mAC} and the aerodynamic center are even more accurate, but stall and drag information are best obtained by wind tunnel or flight test. In supersonic flow the theory predicts a wave drag, associated with the generation of disturbances that are propagated to infinity. While this often represents a substantial fraction of the total resistance, it is always smaller than the total drag experienced in real gases, which exert tangential stresses on the surface of the wing. Drag forces are ordinarily much smaller than lift forces, however, so that the large chordwise stiffnesses of wing structures usually permit the former to be neglected completely in the treatment of aeroelastic problems.

5–5 Finite wings in steady motion. As a consequence of our approach to the lifting airfoil in Section 5–4, we naturally think of representing a three-dimensional planform with a vortex sheet covering its projection R_a onto the xy-plane. Thickness effects are handled by the solution given in Eq. (5–85), so we replace the actual wing with its infinitely thin mean surface $z = z_a(x,y)$, and the pattern of velocities induced by the vortex sheet must then meet the linearized boundary condition

$$w = \frac{\partial \phi'}{\partial z} = U \frac{\partial z_a}{\partial x}; \qquad \text{for } z = 0, \ (x,y) \text{ in } R_a. \qquad (5\text{–}120)$$

There is nothing in the derivations of Eqs. (5–92) and (5–93) that limits their applicability just to two-dimensional flow. In fact, the direct proportionality among local wing loading $(p_U - p_L)$, tangential velocity discontinuity $(u_U' - u_L')$, and bound circulation per unit chord γ_a is a key fact in developing our theory. It tells us that, since $(p_U - p_L)$ is a function of both chordwise and spanwise coordinates on the wing, the circulation must also be $\gamma_a = \gamma_a(x,y)$. But this implies, when we proceed along a particular vortex line in the y-direction, that its strength may be changing. Unless there is to be a violation of the fundamental law of continuity of vortices, there must then also be vortex lines with their axes in the x-direction. Putting it another way, the resultant vortex lines may be oriented in any direction in the xy-plane; γ_a is locally proportional to the magnitude of the spanwise component of the local resultant vorticity vector. (Actually the *vorticity* is infinite when the sheet has zero thickness, but it is a useful concept and has meaning if we imagine the sheet to be a very thin region of uniform shear.) We define the chordwise component of the same vector (Ref. 5–15) by calling $\delta_a(x,y)$ the circulation per unit distance in the y-direction, taken about the *negative* x-direction. The reason for this choice of sign is to make δ_a proportional to the discontinuity of spanwise velocity component v as one passes through the vortex sheet in the positive z-direction. Just as Fig. 5–7 leads to the result

$$\gamma_a = u_U' - u_L', \qquad (5\text{–}121a)$$

so a similar picture taken by cutting the vortex sheet with a yz-plane, replacing dx by dy, γ_a by δ_a, and the velocity components by v_U and v_L, yields

$$\delta_a = v_U - v_L. \qquad (5\text{–}121b)$$

Equations (5–121) also furnish a simple means of expressing quantitatively the continuity of vortex lines. For, if ϕ_U' and ϕ_L' denote the values of the disturbance potential just above and below the sheet, we have

$$\frac{\partial \gamma_a}{\partial y} = \frac{\partial (u_U' - u_L')}{\partial y} = \frac{\partial^2}{\partial y \partial x} (\phi_U' - \phi_L')$$

$$= \frac{\partial (v_U - v_L)}{\partial x} = \frac{\partial \delta_a}{\partial x}. \tag{5–122}$$

Equality between $\partial \gamma_a / \partial y$ and $\partial \delta_a / \partial x$ can be interpreted physically to mean that, if the strength γ_a of a particular spanwise vortex line increases in the y-direction $(\partial \gamma_a / \partial y > 0)$; this must come about as a result of vortices turning away from the x-direction, so that the strength $-\delta_a$ of the adjacent chordwise vortex line must be decreasing in the x-direction $(-\partial \delta_a / \partial x < 0$ or $\partial \delta_a / \partial x > 0)$.

A close examination of the problem we are attempting to solve also convinces us that the vortex sheet cannot be limited to region R_a alone. If it were, it would have to consist of a series of closed rings; but the total circulation about the entire wing at any spanwise station would then have to be zero, and Eq. (5–106) would prove the wing to be developing no lift. The need for a solution giving finite lift requires that vortex lines somehow extend away from R_a to infinity. Since they have their source at the wing, the only direction in which they can move while still obeying the rule that they must always be attached to the same particles in the open flow, is downstream along the wake. To be consistent with the idea of small disturbances, we therefore assume a wake region R_w which lies in the xy-plane between the downstream projections of the wing tips, and we fill it with a vortex sheet having component circulations per unit length γ_w and δ_w. For vortices in R_w, Eq. (5–122) reads

$$\frac{\partial \gamma_w}{\partial y} = \frac{\partial \delta_w}{\partial x}. \tag{5–123}$$

Its implications for steady and unsteady flows are discussed below. Boundaries of R_a and R_w are shown in Fig. 5–9.

In view of Eqs. (5–122) and (5–123), we may regard either circulation component as the principal unknown of the problem and compute the other one afterward by a simple differentiation and integration. The direct relationship between the γ-component and the load distribution over the wing singles it out immediately as the one to use. Combination of the Biot-Savart law with boundary condition (5–120) will be shown to produce an integral equation for γ_a which is quite analogous to Eq. (5–95) in the two-dimensional case. Let the regions R_a and R_w of the vortex sheet be fixed relative to the coordinate system, as illustrated in Fig. 5–9. We

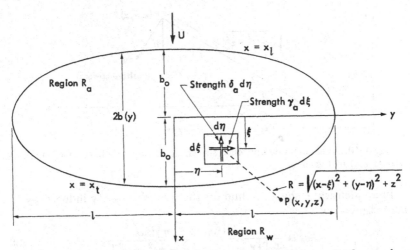

Fig. 5–9. Dimensions of the projection of a finite wing onto the xy-plane, with wing region R_a and wake region R_w. Also shown are the vortices contained within an elementary area $d\xi\,d\eta$ of the vortex sheet representing the wing.

concentrate our attention upon the vortices contained within a small rectangular element $d\xi\,d\eta$ of the sheet, centered at $x = \xi$, $y = \eta$, and ask what is the vertical velocity w induced by them at an arbitrary spatial point $p(x,y,z)$. First we apply Eq. (5–62b) to find the total velocity dq due to the spanwise vortex, which has a circulation $\Gamma = \gamma_a d\xi$. Here the resultant distance R is $\sqrt{(x - \xi)^2 + (y - \eta)^2 + z^2}$, and the perpendicular length r from p to the axis of the vortex is $\sqrt{(x - \xi)^2 + z^2}$. Substituting all these into Eq. (5–62b), we obtain

$$dq = \frac{\sqrt{(x - \xi)^2 + z^2}\, \gamma_a(\xi,\eta)d\xi\,d\eta}{4\pi[(x - \xi)^2 + (y - \eta)^2 + z^2]^{\frac{3}{2}}}. \tag{5–124}$$

According to the right-hand rule, dq lies in the xz-plane through point p, pointing generally downward perpendicular to the length r. Its vertical component dw_γ (positive upward) is found by multiplying it with the cosine of its inclination μ to the z-direction. This angle μ has sides perpendicular to r and to the length $(x - \xi)$ from the vortex line to the projection of (x,y,z) on the xy-plane, as shown in Fig. 5–10. Hence $\cos \mu$ is $(x - \xi)/\sqrt{(x - \xi)^2 + z^2}$, and

$$dw_\gamma = -\,dq\,\frac{(x - \xi)}{\sqrt{(x - \xi)^2 + z^2}}$$

$$= \frac{-\gamma_a(\xi,\eta)[x - \xi]d\xi\,d\eta}{4\pi[(x - \xi)^2 + (y - \eta)^2 + z^2]^{\frac{3}{2}}}. \tag{5–125}$$

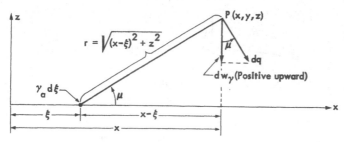

Fig. 5–10. Velocities dq and dw_γ, shown in an xz-plane passing through the point (x,y,z) of Fig. 5–9.

In an identical fashion, we find for the vertical velocity induced by the chordwise vortex

$$dw_\delta = \frac{- \delta_a(\xi,\eta)[y - \eta]d\xi d\eta}{4\pi[(x - \xi)^2 + (y - \eta)^2 + z^2]^{\frac{3}{2}}}.$$ (5–126)

The total velocity due to vortices within $d\xi d\eta$ is

$$dw = dw_\gamma + dw_\delta.$$ (5–127)

To calculate the effect of the entire sheet, we simply sum the elementary contributions by integration over wing and wake regions,

$$w(x,y,z) = -\frac{1}{4\pi}\iint_{R_a} \frac{\gamma_a(\xi,\eta)[x - \xi] + \delta_a(\xi,\eta)[y - \eta]}{[(x - \xi)^2 + (y - \eta)^2 + z^2]^{\frac{3}{2}}} d\xi d\eta$$
$$-\frac{1}{4\pi}\iint_{R_w} \frac{\gamma_w(\xi,\eta)[x - \xi] + \delta_w(\xi,\eta)[y - \eta]}{[(x - \xi)^2 + (y - \eta)^2 + z^2]^{\frac{3}{2}}} d\xi d\eta.$$ (5–128)

Before proceeding to manipulate Eq. (5–128) into a form suitable for our purposes, we mention several points of general importance. To begin with, it is not necessary from a mathematical standpoint to derive Eq. (5–128) using the concept of the vortex sheet. The Biot-Savart law assures, of course, that what we have obtained satisfies Laplace's equation. We could have started, however, from the stipulation that we wished a harmonic function (i.e., a function satisfying Laplace's equation) corresponding to a flow pattern antisymmetrical in z and having discontinuities of ϕ' over R_a and R_w. Or we could have sought a representation of the acceleration potential ψ which had a discontinuity over R_a only. In either case, we should have used superpositions of elementary solutions known as doublets,

$$\phi_D = \frac{\partial \phi_s}{\partial z} = \frac{H[z - \zeta]}{4\pi[(x - \xi)^2 + (y - \eta)^2 + (z - \zeta)^2]^{\frac{3}{2}}},$$ (5–129)

and the integral relation called Green's theorem. Such methods all lead effectively to Eq. (5–128) or some equivalent form. Their elegance is not believed to outweigh the superior physical clarity of the vortex concept. Nevertheless, the acceleration potential has been used to advantage in dealing with certain wing problems. Examples of this will be seen in the chapters on compressible flow. Because the wing is the only surface of pressure discontinuity, ψ is also a powerful tool for an incompressible fluid when the region R_a can be made to cover the entire area of some coordinate surface in a suitably chosen system. Kinner's solution (Ref. 5–17) for the load distribution on a circular wing in prolate spheroidal coordinates is the classical example of this approach.

The condition of zero pressure discontinuity across R_w, together with Eq. (5–93), shows that for steady flow γ_w must vanish. It follows from Eq. (5–123) that

$$\frac{\partial \delta_w}{\partial x} = 0 \qquad (5\text{–}130)$$

everywhere in R_w, or that δ_w is a function of y only throughout the wake. Since the forward boundary of R_w is the trailing edge $x = x_t$ of the wing, where Kutta's hypothesis requires the liquid motion to be smooth and continuous, we can follow any line $y = $ const. up across x_t and evaluate δ_w as

$$\delta_w(x,y) = \delta_w(x_t,y) = \delta_a(x_t,y). \qquad (5\text{–}131)$$

This is put into more familiar terms by observing that there are no vortices ahead of the leading edge, so Eq. (5–122) can be integrated to give*

$$\delta_a(x_t,y) = \int_{x_l}^{x_t} \frac{\partial \gamma_a}{\partial y} \, dx = \frac{d}{dy} \int_{x_l}^{x_t} \gamma_a dx = \frac{d\Gamma}{dy}, \qquad (5\text{–}132)$$

where $\Gamma(y)$ is the total bound circulation at station y on the wing. Substituting into Eq. (5–131),

$$\delta_w(x,y) = \frac{d\Gamma(y)}{dy}. \qquad (5\text{–}133)$$

Thus the wake is composed of streamwise vortices of constant strength, their running circulation at any station y being determined by the local rate of change of total wing circulation. These are the well-known trailing vortices, which are supposed to extend downstream to the place where steady flight began. There their ends are closed up by a starting vortex parallel to the span. Only in steady flow is the wake composed of trailing vortices alone. We shall see in Chapter 7 that the unsteady wake is a

---

* The reader is cautioned that the second equality implies $\gamma_a(x_t,y) = \gamma_a(x_l,y) = 0$. While γ_a has been assumed to vanish at the trailing edge, the condition of zero $\gamma_a(x_l,y)$ is justified only when we assert that all integrations over R_a extend a short distance beyond the actual leading edge into the region of zero vorticity. This artifice is not inconsistent with other steps of the analysis. We have also made the tacit presumption here that x_t and x_l are single-valued functions of y. That is, the wing can have no re-entrant side edges, a limitation which is met by all practical planforms.

complicated, time-dependent pattern of shed γ_w-vortices and trailers. The condition of zero pressure discontinuity in R_w tells us, however, that the pattern is embedded in the general flow and moves downstream with speed U. As noted in Section 5-3, its entire form is determined by the time history of circulation on the wing that generates it.

We are now in a position to return to Eq. (5-128), insert our information about γ_w and δ_w, and construct the desired integral equation by letting z approach zero. In the limit, as Reissner shows in Ref. 5-16, the continuity of w forces the singular integral to assume the same sort of Cauchy principal value that occurred in the case of the airfoil:

$$
\begin{aligned}
w(x,y,0) = U \frac{\partial z_a}{\partial x} \\
= -\frac{1}{4\pi} \oint\oint_{R_a} \frac{\gamma_a(\xi,\eta)[x - \xi] + \delta_a(\xi,\eta)[y - \eta]}{[(x - \xi)^2 + (y - \eta)^2]^{\frac{3}{2}}} d\xi d\eta \\
- \frac{1}{4\pi} \iint_{R_w} \frac{\delta_a(\xi_t,\eta)[y - \eta]}{[(x - \xi)^2 + (y - \eta)^2]^{\frac{3}{2}}} d\xi d\eta
\end{aligned}
\tag{5-134}
$$

for all (x,y) in R_a.

At this point we introduce, for convenience in our presentation, the limitation that the wing has a rectangular planform of span $2l$ and chord $2b$. Such a step is not as restrictive as it may seem, since several of the results we shall deduce can also be proved valid for differently shaped wings, particularly those with straight trailing edges. The advantage of constant limits of integration could also be obtained, at the sacrifice of orthogonal coordinates, by the interesting variable change

$$
x^* = \frac{2x - x_l - x_t}{x_t - x_l},
\tag{5-135a}
$$

$$
y^* = y/b_0.
\tag{5-135b}
$$

This was first adopted for wing theory by Reissner (Ref. 5-15, page 18), and it has also been used by W. P. Jones (Ref. 5-18). The effect is to transform the leading and trailing edges into the lines $x^* = -1$ and $x^* = 1$, respectively.

We substitute the R_a limits $x = \pm b$ and $y = \pm l$ into Eq. (5-134), and proceed to manipulate the various terms as follows: the γ_a integral is integrated by parts with respect to η, observing that the integrated portion vanishes because the loading, hence γ_a, is zero at both wing tips; the δ_a integral is integrated by parts with respect to ξ, observing that the vanishing of δ_a causes the integrated portion to be zero at its lower limit $x = -b$; Eq. (5-122) is used to replace $\partial\delta_a/\partial\xi$ with $\partial\gamma_a/\partial\eta$, so that the double integrals resulting from the first two steps can be combined; in the wake part, the ξ-integration is carried out, the portion from the lower limit being cancelled by the integrated part of the δ_a integral; and finally, $\delta_a(\xi_t,\eta) =$

$\delta_a(b,\eta)$ is replaced by $d\Gamma/d\eta$ from Eq. (5–133). Having described these steps, we write them out in a single, long series:

$$w(x,y,0) = U\,\frac{\partial z_a}{\partial x}$$

$$= -\frac{1}{4\pi}\oint_{-b}^{b}\oint_{-l}^{l}\frac{\gamma_a(\xi,\eta)[x-\xi]}{[(x-\xi)^2+(y-\eta)^2]^{\frac{3}{2}}}\,d\eta\,d\xi$$

$$-\frac{1}{4\pi}\oint_{-l}^{l}\oint_{-b}^{b}\frac{\delta_a(\xi,\eta)[y-\eta]}{[(x-\xi)^2+(y-\eta)^2]^{\frac{3}{2}}}\,d\xi\,d\eta$$

$$-\frac{1}{4\pi}\int_{-l}^{l}\delta_a(b,\eta)[y-\eta]\left\{\int_{b}^{\infty}\frac{d\xi}{[(x-\xi)^2+(y-\eta)^2]^{\frac{3}{2}}}\right\}d\eta$$

$$= -\frac{1}{4\pi}\oint_{-b}^{b}\left\{\frac{-\gamma_a(\xi,\eta)[y-\eta]}{[x-\xi]\sqrt{(x-\xi)^2+(y-\eta)^2}}\Bigg|_{\eta=-l}^{\eta=l}\right\}d\xi$$

$$-\frac{1}{4\pi}\oint_{-b}^{b}\oint_{-l}^{l}\frac{\partial\gamma_a}{\partial\eta}\frac{[y-\eta]}{[x-\xi]\sqrt{(x-\xi)^2+(y-\eta)^2}}\,d\eta\,d\xi$$

$$-\frac{1}{4\pi}\oint_{-l}^{l}\left\{\frac{-\delta_a(\xi,\eta)[x-\xi]}{[y-\eta]\sqrt{(x-\xi)^2+(y-\eta)^2}}\Bigg|_{\xi=-b}^{\xi=b}\right\}d\eta$$

$$-\frac{1}{4\pi}\oint_{-l}^{l}\oint_{-b}^{b}\frac{\partial\delta_a}{\partial\xi}\frac{[x-\xi]}{[y-\eta]\sqrt{(x-\xi)^2+(y-\eta)^2}}\,d\xi\,d\eta$$

$$-\frac{1}{4\pi}\int_{-l}^{l}\delta_a(b,\eta)\left\{\frac{-[x-\xi]}{[y-\eta]\sqrt{(x-\xi)^2+(y-\eta)^2}}\Bigg|_{\xi=b}^{\xi=\infty}\right\}d\eta$$

$$= 0 - 0 - \frac{1}{4\pi}\oint_{-l}^{l}\oint_{-b}^{b}\frac{\partial\gamma_a}{\partial\eta}\frac{[y-\eta]}{[x-\xi]\sqrt{(x-\xi)^2+(y-\eta)^2}}\,d\xi\,d\eta$$

$$+\frac{1}{4\pi}\oint_{-l}^{l}\frac{\delta_a(b,\eta)[x-b]}{[y-\eta]\sqrt{(x-b)^2+(y-\eta)^2}}\,d\eta - 0$$

$$-\frac{1}{4\pi}\oint_{-l}^{l}\oint_{-b}^{b}\frac{\partial\gamma_a}{\partial\eta}\frac{[x-\xi]}{[y-\eta]\sqrt{(x-\xi)^2+(y-\eta)^2}}\,d\xi\,d\eta$$

$$-\frac{1}{4\pi}\oint_{-l}^{l}\delta_a(b,\eta)$$

$$\times\left\{-\frac{1}{y-\eta}+\frac{[x-b]}{[y-\eta]\sqrt{(x-b)^2+(y-\eta)^2}}\right\}d\eta$$

$$=-\frac{1}{4\pi}\oint_{-l}^{l}\oint_{-b}^{b}\frac{\partial\gamma_a}{\partial\eta}\frac{\sqrt{(x-\xi)^2+(y-\eta)^2}}{(x-\xi)(y-\eta)}\,d\xi\,d\eta$$

$$-\frac{1}{4\pi}\oint_{-l}^{l}\frac{d\Gamma}{d\eta}\frac{d\eta}{(y-\eta)}. \tag{5–136}$$

The various exchanges of order of integration and limiting processes implied in Eq. (5–136) are justified because the singular integrals are given their Cauchy principal values. The important final result is rewritten:

$$
U \frac{\partial z_a}{\partial x} = - \frac{1}{4\pi} \oint_{-l}^{l} \frac{d\Gamma}{d\eta} \frac{d\eta}{(y - \eta)}
$$

$$
- \frac{1}{4\pi} \oint_{-l}^{l} \oint_{-b}^{b} \frac{\partial \gamma_a}{\partial \eta} \frac{\sqrt{(x - \xi)^2 + (y - \eta)^2}}{(x - \xi)(y - \eta)} \, d\xi d\eta \qquad (5\text{–}137\text{a})
$$

for all (x,y) in $-b \leq x \leq b$, $-l \leq y \leq l$.

We remark, without going through the steps in detail, that the derivation of Eq. (5–137a) retains its validity when the planform has a straight trailing edge $x = x_t$ but a leading-edge location $x = x_l(y)$ which varies from one station to another. The only difficult point which arises while modifying the operations in Eq. (5–136) concerns the first integration by parts, when we are forced to evaluate $\gamma_a(\xi,\eta)$ along the leading edge. In accordance with a previous footnote, we recall that all the integrations may be assumed to extend a short distance off the planform, so we set γ_a equal to zero. The resulting equation is

$$
U \frac{\partial z_a}{\partial x} = - \frac{1}{4\pi} \oint_{-l}^{l} \frac{d\Gamma}{d\eta} \frac{d\eta}{(y - \eta)}
$$

$$
- \frac{1}{4\pi} \oint_{-l}^{l} \oint_{x_l(\eta)}^{x_t} \frac{\partial \gamma_a}{\partial \eta} \frac{\sqrt{(x - \xi)^2 + (y - \eta)^2}}{(x - \xi)(y - \eta)} \, d\xi d\eta \qquad (5\text{–}137\text{b})
$$

for all (x,y) in $x_l \leq x \leq x_t$, $-l \leq y \leq l$.

Despite their linearity, equations like (5–134) and (5–137) have been solved without modifications only in a handful of simple special cases. Two methods of tackling Eq. (5–137a) which might seem straightforward in principle are (1) to approximate the exact integrals by means of a numerical integration formula, yielding a set of algebraic equations for the values of γ_a (or some function of γ_a) at discrete points over the planform; or (2) to substitute an appropriate double Fourier series for γ_a, carry out the required definite integrals, and satisfy the resulting algebraic equation at as many points as there are terms taken in the series. Both these schemes encounter practical difficulties. The first is handicapped by the need for dealing specially with the strong singularities of the integrands along lines $\eta = y$ and $\xi = x$ (one successful attack on a similar problem in compressible flow will be discussed in Chapter 7). Some of the definite integrals that are generated by the second method are so unwieldy that no one has deemed it worth the trouble to evaluate them. Nevertheless, we point out for future reference that Eqs. (5–137) are theoretically reducible to algebraic systems of the form

$$\left\{\frac{\partial z_a}{\partial x}\right\} = \frac{1}{U}[A]\{\gamma_a\}\cdot \tag{5–138}$$

Here the column matrices consist of the values of known $\partial z_a/\partial x$ and unknown γ_a at a suitably chosen network of points over the wing. The square aerodynamic matrix $[A]$ contains known, finite elements and can be inverted to give

$$\frac{1}{U}\{\gamma_a\} = [A]^{-1}\left\{\frac{\partial z_a}{\partial x}\right\}\cdot \tag{5–139}$$

We now turn our attention to the various approximate procedures for solving the linearized problem of the lifting wing which have proved useful in the past. Three cases of particular interest to the aeroelastician are covered in the following subsections.

(a) *Lifting-line theory.* Prandtl's classical theory of the loaded line is deduced from Eq. (5–137a) by observing that the length $|y - \eta|$ greatly exceeds the length $|x - \xi|$ over most of the surface of the large-aspect ratio, straight wings for which this method is applicable. Moreover, in any symmetrical rectangular area around the singular point $\xi = x$, $\eta = y$, the coefficient of $\partial\gamma_a/\partial\eta$ in the last integral of Eq. (5–137a) is an odd function of $(x - \xi)$ and $(y - \eta)$. Hence such an area makes relatively small numerical contributions to the complete equation, except in xy-regions where $\partial\gamma_a/\partial\eta$ is a rapidly varying function. These two considerations provide an excuse, if not a fully rigorous mathematical justification, for replacing the radical $\sqrt{(x - \xi)^2 + (y - \eta)^2}$ by the approximation $|y - \eta|$. We can then simplify the integral containing this factor as follows:

$$\oint_{-l}^{l}\oint_{-b}^{b}\frac{\partial\gamma_a}{\partial\eta}\frac{\sqrt{(x - \xi)^2 + (y - \eta)^2}}{(x - \xi)(y - \eta)}\,d\xi d\eta$$

$$\cong \oint_{-b}^{b}\frac{1}{(x - \xi)}\oint_{-l}^{l}\frac{\partial\gamma_a}{\partial\eta}\frac{|y - \eta|}{(y - \eta)}\,d\eta d\xi$$

$$= \oint_{-b}^{b}\frac{1}{(x - \xi)}\left\{\int_{-l}^{y}\frac{\partial\gamma_a}{\partial\eta}\,d\eta - \int_{y}^{l}\frac{\partial\gamma_a}{\partial\eta}\,d\eta\right\}d\xi$$

$$= 2\oint_{-b}^{b}\frac{\gamma_a(\xi,y)}{x - \xi}\,d\xi. \tag{5–140}$$

The last two steps are correct because $|y - \eta|/(y - \eta)$ just equals $+1$ for $\eta < y$ and -1 for $\eta > y$, and because γ_a vanishes at the wingtip limits $\eta = \pm l$. Putting Eq. (5–140) into Eq. (5–137a), we get a relation involving only single integrals, of which the last one is reminiscent of thin-airfoil theory [cf. Eq. (5–95)]:

$$U \frac{\partial z_a}{\partial x} = - \frac{1}{4\pi} \oint_{-l}^{l} \frac{d\Gamma}{d\eta} \frac{d\eta}{(y - \eta)} - \frac{1}{2\pi} \oint_{-b}^{b} \frac{\gamma_a(\xi,y)}{x - \xi} d\xi. \qquad (5\text{–}141)$$

The final operation, which produces Prandtl's equation, is motivated by our wish to be able to calculate the *spanwise* distribution of load on the large-aspect ratio wing. This we accomplish by eliminating x from Eq. (5–141) with a chordwise integration. In so doing we multiply through by a weighting factor $\sqrt{(b + x)/(b - x)}$ (because of the very useful effect Munk's integral formula (5–108) tells us it will have on the term containing γ_a):

$$\oint_{-b}^{b} \sqrt{\frac{b + x}{b - x}} \oint_{-b}^{b} \frac{\gamma_a(\xi,y)}{x - \xi} d\xi dx$$

$$= \oint_{-b}^{b} \gamma_a(\xi,y) \left\{ \oint_{-b}^{b} \sqrt{\frac{b + x}{b - x}} \frac{dx}{(x - \xi)} \right\} d\xi = \oint_{-b}^{b} \gamma_a(\xi,y) \left\{ \oint_{0}^{\pi} \frac{(1 + \cos \theta)d\theta}{(\cos \theta - \xi)} \right\} d\xi$$

$$= \pi \int_{-b}^{b} \gamma_a(\xi,y) d\xi = \pi \Gamma(y). \qquad (5\text{–}142)$$

As in Section 5–4, the inner integral was here simplified by the variable change $x/b = x^* = \cos \theta$, and Eq. (5–100) was used with $r = 1$. The weighted x-integral occurring in the first term on the right of Eq. (5–141) is just πb, so we obtain from the complete integration process

$$U \int_{-b}^{b} \sqrt{\frac{b + x}{b - x}} \frac{\partial z_a(x,y)}{\partial x} dx = - \frac{\Gamma}{2} - \frac{b}{4} \oint_{-l}^{l} \frac{d\Gamma}{d\eta} \frac{d\eta}{(y - \eta)}. \qquad (5\text{–}143)$$

Thin-airfoil theory permits us to interpret the left-hand expression in Eq. (5–143) as being proportional to the angle of attack at station y, measured from the zero-lift attitude of the cross section there. Thus a combination of Eqs. (5–108) and (5–113) yields

$$U \int_{-b}^{b} \sqrt{\frac{b + x}{b - x}} \frac{\partial z_a(x,y)}{\partial x} dx = Ub \int_{-1}^{1} \sqrt{\frac{1 + x^*}{1 - x^*}} \frac{\partial z_a(x^*,y)}{\partial x} dx^*$$

$$= - \frac{L_2}{2\rho U} = - \pi U b \alpha, \qquad (5\text{–}144)$$

where L_2 temporarily denotes the two-dimensional lift per unit span. Substituting Eq. (5–144) into Eq. (5–143), and multiplying by 2, we get

$$\Gamma(y) = 2\pi U b \left[\alpha(y) - \frac{1}{4\pi U} \oint_{-l}^{l} \frac{d\Gamma}{d\eta} \frac{d\eta}{(y - \eta)} \right]. \qquad (5\text{–}145)$$

Equation (5–145) is the lifting-line formula, specialized for a rectangular wing whose sectional lift-curve slopes are all 2π. As is well known to aeronautical engineers, the general formula was originally derived by replacing the actual vortex sheet with a single, concentrated bound vortex of strength $\Gamma(y)$, from which emanates a wake of trailers having circulation $d\Gamma/dy$ per unit spanwise distance. The running lift $\rho U \Gamma$ at each section was equated to the lift $a_0 \frac{1}{2} \rho U^2 c \alpha_{\text{eff}}$ of a two-dimensional airfoil with chord c and lift-curve slope a_0, working at an effective angle of attack determined by the induced flow pattern there. α_{eff} was calculated by subtracting from the geometrical angle α (measured from zero-lift attitude) the contribution of the so-called downwash. This downwash, assumed constant along each airfoil chord, is just the downward velocity at the bound vortex line due to the entire vortex sheet. A simple application of the Biot-Savart law therefore yields

$$\alpha_{\text{eff}} \cong \alpha - \frac{\text{Downwash}}{U} = \alpha - \frac{1}{4\pi U} \oint_{-l}^{l} \frac{d\Gamma}{d\eta} \frac{d\eta}{(y - \eta)} . \qquad (5\text{–}146)$$

The equation between the two expressions for running lift, after substitution of Eq. (5–146) and division by ρU, finally reads

$$\Gamma(y) = a_0 U \frac{c}{2}\left[\alpha(y) - \frac{1}{4\pi U} \oint_{-l}^{l} \frac{d\Gamma}{d\eta} \frac{d\eta}{(y - \eta)} \right]. \qquad (5\text{–}147)$$

Since $c/2 = b$, this is obviously consistent with Eq. (5–145), although here both c and a_0 may be functions of the spanwise coordinate y.*

Two methods of solving Eq. (5–147) are in common use. The best known is Glauert's Fourier series substitution, which leads in such a natural way to the familiar concepts of elliptic loading and minimum induced drag. A convenient angle variable resembling the one introduced in Eq. (5–99) is defined by

$$\eta = l \cos \theta, \qquad y = l \cos \phi. \qquad (5\text{–}148)$$

This puts the wing tips at $\theta = 0$ and $\theta = \pi$. Since Γ is known to vanish at both tips, it is taken in the form of a Fourier sine series:

$$\Gamma(y) = \Gamma(\phi) = Ul \sum_{r=1} \bar{A}_r \sin r\phi. \qquad (5\text{–}149)$$

The general integral formula (5–100) comes into use when we substitute

---

* Symbols c and $2b$ are used interchangeably for wing chord throughout this book. In view of past usages, we feel the reader is best served by becoming accustomed to this practice. We use $2b$ generally when developing fundamental aerodynamic theory, and c when discussing applications to aeroelastic problems on actual aircraft.

Eq. (5–149) into the right side of Eq. (5–146):

$$\oint_{-l}^{l} \frac{d\Gamma}{d\eta} \frac{d\eta}{(y-\eta)} = -\frac{1}{l} \oint_{0}^{\pi} \frac{d\Gamma}{d\theta} \frac{d\theta}{(\cos\phi - \cos\theta)}$$

$$= U \oint_{0}^{\pi} \sum_{r=1} \frac{r\bar{A}_r \cos r\theta}{\cos\theta - \cos\phi} d\theta$$

$$= U \sum_{r=1} r\bar{A}_r \oint_{0}^{\pi} \frac{\cos r\theta d\theta}{\cos\theta - \cos\phi}$$

$$= \pi U \sum_{r=1} r\bar{A}_r \frac{\sin r\phi}{\sin\phi}. \qquad (5\text{–}150)$$

Inserting Eqs. (5–149) and (5–150) into Eq. (5–147), dividing by Ul and rearranging, we obtain the algebraic equality

$$\frac{a_0 c}{2l} \alpha = \sum_{r=1} \bar{A}_r \left[\sin r\phi + \frac{a_0 c}{8l} \frac{r \sin r\phi}{\sin\phi} \right]. \qquad (5\text{–}151)$$

The sine of an odd multiple of ϕ makes a contribution to the spanwise lift distribution which is symmetrical about mid-span, whereas even multiples are antisymmetrical. Hence α may be divided into portions α^s and α^a, the former equaling the sum of odd terms on the right of Eq. (5–151), the latter equaling the sum of even terms. It is customary to solve the two equations thus obtained separately for the constants $\bar{A}_r{}^s$ and $\bar{A}_r{}^a$ by requiring them to be satisfied identically at a number of stations along the wing semispan equal to the number of constants needed for adequate convergence in Eq. (5–149). In most aeroelastic problems, convenience dictates that these should be the same stations for which structural stiffness properties are known, so that aerodynamic and elastic equations can be combined straightforwardly. This is by no means necessary, however.

In matrix notation, the symmetrical part of Eq. (5–151) for such a series of stations is

$$\frac{1}{2l} [a_0 c]\{\alpha^s\} = [\sin r\phi]\{\bar{A}_r{}^s\} + \frac{1}{8l} \left[\frac{a_0 c}{\sin\phi} \right] [r \sin r\phi]\{\bar{A}_r{}^s\}. \qquad (5\text{–}152)$$

At each of these stations (or at any other station of interest) the local sectional lift is

$$\rho U \Gamma = \tfrac{1}{2}\rho U^2 c c_l. \qquad (5\text{–}153)$$

Therefore, Eq. (5–149) leads to

$$c(y)c_l(y) = 2l \sum_{r=1} \bar{A}_r \sin r\phi, \qquad (5\text{–}154)$$

or in matrix form for the symmetrical case,

$$\{cc_l{}^s\} = 2l[\sin r\phi]\{\bar{A}_r{}^s\}. \tag{5–155}$$

An aerodynamic matrix relating lift coefficient and angle of attack is derived by eliminating $\bar{A}_r{}^s$ between (5–152) and (5–155):

$$\frac{1}{2l}[a_0c]\{\alpha^s\} = \frac{1}{2l}\left([\sin r\phi] + \frac{1}{8l}\left[\frac{a_0c}{\sin\phi}\right][r\sin r\phi]\right)[\sin r\phi]^{-1}\{cc_l{}^s\}. \tag{5–156}$$

After cancelling and multiplying by $\left[\dfrac{1}{a_0c}\right]$, we obtain

$$\{\alpha^s\} = \left(\left[\frac{1}{a_0c}\right] + \frac{1}{8l}\left[\frac{1}{\sin\phi}\right][r\sin r\phi][\sin r\phi]^{-1}\right)\{cc_l{}^s\}$$

$$= [A^s]\{cc_l{}^s\}. \tag{5–157}$$

The matrix $[A^s]$ is quite analogous to the more exact relations between local slope and wing loading exemplified by Eq. (5–138), except that the present simpler type of wing undergoes no chordwise deformation and leads to a one- rather than a two-dimensional model. Such systematized representations of the solution of steady-state aerodynamic problems were apparently first suggested by Sears and Pai (Ref. 5–20). The matrix $[A^a]$ has a form identical to $[A^s]$, except that even rather than odd values of r are involved and no terms need be included for the mid-span station, where $\alpha^a = c_l{}^a = 0$. It is also evident that Eq. (5–157) reduces to aerodynamic strip theory (finite-span effects neglected) whenever the size of the wing span $2l$ causes the second term in parentheses to become negligibly small.

When the semispan is divided into n intervals, the sine matrices in Eq. (5–157) are as follows:

$$\left[\frac{1}{\sin\phi}\right] = \begin{bmatrix} \dfrac{1}{\sin\phi_1} & 0 & 0 & \cdots & 0 \\ 0 & \dfrac{1}{\sin\phi_2} & 0 & \cdots & 0 \\ 0 & 0 & \dfrac{1}{\sin\phi_3} & \cdots & 0 \\ \cdot & \cdot & \cdot & & \cdot \\ \cdot & \cdot & \cdot & & \cdot \\ \cdot & \cdot & \cdot & & \cdot \\ 0 & 0 & 0 & \cdots & \dfrac{1}{\sin\phi_n} \end{bmatrix}, \tag{5–158}$$

$[r \sin r\phi] =$

$$
\begin{bmatrix}
\sin \phi_1 & 3 \sin 3\phi_1 & 5 \sin 5\phi_1 & \cdots & (2n-1) \sin (2n-1)\phi_1 \\
\sin \phi_2 & 3 \sin 3\phi_2 & 5 \sin 5\phi_2 & \cdots & (2n-1) \sin (2n-1)\phi_2 \\
\sin \phi_3 & 3 \sin 3\phi_3 & 5 \sin 5\phi_3 & \cdots & (2n-1) \sin (2n-1)\phi_3 \\
\cdot & \cdot & \cdot & & \cdot \\
\cdot & \cdot & \cdot & & \cdot \\
\cdot & \cdot & \cdot & & \cdot \\
\sin \phi_n & 3 \sin 3\phi_n & 5 \sin 5\phi_n & \cdots & (2n-1) \sin (2n-1)\phi_n
\end{bmatrix} . \qquad (5\text{--}159)
$$

Two formulas of interest are found when we examine the total wing lift associated with Glauert's substitution (5–149). This is calculated from the following integration of the load per unit span:

$$
\begin{aligned}
L &= \int_{-l}^{l} \rho U \Gamma \, dy = \rho U l \int_0^\pi \Gamma(\phi) \sin \phi d\phi \\
&= \rho U^2 l^2 \sum_{r=1} \bar{A}_r \int_0^\pi \sin r\phi \sin \phi d\phi \\
&= \pi \frac{\rho}{2} U^2 l^2 \bar{A}_1.
\end{aligned}
\qquad (5\text{--}160)
$$

All terms but the first vanish because of the orthogonality of sines of different integral multiples of ϕ in the interval 0 to π. If S denotes the plan area and AR the aspect ratio,

$$
\mathrm{AR} = \frac{(2l)^2}{S}, \qquad (5\text{--}161)
$$

the coefficient of total lift is

$$
C_L = \frac{L}{\frac{1}{2}\rho U^2 S} = \frac{\pi \mathrm{AR} \bar{A}_1}{4} \qquad (5\text{--}162)
$$

Because lift depends only on the first term in the Fourier series while induced drag ("drag due to lift," cf. Chapter IX of Ref. 5–1, for instance) is contributed by all terms, designers have long recognized the desirability of wings with elliptic load distribution

$$
\Gamma = U l \bar{A}_1 \sin \phi. \qquad (5\text{--}163)
$$

(When plotted vs. y, $\sin \phi$ is shaped like the upper half of an ellipse.) Equation (5–151) shows that an elliptically loaded wing without aerodynamic twist ($\alpha = $ const.) must have an elliptic planform. With c_0 the root chord, Eq. (5–151) gives

$$
\bar{A}_1 = \frac{a_0 c_0 \alpha}{2l[1 + (a_0 c_0/8l)]} = a_0 \alpha \frac{1}{(2l/c_0) + \frac{1}{4}a_0}. \qquad (5\text{--}164)
$$

Since this wing has an aspect ratio

$$AR = \frac{(2l)^2}{\frac{\pi}{2} l c_0} = \frac{8}{\pi} \frac{l}{c_0}, \tag{5-165}$$

Eq. (5–162) yields a lift curve slope or additional loading

$$\frac{dC_L}{d\alpha} = \dot{a}_0 \frac{AR}{AR + (a_0/\pi)} \cong a_0 \frac{AR}{AR + 2}, \tag{5-166}$$

where a_0 has been given its approximate value 2π in the smaller denominator term. Equation (5–166) is known to provide a very accurate estimate even for wings whose additional loading is only roughly elliptical, such as moderately tapered planforms with aspect ratios down to 3.

The problem of calculating spanwise distribution of pitching moment M_y is more or less overlooked in the derivation of one-dimensional methods such as lifting-line theory. It is necessary to take recourse to airfoil theory, which states that the moment on any section is determined by the lift acting at the quarter-chord line plus a constant increment (c_{mAC}) independent of angle of attack. We therefore place the bound vortex along the quarter chord, which we assume straight. The lift associated with additional loading, which comes from a constant α-increment without camber, is assumed to act directly on this vortex line. The lift associated with basic loading (twist and camber) is also placed on this line, but there is added about the line a constant moment equal to the aerodynamic-center moment for the actual camber shape, as predicted by thin-airfoil theory.

The second procedure for solving Eq. (5–147) involves the so-called Gauss mechanical quadrature and was first suggested by Multhopp (Ref. 5–21). Utilizing the same angle variable ϕ as in Glauert's substitution (Eq. 5–148b), the wing is covered by an odd number m of equally spaced ϕ-stations

$$\phi_n = \frac{n\pi}{m+1}, \qquad n = 1, 2, \cdots, m. \tag{5-167}$$

The cosine relationship between ϕ and y causes these stations to concentrate near the tips of the actual physical planform, but this is claimed to be advantageous because more rapid variations of lift occur there. The mechanical quadrature or numerical integration formula states simply that the spanwise integral of any $f(y)$ which vanishes at the wing tips (such as lift) can be computed from

$$\int_{-l}^{l} f(y)dy = l \int_{0}^{\pi} f(\phi) \sin \phi d\phi$$

$$\cong \frac{\pi l}{m+1} \sum_{n=1}^{m} f(\phi_n) \sin \phi_n. \tag{5-168}$$

This formulation proves to be exact whenever $f(\phi)$ is a Fourier sine series with terms up to $\sin 2m\phi$ or less. Multhopp uses the quadrature as an ingenious means of expressing the coefficients in a Glauert series for the dimensionless circulation

$$G(\phi) = \frac{\Gamma(\phi)}{2lU} = \sum_{\mu=1}^{m} 2\bar{A}_\mu \sin \mu\phi. \qquad (5\text{-}169)$$

He notes that if any $2\bar{A}_\mu$ is represented by the standard formula for the μth element in a Fourier sine series, and if the resulting integral is calculated by Eq. (5–168), one obtains

$$2\bar{A}_\mu = \frac{2}{\pi} \int_0^\pi G(\phi) \sin \mu\phi d\phi$$

$$= \frac{2}{\pi} \int_0^\pi \left[\frac{G(\phi) \sin \mu\phi}{\sin \phi} \right] \sin \phi d\phi$$

$$= \frac{2}{m+1} \sum_{n=1}^{m} \frac{G(\phi_n) \sin \mu\phi_n}{\sin \phi_n} \sin \phi_n. \qquad (5\text{-}170)$$

Substituted into Eq. (5–169), Eq. (5–170) produces an expression for G in terms of its values G_n at the particular stations of interest:

$$G(\phi) = \frac{2}{m+1} \sum_{n=1}^{m} G_n \sum_{\mu=1}^{m} \sin \mu\phi_n \sin \mu\phi. \qquad (5\text{-}171)$$

The use of Eq. (5–171) in Eq. (5–147) depends on evaluation of the integral on the right for any particular station $n = \nu$:

$$\frac{1}{4\pi U} \oint_{-l}^{l} \frac{d\Gamma}{d\eta} \frac{d\eta}{(y_\nu - \eta)} = -\frac{1}{2\pi} \oint_0^\pi \frac{dG}{d\theta} \frac{d\theta}{(\cos \phi_\nu - \cos \theta)}$$

$$= \frac{1}{\pi(m+1)} \sum_{n=1}^{m} G_n \sum_{\mu=1}^{m} \mu \sin \mu\phi_n \oint_0^\pi \frac{\cos \mu\theta d\theta}{(\cos \theta - \cos \phi_\nu)}$$

$$= \frac{1}{m+1} \sum_{n=1}^{m} G_n \sum_{\mu=1}^{m} \frac{\mu \sin \mu\phi_n \sin \mu\phi_\nu}{\sin \phi_\nu}$$

$$= b_{\nu\nu}G_\nu - \sum_{n=1}^{m}{}' b_{\nu n}G_n. \qquad (5\text{-}172)$$

Here the prime on the summation indicates that the term with $n = \nu$ is excluded. In Eq. (5–172) the familiar integral formula (5–100) has been used. Also $b_{\nu\nu}$ and $b_{\nu n}$ are shorthand for the summations on μ which are the coefficients of the G_n; these can be identified as closed forms by means of simple operations involving the complex variable and the sum of a geometric series (cf. Ref. 5–22, page 16):

$$b_{\nu\nu} = \frac{1}{m+1} \sum_{\mu=1}^{m} \frac{\mu \sin^2 \mu\phi_\nu}{\sin \phi_\nu} = \frac{m+1}{4 \sin \phi_\nu}, \qquad (5\text{-}173a)$$

$$b_{\nu n} = -\frac{1}{m+1} \sum_{\mu=1}^{m} \frac{\mu \sin \mu\phi_n \sin \mu\phi_\nu}{\sin \phi_\nu}$$

$$= \frac{\sin \phi_n}{(\cos \phi_n - \cos \phi_\nu)^2} \left[\frac{1 - (-1)^{n-\nu}}{2(m+1)} \right]. \tag{5-173b}$$

When Eq. (5–172) is substituted into Eq. (5–147) and all terms are given their values (denoted by subscript ν) at station $y = y_\nu$, we obtain

$$\frac{\Gamma_\nu}{2Ul} = G_\nu = \frac{a_{0\nu}c_\nu}{4l} \left[\alpha_\nu - b_{\nu\nu}G_\nu + \sum_{n=1}^{m}{}' b_{\nu n}G_n \right]. \tag{5-174}$$

This is usually condensed into

$$b_\nu G_\nu = \alpha_\nu + \sum_{n=1}^{m}{}' b_{\nu n}G_n, \tag{5-175a}$$

where

$$b_\nu = b_{\nu\nu} + \frac{4l}{a_{0\nu}c_\nu}. \tag{5-175b}$$

The system of equations (5–175a) at m stations y_ν can be solved simultaneously for m point values of

$$G = \frac{\Gamma}{2lU} = \frac{cc_l}{4l}. \tag{5-176}$$

Multhopp has pointed out, however, that the vanishing of $b_{\nu n}$ when $(n - \nu)$ is an even integer makes it possible explicitly to calculate all the even G_ν from the odd ones, or vice versa, through Eq. (5–175a). He suggests an iteration procedure by which even and odd sets are successively computed, until the process converges, without the need to solve simultaneous equations.

When the wing is symmetrical about mid-span ($G_n = G_{(m+1-n)}$) or antisymmetrical ($G_n = -G_{(m+1-n)}$), Eqs. (5–175a) can be reduced in an obvious way to systems of $(m + 1)/2$ and $(m - 1)/2$ equations, respectively. The details of these reductions are fully covered in Ref. 5–21, which also furnishes tables of the "aerodynamic influence coefficients" $b_{\nu\nu}$ and $b_{\nu n}$ for the commonly used numbers $m = 3$, 7, and 15. We close this section by pointing out that Eq. (5–176) shows Eqs. (5–175a) to be expressible in a matrix form quite similar to Eq. (5–157):

$$\{\alpha\} = [A]\{cc_l\}. \tag{5-177}$$

(b) *Weissinger L-method.* The integral equation proposed by Weissinger in Ref. 5–23 has been widely adopted as the standard tool for calculating spanwise load distribution on wings which fall outside the scope of lifting-line theory, particularly those with sweep-back or sweep-forward exceeding about 15°. When the planform is rectangular, we can derive Weissinger's result from our exact Eq. (5–137a) by replacing $(x - \xi)^2$ under the radical in the last integral with the rough average value b^2. This approximation is justified because $|x - \xi|$ always lies between 0 and one chord length $2b$, being in the range 0 to b a large majority of the time for most points (x,y). It has no more rigorous mathematical foundation, however, than does the omission of $(x - \xi)^2$ which leads to lifting-line theory. We obtain in this way

$$
U \frac{\partial z_a}{\partial x} = - \frac{1}{4\pi} \oint_{-l}^{l} \frac{d\Gamma}{d\eta} \frac{d\eta}{(y - \eta)}
$$

$$
- \frac{1}{4\pi} \oint_{-l}^{l} \oint_{-b}^{b} \frac{\partial \gamma_a}{\partial \eta} \frac{\sqrt{b^2 + (y - \eta)^2}}{(x - \xi)(y - \eta)} \, d\xi d\eta. \qquad (5\text{--}178)
$$

An equation independent of x, with circulation $\Gamma(y)$ as the principal unknown, is constructed just as in the previous subsection by integrating Eq. (5–178) across the chord with the weighting factor $\sqrt{(b + x)/(b - x)}$. The effects of this operation on the first two terms are the same as before [see Eq. (5–144)]. The last term on the right is modified as follows:

$$
\int_{-b}^{b} \sqrt{\frac{b + x}{b - x}} \oint_{-l}^{l} \oint_{-b}^{b} \frac{\partial \gamma_a(\xi, \eta)}{\partial \eta} \frac{\sqrt{b^2 + (y - \eta)^2}}{(x - \xi)(y - \eta)} \, d\xi d\eta dx
$$

$$
= \oint_{-l}^{l} \frac{\sqrt{b^2 + (y - \eta)^2}}{(y - \eta)} \oint_{-b}^{b} \frac{\partial \gamma_a(\xi, \eta)}{\partial \eta} \left\{ \oint_{-b}^{b} \sqrt{\frac{b + x}{b - x}} \frac{dx}{(x - \xi)} \right\} d\xi d\eta
$$

$$
= \pi \oint_{-l}^{l} \frac{\sqrt{b^2 + (y - \eta)^2}}{(y - \eta)} \frac{d}{d\eta} \left[\int_{-b}^{b} \gamma_a(\xi, \eta) d\xi \right] d\eta
$$

$$
= \pi \oint_{-l}^{l} \frac{d\Gamma}{d\eta} \frac{\sqrt{b^2 + (y - \eta)^2}}{(y - \eta)} \, d\eta. \qquad (5\text{--}179)
$$

Weighted integration of the entire Eq. (5–178) accordingly leads to

$$
\pi U b \alpha = \frac{b}{4} \oint_{-l}^{l} \frac{d\Gamma}{d\eta} \frac{d\eta}{(y - \eta)} + \frac{1}{4} \oint_{-l}^{l} \frac{d\Gamma}{d\eta} \frac{\sqrt{b^2 + (y - \eta)^2}}{(y - \eta)} \, d\eta. \qquad (5\text{--}180)
$$

We introduce the dimensionless quantities

$$
G = \frac{\Gamma}{2lU}, \qquad (5\text{--}169)
$$

$$
y^* = \frac{y}{l}, \qquad \eta^* = \frac{\eta}{l}, \qquad (5\text{--}181)
$$

and divide Eq. (5–180) by the coefficient $\pi U b$ of the left-hand side. In the resulting equation, we eliminate the singularity from the integrand of the second integral on the right by adding and subtracting a term $1/(y^* - \eta^*)$. The added part combines with the first integral, so we have finally

$$\alpha(y^*) = \frac{1}{\pi} \oint_{-1}^{1} \frac{dG}{d\eta^*} \frac{d\eta^*}{(y^* - \eta^*)} + \frac{l}{2\pi b} \int_{-1}^{1} \frac{dG}{d\eta^*} \frac{\sqrt{1 + (l/b)^2(y^* - \eta^*)^2} - 1}{(l/b)(y^* - \eta^*)} d\eta^*.$$

$$(5\text{-}182)$$

The second integral has been multiplied and divided by the aspect ratio $l/b = 2l/c$ to show that the coefficient of $dG/d\eta^*$, which is known as Weissinger's L-function

$$L(y^*, \eta^*) = \frac{\sqrt{1 + (l/b)^2(y^* - \eta^*)^2} - 1}{(l/b)(y^* - \eta^*)}, \qquad (5\text{-}183)$$

depends only on the single combination of variables $(l/b)(y^* - \eta^*)$.

Equation (5-182) is the Weissinger formula specialized for a rectangular wing. His original derivation relied on a lifting-line type of representation of the vortex sheet, but it is valid for any planform whose quarter-chord line is rectilinear over each half span. For simplicity, we illustrate his procedure on a straight wing (Fig. 5-11) with its quarter-chord line assumed normal to the flight direction. The vortex sheet here is identical with that used in lifting-line theory, consisting of a concentrated bound vortex of strength $\Gamma(y)$ and a trailing sheet with circulation $d\Gamma/dy$ per unit spanwise distance. In the present theory, however, the boundary conditions are satisfied by requiring the inclination of the streamlines w/U at the three-quarter-chord point of each station to be determined by the angle of attack α, measured from local zero lift attitude. The reason for this particular scheme of introducing α is its exactness in

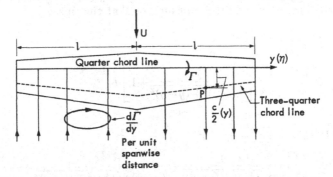

Fig. 5-11. Weissinger's representation of a lifting wing by a bound vortex along the quarter-chord line, with boundary conditions satisfied at the three-quarter-chord line.

the case of a two-dimensional airfoil with lift-curve slope 2π, for then the lift per unit span is given by

$$L_2 = 2\pi \tfrac{1}{2}\rho U^2 c\alpha. \qquad (5\text{-}184)$$

But, if we replace α in Eq. (5-184) with the slope

$$-\frac{w}{U} = \frac{1}{U}\frac{\Gamma_2}{2\pi\tfrac{1}{2}c} \qquad (5\text{-}185)$$

of the streamlines a distance $c/2$ downstream from a concentrated vortex of strength Γ_2, we are led to

$$L_2 = \frac{\pi\rho U^2 c \Gamma_2}{\pi U c} = \rho U \Gamma_2. \qquad (5\text{-}186)$$

This is the correct lift-circulation relation in two dimensions.

We employ the Biot-Savart law (Eq. 5-62b) to calculate the total w/U at a typical point p, located on the three-quarter-chord line at a distance y outboard from mid-span (Fig. 5-11). The element $\Gamma d\eta$ of the bound vortex at station η contributes a vertical velocity

$$dw = -\frac{r}{4\pi R^3}\,\Gamma d\eta = \frac{-(\tfrac{1}{2}c)\,\Gamma d\eta}{4\pi[(\tfrac{1}{2}c)^2 + (y-\eta)^2]^{\frac{3}{2}}}. \qquad (5\text{-}187)$$

The trailer from station η, whose circulation is $(d\Gamma/d\eta)d\eta$ throughout its length, is found by an integration of Eq. (5-62b) to induce

$$dw = \frac{-(d\Gamma/d\eta)d\eta}{4\pi(y-\eta)}\left[\frac{\tfrac{1}{2}c}{\sqrt{(\tfrac{1}{2}c)^2 + (y-\eta)^2}} + 1\right]. \qquad (5\text{-}188)$$

We sum all the contributions of the sheet by integrating Eqs. (5-187) and (5-188) from wing tip to wing tip. It then turns out that the first integral can be conveniently combined with the second by partial integration with respect to η, together with the vanishing of Γ at the tips.

$$
\begin{aligned}
-w = {} & \frac{1}{4\pi}\int_{-l}^{l}\frac{(\tfrac{1}{2}c)\,\Gamma d\eta}{[(\tfrac{1}{2}c)^2 + (y-\eta)^2]^{\frac{3}{2}}} \\
& + \frac{1}{4\pi}\oint_{-l}^{l}\frac{d\Gamma}{d\eta}\frac{1}{(y-\eta)}\left[\frac{\tfrac{1}{2}c}{\sqrt{(\tfrac{1}{2}c)^2 + (y-\eta)^2}} + 1\right]d\eta \\
= {} & -\frac{1}{4\pi}\frac{(y-\eta)\,\Gamma(\eta)}{(\tfrac{1}{2}c)\sqrt{(\tfrac{1}{2}c)^2 + (y-\eta)^2}}\bigg|_{\eta=-l}^{\eta=l} \\
& + \frac{1}{4\pi}\oint_{-l}^{l}\frac{d\Gamma}{d\eta}\left\{\frac{(y-\eta)}{\tfrac{1}{2}c\sqrt{(\tfrac{1}{2}c)^2 + (y-\eta)^2}} + \frac{\tfrac{1}{2}c}{(y-\eta)\sqrt{(\tfrac{1}{2}c)^2 + (y-\eta)^2}} + \frac{1}{(y-\eta)}\right\}d\eta \\
= {} & 0 + \frac{1}{4\pi}\oint_{-l}^{l}\frac{d\Gamma}{d\eta}\frac{d\eta}{(y-\eta)} + \frac{1}{2\pi c}\oint_{-l}^{l}\frac{d\Gamma}{d\eta}\frac{\sqrt{(\tfrac{1}{2}c)^2 + (y-\eta)^2}}{(y-\eta)}d\eta. \qquad (5\text{-}189)
\end{aligned}
$$

Since $-w = U\alpha$ according to the boundary condition, Eq. (5–189) is identical with Eq. (5–180), with the important difference that here the planform is arbitrary and c is the value of the chord at station y. Therefore, the substitutions (5–169) and (5–181) will reduce Eq. (5–189) to

$$\alpha(y^*) = \frac{1}{\pi} \oint_{-1}^{1} \frac{dG}{d\eta^*} \frac{d\eta^*}{(y^* - \eta^*)} + \frac{l}{\pi c(y^*)} \int_{-1}^{1} \frac{dG}{d\eta^*} L(y^*,\eta^*) d\eta^*, \quad (5\text{–}190)$$

where $L(y^*,\eta^*)$ is the expression given by Eq. (5–183), except that $2b = c$ is a function of y^*.

When the wing has sweep, Weissinger still terminates the trailers at a concentrated vortex along the quarter-chord line and satisfies boundary conditions at the three-quarter chord, but the bound vortex on each side of mid-span is now swept through an angle Λ from the normal to the flight direction (Fig. 5–12). Quantities like α and c are still measured in cross sections parallel to the aircraft's plane of symmetry, and the y-direction is normal to this plane. A rather more complicated application of the Biot-Savart law (e.g., Ref. 5–22) gives for the swept-wing $L(y^*,\eta^*)$ function of Eq. (5–190) (when $y^* \geq 0$):

$$L(y^*,\eta^*) = \frac{\sqrt{[1 + (2l/c)(y^* + \eta^*)\tan\Lambda]^2 + (2l/c)^2(y^* - \eta^*)^2}}{(2l/c)(y^* - \eta^*)[1 + 2(2l/c)y^*\tan\Lambda]}$$

$$- \frac{1}{(2l/c)(y^* - \eta^*)}$$

$$+ \frac{2\tan\Lambda\sqrt{[1 + (2l/c)y^*\tan\Lambda]^2 + (2l/c)^2(y^*)^2}}{1 + 2(2l/c)y^*\tan\Lambda} \quad (5\text{–}191a)$$

when $\eta^* \leq 0$; and

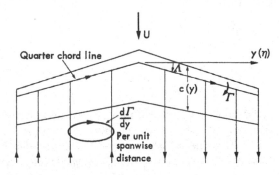

Fig. 5–12. Weissinger's vortex sheet for a wing with quarter-chord line swept back through an angle Λ.

$$L(y^*,\eta^*) = -\frac{1}{(2l/c)(y^* - \eta^*)}$$

$$+ \frac{\sqrt{[1 + (2l/c)(y^* - \eta^*)\tan\Lambda]^2 + (2l/c)^2(y^* - \eta^*)^2}}{(2l/c)(y^* - \eta^*)} \quad (5\text{-}191b)$$

when $\eta^* \geq 0$.

Here $c = c(y^*)$ wherever it appears; clearly $L(y^*,\eta^*)$ is an involved function of y^*, η^*, $2l/c(y^*)$, and Λ separately. Equations (5-191) reduce to Eq. (5-183) when $\Lambda = 0$. Except in this special case, the vortex sheet of Fig. 5-12 induces infinite downwash along its leading edges, which explains the complete failure of ordinary lifting-line theory in the face of sweep.

Solution of Eq. (5-190) for straight or swept wings is carried out systematically by means of Multhopp's Fourier series (5-171), using the quadrature formula (5-168) on the integral containing $L(y^*,\eta^*)$. The relation between Weissinger's spanwise coordinate and the angle variable is

$$\eta^* = \cos\theta, \qquad y^* = \cos\phi. \quad (5\text{-}192)$$

The first integral in Eq. (5-190) differs only by a constant factor from the one occurring in lifting-line theory, so that the coefficients $b_{\nu\nu}$ and $b_{\nu n}$ of Eq. (5-172) make their appearance again. After substitution of the series for G, the second integral is evaluated by a trapezoidal formula in variable ϕ, which is really just an extension of Eq. (5-168) to cover the case where the integrand does not vanish at the wing tips. In the notation of De Young and Harper (Ref. 5-22), the algebraic equation so obtained may be written as follows for station $y_\nu^* = \cos\phi_\nu$:

$$\alpha_\nu = \left[2b_{\nu\nu} + \frac{2l}{c_\nu}g_{\nu\nu}\right]G_\nu + \sum_{n=1}^{m}{}' \left[\frac{2l}{c_\nu}g_{\nu n} - 2b_{\nu n}\right]G_n, \quad (5\text{-}193a)$$

where

$$g_{\nu n} = \frac{-1}{2(M+1)}\left\{\frac{L_{\nu 0}f_{n0} + L_{\nu,M+1}f_{n,M+1}}{2} + \sum_{r=1}^{M}L_{\nu r}f_{nr}\right\}, \quad (5\text{-}193b)$$

$$L_{\nu r} = L(\phi_\nu,\theta_r) = L(y_\nu^*,\eta_r^*), \quad (5\text{-}193c)$$

and

$$f_{nr} = f_n(\phi_r) = \frac{2}{m+1}\sum_{\mu=1}^{m}\mu\sin\mu\phi_n\cos\mu\phi_r. \quad (5\text{-}193d)$$

Here M is an arbitrarily selected odd integer, not necessarily the same as m, used for quadrature calculation of the L-integrals. Inasmuch as $G = cc_l/2l$, with c_l here the lift per unit y-distance divided by $(\rho/2)U^2c$, Eq. (5-193a) for the stations $\nu = 1, 2, 3, \cdots, m$ displays once more the familiar matrix form

$$\{\alpha\} = [A]\{cc_l\}. \quad (5\text{-}194)$$

The aerodynamic matrix $[A]$ reads explicitly

$$[A] = \frac{1}{l}$$

$$\times \begin{bmatrix} \left(b_{11} + \dfrac{l}{c_1}\, g_{11}\right) & \left(\dfrac{l}{c_1}\, g_{12} - b_{12}\right) & \cdots & \left(\dfrac{l}{c_1}\, g_{1m} - b_{1m}\right) \\ \left(\dfrac{l}{c_2}\, g_{21} - b_{21}\right) & \left(b_{22} + \dfrac{l}{c_2}\, g_{22}\right) & & \cdot \\ \cdot & \cdot & & \cdot \\ \cdot & \cdot & & \cdot \\ \cdot & \cdot & & \cdot \\ \left(\dfrac{l}{c_m}\, g_{m1} - b_{m1}\right) & \cdot & \cdots & \left(b_{mm} + \dfrac{l}{c_m}\, g_{mm}\right) \end{bmatrix}. \qquad (5\text{–}195)$$

In Ref. 5–22 an extensive and informative discussion is presented on the application of Weissinger's theory to symmetrically loaded wings, including corrections for compressibility and for deviations of the two-dimensional section lift-curve slope from 2π. The assumption of symmetry $(G_n = G_{(m+1-n)})$ naturally permits the contraction of Eq. (5–195) into a system of $(m + 1)/2$ equations

$$\{\alpha^s\} = [A^s]\{cc_l{}^s\}. \qquad (5\text{–}196)$$

De Young and Harper write Eq. (5–196)

$$\alpha_\nu = \sum_{n=1}^{\frac{1}{2}(m+1)} a_{\nu n} G_n \qquad (5\text{–}197)$$

for $\nu = 1, 2, \cdots, (m + 1)/2$. With $m = 7$, Ref. 5–22 provides complete graphs of the influence coefficients $a_{\nu n}$ vs. the two parameters on which they depend for each of the 16 possible combinations of ν and n. These curves are arranged to handle the Prandtl-Glauert compressibility correction, which we shall discuss in Chapter 6, and to adjust for variations in lift-curve slope. For zero Mach number and slope 2π the parameters of Ref. 5–22 are essentially the sweep angle Λ itself and the "local aspect ratio" $2l/c_\nu$. The authors distinguish between the calculation of basic loading, or lift distribution due to twist at zero total C_L, and additional loading, or lift distribution due to change in angle of attack. The latter problem involves constant values of α across the span and yields information on three-dimensional lift-curve slope and aerodynamic-center location. In aeroelastic applications we are normally interested in elastic bending and twisting of the wing, so the most important results of Weissinger's theory for us are relations like Eqs. (5–194) and (5–196), in which the spanwise variation of α may be arbitrary. Chapter 8 demonstrates the use of these equations in the study of divergence, aileron reversal and associated phenomena.

(c) *Wings of very low aspect ratio*. In view of the increasing number of aircraft with very narrow lifting surfaces, many of them having the triangular or delta planform, we complete our discussion of finite wings in incompressible flow by presenting a theory suitable for vanishingly small aspect ratios. This method is based on a suggestion of R. T. Jones (Ref. 5–24) for extending Munk's momentum theory of airships. Because of the manner in which the exact integral equation is modified, Jones' theory may be said to bear the same relationship to wings of low aspect ratio that lifting-line theory does to large aspect-ratio ones.

Since the results are relatively meaningless for rectangular planforms, we concentrate on the wing shape shown in Fig. 5–13. The trailing edge is straight, and we can apply Eq. (5–137b). However, the length $|x - \xi|$ greatly exceeds the length $|y - \eta|$ over most of the surface of the narrow wing. For this reason, and because of the behavior of the integrand near $\xi = x$, $\eta = y$ described in subsection (a), we are justified in replacing the radical $\sqrt{(x - \xi)^2 + (y - \eta)^2}$ of Eq. (5–137b) by the approximation $|x - \xi|$. The equation may then be simplified as follows.

$$U \frac{\partial z_a}{\partial x} \cong -\frac{1}{4\pi} \oint_{-l}^{l} \frac{d\Gamma}{d\eta} \frac{d\eta}{(y - \eta)}$$

$$-\frac{1}{4\pi} \oint_{-l}^{l} \oint_{x_l(\eta)}^{b_0} \frac{\partial \gamma_a}{\partial \eta} \frac{|x - \xi|}{(x - \xi)(y - \eta)} d\xi d\eta$$

$$= -\frac{1}{4\pi} \oint_{-l}^{l} \frac{1}{(y - \eta)} \frac{\partial}{\partial \eta} \left\{ \int_{x_l(\eta)}^{b_0} \gamma_a(\xi, \eta) d\xi \right\} d\eta$$

$$-\frac{1}{4\pi} \oint_{-l}^{l} \frac{1}{(y - \eta)} \frac{\partial}{\partial \eta} \left\{ \int_{x_l(\eta)}^{b_0} \gamma_a(\xi, \eta) \frac{|x - \xi|}{(x - \xi)} d\xi \right\} d\eta$$

$$= -\frac{1}{4\pi} \oint_{-l}^{l} \frac{1}{(y - \eta)} \frac{\partial}{\partial \eta} \left\{ \int_{x_l(\eta)}^{x} \gamma_a(\xi, \eta) d\xi + \int_{x}^{b_0} \gamma_a(\xi, \eta) d\xi \right.$$

$$\left. + \int_{x_l(\eta)}^{x} \gamma_a(\xi, \eta) d\xi - \int_{x}^{b_0} \gamma_a(\xi, \eta) d\xi \right\} d\eta$$

$$= -\frac{1}{2\pi} \oint_{-l}^{l} \frac{1}{(y - \eta)} \frac{\partial}{\partial \eta} \left\{ \int_{x_l(\eta)}^{x} \gamma_a(\xi, \eta) d\xi \right\} d\eta. \tag{5–198}$$

As discussed early in this section, the interchange of order of differentiation and integration in the second integral is proper because γ_a vanishes both at the trailing edge $\xi = b_0$ and a short distance off the leading edge $\xi = [x_l(\eta) - \epsilon]$, to which point we may extend the integration. Since γ_a is just the discontinuity through the wing of the disturbance velocity component u', we can identify the last bracketed quantity in Eq. (5–198) with

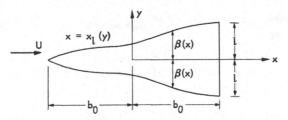

Fig. 5-13. Symmetrical wing of low aspect ratio with straight trailing edge.

the discontinuity $\Delta\phi' = \phi_U' - \phi_L'$ of disturbance velocity potential

$$\int_{x_l(\eta)}^{x} \gamma_a(\xi,\eta)d\xi = \int_{x_l(\eta)}^{x} [u_U' - u_L']d\xi = \int_{x_l(\eta)}^{x} \frac{\partial\Delta\phi'}{\partial\xi} d\xi = \Delta\phi'(x,\eta). \quad (5\text{-}199)$$

$\Delta\phi'$ vanishes at the lower limit, because this point is slightly ahead of the wing. The integral yielding $\Delta\phi'$ goes only from the wing leading edge to chordwise station x, so we can replace the spanwise integration limits $\pm l$ with the extreme dimensions $\pm\beta(x)$ of the planform at that station. Therefore Eq. (5-198) becomes

$$U\frac{\partial z_a}{\partial x} = -\frac{1}{2\pi}\oint_{-\beta(x)}^{\beta(x)} \frac{1}{(y-\eta)} \frac{\partial\Delta\phi'(x,\eta)}{\partial\eta} d\eta \quad (5\text{-}200)$$

for all (x,y) in $x_l \leq x \leq b_0$, $-l \leq y \leq l$.

The right side of Eq. (5-200) just gives the vertical velocity induced at (x, y) by the vortex sheet, but the integral is seen to depend only on conditions ahead of the line x = const. We conclude that, within the assumptions of this theory, the flow is uninfluenced by the presence of the wake. This is, of course, a great simplification, being even more significant in unsteady flow. Its one unfortunate effect is that no mechanism exists by which the liquid can be made aware of an approaching trailing edge, so that Kutta's hypothesis is not, in general, fulfilled. This constitutes a major weakness of the theory, because even on very narrow delta wings the actual pressure discontinuity tends to drop to zero at the rear.

Integral equation (5-200) is of the same form as Eq. (5-95), so it can be solved with Fourier series or inverted directly by Söhngen's methods (Ref. 5-13). We choose the latter for convenience and introduce temporarily the dimensionless variables

$$\eta^* = \frac{\eta}{\beta(x)}, \qquad y^* = \frac{y}{\beta(x)}, \quad (5\text{-}201)$$

which put unit limits onto Eq. (5-200):

$$U\frac{\partial z_a}{\partial x} = -\frac{1}{2\pi}\oint_{-1}^{1} \frac{\partial\Delta\phi'}{\partial\eta} \frac{d\eta^*}{(y^* - \eta^*)}. \quad (5\text{-}202)$$

As an auxiliary condition to make the solution unique, we find that

$$\int_{-1}^{1} \frac{\partial \Delta \phi'}{\partial \eta} \, d\eta^* = \frac{1}{\beta(x)} \int_{-1}^{1} \frac{\partial \Delta \phi'}{\partial \eta^*} \, d\eta^* = \frac{1}{\beta(x)} \Delta \phi' \Big|_{\eta^* = -1}^{\eta^* = 1} = 0, \quad (5\text{-}203)$$

since the velocity potential has no discontinuity just off the leading edges. The appropriate formulas from Ref. 5–13 state that the solution to the integral equation

$$g(y^*) = \frac{1}{2\pi} \oint_{-1}^{1} \frac{f(\eta^*)}{(y^* - \eta^*)} \, d\eta^* \qquad (5\text{-}204)$$

for which

$$\int_{-1}^{1} f(\eta^*) \, d\eta^* = 0 \qquad (5\text{-}205)$$

is

$$f(y^*) = -\frac{2}{\pi \sqrt{1 - y^{*2}}} \oint_{-1}^{1} \frac{g(\eta^*) \sqrt{1 - \eta^{*2}}}{(y^* - \eta^*)} \, d\eta^*. \qquad (5\text{-}206)$$

Here f and g can be identified with $\partial \Delta \phi' / \partial \eta$ and $-U(\partial z_a / \partial x)$, respectively, so Eq. (5–206) yields the result

$$\frac{\partial \Delta \phi'}{\partial y} = \frac{2U}{\pi \sqrt{1 - y^{*2}}} \oint_{-1}^{1} \frac{\partial z_a}{\partial x} \frac{\sqrt{1 - \eta^{*2}}}{(y^* - \eta^*)} \, d\eta^*. \qquad (5\text{-}207\text{a})$$

In terms of the physical variables,

$$\frac{\partial \Delta \phi'(x,y)}{\partial y} = \frac{2U}{\pi \sqrt{\beta^2(x) - y^2}} \oint_{-\beta(x)}^{\beta(x)} \frac{\partial z_a(x,\eta)}{\partial x} \frac{\sqrt{\beta^2(x) - \eta^2}}{(y - \eta)} \, d\eta. \qquad (5\text{-}207\text{b})$$

Once $\Delta \phi'$ is known from integrating Eq. (5–207b), we can find the pressure distribution by means of Eq. (5–93):

$$\frac{p_U - p_L}{(\frac{1}{2}\rho) U^2} = \frac{-2\gamma_a}{U} = -\frac{2}{U} \frac{\partial \Delta \phi'}{\partial x} = -\frac{2}{U} \frac{\partial}{\partial x} \int_{-\beta(x)}^{y} \frac{\partial \Delta \phi'(x,y')}{\partial y'} \, dy'. \qquad (5\text{-}208)$$

A case of special interest among low-aspect ratio wings is the one where slope changes or elastic deformations take place only when one proceeds in the chordwise direction, so that $\partial z_a / \partial x$ is independent of y. We can then integrate Eq. (5–208) easily.

$$\frac{p_U - p_L}{(\frac{1}{2}\rho) U^2} = -\frac{2}{U} \frac{\partial}{\partial x}$$

$$\times \left\{ \frac{2U}{\pi} \frac{\partial z_a(x)}{\partial x} \int_{-\beta(x)}^{y} \frac{1}{\sqrt{\beta^2(x) - y'^2}} \oint_{-\beta(x)}^{\beta(x)} \frac{\sqrt{\beta^2(x) - \eta^2}}{(y' - \eta)} \, d\eta \, dy' \right\}$$

$$= 4 \frac{\partial}{\partial x} \left\{ \sqrt{\beta^2(x) - y^2} \frac{dz_a}{dx} \right\}$$

$$= 4 \sqrt{\beta^2(x) - y^2} \frac{d^2 z_a}{dx^2} + \frac{4\beta(x)}{\sqrt{\beta^2(x) - y^2}} \frac{d\beta}{dx} \frac{dz_a}{dx}. \qquad (5\text{-}209)$$

It is evident that the Kutta condition of zero $p_U - p_L$ would be satisfied at the trailing edge $x = b_0$ only if the curvature of the planform d^2z_a/dx^2 vanished there, and simultaneously either the slope dz_a/dx or the rate of growth of the span $d\beta/dx$ were zero.

It is instructive to integrate Eq. (5–209) and find the spanwise and chordwise lift distributions. The lift per unit span at station y comes from an x-integration:

$$L(y) = -\int_{x_l(y)}^{b_0} [p_U - p_L]dx = -2\rho U^2 \int_{x_l(y)}^{b_0} \frac{\partial}{\partial x}\left[\sqrt{\beta^2(x) - y^2}\,\frac{dz_a}{dx}\right]dx$$

$$= -2\rho U^2 \sqrt{l^2 - y^2}\left(\frac{dz_a}{dx}\right)_{\text{T.E.}}, \tag{5–210}$$

the integrand vanishing at the leading edge, where $y = \beta(x_l)$. Equation (5–210) shows that any low-aspect ratio wing with rigid spanwise cross sections is elliptically loaded, with the lift magnitude fixed *only* by the slope at the trailing edge. The total lift is just

$$L = \int_{-l}^{l} L(y)dy = -\pi\rho U^2 l^2 \left(\frac{dz_a}{dx}\right)_{\text{T.E.}} \tag{5–211}$$

For the flat plate or the calculation of additional loading, this trailing-edge slope is the negative of the angle of attack α, so that

$$L = \pi\rho U^2 l^2 \alpha \tag{5–212a}$$

and

$$\frac{dC_L}{d\alpha} = \frac{2\pi l^2}{S} = \frac{\pi}{2}\,\text{AR}. \tag{5–212b}$$

Equation (5–212b) is an especially well-known formula.

For some purposes, an expression of greater interest is the lift per unit chord at station x, which we find by y-integration:

$$L(x) = -\int_{-\beta(x)}^{\beta(x)} [p_U - p_L]dy$$

$$= -2\rho U^2 \frac{\partial}{\partial x}\left\{\frac{dz_a}{dx}\int_{-\beta(x)}^{\beta(x)} \sqrt{\beta^2(x) - y^2}\,dy\right\}$$

$$= -U^2 \frac{d}{dx}\left[\pi\rho\beta^2(x)\frac{dz_a}{dx}\right]. \tag{5–213}$$

The interchange of differentiation and integration is justified here by the fact that the integrand vanishes at the limits $y = \pm\beta(x)$.

All of the results (5–209)–(5–213) can be derived equally well from Jones' physical model of the flow over the narrow wing. He assumed the motion to be two-dimensional in planes normal to the flight direction. It is therefore possible to neglect the disturbance velocities u' and to follow the history of the liquid slab contained between yz-planes separated a distance Δx as it passes over the wing with velocity U. Momentum is imparted to this slab in the z-direction, the reactions to the forces producing the momentum being the forces exerted by the flow on the wing. This momentum is calculated by observing that the motion in the slab is the same as that which would be produced by section Δx of a two-dimensional flat plate of width $2\beta(x)$, translating normal to itself in the positive z-direction with a velocity $U dz_a/dx$ and simultaneously widening at a rate

$$\frac{d[2\beta(x)]}{dx}\frac{dx}{dt} = 2\frac{d\beta}{dx}U. \tag{5-214}$$

In the sense of Section 5–2, the impulsive force required to give such a plate a z-velocity equal to $U dz_a/dx$ is

$$\Delta I_x = [\pi\rho\beta^2(x)\Delta x]U\frac{dz_a}{dx}. \tag{5-215}$$

This follows from the fact that the virtual mass of the plate is $\pi\rho\beta^2(x)\Delta x$, as we shall see in Section 5–6. The reaction to the force per unit chord needed to bring about the actual momentum change is just the running chordwise lift:

$$L(x) = -\frac{d}{dt}\left[\frac{\Delta I_x}{\Delta x}\right] = -\frac{d}{dx}\left[\frac{\Delta I_x}{\Delta x}\right]\frac{dx}{dt}$$

$$= -U^2\frac{d}{dx}\left[\pi\rho\beta^2(x)\frac{dz_a}{dx}\right], \tag{5-216}$$

where we have divided by Δx to get lift per unit chord and replaced dx/dt by U, the time rate at which the liquid slab moves along the wing. Equation (5–216) gives the same final result as Eq. (5–213); a chordwise integration will obviously lead to the same expression for total lift.

Jones' method also shows us why the total lift should be determined only by conditions at the trailing edge. Considering the flow once more as contained in a series of slabs between xz-planes fixed in the liquid at rest, we see that its steadiness assures that the total z-momentum in all slabs instantaneously between the vertex and the trailing edge is constant. Therefore, the rate at which momentum is being added to the entire liquid mass equals that at which momentum is shed from the trailing edge. The momentum in each slab of thickness Δx, as it passes off this edge, is $-\pi\rho l^2 U(dz_a/dx)_{\text{T.E.}}$, which is fixed by the span $2l$ and the trailing-edge angle of attack. These slabs are assumed to proceed to infinity without further alterations in their flow patterns, a situation that can be interpreted as giving rise to a uniform wake of trailing vortices with the same width as the wing. This picture also explains Jones' conclusion that no further lift can be expected on a narrow wing from chordwise stations aft of the position of maximum span; these are in the wake of the widest section and cannot be presumed to generate the same two-dimensional lateral motion as stations on the widening part of the wing.

Finally Jones points out that there is no reason to anticipate appreciable changes in the lateral flow when the flight Mach number is so large that compressibility effects would be important on wider wings. Reference 5–24 presents an experimental comparison which shows the lift-curve slope of a narrow delta to be substantially independent of M even into the supersonic range.

Besides those described in the preceding subsections, there exist numerous other approximate theories for the finite lifting surface in steady, incompressible flow. None of these has yet proved itself as generally useful to the aeroelastician as the methods of Prandtl and Weissinger. However, the advent of wings radically different from those that were in general favor during the first 40 years of powered flight suggests that this may not always be the case.

We call attention particularly to the theory of Lawrence (Ref. 5–25), which seems most suited to aspect ratios in the neighborhood of unity. His procedure involves replacement of the radical $\sqrt{(x-\xi)^2 + (y-\eta)^2}$ in Eqs. (5–137) by an approximation of the form $\frac{1}{2}[|x-\xi| + \sqrt{(x-\xi)^2 + l^2}]$, and solution for chordwise load distribution by modified Fourier series. Although some rather tedious integrations are needed to reduce Lawrence's formulas to algebraic equations, the theory is well systematized and shows excellent agreement both with experimental data and with other theories in areas where their applicability overlaps.

Another promising scheme for rectangular planforms (Laidlaw, Ref. 5–26) simplifies the unwieldy integral of Eq. (5–137a) by an approximation

$$\sqrt{(x-\xi)^2 + (y-\eta)^2} \cong \lambda_0(\text{AR})|x-\xi| + \lambda_\infty(\text{AR})|y-\eta|. \quad (5\text{–}217)$$

The parameters λ_0 and λ_∞ are calculated, as functions of aspect ratio, by the requirement that the quadruple integral of the square of the difference between the two sides of Eq. (5–217) with respect to x, y, ξ, and η over the planform shall be a minimum relative to variations in their values. As might be expected, λ_0 has the value unity in the limit of zero AR and gradually approaches zero as AR increases, while λ_∞ rises from zero to unity going through the AR range. Because of the complete symmetry, they coincide at 0.771 when the aspect ratio is one. Laidlaw's method makes single integrals out of all those in Eq. (5–137a), producing an equation that is adapted to computing the over-all pressure distribution by means of an appropriate double Fourier series for γ_a. The definite integrals to be carried out offer no more serious difficulties than those arising in Ref. 5–25, and a matrix equation exactly like (5–138) is the end product. This theory has the potential advantages that it applies without question to the entire aspect-ratio range and yields the pressures directly, without the need for supplementary assumptions and calculations. If it can be extended to other planforms, it should be well suited to the needs of the aeroelastician who wishes to predict the plate-like deformations of low-aspect ratio wing structures.

We complete this section by presenting certain semi-empirical formulas that are useful in estimating the general effects of finite span on loading in incompressible flow. For straight wings with ordinary planforms, Eq. (5–166),

$$\frac{dC_L}{d\alpha} = a_0 \frac{AR}{AR + (a_0/\pi)} \cong a_0 \frac{AR}{AR + 2}, \qquad (5\text{–}166)$$

is accurate when AR exceeds about 4 but gives too high numerical values below that. Diederich (Ref. 5–27) points out that the formula

$$\frac{dC_L}{d\alpha} = a_0 \frac{AR}{AR\sqrt{1 + (a_0/\pi AR)^2} + a_0/\pi} \qquad (5\text{–}218)$$

covers the whole range rather well, approaching Eq. (5–166) in the limit of large AR and Eq. (5–212b) as AR goes to zero. For the special purpose of correcting torsional divergence speed for the influence of aspect ratio, a problem in which C_L variations are definitely not accompanied by constant α changes over the entire span, Diederich and Budiansky (Ref. 5–28) introduce an alternative formula to Eq. (5–166):

$$\frac{dC_L}{d\alpha} = a_0 \frac{AR}{AR + 4}. \qquad (5\text{–}219)$$

When a wing is swept at an angle Λ, but the aspect ratio is so great that the presence of tips can be overlooked, the effect can be accurately estimated from thin-airfoil theory. We imagine an infinite swept wing of uniform properties in a flow U and discover that we can eliminate the sweep entirely by adopting a coordinate system which slides backward along the span with velocity $U \sin \Lambda$. In fact, the spanwise component of the flow can have no influence whatever on the pressure distribution, so long as we forget the presence of viscosity and boundary layer. In the new system we see a straight wing of chord $\bar{c}$ and angle of attack $\bar{\alpha}$ in a stream $U \cos \Lambda$. Inserting the two-dimensional experimental value a_0 for the theoretical lift-curve slope 2π, we get from Eq. (5–111) the lift per unit spanwise distance:

$$\bar{L} = a_0(\tfrac{1}{2}\rho)(U \cos \Lambda)^2 \bar{c}\bar{\alpha}. \qquad (5\text{–}220)$$

But the lift per unit distance normal to the flight direction of the swept wing is $\bar{L}/\cos \Lambda$; the chord is

$$c = \frac{\bar{c}}{\cos \Lambda}; \qquad (5\text{–}221a)$$

and the angle of attack in a plane containing the flight direction is

$$\alpha = \bar{\alpha} \cos \Lambda. \tag{5–221b}$$

Making these substitutions in Eq. (5–220), we obtain

$$L = \frac{\bar{L}}{\cos \Lambda} = a_0 (\tfrac{1}{2}\rho) U^2 c \cos \Lambda \alpha, \tag{5–222a}$$

so that

$$\frac{dc_l}{d\alpha} = \frac{L}{(\tfrac{1}{2}\rho) U^2 c \alpha} = a_0 \cos \Lambda. \tag{5–222b}$$

The important conclusion expressed by Eq. (5–222b) has prompted Diederich (Ref. 5–27) to account for sweep by modifying Eq. (5–166) to

$$\frac{dC_L}{d\alpha} = \frac{AR}{AR + (a_0/\pi) \cos \Lambda} a_0 \cos \Lambda. \tag{5–223}$$

Similarly, he generalizes Eq. (5–218) into

$$\frac{dC_L}{d\alpha} = \frac{AR a_0 \cos \Lambda}{AR \sqrt{1 + (a_0 \cos \Lambda / \pi AR)^2} + (a_0/\pi) \cos \Lambda}. \tag{5–224}$$

The ratio $(dC_L/d\alpha)/a_0 \cos \Lambda$ thus depends on the single planform parameter $\pi AR/a_0 \cos \Lambda$. Measurements on a wide variety of wings agree excellently with Eq. (5–224).

5–6 Thin airfoils oscillating in incompressible flow. The problem of the thin airfoil performing small lateral oscillations in a uniform stream of incompressible fluid was for many years at the heart of all flutter prediction. Its results still provide a source of relatively reliable reference data on large-aspect ratio wings, even though their actual flutter speeds may be transonic or supersonic. The case of simple harmonic motion was partially solved by Glauert in 1929 (Ref. 5–30), but the complete solution was first published in the United States by Theodorsen (Ref. 5–29). Other European aerodynamicists reached essentially the same conclusions independently and presented them during the same period (e.g., Cicala, Ref. 5–31; Ellenberger, Ref. 5–32; von Borbely, Ref. 5–33; Kassner and Fingado, Ref. 5–34; and Küssner, Ref. 5–35). Another interesting discussion appears in Chapter V, Sects. 6–8, pp. 293–304 of von Kármán and Burgers' general treatment of wing theory (Ref. 5–9). This seems to be the first place where, for special motions of the airfoil, the important integrals arising from the wake circulation were identified as Bessel functions.

We commence with an interpretation of Theodorsen's method, which is familiar to many workers in aeroelasticity and permits an illuminating separation of the circulatory and noncirculatory portions of the flow.

Adopting the assumptions of small-disturbance theory, we must solve Laplace's equation

$$\nabla^2 \phi' = 0 \tag{5-65}$$

for the disturbance potential, subject to the two-dimensional boundary condition [cf. Eqs. (5-36), (5-75), and (5-88)] along the airfoil of chord $2b$:

$$w = \frac{\partial z_a}{\partial t} + U \frac{\partial z_a}{\partial x}$$

$$= w_a(x,t); \quad \text{for } z = 0, \ -b \leq x \leq b, \tag{5-225}$$

where $w_a(x,t)$ is convenient shorthand for Eq. (5-225), and may be regarded as a known function. Overlooking any possibility of lag in the adjustment of flow at the trailing edge, we apply Kutta's hypothesis of finite, continuous velocities and pressures at $x = b$.

Theodorsen divides his solution into two parts: Eq. (5-225) is satisfied with an appropriate distribution of sources and sinks just above and below the line $z = 0$; a pattern of vortices is then put on this line, with counter-vortices along the wake to infinity, in such a way that Kutta's hypothesis is fulfilled without disturbing the boundary conditions at the airfoil. Each piece of the flow is most conveniently obtained by using Joukowski's conformal transformation to map a circle of radius $b/2$ onto the airfoil's projection.

For a discussion of transformations in general, and Joukowski's in particular, the reader is referred to Chapters VII and VIII of Ref. 5-1. The details which are important to us here are as follows. The circle

$$r = \tfrac{1}{2}b \tag{5-226}$$

in the XZ-plane is transformed into the line or "slit"

$$-b \leq x \leq b, \quad z = 0, \tag{5-227}$$

in the xz-plane by means of the formula

$$x + iz = (X + iZ) + \frac{b^2}{4(X + iZ)}, \tag{5-228}$$

where $i = \sqrt{-1}$. The two planes are illustrated in Fig. 5-14, which also shows the polar coordinates r and θ, related to X and Z by

$$X = r \cos \theta, \quad Z = r \sin \theta, \tag{5-229a,b}$$

$$X + iZ = r(\cos \theta + i \sin \theta) = re^{i\theta}. \tag{5-229c}$$

To find the correspondence between points on the circle and on the slit, we substitute Eq. (5-229c) into Eq. (5-228) and set $r = b/2$. Equating reals and imaginaries, we get

$$x = b \cos \theta, \quad z = 0. \tag{5-230}$$

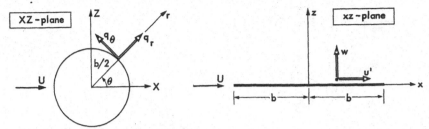

Fig. 5–14. The XZ- and xz-planes, related by Joukowski's conformal transformation.

In connection with the wake countervortices, we need the relation between positions on the X-axis outside the circle and those on the x-axis. This is derived by setting $Z = 0$ in Eq. (5–228), and equating reals and imaginaries:

$$x = X + b^2/4X, \qquad z = Z = 0. \tag{5–231}$$

Flow velocities in the two planes are obtained from one another through the complex formula, containing the derivative of the transformation function, which is a property of all conformal transformations:

$$u' - iw = \frac{q_X - iq_Z}{[d(x + iz)/d(X + iZ)]}. \tag{5–232}$$

Here q_X and q_Z are the Cartesian disturbance-velocity components in the XZ-plane. We need to compute velocities on the slit from those on the circle, and vice versa, so we substitute

$$\left[\frac{d(x + iz)}{d(X + iZ)}\right]_{r=b/2} = \left[1 - \frac{b^2}{4(X + iZ)^2}\right]_{r=b/2}$$

$$= 1 - \frac{b^2}{b^2 e^{i2\theta}} = 2\sin\theta e^{i\theta} \tag{5–233}$$

and take the absolute magnitudes of both sides of Eq. (5–232):

$$|u' - iw| = \sqrt{u'^2 + w^2} = \frac{\sqrt{q_X^2 + q_Z^2}}{|2\sin\theta|} = \frac{\sqrt{q_\theta^2 + q_r^2}}{|2\sin\theta|}, \tag{5–234}$$

where q_r and q_θ are the radial and tangential components in the XZ-plane, illustrated in Fig. 5–14.

Equation (5–234) may also be used to relate the magnitudes of the components separately. Conformal transformation preserves the angle at which two lines meet. In particular, the angle at which a streamline or local velocity vector is inclined to the circle is the same as the angle at which the corresponding velocity vector meets the slit at the correspond-

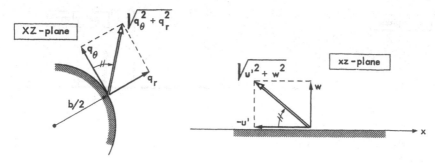

Fig. 5-15. Resultant velocity vectors at corresponding points on circle and slit meet the surfaces at the same angle.

ing point (Fig. 5-15). Hence the radial and tangential components are in the same proportions to u' and w, respectively, as the resultants are to one another:

$$|u'| = \frac{|q_\theta|}{|2\sin\theta|},\qquad\qquad (5\text{-}235a)$$

$$|w| = \frac{|q_r|}{|2\sin\theta|}.\qquad\qquad (5\text{-}236a)$$

On the upper surface, where $\sin\theta \geq 0$, the positive directions of r and z are the same, while those of θ and x are opposite, so that

$$\left.\begin{aligned} q_\theta &= -2u'\sin\theta, \\ q_r &= 2w\sin\theta, \end{aligned}\right\} \; 0 \leq \theta \leq \pi.\qquad \begin{aligned} (5\text{-}235b) \\ (5\text{-}236b) \end{aligned}$$

Before proceeding with Theodorsen's solution, we must know how the disturbance velocity potentials in the two planes are connected. Their general dependence can be described by stating that the differences between the values of ϕ' at corresponding pairs of points in the two planes are equal. This follows from Eq. (5-232), which may be written

$$(u' - iw)d(x + iz) = (q_X - iq_Z)d(X + iZ).\qquad (5\text{-}237)$$

Equating reals,

$$u'dx + wdz = q_X dX + q_Z dZ.\qquad (5\text{-}238)$$

Equation (5-238) is equivalent to

$$d\phi'(x,y) = \frac{\partial\phi'}{\partial x}dx + \frac{\partial\phi'}{\partial z}dz = \frac{\partial\phi'}{\partial X}dX + \frac{\partial\phi'}{\partial Z}dZ = d\phi'(X,Y)\qquad (5\text{-}239)$$

and can be integrated between corresponding points 1 and 2 to give

$$\phi'(x_2,y_2) - \phi'(x_1,y_1) = \int_1^2 d\phi'(x,y) = \int_1^2 d\phi'(X,Y)$$

$$= \phi'(X_2,Y_2) - \phi'(X_1,Y_1). \qquad (5\text{-}240)$$

For corresponding integration paths along the circle and slit

$$d\phi'(X,Y) = q_\theta(\tfrac{1}{2}b)d\theta, \qquad (5\text{-}241)$$

$$d\phi'(x,y) = u'dx. \qquad (5\text{-}242)$$

Hence we can specialize Eq. (5–240) for two such paths, between θ_1 and θ_2, and x_1 and x_2 (noting that the integration runs in the negative x-direction when $\theta_2 > \theta_1$, $x_2 < x_1$):

$$\phi_2' - \phi_1' = \int_{\theta_1}^{\theta_2} q_\theta \frac{b}{2} d\theta = -\int_{x_1}^{x_2} u'dx. \qquad (5\text{-}243)$$

To satisfy condition (5–225), Theodorsen puts a sheet of two-dimensional sources on the upper half of the circle and sinks of equal strength on the lower half. When this flow pattern is transformed into the xz-plane, the sources and sinks appear to merge along the slit, and one might think they would cancel each other identically. That this is by no means the case we realize when we see that the upper and lower surfaces of the slit are not in contact at all. Equation (5–228) transforms all points outside the circle into the whole exterior of the slit, thus filling up the entire xz-plane. However, points inside the circle are transformed so as to fill the entire plane a second time. This is evident because very small values of X and Z around the center, when substituted into Eq. (5–228), yield very large values of x or z, or both. The XZ-origin transforms to infinity in the xz-plane. Thus the whole XZ-plane corresponds to two sheets, or Riemann surfaces, on its transformed counterpart. We pass from one of these to the other only when we try to cross the slit from above to below, or vice versa, which is analogous to going from the exterior to the interior of the circle. In mathematical terms, we put along the slit between $x = .-b$ and $x = b$ a barrier or "cut," which we are forbidden to pass. But we must always remember that the liquid forced downward by the sources does not go directly into the sinks, but creates another, wholly separate, streamline pattern on the second Riemann surface, and travels a long path before ultimately coming around and supplying from above a portion of the liquid required by the sinks. The streamlines of the flows in both surfaces due to sources and sinks of constant strength are illustrated schematically in Fig. 5–16, along with the equivalent streamlines in the XZ-plane.

Being point functions with respect to vertical velocity, these sources and sinks can be made to fulfill Eq. (5–225) just as they were used to represent the thickness of a wing in Section 5–4. Considering first only the

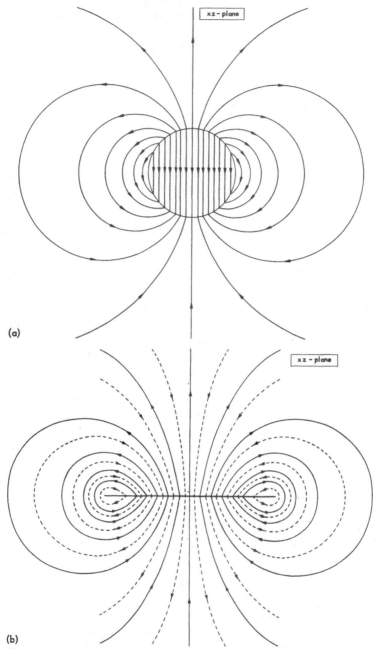

(a)

(b)

Fig. 5–16. Streamlines of the flow due to constant-strength sources and sinks above and below the slit in the xz-plane. The dashed streamlines are those on the second Riemann surface.

source sheet, we know that a single two-dimensional source of strength H, located at $x = \xi$, $z = \zeta$, is represented by

$$\phi_{s2} = \frac{H}{4\pi} \ln \left[(x - \xi)^2 + (z - \zeta)^2 \right]. \tag{5–244}$$

A continuous distribution of these over the upper side of the slit, with strength $H^+(\xi,t)$ per unit x-distance, therefore has the disturbance potential

$$\phi'(x,z,t) = \frac{1}{4\pi} \int_{-b}^{b} H^+(\xi,t) \ln \left[(x - \xi)^2 + z^2 \right] d\xi. \tag{5–245}$$

To remain on the correct Riemann surface we limit ourselves to $z > 0$, and we proceed to calculate

$$w(x,0^+,t) = \frac{\partial \phi'}{\partial z} (x,0^+,t)$$

$$= \frac{1}{4\pi} \lim_{z \to 0^+} \frac{\partial}{\partial z} \int_{-b}^{b} H^+(\xi,t) \ln \left[(x - \xi)^2 + z^2 \right] d\xi$$

$$= \frac{1}{2\pi} \lim_{z \to 0^+} z \int_{-b}^{b} \frac{H^+(\xi,t)d\xi}{\left[(x - \xi)^2 + z^2 \right]}. \tag{5–246}$$

As z assumes smaller and smaller positive values, the integral is caused to vanish by its multiplying factor, except in the vicinity of point $\xi = x$, where the integrand tends to infinity. We isolate this point with a short line of length 2ϵ, centered on it, obtaining

$$w(x,0^+,t) = \frac{1}{2\pi} \lim_{z \to 0^+} z \int_{x-\epsilon}^{x+\epsilon} \frac{H^+(\xi,t)d\xi}{\left[(x - \xi)^2 + z^2 \right]}. \tag{5–247}$$

Being a continuous function, $H^+(\xi,t)$ differs from its central value $H^+(x,t)$ by an amount of order ϵ. Since time t is just carried along as a parameter in the spatial integration, we can take H outside the integral sign. Introducing the temporary integration variable $\xi' = (x - \xi)$, we find

$$w(x,0^+,t) = \frac{H^+(x,t)}{2\pi} \lim_{z \to 0^+} \left[z \int_{-\epsilon}^{\epsilon} \frac{d\xi'}{[\xi'^2 + z^2]} \right]$$

$$= \frac{H^+(x,t)}{2\pi} \lim_{z \to 0^+} \left[\tan^{-1} \left(\frac{\epsilon}{z} \right) - \tan^{-1} \left(-\frac{\epsilon}{z} \right) \right]. \tag{5–248}$$

The limiting process called for here is identical with the one discussed in Section 5–4, leading to Eq. (5–83a). We let z and ξ approach zero in such a way that the ratio ϵ/z becomes indefinitely large and obtain

$$w(x,0^+,t) = \tfrac{1}{2}H^+(x,t). \tag{5–249}$$

Boundary condition (5–225) therefore shows that the strength per unit length of the source sheet must be

$$H^+(x,t) = 2w_a(x,t). \tag{5–250}$$

As in Section 5–4, we might also have reached this conclusion physically by arguing that the sources discharge H^+ ft^3 of liquid per unit time per unit area of the sheet (the spanwise dimension is unity). This liquid can escape only by moving away from the sheet, half of it going upward with normal velocity $H^+/2$, the other half disappearing downward onto the other Riemann surface. In a similar way, the strength H^- of sinks just below the slit needed to produce there an equal upward velocity w_a should be

$$H^-(x,t) = -2w_a(x,t). \tag{5–251}$$

The point-function character of the source and sink sheets makes it possible to calculate the strength of either one, as we have done, without reference to the other. We can demonstrate more rigorously that the sinks have no influence on the vertical velocity at the sources, and vice versa, by examining the streamlines due to any particular source-sink pair. In the XY-plane the two are centered on the circle $r = b/2$ at points equidistant above and below the X-axis. However, any book on elementary fluid dynamics (e.g., Ref. 5–1, Chapter VI) shows that the streamlines consist of all circles which pass through the centers of both the source and the sink. The circle $r = b/2$ coincides with one of these, so is itself a streamline. But this circle transforms into the slit in the xy-plane, making it one of the streamlines there. Thus no sink is able to produce any vertical velocity along the source sheet, and conversely. Of course, each source does generate upward motion just above itself, but its immediate neighborhood is dominated completely by its own outflow, and this motion is the w predicted by Eq. (5–249).

To compute the velocity potential ϕ' of the disturbance flow, it is also convenient to begin with the surface of the circle. The normal velocity there due to the sources is found from Eq. (5–236b) to be

$$q_r = 2w \sin \theta = 2w_a \sin \theta \tag{5–252}$$

on the upper surface $0 \le \theta \le \pi$. The curved source sheet still has its local strength proportional to this normal velocity, so by analogy with Eq. (5–250),

$$H^+(r = b/2,\theta,t) = 2q_r = 4w_a \sin \theta. \tag{5–253}$$

For the sinks on the lower surface

$$H^-(r = b/2,\theta,t) = -4w_a \sin \theta. \tag{5–254}$$

We consider first just the particular element of the source sheet with length $(\frac{1}{2}b)d\phi$ at position $\theta = \phi$ on the circle, together with the corresponding sink centered around $-\phi$. At an arbitrary point p, defined by the angle θ (Fig. 5–17), we calculate the resultant tangential velocity due to this pair. The velocity induced anywhere by a source or sink is along the radial line from its center and has a magnitude equal to the strength divided by 2π times the distance. Hence the source gives rise at P to

$$|dq^{+}| = \frac{H^{+}(\frac{1}{2}b)d\phi}{2\pi b \sin\left[\frac{1}{2}(\phi - \theta)\right]}$$

$$= \frac{[4w_a \sin \phi](\frac{1}{2}b)d\phi}{2\pi b \sin\left[\frac{1}{2}(\phi - \theta)\right]} = \frac{w_a \sin \phi d\phi}{\pi \sin\left[\frac{1}{2}(\phi - \theta)\right]}. \quad (5\text{–}255\text{a})$$

Similarly, the sink induces

$$|dq^{-}| = \frac{w_a \sin \phi d\phi}{\pi \sin\left[\frac{1}{2}(\phi + \theta)\right]}. \quad (5\text{–}255\text{b})$$

The lengths and angles involved are illustrated in Fig. 5–17. For positive w_a, both $|dq^{+}|$ and $|dq^{-}|$ have negative θ-components. If we add these, we get for the combined effect

$$dq_\theta = -|dq^{+}| \cos\left[\tfrac{1}{2}(\phi - \theta)\right] - |dq^{-}| \cos\left[\tfrac{1}{2}(\phi + \theta)\right]$$

$$= -\frac{w_a \sin \phi d\phi}{\pi}\left[\frac{\cos\left[\tfrac{1}{2}(\phi - \theta)\right]}{\sin\left[\tfrac{1}{2}(\phi - \theta)\right]} + \frac{\cos\left[\tfrac{1}{2}(\phi + \theta)\right]}{\sin\left[\tfrac{1}{2}(\phi + \theta)\right]}\right]$$

$$= \frac{2w_a \sin^2 \phi d\phi}{\pi(\cos \phi - \cos \theta)}. \quad (5\text{–}256)$$

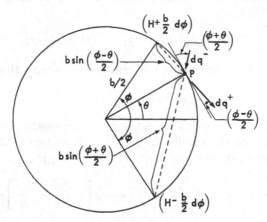

Fig. 5–17.   Velocities dq^{+} and dq^{-} induced at a point p on the circle by a source-sink pair located at ϕ above and below the horizontal.

Here use has been made of elementary trigonometric formulas for sums and products of cosines and sines. It is also easily proved that the normal velocity dq_r at θ is zero, which confirms our previous statement that the circle is a streamline.

Having the tangential velocity due to a differential element of the source-sink sheet, we find its aggregate effect by integrating Eq. (5–256) over all possible pairs. This is accomplished by varying ϕ from 0 to π.

$$q_\theta(\theta,t) = \frac{2}{\pi} \int_0^\pi \frac{w_a \sin^2 \phi\, d\phi}{(\cos \phi - \cos \theta)} \,. \tag{5–257}$$

The disturbance velocity potential ϕ_U' at an arbitrary point on the upper half of the circle, and at the corresponding point on top of the slit, is obtained by substituting Eq. (5–257) into Eq. (5–243). In so doing, we assume that ϕ' is zero at the leading edge $\theta = \pi$, which is permissible since an arbitrary function of time can be added without changing the flow.

$$\phi'(\pi,t) - \phi_U'(\theta,t) = \int_\theta^\pi q_\theta \frac{b}{2}\, d\theta, \tag{5–258}$$

whence

$$\phi_U'(\theta,t) = -\frac{b}{\pi} \int_\theta^\pi \int_0^\pi \frac{w_a \sin^2 \phi\, d\phi\, d\theta}{(\cos \phi - \cos \theta)} \,. \tag{5–259}$$

Because of the antisymmetry with respect to the X-axis of the flow pattern due to the source-sink sheet, the value of q_θ is the same at symmetrically located points on the upper and lower halves of the circle. Therefore, Eq. (5–243) shows that the change in ϕ' going from any upper point around to $\theta = \pi$ must equal the change going from $\theta = \pi$ to the corresponding lower point. That is,

$$\phi'(\pi,t) - \phi_U'(\theta,t) = \phi_L'(-\theta,t) - \phi'(\pi,t) \tag{5–260}$$

from which it follows, because of the vanishing of $\phi'(\pi,t)$, that

$$\phi_L'(-\theta,t) = -\phi_U'(\theta,t). \tag{5–261}$$

Equation (5–261) is useful for calculating the pressure distribution on the slit representing the airfoil. In Section 5–4 we derived the linearized Bernoulli equation for unsteady flow [cf. Eq. (5–73)]:

$$p - p_\infty = -\rho U u' - \rho \frac{\partial \phi'}{\partial t} = -\rho \left[U \frac{\partial \phi'}{\partial x} + \frac{\partial \phi'}{\partial t} \right]. \tag{5–262}$$

Since we can replace the lower limit θ in Eq. (5–259) by the equivalent value of $x = b \cos \theta$, Eqs. (5–262) and (5–261) yield for the pressure difference between the upper and lower surfaces

$$p_U - p_L = -\rho\left[U\left(\frac{\partial\phi_U{}'}{\partial x} - \frac{\partial\phi_L{}'}{\partial x}\right) + \left(\frac{\partial\phi_U{}'}{\partial t} - \frac{\partial\phi_L{}'}{\partial t}\right)\right]$$

$$= -2\rho\left[U\frac{\partial\phi_U{}'}{\partial x} + \frac{\partial\phi_U{}'}{\partial t}\right] = -2\rho\left[\frac{\partial\phi_U{}'}{\partial t} - \frac{U}{b\sin\theta}\frac{\partial\phi_U{}'}{\partial\theta}\right]. \quad (5\text{-}263)$$

By integrating Eq. (5–263) the lift and moment (about an axis at $x = ba$) per unit span, due to the source-sink part of the flow, are found to be

$$L_{NC} = -\int_{-b}^{b}(p_U - p_L)dx$$

$$= 2\rho\int_{-b}^{b}\frac{\partial\phi_U{}'}{\partial t}\,dx + 2\rho U\int_{-b}^{b}\frac{\partial\phi_U{}'}{\partial x}\,dx$$

$$= 2\rho\frac{\partial}{\partial t}\int_{-b}^{b}\phi_U{}'dx = 2\rho b\frac{\partial}{\partial t}\int_{0}^{\pi}\phi_U{}'\sin\theta d\theta \quad (5\text{-}264)$$

$$M_{yNC} = \int_{-b}^{b}(p_U - p_L)[x - ba]dx$$

$$= -2\rho U\int_{-b}^{b}\frac{\partial\phi_U{}'}{\partial x}x dx - 2\rho\int_{-b}^{b}\frac{\partial\phi_U{}'}{\partial t}x dx + baL_{NC}$$

$$= -2\rho U\left[\phi_U{}'x\Big|_{x=-b}^{x=b} - \int_{-b}^{b}\phi_U{}'dx\right]$$

$$-2\rho\frac{\partial}{\partial t}\int_{-b}^{b}\phi_U{}'x dx + 2\rho ba\frac{\partial}{\partial t}\int_{-b}^{b}\phi_U{}'dx$$

$$= 2\rho U\int_{-b}^{b}\phi_U{}'dx - 2\rho\frac{\partial}{\partial t}\int_{-b}^{b}\phi_U{}'[x - ba]dx$$

$$= 2\rho Ub\int_{0}^{\pi}\phi_U{}'\sin\theta d\theta - 2\rho b^2\frac{\partial}{\partial t}\int_{0}^{\pi}\phi_U{}'[\cos\theta - a]\sin\theta d\theta. \quad (5\text{-}265)$$

The last line in each of these formulas makes it possible to work directly from $\phi_U{}'$ as a function of θ, which is the form given by Eq. (5–259). In deriving them, we have used the fact that $\phi_U{}'$ vanishes both at the leading and trailing edges. Equation (5–261) establishes this fact for the latter point ($\theta = 0$), since the sources and sinks introduce no circulation around the circle or slit, ϕ' is single valued, and

$$\phi_U{}'(\theta = 0,t) = \phi_L{}'(\theta = 0,t) = 0. \quad (5\text{-}266)$$

The subscript NC on L and M_y indicates the noncirculatory character of the flow. Incidentally, Eq. (5–264) shows there would be no lift in any steady flow, which we might also conclude from Eq. (5–106) in the absence of circulation.

In view of their derivation, Eqs. (5–259) through (5–265) are valid for arbitrary unsteady motion of the airfoil. It is instructive to apply them to a few cases of practical interest. Following Theodorsen, we consider a chordwise-rigid airfoil with an aerodynamically unbalanced trailing-edge flap or control surface hinged at $x = bc$. The airfoil may move in vertical translation $h(t)$ and rotate about an axis at $x = ba$ through an angle $\alpha(t)$; $\beta(t)$ denotes the angular displacement of the flap relative to the chordline of the airfoil. The positive directions of these variables are pictured in Fig. 5–18. We omit the effects of camber and constant initial angle of attack, since these can be handled by steady-flow theory and the results afterward superimposed on what we calculate. The function $z_a(x,t)$ representing the instantaneous small displacement of the chordline is

$$z_a(x,t) = \begin{cases} -h - \alpha[x - ba]; & \text{for } -b \le x \le bc, \\ -h - \alpha[x - ba] - \beta[x - bc]; & \text{for } bc \le x \le b. \end{cases} \tag{5–267}$$

Substituting this into Eq. (5–225), we find

$$w_a(x,t) =$$
$$\begin{cases} -\dot{h} - U\alpha - \dot{\alpha}[x - ba]; & \text{for } -b \le x \le bc, \\ -\dot{h} - U\alpha - \dot{\alpha}[x - ba] - U\beta - \dot{\beta}[x - bc]; & \text{for } bc \le x \le b, \end{cases} \tag{5–268}$$

where the dot indicates the derivative with respect to t.

We evaluate here the velocity potentials, pressures, and forces due to the h- and α-motions. Results for the flap will simply be stated, since their rather tedious derivation is elaborated in Appendix III of Ref. 5–29. Using $x = b \cos \theta$ and Eq. (5–100) to assist the integrations, we find by inserting Eqs. (5–268), with $\beta = 0$, into Eq. (5–259),

$$\phi_U{}'(\theta,t) = \frac{b}{\pi}[\dot{h} + U\alpha] \int_\theta^\pi \int_0^\pi \frac{\sin^2 \phi\, d\phi\, d\theta}{(\cos \phi - \cos \theta)}$$
$$+ \frac{b^2 \dot{\alpha}}{\pi} \int_\theta^\pi \int_0^\pi \frac{\sin^2 \phi[\cos \phi - a]\, d\phi\, d\theta}{(\cos \phi - \cos \theta)}$$
$$= b[\dot{h} + U\alpha] \sin \theta + b^2 \dot{\alpha} \sin \theta[\tfrac{1}{2} \cos \theta - a]. \tag{5–269}$$

(These formulas are easily compared with Theodorsen's, because he employs a dimensionless variable defined by $x = \cos \theta$.) For the pressure distribution, Eq. (5–263) gives

$$(p_U - p_L)_{NC} = -2\rho\{-\dot{h}U \cot \theta + \ddot{h}b \sin \theta - \alpha U^2 \cot \theta$$
$$+\dot{\alpha}Ub[\sin \theta - \cot \theta(\tfrac{1}{2} \cos \theta - a) + \tfrac{1}{2} \sin \theta]$$
$$+\ddot{\alpha}b^2 \sin \theta[\tfrac{1}{2} \cos \theta - a]\}, \tag{5–270}$$

and for lift and moment we get from Eqs. (5–269), (5–264), and (5–265)

$$L_{NC} = \pi\rho b^2[\ddot{h} + U\dot{\alpha} - ba\ddot{\alpha}], \tag{5–271}$$

$$M_{yNC} = \pi\rho b^2[U\dot{h} + ba\ddot{h} + U^2\alpha - b^2(\tfrac{1}{8} + a^2)\ddot{\alpha}]. \tag{5–272}$$

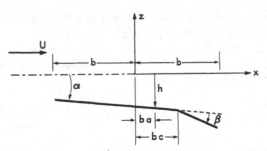

Fig. 5–18. Mean line of chordwise-rigid airfoil with trailing-edge flap having zero gap and aerodynamic balance. $h(t)$ is the downward displacement of the axis $(x = ba)$ of rotation, $\alpha(t)$; $\beta(t)$ is the angle between chordline and flap position. All quantities positive, as shown.

There is much to be learned from Eqs. (5–270)–(5–272) about the detailed effects of the noncirculatory flow. For example, we see that the problems associated with the $\dot{h}$ and $U\alpha$ terms in Eq. (5–268) are analogous to one another; each corresponds to constant vertical velocity over the airfoil at any instant. This is sometimes called the angle-of-attack or plunging type of motion, in contrast with pure angular-velocity motion, whose effects stem from the $\dot{\alpha}$ term in Eq. (5–268). The contributions from $\dot{h}$ or $U\alpha$ can be traced side by side throughout Eqs. (5–269)–(5–272). We note that $\dot{h}$ produces two distinct pressure distributions: $2\rho U\dot{h} \cot \theta$, which is distributed antisymmetrically along the chord, going from zero at midchord to positive and negative infinities at the trailing and leading edges, respectively; and $-2\rho b\ddot{h} \sin \theta$, which is distributed elliptically as a function of x. The former produces no resultant force but only the destabilizing couple $\pi\rho b^2 U\dot{h}$. The latter exerts a lift $\pi\rho b^2 \ddot{h}$ in direct opposition to $\ddot{h}$ and proportional thereto; this force acts at midchord, as evidenced by the vanishing of $\pi\rho b^3 a\ddot{h}$ in the moment when $a = 0$. Recalling some of our generalizations about noncirculatory liquid motion in Section 5–2, we can readily relate them to what we find here. Thus the virtual mass associated with lateral acceleration $\ddot{h}$ is $\rho(\pi b^2)$ per unit span. Someone has pointed out that this is just the mass of a cylinder of liquid with diameter equal to the wing chord. The couple $\pi\rho b^2 U\dot{h}$, which is present even when the "angle of attack" is constant, is similar to the couple [Eq. (5–58)] experienced by a three-dimensional body in steady flight and has the same physical origin. Turning to the angular-velocity terms, we discover that angular acceleration causes a moment $-\pi\rho b^4(\frac{1}{8} + a^2)\ddot{\alpha}$, plus a lift $-\pi\rho b^3 a\ddot{\alpha}$ when the rotational axis is displaced from midchord. The former may be thought of as due to a virtual moment of inertia $\rho[\pi b^4(\frac{1}{8} + a^2)]$. From this result we see the fallacy of associating the idea of virtual mass too closely with particular volumes of liquid: whereas a cylinder with chord $2b$ as diameter would contain the liquid mass connected with linear accelerations, the corresponding cylinder for angular accelerations about a mid-chord axis would have a diameter $\sqrt{2}\,b$.

To conclude the treatment of the noncirculatory flow, the velocity potential, lift, and moment due to rotation of the unbalanced flap are reproduced from Ref. 5–29.

$$(\phi_U')_\beta = \frac{Ub}{\pi}\beta\left\{\sin\theta\cos^{-1}c - (\cos\theta - c)\ln\left[\frac{1 - c\cos\theta - \sin\theta\sqrt{1-c^2}}{\cos\theta - c}\right]\right\}$$

$$+ \frac{b^2}{2\pi}\dot{\beta}\left\{\sin\theta\sqrt{1-c^2} + (\cos\theta - 2c)\sin\theta\cos^{-1}c\right.$$

$$\left. - (\cos\theta - c)^2\ln\left[\frac{1 - c\cos\theta - \sin\theta\sqrt{1-c^2}}{\cos\theta - c}\right]\right\}, \tag{5-273}$$

$$(L_{NC})_\beta = -\rho b^2 U\dot{\beta}[c\sqrt{1-c^2} - \cos^{-1}c] - \rho b^3\ddot{\beta}[c\cos^{-1}c - \tfrac{1}{3}(2+c^2)\sqrt{1-c^2}], \tag{5-274}$$

$$(M_{yNC})_\beta = -\rho b^2 U^2\beta[c\sqrt{1-c^2} - \cos^{-1}c]$$

$$- \rho b^3 U\dot{\beta}\{\tfrac{1}{3}\sqrt{1-c^2}\,(c^2-1) - (c-a)[c\sqrt{1-c^2} - \cos^{-1}c]\}$$

$$- \rho b^4\ddot{\beta}\{(\tfrac{1}{8} + c^2)\cos^{-1}c - \tfrac{1}{8}c\sqrt{1-c^2}\,(7+2c^2)$$

$$+ (c-a)[\tfrac{1}{3}\sqrt{1-c^2}\,(2+c^2) - c\cos^{-1}c]\}. \tag{5-275}$$

When contributions of all degrees of freedom are included, the hinge moment per unit span acting on the flap, positive in a sense to depress the trailing edge, is

$$M_{\beta NC} = \int_{bc}^{b} [p_U - p_L][x - bc]\,dx\ .$$

$$= -\rho b^2 U\dot{h}[c\sqrt{1-c^2} - \cos^{-1}c]$$

$$+ \rho b^3\ddot{h}[c\cos^{-1}c - \tfrac{1}{3}(2+c^2)\sqrt{1-c^2}] - \rho b^2 U^2\alpha[c\sqrt{1-c^2} - \cos^{-1}c]$$

$$+ \rho b^3 U\dot{\alpha}\{a[c\sqrt{1-c^2} - \cos^{-1}c] + \tfrac{1}{3}(\sqrt{1-c^2})^3$$

$$- \tfrac{1}{3}(2+c^2)\sqrt{1-c^2} + c\cos^{-1}c\}$$

$$- \rho b^4\ddot{\alpha}\{(\tfrac{1}{8}+c^2)\cos^{-1}c - \tfrac{1}{8}c\sqrt{1-c^2}\,(7+2c^2)$$

$$+ (c-a)[\tfrac{1}{3}(2+c^2)\sqrt{1-c^2} - c\cos^{-1}c]\}$$

$$- \frac{\rho b^2 U^2}{\pi}\beta[2c\sqrt{1-c^2}\cos^{-1}c - (1-c^2) - (\cos^{-1}c)^2]$$

$$+ \frac{\rho b^4}{\pi}\ddot{\beta}[\tfrac{1}{4}c\sqrt{1-c^2}\cos^{-1}c\,(7+2c^2) - (\tfrac{1}{8}+c^2)(\cos^{-1}c)^2$$

$$- \tfrac{1}{8}(1-c^2)(5c^2+4)]. \tag{5-276}$$

By itself, the noncirculatory solution is incapable of fulfilling Kutta's hypothesis. This can easily be seen by examining Eq. (5–270) or calculating the disturbance velocity component u', on which the pressure depends, at the trailing edge $x = b$, $\theta = 0$. In general,

$$|u'| = \frac{|q_\theta|}{|2 \sin \theta|}, \tag{5-235a}$$

but this will have an infinite value where $\sin \theta = 0$, unless q_θ also vanishes there. However, the q_θ given by Eq. (5–257) is zero only for that very special motion of the airfoil which satisfies the integral equation

$$\int_0^\pi \frac{w_a(\phi,t) \sin^2 \phi\, d\phi}{(\cos \phi - 1)} = 0, \tag{5-277}$$

and this has purely academic interest. It is therefore necessary to superimpose some additional flow pattern that just cancels the noncirculatory $q_\theta(0,t)$. By analogy with techniques proven for the airfoil in steady motion, Theodorsen accomplishes this by means of bound vortices plus a wake of shed countervortices continually moving away from the airfoil at the free-stream velocity. Since they originate at the trailing edge, the centers of the wake vortices lie along the positive x-axis beyond $x = b$. Their counterparts in the XY-plane are therefore on the X-axis, and corresponding positions can be determined from Eqs. (5–231). The law of zero total circulation is met, and simultaneously the circle and slit are made streamlines of the vortex flow so as not to disturb the boundary conditions, by pairing with each wake vortex a bound one of opposite circulation at the "image" position inside the circle. As proved in books on fluid mechanics, the image of a vortex at point χ on the X-axis lies at $X = b^2/4\chi$, when the circle $r = b/2$ is to be preserved as a streamline.

We start by deriving the properties of the flow due to a single bound vortex of positive strength Γ_0 and its image $-\Gamma_0$. Figure 5–19 pictures these two, together with the resultant velocities $|q^+|$ and $|q^-|$ induced by them at an arbitrary point p on the circle. As with any two-dimensional vortex, each of these velocities is normal to the line from the vortex inducing it to p and has a magnitude equal to the strength divided by 2π times the distance. Using symbols defined in Fig. 5–19, we sum the tangential components of $|q^+|$ and $|q^-|$ and obtain the velocity q_θ at p.*

$$q_\theta = |q^-| \cos(\theta_2 - \theta) - |q^+| \cos(\theta_1 - \theta)$$

$$= \frac{\Gamma_0}{2\pi} \left[\frac{r_2 \cos(\theta_2 - \theta)}{r_2^2} - \frac{r_1 \cos(\theta_1 - \theta)}{r_1^2} \right]. \tag{5-278}$$

---

* A short calculation shows that q_r due to the combination is zero, thus proving that the circle $r = b/2$ is a streamline.

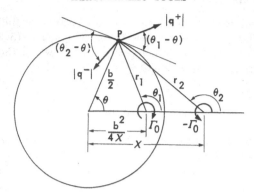

Fig. 5–19. Bound vortex of strength Γ_0 on the circle with its image $-\Gamma_0$ at point $X = \chi$. Also shown are the resultant velocities $|q^+|$ and $|q^-|$ induced by the two at an arbitrary point on the circle $r = b/2$.

By application of the cosine-law to two triangles in Fig. 5–19,

$$r_2{}^2 = \chi^2 + \left(\frac{b}{2}\right)^2 - 2\chi\frac{b}{2}\cos\theta, \tag{5–279a}$$

$$r_1{}^2 = \left[\frac{b^2}{4\chi}\right]^2 + \left(\frac{b}{2}\right)^2 - 2\frac{b^2}{4\chi}\cdot\frac{b}{2}\cos\theta. \tag{5–279b}$$

Also, we can drop perpendiculars from the centers of the two vortices onto the line from p to the center of the circle and find by simple trigonometry

$$r_2\cos(\theta_2 - \theta) = \frac{b}{2} - \chi\cos\theta, \tag{5–280a}$$

$$r_1\cos(\theta_1 - \theta) = \frac{b}{2} - \frac{b^2}{4\chi}\cos\theta. \tag{5–280b}$$

Equations (5–279) and (5–280) are substituted into Eq. (5–278) and the result is put over a common denominator.

$$q_\theta = \frac{\Gamma_0}{2\pi}\left\{\frac{\frac{1}{2}b - \chi\cos\theta}{\chi^2 + (\frac{1}{2}b)^2 - \chi b\cos\theta} - \frac{\frac{1}{2}b - (b^2/4\chi)\cos\theta}{(b^2/4\chi^2)[(\frac{1}{2}b)^2 + \chi^2 - \chi b\cos\theta]}\right\}$$

$$= -\frac{\Gamma_0}{\pi b}\left[\frac{\chi^2 - (\frac{1}{2}b)^2}{\chi^2 + (\frac{1}{2}b)^2 - \chi b\cos\theta}\right]. \tag{5–281}$$

Assuming again that ϕ' vanishes at the leading edge, we calculate the velocity potential on the upper surface of the circle or slit from Eq. (5–258):

$$\phi_U{}'(\theta,t) = -\int_\theta^\pi q_\theta \frac{b}{2}\, d\theta$$

$$= \frac{\Gamma_0[\chi^2 - (\tfrac{1}{2}b)^2]}{2\pi} \int_\theta^\pi \frac{d\theta}{\chi^2 + (\tfrac{1}{2}b)^2 - \chi b \cos\theta}$$

$$= \frac{\Gamma_0}{\pi} \tan^{-1}\left[\frac{\chi - \tfrac{1}{2}b}{\chi + \tfrac{1}{2}b}\sqrt{\frac{1 + \cos\theta}{1 - \cos\theta}}\right]. \tag{5-282}$$

We next determine the pressure distribution from Eq. (5-263), which still applies to the circulatory flow because of the antisymmetry of the disturbance velocities with respect to the X-axis. When this is done, the time variable enters indirectly through the location χ of the wake vortex, whose position is dependent on t. Although $\chi(t)$ is not known directly, we have assumed that the location $x = \xi$ in the xz-plane moves downstream with velocity U, so that

$$\frac{d\xi}{dt} = U. \tag{5-283}$$

ξ is related to χ through the first of Eqs. (5-231):

$$\xi = \chi + \frac{b^2}{4\chi}. \tag{5-284}$$

Equation (5-284) can be put into the equivalent form

$$\sqrt{\frac{\xi - b}{\xi + b}} = \frac{x - (b/2)}{x + (b/2)}, \tag{5-285}$$

which allows us to write Eq. (5-282) as

$$\phi_U{}'(\theta,t) = \frac{\Gamma_0}{\pi} \tan^{-1}\left[\sqrt{\frac{(\xi - b)(1 + \cos\theta)}{(\xi + b)(1 - \cos\theta)}}\right]. \tag{5-286}$$

Hence the pressure distribution due to the vortex pair is, after some manipulation,

$$(p_U - p_L)_{\Gamma_0} = -2\rho\left[\frac{\partial\phi_U{}'}{\partial t} - \frac{U}{b \sin\theta}\frac{\partial\phi_U{}'}{\partial\theta}\right]$$

$$= -2\rho\left[\frac{\partial\phi_U{}'}{\partial\xi}\frac{d\xi}{dt} - \frac{U}{b \sin\theta}\frac{\partial\phi_U{}'}{\partial\theta}\right]$$

$$= \frac{-\rho U\Gamma_0[\xi + b \cos\theta]}{\pi b \sin\theta\sqrt{\xi^2 - b^2}}. \tag{5-287}$$

We cannot use the final expressions in Eqs. (5-264) and (5-265) to compute lift and moment, because $\phi_U{}'$ no longer vanishes at the trailing edge when there is circulation. In fact, Eq. (5–286) yields

$$\phi_U{}'(0,t) = \Gamma_0/2, \qquad (5\text{–}288)$$

which confirms the known fact that the velocity potential should increase by an amount equal to the circulation for each 360-degree clockwise circuit around a vortex. We can find the resultant aerodynamic loads, however, either by integration of $(p_U - p_L)$ or indirectly, using the first line of Eq. (5–287). They turn out to be

$$L_{\Gamma_0} = \frac{\rho U \Gamma_0 \xi}{\sqrt{\xi^2 - b^2}}, \qquad (5\text{–}289)$$

$$M_{y\Gamma_0} = \frac{\rho U \Gamma_0 b^2}{\sqrt{\xi^2 - b^2}} \left[\frac{\xi}{b} a - \tfrac{1}{2} \right], \qquad (5\text{–}290)$$

$$M_{\beta\Gamma_0} = -\frac{\rho U \Gamma_0 b}{\pi} \left\{ \frac{\xi}{\sqrt{\xi^2 - b^2}} \left[\left(1 + \frac{c}{2}\right) \sqrt{1 - c^2} - (c + \tfrac{1}{2}) \cos^{-1} c \right] \right.$$
$$\left. + \tfrac{1}{2} \sqrt{\frac{\xi + b}{\xi - b}} \left[\cos^{-1} c - c\sqrt{1 - c^2} \right] \right\}. \qquad (5\text{–}291)$$

As ξ becomes very large, the flow approaches that of a single bound vortex Γ_0. The lift then becomes $\rho U \Gamma_0$, and Eq. (5–290) shows it to act at midchord.

The effects of the entire circulatory flow follow from Eqs. (5–286)–(5–291) when we identify the concentrated vortex $-\Gamma_0$ with a particular element $\gamma_w d\xi$ of the shed vortex sheet. As in Section 5–5, $\gamma_w = \gamma_w(\xi,t)$ denotes the circulation per unit length of the wake, positive in the customary sense. With

$$\Gamma_0 = -\gamma_w d\xi, \qquad (5\text{–}292)$$

we find the aggregate pressure distribution and aerodynamic loading by integrating over the complete wake from $\xi = b$ to $\xi = \infty$, bearing in mind that the upper limit might be replaced by $b + Ut_0$ if we were dealing with a motion that had gone on only for a finite time interval t_0.

$$p_U - p_L = \frac{\rho U}{\pi b \sin \theta}$$
$$\times \int_b^\infty \left[\frac{\xi}{\sqrt{\xi^2 - b^2}} (1 - \cos \theta) + \sqrt{\frac{\xi + b}{\xi - b}} \cos \theta \right] \gamma_w(\xi,t) d\xi, \qquad (5\text{–}293)$$

$$L_C = -\rho U \int_b^\infty \frac{\xi}{\sqrt{\xi^2 - b^2}} \gamma_w(\xi,t) d\xi, \qquad (5\text{–}294)$$

$$M_{yC} = \rho Ub \int_b^\infty \left[\frac{1}{2} \sqrt{\frac{\xi + b}{\xi - b}} - (a + \tfrac{1}{2}) \frac{\xi}{\sqrt{\xi^2 - b^2}} \right] \gamma_w(\xi,t)d\xi, \quad (5\text{--}295)$$

$$M_{\beta C} = \frac{\rho Ub}{\pi} \left\{ \left[\left(1 + \frac{c}{2} \right) \sqrt{1 - c^2} - (c + \tfrac{1}{2}) \cos^{-1} c \right] \right.$$

$$\times \int_b^\infty \frac{\xi}{\sqrt{\xi^2 - b^2}} \gamma_w(\xi,t)d\xi$$

$$\left. + \tfrac{1}{2}[\cos^{-1} c - c\sqrt{1 - c^2}] \int_b^\infty \sqrt{\frac{\xi + b}{\xi - b}} \gamma_w(\xi,t)d\xi \right\} \cdot \quad (5\text{--}296)$$

We have rearranged terms here to indicate that the circulatory pressure, lift, pitching moment, and hinge moment depend only on two different weighted integrals of the wake circulation.

It remains to assemble the source and vortex flows by applying Kutta's hypothesis to their combination. Equation (5–235a) shows how this can conveniently be done by making $q_\theta(0,t)$ from the two parts add to zero. The noncirculatory q_θ appears in Eq. (5–257); the circulatory one is calculated by replacing χ with ξ in Eq. (5–281) and integrating over the wake. Summing the two results with $\theta = 0$, we find

$$\frac{2}{\pi} \int_0^\pi \frac{w_a \sin^2 \phi d\phi}{(\cos \phi - 1)} + \frac{1}{\pi b} \int_b^\infty \sqrt{\frac{\xi + b}{\xi - b}} \gamma_w(\xi,t)d\xi = 0. \quad (5\text{--}297)$$

Since w_a is given, Eq. (5–297) is an integral equation for the wake circulation γ_w, on which all the important properties of the circulatory flow depend. Theodorsen symbolized one-half the first integral by the single letter Q. For the airfoil motion described by Eq. (5–268), Eq. (5–297) therefore reads

$$-\frac{1}{2\pi b} \int_b^\infty \sqrt{\frac{\xi + b}{\xi - b}} \gamma_w(\xi,t)d\xi = Q$$

$$= \frac{1}{\pi} \int_0^\pi [1 + \cos \phi][\dot{h} + U\alpha + \dot{\alpha}b(\cos \phi - a)]d\phi$$

$$+ \frac{1}{\pi} \int_0^{\cos^{-1} c} [1 + \cos \phi][U\beta + \dot{\beta}b(\cos \phi - c)]d\phi$$

$$= \dot{h} + U\alpha + b(\tfrac{1}{2} - a)\dot{\alpha} + \frac{U\beta}{\pi}[\sqrt{1 - c^2} + \cos^{-1} c]$$

$$+ \frac{b\dot{\beta}}{2\pi}[(1 - 2c) \cos^{-1} c + (2 - c) \sqrt{1 - c^2}]. \quad (5\text{--}298)$$

We add the circulatory and noncirculatory pressure distributions and aerodynamic loads, taking advantage of the identity between the known Q

and one of the wake integrals which constitutes the first line of Eq. (5–298).

$$(p_U - p_L) = (p_U - p_L)_{NC}$$

$$- 2\rho U Q \left\{ \cot \theta + \left[\frac{1 - \cos \theta}{\sin \theta} \right] \frac{\displaystyle\int_b^\infty \frac{\xi}{\sqrt{\xi^2 - b^2}} \gamma_w(\xi,t)d\xi}{\displaystyle\int_b^\infty \sqrt{\frac{\xi + b}{\xi - b}} \gamma_w(\xi,t)d\xi} \right\} \cdot \quad (5\text{–}299)$$

$$L = L_{NC} + 2\pi\rho U b Q \frac{\displaystyle\int_b^\infty \frac{\xi}{\sqrt{\xi^2 - b^2}} \gamma_w(\xi,t)d\xi}{\displaystyle\int_b^\infty \sqrt{\frac{\xi + b}{\xi - b}} \gamma_w(\xi,t)d\xi} \cdot \quad (5\text{–}300)$$

$$M_y = M_{yNC} - 2\pi\rho U b^2 Q \left\{ \tfrac{1}{2} - (a + \tfrac{1}{2}) \frac{\displaystyle\int_b^\infty \frac{\xi}{\sqrt{\xi^2 - b^2}} \gamma_w(\xi,t)d\xi}{\displaystyle\int_b^\infty \sqrt{\frac{\xi + b}{\xi - b}} \gamma_w(\xi,t)d\xi} \right\} \cdot \quad (5\text{–}301)$$

$$M_\beta = M_{\beta NC} - 2\rho U b^2 Q \left\{ \tfrac{1}{2}[\cos^{-1} c - c\sqrt{1 - c^2}] \right.$$

$$\left. + \left[\left(1 + \frac{c}{2}\right)\sqrt{1 - c^2} - (c + \tfrac{1}{2}) \cos^{-1} c \right] \frac{\displaystyle\int_b^\infty \frac{\xi}{\sqrt{\xi^2 - b^2}} \gamma_w(\xi,t)d\xi}{\displaystyle\int_b^\infty \sqrt{\frac{\xi + b}{\xi - b}} \gamma_w(\xi,t)d\xi} \right\} \cdot \quad (5\text{–}302)$$

The influence of the wake circulation is seen to enter always as a particular ratio of two integrals.

To this point in the derivation, no restriction whatever has been placed on the time dependence of the motion except that it must be sufficiently limited to admit the small-disturbance assumption. We now restrict ourselves to simple harmonic oscillations and ascribe the following forms to z_a and w_a:

$$z_a(x,t) = \bar{z}_a(x)e^{i\omega t}, \quad\quad (5\text{–}303)$$

$$w_a(x,t) = \bar{w}_a(x)e^{i\omega t}, \quad\quad (5\text{–}304)$$

where $\bar{w}_a$ and $\bar{z}_a$ may be complex numbers to allow phase differences from point to point. The actual vibration of the physical system is, of course, represented by the real parts of the complex z_a; the linearity of the relations among all the dependent variables permits us to benefit from the great mathematical convenience of the $e^{i\omega t}$ notation. If the motion has gone on for an indefinite period, the linear Eq. (5–297) shows that γ_w must also be proportional to $e^{i\omega t}$. This fact establishes its functionality in both ξ and t, because in our simplified model γ_w is constant at any point which drifts downstream with velocity U. Such a point is described by the equation

$$\xi = \xi_0 + Ut, \tag{5–305}$$

where ξ_0 is its location at $t = 0$. Hence γ_w must be constant when $\xi_0 = \xi - Ut$ is constant, and it can depend on ξ and t only in the form $(\xi - Ut)$ or in some combination (e.g., $(t - \xi/U)$) which has the same property. Accordingly, we describe the simple harmonic wake by assuming

$$\gamma_w(\xi,t) = \bar{\gamma}_w e^{i\omega[t-(\xi/U)]} = \bar{\gamma}_w e^{i(\omega t - k\xi^*)}, \tag{5–306}$$

where

$$k = \frac{\omega b}{U} \tag{5–307}$$

is known as the reduced frequency of oscillation, and

$$\xi^* = \xi/b \tag{5–96}$$

is the dimensionless variable used in Section 5–4. The ratio of integrals through which the wake enters the problem becomes*

$$\frac{\displaystyle\int_b^\infty \frac{\xi}{\sqrt{\xi^2 - b^2}}\,\gamma_w(\xi,t)d\xi}{\displaystyle\int_b^\infty \sqrt{\frac{\xi + b}{\xi - b}}\,\gamma_w(\xi,t)d\xi} = \frac{\bar{\gamma}_w e^{i\omega t}b \displaystyle\int_1^\infty \frac{\xi^*}{\sqrt{\xi^{*2}-1}}\,e^{-ik\xi^*}d\xi^*}{\bar{\gamma}_w e^{i\omega t}b \displaystyle\int_1^\infty \sqrt{\frac{\xi^* + 1}{\xi^* - 1}}\,e^{-ik\xi^*}d\xi^*}$$

$$= \frac{\displaystyle\int_1^\infty \frac{\xi^*}{\sqrt{\xi^{*2}-1}}\,e^{-ik\xi^*}d\xi^*}{\displaystyle\int_1^\infty \sqrt{\frac{\xi^* + 1}{\xi^* - 1}}\,e^{-ik\xi^*}d\xi^*} = C(k), \tag{5–308}$$

a complex function of reduced frequency k only. Making use of formula (34), p. 51, of Ref. 5–36, Theodorsen identified the integrals in Eq. (5–308) as Hankel functions of the second kind,

$$C(k) = F(k) + iG(k) = \frac{H_1^{(2)}(k)}{H_1^{(2)}(k) + iH_0^{(2)}(k)}, \tag{5–309}$$

where $H_n^{(2)}$ is a combination, useful in radiation problems, of Bessel functions of the first and second kinds:

$$H_n^{(2)} = J_n - iY_n. \tag{5–310}$$

---

*The reader's attention is called to subsequent comments regarding the mathematical existence of the integrals occurring in this equation.

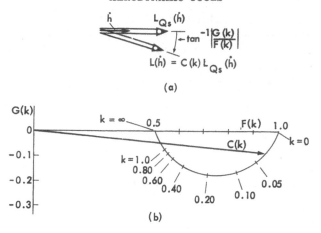

(a)

(b)

Fig. 5-20.   (a) The function of $C(k)$ in reducing amplitude and shifting phase of the circulatory lift due to a vertical oscillation h.   (b) A complex polar plot of $C(k) = F(k) + iG(k)$.

$C(k)$ is plotted as a complex vector in Fig. 5-20(b).   Substitution of Eqs. (5-308) and (5-309) into Eqs. (5-299)-(5-302) completes the closed-form solution.   Certain cancellations occur between circulatory and non-circulatory portions of the moment expressions.   The final lift and pitching moment due to vertical translation and pitching, for example, can be written

$$L = \pi \rho b^2 [\ddot{h} + U\dot{\alpha} - ba\ddot{\alpha}] + 2\pi \rho U b C(k)[\dot{h} + U\alpha + b(\tfrac{1}{2} - a)\dot{\alpha}], \quad (5\text{-}311)$$

$$M_y = \pi \rho b^2 [ba\ddot{h} - Ub(\tfrac{1}{2} - a)\dot{\alpha} - b^2(\tfrac{1}{8} + a^2)\ddot{\alpha}]$$
$$+ 2\pi \rho U b^2 (a + \tfrac{1}{2}) C(k)[\dot{h} + U\alpha + b(\tfrac{1}{2} - a)\dot{\alpha}]. \quad (5\text{-}312)$$

Speaking in a strict mathematical sense, the two integrals leading to Theodorsen's function $C(k)$ are not convergent.   As ξ^* tends to the infinite upper limit, both integrands approach $(\cos k\xi^* - i \sin k\xi^*)$, which oscillates continuously with absolute magnitude unity.   We can show that Eq. (5-309) is a correct result, however, by thinking first of a very slightly divergent oscillation, for which k is a complex number with a negative imaginary part.   The integrands then converge at infinity, and Ref. 5-36 proves that the Hankel-function representation is exact.   On physical grounds, we should expect the aerodynamic loads to be continuous functions of the degree of damping, and their limiting expressions as the damping goes to zero contain just the $C(k)$ function.   Since this reasoning fails when we try to cross the line to the case of convergent oscillations, it is fortunate that other solutions for simple harmonic motion have been presented which avoid the questionable step.   The most compact and inclusive is that of Schwarz (Ref. 5-37), who applies Söhngen's inversion formula (5–103) and (5–104) to the integral equation of a vortex sheet representing the airfoil and its wake. We shall review here the significant portions of his work.

Consider a vortex sheet like that described by Eq. (5–95), except that the time dependence of the total circulation around the airfoil calls for a wake of two-dimensional shed vortices along the x-axis from the trailing edge to infinity. With γ_a and γ_w the running circulations on the wing and wake, respectively, the integral equation replacing (5–95) is

$$w(x,0,t) = w_a(x,t)$$

$$= -\frac{1}{2\pi} \oint_{-b}^{b} \frac{\gamma_a(\xi,t)}{x - \xi}\, d\xi - \frac{1}{2\pi} \int_{b}^{\infty} \frac{\gamma_w(\xi,t)}{x - \xi}\, d\xi. \quad (5\text{–}313a)$$

In accordance with the assumption of simple harmonic motion,

$$\gamma_a(\xi,t) = \bar{\gamma}_a(\xi)e^{i\omega t}, \quad (5\text{–}314)$$

$$\gamma_w(\xi,t) = \bar{\gamma}_w(\xi)e^{i\omega t}, \quad (5\text{–}315)$$

$$w_a(x,t) = \bar{w}_a(x)e^{i\omega t} = \left[i\omega\bar{z}_a(x) + U\frac{d\bar{z}_a}{dx} \right]e^{i\omega t}. \quad (5\text{–}316)$$

These transform Eq. (5–313a) into

$$\bar{w}_a(x) = -\frac{1}{2\pi} \oint_{-b}^{b} \frac{\bar{\gamma}_a(\xi)\,d\xi}{x - \xi} - \frac{1}{2\pi} \int_{b}^{\infty} \frac{\bar{\gamma}_w(\xi)}{x - \xi}\, d\xi. \quad (5\text{–}313b)$$

The quantity γ_w can be written in terms of the airfoil circulation Γ. A physically plausible derivation of their relationship consists of saying that the wake vortex element shed from the trailing edge during any small time interval dt has a circulation equal and opposite to the corresponding change of wing circulation:

$$\gamma_w(b,t)dx = -\frac{d\Gamma(t)}{dt}\, dt. \quad (5\text{–}317)$$

Since the fluid is all moving at approximately U, we assume $dx = U\,dt$ and

$$U\gamma_w(b,t) = -\frac{d\Gamma(t)}{dt}. \quad (5\text{–}318)$$

The vortex element at a general point ξ of the wake was shed in the past at a moment determined by the time interval $(\xi - b)/U$ required for it to reach ξ. Hence,

$$U\gamma_w\left(\xi, t + \frac{\xi - b}{U}\right) = -\frac{d\Gamma(t)}{dt}. \quad (5\text{–}319)$$

Since all variables are proportional to $e^{i\omega t}$, this can be written

$$U\bar{\gamma}_w e^{i\omega\left(t + \frac{\xi - b}{U}\right)} = -i\omega\bar{\Gamma}e^{i\omega t} \quad (5\text{–}320a)$$

or

$$\bar{\gamma}_w(\xi) = -\frac{i\omega}{U}\,\bar{\Gamma}e^{ik}e^{-i(\omega\xi/U)}. \qquad (5\text{--}320\text{b})$$

We can derive Eq. (5–320b) more rigorously starting from the condition that no pressure discontinuity exists across the wake. The very generally applicable relation (5–92) between the tangential disturbance velocity and the circulation,

$$u_U' = \tfrac{1}{2}\gamma_a, \qquad (5\text{--}92)$$

permits Eq. (5–263) to be rewritten entirely in terms of γ.

$$
\begin{aligned}
p_U - p_L &= -2\rho\left[U\frac{\partial\phi_U'}{\partial x} + \frac{\partial\phi_U'}{\partial t}\right] \\
&= -2\rho\left[Uu_U' + \frac{\partial}{\partial t}\int_{-b}^{x} u_U'\,d\xi\right] \\
&= -\rho\left[U\gamma_a + \frac{\partial}{\partial t}\int_{-b}^{x}\gamma_a\,d\xi\right]. \qquad (5\text{--}321)
\end{aligned}
$$

Equation (5–321) applies to points on the airfoil. When it is extended to the wake, the last term must be replaced by

$$\frac{\partial}{\partial t}\left[\int_{-b}^{b}\gamma_a\,d\xi + \int_{b}^{x}\gamma_w\,d\xi\right] = \frac{d\Gamma}{dt} + \frac{\partial}{\partial t}\int_{b}^{x}\gamma_w\,d\xi, \qquad (5\text{--}322)$$

where we have appealed to Kutta's hypothesis to ensure no singularity at the trailing edge. Substituting Eq. (5–322) into (5–321) and assuming simple harmonic motion, we obtain for the condition of zero pressure discontinuity

$$U\gamma_w + i\omega\Gamma + i\omega\int_{b}^{x}\gamma_w\,d\xi = 0. \qquad (5\text{--}323)$$

We take the derivative with respect to x and construct a differential equation for γ_w:

$$U\frac{\partial\gamma_w}{\partial x} + i\omega\gamma_w = 0, \qquad (5\text{--}324)$$

which is solved by

$$\bar{\gamma}_w(x,t) = F(t)e^{-i(\omega x/U)}. \qquad (5\text{--}325)$$

The function of integration $F(t)$ is evaluated by putting Eq. (5–325) into Eq. (5–323) and letting $x = b$:

$$F(t) = -\frac{i\omega}{U}\,\bar{\Gamma}e^{i(\omega b/U)} = -\frac{i\omega}{U}\,\bar{\Gamma}e^{ik}. \qquad (5\text{--}326)$$

When we replace $F(t)$ in Eq. (5–325) and cancel out the time factor $e^{i\omega t}$, we reproduce Eq. (5–320b).

We may use a reduced circulation (Ref. 5–15)

$$\bar{\Omega} = \frac{\bar{\Gamma}}{b} e^{ik}, \qquad (5\text{–}327)$$

which simplifies Eq. (5–320b) to

$$\bar{\gamma}_w(\xi) = -ik\bar{\Omega}e^{-i(\omega\xi/U)}. \qquad (5\text{–}320c)$$

Substituting Eq. (5–320c) into Eq. (5–313b) yields

$$\bar{w}_a(x) = -\frac{1}{2\pi} \oint_{-b}^{b} \frac{\bar{\gamma}_a d\xi}{x - \xi} + \frac{ik\bar{\Omega}}{2\pi} \int_{b}^{\infty} \frac{e^{-i(\omega\xi/U)}}{x - \xi}\, d\xi. \qquad (5\text{–}328)$$

In terms of the dimensionless chordwise variables x^* and ξ^*,

$$-\bar{w}_a(x^*) + \frac{ik\bar{\Omega}}{2\pi} \int_{1}^{\infty} \frac{e^{-ik\xi^*}}{x^* - \xi^*}\, d\xi^* = \frac{1}{2\pi} \oint_{-1}^{1} \frac{\bar{\gamma}_a}{x^* - \xi^*}\, d\xi^*. \qquad (5\text{–}329)$$

Schwarz (Ref. 5–37) discovered that, if one temporarily regards $\bar{\Omega}$ as a known constant, Eq. (5–329) can be solved by Söhngen's relations (5–103) and (5–104). As discussed in Section 5–4, these simultaneously furnish an integral expression for the unknown $\bar{\gamma}_a$ on the right and apply Kutta's hypothesis. Here they lead to

$$\bar{\gamma}_a(x^*) = -\frac{2}{\pi} \sqrt{\frac{1 - x^*}{1 + x^*}} \left\{ -\oint_{-1}^{1} \sqrt{\frac{1 + \xi^*}{1 - \xi^*}}\, \frac{\bar{w}_a(\xi^*)}{(x^* - \xi^*)}\, d\xi^* \right.$$

$$\left. + \frac{ik\bar{\Omega}}{2\pi} \oint_{-1}^{1} \sqrt{\frac{1 + \xi^*}{1 - \xi^*}} \left[\int_{1}^{\infty} \frac{e^{-ik\lambda}d\lambda}{\xi^* - \lambda} \right] \frac{d\xi^*}{(x^* - \xi^*)} \right\}, \qquad (5\text{–}330)$$

where λ is a dummy integration variable running along the wake. In the second term on the right, interchange of order of integration, substitution of $\xi^* = \cos\theta$, and use of the integrals (5–100) yield

$$\bar{\gamma}_a(x^*) = \frac{2}{\pi} \sqrt{\frac{1 - x^*}{1 + x^*}} \left\{ \oint_{-1}^{1} \sqrt{\frac{1 + \xi^*}{1 - \xi^*}}\, \frac{\bar{w}_a(\xi^*)}{(x^* - \xi^*)}\, d\xi^* \right.$$

$$\left. + \frac{ik\bar{\Omega}}{2} \int_{1}^{\infty} \sqrt{\frac{\lambda + 1}{\lambda - 1}}\, \frac{e^{-ik\lambda}}{(x^* - \lambda)}\, d\lambda \right\}. \qquad (5\text{–}331)$$

Equation (5–331) can be solved for the circulation amplitude $\bar{\Omega}$. After integrating it between $x^* = -1$ and 1, and interchanging the orders of both double integrals, we obtain

$$\int_{-1}^{1} \bar{\gamma}_a(x^*)dx^* = \bar{\Omega}e^{-ik}$$

$$= -2\int_{-1}^{1} \sqrt{\frac{1+\xi^*}{1-\xi^*}} \bar{w}_a(\xi^*)d\xi^*$$

$$-ik\bar{\Omega}\int_{1}^{\infty} \left[\sqrt{\frac{\lambda+1}{\lambda-1}} - 1 \right] e^{-ik\lambda}d\lambda. \quad (5\text{-}332)$$

The last integral on the right is convergent, and application of the same formulas that produced Eq. (5–309) leads to

$$\int_{1}^{\infty} \left[\sqrt{\frac{\lambda+1}{\lambda-1}} - 1 \right] e^{-ik\lambda}d\lambda = -\frac{\pi}{2}[H_1^{(2)}(k) + iH_0^{(2)}(k)] - \frac{e^{-ik}}{ik}. \quad (5\text{-}333)$$

Using Eq. (5–333), and dividing Eq. (5–332) by the coefficient of $\bar{\Omega}$,

$$\bar{\Omega} = \frac{4\int_{-1}^{1} \sqrt{\frac{1+\xi^*}{1-\xi^*}} \bar{w}_a(\xi^*)d\xi^*}{\pi ik[H_1^{(2)}(k) + iH_0^{(2)}(k)]}. \quad (5\text{-}334)$$

The pressure distribution over the airfoil can be calculated from Eq. (5–321). If we define

$$p_U - p_L = \Delta\bar{p}_a(x)e^{i\omega t}, \quad (5\text{-}335)$$

then the simple harmonic form of Eq. (5–321), with $e^{i\omega t}$ divided out and $\xi^* = \xi/b$ substituted, reads

$$\frac{\Delta\bar{p}_a(x^*)}{\rho U} = -\bar{\gamma}_a(x^*) - ik\int_{-1}^{x^*} \bar{\gamma}_a(\xi^*)d\xi^*. \quad (5\text{-}336)$$

This result calls for another integration of Eq. (5–331). If we do this, inverting the order of both integrals and making use of well-known integration formulas, we get

$$\int_{-1}^{x^*} \bar{\gamma}_a(\xi^*)d\xi^*$$

$$= -\frac{2}{\pi}\int_{-1}^{1} \left[\sqrt{\frac{1+\xi^*}{1-\xi^*}} \left(\frac{\pi}{2} + \sin^{-1}x^* \right) + \Lambda_1(x^*,\xi^*) \right] \bar{w}_a d\xi^*$$

$$-\frac{ik\bar{\Omega}}{\pi}\int_{1}^{\infty} \left[\sqrt{\frac{\lambda+1}{\lambda-1}} \left(\frac{\pi}{2} + \sin^{-1}x^* \right) + \Lambda_2(x^*,\lambda) \right] e^{-ik\lambda}d\lambda, \quad (5\text{-}337a)$$

where

$$\Lambda_1(x^*,\xi^*) = \frac{1}{2}\ln\left[\frac{1 - x^*\xi^* + \sqrt{1-\xi^{*2}}\sqrt{1-x^{*2}}}{1 - x^*\xi^* - \sqrt{1-\xi^{*2}}\sqrt{1-x^{*2}}} \right] \quad (5\text{-}337b)$$

and

$$\Lambda_2(x^*,\lambda) = 2 \tan^{-1}\left[\sqrt{\frac{(1 - x^*)(\lambda + 1)}{(1 + x^*)(\lambda - 1)}}\right] - \pi. \qquad (5\text{-}337c)$$

Equation (5-337a) can be put into a simpler form by multiplying Eq. (5-332) by $(\frac{1}{2} + \pi^{-1}\sin^{-1} x^*)$ and subtracting.

$$\int_{-1}^{x^*} \bar{\gamma}_a(\xi^*)d\xi^* - \frac{e^{-ik\bar{\Omega}}}{\pi}\left(\frac{\pi}{2} + \sin^{-1} x^*\right) = -\frac{2}{\pi}\int_{-1}^{1} \Lambda_1(x^*,\xi^*)\bar{w}_a(\xi^*)d\xi^*$$

$$-\frac{ik\bar{\Omega}}{\pi}\int_{1}^{\infty}\left[\frac{\pi}{2} + \sin^{-1} x^* + \Lambda_2(x^*,\lambda)\right]e^{-ik\lambda}d\lambda. \qquad (5\text{-}338)$$

Integration of the last term by parts and cancelling between the two sides gives

$$\int_{-1}^{x^*} \bar{\gamma}_a(\xi^*)d\xi^* = -\frac{2}{\pi}\int_{-1}^{1} \Lambda_1(x^*,\xi^*)\bar{w}_a(\xi^*)d\xi^* - \frac{\bar{\Omega}}{\pi}\int_{1}^{\infty}\frac{\partial\Lambda_2}{\partial\lambda}e^{-ik\lambda}d\lambda. \qquad (5\text{-}339)$$

Equations (5-331) and (5-339) are now substituted into Eq. (5-336). Another important cancellation occurs because

$$\frac{\partial\Lambda_2}{\partial\lambda} = \sqrt{\frac{1 - x^*}{1 + x^*}}\left[\frac{1}{\sqrt{\lambda^2 - 1}} + \sqrt{\frac{\lambda + 1}{\lambda - 1}}\frac{1}{(x^* - \lambda)}\right]. \qquad (5\text{-}340)$$

The simplified result is

$$\frac{-\Delta\bar{p}_a(x^*)}{\rho U} = \frac{-ik\bar{\Omega}}{\pi}\sqrt{\frac{1 - x^*}{1 + x^*}}\int_{1}^{\infty}\frac{e^{-ik\lambda}}{\sqrt{\lambda^2 - 1}}d\lambda$$

$$+ \frac{2}{\pi}\oint_{-1}^{1}\left[\sqrt{\frac{1 - x^*}{1 + x^*}}\sqrt{\frac{1 + \xi^*}{1 - \xi^*}}\frac{1}{(x^* - \xi^*)}\right.$$

$$\left.- ik\Lambda_1(x^*,\xi^*)\right]\bar{w}_a(\xi^*)d\xi^*. \qquad (5\text{-}341)$$

The last steps in finding the pressure distribution are to insert $\bar{\Omega}$ from Eq. (5-334) and replace the first integral on the right of Eq. (5-341) by its value $-i(\pi/2)H_0^{(2)}(k)$; these substitutions lead to

$$\frac{-\Delta\bar{p}_a(x^*)}{\rho U} = \frac{2}{\pi}[1 - C(k)]\sqrt{\frac{1 - x^*}{1 + x^*}}\int_{-1}^{1}\sqrt{\frac{1 + \xi^*}{1 - \xi^*}}\bar{w}_a(\xi^*)d\xi^*$$

$$+ \frac{2}{\pi}\oint_{-1}^{1}\left[\sqrt{\frac{1 - x^*}{1 + x^*}}\sqrt{\frac{1 + \xi^*}{1 - \xi^*}}\frac{1}{(x^* - \xi^*)}\right.$$

$$\left.- ik\Lambda_1(x^*,\xi^*)\right]\bar{w}_a(\xi^*)d\xi^* \qquad (5\text{-}342)$$

Here $C(k)$ is again Theodorsen's function, defined by Eq. (5-309). The only differences between Eq. (5-342) and Eq. (5-299) (with the ratio of integrals replaced by $C(k)$ and simple harmonic motion specified) consist of the more rigorous derivation of Eq. (5-342) and the fact that Schwarz and Söhngen have proved their form to be valid for $\bar{w}_a(x^*)$ with discontinuities and integrable singularities. For given types of airfoil oscillation, the resultant lifts and moments per unit span deduced by appropriate integrations of Eq. (5-342) are identical with those found by Theodorsen.

We draw the interesting conclusion from Eq. (5-342) that the chordwise pressure variation associated with the wake (the $1 - C(k)$ term) is the same as the flat-plate loading given by Eq. (5-110), regardless of the motion; the corresponding total force must always act at the quarter-chord point. We can also identify circulatory and noncirculatory parts of the distribution in Eq. (5-342). The latter is produced by a wing vortex sheet that satisfies the instantaneous boundary conditions but has zero total circulation; it is calculated by setting $\bar{\Omega} = 0$ and inverting what is left of Eq. (5-329) with another of Söhngen's formulas, this one implying the auxiliary requirement

$$\int_{-1}^{1} \bar{\gamma}_a(\xi^*)d\xi^* = 0. \qquad (5\text{-}343)$$

The result is

$$\frac{-(\Delta \bar{p}_a)_{\mathrm{NC}}}{\rho U} = \frac{2}{\pi} \oint_{-1}^{1} \left[\frac{\sqrt{1 - \xi^{*2}}}{\sqrt{1 - x^{*2}}} \frac{1}{(x^* - \xi^*)} - ik\Lambda_1(x^*,\xi^*) \right] \bar{w}_a(\xi^*)d\xi^*.$$
$$(5\text{-}344)$$

The difference between the right sides of Eqs. (5-342) and (5-344) takes care of Kutta's hypothesis and the wake vortices.

In the solutions for simple harmonic motion derived in this section, we see that the function $C(k)$ plays a unique part. The argument of all the Bessel functions which make up $C(k)$ is the reduced frequency itself, and for this reason we regard k as the best measure of the "unsteadiness" of an incompressible flow and assign it an important role in our discussions. This fact also explains the adoption of the rather unnatural semichord b in place of the chord $c = 2b$, and the relatively infrequent appearance in aeroelasticity of reduced frequencies and velocities based on c. A clear idea of the significance of $C(k)$ comes from an examination of the simplified lift and moment expressions employed by aeronautical engineers when dealing with low-reduced-frequency unsteady motions like dynamic stability modes. At least two such simplifications are in common use. For the most approximate of these it is assumed that all aerodynamic loads can be calculated from steady-state formulas like Eqs. (5-111) and (5-112), except that the angle of attack α is replaced by the instantaneous inclination between the resultant velocity vector and the chordline. In the case of the h and α motions described in Fig. 5-18, this procedure would lead to

$$L \cong 2\pi\rho Ub[\dot{h} + U\alpha], \tag{5-345}$$

$$M_y \cong -L[ba + (b/2)]. \tag{5-346}$$

A more sophisticated approximation is the so-called quasi steady-state assumption, which neglects only the influence of the wake vortices on the flow. This is equivalent to replacing $C(k)$, or the ratio of integrals in Eqs. (5-299)-(5-302), with the value 1 approached when k goes to zero. Equations (5-311) and (5-312) give for the quasi steady-state lift and moment due to vertical translation and pitching oscillations:

$$L_{\mathrm{QS}} = \pi\rho b^2[\ddot{h} + U\dot{\alpha} - ba\ddot{\alpha}] + 2\pi\rho Ub[\dot{h} + U\alpha + b(\tfrac{1}{2} - a)\dot{\alpha}], \tag{5-347}$$

$$M_{y\mathrm{QS}} = \pi\rho b^2[ba\ddot{h} - Ub(\tfrac{1}{2} - a)\dot{\alpha} - b^2(\tfrac{1}{8} + a^2)\ddot{\alpha}]$$
$$+ 2\pi\rho Ub^2(a + \tfrac{1}{2})[\dot{h} + U\alpha + b(\tfrac{1}{2} - a)\dot{\alpha}]. \tag{5-348}$$

The terms containing the accelerations $\ddot{h}$ and $\ddot{\alpha}$ can be lumped with inertial forces due to the aircraft, and they are often justifiably omitted when the relative density is very large. However, the term with $\dot{\alpha}$ in the first bracket of Eq. (5-348) describes a damping in pitch which comes from the circulatory part of the solution. On conventional airplanes its contribution is rather small compared with damping about the center of gravity from the stabilizer, but neglecting it may occasionally lead to a marked underestimation of stability.

We get a physical interpretation of $C(k)$ by comparing the circulatory lifts, in the exact Eq. (5-311) and the quasi-steady Eq. (5-347), produced by angle-of-attack motion $\dot{h}$. The two forms are related by

$$2\pi\rho UbC(k)\dot{h} = L_C(\dot{h}) = C(k) \cdot L_{\mathrm{QS}}(\dot{h}), \tag{5-349}$$

where $\dot{h}$ may be regarded as a complex vector rotating with angular velocity ω. $L_{\mathrm{QS}}(\dot{h})$ is a vector proportional to $\dot{h}$ and therefore in phase with it; since the directions of L and h are opposite, $L_{\mathrm{QS}}(\dot{h})$ is a damping force directly resisting the vertical oscillation. When unsteadiness is important, as evidenced by an appreciable value of k, the lift vector $L_{\mathrm{QS}}(\dot{h})$ is modified by multiplication with the complex number $C(k)$. This brings about a reduction in magnitude and a negative rotation or phase shift, so that $L_C(\dot{h})$ always lags behind its quasi-steady counterpart. Figure 5-20 illustrates the two lift vectors and presents a complex polar plot of $C(k)$. The lift-reduction factor $C(k)$ falls gradually to $\tfrac{1}{2}$ as k goes from zero to infinity. However, its effect as a producer of phase lag takes over very quickly, since its imaginary part $G(k)$ starts out from zero with a negatively infinite slope as k increases, and the maximum rotation of the vector $L_{\mathrm{QS}}(\dot{h})$ occurs at about $k = 0.2$. Comparable phase shifts are, of course, produced by $C(k)$ on all the circulatory parts of the quasi steady-state aerodynamic loads. Since the amount of energy extracted from or added to an oscillatory motion by the airstream is very sensitive to certain of these phase angles, we can see how important unsteady flow is in controlling an instability like flutter, even though the reduced frequency may be of the order of 0.1 or less.

Probably the most accurate table of $C(k)$ appears in Appendix V–J of Ref. 5–38. Tables of the aerodynamic loads and extensions of the soultion to cover other airfoil motions abound. For example, Theodorsen and Garrick (Ref. 5–39) have dealt with flaps and tabs having aerodynamic balance and sealed gaps. Effects of both open and sealed gap are considered by Küssner and Schwarz (Ref. 5–40). The tables most commonly employed in the United States are those of Smilg and Wasserman (Ref. 5–41). These authors present the quantities listed in Eqs. (5–300)–(5–302) in the following notation (the gap is assumed to be open, so the Küssner-Schwarz solution is actually used):

$$L = -\pi\rho b^3 \omega^2 \left\{ L_h \frac{h}{b} + [L_\alpha - L_h(\tfrac{1}{2} + a)]\alpha + L_\beta \beta \right\}. \quad (5\text{–}350)$$

$$\begin{aligned} M_y = \pi\rho b^4 \omega^2 \{ & [M_h - L_h(\tfrac{1}{2} + a)]\frac{h}{b} \\ & + [M_\alpha - (L_\alpha + M_h)(\tfrac{1}{2} + a) + L_h(\tfrac{1}{2} + a)^2]\alpha \\ & + [M_\beta - L_\beta(\tfrac{1}{2} + a)]\beta \}. \end{aligned} \quad (5\text{–}351)$$

$$M_\beta = \pi\rho b^4 \omega^2 \left\{ T_h \frac{h}{b} + [T_\alpha - T_h(\tfrac{1}{2} + a)]\alpha + T_\beta \beta \right\}. \quad (5\text{–}352)$$

Here

$$h = \bar{h}_0 e^{i\omega t}, \quad (5\text{–}353)$$

$$\alpha = \bar{\alpha}_0 e^{i\omega t}, \quad (5\text{–}354)$$

$$\beta = \bar{\beta}_0 e^{i\omega t}, \quad (5\text{–}355)$$

and L_h, L_α, M_h, and M_α are tabulated dimensionless functions of the reduced frequency; the remaining coefficients in the brackets are tabulated vs. reduced frequency and location of the flap's leading edge. The tables in Ref. 5–41 also include aerodynamic balance of the flap and the effects of a tab hinged at its nose. More elaborate tables of aerodynamic coefficients for control surfaces appear in Ref. 5–42. Spielberg (Ref. 5–46) presents some values of the coefficients needed for a flutter analysis including a parabolic mode of chordwise deformation.

The accuracy of the aerodynamic theory for oscillations of a two-dimensional airfoil has been explored in a multitude of experimental investigations. Probably the most extensive study of vertical translation and pitching motions is Halfman's (Ref. 5–43). All his data were taken at uniformly high Reynolds numbers, but this fact, together with mechanical limitations on vibration frequencies, made it impossible to work at greater than about $k = 0.4$. Below this value, the amplitudes and phase angles of all force and moment components differ from theory by less than experimental error, with one exception. This is the lift due to pitching motion, which, as the reduced frequency approaches zero, falls gradually toward a value about 0.85 to 0.9 of the theoretical. This result might be anticipated, in view of the fact that the steady-state lift-curve slope of an airfoil is roughly the same multiple of 2π. Greidanus, van de Vooren, and Bergh (Ref. 5–44) present measurements comparable to Halfman's, but their reduced frequency range extends up to 1.0, at the expense of wide Reynolds number variations. Tripper wires were used over the front of the airfoil to compensate for Reynolds-number effects on the boundary layer. Their data show much the same trends as Halfman's for the same values of k. Above $k = 0.4$, several aerodynamic terms gradually diverge from their theoretical counterparts. These discrepancies are ascribed to a lag in fulfillment of Kutta's hypothesis, and to errors in the assumption that the wake vortices move downstream in a plane, both of which would be expected to influence the flow more severely at higher reduced frequencies.

Andreopolis, Cheilek, and Donovan (Ref. 5–45) measured certain airloads on an oscillating flap-tab combination behind a fixed airfoil. Their observations confirm Theodorsen's predictions of flap hinge moments quite satisfactorily, but large inaccuracies occur in the moments associated with motion of the tab.

5–7 Arbitrary motion of thin airfoils in incompressible flow; the gust problem. Just as the solutions presented in Section 5–6 are valuable for flutter analysts, an aeroelastician dealing with unsteady phenomena like rapid maneuvers and gust entry needs a theory for more general small motions of airfoils. Two methods of approach have been used extensively in solving this problem. The first of these (from a historical standpoint) consists of direct, numerical attack on some form of the integral equation of the vortex sheet. The other employs Fourier-integral superposition of the linear results for simple harmonic motion. We describe the latter method in detail for incompressible flow, since it builds on the material given in Section 5–6.

By way of introducing the subject, we point out that Eqs. (5–299)–(5–302), with $C(k)$ substituted from Eq. (5–308), are applicable for divergent oscillations proportional to $e^{\lambda t}$, where $\lambda = \mu + i\omega$ and $\mu > 0$. To make this extension, we simply give h, α, etc. the appropriate time dependence and regard k in $C(k)$ as a complex argument:

$$k = \frac{\omega b}{U} - i\frac{\mu b}{U}. \tag{5–356}$$

Values of $C(k)$ for complex k are tabulated in Ref. 5–47. This important generalization of Theodorsen's formulas was implied in one of Söhngen's papers (Ref. 5–48) and developed by W. P. Jones (Ref. 5–49). For convergent oscillations, when $\mu < 0$, the same straightforward conclusion is obviously not valid, because the integrals in Eq. (5–308) diverge. In fact, Söhngen proves (Ref. 5–38) that the aerodynamic loads due to a decaying oscillation which was suddenly started at $t = 0$ consist of two parts, one containing the factor $e^{\lambda t}$, the other proportional to $1/t$ as t becomes large. For $\mu < 0$, the second part dies out more slowly than the first. Fraeys de Veubeke, in his comprehensive study of unsteady airfoil motion in low-speed flow (Ref. 5–50), shows that the $e^{\lambda t}$ portion of the loading due to convergent oscillations is analogous to Theodorsen's and suggests a procedure for finding the $1/t$ remainder. The tables of Ref. 5–47 include $C(k)$ for $\mu < 0$, so they are useful both in calculating the exponential loads and in estimating the resultants during any part of the transient regime when the $1/t$ loads are negligibly small by comparison.

The noncirculatory results of Section 5–6 are correct regardless of the nature of unsteady motion, so long as it is small. We shall concentrate here on finding lift and moment. Considering first only vertical translation and pitching, we can state that the noncirculatory part of Eq. (5–311) and the part of Eq. (5–312) not containing $C(k)$ go unchanged. The latter does have a term proportional to $\dot{\alpha}$ which is known to be circulatory, but the coefficient is a constant independent of frequency. Therefore, we can state, by the principle of superposition of simple harmonic components, that if its form is identical for every harmonic it must carry through and be the same for a more general time function constructed by summing them.

The remaining terms in Eqs. (5–311) and (5–312) are related to the oscillation only through the quantity

$$w_{\frac{3}{4}c}(t) = -[\dot{h} + U\alpha + b(\tfrac{1}{2} - a)\dot{\alpha}].\qquad(5\text{–}357)$$

This is the instantaneous vertical velocity of the liquid particle in contact with the three-quarter chord point of the airfoil. We are able to represent $w_{\frac{3}{4}c}(t)$ for arbitrary motion by the Fourier integral*

$$w_{\frac{3}{4}c}(t) = \frac{1}{2\pi}\int_{-\infty}^{\infty} f(\omega)e^{i\omega t}d\omega,\qquad(5\text{–}358\mathrm{a})$$

which can be inverted into

$$f(\omega) = \int_{-\infty}^{\infty} w_{\frac{3}{4}c}(t)e^{-i\omega t}dt.\qquad(5\text{–}358\mathrm{b})$$

---

* For discussions of the theory of the Fourier integral, the reader is referred to books like Hildebrand's (Ref. 5–51) and Sneddon's (Ref. 5–52).

Since the circulatory part of the lift per unit span for any Fourier component with unit amplitude of $w_{\frac{3}{4}c}$ is

$$\Delta L_c = -2\pi\rho Ub C(k)e^{i\omega t}, \tag{5-359}$$

the resultant must be

$$L_c = \frac{1}{2\pi}\int_{-\infty}^{\infty}f(\omega)\Delta L_c d\omega = -\frac{2\pi\rho Ub}{2\pi}\int_{-\infty}^{\infty}C(k)f(\omega)e^{i\omega t}d\omega. \tag{5-360}$$

The total running lift and moment (about the axis at $x = ba$) are

$$L = \pi\rho b^2[\ddot{h} + U\dot{\alpha} - ba\ddot{\alpha}] - \rho Ub\int_{-\infty}^{\infty}C\left(\frac{\omega b}{U}\right)f(\omega)e^{i\omega t}d\omega, \tag{5-361}$$

$$M_y = \pi\rho b^2[ba\ddot{h} - Ub(\tfrac{1}{2} - a)\dot{\alpha} - b^2(\tfrac{1}{8} + a^2)\ddot{\alpha}]$$
$$- \rho Ub^2(\tfrac{1}{2} + a)\int_{-\infty}^{\infty}C\left(\frac{\omega b}{U}\right)f(\omega)e^{i\omega t}d\omega. \tag{5-362}$$

We observe that the lift associated with $C(k)$ acts at the quarter-chord point $a = -\frac{1}{2}$. The quarter and three-quarter chord points play such a distinctive role in this theory that Küssner has named them the forward and rear neutral points of the airfoil.

Fourier integral superposition can be used with any function that has a finite number of finite discontinuities and whose absolute value has a finite integral in the range $t = -\infty$ to $+\infty$. This should cover a majority of functions of engineering interest. A case that does not quite fall within the limitations, but can still be handled, is Wagner's problem of the step change in angle of attack:

$$w_{\frac{3}{4}c} = \begin{cases} 0, & t < 0, \\ -U\alpha_0, & t > 0. \end{cases} \tag{5-363}$$

This has the Fourier integral formula

$$w_{\frac{3}{4}c} = -\frac{U\alpha_0}{2\pi}\int_{-\infty}^{\infty}\frac{e^{i\omega t}}{i\omega}d\omega, \tag{5-364}$$

where the path of integration makes an infinitesimal loop below the pole at the origin. Equation (5-360) yields the circulatory lift due to this motion (sometimes called the indicial lift).

$$L = \rho Ub U\alpha_0\int_{-\infty}^{\infty}\frac{C(k)}{ik}e^{iks}dk. \tag{5-365}$$

Here

$$s = \frac{Ut}{b} \tag{5-366}$$

is the distance in semichords traveled by the airfoil after the step. Equation (5–365) is often written in terms of Wagner's function $\phi(s)$:

$$L = 2\pi\rho U^2 b\alpha_0 \phi(s). \tag{5-367}$$

Since its coefficient is the steady-state lift, this function approaches unity for large values of s. It was originally computed in Ref. 5–53 by the vortex-sheet method. The present formula is

$$\phi(s) = \frac{1}{2\pi i} \int_{-\infty}^{\infty} \frac{C(k)}{k} e^{iks} dk$$

$$= 1(s) + \frac{1}{2\pi i} \int_{-\infty}^{\infty} \frac{C(k) - 1}{k} e^{iks} dk, \tag{5-368a}$$

where $1(s)$ is the step function that jumps from zero to unity at $s = 0$.

Separating e^{iks} and $C(k)$ into their real and imaginary parts, we rewrite the last integral in Eq. (5–368a), as follows:

$$\phi(s) = 1(s) + \frac{1}{2\pi} \int_{-\infty}^{\infty} [F(k) - 1] \frac{\sin ks}{k} dk$$

$$+ \frac{1}{2\pi} \int_{-\infty}^{\infty} G(k) \frac{\cos ks}{k} dk$$

$$+ \frac{i}{2\pi} \int_{-\infty}^{\infty} \left\{ G(k) \frac{\sin ks}{k} - [F(k) - 1] \frac{\cos ks}{k} \right\} dk$$

$$= 1(s) + \frac{1}{\pi} \int_{0}^{\infty} \left\{ [F(k) - 1] \frac{\sin ks}{k} + G(k) \frac{\cos ks}{k} \right\} dk. \tag{5-368b}$$

Here we have used the facts that $F(k)$ and $\cos ks$ are even functions of the variable of integration, whereas $G(k)$ and $\sin ks$ are odd. The symmetries of $F(k)$ and $G(k)$ can be demonstrated either by studying their forms as combinations of Bessel functions, or by observing that the imaginary part of $\phi(s)$ in Eq. (5–368b) must be zero for all values of s. Moreover, $C(k)$ is the complex admittance function for steady-state oscillations of a linear system, and any such function is known to have an even real part and an odd imaginary part.

Equation (5–368b) can be simplified further through the vanishing of $\phi(s)$ and $1(s)$ when s is negative. For $s < 0$, it gives

$$-\frac{1}{\pi} \int_{0}^{\infty} [F(k) - 1] \frac{\sin ks}{k} dk = \frac{1}{\pi} \int_{0}^{\infty} G(k) \frac{\cos ks}{k} dk. \tag{5-368c}$$

Since $\sin ks$ is an odd function, Eq. (5–368c) can be written in terms of absolute values of s, as follows:

$$\frac{1}{\pi} \int_{0}^{\infty} [F(k) - 1] \frac{\sin k|s|}{k} dk = \frac{1}{\pi} \int_{0}^{\infty} G(k) \frac{\cos k|s|}{k} dk. \tag{5-368d}$$

However, Eq. (5–368d) holds for all *positive* numbers $|s|$ and shows that the two parts of the last integral in Eq. (5–368b) are equal. Hence Eq. (5–368b) reduces finally, for computational purposes, to

$$\phi(s) = \frac{2}{\pi}\int_0^\infty \frac{F(k)}{k}\sin ksdk = 1 + \frac{2}{\pi}\int_0^\infty \frac{G(k)}{k}\cos ksdk. \quad (5\text{–}369)$$

This is obviously limited to $s > 0$; the first formula in Eq. (5–369) is a more practical one, since its integrand is finite as k approaches zero. A plot of $\phi(s)$ is shown in Fig. 5–21.

In many applications it is convenient to use $\phi(s)$ itself when writing the circulatory lift and moment due to arbitrary motion. Thus, for the rigid airfoil starting from rest at $t = 0$, use of the Duhamel or superposition integral leads to

$$L = \pi\rho b^2[\ddot{h} + U\dot{\alpha} - ba\ddot{\alpha}]$$

$$-2\pi\rho Ub\left[w_{\frac{3}{4}c}(0)\phi(s) + \int_0^s \frac{dw_{\frac{3}{4}c}(\sigma)}{d\sigma}\phi(s-\sigma)d\sigma\right] \quad (5\text{–}370)$$

per unit span. Although $\phi(s)$ has a relatively simple form, it is not expressible in terms of well-known functions. Therefore we often resort to certain convenient approximate representations, such as

$$\phi(s) \cong 1 - 0.165e^{-0.0455s} - 0.335e^{-0.3s} \quad (5\text{–}371)$$

and

$$\phi(s) \cong \frac{s+2}{s+4}. \quad (5\text{–}372)$$

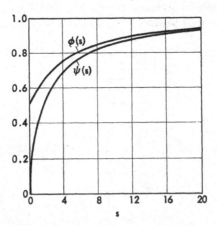

Fig. 5–21. Wagner's function $\phi(s)$ for indicial lift and Küssner's function $\psi(s)$ for lift due to a sharp-edged gust, plotted as functions of distance traveled in semichordlengths.

The former possesses an elementary Laplace transform, making it handy when operational techniques are employed to analyze the motion of linear systems. It has been criticized because it gives incorrect limiting behavior for large values of s. Since the difference between $\phi(s)$ and unity actually dies out like $1/s$, Eq. (5-372) is a better approximation for this limit.

Another useful application of the Fourier integral to unsteady liquid motion around wings is in the field of gust loading. When a thin airfoil enters a region of atmospheric turbulence with normal velocity distribution $w_G(w_G \ll U)$, we can derive a boundary condition analogous to Eq. (5-225). This situation is illustrated in Fig. 5-22. We are interested only in forces due to the gust structure itself, since we can deal with motion of the rigid airfoil in response to these loads by methods outlined above and superimpose the results. Accordingly, the boundary condition specifies that the total vertical velocity due to the gust and the vortex sheet simulating the airfoil must vanish:

$$w_G + w_a = 0 \tag{5-373a}$$

or

$$w_a(x,t) = -w_G; \quad \text{for } z = 0, \ -b \le x \le b. \tag{5-373b}$$

We assume the turbulence to be embedded in the atmosphere in such a way that it moves past the airfoil at velocity U without an appreciable change in w_G for any particular particle during its encounter. As in the case of γ_w [cf. Eq. (5-306)], we express this fact by making w_G a function of x and t in the combination $(x - Ut)$.

To permit Fourier superposition, we must first apply the theory of Section 5-6 to a simple harmonic gust problem. Such a one is the so-called sinusoidal gust, described physically by the real part of

$$w_G(x - Ut) = \bar{w}_G e^{i\omega[t - (x/U)]} = \bar{w}_G e^{ik(s - x^*)}. \tag{5-374}$$

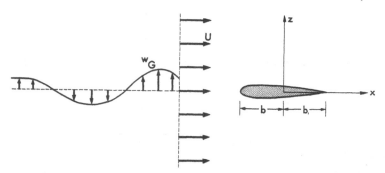

Fig. 5-22. Thin airfoil encountering a region of atmospheric turbulence with vertical gust velocity $w_G(x - Ut)$.

Substituted into Eq. (5–373b), Eq. (5–374) leads to

$$\overline{w}_a(x^*) = -\overline{w}_G e^{-ikx^*} \tag{5–375}$$

This, in turn, is suitable for use with Schwarz' solution (5–342) to calculate the pressure distribution, lift, and moment per unit span. The last two are

$$L = 2\pi\rho U b \overline{w}_G \{C(k)[J_0(k) - iJ_1(k)] + iJ_1(k)\}e^{i\omega t}, \tag{5–376}$$

$$M_y = b(\tfrac{1}{2} + a)L. \tag{5–377}$$

Thus the entire aerodynamic force due to a sinusoidal gust acts at the quarter-chord point. By superposition, the same must be true for any other gust which is a function of $(x - Ut)$ only.

The Fourier integral makes it possible to calculate from Eqs. (5–376) and (5–377) the airloads caused by arbitrary w_G. For example, we consider a sharp-edged gust striking the leading edge of the airfoil at $t = 0$ (Küssner's problem, Ref. 5–35, etc.):

$$w_G = \begin{cases} 0, & x > Ut - b, \\ w_0, & x < Ut - b. \end{cases} \tag{5–378}$$

Following Garrick (Ref. 5–54), we can represent this as

$$w_G = \frac{w_0}{2\pi} \int_{-\infty}^{\infty} \frac{e^{i\omega\left(t - \frac{b}{U} - \frac{x}{U}\right)}}{i\omega} \, d\omega = \frac{w_0}{2\pi} \int_{-\infty}^{\infty} \frac{e^{ik(s - x^* - 1)}}{ik} \, dk, \tag{5–379}$$

where again the path of integration must take an infinitesimal loop below the origin. Therefore, the lift per unit span works out to be

$$L = \rho U b w_0 \int_{-\infty}^{\infty} \frac{\{C(k)[J_0(k) - iJ_1(k)] + iJ_1(k)\}e^{ik(s-1)}}{ik} \, dk, \tag{5–380}$$

while the moment is again given by Eq. (5–377). This can, of course, be simplified for computing by the same sort of symmetry considerations that led to Eq. (5–369).

From Eq. (5–380) we determine Küssner's function representing dimensionless lift development due to a sharp-edged gust:

$$L(s) = 2\pi\rho U b w_0 \psi(s), \tag{5–381a}$$

with

$$\psi(s) = \frac{1}{2\pi i} \int_{-\infty}^{\infty} \frac{[F_G(k) + iG_G(k)]e^{ik(s-1)}}{k} \, dk$$

$$= \frac{2}{\pi} \int_0^{\infty} \frac{[F_G(k) - G_G(k)] \sin ks \sin k}{k} \, dk \tag{5–381b}$$

for $s > 0$.

Here $F_G(k)$ and $G_G(k)$ are the real and imaginary parts of the expression in braces in Eq. (5–376), and we have reduced the integral as in the case of $\phi(s)$. Like Wagner's function, $\psi(s)$ cannot be written in simple algebraic terms. It increases from 0 at $t = 0$ to 1 at $t = \infty$, as illustrated in Fig. 5–21. Both of the basic indicial functions for incompressible flow have been tabulated by many authors, including Sears (Ref. 5–55).

If an arbitrary $w_G(s)$ is defined, for $s \geq 0$, as the gust velocity encountered by the airfoil's leading edge at the instant $t = sb/U$, Duhamel's integral gives

$$L = 2\pi\rho Ub \left\{ w_G(0)\psi(s) + \int_0^s \frac{dw_G(\sigma)}{d\sigma} \psi(s - \sigma)d\sigma \right\}. \qquad (5\text{–}382)$$

Like $\phi(s)$, $\psi(s)$ is frequently replaced by simple algebraic approximations, such as

$$\psi(s) \cong 1 - 0.500e^{-0.130s} - 0.500e^{-s} \qquad (5\text{–}383)$$

and

$$\psi(s) \cong \frac{s^2 + s}{s^2 + 2.82s + 0.80}. \qquad (5\text{–}384)$$

The former has a convenient Laplace transform but, as with $\phi(s)$, the latter gives a better asymptotic representation for large values of s.

It is an unfortunate fact that no very accurate or extensive experimental research has yet been carried out with the objective of establishing the accuracy of Wagner's and Küssner's functions. This lack can be attributed to difficulties with measuring equipment in the presence of the sudden changes of load that are encountered. It is hoped that the appearance of pressure pickups with very rapid dynamic response will before long provide the tools for successful measurements. Meanwhile we can assert that, if we assume the validity of the superposition principle, tests on oscillating airfoils provide us with indirect verification of Wagner's result.

We conclude the topic of arbitrary airfoil motion with a review of important formulas from the vortex-sheet approach to the problem. Among the earliest publications on this subject are those of Wagner and Küssner (Refs. 5–53 and 5–35). However, papers by von Kármán and Sears (Refs. 5–55 and 5–56) laid the foundations of most research in the United States, and we follow their method here. Our notation is the same as that used with Theodorsen's work in Section 5–6. The total instantaneous circulation on the airfoil is

$$\Gamma = \int_{-b}^{b} \gamma_a(x,t)dx. \qquad (5\text{–}385)$$

Moreover, the condition of zero total circulation in the entire flow can be written

$$\Gamma + \int_{b}^{\infty} \gamma_w(x,t)dx = 0. \qquad (5\text{–}386)$$

We separate the running bound circulation into two parts:

$$\gamma_a(x,t) = \gamma_0(x,t) + \gamma_1(x,t). \tag{5-387}$$

Here γ_0 is the quasi steady-state result and can be readily calculated. For instance, Eq. (5-105) with $U dz_a/dx$ replaced by w_a reads

$$\gamma_0(x,t) = \frac{2}{\pi} \sqrt{\frac{1-x^*}{1+x^*}} \oint_{-1}^{1} \sqrt{\frac{1+\xi^*}{1-\xi^*}} \frac{w_a(\xi^*,t)}{(x^*-\xi^*)} d\xi^*, \tag{5-388}$$

whence, by integration,

$$\Gamma_0(t) = \int_{-b}^{b} \gamma_0(x,t)dx = -2b \int_{-1}^{1} \sqrt{\frac{1+\xi^*}{1-\xi^*}} w_a(\xi^*,t)d\xi^*. \tag{5-389}$$

γ_1 is the additional circulation due to unsteadiness of the flow. Each element of γ_1 is paired with an opposite γ_w-element in the wake. The combined $\gamma_1 - \gamma_w$ flow must satisfy Kutta's hypothesis, since γ_0 does. It must also leave undisturbed the boundary condition (5-225) satisfied at the airfoil surface by γ_0. But these are exactly the properties of the pattern of vortices and countervortices used by Theodorsen in the second part of his solution. We may therefore adopt for our purposes any of Theodorsen's formulas that are useful. Equation (5-297), in particular, helps us to find γ_w and

$$\Gamma_1(t) = \int_{-b}^{b} \gamma_1(x,t)dx = \Gamma(t) - \Gamma_0(t). \tag{5-390}$$

We rewrite Eq. (5-297) with the variable $\xi^* = \cos \phi$ replacing ϕ in the left-hand integral:

$$\frac{2}{\pi} \int_{-1}^{1} \frac{\sqrt{1-\xi^{*2}}}{(\xi^*-1)} w_a(\xi^*,t)d\xi^* + \frac{1}{\pi b} \int_{b}^{\infty} \sqrt{\frac{\xi+b}{\xi-b}} \gamma_w(\xi,t)d\xi = 0. \tag{5-391}$$

If this is multiplied by πb, and $\sqrt{1-\xi^{*2}}/(\xi^*-1)$ is reduced to $-\sqrt{(1+\xi^*)/(1-\xi^*)}$, the left-hand integral becomes equal to the quasi-steady circulation in Eq. (5-389).

$$\Gamma_0(t) + \int_{b}^{\infty} \sqrt{\frac{\xi+b}{\xi-b}} \gamma_w(\xi,t)d\xi = 0. \tag{5-392}$$

In view of Eqs. (5-386) and (5-390), this can be modified to read

$$\Gamma_1(t) = \int_{b}^{\infty} \left[\sqrt{\frac{\xi+b}{\xi-b}} - 1 \right] \gamma_w(\xi,t)d\xi. \tag{5-393}$$

Since Γ_0 is known, Eq. (5-392) is an integral equation for the wake circulation distribution, from which Γ_1 can be computed by Eq. (5-393).

Assuming the circulation pattern to be known, it remains to find the aerodynamic loads on the airfoil. This might be done by Eqs. (5–299)–(5–302), but Ref. 5–56 presents an interesting alternative set of formulas. We can derive these by an extensive manipulation of Bernoulli's equation or, following Ref. 5–56, by considering the rates of change of the impulses [cf. Eq. (5–60)] of all the vortices in the pattern. The results are

$$L = \rho U \Gamma_0 - \rho \frac{d}{dt} \int_{-b}^{b} \gamma_0(x,t) x \, dx + \rho U b \int_{b}^{\infty} \frac{\gamma_w(x,t) \, dx}{\sqrt{x^2 - b^2}}, \quad (5\text{–}394)$$

$$M_y = -\rho U \int_{-b}^{b} \gamma_0(x,t) x \, dx + \frac{\rho}{2} \frac{d}{dt} \int_{-b}^{b} \gamma_0(x,t) \left[x^2 - \left(\frac{b}{2}\right)^2 \right] dx$$

$$+ \frac{\rho}{2} U b^2 \int_{b}^{\infty} \frac{\gamma_w(x,t) \, dx}{\sqrt{x^2 - b^2}}. \quad (5\text{–}395)$$

Here M_y is taken about an axis at midchord $x = 0$; it can be transferred to $x = ba$ by adding the product of L and the axis shift ba. In each of Eqs. (5–394) and (5–395) the lead term is evidently the quasi steady-state circulatory effect. The second term is the noncirculatory contribution, while the third term represents the unsteady circulatory influence of the wake, vanishing whenever γ_w is negligible or the distance $\sqrt{x^2 - b^2}$ in the denominator becomes very large. As mentioned previously, we can see from the final integrals in Eqs. (5–394) and (5–395) that the lift due to the wake acts always at the quarter-chord point.

Neumark (Ref. 5–57) has derived an analogous expression for the chordwise pressure distribution

$$p_U - p_L = -\rho U \gamma_0 - \rho \frac{\partial}{\partial t} \int_{-b}^{x} \gamma_{0n}(x,t) \, dx - \frac{\rho U}{\pi} \sqrt{\frac{b - x}{b + x}} \int_{b}^{\infty} \frac{\gamma_w(\xi,t)}{\sqrt{\xi^2 - b^2}} \, d\xi,$$
$$(5\text{–}396)$$

where γ_{0n} is the noncirculatory part of the airfoil vortex sheet, which can be shown by Söhngen's formulas (Ref. 5–13) to be

$$\gamma_{0n}(x,t) = -\frac{2}{\pi \sqrt{b^2 - x^2}} \int_{-b}^{b} \frac{\sqrt{b^2 - \xi^2}}{(x - \xi)} w_a(\xi,t) \, d\xi. \quad (5\text{–}397)$$

There are two steps in solving a practical problem by the method of von Kármán and Sears. Equation (5–392) is first used to calculate $\gamma_w(x,t)$. Equations (5–394) and (5–395) can then be integrated directly to find lift and moment. Although this has often been done by numerical integration, Sears (Ref. 5–55) points out the remarkable simplification that is gained by introducing Laplace transformation on the time variable.

Thus, if we put Eq. (5–392) into dimensionless form and recall that γ_w depends only on $(x - Ut)$,

$$\frac{\Gamma_0(s)}{Ub} + \int_1^\infty \sqrt{\frac{\xi^* + 1}{\xi^* - 1}} \frac{\gamma_w(\xi^* - s)}{U} d\xi^* = 0. \tag{5–398}$$

For the indicial type of problem, where the unsteady motion begins at $s = t = 0$, the wake only extends up to $\xi^* = 1 + s$. We make the variable change

$$\xi^* = 1 + s - \sigma$$

and observe that γ_w is a function of σ only, since this differs only by a constant from $(\xi^* - s)$.

$$\int_0^s \sqrt{\frac{2 + (s - \sigma)}{(s - \sigma)}} \frac{\gamma_w(\sigma)}{U} d\sigma = -\frac{\Gamma_0(s)}{Ub}. \tag{5–399}$$

In a similar fashion, the last term in Eq. (5–394), which is the only part of the lift that is difficult to calculate, can be made to read

$$\frac{L_2(s)}{\rho U^2 b} = \int_0^s \frac{\gamma_w(\sigma)}{U} \frac{d\sigma}{\sqrt{(s - \sigma)^2 + 2(s - \sigma)}}. \tag{5–400}$$

In the sense of operational mathematics (see, for example, Churchill, Ref. 5–58), the right sides of Eqs. (5–399) and (5–400) are convolution integrals. That is, if we apply the Laplace transformation

$$\mathcal{L}\{f(s)\} = \int_0^\infty e^{-ps} f(s) ds, \tag{5–401}$$

these integrals reduce to products of the transforms of their integrands, regarded as functions of s. We therefore employ the relations

$$\mathcal{L}\left\{\frac{1}{\sqrt{s^2 + 2s}}\right\} = e^p K_0(p), \tag{5–402}$$

$$\mathcal{L}\left\{\sqrt{\frac{2 + s}{s}}\right\} = e^p[K_0(p) + K_1(p)] \tag{5–403}$$

(where K_0 and K_1 are Bessel functions of the second kind and imaginary argument) and transform Eqs. (5–399) and (5–400) into

$$\mathcal{L}\left\{\frac{\Gamma_0}{Ub}\right\} = -e^p[K_0(p) + K_1(p)]\mathcal{L}\left\{\frac{\gamma_w}{U}\right\}, \tag{5–404}$$

$$\mathcal{L}\left\{\frac{L_2}{\rho U^2 b}\right\} = e^p K_0(p)\mathcal{L}\left\{\frac{\gamma_w}{U}\right\}. \tag{5–405}$$

Eliminating the transform of the unknown wake circulation,

$$\mathcal{L}\left\{\frac{L_2}{\rho U^2 b}\right\} = -\frac{K_0(p)}{K_0(p) + K_1(p)} \mathcal{L}\left\{\frac{\Gamma_0}{Ub}\right\}. \tag{5-406}$$

Sears applies Eq. (5-406) to several special cases in Ref. 5-55. For instance, in Wagner's problem with α_0 replaced by $1/2\pi$ to give unit steady-state lift coefficients,

$$\frac{\Gamma_0}{Ub} = 1 \tag{5-407}$$

and

$$\mathcal{L}\{1\} = \frac{1}{p}. \tag{5-408}$$

Therefore, Eqs. (5-406) and (5-394) give for the resultant lift

$$\frac{L}{\rho U^2 b} = \frac{\Gamma_0}{Ub} + \frac{L_2}{\rho U^2 b} = 1 - \mathcal{L}^{-1}\left\{\frac{K_0(p)}{p[K_0(p) + K_1(p)]}\right\} = \phi(s). \tag{5-409}$$

There is no noncirculatory lift after $s = 0$. Although this transform cannot be inverted in closed form, it does provide a very efficient numerical scheme for calculating $\phi(s)$.

As a second example, we use Eq. (5-406) to reproduce our previous results for simple harmonic motion. We recall the fact that the response of any linear system to simple harmonic variation of an input quantity is the Laplace transform of its response to an equal-amplitude impulse of the same input, with p replaced by $\sqrt{-1}$ times the frequency of oscillation. Consider angle-of-attack motion $\dot{h}$. The quasi-steady circulatory lift is

$$L_0 = 2\pi \rho b U \dot{h} \tag{5-410}$$

and the noncirculatory part is

$$L_1 = \pi \rho b^2 \ddot{h}. \tag{5-411}$$

Hence,

$$\frac{\Gamma_0}{Ub} = \frac{L_0}{\rho U^2 b} = \frac{2\pi \dot{h}}{U}. \tag{5-412}$$

When this is multiplied by a *unit* impulse, the Laplace transform is

$$\mathcal{L}\left\{\frac{\Gamma_0}{Ub}\right\} = 2\pi \frac{\dot{h}}{U} \cdot 1. \tag{5-413}$$

Replacing p by ik in Eq. (5-406), and using the aforementioned principle for simple harmonic motion, we get

$$\frac{L_2}{\rho U^2 b} = \frac{-K_0(ik)}{K_0(ik) + K_1(ik)} 2\pi \frac{\dot{h}}{U}. \tag{5-414}$$

Therefore

$$L = L_1 + L_0 + L_2 = \pi\rho b^2 \ddot{h} + 2\pi\rho U b \dot{h} \left[1 - \frac{K_0(ik)}{K_0(ik) + K_1(ik)} \right]$$

$$= \pi\rho b^2 \ddot{h} + 2\pi\rho U b \dot{h} \left[\frac{K_1(ik)}{K_0(ik) + K_1(ik)} \right]. \tag{5-415}$$

However,

$$K_1(ik) = \frac{\pi}{2} H_1^{(2)}(k), \qquad K_0(ik) = i\frac{\pi}{2} H_0^{(2)}(k), \tag{5-416}$$

so that

$$L = \pi\rho b^2 \ddot{h} + 2\pi\rho U b \dot{h} \frac{H_1^{(2)}(k)}{[H_1^{(2)}(k) + i H_0^{(2)}(k)]}$$

$$= \pi\rho b^2 \ddot{h} + 2\pi\rho U b C(k) \dot{h}, \tag{5-417}$$

which confirms Eqs. (5–309) and (5–311). Similar checks on other types of airfoil oscillation can be carried through without difficulty.

CHAPTER 6

AERODYNAMIC TOOLS: COMPRESSIBLE FLOW

6–1 Introduction. Problems involving flight at high subsonic, transonic, and supersonic speeds are now encountered routinely by the practicing aeroelastician. He is fortunate that among the three areas of applied mechanics which his work encompasses—dynamics, elasticity, and aerodynamics—only the last is seriously affected by the transition to these speeds. At least this is true until the effects of heat transfer begin appreciably to alter the stresses and properties of structural materials. This fact makes it possible always to employ the same static-deformation relations and equations of motion, simply leaving certain brackets or aerodynamic matrices to be filled with coefficients appropriate to the Mach number range in which the aircraft is flying. A similar remark can be made about the representation of aeroelastic systems on analog computing machines, for this process is, of course, just another way of setting up the aforementioned equations. Aeroelastic wind-tunnel and flight testing may require a complete overhaul of techniques when speeds go up, however, as we shall see in chapters on those subjects.

We make these opening remarks because some aeroelasticians have tended to take rather lightly the problems of what to put in their aerodynamic matrices or how to design aerodynamic circuits for computers, and they have occasionally chafed when aerodynamicists were unable to furnish accurate, linear theories in simple mathematical forms. By no means is this always an easy matter when compressibility must be accounted for, as one might guess from the difficulties met in unsteady incompressible flow. Luckily, considerable progress has been made in some areas. Our purposes in this chapter and parts of the next are to delineate these areas and suggest a few lines of approach that may prove fruitful in the future.

The concept of small disturbances, expressed in Eqs. (5–31) through (5–33), underlies most compressible flow theory that is useful in aeroelasticity. In Section 5–1 we have shown its principal consequence to be the linearized partial differential equation (5–44), which can be rewritten

$$\nabla^2 \phi' - \frac{1}{a_\infty^2} \frac{\partial^2 \phi'}{\partial t^2} - \frac{2M}{a_\infty} \frac{\partial^2 \phi'}{\partial x \partial t} - M^2 \frac{\partial^2 \phi'}{\partial x^2} = 0, \qquad (6\text{–}1)$$

with

$$M = U/a_\infty, \qquad (6\text{–}2)$$

the Mach number of the free stream. Equation (6–1) is known to be a valid deduction from the small-disturbance hypothesis, except in certain

294

steady or nearly steady transonic flows over large aspect-ratio wings and airfoils. When flight occurs at hypersonic speed, that is,

$$M \gg 1, \tag{6-3}$$

it is no longer reasonable to regard u', v, w as small just because they are negligible compared with U. They can then be of the same order of magnitude as the speed of sound, and the linearity of the theory breaks down.

When Eq. (6–1) does hold, we do not expect the boundary conditions at the wing, which are purely geometrical, to be changed by compressibility. For the symmetrical problem of thickness distribution in steady flow, shown in Section 5–4 to be the only symmetrical case of major interest, Eqs. (5–76) and (5–77) are appropriate.

$$w = \frac{\partial \phi'}{\partial z} = U \frac{\partial z_t}{\partial x}; \qquad \text{for } z = 0^+, (x,y) \text{ on } R_a, \tag{5-76}$$

$$w = \frac{\partial \phi'}{\partial z} = -U \frac{\partial z_t}{\partial x}; \qquad \text{for } z = 0^-, (x,y) \text{ on } R_a. \tag{5-77}$$

For antisymmetrical flow over lifting wings, with angle of attack and camber, in unsteady motion, we modify Eq. (5–225) to

$$w = \frac{\partial \phi'}{\partial z} = \frac{\partial z_a}{\partial t} + U \frac{\partial z_a}{\partial x} = w_a(x,y,t); \qquad \text{for } z = 0, (x,y) \text{ on } R_a. \tag{6-4}$$

As we shall see later, Kutta's hypothesis is logical on physical grounds for compressible flow, but only when the trailing edge is of the type known as subsonic. Moreover, the condition of vanishing disturbances at infinity must be interpreted in accordance with the form of Eq. (6–1) that governs the problem.

We might imagine that the pressure-velocity relation, or Bernoulli's equation,

$$\frac{\partial \phi}{\partial t} + \frac{q^2}{2} + \int \frac{dp}{\rho} = F(t), \tag{5-13}$$

would change because of compressibility, but this is not so within the small-disturbance framework. In Section 5–1, it was seen that Eqs. (5–31) and (5–33) lead to

$$\frac{\partial \phi}{\partial t} + \frac{q^2}{2} \cong \frac{\partial \phi'}{\partial t} + \frac{U^2}{2} + Uu'. \tag{6-5}$$

Furthermore, we found $F(t) = U^2/2$ when there is uniform, parallel fluid motion far upstream from the wing, and this relation holds when the integral of $1/\rho(p)$ is taken between the free stream state $p = p_\infty$ and local conditions where the speed is q. However, because there exists an intimate, continuous relationship among pressure, density, velocity, and velocity potential, small perturbations in two of them imply the same for the others. We can therefore assume

$$(p - p_\infty) \ll p_\infty \qquad (6\text{--}6)$$

and deduce therefrom

$$\rho(p) = \rho_\infty + \left(\frac{d\rho}{dp}\right)_\infty [p - p_\infty] + \frac{1}{2!}\left(\frac{d^2\rho}{dp^2}\right)_\infty [p - p_\infty]^2 + \cdots, \qquad (6\text{--}7)$$

where all subsequent terms on the right are small compared with the first. It follows that we can develop an approximation for the integral in Eq. (5–13):

$$\int_{p_\infty}^{p} \frac{dp}{\rho(p)} = \int_{p_\infty}^{p} \frac{dp}{\rho_\infty\left\{1 + \left(\dfrac{d\rho}{dp}\right)_\infty \dfrac{[p - p_\infty]}{\rho_\infty} + \cdots\right\}}$$

$$= \frac{1}{\rho_\infty}\int_{p_\infty}^{p}\left\{1 - \frac{p - p_\infty}{\gamma p_\infty} - \cdots\right\} dp \cong \frac{p - p_\infty}{\rho_\infty}. \qquad (6\text{--}8)$$

Here we have made use of the isentropic pressure-density formula, which can be written

$$\left(\frac{dp}{d\rho}\right)_\infty = \frac{\gamma p_\infty}{\rho_\infty}. \qquad (6\text{--}9)$$

When Eqs. (6–8), (6–5), and (5–14) are substituted into Eq. (5–13), and the result is solved for the pressure, we get

$$p - p_\infty = -\rho_\infty\left[\frac{\partial \phi'}{\partial t} + Uu'\right] \qquad (6\text{--}10a)$$

or

$$C_p = \frac{p - p_\infty}{\frac{1}{2}\rho_\infty U^2} = -\frac{2}{U^2}\frac{\partial \phi'}{\partial t} - \frac{2}{U}\frac{\partial \phi'}{\partial x}. \qquad (6\text{--}10b)$$

Equation (6–10b) closely resembles Eq. (5–73), but the latter was restricted to incompressible flow.

6–2 Wings and airfoils in steady subsonic flow; the Prandtl-Glauert transformation. The differential equation describing any steady subsonic flow with small disturbances is found by eliminating two time-derivative terms from Eq. (6–1) and inserting the Mach number of the free stream:

$$(1 - M^2)\frac{\partial^2 \phi'}{\partial x^2} + \frac{\partial^2 \phi'}{\partial y^2} + \frac{\partial^2 \phi'}{\partial z^2} = 0. \qquad (6\text{--}11)$$

When we are dealing with a nearly plane wing or body, the boundary conditions are given by Eq. (5–76) or Eq. (6–4). For present purposes, it suffices to discuss just one of these types; we concentrate on the more important lifting case (6–4), dropping the unsteady term.

$$w = \frac{\partial \phi'}{\partial z} = U \frac{\partial z_a}{\partial x} = w_a(x,y); \quad \text{for } z = 0, \ (x,y) \text{ on } R_a. \quad (6\text{--}12)$$

One of the early fundamental discoveries in the linearized theory of compressible fluid motion was that for every problem governed by differential equation (6–11) there is an equivalent incompressible flow. This is because (6–11) differs from Laplace's equation only by a scale distortion in the x-direction. We introduce an auxiliary set of independent and dependent variables through the so-called Prandtl-Glauert transformation:

$$x_0 = \frac{x}{\sqrt{1 - M^2}}, \quad (6\text{--}13a)$$

$$y_0 = y, \quad (6\text{--}13b)$$

$$z_0 = z, \quad (6\text{--}13c)$$

$$\phi_0'(x_0,y_0,z_0) = \phi'(x,y,z). \quad (6\text{--}14)$$

Equation (6–14) means that the numerical value of the function ϕ_0' at point (x_0,y_0,z_0) equals that of the disturbance potential ϕ' in the original Cartesian coordinates at point (x,y,z), which is related to the corresponding transformed point through Eqs. (6–13). When Eqs. (6–13) and (6–14) are substituted into Eqs. (6–11) and (6–12), the following new mathematical problem emerges:

$$\frac{\partial^2 \phi_0'}{\partial x_0{}^2} + \frac{\partial^2 \phi_0'}{\partial y_0{}^2} + \frac{\partial^2 \phi_0'}{\partial z_0{}^2} = 0, \quad (6\text{--}15)$$

$$\frac{\partial \phi_0'}{\partial z_0} = w_{a0}(x_0,y_0); \quad \text{for } z_0 = 0, \ (x_0,y_0) \text{ on } R_{a0}. \quad (6\text{--}16)$$

Here w_{a0} is a given function, equal to w_a at corresponding points. R_{a0} is the region in the $x_0 y_0$-plane bounded by the line obtained when the xy-boundary of R_a is transformed. In transforming Eqs. (6–11) and (6–12), we have been careful to work only with the single dependent variable ϕ', from which all properties of the compressible flow can be calculated by differentiation. Most interpretational difficulties that arise in connection with this transformation seem to come from the attempt to transform more than one dependent variable at a time (such as ϕ' and w). In fact, ϕ' alone contains the complete solution.

Since (6–15) is just Laplace's equation, it is instructive to provide (6–15)–(6–16) with a physical interpretation. ϕ_0' can be regarded as the disturbance velocity potential for steady flow of incompressible liquid over a wing whose planform has the projection R_{a0} on the $x_0 y_0$-plane. The new wing has such a shape that it gives particles in contact with its upper and lower surfaces the vertical velocity distribution $w_{a0}(x_0,y_0)$. If we make the convenient assumption that both uniform streams have the same speed U, that is, the total velocity potential is

$$\phi_0 = \phi_0' + Ux_0, \tag{6-17}$$

we deduce from (6-12) and (6-16) that the two wings have the same slopes at corresponding points. Because the speed of sound in the original gas is so low as to make compressibility effects important, it is best to imagine that the two flows involve different fluids, such as air and water, or Freon and air. The two wing planforms obviously differ, Eq. (6-13a) showing that x_0-dimensions in the incompressible flow have been stretched by $1/\sqrt{1 - M^2}$. This increases both the tangent of the sweep angle and the streamwise chord at every station in these proportions. The wingspan is unchanged, so the enlarged plan area causes the aspect ratio to be reduced by $\sqrt{1 - M^2}$. Two pairs of equivalent planforms are illustrated in Fig. 6-1.

The significance of the discovery expressed in Eqs. (6-13)-(6-16) is that it brings to bear all the powerful techniques of potential theory onto subsonic flow problems. We can use any of the methods for airfoils and wings

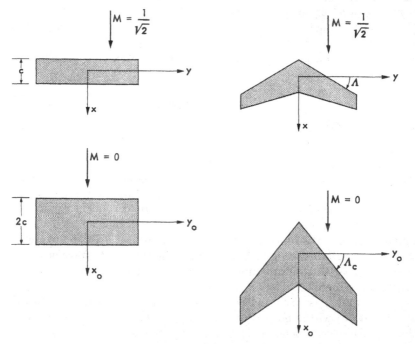

Fig. 6-1. Equivalent straight and sweptback planforms obtained by the Prandtl-Glauert transformation, Eq. (6-13).

described in Sections 5–4 and 5–5, or in the references given there, which happen to apply to the particular incompressible flow under study. In addition to Eq. (6–16), the subsidiary boundary conditions are the same. Since both differential equations (6–11) and (6–15) are of elliptic type (cf. Ref. 6–1, Chap. 7), all that is required at infinity is that ϕ_0' die out to zero everywhere except near the wake. For incompressible flow we have applied Kutta's hypothesis by making the pressure discontinuity through the wing drop to zero at the trailing edge. When the fluid is compressible but the main stream is subsonic, the same indeterminacy with regard to total circulation at any spanwise station is encountered, and we have experimental evidence to prove that the uniquely correct circulation is again given by the condition of smooth flow-off. Moreover, the trailing edge of region R_a transforms into that of R_{a0}.

Assuming we have solved Eqs. (6–15) and (6–16), subject to the additional conditions just mentioned, the last step in the transformation process is to determine the aerodynamic loading on the original wing. This we do by transforming the steady-flow form of Bernoulli's equation (6–10b):

$$C_p = \frac{p - p_\infty}{\frac{1}{2}\rho_\infty U^2} = -\frac{2}{U}\frac{\partial \phi'}{\partial x}. \tag{6–18}$$

The local pressure coefficient C_{p0} at any point (x_0, y_0) on the transformed wing is

$$C_{p0} = \frac{p_0 - p_\infty}{\frac{1}{2}\rho U^2} = -\frac{2}{U}\frac{\partial \phi_0'}{\partial x_0}. \tag{6–19}$$

Equations (6–13) and (6–14) yield

$$\frac{\partial \phi'}{\partial x} = \frac{1}{\sqrt{1 - M^2}}\frac{\partial \phi_0'}{\partial x_0}, \tag{6–20}$$

and we can eliminate the derivatives between Eqs. (6–18) and (6–19), thus obtaining

$$C_p = \frac{C_{p0}}{\sqrt{1 - M^2}}. \tag{6–21}$$

If we regard the free-stream densities as equal, we conclude that the disturbance pressure at any point on a wing in compressible fluid is increased over that at a corresponding point on the equivalent wing in incompressible fluid by the factor $1/\sqrt{1 - M^2}$.

Since they are dimensionless integrals of C_p over the same dimensionless surface, the coefficients of running or total lift and moment on the two wings are also related as in Eq. (6–21). However, the actual lifts and lifts per unit span, which are especially important because they are often calculated by means of lifting-line theory [Eq. (5–147)] or Weissinger theory

[Eq. (5–190)], are equal to one another. This we prove as follows. The running lift on the original wing is

$$L(y) = -\int_{x_l(y)}^{x_t(y)} [p_U - p_L]dx, \tag{6–22}$$

where x_l and x_t are the coordinates of the local leading and trailing edges. Similarly, the corresponding lift on the transformed wing is

$$L_0(y_0) = -\int_{x_{l0}(y_0)}^{x_{t0}(y_0)} [p_{0U} - p_{0L}]dx_0. \tag{6–23}$$

The dynamic pressures are equal in the two flows, so Eq. (6–21) implies

$$p_U - p_L = \frac{p_{0U} - p_{0L}}{\sqrt{1 - M^2}}. \tag{6–24}$$

If we substitute Eq. (6–24) into Eq. (6–23) and make the transformation (6–13a), we find

$$L_0(y_0) = -\int_{x_l(y)}^{x_t(y)} \sqrt{1 - M^2}\,[p_U - p_L]\frac{dx}{\sqrt{1 - M^2}} = L(y). \tag{6–25}$$

Therefore, the chordwise shrinking of the wing just compensates for the pressure rise in returning from the incompressible to the compressible flow. The total lifts are also the same, because the transformation (6–13b) leaves a spanwise integration unchanged.

Equation (6–25) enables us to solve for the spanwise loading, or the cc_l-distribution, on the distorted planform and transfer it without modification to the original one. This is, for example, the procedure recommended by DeYoung and Harper (Ref. 6–2) for the application of Weissinger's integral equation to straight and swept wings. Pitching moment, aerodynamic-center location, induced drag, and the like are most easily computed after the loading has been put back on the actual wing.

When conditions are three-dimensional, there is no way of avoiding the inconvenience of analyzing a new planform for each different Mach number M. However, the simple properties of two-dimensional flows enable us to derive relations between pressures on the same airfoil with and without compressibility. When we transform such an airfoil with (6–13a) and (6–13b), its chord is elongated by $1/\sqrt{1 - M^2}$, but we have seen that its chordwise slopes dz_a/dx come through unaltered. This means that the angle of attack, shape, and dimensionless camber are preserved in the equivalent incompressible flow. No span effect is present, however, and dimensionless properties of an airfoil in a particular orientation, such as lift coefficient, moment coefficient, and pressure-coefficient distribution, are not affected by a simple scale change like a $1/\sqrt{1 - M^2}$ increase of all dimensions.   Hence we can state that Eq. (6–21) is valid for identical airfoils.   It follows that

$$c_l = \frac{c_{l0}}{\sqrt{1 - M^2}},$$

(6–26)

$$c_m = \frac{c_{m0}}{\sqrt{1 - M^2}},$$

(6–27)

$$\frac{dc_l}{d\alpha} = \frac{1}{\sqrt{1 - M^2}} \left(\frac{dc_l}{d\alpha}\right)_0 = \frac{a_0}{\sqrt{1 - M^2}},$$

(6–28)

all relate properties of the same two-dimensional profile under compressible and incompressible flight conditions.   The aerodynamic center remains at the quarter-chord point, and the invariance of boundary conditions shows also that Eq. (6–21) holds for the pressures due to airfoil thickness. The compressibility effect on lift-curve slope, Eq. (6–28), is an especially well-known formula.   It has been quite satisfactorily verified by experiments on thin wings at low angles of attack up to within a few hundredths of their critical Mach numbers.

The result expressed by Eqs. (6–21) and (6–26)–(6–28) is called the Prandtl-Glauert compressibility correction.   It applies strictly to two-dimensional flows and its unrestricted use for finite wings can be misleading, especially when the aspect ratio is small.   The correction also has no theoretical validity with respect to drag, which is caused mainly by the viscous shear forces on thin airfoils.   In fact, the drag coefficient c_D tends to stay constant, independent of Mach number, until its sudden sharp rise heralds the appearance of shocks on the wing surface.   Other more accurate two-dimensional flow theories have been employed to derive nonlinear compressibility correction formulas.   Perhaps the best known and most accurate of these is the one due to von Kármán and Tsien (see Ref. 6–3, Chap. 11),

$$C_p = \frac{C_{p0}}{\sqrt{1 - M^2} + \dfrac{M^2 C_{p0}}{2\sqrt{1 - M^2} + 2}}.$$

(6–29)

Because C_{p0} appears in the denominator of Eq. (6–29), it cannot be integrated straightforwardly to find lift and moment corrections.   The various linear relations that can be derived by the Prandtl-Glauert transformation have been used almost exclusively in solving the steady, subsonic problems of aeroelasticity.

The boundary conditions must be transformed in a different way when one treats the flow over a thicker object, like a body of revolution.   They can no longer be satisfied on the plane $z = 0$ but must follow the body's contours.   One possibility then is to use Goethert's form of the transformation, which is well explained in Ref. 6–4.

We naturally expect all these corrections to become invalid at flight speeds above the critical Mach number of the wing or body, beyond which supersonic regions begin to appear locally on the surface and shocks form. Formulas like (6–28) are meaningless in the transonic range, where the denominator approaches zero. In two-dimensional flow, this phenomenon reflects the essential nonlinearity of the governing differential equation, present in spite of the small-disturbance hypothesis. For three-dimensional wings, however, an uncritical examination of transformation Eq. (6–13) suggests that, when $\sqrt{1 - M^2}$ is close to zero, any planform has a very low aspect-ratio incompressible equivalent. The theory described in Subsection 5–5(c) can be employed to calculate its loading. Heaslet, Lomax, and Spreiter (Ref. 6–5) have thoroughly discussed the limiting theory that is obtained at $M = 1$.

At the lower end of the Mach-number range, Eq. (6–21) explains why incompressible flow theory can be used successfully up to fairly high subsonic speeds. The factor $1/\sqrt{1 - M^2}$ approaches unity with zero slope as M goes to zero. Even at $M = 0.3$, its numerical value is 1.048, so that formulas based on constant density are in error by less than 5 percent.

We conclude this section by extending the semiempirical formulas given at the end of Section 5–5 into the subsonic compressible-flow range. Equation (5–166), for example, follows from lifting-line theory and is therefore accurate only for straight wings of large-aspect ratio. Since lift coefficients and lift-curve slopes of equivalent wings are related through Eq. (6–21), we must take the lift-curve slope of the lower-aspect ratio incompressible wing and increase it by $1/\sqrt{1 - M^2}$. This leads to

$$\frac{dC_L}{d\alpha} = \frac{a_0}{\sqrt{1 - M^2}} \frac{\sqrt{1 - M^2}\ \mathrm{AR}}{\left[\sqrt{1 - M^2}\ \mathrm{AR} + (a_0/\pi)\right]}$$

$$\cong a_0 \frac{\mathrm{AR}}{\left[\sqrt{1 - M^2}\ \mathrm{AR} + 2\right]}, \tag{6–30}$$

where a_0 is the two-dimensional, *incompressible* value. The same process, applied to the more accurate Eq. (5–218), produces

$$\frac{dC_L}{d\alpha} = \frac{a_0}{\sqrt{1 - M^2}} \frac{\mathrm{AR}\sqrt{1 - M^2}}{\left[\mathrm{AR}\sqrt{1 - M^2}\sqrt{1 + \left(\dfrac{a_0}{\pi \mathrm{AR}\sqrt{1 - M^2}}\right)^2} + \dfrac{a_0}{\pi}\right]}. \tag{6–31}$$

On a two-dimensional swept wing, we recall, all the aerodynamic loads are determined by the component of flow velocity normal to the leading edge. Since the same reasoning holds when the fluid is compressible, the Mach number governing compressibility effects is based on this normal component; it equals $M \cos \Lambda$, where M is the resultant flight Mach number. We can therefore say that the lift-curve slope per unit distance along the swept span is

$$\frac{d\bar{c}_l}{d\bar{\alpha}} = \frac{a_0}{\sqrt{1 - M^2 \cos^2 \Lambda}}, \qquad (6\text{-}32)$$

where a_0 is the two-dimensional slope of the profile taken normal to the span. Because of Eq. (5–222b), the two-dimensional lift for sections and angles of attack parallel to the flight direction is

$$\frac{dc_l}{d\alpha} = \frac{a_0 \cos \Lambda}{\sqrt{1 - M^2 \cos^2 \Lambda}}. \qquad (6\text{-}33)$$

Equations (6–32) and (6–33) illustrate the important property of large aspect-ratio swept wings that the onset of compressibility effects tends to be delayed until $M \cos \Lambda$ approaches critical values.

Finite swept wings have their incompressible equivalents just like straight ones. The effective sweep angle (Fig. 6–1) is given by

$$\tan \Lambda_e = \frac{\tan \Lambda}{\sqrt{1 - M^2}}, \qquad (6\text{-}34a)$$

$$\cos \Lambda_e = \frac{\sqrt{1 - M^2}}{\sqrt{1 - M^2 \cos^2 \Lambda}} \cos \Lambda. \qquad (6\text{-}34b)$$

Accordingly, the natural extension of Eq. (5–224) is Diederich's very general formula (Ref. 6–6)

$$\frac{dC_L}{d\alpha} = \frac{a_0 \cos \Lambda_e}{\sqrt{1 - M^2}} \frac{AR\sqrt{1 - M^2}}{AR\sqrt{1 - M^2}\sqrt{1 + \left(\frac{a_0 \cos \Lambda_e}{\pi AR\sqrt{1 - M^2}}\right)^2} + \frac{a_0 \cos \Lambda_e}{\pi}}. \qquad (6\text{-}35)$$

Measured data in Ref. 6–6 up to $M = 0.7$ and down to $AR = 1.5$ verify the accuracy of Eq. (6–35) remarkably well.

6–3 Airfoils and wings in steady supersonic flow. Along with all unsteady compressible flows, steady supersonic flow exhibits a whole series of new phenomena, distinct from those with which we are familiar for fluid moving steadily at lower speeds. In simple physical terms, the character of the motion is dominated by the fact that any disturbance takes a finite time to be propagated from one point to another, and no signals whatever are able to make forward progress against a general stream in excess of the speed of sound. We can prove that all flow perturbations due to a slender wing or body are confined behind certain rearward-leaning cones and wedges which emanate from its leading edge.

The linearized theory derived in Section 5–1 contains most of the qualitatively important elements actually observed in supersonic flight of aircraft, but there are a few exceptions. The most serious of these is that the assumption of constant entropy throughout the entire gas rules out all shocks of appreciable strength. Neglecting viscosity and heat conductivity also eliminates mechanical and thermal boundary layers. Of itself, this is no more serious an oversight than in subsonic flow, until heat transfer begins to assume primary importance for the designer beyond about $M = 3$. But viscosity also has a part in forming shocks; and boundary layer thickening or separation produced in regions where shocks come close to body surfaces may, even on very slender wings, extend the influence of viscosity into much larger volumes of flow than would be affected in subsonic flight.

Supersonic small-disturbance theory contains one curious internal inconsistency. Whereas the governing differential equation implies that various space and time derivatives of the velocity field must everywhere be relatively small, finite velocity discontinuities (infinite derivatives) are found to occur across certain surfaces in the flow solutions. We are forced to regard these, like pressure singularities in subsonic solutions, as part of the price paid for linearization. The only real justification comes from good agreement between theory and experiment, whenever such is obtained.

Before undertaking any mathematical treatment, we examine a few important physical facts deducible from the assumption of small disturbances. Section 5–1 proves that deviations of the speed of sound from its free-stream value a_∞ must be neglected. It follows from this and Eq. (5–33) that any signal emitted from a point (x,y,z), in a coordinate system fixed to a body in the flow, can be felt only inside or on the surface of a right circular cone whose axis points downstream from (x,y,z). As seen by an observer moving with the fluid, a particular pulse emitted at $t = 0$ expands on a spherical surface of instantaneous radius $a_\infty t$, but the center of this sphere goes downstream relative to our coordinate system with velocity U. As illustrated in Fig. 6–2, the envelope of positions of the sphere is a cone of semivertex angle

$$\mu = \sin^{-1} \frac{a_\infty t}{Ut} = \sin^{-1} \frac{1}{M} = \tan^{-1} \frac{1}{\sqrt{M^2 - 1}}, \qquad (6\text{–}36)$$

where μ is the Mach angle. Any train of disturbances from (x,y,z) lies within the cone, which, when the x-direction coincides with the free stream,

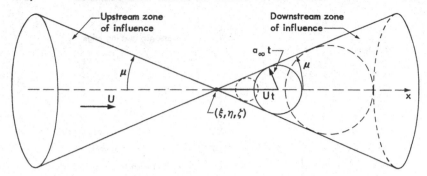

Fig. 6–2. Generation of the Mach cone by a disturbance starting from point (ξ, η, ζ) at $t = 0$. Also shown are the upstream and downstream zones of influence of (ξ, η, ζ).

has the equation

$$(\xi - x)^2 - (M^2 - 1)[(\eta - y)^2 + (\zeta - z)^2] = 0; \quad \text{for } x \leq \xi. \quad (6\text{–}37)$$

This is known as the Mach cone or downstream zone of influence of the point (x,y,z).

Such a point can itself be affected only by those sources of disturbance in whose downstream zone it falls. The locus of all these is evidently the Mach cone directed forward from (x,y,z). It is designated the forecone or upstream zone of influence of (x,y,z) and has the same Eq. (6–37), except that ξ is limited by $x \geq \xi$. Both influence zones are shown in Fig. 6–2.

When the body giving rise to disturbances in supersonic flow is a wing lying nearly on the xy-plane, its entire downstream zone of influence is bounded by the envelope of rearward Mach cones from its leading edge. If this edge is straight or the wing is a two-dimensional airfoil, for example, the cones coalesce into the so-called Mach wedge. The problem of calculating the fluid motion at any particular point due to an arbitrary supersonic wing is simplified because conditions at such a point are influenced only by the portion of the xy-plane intercepted by the forecone. For a general point (x,y,z), this area of influence lies within the hyperbola found by setting $\zeta = 0$ in Eq. (6–37) and limiting ξ to be less than x.

$$\xi = x - \sqrt{(M^2 - 1)[(y - \eta)^2 + z^2]} \quad (6\text{–}38a)$$

or

$$\eta = y \pm \sqrt{\frac{(x - \xi)^2}{M^2 - 1} - z^2}. \quad (6\text{–}38b)$$

Here ξ and η are running coordinates on the plane of the wing. For a point $(x,y,0^+)$ at the surface, Eq. (6–38a) reduces to the pair of straight lines

$$\xi - x = \pm \sqrt{M^2 - 1}\,(y - \eta); \quad \text{for } \xi \leq x. \quad (6\text{–}39)$$

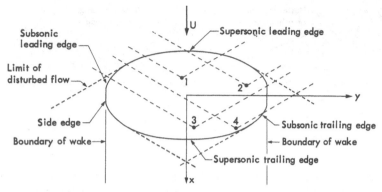

Fig. 6-3. Generalized supersonic planform, showing various points on the surface and their upstream regions of influence.

In Fig. 6-3 we illustrate a generalized supersonic planform and four points $(x,y,0^+)$ in different positions relative to the edges, together with their upstream boundaries of influence as given by Eq. (6-39). The various dashed lines in Fig. 6-3 are Mach lines in the plane of the wing; like the influence boundaries, they all make an angle μ with the x-axis. We see from this figure that the nature of the region which affects any point depends on both the location of the point and the kinds of leading and trailing edges the wing has. To begin with, any point ahead of the leading edge and ahead of the Mach lines from this edge is undisturbed, and we can set $\phi' = 0$ $(\phi = Ux)$ there; this is also true of any part of the entire flow in front of the aforementioned envelope of leading-edge Mach cones.

Point 1 in the figure is influenced only by points on its own side of the wing surface between it and the leading edge. The flow ahead of this edge does not contribute to ϕ'. That portion of the leading edge which is in the upstream zone of point 1 has the property that the component of the free-stream velocity normal to it exceeds a_∞; it is therefore called a supersonic leading edge. In regions where the upstream zones intercept only supersonic edges (e.g., the entire planform of a two-dimensional airfoil), the upper and lower surfaces of the wing are said to be independent. The flow over either the top or bottom can be calculated without reference to the shape of the opposite side.

Point 2 is affected by a disturbed flow region off the wing, because there occurs in its upstream zone a portion of subsonic leading edge, where the normal component of U is less than a_∞. The shape of the wing surface on the side opposite to point 2 affects conditions off the wing, and these, in turn, contribute to ϕ' at point 2. Hence the surfaces can no longer be regarded as independent.

Point 3 is influenced by two opposite sections of subsonic leading edge, with their corresponding off-the-wing disturbed areas. Point 4 receives signals from supersonic and subsonic leading edges, a side edge, and a sub-sonic trailing edge, from which we expect some sort of wake to be emanating. Obviously, no supersonic trailing edge could be contained in the upstream zone of influence of any point on a wing of practical importance.

Mathematical solution of the linearized problem grows progressively more difficult as we go from points like 1 to 2 to 3 to 4. Quite different techniques prove successful for different sorts of influence regions; this leads to the unfortunate situation that each new class of supersonic planforms usually calls for another method of approach. No such broadly applicable scheme as Weissinger's integral equation has been discovered, and none seems to be in prospect.

The possibility of two different kinds of trailing edge raises a question about Kutta's hypothesis in supersonic theory. No smooth flow-off condition is imposed when an edge is supersonic, because there is no mechanism by which the approaching fluid particles can be apprised of its imminence in time for the flow to adjust itself. In fact, the load distribution on a wing with all supersonic rearward boundaries is calculated just as if the planform continued downstream to infinity. A subsonic trailing edge can influence the flow over the wing in its neighborhood, however, and the custom has been to assume zero pressure discontinuity there. Mathematical solutions turn out to have the same sort of indeterminacy which is also removed by Kutta's hypothesis in subsonic flow, and whatever experimental evidence there is justifies using this condition. In view of the difficulties involved, it is fortunate that only a few sweptback wings encountered on actual supersonic aircraft have trailing edges that are subsonic.

As an illustration of a very simple solution, which is extensively employed in the strip-theory analysis of static aeroelastic properties of supersonic wings, we consider a two-dimensional airfoil with arbitrary camber and thickness distribution. Differential equation (6–1) is most appropriately written

$$(M^2 - 1)\frac{\partial^2 \phi'}{\partial x^2} - \frac{\partial^2 \phi'}{\partial z^2} = 0. \qquad (6\text{–}40)$$

If the chord is $2b$, we can follow Sections 5–1 and 5–4 in prescribing the upper and lower surface boundary conditions:

$$w = \frac{\partial \phi'}{\partial z} = U\frac{dz_\mathrm{U}}{dx}\ ; \qquad \text{for } z = 0^+, -b \le x \le b, \qquad (6\text{–}41\mathrm{a})$$

$$w = \frac{\partial \phi'}{\partial z} = U\frac{dz_\mathrm{L}}{dx}\ ; \qquad \text{for } z = 0^-, -b \le x \le b. \qquad (6\text{–}41\mathrm{b})$$

Equation (6–40) is recognized as the wave equation in two dimensions, and any textbook on partial differential equations proves that it is satisfied by

$$\phi' = f(x - z\sqrt{M^2 - 1}) + g(x + z\sqrt{M^2 - 1}), \qquad (6\text{-}42)$$

where f and g are arbitrary functions of the prescribed combinations of independent variables. The present problem can be handled by means of Eq. (6-42) (e.g., Ref. 6-3, Chap. 9), but a different method, which corresponds better with what we use elsewhere in this book, consists of replacing the airfoil by a sheet of supersonic sources.

The disturbance potential giving the effect at point (x,y,z) of a steady, three-dimensional supersonic source of strength H, centered at (ξ,η,ζ), is

$$\phi_s' = \frac{-H}{2\pi\sqrt{(x - \xi)^2 - (M^2 - 1)[(y - \eta)^2 + (z - \zeta)^2]}} . \qquad (6\text{-}43)$$

Equation (6-43) is a solution of

$$(M^2 - 1)\frac{\partial^2\phi'}{\partial x^2} - \frac{\partial^2\phi'}{\partial y^2} - \frac{\partial^2\phi'}{\partial z^2} = 0, \qquad (6\text{-}44)$$

of which (6-40) is the two-dimensional case. ϕ_s' must be assigned the value zero outside the downstream zone of influence of (ξ,η,ζ), whose boundary is given by the vanishing of the quantity under the radical in (6-43), with $x \geq \xi$. It is singular all over this boundary, because the signals sent out by a *concentrated* source reinforce one another there to an infinite degree. A sheet of supersonic sources over the $\xi\eta$-plane, whose strength per unit area, $H(\xi)$, is independent of the spanwise coordinate, has the potential

$$\phi'(x,z) = \frac{-1}{2\pi}\int_{-b}^{x - z\sqrt{M^2 - 1}}\int_{-\sqrt{\frac{(x - \xi)^2}{M^2 - 1} - z^2}}^{\sqrt{\frac{(x - \xi)^2}{M^2 - 1} - z^2}}\frac{H(\xi)d\eta d\xi}{\sqrt{(x - \xi)^2 - (M^2 - 1)[\eta^2 + z^2]}} .$$
$$(6\text{-}45)$$

Here y has been set equal to zero for convenience, since the two-dimensional flow is expected to be the same in any plane $y = $ const. The integration limits of ξ and η lie on hyperbola (6-38) generated by the intersection of the forecone from $(x,0,z)$ with the plane of the wing, behind the leading edge $\xi = -b$.

We concentrate on the upper surface of the airfoil, characterized by Eq. (6-41a), and restrict $z \geq 0$. The flow is uninfluenced by the shape of the lower surface and can be found independently thereof, which fact justifies our using elementary solutions symmetrical in z although we know the actual motion has no such symmetry. Equation (6-45) can be integrated once, since the unknown function H does not affect the η-integral:

$$\phi'(x,z) = -\frac{1}{2\pi} \int_{-b}^{x-z\sqrt{M^2-1}} \frac{H(\xi)}{\sqrt{M^2-1}}$$

$$\times \left\{ \sin^{-1} \left[\frac{\sqrt{M^2-1}\,\eta}{\sqrt{(x-\xi)^2 - (M^2-1)z^2}} \right] \Bigg|_{-\sqrt{\frac{(x-\xi)^2}{M^2-1}-z^2}}^{\sqrt{\frac{(x-\xi)^2}{M^2-1}-z^2}} \right\} d\xi$$

$$= -\frac{1}{2\sqrt{M^2-1}} \int_{-b}^{x-z\sqrt{M^2-1}} H(\xi)d\xi. \tag{6-46}$$

When we substitute Eq. (6–46) into the boundary condition (6–41a), using the formula for differentiating a definite integral with respect to a parameter, we find

$$\frac{\partial \phi'(x,0^+)}{\partial z} = \lim_{z\to 0^+} \left\{ \frac{-1}{2\sqrt{M^2-1}} \frac{\partial(x - z\sqrt{M^2-1})}{\partial z} H(x - z\sqrt{M^2-1}) \right\}$$

$$= \frac{H(x)}{2} = U\frac{dz_U}{dx}. \tag{6-47}$$

The local value of the strength of the source sheet is proportional to the vertical velocity there, and we can say the sources in such a sheet are point functions with respect to w.

When Eq. (6–47) is substituted into Eq. (6–46), we are led to a simple expression for ϕ':

$$\phi'(x,z) = \frac{-U}{\sqrt{M^2-1}} \int_{-b}^{x-z\sqrt{M^2-1}} \frac{dz_U}{d\xi} d\xi$$

$$= \frac{-U}{\sqrt{M^2-1}} [z_U(x - z\sqrt{M^2-1}) - z_U(-b)]. \tag{6-48}$$

The disturbance potential ϕ_U' just above the upper surface is

$$\phi_U'(x,0^+) = \frac{-U}{\sqrt{M^2-1}} [z_U(x) - z_U(-b)]. \tag{6-49}$$

This depends only on the difference in height between the leading edge and the point in question. Moreover, as we move upward away from the surface, ϕ' remains constant along the rearward-leaning Mach line

$$x - z\sqrt{M^2-1} = \text{const}, \tag{6-50}$$

its value at (x,z) being equal to that at the point where the Mach line through (x,z) strikes the upper surface. ϕ' in Eq. (6–48) has the form of the function f in Eq. (6–42), something we could have anticipated by observing that any solution for $z > 0$ which remained constant along $x + z\sqrt{M^2-1} = \text{const.}$ Mach lines would project forward into the forbidden region beyond the leading-edge Mach wedge, so that g must vanish.

Conditions are just reversed on the lower half of the airfoil. A repetition of the foregoing steps with $z < 0$ yields

$$\phi'(x,z) = \frac{U}{\sqrt{M^2 - 1}} \left[z_L(x + z\sqrt{M^2 - 1}) - z_L(-b) \right] \quad (6\text{-}51)$$

and

$$\phi_L'(x,0^-) = \frac{U}{\sqrt{M^2 - 1}} \left[z_L(x) - z_L(-b) \right]. \quad (6\text{-}52)$$

Equation (6-51) contains only the g type of function from Eq. (6-42), the solution being constant along rearward-leaning Mach lines projecting downward from the surface.

The pressure distribution and aerodynamic loads follow from Eq. (6-10), with the time derivative removed.

$$\frac{p_U - p_\infty}{(\frac{1}{2}\rho_\infty)U^2} = -\frac{2}{U}\frac{\partial \phi_U'}{\partial x} = \frac{2}{\sqrt{M^2 - 1}}\frac{dz_U}{dx}, \quad (6\text{-}53)$$

$$\frac{p_L - p_\infty}{(\frac{1}{2}\rho_\infty)U^2} = -\frac{2}{\sqrt{M^2 - 1}}\frac{dz_L}{dx}. \quad (6\text{-}54)$$

Equations (6-53) and (6-54) express a familiar fact about linearized, two-dimensional supersonic flow: the disturbance pressure is directly proportional to the angle through which a streamline is turned from the free-stream direction. Rotation in a sense to spread the streamlines causes expansion, whereas forcing them together brings about compression. As in Section 5-4, we can separate the airfoil contour into symmetrical and antisymmetrical parts. The antisymmetrical pressure distribution, responsible for lift and pitching moment, is determined by replacing both z_U and z_L with the mean line $z_a(x)$. If we make this substitution and subtract Eq. (6-54) from Eq. (6-53),

$$\frac{p_U - p_L}{(\frac{1}{2}\rho_\infty)U^2} = \frac{4}{\sqrt{M^2 - 1}}\frac{dz_a}{dx}. \quad (6\text{-}55)$$

Since the lift and pitching moment (about an axis at $x = ba$) per unit span are

$$L = -\int_{-b}^{b} [p_U - p_L]dx, \quad (5\text{-}106)$$

$$M_y = \int_{-b}^{b} [p_U - p_L][x - ba]dx, \quad (5\text{-}107)$$

the corresponding coefficients, based on Eq. (6-55), turn out to be

$$c_l = \frac{L}{(\frac{1}{2}\rho_\infty)U^2(2b)} = \frac{-2}{b\sqrt{M^2 - 1}} [z_a(b) - z_a(-b)], \quad (6\text{-}56)$$

$$c_m = \frac{M_y}{(\frac{1}{2}\rho_\infty)U^2(2b)^2} = \frac{1}{b\sqrt{M^2-1}} \int_{-b}^{b} \frac{dz_a}{dx} \left[\frac{x}{b} - a\right] dx$$

$$= \frac{1}{b\sqrt{M^2-1}} \left[z_a(b) + z_a(-b) - \int_{-b}^{b} \frac{z_a}{b} dx\right] + \frac{a}{2} c_l. \quad (6\text{-}57)$$

The problem of additional loading, associated with the flat plate at angle of attack α, is treated by inserting

$$z_a = -\alpha x \quad (6\text{-}58)$$

into Eqs. (6-56) and (6-57). Thus we obtain

$$c_l = \frac{4\alpha}{\sqrt{M^2-1}}, \quad (6\text{-}59)$$

$$c_m = \frac{2a\alpha}{\sqrt{M^2-1}} = \frac{a}{2} c_l, \quad (6\text{-}60)$$

and, for the lift-curve slope,

$$\frac{dc_l}{d\alpha} = \frac{4}{\sqrt{M^2-1}}. \quad (6\text{-}61)$$

Equation (6-60) shows the theoretical aerodynamic center of all supersonic airfoils to be at midchord $a = 0$. Nonlinear effects of thickness tend to move it forward, so that it is measured to be somewhere between 40% and 50% of the chord behind the leading edge on profiles in actual use. Figure 6-4 presents a plot of the lift-curve slope predicted by linearized theory [Eqs. (6-28) and (6-61)] throughout the Mach-number range for which these formulas are valid. At supersonic speeds tests like those published by Vincenti (Ref. 6-7) tend. to verify Eq. (6-61) rather closely, except when M is too near unity.

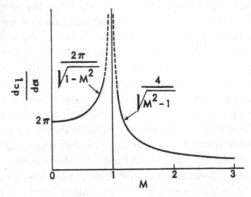

Fig. 6-4. Lift-curve slope of a two-dimensional airfoil, as predicted by small-disturbance theory at subsonic speeds.

It is well known that small-disturbance theory predicts a drag in two-dimensional supersonic flow. Although it is not very important in aero-elastic applications, we can readily calculate it by summing all downstream components of the pressures given by Eqs. (6–53) and (6–54). The constant p_∞ exerts no resultant force, so we write

$$D = \int_{-b}^{b} [p_U - p_\infty] \frac{dz_U}{dx} dx - \int_{-b}^{b} [p_L - p_\infty] \frac{dz_L}{dx} dx$$

$$= \frac{\rho_\infty U^2}{\sqrt{M^2 - 1}} \int_{-b}^{b} \left[\left(\frac{dz_U}{dx} \right)^2 + \left(\frac{dz_L}{dx} \right)^2 \right] dx, \qquad (6\text{–}62)$$

where D is obviously always positive. Both the mean-line shape and the thickness contribute to drag. Despite the squared terms, their contributions can be separated, because

$$z_U = z_a + z_t, \qquad z_L = z_a - z_t. \qquad (5\text{–}75)$$

This substitution in Eq. (6–62) causes a cancellation of cross-product terms, leaving

$$D = \frac{2\rho_\infty U^2}{\sqrt{M^2 - 1}} \int_{-b}^{b} \left[\left(\frac{dz_a}{dx} \right)^2 + \left(\frac{dz_t}{dx} \right)^2 \right] dx. \qquad (6\text{–}63)$$

The physical explanation for this particular type of supersonic drag is that the disturbance potentials (6–49) and (6–51) do not die out as we proceed to infinity between the Mach lines from the leading and trailing edges. There takes place a continual outward propagation of mechanical energy of fluid motion at infinity, which, incidentally, cannot be dissipated there because viscous friction and entropy changes are ruled out of the theory. The work needed to produce this energy is done by the forward force D, per unit spanwise distance, exerted by the wing as it moves through the gas at rest. Since it has its explanation in a wavelike propagation of energy, this type of fluid resistance is called wave drag.

The drag of finite lifting surfaces is substantially higher than would be predicted by Eq. (6–63) because of two additional phenomena. The first of these, which is encompassed by linearized theory, is the wake of vortices with running circulation proportional to the spanwise lift variation which trails to infinity behind the wing, steadily withdrawing mechanical energy from the system. The second includes direct shear forces and distortion of pressure distribution due to viscosity, whose influence is felt at high speeds both through shocks and the boundary layer.

All the foregoing formulas for supersonic airfoils are associated with the name of Ackeret, who pioneered the application of the small-disturbance idea in this Mach-number range (Ref. 6–8).

The theory of three-dimensional flow over thin supersonic wings has progressed along several fairly distinct lines. Those in most widespread use are source methods (Refs. 6–9 and 6–10), conical-flow theory (e.g., Lagerstrom, Ref. 6–11), and the superposition of other elementary solutions of the steady-flow wave equation, such as doublets and vortices (Refs. 6–12, 6–13, 6–14, and 6–15). Green's theorem and the concept of the finite part of an integral have done much to unify superposition techniques and broaden their applicability.

We illustrate here only the simplest of these methods, which is essentially the representation of flow due to independent upper and lower wing surfaces by three-dimensional source sheets. The differential equation to be solved is Eq. (6–44), while the linearized boundary conditions are just those given in Eqs. (6–41), with allowance made for the dependence of z_U and z_L on the spanwise coordinate y.

$$w = \frac{\partial \phi'}{\partial z} = U \frac{\partial z_U}{\partial x} ; \qquad \text{for } z = 0^+, \ (x,y) \text{ on } R_a, \qquad \text{(6–64a)}$$

$$w = \frac{\partial \phi'}{\partial z} = U \frac{\partial z_L}{\partial x} ; \qquad \text{for } z = 0^-, \ (x,y) \text{ on } R_a. \qquad \text{(6–64b)}$$

As in previous work, R_a denotes the projection of the planform onto the xy-plane. Concentrating on the upper half of the wing, we make the restriction $z \geq 0$. We start from a source distribution like that in Eq. (6–45), except that we must now allow for arbitrary values of y, and the strength per unit area H varies both spanwise and chordwise:

$$\phi'(x,y,z) = -\frac{1}{2\pi} \iint_{R_a'} \frac{H(\xi,\eta)\,d\xi d\eta}{\sqrt{(x-\xi)^2 - (M^2-1)[(y-\eta)^2 + z^2]}} . \qquad \text{(6–65)}$$

The limits on these integrals are the boundaries of R_a', which is the intersection of the forecone from (x,y,z) with the planform R_a (Fig. 6–5). The forward extent of this area is unimportant, however, since the sources turn out to have zero strength in any undisturbed region ahead of the leading edge. Hence we can use $-C$ (C a large enough positive number) and $[x - z\sqrt{M^2 - 1}]$ for the outer ξ-integration, and insert the two values of

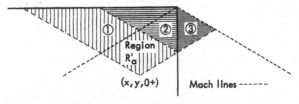

Fig. 6–5.   Rectangular wing tip in supersonic flow, showing the integration areas used in Evvard's theory.

η given in Eq. (6–38b) on the inner η-integration:

$$\phi'(x,y,z) = -\frac{1}{2\pi}$$

$$\times \int_{-c}^{x-z\sqrt{M^2-1}} \int_{y-\sqrt{\frac{(x-\xi)^2}{M^2-1}-z^2}}^{y+\sqrt{\frac{(x-\xi)^2}{M^2-1}-z^2}} \frac{H(\xi,\eta)d\eta d\xi}{\sqrt{(x-\xi)^2-(M^2-1)[(y-\eta)^2+z^2]}}. \quad (6\text{–}66)$$

Many schemes have been suggested for substituting Eq. (6–66) into boundary condition (6–64), the first published in the United States being Puckett's (Ref. 6–9). We employ one here which closely parallels the corresponding step for two-dimensional, unsteady supersonic flow. We observe that the shorthand notation

$$Y_1 = y - \sqrt{\frac{(x-\xi)^2}{M^2-1} - z^2}, \qquad Y_2 = y + \sqrt{\frac{(x-\xi)^2}{M^2-1} - z^2} \quad (6\text{–}67)$$

enables us to write Eq. (6–66) as

$$\phi'(x,y,z) = -\frac{1}{2\pi\sqrt{M^2-1}} \int_{-c}^{x-z\sqrt{M^2-1}} \int_{Y_1}^{Y_2} \frac{H(\xi,\eta)d\eta d\xi}{\sqrt{(\eta-Y_1)(Y_2-\eta)}}. \quad (6\text{–}68)$$

Explicit factorization of the denominator radical is possible, because we know it vanishes at both limits of integration. The variable η is replaced by an especially convenient dummy integration variable θ, defined by

$$\eta = \frac{Y_2+Y_1}{2} - \frac{Y_2-Y_1}{2}\cos\theta = y - \sqrt{\frac{(x-\xi)^2}{M^2-1} - z^2}\cos\theta \quad (6\text{–}69a)$$

or

$$\theta = \cos^{-1}\left[\frac{Y_2+Y_1-2\eta}{Y_2-Y_1}\right]. \quad (6\text{–}69b)$$

It is easy to show that the limits Y_1 and Y_2 in Eq. (6–68) correspond, respectively, to 0 and π in θ. Moreover,

$$\frac{d\eta}{\sqrt{(\eta-Y_1)(Y_2-\eta)}} = \frac{\dfrac{Y_2-Y_1}{2}\sin\theta d\theta}{\sqrt{\left(\dfrac{Y_2-Y_1}{2}\right)^2(1-\cos\theta)(1+\cos\theta)}} = d\theta. \quad (6\text{–}70)$$

Hence Eq. (6–68) reads

$$\phi'(x,y,z)$$

$$= -\frac{1}{2\pi\sqrt{M^2-1}} \int_{-c}^{x-z\sqrt{M^2-1}} \int_0^\pi H\left(\xi, y-\sqrt{\frac{(x-\xi)^2}{M^2-1}-z^2}\cos\theta\right)d\theta d\xi. \quad (6\text{–}71)$$

To apply the boundary condition, we need the derivative with respect to z, which is obtained by the rule for differentiating a definite integral with respect to a parameter. In the term containing the derivative of the upper limit, the dependence of H on θ disappears, as follows:

$$\frac{\partial \phi'}{\partial z} = -\frac{\dfrac{\partial (x - z\sqrt{M^2 - 1})}{\partial z}}{2\pi\sqrt{M^2 - 1}}$$

$$\times \int_0^\pi H\left[(x - z\sqrt{M^2 - 1}),\ y - \sqrt{\frac{(z\sqrt{M^2 - 1})^2}{M^2 - 1} - z^2}\cos\theta \right] d\theta$$

$$- \frac{1}{2\pi\sqrt{M^2 - 1}} \int_{-C}^{x - z\sqrt{M^2 - 1}} \int_0^\pi \frac{\partial H\left(\xi, y - \sqrt{\dfrac{(x - \xi)^2}{M^2 - 1} - z^2}\cos\theta \right)}{\partial z}\, d\theta\, d\xi$$

$$= \frac{1}{2\pi} H(x - z\sqrt{M^2 - 1},\ y) \int_0^\pi d\theta$$

$$- \frac{1}{2\pi\sqrt{M^2 - 1}} \int_{-C}^{x - z\sqrt{M^2 - 1}} \int_0^\pi \frac{\partial H}{\partial\left(y - \sqrt{\dfrac{(x - \xi)^2}{M^2 - 1} - z^2}\cos\theta \right)}$$

$$\times \frac{\partial\left(y - \sqrt{\dfrac{(x - \xi)^2}{M^2 - 1} - z^2}\cos\theta \right)}{\partial z}\, d\theta\, d\xi = \frac{H(x - z\sqrt{M^2 - 1},\ y)}{2}.$$

$$+ \frac{z}{2\pi} \int_{-C}^{x - z\sqrt{M^2 - 1}} \int_0^\pi \frac{\dfrac{\partial H}{\partial\left(y - \sqrt{\dfrac{(x - \xi)^2}{M^2 - 1} - z^2}\cos\theta \right)}\cos\theta}{\sqrt{(x - \xi)^2 - z^2(M^2 - 1)}}\, d\theta\, d\xi. \quad (6\text{–}72)$$

We wish to take the limit as z approaches 0^+, in order to apply Eq. (6–64a). The first term of the final expression in Eq. (6–72) clearly becomes one-half of $H(x,y)$. The factor z outside causes the second term to vanish, because H and its derivatives are continuous functions, so that the integral multiplied by z is no greater than the largest value of the derivative of H times, say, the quantity

$$\int_{-C}^{x - z\sqrt{M^2 - 1}} \frac{d\xi}{\sqrt{(x - \xi)^2 - z^2(M^2 - 1)}} \int_0^\pi |\cos\theta|\, d\theta = 2\cosh^{-1}\left[\frac{x + C}{z\sqrt{M^2 - 1}} \right].$$

$$(6\text{–}73)$$

As z goes to zero, the product $z \cosh^{-1}(1/z)$ also vanishes. Therefore, Eq. (6–72) and boundary condition (6–64a) yield

$$\frac{\partial \phi'(x,y,0^+)}{\partial z} = \frac{H(x,y)}{2} = U \frac{\partial z_U}{\partial x}. \tag{6–74}$$

We discover that the source sheet remains a point function with respect to the normal velocity component, even when its strength varies in both coordinate directions. This result can, of course, be proved in many different ways (Ref. 6–10). The final solution for the upper wing surface is found by substituting Eq. (6–74) into Eq. (6–65):

$$\phi'(x,y,z) = -\frac{U}{\pi} \iint_{R_a'} \frac{[\partial z_U(\xi,\eta)/\partial x]\,d\xi\,d\eta}{\sqrt{(x-\xi)^2 - (M^2-1)[(y-\eta)^2 + z^2]}} \tag{6–75}$$

for $z \geq 0^+$.

A corresponding analysis of the lower wing surface leads to

$$\phi'(x,y,z) = \frac{U}{\pi} \iint_{R_a'} \frac{[\partial z_L(\xi,\eta)/\partial x]\,d\xi\,d\eta}{\sqrt{(x-\xi)^2 - (M^2-1)[(y-\eta)^2 + z^2]}} \tag{6–76}$$

for $z \leq 0^-$.

Equations (6–75) and (6–76) bear a close resemblance to Eq. (5–85) for incompressible flow. The analogy is perfect when we are dealing with the problem of symmetrical thickness distribution; then z_U and z_L are replaced with $\pm z_t$, and the same source sheet is suitable for representing the entire flow:

$$\phi'(x,y,z) = -\frac{U}{\pi} \iint_{R_a'} \frac{[\partial z_t(\xi,\eta)/\partial x]\,d\xi\,d\eta}{\sqrt{(x-\xi)^2 - (M^2-1)[(y-\eta)^2 + z^2]}}. \tag{6–77}$$

The antisymmetrical lifting problem is solved by

$$\phi'(x,y,z) = -\frac{U}{\pi} \iint_{R_a'} \frac{[\partial z_a(\xi,\eta)/\partial x]\,d\xi\,d\eta}{\sqrt{(x-\xi)^2 - (M^2-1)[(y-\eta)^2 + z^2]}} \tag{6–78}$$

for $z \geq 0^+$. In calculating aerodynamic loads, we employ symmetry considerations to show that the pressure difference through the wing equals twice the disturbance pressure on the upper surface, which we get, in turn, by substituting Eq. (6–78) into Eq. (6–10).

Implicit in all that we have done so far on supersonic wings are the assumptions that the upper and lower surfaces are independent of each other and that all slopes $\partial z_U/\partial x$, etc. in Eqs. (6–75)–(6–78) are known over the whole area of integration. Strictly speaking, these are true only for planforms whose leading and trailing edges are entirely supersonic, like the triangular (delta) wing of large aspect ratio. Since such wings are in the minority, it is fortunate that certain devices exist to broaden the applicability of Eqs. (6–75)–(6–78) enormously. For example, Eq. (6–77) is valid for any nonlifting planform whatever. This is because, by symmetry, w is an odd function of z throughout the flow, but it must be continuous through the plane $z = 0$ and vanishes there everywhere except on the portion of the plane occupied by the wing. We can imagine the disturbed flow regions off subsonic leading and trailing edges (cf. Fig. 6–3) to be replaced by a sheet of sources which produces the same flow pattern, and Eq. (6–74) proves that such a sheet would have zero strength except on R_a. Accordingly, Eq. (6–77) may be used to estimate pressure distribution and wave drag due to thickness in any case of interest, as was first pointed out by Puckett (Ref. 6–9).

Lifting wings cause substantially more difficulty. However, Eq. (6–78) can be used without modification to find the loading at any point like 1 in Fig. 6–3, for which $\partial z_a/\partial x$ is known over the whole integration area $R_a{}'$. On many planforms this leaves only certain tip and trailing-edge regions to be handled by more powerful techniques. In Ref. 6–10 and elsewhere, Evvard discovered and exploited a theorem which does much to enhance the usefulness of the source method. By representing the disturbed area off a subsonic leading or side edge with an artificial, equivalent source sheet, he proved for the lifting problem that the contribution to $\phi'(x,y,0^+)$ from the portion of $R_a{}'$ beyond the edge is just cancelled by the contribution from a region forward of the other Mach line passing through the intersection of that edge with the boundary of $R_a{}'$. On the rectangular wing tip of Fig. 6–5, for instance, areas ② and ③ nullify each other, so that Eq. (6–78) can be integrated over area ① only, where the wing slope is known.

Evvard's theorem is readily extended to cover points like 3 in Fig. 6–3 and even regions influenced by trailing edges. His results are especially valuable in aeroelastic applications, where we often wish to find the load distribution due to an arbitrary elastic deformation of the wing. The problem is reduced to one of integrating Eq. (6–78) with some relatively easily integrated functions, like power series, approximating the slopes of the deflected structure. Although such calculations are by no means always simple, the general method of approach is much more fruitful than conical-flow theory, for example, which is essentially limited to cases of constant angle of attack.

The results of steady-flow supersonic wing theories have been published at length in many places, so we give no further examples here. One useful collection of lift, moment, and drag formulas for a variety of practically interesting planforms has been assembled by Lapin (Ref. 6–16).

6–4 Oscillating airfoils in subsonic flow. In the study of oscillating airfoils by small-disturbance theory, the airspeed range which has presented the greatest analytical obstacles is the higher subsonic. Two general methods of attack on this problem have been pursued. The first involves solution of an integral equation for the intensity distribution of acceleration potential (pressure function) doublets; it was proposed by Possio (Ref.

6–17).   Dietze (Ref. 6–18) and Fettis. (Ref. 6–19), among others, have refined Possio's techniques and furnished extensive computations.

The second method consists of transforming the differential equation for the velocity or acceleration potential into elliptical coordinates and expanding the solution as an infinite series of Mathieu functions. Investigations along this line were published by Reissner and Sherman (Ref. 6–20), Haskind (Ref. 6–21), and Reissner (Ref. 6–22, among others). Use of the acceleration potential in this manner was first suggested in a doctoral thesis by Timman.   Applications and extensions of his work appear in Refs. 6–23, 6–24, and 6–25. The Mathieu function procedure gives a mathematically exact solution of the problem and may ultimately lead to more convenient calculations than Possio's. The most accurate existing tables are based on the integral equation, however.

In this section we devote principal attention to Possio's theory, since it does not demand of the reader an extensive knowledge of the properties of the relatively unfamiliar Mathieu functions. A brief discussion of the acceleration potential ψ is given at the end of Section 5–1, where we show how, in incompressible flow, it is proportional to the disturbance pressure. Equation (6–8) demonstrates that, because of the assumption of small disturbances, the relationship is unaltered when we allow for compressibility. That is,

$$p - p_\infty = -\rho_\infty \psi. \tag{6–79}$$

The equation between ψ and the disturbance velocity potential is also unchanged. It reads

$$\psi = \frac{\partial \phi'}{\partial t} + U \frac{\partial \phi'}{\partial x}. \tag{5–50b}$$

Since the operator on the right is the linearized substantial derivative, Eq. (5–50b) can be interpreted to mean that ψ is the substantial derivative of ϕ', just as the acceleration, whose potential ψ is, represents the substantial derivative of the velocity.

In view of the linear expression (5–50b) by which ψ is obtained from ϕ', we can take the substantial derivative of Eq. (5–44), interchange the orders of the two sets of linear operators, and thus prove that ψ is governed by exactly the same partial differential equation as ϕ'. Therefore, for any two-dimensional problem,

$$\frac{\partial^2 \psi}{\partial x^2} + \frac{\partial^2 \psi}{\partial z^2} - \left[\frac{1}{a_\infty^2} \frac{\partial^2 \psi}{\partial t^2} + \frac{2M}{a_\infty} \frac{\partial^2 \psi}{\partial x \partial t} + M^2 \frac{\partial^2 \psi}{\partial x^2} \right] = 0. \tag{6–80}$$

To apply boundary conditions, we must know how ψ is related to the velocity components. For instance, we can start from

$$Dw = \frac{\partial \psi}{\partial z} dt \tag{6–81}$$

and integrate along the path of an individual fluid particle. Except for nonlinear terms, such a trajectory can be approximated by a line $z = $ const, running from some remote starting point $(-\infty, z)$ up to the spot (x,z) occupied by the particle at time t. If τ is a dummy integration variable representing all instants of time between $-\infty$ and t, we observe that at any τ the particle was located at $(x - U(t - \tau), z)$, so that

$$w(x,z,t) = \int_{\tau=-\infty}^{\tau=t} Dw = \int_{-\infty}^{t} \frac{\partial \psi(x - U(t - \tau), z, \tau)}{\partial z} \, d\tau. \quad (6\text{-}82)$$

Changing variable in the integral from τ to $\xi' = x - U(t - \tau)$,

$$w(x,z,t) = \int_{-\infty}^{x} \frac{\partial \psi \left(\xi', z, t - \dfrac{x - \xi'}{U} \right)}{\partial z} \frac{d\xi'}{U}. \quad (6\text{-}83)$$

As in previous work, we place the airfoil projection between $x = -b$ and $x = b$. The primary boundary condition of the lifting problem, with arbitrary two-dimensional motion, is

$$w = \frac{\partial z_a}{\partial t} + U \frac{\partial z_a}{\partial x} = w_a(x,t); \quad \text{for } z = 0, \ -b \leq x \leq b. \quad (5\text{-}225)$$

Specialized to simple harmonic motion, when all quantities contain the time-dependence factor $e^{i\omega t}$, Eq. (5-225) reads

$$\bar{w}(x,0) = i\omega \bar{z}_a(x) + U \frac{d\bar{z}_a}{dx} = \bar{w}_a(x); \quad \text{for } z = 0, \ -b \leq x \leq b, \quad (6\text{-}84)$$

where $\bar{w}_a(x)$ may be regarded as a known function, since it is shorthand for the right side.

We list the auxiliary boundary conditions as follows:

(1) There is zero pressure discontinuity across the x-axis, both ahead of the airfoil leading edge $(x < -b)$ and in the wake $(x > b)$. In view of the antisymmetry of the entire pressure field and Eq. (6-79), this means no discontinuity of ψ across $z = 0$ for $|x| > b$, so that $\psi = 0$ there.

(2) All signals set up in the gas must propagate outward from the airfoil toward infinity, without reflection. The presence of time derivatives in Eq. (6-80) makes it hyperbolic, so the simple condition of vanishing disturbances at infinity is not sufficient here, as it was in the case of the elliptic Eq. (6-11).

(3) The flow still being subsonic, Kutta's hypothesis must be fulfilled at the trailing edge. This is assured straightforwardly by specifying that ψ can have no discontinuity at $x = b$.

We can derive Possio's integral equation by representing the airfoil with a sheet of acceleration potential doublets along the projection of the airfoil.

The doublet is obtained from a simple solution of Eq. (6–80) known as a source pulse:

$$\psi_{sp} = \frac{A(\xi,\zeta,T)}{\sqrt{a_\infty^2(t-T)^2 - [(x-\xi) - U(t-T)]^2 - (z-\zeta)^2}}, \quad (6\text{–}85)$$

where ψ_{sp} may be interpreted as a two-dimensional pressure disturbance, which originates at the point (ξ,ζ) at the time T and propagates outward in a cylindrical fluid region, while being carried downstream at velocity U by the main flow. The right side of Eq. (6–85) is taken to be zero whenever $(t-T)$ is so small that the expression under the radical is negative. The positive value of $(t-T)$ which just causes the expression to vanish is the interval of time it takes the disturbance to get from (ξ,ζ) to (x,z) by the shortest path. Figure 6–6 shows the expanding cylindrical envelope of the disturbance at just the instant when it first hits (x,z). From the geometry of this figure, it is easily shown that $\{a_\infty^2(t-T)^2 - [(x-\xi) - U(t-T)]^2 - (z-\zeta)^2\}$ equals zero. Since the source is actually distributed along a line extending to infinity in both y-directions, the signal originating at a particular moment T continues to be felt at (x,z) forever after the first instant of arrival.

If the source point (ξ,ζ) has been pulsing sinusoidally for a long time, its effect is found by assuming simple harmonic dependence of A on T:

$$A(\xi,\zeta,T) = -\frac{\bar{A}(\xi,\zeta)U^2 a_\infty e^{i\omega T}dT}{2\pi} \quad (6\text{–}86)$$

and integrating over T:

$$\psi_s(x,z,t)$$
$$= -\frac{\bar{A}(\xi,\zeta)U^2 a_\infty}{2\pi}\int_{-\infty}^{g} \frac{e^{i\omega T}dT}{\sqrt{a_\infty^2(t-T)^2 - [(x-\xi) - U(t-T)]^2 - (z-\zeta)^2}}, \quad (6\text{–}87)$$

where $\bar{A}$ is a complex constant associated with the center (ξ,ζ). In Eq. (6–87), g denotes the most recent value of T for which a disturbance

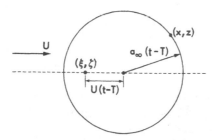

Fig. 6–6. The disturbance produced by a subsonic source pulse, centered at (ξ,ζ), shown when it first reaches the point (x,z).

originating at (ξ, ζ) can be felt at (x,y) at time t. As discussed above, it is given by the positive root of the radical:

$$g = t - \frac{\sqrt{(x - \xi)^2 + (z - \zeta)^2(1 - M^2)} - M(x - \xi)}{a_\infty(1 - M^2)}. \qquad (6\text{--}88)$$

If we substitute into Eq. (6–87) the new variables

$$w' = \frac{kM}{1 - M^2}\sqrt{\left(\frac{x - \xi}{b}\right)^2 + (1 - M^2)\left(\frac{z - \zeta}{b}\right)^2}, \qquad (6\text{--}89)$$

$$\lambda = \frac{1}{w'}\left[\omega(t - T) + \frac{kM^2}{1 - M^2}\left(\frac{x - \xi}{b}\right)\right], \qquad (6\text{--}90)$$

where k is the reduced frequency $\omega b/U$, the denominator of the integral simplifies considerably, leading to

$$\psi_s(x,z,t) = \frac{-\bar{A}(\xi,\zeta)U^2 e^{i\left[\omega t + \frac{kM^2(x-\xi)}{(1-M^2)b}\right]}}{2\pi\sqrt{1 - M^2}}\int_1^\infty \frac{e^{-iw'\lambda}}{\sqrt{\lambda^2 - 1}}\,d\lambda. \qquad (6\text{--}91)$$

The integral in Eq. (6–91) is the same as one involved in the definition of Theodorsen's $C(k)$ function (Section 5–6). It is equal to $-\pi i/2$ times the Hankel function of the second kind and zero order, so that

$$\psi_s(x,z,t) = \frac{\bar{A}(\xi,\zeta)U^2 e^{i\left[\omega t + \frac{kM^2(x-\xi)}{(1-M^2)b}\right]}}{4\sqrt{1 - M^2}}\,iH_0^{(2)}(w'). \qquad (6\text{--}92)$$

A sinusoidally pulsating doublet, with its axis vertical, is generated from Eq. (6–92) by placing a positive source ψ_s just above an equal negative source $-\psi_s$ and making the separation between them approach zero while increasing their strengths in inverse proportion to that separation. This operation is mathematically equivalent to taking the derivative with respect to the variable ζ, which represents the vertical location of the source. Since we want doublets on the x-axis, we simultaneously perform the differentiation and set ζ equal to zero:

$$\psi_D(x,z,t) = \left(\frac{\partial\psi_s}{\partial\zeta}\right)_{\zeta=0}$$

$$= \frac{\bar{A}(\xi,0)U^2 e^{i\left[\omega t + \frac{kM^2(x-\xi)}{(1-M^2)b}\right]}}{4\sqrt{1 - M^2}}\,i\,\frac{dH_0^{(2)}}{dw'}\,\frac{\partial w'}{\partial\zeta}$$

$$= \frac{\bar{A}(\xi)\omega^2 M^2 z}{4(1 - M^2)^{\frac{3}{2}}}\,i\,\frac{H_1^{(2)}(w')}{w'}\,e^{i\left[\omega t + \frac{kM^2(x-\xi)}{(1-M^2)b}\right]}. \qquad (6\text{--}93)$$

It can be proved that any doublet in a horizontal sheet obtained by integrating Eq. (6–92) along the x-axis is a point function with respect to the value of ψ itself. That is, the local disturbance pressure and doublet strength are directly proportional to each other. The factor z in Eq. (6–93) makes all the doublets odd functions of z, so the sheet produces a pressure discontinuity. Hence we need only put the sheet along the airfoil chord, letting its strength drop to zero in the wake and thus automatically satisfying boundary condition (1). The acceleration potential of the entire sheet is

$$\psi(x,z,t) = \int_{-b}^{b} \psi_D d\xi = \frac{\omega^2 M^2 e^{i\omega t}}{4(1 - M^2)^{\frac{3}{2}}} iz \int_{-b}^{b} \bar{A}(\xi) \frac{H_1^{(2)}(w')}{w'} e^{i\frac{kM^2(x-\xi)}{(1-M^2)b}} d\xi. \quad (6\text{–}94)$$

Letting z approach zero from the positive half-plane in order to study the point-function character of Eq. (6–94), we observe that the z-factor causes the right side to vanish, except in the immediate neighborhood of the point $\xi = x$, where the integrand is infinite in the limit. We isolate this point with a strip of length 2ϵ, sufficiently small so that the continuous functions $\bar{A}(\xi)$ and $e^{i\frac{kM^2(x-\xi)}{(1-M^2)b}}$ can be replaced by their values $\bar{A}(x)$ and 1 at the center of the interval. This produces

$$\psi(x,0^+,t) = \frac{\omega^2 M^2 e^{i\omega t}}{4(1 - M^2)^{\frac{3}{2}}} \bar{A}(x) \lim_{z \to 0^+} \left\{ iz \int_{x-\epsilon}^{x+\epsilon} \frac{H_1^{(2)}(w')}{w'} d\xi \right\}. \quad (6\text{–}95)$$

The value of w' is given by Eq. (6–89) with $\zeta = 0$, and it is a very small quantity throughout the integration range. We can therefore take the first term in the series for the Hankel function (Ref. 6–1, Chapter 4):

$$H_1^{(2)}(w') \cong \frac{2i}{\pi w'}. \quad (6\text{–}96)$$

This leads to

$$\lim_{z \to 0^+} \left\{ iz \int_{x-\epsilon}^{x+\epsilon} \frac{H_1^{(2)}(w')}{w'} d\xi \right\} = \lim_{z \to 0^+} \left\{ -\frac{2z}{\pi} \int_{x-\epsilon}^{x+\epsilon} \frac{d\xi}{w'^2} \right\}$$

$$= \frac{-2b^2(1 - M^2)^2}{\pi M^2 k^2} \lim_{z \to 0^+} \left\{ z \int_{-\epsilon}^{\epsilon} \frac{d\xi'}{[\xi'^2 + (1 - M^2)z^2]} \right\}$$

$$= \frac{-2b^2(1 - M^2)^2}{\pi M^2 k^2} \lim_{z \to 0^+} \left\{ \frac{z}{z\sqrt{1 - M^2}} \tan^{-1}\left(\frac{\epsilon}{z\sqrt{1 - M^2}} \right) \right.$$

$$\left. - \frac{z}{z\sqrt{1 - M^2}} \tan^{-1}\left(\frac{-\epsilon}{z\sqrt{1 - M^2}} \right) \right\}$$

$$= \frac{-2b^2(1 - M^2)^{\frac{3}{2}}}{M^2 k^2}. \quad (6\text{–}97)$$

As in other applications of this same process (cf. Section 5–4), we go to the limit in such a way that ϵ/z becomes indefinitely large. This ensures that any contributions from the doublet sheet outside the strip 2ϵ are negligible. Substituting Eq. (6–97) into Eq. (6–95), we find after several cancellations

$$\psi(x,0^+,t) = -\frac{U^2\bar{A}(x)}{2}e^{i\omega t}.$$ (6–98)

Identical steps carried out for negative values of z lead to

$$\psi(x,0^-,t) = \frac{U^2\bar{A}(x)}{2}e^{i\omega t}.$$ (6–99)

In view of Eq. (6–79),

$$p_U - p_\infty = -\rho_\infty\psi(x,0^+,t), \qquad p_L - p_\infty = -\rho_\infty\psi(x,0^-,t).$$ (6–100)

The simple harmonic discontinuity of pressure

$$p_U - p_L = \Delta\bar{p}_a(x)e^{i\omega t}$$ (6–101)

across the airfoil is therefore proportional to $\bar{A}(x)$ through Eqs. (6–98) and (6–99).

$$\Delta\bar{p}_a(x)e^{i\omega t} = -\rho_\infty[\psi(x,0^+,t) - \psi(x,0^-,t)] = \rho_\infty U^2\bar{A}(x)e^{i\omega t}$$ (6–102)

We can rewrite Eq. (6–102)

$$\bar{A}(x) = \frac{\Delta\bar{p}_a(x)}{\rho_\infty U^2}$$ (6–103)

and substitute it back into the integral of Eq. (6–94) for the doublet sheet.

$$\psi(x,z,t) = \frac{\omega^2 M^2 ize^{i\omega t}}{4\rho_\infty U^2(1 - M^2)^{\frac{3}{2}}} \int_{-b}^{b} \frac{H_1^{(2)}(w')}{w'} e^{i\frac{kM^2(x-\xi)}{(1-M^2)b}} \Delta\bar{p}_a(\xi)d\xi$$

$$= \frac{-ie^{i\omega t}}{4\rho_\infty\sqrt{1 - M^2}} \int_{-b}^{b} \left[\frac{\partial H_0^{(2)}(w')}{\partial z}\right]_{\zeta=0} e^{i\frac{kM^2(x-\xi)}{(1-M^2)b}} \Delta\bar{p}_a(\xi)d\xi.$$ (6–104)

In order to satisfy boundary condition (6–84), we must calculate $\bar{w}(x,0)$ by inserting Eq. (6–104) into Eq. (6–83) and letting z approach zero. For simple harmonic motion, Eq. (6–83) reads

$$w(x,0,t) = \frac{1}{U}e^{-i(\omega x/U)} \int_{-\infty}^{x} \left[\frac{\partial\psi}{\partial z}\right]_{z\to 0^+} e^{i(\omega\xi'/U)}d\xi'.$$ (6–105)

If we substitute the last line of Eq. (6–104), cancel $e^{i\omega t}$, and invert the order of integration, we obtain

$$\bar{w}(x,0) = \bar{w}_a(x)$$

$$= \frac{-ie^{-i(kx/b)}}{4\rho_\infty U\sqrt{1-M^2}} \int_{-b}^{b} \Delta\bar{p}_a(\xi)e^{i(k\xi/b)} \int_{-\infty}^{x} \left[\frac{\partial^2 H_0^{(2)}}{\partial z^2}\right]_{\substack{z\to 0^+ \\ \zeta\to 0}} e^{i\frac{k(\xi'-\xi)}{(1-M^2)b}} d\xi'd\xi.$$

(6–106)

In this form, the limit of the second derivative of $H_0^{(2)}$ at $z = 0$ has such a strong singularity that Eq. (6–106) cannot be used. One way around this difficulty is to observe that $H_0^{(2)}(w')$ is a solution of the partial differential equation

$$(1 - M^2) \frac{\partial^2 H_0^{(2)}}{\partial \xi'^2} + \frac{\partial^2 H_0^{(2)}}{\partial z^2} + \frac{\omega^2}{a_\infty^2(1 - M^2)} H_0^{(2)} = 0. \quad (6\text{–}107)$$

We can therefore rewrite the inner integral of Eq. (6–106) (omitting temporarily the brackets indicating the limits):

$$I = \int_{-\infty}^x \frac{\partial^2 H_0^{(2)}}{\partial z^2} e^{i\frac{k(\xi'-\xi)}{(1-M^2)b}} d\xi' = -(1 - M^2) \int_{-\infty}^x \frac{\partial^2 H_0^{(2)}}{\partial \xi'^2} e^{i\frac{k(\xi'-\xi)}{(1-M^2)b}} d\xi'$$

$$- \frac{\omega^2}{a_\infty^2(1 - M^2)} \int_{-\infty}^x H_0^{(2)} e^{i\frac{k(\xi'-\xi)}{(1-M^2)b}} d\xi'$$

$$= -(1 - M^2) \int_{-\infty}^{x-\xi} \frac{\partial^2 H_0^{(2)}}{\partial \xi_0^2} e^{i\frac{k\xi_0}{(1-M^2)b}} d\xi_0$$

$$- \frac{\omega^2}{a_\infty^2(1 - M^2)} \int_{-\infty}^{x-\xi} H_0^{(2)} e^{i\frac{k\xi_0}{(1-M^2)b}} d\xi_0, \quad (6\text{–}108)$$

where $\xi_0 = \xi' - \xi$. A pair of integrations by parts applied to the first integral produces, after setting $z = 0^+$, the following integrable representation of I.

$$I = \frac{\omega M}{U} \frac{|x - \xi|}{(x - \xi)} e^{i\frac{k(x-\xi)}{(1-M^2)b}} H_1^{(2)} \left(\frac{kM|x - \xi|}{(1 - M^2)b} \right)$$

$$+ i \frac{\omega}{U} e^{i\frac{k(x-\xi)}{(1-M^2)b}} H_0^{(2)} \left(\frac{kM|x - \xi|}{(1 - M^2)b} \right)$$

$$+ \frac{\omega^2}{U^2} \int_{-\infty}^{x-\xi} H_0^{(2)} \left(\frac{kM|\xi_0|}{(1 - M^2)b} \right) e^{i\frac{k\xi_0}{(1-M^2)b}} d\xi_0. \quad (6\text{–}109)$$

Equation (6–109) can be simplified slightly by means of the integral (see Ref. 6–26)

$$\int_{-\infty}^0 H_0^{(2)}(M|u|)e^{iu} du = \frac{2}{\pi\sqrt{1 - M^2}} \ln \frac{1 + \sqrt{1 - M^2}}{M}. \quad (6\text{–}110)$$

When Eqs. (6–109) and (6–110) are substituted into Eq. (6–106) and certain algebraic manipulations are carried out, we find

$$\bar{w}_a(x) = -\frac{\omega}{\rho_\infty U^2} \oint_{-b}^b \Delta \bar{p}_a(\xi) K \left(M, \frac{k(x-\xi)}{b} \right) d\xi; \quad \text{for } -b \le x \le b. \quad (6\text{–}111)$$

Here the kernel function K is defined by

$$K\left(M, \frac{k(x - \xi)}{b}\right) = \frac{1}{4\sqrt{1 - M^2}}\left\{e^{i\frac{kM^2(x-\xi)}{(1-M^2)b}}\left[iM\frac{|x - \xi|}{(x - \xi)}H_1^{(2)}\left(\frac{kM|x - \xi|}{(1 - M^2)b}\right)\right.\right.$$

$$-H_0^{(2)}\left(\frac{kM|x - \xi|}{(1 - M^2)b}\right)\right] + i(1 - M^2)e^{-i\frac{k(x-\xi)}{b}}\left[\frac{2}{\pi\sqrt{1 - M^2}}\ln\frac{1 + \sqrt{1 - M^2}}{M}\right.$$

$$+\int_0^{\frac{k(x-\xi)}{(1-M^2)b}} e^{iu}H_0^{(2)}(M|u|)du\left.\right]\right\}, \tag{6–112}$$

where u is shorthand for the integration variable

$$u = \frac{k(\xi' - \xi)}{(1 - M^2)b}. \tag{6–113}$$

Equation (6–111) is Possio's integral equation. Although not a simple function, the kernel K depends on only two unrelated variables, and it has been fully tabulated by Dietze (Ref. 6–18). Following a well-known procedure from steady-state thin-airfoil theory, Possio attempted to solve Eq. (6–111) by introducing the angle variable

$$\frac{\xi}{b} = \cos \theta . \tag{6–114}$$

His Fourier series for the unknown $\Delta\bar{p}_a$, containing the appropriate leading-edge singularity and satisfying Kutta's hypothesis, was of the general form

$$\Delta\bar{p}_a(\theta) = A_0 \cot\frac{\theta}{2} + \sum_{n=1}^{\infty} A_n \sin n\theta. \tag{6–115}$$

This substitution does not permit a straightforward inversion, and it is necessary to compute a fixed number of coefficients approximately by satisfying Eq. (6–111) at an equal number of points along the airfoil chord. A numerical integration must be made of each term. Convergence of the series (6–115) is not always rapid, and discontinuities of $\bar{w}_a$, like that at the leading edge of a hinged flap, cannot be accounted for accurately.

Other means of solving the integral equation have proved more fruitful. The scheme suggested by Dietze (Ref. 6–18) is based on development of the kernel function (6–112) in terms of its relatively simple counterpart from incompressible flow:

$$K\left(0, \frac{k(x - \xi)}{b}\right) = \frac{1}{2\pi}\left\{ie^{-i\frac{k(x-\xi)}{b}}\right.$$

$$\times\left[Ci\left(\frac{k(x - \xi)}{b}\right) + i\left(\frac{\pi}{2} + Si\left(\frac{k(x - \xi)}{b}\right)\right)\right] - \frac{1}{\frac{k(x - \xi)}{b}}\left.\right\}. \tag{6–116}$$

Here Ci and Si represent the cosine and sine integral functions, tabulated by Jahnke and Emde (Ref. 6–27). Starting from the known solution for incompressible fluid, Dietze derives an iteration procedure, consisting of successive approximate solutions to an integral equation whose kernel is

$$\Delta K = K\left(0, \frac{k(x - \xi)}{b}\right) - K\left(M, \frac{k(x - \xi)}{b}\right). \tag{6–117}$$

The convergence is good in the practical range of reduced frequencies. Dietze's calculations for the airfoil oscillating in vertical translation and pitch, with rotation of an aerodynamically unbalanced flap, at Mach numbers of 0.5 and 0.7, form the basis of useful tables. It appears for $M < 0.5$ that the effects of compressibility are small enough to be omitted from flutter analyses.

The weakness of the Dietze integral equation is that its kernel still has a singularity at $x = \xi$. Fettis overcame this difficulty (Ref. 6–19) by subtracting out the entire singular part of Eq. (6–112) and approximating the remainder with an algebraic polynomial. His method makes use of the known incompressible solution but does not require iteration.

Quite complete tables of aerodynamic coefficients, found by Fettis' method for $M = 0.7$, appear in Ref. 6–28. These can be used for lift, moment, and hinge moment on an oscillating wing-flap combination with aerodynamic balance. The notation is that of Ref. 6–29* [cf. Eqs. (5–350)–(5–352)]. Hence this tabulation is particularly convenient for compressible-flow flutter analyses and similar calculations on systems whose equations of motion are already available in terms of the same symbols. Dietze's tables have been supplemented somewhat and put into the Ref. 6–69 notation by Luke (Ref. 6–30); extended computations based on Dietze's procedure are given by Turner and Rabinowitz (Ref. 6–31).

To complete our discussion of the oscillating airfoil in subsonic flight, we present a resumé of the alternative approach to this problem, which involves transformation to elliptical coordinates and expansion of the dependent variable in series of Mathieu functions. Following Reissner (Ref. 6–22), we assume a two-dimensional disturbance potential in simple harmonic form:

$$\phi'(x,z,t) = \bar{\phi}'(x,z)e^{i\omega t}. \tag{6–118}$$

---

* In compressible flow, the coefficients L_h, L_α, M_h, and M_α are functions of M and reduced frequency k. All other coefficients depend on M, k, and the dimensionless coordinate of the flap leading edge. The notation of Ref. 6–29 is sufficiently general to be suitable for a similar tabulation of any linear theory, subsonic or supersonic. However, the coefficients for Mach numbers 1 and greater are extensively published in a different notation, which we list in Section 6–6.

Equation (6–118) modifies the differential equation (6–1) into

$$(1 - M^2) \frac{\partial^2 \bar{\phi}'}{\partial x^2} + \frac{\partial^2 \bar{\phi}'}{\partial z^2} - \frac{2Mi\omega}{a_\infty} \frac{\partial \bar{\phi}'}{\partial x} + \frac{\omega^2}{a_\infty^2} \bar{\phi}' = 0. \quad (6\text{–}119)$$

The transformation of variables

$$\chi = e^{-i\frac{kM^2 x^*}{1-M^2}} \bar{\phi}', \quad (6\text{–}120)$$

$$x^* = \frac{x}{b}, \quad z^* = \sqrt{1 - M^2}\, \frac{z}{b} \quad (6\text{–}121)$$

simplifies Eq. (6–119) as follows:

$$\frac{\partial^2 \chi}{\partial x^{*2}} + \frac{\partial^2 \chi}{\partial z^{*2}} + \kappa^2 \chi = 0, \quad (6\text{–}122)$$

with

$$\kappa = \frac{kM}{1 - M^2}. \quad (6\text{–}123)$$

In the same terms, Eq. (6–10a) for the amplitude of the simple harmonic disturbance pressure reads

$$\overline{p - p_\infty} = - \frac{\rho_\infty U}{b} e^{i\frac{kM^2 x^*}{1-M^2}} \left[\frac{ik\chi}{1 - M^2} + \frac{\partial \chi}{\partial x^*} \right]. \quad (6\text{–}124)$$

The principal boundary condition (6–4) becomes

$$\frac{\partial \chi}{\partial z^*} = \frac{Ue^{-i\frac{kM^2 x^*}{1-M^2}}}{\sqrt{1 - M^2}} \left[ik\bar{z}_a + \frac{\partial \bar{z}_a}{\partial x^*} \right]; \quad \text{for } z^* = 0, \ -1 \le x^* \le 1. \quad (6\text{–}125)$$

Moreover, the disturbance pressure given by Eq. (6–124) can have no discontinuities across the x^*-axis for $|x^*| > 1$, and Kutta's hypothesis requires it to be continuous at the trailing edge $x^* = 1$.

The modified disturbance potential χ is split into two parts:

$$\chi = \chi_1 + \chi_2 \quad (6\text{–}126)$$

such that χ_1 represents a noncirculatory flow but satisfies boundary condition (6–125), while χ_2 applies a circulation about the airfoil, adjusts χ_1 to meet Kutta's hypothesis, and leaves Eq. (6–125) unaffected. That is,

$$\frac{\partial \chi_1}{\partial z^*} = \frac{Ue^{-i\frac{kM^2 x^*}{1-M^2}}}{\sqrt{1 - M^2}} \left[ik\bar{z}_a + \frac{\partial \bar{z}_a}{\partial x^*} \right],$$

$$\frac{\partial \chi_2}{\partial z^*} = 0; \quad \text{for } z^* = 0, \ -1 \le x^* \le 1. \quad (6\text{–}127)$$

Equation (6–122) is transformed into elliptical coordinates by means of

$$x^* = \cosh \xi \cos \zeta, \quad z^* = \sinh \xi \sin \zeta, \quad (6\text{–}128)$$

where ξ has the property that the figures $\xi = $ const are confocal ellipses on the foci $x^* = \pm 1$. As ξ increases from 0 to ∞, these ellipses grow from the slit $(-1 \leq x^* \leq 1, z^* = 0)$ to a circle of infinite radius. The figures $\zeta = $ const are hyperbolas with the same foci as the ellipses. ζ may be taken to vary from 0 to 2π; $\zeta = 0$ is the upper side of the x^*-axis from $x^* = 1$ to ∞, $\zeta = \pi/2$ is the upper half of the z^*-axis, etc. This coordinate system is especially suited to the present problem, since the entire surface $\xi = 0$ coincides with the airfoil projection. We illustrate a few of the curves $\xi = $ const and $\zeta = $ const in Fig. 6–7.

Substituting Eqs. (6–128) into Eq. (6–122) yields the transformed differential equation

$$\frac{\partial^2 \chi}{\partial \xi^2} + \frac{\partial^2 \chi}{\partial \zeta^2} + \kappa^2 [\cosh^2 \xi - \cos^2 \zeta] \chi = 0. \qquad (6\text{–}129)$$

The principal boundary condition (6–127) is simply

$$\frac{1}{\sin \zeta} \frac{\partial \chi_1}{\partial \xi} = \frac{U e^{-i\frac{kM^2 \cos \zeta}{1 - M^2}}}{\sqrt{1 - M^2}} \left[ik\bar{z}_a + \frac{\partial \bar{z}_a}{\partial x^*} \right] = g(\zeta), \qquad (6\text{–}130)$$

$$\frac{\partial \chi_2}{\partial \xi} = 0; \qquad \text{for } \xi = 0.$$

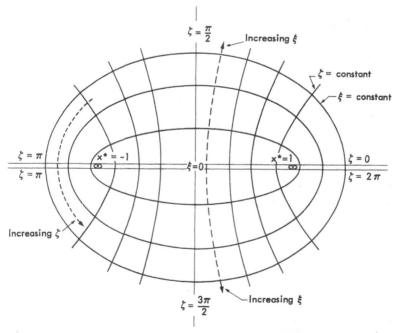

Fig. 6–7. Coordinate surfaces associated with the system of elliptic-cylinder coordinates ξ and ζ.

The disturbance pressure, Eq. (6–124), may be written, for $\xi = 0$,

$$\overline{p - p_\infty} = -\frac{\rho_\infty U}{b} e^{i\frac{kM^2 \cos \zeta}{1 - M^2}} \left[\frac{ik\chi(0,\zeta)}{1 - M^2} - \frac{1}{\sin \zeta}\frac{\partial \chi(0,\zeta)}{\partial \zeta}\right]. \quad (6\text{–}131)$$

If we assume a separated solution

$$\chi = F(\xi)G(\zeta), \qquad (6\text{–}132)$$

the differential equation (6–129) splits into two Mathieu equations:

$$\frac{d^2 F}{d\xi^2} + (\kappa^2 \cosh^2 \xi - c)F = 0, \qquad (6\text{–}133a)$$

$$\frac{d^2 G}{d\zeta^2} - (\kappa^2 \cos^2 \zeta - c)G = 0. \qquad (6\text{–}133b)$$

Possible values of the separation constant c are determined by the requirement that, for a given κ, $G(\zeta)$ must be periodic of period 2π to prevent discontinuities along the positive x^*-axis. In McLachlan's book (Ref. 6–32) we find two types of eigenfunctions of Eq. (6–133b) which have this periodicity, known as $ce_m(\zeta)$ and $se_m(\zeta)$.† The former has the same symmetry with respect to $\zeta = 0$ as $\cos n\zeta$, can be expanded as an infinite series of such cosines, and reduces to $\cos m\zeta$ for $\kappa = 0$. The latter is similarly related to $\sin n\zeta$.

The solutions of Eq. (6–133a) which correspond to $ce_m(\zeta)$ and $se_m(\zeta)$, and have the important property that they represent outgoing disturbance propagation at infinity, are designated $Ce_m(\xi)$ and $Se_m(\xi)$. In their asymptotic behavior for large values of ξ, they have a wavelike factor in common with the Hankel functions $H_n^{(2)}$.

The noncirculatory part of the flow χ_1 must be an odd function of ζ, since $(1/\sin \zeta)(\partial \chi_1/\partial \xi)$, which is proportional to the vertical velocity, must be even in ζ. Hence we select a series of the $se_m(\zeta)$ for χ_1:

$$\chi_1 = \sum_{m=0}^{\infty} b_m se_m(\zeta)Se_m(\xi). \qquad (6\text{–}134)$$

Because of the orthogonality of these Mathieu functions in the interval 0 to π,

$$\int_0^\pi se_n(\zeta)se_n(\zeta)d\zeta = 0, \qquad m \neq n, \qquad (6\text{–}135)$$

† The notation we use here for the Mathieu functions strictly follows Ref. 6–22, which differs in certain respects from both Ref. 6–32 and Ref. 6–33. For instance, ce_m and se_m are proportional to the functions called Se_{2r} (or Se_{2r+1}) and So_{2r} (or So_{2r+1}), respectively, by the National Bureau of Standards. However, the latter are normalized in a different way. There is no general agreement on symbols for the functions which Reissner denotes Ce_m and Se_m.

and the assumed normalization

$$\frac{dSe_m(0)}{d\xi} = 1, \tag{6-136}$$

we can compute individual coefficients in (6-134) from Eq. (6-130) as follows:

$$b_m = \int_0^\pi \sin \zeta g(\zeta) se_m(\zeta) d\zeta. \tag{6-137}$$

Thus χ_1 is completely determined, and the noncirculatory pressure, lift, and moment are obtained through Eq. (6-131). Incidentally, the noncirculatory character of this part of the flow can be confirmed from the fact that χ_1 (hence ϕ') does not change when the variable ζ is increased by 2π, corresponding to a complete circuit around the airfoil.

The most difficult part of the Mathieu-function approach is to find a suitable form for the function χ_2. In Ref. 6-22, Reissner states that Haskind's representation (Ref. 6-21) appears to be the most convenient. χ_2 must satisfy condition (6-130), but the more awkward limitations on it come from the continuity of pressure on $z^* = 0$ ahead of the airfoil and in the wake. Since $\chi_1 = 0$ and thus gives rise to no disturbance pressure along either of these lines, Reissner shows that the requirements on χ_2 may be written

$$\chi_2 = 0; \quad \text{for } z^* = 0, x^* \leq -1, \tag{6-138a}$$

$$\frac{ik\chi_2}{1 - M^2} + \frac{\partial \chi_2}{\partial x^*} = 0; \quad \text{for } z^* = 0, x^* \geq 1. \tag{6-138b}$$

The second of these is equivalent to letting

$$\chi_2 = Ce^{-i(kx^*/1 - M^2)} \tag{6-139}$$

along the wake, with C an arbitrary constant. Simply putting $\chi_2 = 0$ in the wake is not enough, since it eliminates circulation. The Haskind approach is to introduce another solution of Mathieu's equations $W(\xi,\zeta)$, such that its z^*-derivative is proportional to the disturbance pressure.

$$\frac{\partial W}{\partial z^*} = \frac{ik}{1 - M^2} \chi_2 + \frac{\partial \chi_2}{\partial x^*}. \tag{6-140}$$

Since χ_2, like χ_1, is an odd function of z, W must be even in z (and ζ). Hence it is written

$$W = \sum_{m=0}^\infty a_m ce_m(\zeta) Ce_m(\xi). \tag{6-141}$$

Equations (6–140) and (6–141) meet all conditions on pressure, except Kutta's hypothesis. The vanishing of the vertical velocity at the airfoil due to χ_2 leads first to

$$W = A \sum_{m=0}^{\infty} \alpha_m ce_m(\zeta)Ce_m(\xi) + B \sum_{m=0}^{\infty} \beta_m ce_m(\zeta)Ce_m(\xi). \quad (6\text{–}142)$$

Here A and B are as yet undetermined constants and

$$\alpha_m = \frac{\displaystyle\int_0^{\pi} \cos(\kappa \cos \zeta)ce_m(\zeta)d\zeta}{Ce_m(0)}, \quad (6\text{–}143a)$$

$$\beta_m = \frac{\displaystyle\int_0^{\pi} \frac{\sin(\kappa \cos \zeta)ce_m(\zeta)}{\kappa}d\zeta}{Ce_m(0)}. \quad (6\text{–}143b)$$

A refinement of the zero-vertical-velocity requirement then produces the following relation between A and B:

$$A\left[e^{-i(\kappa/M)}\kappa^2\left(i\frac{1}{M}\sin \kappa + \cos \kappa \right) \right.$$
$$\left. + \kappa^2\left(1 - \frac{1}{M^2} \right) \sum_{n=0}^{\infty} \alpha_n ce_n(\pi)\int_0^{\infty} e^{-i(\kappa/M)\cosh \xi}Ce_n{}'(\xi)d\xi \right]$$
$$+ B\left[e^{-i(\kappa/M)}\left(i\frac{\kappa}{M}\cos \kappa - \kappa \sin \kappa \right) \right.$$
$$\left. + \kappa^2\left(1 - \frac{1}{M^2} \right) \sum_{n=0}^{\infty} \beta_n ce_n(\pi)\int_0^{\infty} e^{-i(\kappa/M)\cosh \xi}Ce_n{}'(\xi)d\xi \right] = 0, \quad (6\text{–}144)$$

where the prime on $Ce_n(\xi)$ indicates the derivative with respect to ξ.

Finally, the finite pressure at the trailing edge brings the two parts of the solution together and furnishes a second equation between A and B:

$$A \sum_{n=0}^{\infty} \alpha_n ce_n(0) + B \sum_{n=0}^{\infty} \beta_n ce_n(0) = \sum_{n=0}^{\infty} b_n se_n{}'(0)Se_n(0). \quad (6\text{–}145)$$

We see from the foregoing summary that numerical computations associated with the Mathieu-function approach are certainly not rapid or straightforward. Infinite sums occur even in the case of simple motions like vertical translation and pitching; the number of terms which must be taken to assure convergence varies widely with the reduced frequency and Mach number. Moreover, certain sets of integrals, like

$$\int_0^{\infty} e^{-i(\kappa/M)\cosh \xi}Ce_n{}'(\xi)d\xi,$$

cannot even be expressed as infinite sums of tabulated coefficients but must be evaluated numerically. All these operations will certainly be made more accurate and efficient because of tables of Mathieu functions issued by the National Bureau of Standards (Ref. 6–33).

Aerodynamic coefficients for the oscillating chordwise-rigid airfoil have been calculated using Mathieu functions by Timman, van de Vooren, and Greidanus (Ref. 6–24). They are based on the acceleration potential method and are given for Mach numbers of 0.35, 0.5, 0.6, 0.7, and 0.8.

There has been no complete experimental confirmation for aerodynamic loading of the oscillating airfoil in subsonic compressible flow. At best, flutter tests can furnish only an indirect check, since flutter equations contain assumed mode shapes, and there can be errors in estimating elastic and inertia as well as aerodynamic properties of a wing. Nevertheless, there exist measurements which show the trend of flutter behavior with Mach number to follow the theory. Discrepancies that have been reported between tests and theoretical results are isolated rather than systematic.

6–5 Arbitrary small motions of airfoils in subsonic flow. The only fully workable way of predicting aerodynamic loads due to arbitrary small motions of an airfoil in subsonic flight is by means of Fourier-integral superposition of theoretical results for simple harmonic oscillations. In order to avoid difficulties with convergence, indicial-admittance functions are first determined for the important airfoil degrees of freedom. Duhamel's superposition integral is then used to give lift, moment, etc. for more general motions.

In this section we emphasize vertical translation (plunging) and pitching of a wing whose chordwise sections do not deform, inasmuch as existing tables cover only these cases. In principle, however, the expressions we shall derive have obvious counterparts for rotation of a leading- or trailing-edge flap, prescribed modes of chordwise bending, and the like.

Certain effects of compressibility complicate the employment of the Fourier integral for indicial motions. In fluid with a finite speed of sound, noncirculatory flow patterns do not adjust themselves immediately to changing boundary conditions. Any concept of virtual mass is meaningless. The noncirculatory lift and moment, which can be calculated only at some pains, do not depend on instantaneous accelerations and velocities of the wing but are affected by their time history. Accordingly, we gain no advantage by separating out circulatory and noncirculatory parts of the flow. Indeed, the principal justification for doing so in the incompressible case is that the noncirculatory loads become infinite at the start of impulsive motion, but this singularity is eliminated when compressibility cushions the impact.

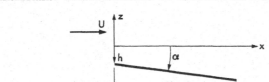

Fig. 6–8. Wing performing vertical translation and pitching about an axis through the leading edge at midspan.

In contrast to what we found in Section 5–7, the circulation around a chordwise-rigid airfoil in compressible flow is not determined just by the vertical velocity at the three-quarter chord point. This means that two indicial-admittance functions, one for vertical translation and another for pitching velocity, replace the single Wagner function. In order to see why, let us consider the boundary condition on the vertical velocity of fluid particles in contact with a wing which is executing the h and α motions described in Fig. 6–8. We temporarily place the axis of rotation through the leading edge at midspan, which point we assume projects onto the origin of our coordinate system. The boundary condition states that the derivative of ϕ' with respect to z equals

$$w_a(x,t) = -[\dot{h} + U\alpha + \dot{\alpha}x]. \tag{6–146}$$

As pointed out in Section 5–6, both the h and $U\alpha$ terms in Eq. (6–146) give constant w_a along the chord and can be treated interchangeably. The $\dot{\alpha}$-term produces a linear variation of w_a with x, so that it must be handled separately. Wing motions in which each of the two types of vertical velocity distribution is generated, without giving rise to the other, are illustrated in Figs. 6–9(a) and 6–9(b).

Also shown in Fig. 6–9 are indicial motions of the two types. Let us suppose that, at a given flight Mach number, we can solve for the indicial lift and pitching moment (about the axis of rotation) in each case. For a vertical velocity $\dot{h}_0$, starting at $t = 0$, we write

$$L_T{}'(s) = 2\pi \frac{\rho_\infty}{2} U^2 S \frac{\dot{h}_0}{U} \phi_c(s), \tag{6–147}$$

$$M_{yT}{}'(s) = 2\pi \frac{\rho_\infty}{2} U^2 S(2b) \frac{\dot{h}_0}{U} \phi_{cM}(s), \tag{6–148}$$

where

$$s = \frac{Ut}{b} \tag{5–365}$$

is the distance traveled by the wing after the start of plunging, measured in reference semichords. S denotes the wing area. The factor 2π has been chosen by comparison with the incompressible problem, so that for a two-dimensional airfoil flying at low speed the dimensionless indicial func-

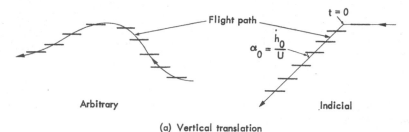

(a) Vertical translation

(b) Pitching

Fig. 6-9. Wings performing arbitrary and indicial motions (a) in vertical translation (plunging), and (b) in pure pitch, without angle-of-attack changes.

tion $\phi_c(s)$ will approach unity when s becomes very large. The ratio $-\phi_{cM}(s)/\phi_c(s)$ is adjusted to measure how far the instantaneous center of pressure lies behind the leading edge, in terms of the reference chord $2b$.

For an indicial angular velocity q_0 about the leading-edge axis, we write*

$$L_q'(s) = 4\pi \frac{\rho_\infty}{2} U^2 S \left(\frac{q_0 b}{U}\right) \phi_{cq}(s), \qquad (6\text{--}149)$$

$$M_{yq}'(s) = 4\pi \frac{\rho_\infty}{2} U^2 S(2b) \left(\frac{q_0 b}{U}\right) \phi_{cMq}(s). \qquad (6\text{--}150)$$

This definition of $\phi_{cq}(s)$ leads to the asymptotic value $\frac{3}{4}$ for the two-dimensional airfoil in incompressible flow, as can be proved by equating Eqs. (6-150) and (5-366), and replacing α_0 in the latter by the effective angle of attack $3q_0 b/2U$ due to q_0 at the three-quarter chord point. As before, $-\phi_{cMq}(s)/\phi_{cq}(s)$ measures the dimensionless location of the center of pressure.

---

* The dimensionless indicial functions here are chosen to correspond with those calculated by Mazelsky (Refs. 6-34 and 6-35), although the need for preserving notational homogeneity throughout the book prevents us from using exactly his symbols. $\phi_c(s)$ and $\phi_{cq}(s)$ are numerically equal to Mazelsky's $k_1(s)$ and $k_{1q}(s)$ for the two-dimensional airfoil. The moment functions are the same as $m_1(s)$ and $m_{1q}(s)$, except for a shift of axis.

Given the four indicial functions defined in Eqs. (6–147)–(6–150), and recognizing the identity between $\dot{h}/U$ and α as far as their association with $\phi_c(s)$ and $\phi_{cM}(s)$ is concerned, we can use Duhamel's integral to describe the lift and moment produced by any small motion of the wing. The appropriate expressions, when the transient starts at $t = 0$, are

$$L'(s) = 2\pi \frac{\rho_\infty}{2} U^2 S \left\{ \left[\alpha(0) + \frac{\dot{h}(0)}{U} \right] \phi_c(s) \right.$$

$$\left. + \int_0^s \frac{d}{d\sigma} \left[\alpha(\sigma) + \frac{\dot{h}(\sigma)}{U} \right] \phi_c(s - \sigma)d\sigma \right\}$$

$$+ 4\pi \frac{\rho_\infty}{2} U^2 S \left\{ \frac{b}{U} \dot{\alpha}(0)\phi_{cq}(s) + \int_0^s \frac{b}{U} \frac{d\dot{\alpha}(\sigma)}{d\sigma} \phi_{cq}(s - \sigma)d\sigma \right\}, \quad (6\text{–}151)$$

$$M_y'(s) = 2\pi \frac{\rho_\infty}{2} U^2 S(2b) \left\{ \left[\alpha(0) + \frac{\dot{h}(0)}{U} \right] \phi_{cM}(s) \right.$$

$$\left. + \int_0^s \frac{d}{d\sigma} \left[\alpha(\sigma) + \frac{\dot{h}(\sigma)}{U} \right] \phi_{cM}(s - \sigma)d\sigma \right\}$$

$$+ 4\pi \frac{\rho_\infty}{2} U^2 S(2b) \left\{ \frac{b}{U} \dot{\alpha}(0)\phi_{cMq}(s) + \int_0^s \frac{b}{U} \frac{d\dot{\alpha}(\sigma)}{d\sigma} \phi_{cMq}(s - \sigma)d\sigma \right\}.$$

$$(6\text{–}152)$$

Here the superscript dot indicates the derivative with respect to physical time; it can be replaced by an additional derivative with respect to s or σ, as the case may be, if the factor U/b is appended to each term so modified.

Equations (6–151) and (6–152) are perfectly general, applying within the theoretical limitations to any rigid wing in any flight range where the indicial functions are available. When the lift and moment are those per unit span of a two-dimensional airfoil, we need only to substitute the chord $2b$ for wing area S wherever it occurs. We retain the primes on L' and M_y' as a warning that they were calculated for a pitching axis through the leading edge. If we follow Theodorsen's notation and introduce an axis for both moment and pitching which is located at $b(a + 1)$ behind the leading edge, we can write a new set of indicial lifts and moments, in terms of Eqs. (6–147)–(6–150). When so doing, we employ the general law for transferring the axis of a moment. We also observe that an angular velocity q_0 about $x = b(a + 1)$ is equivalent to the same angular velocity about $x = 0$, plus a vertical translation $h_0 = -q_0 b(a + 1)$ of the whole wing.

$$L_T(s) = L_T'(s) = 2\pi \frac{\rho_\infty}{2} U^2 S \frac{h_0}{U} \{\phi_c(s)\}. \qquad (6\text{–}153)$$

$$M_{yT}(s) = M_{yT}'(s) + L_T(s)b[a + 1]$$

$$= 2\pi \frac{\rho_\infty}{2} U^2 S(2b) \frac{\dot{h}_0}{U} \left\{ \phi_{cM}(s) + \left[\frac{a}{2} + \frac{1}{2} \right] \phi_c(s) \right\}. \tag{6-154}$$

$$L_q(s) = L_q'(s) - 2\pi \frac{\rho_\infty}{2} U^2 S \frac{q_0 b[a + 1]}{U} \phi_c(s)$$

$$= 4\pi \frac{\rho_\infty}{2} U^2 S \left(\frac{q_0 b}{U} \right) \left\{ \phi_{cq}(s) - \left[\frac{a}{2} + \frac{1}{2} \right] \phi_c(s) \right\}. \tag{6-155}$$

$$M_{yq}(s) = M_{yq}'(s) - 2\pi \frac{\rho_\infty}{2} U^2 S(2b) \frac{q_0 b[a + 1]}{U} \phi_{cM}(s) + L_q(s)b[a + 1]$$

$$= 4\pi \frac{\rho_\infty}{2} U^2 S(2b) \left(\frac{q_0 b}{U} \right) \left\{ \phi_{cMq}(s) \right.$$

$$\left. + \left[\frac{a}{2} + \frac{1}{2} \right] [\phi_{cq}(s) - \phi_{cM}(s)] - \left[\frac{a}{2} + \frac{1}{2} \right]^2 \phi_c(s) \right\}. \tag{6-156}$$

The four quantities in braces in Eqs. (6-153)-(6-156) are the new indicial functions. They can be used in place of the original ones in Eq. (6-151) and (6-152) to write $L(s)$ and $M_y(s)$ for arbitrary translation and pitching about the axis a distance ba back of the reference midchord point.

In connection with the application of these formulas, we mention that in problems involving motion of the complete aircraft, such as dynamic stability, confusion sometimes arises between the coordinates employed here and so-called body axes, which are fixed with respect to the machine. The quantities α and h are the displacements of the wing relative to wind-tunnel axes. In other words, they are measured from a Newtonian coordinate system which translates with the forward speed U (assumed constant) but does not partake of any transient lateral disturbances. By contrast, the quantities θ and w, familiar in dynamic stability work, represent the rotation of the chordline in pitch from some direction fixed in space and the absolute velocity (positive downward) of the airplane relative to the air, taken in a direction always normal to the chordline. Thus w/U is the resultant aerodynamic angle of attack of the chordline, while θ is an angular displacement of that chordline without angle-of-attack change. When $w = 0$, θ (which is usually called q) is just the pure pitching motion pictured in Fig. 6-9(b). The transformation by which θ and w may be calculated from α and h reads

$$\frac{w}{U} = \frac{\dot{h}}{U} + \alpha, \tag{6-157}$$

$$\theta = \alpha, \tag{6-158}$$

so long as the origins of coordinates project onto one another.

To give the correct results, both these equations must be substituted simultaneously into formulas like Eqs. (6–151) and (6–152).

We next examine the determination of $\phi_c(s)$, $\phi_{cM}(s)$, $\phi_{cq}(s)$, and $\phi_{cMq}(s)$ for an airfoil in subsonic flight. None of them can be found exactly in terms of known functions throughout the entire range of s. However, the initial and asymptotic magnitudes are easily specified, and Lomax *et al.* (Ref. 6–37) suggest an interesting analogy with steady flow by which the functions are simply expressed when s lies between 0 and $2M/(M + 1)$.

The asymptotic values of the indicial quantities are simply the lifts and moments on airfoils in steady motion. These are readily computed by multiplying their counterparts in incompressible flow by the Prandtl-Glauert factor $1/\sqrt{1 - M^2}$. The incompressible $\phi(\infty)$ and $\phi_M(\infty)$ are $1/2\pi$ times the lift- and moment-curve slopes from Eq. (5–115) and Eq. (5–114) with $a = -1$, whence we obtain

$$\phi_c(\infty) = \frac{1}{\sqrt{1 - M^2}}, \tag{6–159}$$

$$\phi_{cM}(\infty) = \frac{-1}{4\sqrt{1 - M^2}}. \tag{6–160}$$

As expected, the ratio of Eq. (6–160) to Eq. (6–159) shows the final lift to be centered at the quarter-chord point. Since the incompressible indicial lift is known to act at this point throughout the maneuver, the deviation of the compressible center of lift from $x = b/2$ measures the combined effects of the noncirculatory flow and compressibility on the development of the wake. Equations (5–117) and (5–118) for the parabolically cambered airfoil, which produces the same chordwise w_a distribution as pure pitching about a midchord axis, lead to the incompressible counterparts of $\phi_{cq}(\infty)$ and $\phi_{cMq}(\infty)$. The Prandtl-Glauert factor yields

$$\phi_{cq}(\infty) = \frac{3}{4\sqrt{1 - M^2}}, \tag{6–161}$$

$$\phi_{cMq}(\infty) = \frac{-1}{4\sqrt{1 - M^2}}. \tag{6–162}$$

All starting values $\phi_c(0)$, etc., can be found by means of the Fourier integral, since they are equal to the corresponding quantities for simple harmonic motion at infinite frequency (cf. Ref. 6–34, Appendix A). There is an equally easy way of determining them by physical reasoning, however, which can be applied to any wing whatever at any Mach number greater than zero. It proceeds as follows. When a piston moves impulsively from rest into any undisturbed mass of gas with a velocity u_0, small compared with a_∞, it first experiences a disturbance pressure

$$p - p_\infty = \rho_\infty a_\infty u_0. \tag{6–163}$$

This is a result of acoustics or simple-wave theory. It can also be derived starting from Section 6-1. Let the piston move in the x-direction. So long as the flow remains one-dimensional, it is governed by the acoustic differential equation [Eq. (6-1) with $U = 0$]

$$\frac{\partial^2 \phi'}{\partial x^2} = \frac{1}{a_\infty^2} \frac{\partial^2 \phi'}{\partial t^2}, \tag{6-164}$$

and subject to the boundary condition

$$\frac{\partial \phi'}{\partial x} = u_0; \qquad \text{for } x \cong 0, t > 0. \tag{6-165}$$

Any $f(x - a_\infty t)$ is a solution of Eq. (6-164), representing waves propagating parallel to the positive x-axis. A particular f which satisfies

$$\frac{\partial \phi'}{\partial x} = \frac{\partial f}{\partial x} = \frac{df(x - a_\infty t)}{d(x - a_\infty t)} = u_0; \qquad \text{for } x \cong 0, t > 0 \tag{6-166}$$

is the integral of the step function $u_0 1(a_\infty t - x)$. That is,

$$\frac{df}{d(x - a_\infty t)} = u_0 1(a_\infty t - x). \tag{6-167}$$

However, the pressure on the piston can be computed from the linearized form of Bernoulli's equation (5-13), with $q^2/2$ neglected and $F(t)$ set equal to zero because of no disturbances at infinity:

$$\frac{p - p_\infty}{\rho_\infty} = -\frac{\partial \phi'}{\partial t} = -\frac{df}{d(x - a_\infty t)} \frac{d(x - a_\infty t)}{dt} = a_\infty \frac{df}{d(x - a_\infty t)}. \tag{6-168}$$

For any positive value of t, the step function is unity at $x = 0$. We can substitute Eq. (6-167) into Eq. (6-168) and get

$$p - p_\infty = \rho_\infty a_\infty u_0. \tag{6-163}$$

In general, there is no resemblance between one-dimensional motion of a piston and indicial pitching or plunging of a wing; but at the very first instant $t = 0^+$ each little element of wing surface does start off like a little piston with normal velocity w_a into gas at rest. Hence Eq. (6-163) shows that the initial pressure just above a wing starting at $\dot{h}_0$ is

$$p_U - p_\infty = \rho_\infty a_\infty [-\dot{h}_0] = -\rho_\infty \frac{U^2}{M} \frac{\dot{h}_0}{U}. \tag{6-169}$$

The pressure for a wing beginning to pitch at angular velocity q_0 about $x = 0$ is

$$(p_U - p_\infty)_q = \rho_\infty a_\infty [-q_0 x] = -\rho_\infty \frac{U^2}{M} \frac{q_0 x}{U}. \tag{6-170}$$

Because of the antisymmetry of the flow pattern, the disturbance pressures $(p_L - p_\infty)$ on the lower surface are equal and opposite. We can therefore integrate to find the resultant lifts and moments on the two-dimensional airfoil.

$$L_T'(0) = -\int_0^{2b} [p_U - p_L]dx = 2\rho_\infty \frac{U^2(2b)}{M} \frac{\dot{h}_0}{U}$$

$$= 2\pi \frac{\rho_\infty}{2} U^2(2b) \frac{\dot{h}_0}{U} \phi_c(0), \tag{6-171}$$

$$M_{yT}'(0) = \int_0^{2b} [p_U - p_L]xdx = -\rho_\infty \frac{U^2(2b)^2}{M} \frac{\dot{h}_0}{U}$$

$$= 2\pi \frac{\rho_\infty}{2} U^2(2b)^2 \frac{\dot{h}_0}{U} \phi_{cM}(0). \tag{6-172}$$

$$L_q'(0) = -\int_0^{2b} [p_U - p_L]_q dx = 2\rho_\infty \frac{U^2(2b)}{M} \frac{q_0 b}{U}$$

$$= 4\pi \frac{\rho_\infty}{2} U^2(2b) \left(\frac{q_0 b}{U}\right) \phi_{cq}(0), \tag{6-173}$$

$$M_{yq}'(0) = \int_0^{2b} [p_U - p_L]_q xdx = -\tfrac{4}{3}\rho_\infty \frac{U^2(2b)^2}{M} \frac{q_0 b}{U}$$

$$= 4\pi \frac{\rho_\infty}{2} U^2(2b)^2 \left(\frac{q_0 b}{U}\right) \phi_{cMq}(0). \tag{6-174}$$

Whence, finally,

$$\phi_c(0) = \frac{2}{\pi M}, \tag{6-175}$$

$$\phi_{cM}(0) = -\frac{1}{\pi M}, \tag{6-176}$$

$$\phi_{cq}(0) = \frac{1}{\pi M}, \tag{6-177}$$

$$\phi_{cMq}(0) = -\frac{2}{3\pi M}. \tag{6-178}$$

As discussed in Part III of Ref. 6–37, formula (6–163) for the starting pressure leads to many other interesting conclusions. For instance, we can show from Eq. (6–169) that the initial lift coefficient per unit angle of attack $\alpha_0 = \dot{h}_0/U$ for any plane wing at any Mach number is

$$\frac{dc_l(0)}{d\alpha_0} = \frac{4}{M}. \tag{6-179}$$

Equations (6–175)–(6–178) imply that at $s = 0$ the centers of pressure for vertical translation and pitching are at midchord and two-thirds chord, respectively. This we could also deduce from the linear relationship between disturbance pressure and local starting velocity.

During the starting phase of the motion, we can compute the indicial functions by analogy with our steady-flow theory in Section 6–3. We observe the motion in a coordinate system fixed with respect to the fluid at rest, in which the airfoil occupies the position between $x = 0$ and $x = 2b$ at $t = 0$ and afterward translates steadily in the negative x-direction with velocity U. This is still a two-dimensional unsteady problem. The governing differential equation is obtained by linearizing Eq. (5–24), under the assumption that all velocity components are small compared with a_∞, or, alternatively, by setting $U = 0$ in Eq. (6–1):

$$\frac{\partial^2 \phi'}{\partial x^2} + \frac{\partial^2 \phi'}{\partial z^2} - \frac{1}{a_\infty{}^2} \frac{\partial^2 \phi'}{\partial t^2} = 0. \tag{6–180a}$$

The principal boundary condition must be satisfied at the instantaneous position of the projection of the surface of the airfoil, which lies between $x = -Ut$ and $x = -Ut + 2b$. That is,

$$\frac{\partial \phi'}{\partial z} = w_a(x,t); \qquad \text{for } z = 0, -Ut \le x \le -Ut + 2b. \tag{6–180b}$$

To facilitate the analogy, we make the transformation

$$t_1 = a_\infty t \tag{6–181}$$

in Eqs. (6–180a) and (6–180b), finding

$$\frac{\partial^2 \phi'}{\partial t_1{}^2} - \frac{\partial^2 \phi'}{\partial x^2} - \frac{\partial^2 \phi'}{\partial z^2} = 0 \tag{6–182a}$$

and

$$\frac{\partial \phi'}{\partial z} = w_a(x, t_1); \qquad \text{for } z = 0, -Mt_1 \le x \le -Mt_1 + 2b. \tag{6–182b}$$

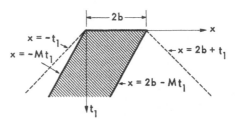

Fig. 6–10. Trace of an impulsively starting airfoil in the t_1 x-plane.

In Fig. 6–10, the shaded area illustrates the portion of the t_1 x-plane on which boundary condition (6–182b) holds. We also show in this figure, as dashed lines, the limits of the region which is disturbed by the airfoil's motion. These limits are determined by observing the actual flow in a particular xz-plane as time progresses. At $t = t_1 = 0$, the initial disturbances start from the wing chord and expand as a series of circles, with centers fixed along the line $0 \leq x \leq 2b$ and instantaneous radii $a_\infty t = t_1$. The foremost of these circles cuts the plane $z = 0$ at the point $x = -t_1$, while the rearmost extends behind the airfoil to $x = 2b + t_1$, so that their traces in the t_1 x-plane are lines through $x = 0$ and $x = 2b$ with slopes $-45°$ and $+45°$, respectively. Moreover, in the t_1 xz-space, these expanding circles generate "Mach cones" with axes parallel to the positive t_1 direction and vertices along the dashed "Mach lines" in Fig. 6–10.

Bearing these facts in mind, we compare differential equation (6–182a) with Eq. (6–44) and discover that the two are identical, except that t_1 and x have assumed the roles of x and y, respectively, and the Mach number is $\sqrt{2}$. Regarding t_1 as the new flight direction for a wing in steady, three-dimensional supersonic flow, we turn to Fig. 6–10 and see that the boundary conditions are exactly those associated with a swept-forward wing tip, having sweep angle

$$\Lambda = -\tan^{-1}\left(\frac{1}{M}\right) \tag{6–183}$$

and cut off perpendicular to the flight direction. *M no longer* represents the analogous steady-flow Mach number, which is $\sqrt{2}$. All auxiliary boundary conditions are in complete accord between the two problems. The disturbed portion of t_1 xz-space is contained within the appropriate Mach cones with 90° vertices; the pressure discontinuity must be zero off the planform; and Kutta's hypothesis is needed at the subsonic trailing edge.

We take the equivalent steady-flow solution from Eq. (6–78), replacing local angle of attack $\partial z_a/\partial x$ with w_a/U and interchanging the necessary independent variables.

$$\phi'(t_1, x, z) = -\frac{1}{\pi} \iint_{R_a'} \frac{w_a(\tau_1, \xi) d\tau_1 d\xi}{\sqrt{(t_1 - \tau_1)^2 - [(x - \xi)^2 + z^2]}} \cdot \tag{6–184}$$

Equation (6–184) is valid only for integration regions R_a' where the source method, or Evvard's extensions of it, can be used.

As an example, we apply this analogy to calculate the pressure distribution, lift, and moment due to indicial plunging, for which

$$w_a = -\dot{h}_0. \tag{6–185}$$

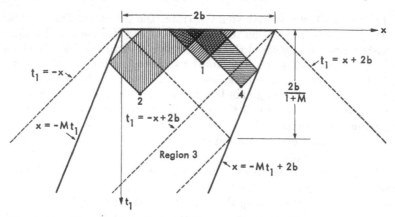

Fig. 6–11. Four regions of the analogous wing planform in the t_1 x-plane, showing typical points in three of them, with corresponding areas of integration (shaded).

Setting $z = 0$, we find $\phi'(t_1,x,0)$ for the regions of the planform pictured in Fig. 6–11.

Typical points are shown in three of these regions, together with solid lines bounding the areas of integration given by Evvard's theorem. For point 1, on the two-dimensional part of the wing, Eq. (6–184) reads

$$\phi_1'(t_1,x,0) = \frac{\dot{h}_0}{\pi} \int_0^{t_1} \int_{x-(t_1-\tau_1)}^{x+(t_1-\tau_1)} \frac{d\xi d\tau_1}{\sqrt{(t_1-\tau_1)^2 - (x-\xi)^2}} = \dot{h}_0 t_1. \quad (6\text{–}186)$$

For point 2, influenced by the subsonic leading edge, we find

$$\phi_2'(t_1,x,0) = \frac{\dot{h}_0}{\pi} \left\{ \int_{(t_1-x)/(1+M)}^{t_1} \int_{x-(t_1-\tau_1)}^{x+(t_1-\tau_1)} \frac{d\xi d\tau_1}{\sqrt{(t_1-\tau_1)^2 - (x-\xi)^2}} \right.$$

$$\left. + \int_0^{(t_1-x)/(1+M)} \int_{-\tau_1-(t_1-x)\left(\frac{M-1}{M+1}\right)}^{x-(t_1-\tau_1)} \frac{d\xi d\tau_1}{\sqrt{(t_1-\tau_1)^2 - (x-\xi)^2}} \right\}$$

$$= \dot{h}_0 \left[t_1 - \frac{2t_1}{\pi} \tan^{-1} \sqrt{\frac{t_1-x}{x+Mt_1}} \right]$$

$$+ \frac{2\dot{h}_0}{\pi(1+M)} \sqrt{(x+Mt_1)(t_1-x)} \cdot \quad (6\text{–}187)$$

The disturbance pressure on the wing surface is derived by setting $U = 0$ in Eq. (6–10) and substituting t_1 in place of t:

$$p - p_\infty = -\rho_\infty \frac{\partial \phi'}{\partial t} = -\rho_\infty a_\infty \frac{\partial \phi'}{\partial t_1} \cdot \quad (6\text{–}188a)$$

We divide by the actual flight dynamic pressure and get

$$\frac{p - p_\infty}{(\frac{1}{2}\rho_\infty)U^2} = -\frac{2}{UM}\frac{\partial\phi'}{\partial t_1}. \tag{6–188b}$$

Because of the antisymmetry of the flow, it follows that

$$\frac{p_U - p_L}{(\frac{1}{2}\rho_\infty)U^2} = -\frac{4}{UM}\frac{\partial\phi'(t_1,x,0)}{\partial t_1}. \tag{6–189}$$

When applied to Eq. (6–186), Eq. (6–189) yields

$$\frac{(p_U - p_L)_1}{(\frac{1}{2}\rho_\infty)U^2} = -\frac{4}{M}\frac{\dot{h}_0}{U}. \tag{6–190}$$

We note that over the entire two-dimensional part of the equivalent swept-forward wing, the pressure has the impulsive starting value given by Eq. (6–169). We can see on physical grounds that this should be true, because no signals from the leading or trailing edges are able to reach the gas in this region, so the flow is exactly what would be produced by a piston of infinite extent.

In the region of point 2, Eqs. (6–189) and (6–187) lead to

$$\frac{(p_U - p_L)_2}{(\frac{1}{2}\rho_\infty)U^2} = -\frac{8}{\pi M}\frac{\dot{h}_0}{U}\left[\frac{M}{1+M}\sqrt{\frac{t_1-x}{x+Mt_1}} + \tan^{-1}\sqrt{\frac{x+Mt_1}{t_1-x}}\right]. \tag{6–191}$$

By the use of extensions of Evvard's technique not covered in Section 6–3, Lomax *et al.* (Ref. 6–37) have determined the pressures in regions 3 and 4 to be

$$\frac{(p_U - p_L)_3}{(\frac{1}{2}\rho_\infty)U^2} = -\frac{\dot{h}_0}{U}\left\{\frac{8}{\pi(1+M)}\sqrt{\frac{t_1-x}{x+Mt_1}}\right.$$

$$+ \frac{4}{\pi M}\left[\sin^{-1}\left(\frac{2x - t_1(1-M)}{t_1(1+M)}\right)\right.$$

$$\left.\left. + \sin^{-1}\left(\frac{2(2b-x) - t_1(1+M)}{t_1(1-M)}\right)\right]\right\}, \tag{6–192}$$

$$\frac{(p_U - p_L)_4}{(\frac{1}{2}\rho_\infty)U^2} = -\frac{8}{\pi M}\frac{\dot{h}_0}{U}\sin^{-1}\sqrt{\frac{x + Mt_1 - 2b}{t_1(M-1)}}. \tag{6–193}$$

Given Eqs. (6–190)–(6–193), we can compute lift and moment per unit span of the airfoil as functions of time. Strictly speaking, there are two ranges of t_1 in which this must be done separately. The first, extending from $t_1 = 0$ to b, involves integration across regions 2, 1, and 4 with respect to x; the second, extending from $t_1 = b$ to $2b/(1 + M)$, involves integration across regions 2, 3, and 4. Fortunately, the expressions turn out to be the same for both ranges and have the forms (Ref. 6–37)

$$L_T'(t_1) = -\int_{-Mt_1}^{2b-Mt_1} [p_U - p_L]dx$$

$$= \frac{2\rho_\infty U^2(2b)}{M} \frac{\dot{h}_0}{U} \left[1 - \frac{t_1}{2b}(1 - M)\right], \qquad (6\text{–}194)$$

$$M_{yT}'(t_1) = \int_{-Mt_1}^{2b-Mt_1} [p_U - p_L][x + Mt_1]dx$$

$$= -\frac{\rho_\infty U^2(2b)^2}{M} \frac{\dot{h}_0}{U} \left[1 - \frac{t_1}{2b}(1 - M) + \frac{t_1^2 M}{8b^2}(M - 2)\right], \qquad$$
$$(6\text{–}195)$$

when

$$0 \le t_1 \le \frac{2b}{1 + M}.$$

We compare Eqs. (6–194) and (6–195) with Eqs. (6–147) and (6–148), making use of the fact that

$$s = \frac{Ut}{b} = \frac{Mt_1}{b}, \qquad (6\text{–}196)$$

and find for the dimensionless indicial functions

$$\phi_c(s) = \frac{2}{\pi M}\left[1 - \frac{s}{2M}(1 - M)\right], \qquad (6\text{–}197)$$

$$\phi_{cM}(s) = -\frac{1}{\pi M}\left[1 - \frac{s}{2M}(1 - M) + \frac{s^2}{8M}(M - 2)\right], \quad (6\text{–}198)$$

when

$$0 \le s \le \frac{2M}{1 + M}.$$

Reference 6–37 shows how this same procedure leads to indicial lift and moment for pitching motion about the airfoil's leading edge. In our notation, the final expressions read

$$L_q'(t_1) = \frac{2\rho_\infty U^2(2b)}{M}\left(\frac{q_0 b}{U}\right)\left[1 + \frac{t_1}{2b}(M - 1) + \frac{t_1^2}{4b^2}\left(M - \frac{M^2}{2}\right)\right], \quad (6\text{–}199)$$

$$M_{yq}'(t_1) = -\frac{4\rho_\infty U^2(2b)^2}{3M}\left(\frac{q_0 b}{U}\right)$$

$$\times \left\{1 - \frac{3t_1}{4b}(1 - M) + \frac{3t_1^2}{32b^2}(1 - M)^2 + \frac{t_1^3}{16b^3}\left[M + \tfrac{1}{4}(1 - M)^3\right]\right\}, \qquad$$
$$(6\text{–}200)$$

when

$$0 \le t_1 \le \frac{2b}{1 + M}.$$

Comparisons with Eqs. (6–149) and (6–150) give the dimensionless indicial functions

$$\phi_{cq}(s) = \frac{1}{\pi M}\left[1 - \frac{s}{2M}(1-M) + \frac{s^2}{4M}\left(1 - \frac{M}{2}\right)\right], \quad (6\text{–}201)$$

$$\phi_{cMq}(s) = -\frac{2}{3\pi M}\left\{1 - \frac{3s}{4M}(1-M) + \frac{3s^2}{32M^2}(1-M)^2\right.$$

$$\left. + \frac{s^3}{16M^3}\left[M + \tfrac{1}{4}(1-M)^3\right]\right\}, \quad (6\text{–}202)$$

when

$$0 \le s \le \frac{2M}{1+M}.$$

Although it is rather limited in the values of s for which it has been employed, we have described the method of supersonic analogy here for two reasons. In the first place it can, with sufficient labor, be extended up to $s = 2M/(1 - M)$, which may cover the entire range of validity of linearized theory when the Mach number is close to unity. Goodman's integral-equation technique (Ref. 6–38) actually allows it to be carried as far as the patience of the person performing the integrations endures. However, integrals appear even in the second interval of s which cannot be expressed in terms of known functions (Ref. 6–37), so that we naturally turn to the Fourier integral for solving subsonic indicial problems. This does not mean that the supersonic analogy has no other more fruitful applications. For instance, two-dimensional impulsive motion at supersonic speeds is analogous to a swept-forward wingtip with supersonic edges. The straightforward source method generates indicial functions in simple form for the entire range of s. Until more extensive tables of aerodynamic coefficients have been computed, it also appears that the analogy scheme furnishes the only rigorous way of treating certain other unsteady subsonic phenomena, like entry into a sharp-edged gust and impulsive rotation of a flap.

Beyond roughly one semichord length from the starting point, the functions $\phi_c(s)$, $\phi_{cM}(s)$, $\phi_{cq}(s)$ and, $\phi_{cMq}(s)$ for subsonic airfoils are predicted most efficiently by Fourier-integral superposition. Since the general relationships between input (motion) and output (aerodynamic load) are the same for any linear system, we can derive appropriate formulas for the airfoil in compressible flow by direct comparison with the incompressible case. In Section 5–7, we showed that if the lift for simple harmonic variation of the velocity at the three-quarter chord point is

$$L(k) = 2\pi\,\frac{\rho}{2}\,U^2(2b)\left(\frac{-w_{\frac{3}{4}c}}{U}\right)[F(k) + iG(k)], \quad (6\text{–}203)$$

then the corresponding force due to an impulsive motion

$$\frac{-w_{\frac{3}{4}c}}{U} = \alpha_0 1(s) \tag{6-204}$$

is

$$L(s) = 2\pi \frac{\rho}{2} U^2 (2b) \alpha_0 \phi(s). \tag{5-367}$$

Here

$$\phi(s) = \frac{2}{\pi} \int_0^\infty \frac{F(k)}{k} \sin ks \, dk = 1 + \frac{2}{\pi} \int_0^\infty \frac{G(k)}{k} \cos ks \, dk \tag{5-369}$$

when $s > 0$. This lift excludes, of course noncirculatory contributions, which are infinite at $s = 0$.

Without further reference to aerodynamic theory, we can assert by analogy a similar result for compressible flow. We assume the total lift per unit span to be expressed by

$$L_T'(k) = 2\pi \frac{\rho_\infty}{2} U^2 (2b) \left(\frac{\dot{h}}{U}\right) [F_c(k) + iG_c(k)]$$

$$= 2\pi \frac{\rho_\infty}{2} U^2 2ik\bar{h}e^{iks} [F_c(k) + iG_c(k)] \tag{6-205}$$

for simple harmonic plunging

$$h = \bar{h}e^{i\omega t} = \bar{h}e^{iks}. \tag{6-206}$$

It follows that an impulsive plunging of magnitude $\dot{h}_0/U$ must give rise to

$$L_T'(s) = 2\pi \frac{\rho_\infty}{2} U^2 (2b) \left(\frac{\dot{h}_0}{U}\right) \phi_c(s), \tag{6-147}$$

where

$$\phi_c(s) = \frac{2}{\pi} \int_0^\infty \frac{F_c(k)}{k} \sin ks \, dk = 1 + \frac{2}{\pi} \int_0^\infty \frac{G_c(k)}{k} \cos ks \, dk. \tag{6-207}$$

Similarly, if the other moments and lift for plunging and pitching about the leading edge have as their counterparts in simple harmonic motion

$$M_{yT}'(k) = 2\pi \frac{\rho_\infty}{2} U^2 (2b)^2 \left(\frac{\dot{h}}{U}\right) [M_c(k) + iN_c(k)], \tag{6-208}$$

$$L_q'(k) = 4\pi \frac{\rho_\infty}{2} U^2 (2b) \left(\frac{qb}{U}\right) [F_{cq}(k) + iG_{cq}(k)], \tag{6-209}$$

$$M_{yq}'(k) = 4\pi \frac{\rho_\infty}{2} U^2 (2b)^2 \left(\frac{qb}{U}\right) [M_{cq}(k) + iN_{cq}(k)], \tag{6-210}$$

the corresponding indicial functions in Eqs. (6-148) through (6-150) must be

$$\phi_{cM}(s) = \frac{2}{\pi} \int_0^\infty \frac{M_c(k)}{k} \sin ksdk = 1 + \frac{2}{\pi} \int_0^\infty \frac{N_c(k)}{k} \cos ksdk, \qquad (6\text{-}211)$$

$$\phi_{cq}(s) = \frac{2}{\pi} \int_0^\infty \frac{F_{cq}(k)}{k} \sin ksdk = 1 + \frac{2}{\pi} \int_0^\infty \frac{G_{cq}(k)}{k} \cos ksdk, \qquad (6\text{-}212)$$

$$\phi_{cMq}(s) = \frac{2}{\pi} \int_0^\infty \frac{M_{cq}(k)}{k} \sin ksdk = 1 + \frac{2}{\pi} \int_0^\infty \frac{N_{cq}(k)}{k} \cos ksdk. \qquad (6\text{-}213)$$

Equations (6-211)-(6-213) are all limited to $s > 0$.

Mazelsky and Drischler (Refs. 6-34, 6-35, and 6-36) have integrated Eqs. (6-207) and (6-211)-(6-213) numerically for Mach numbers 0.5, 0.6, and 0.7. In each case the integral containing F or M is more suitable, since its integrand has no singularity at $k = 0$, whereas the other is logarithmically infinite. The principal difficulty in these evaluations arises from the incompleteness of existing tables for simple harmonic motion, especially at the two lower Mach numbers. In Ref. 6-36 this is overcome by a process of estimating values of $F_c(k)$ for large k. It is possible to invert the Fourier integral representation of $\phi_c(s)$, so that $\phi_c(s)$ appears under the integral sign and $F_c(k)$ outside. The exact solution for the indicial function is known between $s = 0$ and $2M/(1 + M)$, and an important theorem from the theory of Fourier transforms then allows $F_c(k)$ to be approximated, when its argument is large, by a numerical integration of values of $\phi_c(s)$ near $s = 0$.

The tables of Dietze (Ref. 6-18), which form the basis of the calculations, are not in the same notation as Eq. (6-205). However, corresponding quantities are easily identified; e.g.,

$$F_c(k) = \frac{Z_2(k)}{2k}, \qquad (6\text{-}214)$$

where Z_2 is the imaginary part of Dietze's symbol for dimensionless lift due to vertical translation.

Figures 6-12(a) and 6-12(c) plot the two indicial lift functions $\phi_c(s)$ and $\phi_{cq}(s)$ for both incompressible flow and three subsonic Mach numbers. In Fig. 6-12(b) the indicial moment function for plunging motion is shown, but the moment is taken about the quarter-chord axis to demonstrate the rapidity of approach to its final value of zero. The ordinate in Fig. 6-12(b) is

$$(\phi_{cM})_{c/4} = \phi_{cM}(s) + \tfrac{1}{4}\phi_c(s). \qquad (6\text{-}215)$$

Figure 6-12(d) illustrates the indicial moment function for pitching about

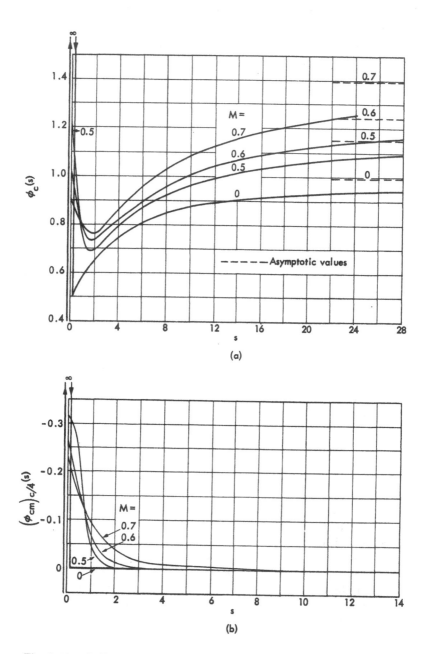

Fig. 6–12. Indicial lift and moment functions from plunging and pitching of an airfoil in subsonic compressible flow. All quantities are defined in the text.

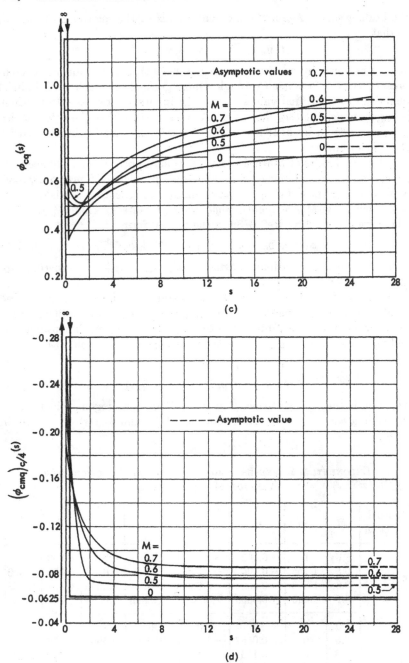

Fig. 6–12.   (Cont'd.)

the leading edge. Again the moment axis is at the quarter-chord point, so that

$$(\phi_{cMq})_{c/4} = \phi_{cMq}(s) + \tfrac{1}{4}\phi_{cq}(s). \tag{6-216}$$

It may be generally concluded from Fig. 6–12 that the influence of compressibility manifests itself through a more gradual approach of all indicial quantities to their final values. Once the impulsive starting effect, which is largely noncirculatory, has died out, we can state that the defect of lift or moment from its final magnitude, at any particular s, increases with the Mach number.

All these results of Mazelsky and Drischler appear to be fully accurate enough for use in solving routine aeroelastic problems. Since many such applications involve Laplace transformation on the time variable, it is desirable to have approximate exponential representations like Eq. (5–370). These are provided by Ref. 6–36 in the typical form

$$\phi(s) = b_0 + b_1 e^{-\beta_1 s} + b_2 e^{-\beta_2 s} + b_3 e^{-\beta_3 s}. \tag{6-217}$$

Table 6–1 lists the various constants associated with the functions for plunging motion.

In view of the importance of gust loading on aircraft in high-speed flight, we need information concerning compressible-flow indicial functions for entry of wings into sharp-edged gusts. The two-dimensional lift and moment per unit span can be expressed, as in Eq. (5–381a) by

$$L(s) = 2\pi\rho_\infty U b w_0 \psi_c(s), \tag{6-218}$$

$$M(s) = 2\pi\rho_\infty U (2b^2) w_0 \psi_{cM}(s), \tag{6-219}$$

where w_0 is the vertical velocity in the uniform gust region, which meets the leading edge at $t = 0$. Although both $\psi_c(s)$ and $\psi_{cM}(s)$ are known in

TABLE 6–1

EXPONENTIAL REPRESENTATIONS OF INDICIAL FUNCTIONS FOR
VARIOUS MACH NUMBERS

Indicial Function	M	b_0	b_1	b_2	b_3	β_1	β_2	β_3
$\phi_c(s)$	0	1	-0.165	-0.335	0	0.0455	0.300	—
	0.5	1.155	-0.406	-0.249	0.773	0.0754	0.372	1.890
	0.6	1.250	-0.452	-0.630	0.893	0.0646	0.481	0.958
	0.7	1.400	-0.5096	-0.567	-0.5866	0.0536	0.357	0.902
$(\phi_{cM})_{c/4}$	0	0	0	0	0	—	—	—
	0.5	0	0.0557	-1.000	0.6263	2.555	3.308	6.09
	0.6	0	-0.100	-1.502	1.336	1.035	4.040	5.022
	0.7	0	-0.2425	0.084	-0.069	0.974	0.668	0.438

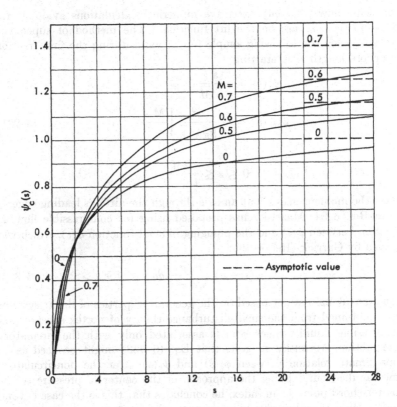

Fig. 6–13.   Indicial lift function for entry into a sharp-edged gust in subsonic compressible flow.   Symbols are defined in the text.

TABLE 6–2

EXPONENTIAL REPRESENTATION OF THE INDICIAL FUNCTION $\psi_c(s)$
FOR SHARP-EDGED GUST ENTRY AT VARIOUS MACH NUMBERS

M	b_0	b_1	b_2	b_3	β_1	β_2	β_3
0	1	−0.500	−0.500	0	0.130	1.000	—
0.5	1.155	−0.450	−0.470	−0.235	0.0716	0.374	2.165
0.6	1.250	−0.410	−0.538	−0.302	0.0545	0.257	1.461
0.7	1.400	−0.563	−0.645	−0.192	0.0542	0.3125	1.474

supersonic flow, the only extensive numerical calculations available for the subsonic case are on the lift function. The method of supersonic analogy yields the following simple forms, valid during the first fraction of a chord length after starting.

$$\psi_c(s) = \frac{s}{\pi\sqrt{M}}, \tag{6-220}$$

$$\psi_{cM}(s) = -\frac{s^2}{8\pi\sqrt{M}}\left[\frac{M+1}{M}\right], \tag{6-221}$$

for

$$0 \leq s \leq \frac{2M}{1+M},$$

where the moment acts about an axis through the airfoil's leading edge.

In Ref. 6–34, Mazelsky has proposed using for compressible flow an equation between $\psi(s)$ and the plunging indicial function $\phi(s)$, which was derived by Garrick (Ref. 6–39):

$$\psi(s) = \frac{1}{\pi}\int_0^2 \phi(s-\sigma)\sqrt{\frac{\sigma}{2-\sigma}}\,d\sigma; \qquad \text{for } s > 2. \tag{6-222}$$

Equation (6–222) is not based on the general properties of linear systems but is obtained from the small-disturbance theory of vortex sheets in incompressible liquid. Since $\phi(s)$ is associated only with the circulatory fluid motion, Mazelsky suggests that Eq. (6–222) might be used as an approximate relation between $\phi_c(s)$ and $\psi_c(s)$ after the noncirculatory flow has died out. Using the approach of the center of pressure to the quarter-chord point as an index, he concludes that this is the case beyond $s = 4$. He is therefore able to estimate $\psi_c(s)$ for $s \geq 6$.

Figure 6–13 shows the curves of $\psi_c(s)$ (from Ref. 6–36) for Mach numbers 0, 0.5, 0.6, and 0.7, calculated by a combination of Eq. (6–220), Eq. (6–222), and interpolation. There are no similar data on $\psi_{cM}(s)$. However, it seems reasonable to estimate that the indicial lift acts at the quarter-chord point throughout almost the whole maneuver, since Eq. (6–221) shows that it is already there at $s = 2M/(M+1)$. Moreover, the asymptotic lift due to the gust is equal to that developed by a flat plate at angle of attack w_0/U, so it also is centered at the quarter chord.

Table 6–2 gives the constants in exponential approximations to $\psi_c(s)$ for various Mach numbers. The literal formula for which they are designed is Eq. (6–217). All asymptotic values are obviously the same as those for $\phi_c(s)$.

Any more rigorous evaluation of $\psi_c(s)$ and $\psi_{cM}(s)$ must be based on Fourier-integral superposition of lifts and moments due to a sinusoidal gust. As shown in Section 5–7, this problem is associated with the boundary condition along the airfoil chord:

$$\bar{w}_a(x^*) = -\bar{w}_G e^{-ikx^*}, \tag{5-375}$$

where $x^* = x/b$. Equation (5–375) can be substituted into Possio's integral equation (6–111) and the pressure distribution determined by one of the standard methods of solution. Alternatively, if Mathieu functions are to be employed, Reissner's boundary condition (6–130) becomes

$$\frac{1}{\sin \zeta} \frac{\partial \chi_1}{\partial \xi} = - \frac{e^{-i\frac{k \cos \zeta}{1 - M^2}}}{\sqrt{1 - M^2}} b \bar{w}_G = g(\zeta); \qquad \text{for } \xi = 0. \qquad (6\text{–}223)$$

Since this condition involves only the relatively simple noncirculatory part of the flow, and no more difficulty will be encountered in evaluating the integrals in Eq. (6–137) than for motions of the rigid airfoil, it appears that this approach is more efficient than Possio's. In fact, calculations for plunging oscillation furnish directly the velocity potential due to a sinusoidal gust at a modified reduced frequency.

Regardless of how they are calculated, the lift and moment in the simple harmonic case can be written

$$L(k) = 2\pi\rho_\infty U b \bar{w}_G [F_G(k) + i G_G(k)] e^{iks}, \qquad (6\text{–}224)$$

$$M(k) = 2\pi\rho_\infty U (2b^2) \bar{w}_G [M_G(k) + i N_G(k)] e^{iks}. \qquad (6\text{–}225)$$

The indicial loads are determined numerically from Fourier integrals like Eq. (5–381b), which may be carried over unchanged to the compressible case.

6–6 Oscillating airfoils at supersonic speeds. The problem of small lateral oscillations of wings in supersonic flight was first solved (Ref. 6–40) by a complicated approach involving "hook" integration. In Ref. 6–41 Garrick demonstrates how the two-dimensional case can be treated much more simply using the method of source pulses. It is his technique that we first discuss here. The differential equation and principal boundary condition are, of course, just Eqs. (6–1) and (5–225), which we reproduce in the most suitable forms for supersonic flow.

$$(M^2 - 1) \frac{\partial^2 \phi'}{\partial x^2} - \frac{\partial^2 \phi'}{\partial z^2} + \frac{1}{a_\infty^2} \frac{\partial^2 \phi'}{\partial t^2} + \frac{2M}{a_\infty} \frac{\partial^2 \phi'}{\partial x \partial t} = 0, \qquad (6\text{–}1)$$

$$w = \frac{\partial \phi'}{\partial z} = \frac{\partial z_a}{\partial t} + U \frac{\partial z_a}{\partial x} = w_a(x,t); \qquad \text{for } z = 0, \ -b \leq x \leq b. \qquad (5\text{–}225)$$

We leave the time dependence of w_a arbitrary, since the solution proves to be valid for any small motion whatever. As auxiliary boundary conditions for Mach numbers greater than unity, we require that there be no disturbances ($\phi' = 0$) ahead of the rearward-leaning Mach lines from the leading edge, and that signals propagate outward from the airfoil at infinity.

We have already shown in Section 6–4 that the upper and lower airfoil surfaces are independent, so that we can find our answers for $z \geq 0^+$ and rely on the antisymmetry of the pressure distribution when calculating aerodynamic loads. Because of the success of source methods for steady-flow problems of this type, we are led to seek a similar procedure in the present case. The source pulse, an exact counterpart of the same elementary solution for the acceleration potential, given by Eq. (6–85), proves to be the necessary tool. Regarded here as a velocity potential which satisfies the two-dimensional Eq. (6–1), this can be written

$$\phi_{sp}{}'(x,z,t) = \frac{A(\xi,\zeta,T)}{\sqrt{a_\infty{}^2(t-T)^2 - [(x-\xi) - U(t-T)]^2 - (z-\zeta)^2}} \qquad (6\text{–}226)$$

The disturbance described by Eq. (6–226) has many of the same properties as its subsonic analog. It gives the effect at point (x,z) at time t of an impulsive signal starting from (ξ,ζ) at T. In supersonic flow we anticipate that the source pulse can be felt just in the downstream zone of influence of (ξ,ζ), which is a wedge with faces inclined at the Mach angle above and below $z = \zeta$. Even in this zone there is only a finite interval of the time delay

$$\tau = t - T \qquad (6\text{–}227)$$

within which the impulse is sensed at (x,z). Before some lower limit $\tau = \tau_1$ the pulse has not had time to reach (x,z); after some $\tau = \tau_2$, the expanding cylinder containing the disturbance gets swept past (x,z) by a uniform stream that exceeds the speed of sound. These two extreme positions of the cylinder are illustrated in Fig. 6–14.

All this behavior is implied by Eq. (6–226), because the quantity in the radical is negative except between the two time delays

$$\tau_1 = \frac{M(x-\xi) - \sqrt{(x-\xi)^2 - (M^2-1)(z-\zeta)^2}}{a_\infty(M^2-1)} \qquad (6\text{–}228a)$$

and

$$\tau_2 = \frac{M(x-\xi) + \sqrt{(x-\xi)^2 - (M^2-1)(z-\zeta)^2}}{a_\infty(M^2-1)}. \qquad (6\text{–}228b)$$

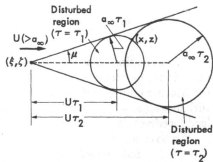

Fig. 6–14. Two positions of the disturbed region produced by a source pulse at point (ξ,ζ) at time $t = 0$ in two-dimensional supersonic flow.

These are both positive inside the downstream zone of influence; they are the two roots of the denominator of Eq. (6–226). Outside this zone, before τ_1, and after τ_2, we set ϕ_{sp}' equal to zero. On the Mach wedge from (ξ,ζ) the radical in Eqs. (6–228) vanishes, so that τ_1 and τ_2 are equal. This is as expected, for a particular pulse should be sensed there only at one instant.

Arbitrary unsteady motion of a thin airfoil can be represented by an appropriate sheet of sources along the projection $z = 0$, $-b \leq x \leq b$ of the airfoil chord. They must also be distributed in time, which we can accomplish by integrating with respect either to T or the dummy variable τ defined in Eq. (6–227). Accounting for the restrictions on what pulses can affect a point (x,z) at a particular instant, we can write the disturbance potential of the sheet:

$$\phi'(x,z,t) = \int_{-b}^{\xi_1} \int_{\tau_1}^{\tau_2} \frac{A(\xi, t - \tau)d\tau d\xi}{\sqrt{a_\infty^2 \tau^2 - [(x - \xi) - U\tau]^2 - z^2}}. \quad (6–229)$$

In Eq. (6–229) ξ_1 is obtained from the fact that no signal reaches (x,z) which originates behind the point $(\xi_1,0)$, where the upstream Mach wedge of (x,z) hits the sheet. That is,

$$\frac{z}{x - \xi_1} = \tan \mu = \frac{1}{\sqrt{M^2 - 1}} \quad (6–230a)$$

or

$$\xi_1 = x - z\sqrt{M^2 - 1}. \quad (6–230b)$$

Since τ_1 and τ_2 are roots of the denominator radical, we remove the coefficient $(U^2 - a_\infty^2)$ of the squared term and rewrite Eq. (6–229):

$$\phi'(x,z,t) = \frac{1}{\sqrt{U^2 - a_\infty^2}} \int_{-b}^{\xi_1} \int_{\tau_1}^{\tau_2} \frac{A(\xi, t - \tau)d\tau d\xi}{\sqrt{(\tau - \tau_1)(\tau_2 - \tau)}}. \quad (6–231)$$

There is no essential difference between Eq. (6–231) and the integral equation (6–68) for steady, three-dimensional supersonic flow. We therefore achieve the same simplification as in Section 6–3 by defining an angle variable

$$\tau = \frac{\tau_2 + \tau_1}{2} - \frac{\tau_2 - \tau_1}{2} \cos \theta. \quad (6–232)$$

By direct analogy with Eq. (6–71), this transforms Eq. (6–231) into

$$\phi'(x,z,t) = \frac{1}{\sqrt{U^2 - a_\infty^2}}$$

$$\times \int_{-b}^{\xi_1} \int_0^{\pi} A\left(\xi, t - \frac{M(x-\xi)}{a_\infty(M^2-1)} + \frac{\sqrt{(x-\xi)^2 - z^2(M^2-1)}}{a_\infty(M^2-1)} \cos \theta\right) d\theta d\xi.$$

$$(6–233)$$

Moreover, it follows by comparison with Eq. (6–74) that

$$\frac{\partial \phi'(x,0^+,t)}{\partial z} = -\frac{\pi}{a_\infty} A(x,t). \qquad (6\text{–}234)$$

Taking advantage of this point-function character of the source pulse, we apply boundary condition (5–225) and get

$$A(x,t) = -\frac{a_\infty}{\pi} w_a(x,t). \qquad (6\text{–}235)$$

Equation (6–235) holds for the upper surface of the airfoil. In the lower half-plane, $\xi_1 = x + z\sqrt{M^2 - 1}$, which leads to the conclusion that the sign in Eqs. (6–234) and (6–235) must be reversed. We do not have to use this fact in applications, because the pressure difference across the airfoil is twice the disturbance pressure $(p_U - p_\infty)$ calculated for $z = 0^+$. Substitution of Eq. (6–235) back into Eq. (6–231) inverts the integral equation and reduces our problem to one of integration:

$$\phi'(x,z,t) = -\frac{1}{\pi\sqrt{M^2 - 1}} \int_{-b}^{\xi_1} \int_{\tau_1}^{\tau_2} \frac{w_a(\xi, t - \tau)d\tau d\xi}{\sqrt{(\tau - \tau_1)(\tau_2 - \tau)}}. \qquad (6\text{–}236)$$

Of the various physical situations that can be handled by Eq. (6–236), the first to be fully explored was simple harmonic vibration of the airfoil (Ref. 6–41). In this case,

$$w_a(\xi, t - \tau) = \bar{w}_a(\xi)e^{i\omega(t-\tau)} = w_a(\xi,t)e^{-i\omega\tau}. \qquad (6\text{–}237)$$

When Eq. (6–237) is put into Eq. (6–236) and the variable θ is reintroduced, we find

$$\phi'(x,z,t) = -\frac{1}{\pi\sqrt{M^2 - 1}} \int_{-b}^{\xi_1} w_a(\xi,t) \int_{\tau_1}^{\tau_2} \frac{e^{-i\omega\tau}}{\sqrt{(\tau - \tau_1)(\tau_2 - \tau)}} d\tau d\xi$$

$$= -\frac{1}{\pi\sqrt{M^2 - 1}} \int_{-b}^{\xi_1} w_a(\xi,t)e^{-i\omega\left(\frac{\tau_2+\tau_1}{2}\right)} \int_0^\pi e^{i\omega\left(\frac{\tau_2-\tau_1}{2}\right)\cos\theta} d\theta d\xi. \qquad (6\text{–}238)$$

After the values of τ_1 and τ_2 for the upper airfoil surface $z = 0$ are substituted, Eq. (6–238) becomes

$$\phi'(x,0^+,t) = -\frac{1}{\pi\sqrt{M^2 - 1}} \int_{-b}^x w_a(\xi,t)e^{-i\omega\frac{(x-\xi)M}{a_\infty(M^2-1)}} \int_0^\pi e^{i\omega\frac{(x-\xi)}{a_\infty(M^2-1)}\cos\theta} d\theta d\xi. \qquad (6\text{–}239)$$

The inner integral in Eq. (6–239) is identical with a well-known representation of the Bessel function:

$$\pi J_0(q) = \int_0^\pi e^{iq\cos\theta} d\theta. \qquad (6\text{–}240)$$

Hence we can employ Eq. (6–240) and the dimensionless variables

$$x^* = \frac{x}{b}, \qquad \xi^* = \frac{\xi}{b} \tag{5–96}$$

to rewrite Eq. (6–239) in its final form

$$\phi'(x^*,0^+,t) = -\frac{b}{\sqrt{M^2-1}} \int_{-1}^{x^*} w_a(\xi^*,t) e^{-i(\bar{\omega}/2)(x^*-\xi^*)} J_0\left(\frac{\bar{\omega}}{2M}(x^*-\xi^*)\right) d\xi^*. \tag{6–241}$$

Here $\bar{\omega}$ is a combination of reduced frequency and Mach number, which plays a role in supersonic theory somewhat similar to that of k in the incompressible problem,

$$\bar{\omega} = \frac{2kM^2}{M^2-1}. \tag{6–242}$$

The pressure distribution and aerodynamic loads on the airfoil follow from previously derived relations.

$$\begin{aligned} p_U - p_L &= -2\rho_\infty \left(\frac{\partial\phi'}{\partial t} + U\frac{\partial\phi'}{\partial x}\right)\bigg|_{z=0^+} \\ &= -2\rho_\infty \frac{U}{b}\left(ik\phi' + \frac{\partial\phi'}{\partial x^*}\right)\bigg|_{z^*=0^+} \end{aligned} \tag{6–243}$$

$$L(t) = -\int_{-b}^{b} [p_U - p_L]dx = -b\int_{-1}^{1} [p_U - p_L]dx^*. \tag{5–106}$$

$$\begin{aligned} M_y(t) &= \int_{-b}^{b} [p_U - p_L][x - ba]dx \\ &= b^2 \int_{-1}^{1} [p_U - p_L][x^* - a]dx^*. \end{aligned} \tag{5–107}$$

As an illustration of the application of these formulas, we consider vertical translation of the airfoil, which is described by

$$w_a = -\dot{h}(t) = -i\omega h_0 e^{i\omega t}. \tag{6–244}$$

Equation (6–241) gives for the disturbance potential

$$\begin{aligned} \phi_h'(x^*,0^+,t) &= \frac{b\dot{h}}{\sqrt{M^2-1}} \int_{-1}^{x^*} e^{-i(\bar{\omega}/2)(x^*-\xi^*)} J_0\left(\frac{\bar{\omega}}{2M}(x^*-\xi^*)\right) d\xi^* \\ &= \frac{2b\dot{h}}{\sqrt{M^2-1}} \int_{0}^{(x^*+1)/2} e^{-i\bar{\omega}u} J_0\left(\frac{\bar{\omega}}{M}u\right) du, \end{aligned} \tag{6–245}$$

where

$$u = \frac{x^* - \xi^*}{2}. \tag{6–246}$$

The pressures from Eq. (6–243) are

$$p_U - p_L = \frac{-4\rho_\infty b}{\sqrt{M^2 - 1}} \left\{ \ddot{h} \int_0^{(x^*+1)/2} e^{-i\bar{\omega}u} J_0\left(\frac{\bar{\omega}}{M} u\right) du \right.$$

$$\left. + \frac{hU}{2b} e^{-i\bar{\omega}[(x^*+1)/2]} J_0\left(\frac{\bar{\omega}}{M}\left(\frac{x^*+1}{2}\right)\right) \right\}. \qquad (6\text{–}247)$$

Lift and moment are most easily calculated by working directly with ϕ_h'. Thus,

$$L_h(t) = 2\rho_\infty b \int_{-1}^{1} \left[\frac{\partial \phi_h'}{\partial t} + \frac{U}{b} \frac{\partial \phi_h'}{\partial x^*} \right] dx^*$$

$$= 2\rho_\infty b \frac{\partial}{\partial t} \int_{-1}^{1} \phi_h' dx^* + 2\rho_\infty U \phi_h'(1,0^+,t), \qquad (6\text{–}248)$$

where we have used the fact that ϕ_h' vanishes at the leading edge. If we make the change of variable

$$u' = \frac{x^* + 1}{2} \qquad (6\text{–}249)$$

in the integral involving x^*, we can substitute Eq. (6–245) into Eq. (6–248) and obtain the final expression

$$L_h(t) = \frac{4\rho_\infty b}{\sqrt{M^2 - 1}} [Ur_1 h + 2br_2 \ddot{h}], \qquad (6\text{–}250)$$

where

$$r_1(M,\bar{\omega}) = \int_0^1 e^{-i\bar{\omega}u} J_0\left(\frac{\bar{\omega}}{M} u\right) du \qquad (6\text{–}251)$$

and

$$r_2(M,\bar{\omega}) = \int_0^1 \int_0^{u'} e^{-i\bar{\omega}u} J_0\left(\frac{\bar{\omega}}{M} u\right) du\, du'. \qquad (6\text{–}252)$$

A similar series of operations leads to the moment

$$M_{yh}(t) = b[a + 1]L_h(t) - \frac{8\rho_\infty b^2}{\sqrt{M^2 - 1}} [Uq_1 h + bq_2 \ddot{h}], \qquad (6\text{–}253)$$

where

$$q_1(M,\bar{\omega}) = \int_0^1 u e^{-i\bar{\omega}u} J_0\left(\frac{\bar{\omega}}{M} u\right) du \qquad (6\text{–}254)$$

and

$$q_2(M,\bar{\omega}) = 2\int_0^1 \int_0^{u'} u' e^{-i\bar{\omega}u} J_0\left(\frac{\bar{\omega}}{M} u\right) du\, du'. \qquad (6\text{–}255)$$

Lift, pitching moment, and hinge moment due to simple harmonic motions of the airfoil-flap combination described by Eq. (5–268) can all be expressed in terms of a series of integrals like r_1 and r_2, which are functions of the Mach number and supersonic reduced-frequency parameter $\bar{\omega}$. As shown by Garrick and Rubinow (Ref. 6–41), these integrals are related algebraically to four simpler ones, given by the equation

$$f_\lambda(M,\bar{\omega}) = \int_0^1 u^\lambda e^{-i\bar{\omega}u} J_0\left(\frac{\bar{\omega}}{M} u\right) du; \qquad \text{for } \lambda = 0, 1, 2, \text{ and } 3. \quad (6\text{–}256)$$

For instance, the definitions (6–251) and (6–254) make it clear that

$$r_1(M,\bar{\omega}) = f_0(M,\bar{\omega}) \qquad (6\text{–}257)$$

and

$$q_1(M,\bar{\omega}) = f_1(M,\bar{\omega}). \qquad (6\text{–}258)$$

Moreover, if we interchange the orders of integration in Eq. (6–252) and Eq. (6–255), we are led to

$$\begin{aligned}
r_2(M,\bar{\omega}) &= \int_0^1 e^{-i\bar{\omega}u} J_0\left(\frac{\bar{\omega}}{M} u\right)\left[\int_u^1 du'\right] du \\
&= \int_0^1 [1 - u]e^{-i\bar{\omega}u} J_0\left(\frac{\bar{\omega}}{M} u\right) du \\
&= f_0(M,\bar{\omega}) - f_1(M,\bar{\omega}), \qquad (6\text{–}259)
\end{aligned}$$

$$\begin{aligned}
q_2(M,\bar{\omega}) &= \int_0^1 e^{-i\bar{\omega}u} J_0\left(\frac{\bar{\omega}}{M} u\right)\left[2\int_u^1 u'du'\right] du \\
&= \int_0^1 e^{-i\bar{\omega}u} J_0\left(\frac{\bar{\omega}}{M} u\right)[1 - u^2]du \\
&= f_0(M,\bar{\omega}) - f_2(M,\bar{\omega}). \qquad (6\text{–}260)
\end{aligned}$$

The integrals f_1, f_2, and f_3 can all be computed from f_0 by means of a recursion formula. f_0 is not expressible in terms of elementary functions, but it has been evaluated by Schwarz (Ref. 6–42), using the series

$$f_0(M,\bar{\omega}) = e^{-i\bar{\omega}} \sum_{n=0}^{\infty} \left(\frac{M^2 - 1}{M^2}\,\bar{\omega}\right)^n \frac{[J_n(\bar{\omega}) + iJ_{n+1}(\bar{\omega})]}{2^n n!\,(2n + 1)}. \quad (6\text{–}261)$$

The appearance of $\bar{\omega}$ as the argument of the Bessel functions here explains its special significance as a supersonic reduced frequency. The tables of Schwarz cover Mach numbers 10/9, 5/4, 10/7, 5/3, 2, 5/2, 10/3, and 5, which accounts for the recurrent appearance of these values in listings of flutter coefficients.

References 6–41 and 6–43 present tables of the f_λ integrals and of the aerodynamic loads on the oscillating airfoil-flap combination without aerodynamic balance or gap. Lift and moments are related to the tabulated quantities as follows:

$$L = 4\rho_\infty U^2 bk^2 \left\{ [L_1 + iL_2]\frac{h}{b} + [L_3 + iL_4]\alpha + [L_5 + iL_6]\beta \right\},$$

$$(6\text{-}262)$$

$$M_y = -4\rho_\infty U^2 b^2 k^2 \left\{ [M_1 + iM_2]\frac{h}{b} + [M_3 + iM_4]\alpha + [M_5 + iM_6]\beta \right\},$$

$$(6\text{-}263)$$

$$M_\beta = -4\rho_\infty U^2 b^2 k^2 \left\{ [N_1 + iN_2]\frac{h}{b} + [N_3 + iN_4]\alpha + [N_5 + iN_6]\beta \right\}.$$

$$(6\text{-}264)$$

Here all the dimensionless real numbers L_1, L_2, L_3, etc., are calculable from the f_λ through intermediate integrals like q_1, q_2, q_3, and q_4. L_1 and L_2 are functions only of M and $\bar{\omega}$, but the remaining coefficients depend also on the axis of rotation a, on the dimensionless coordinate of the flap hingeline e, or on both. For purposes of tabulation, their dependence on a is easily separated out. Thus, those associated with vertical translation and pitching oscillation are written

$$L_3 = L_3{}' - (a + 1)L_1, \qquad (6\text{-}265)$$

$$L_4 = L_4{}' - (a + 1)L_2, \qquad (6\text{-}266)$$

$$M_1 = M_1{}' - (a + 1)L_1, \qquad (6\text{-}267)$$

$$M_2 = M_2{}' - (a + 1)L_2, \qquad (6\text{-}268)$$

$$M_3 = M_3{}' - (a + 1)[(M_1{}' + L_3{}') - (a + 1)L_1], \qquad (6\text{-}269)$$

$$M_4 = M_4{}' - (a + 1)[(M_2{}' + L_4{}') - (a + 1)L_2], \qquad (6\text{-}270)$$

where the various coefficients on the right are independent of a. Incidentally, these latter coefficients have been translated into the notation of Ref. 6–29 and published by Luke (Ref. 6–30). Tables for other Mach numbers are given by Temple and Jahn (Ref. 6–47).

The calculation and tabulation of forces and moments due to rotation of the flap can be much simplified by recognizing their close relationship with those for the airfoil as a whole. The flap's effects are determined by suppressing any motion of the forward portion of the airfoil and finding the disturbance potential and pressure distribution set up by it alone. Within the limits of linearized theory, however, the remainder of the airfoil produces no disturbance when it does not move, and the flap cannot affect the supersonic flow ahead of its own hingeline. Hence the flap behaves just like a smaller airfoil, of chord $b(1 - e)$, rotating about its leading edge. The Mach number of the flow is unchanged, but k and $\bar{\omega}$ are reduced in the ratio $(1 - e)/2$ of the chords. By comparison with Eq. (6–262), we should be able to write the lift on the flap (and thus the force applied by the flap to the entire airfoil)

$$L_\beta = 4\rho_\infty U^2 \left[\frac{b(1 - e)}{2}\right]\left[\frac{\omega b(1 - e)}{2}\right]^2$$

$$\times \left[L_3\left(M,\bar{\omega}\,\frac{1 - e}{2}, a = -1\right) + iL_4\left(M,\bar{\omega}\,\frac{1 - e}{2}, a = -1\right)\right]\beta$$

$$= 4\rho_\infty U^2 \left[\frac{b(1 - e)}{2}\right]\left[\frac{\omega b(1 - e)}{2}\right]^2$$

$$\times \left[L_3'\left(M,\bar{\omega}\,\frac{1 - e}{2}\right) + iL_4'\left(M,\bar{\omega}\,\frac{1 - e}{2}\right)\right]\beta. \qquad (6\text{–}271)$$

Since it is also true, by definition, that

$$L_\beta = 4\rho_\infty U^2 b \left(\frac{\omega b}{U}\right)^2 [L_5(M,\bar{\omega},e) + iL_6(M,\bar{\omega},e)]\beta, \qquad (6\text{–}272)$$

it must follow that*

$$L_5(M,\bar{\omega},e) = \left[\frac{1 - e}{2}\right]^3 L_3'\left(M,\bar{\omega}\,\frac{1 - e}{2}\right), \qquad (6\text{–}273\text{a})$$

$$L_6(M,\bar{\omega},e) = \left[\frac{1 - e}{2}\right]^3 L_4'\left(M,\bar{\omega}\,\frac{1 - e}{2}\right). \qquad (6\text{–}273\text{b})$$

Similar relations can be derived for M_5, M_6, N_5, and N_6. The other coefficients in Eq. (6–264) are found from weighted integrals of the pressure distribution over the flap due to plunging and pitching of the airfoil. Although their connection with the airfoil coefficients cannot be so easily reasoned on physical grounds, simple mathematical relations are derived by Garrick and Rubinow.

In connection with aerodynamic loads on oscillating supersonic airfoils, we should mention that very little experimental data is generally available, especially measurements taken at Reynolds numbers high enough to ensure that the effects of viscosity do not jeopardize comparisons with theory. Several authors, including Van Dyke (Ref. 6–44) and W. P. Jones (Ref. 6–45), have examined in various approximate ways the nonlinear influence

---

* Since reduced frequency k appears explicitly, along with $\bar{\omega}$, in the literal expressions for L_3' and L_4' (Ref. 6–41), it must also be replaced by $k(1 - e)/2$ when making the comparison.

of thickness on the oscillatory pressure distribution. Their principal discovery of importance to the flutter analyst is an appreciable reduction of the damping in pitch of the airfoil, which is associated with the coefficient M_4. The effect is most pronounced for lower Mach numbers and reduced frequencies, and for forward locations of the rotation axis of the airfoil. There results an increase in the range of instability of pitching oscillations over that which is predicted by linear theory. For very high reduced frequencies, a slight improvement of stability may be produced by the nonlinearity, however.

When both Mach number and reduced frequency are large enough, an elementary method of approach proposed by Lighthill (Ref. 6–46) can be applied to the vibrating airfoil. The flow is assumed to consist of a series of fluid slabs, contained between adjacent pairs of yz-planes, within each of which the motion is one-dimensional in the z-direction. Any such slab can be treated like a uniformly translating walled container full of gas; at one end a piston (the airfoil surface) moves in and out, sending forth a train of simple compression and expansion waves. The instantaneous piston velocity is $w_a(x,t)$, the rate at which particles in contact with the airfoil are forced to move parallel to the z-axis. Lighthill's theory is not restricted to motions with small amplitude, but we can linearize it for aeroelastic application by the limitation

$$w_a \ll a_\infty. \tag{6–274}$$

Condition (6–274) permits us to use Eq. (6–163), with u_0 replaced by $w_a(x,t)$, for the instantaneous pressure developed in the one-dimensional flow over the upper surface of the airfoil. Because of the antisymmetry in z, it follows that

$$p_U - p_L = 2\rho_\infty a_\infty w_a(x,t). \tag{6–275}$$

Starting from this elementary point-function relationship, we can derive approximate values for the aerodynamic coefficients, such as

$$L_1 \cong M_1' \cong 0, \tag{6–276a}$$

$$L_2 \cong L_4' \cong M_2' \cong \frac{1}{Mk}, \tag{6–276b}$$

$$L_3' \cong M_3' \cong \frac{1}{Mk^2}, \tag{6–276c}$$

$$M_4' \cong \frac{4}{3Mk}, \tag{6–276d}$$

for $Mk \gg 1$. By an asymptotic examination of the f_λ functions, it is readily proved that these expressions are correct according to the more exact theory, when terms are retained only up to order $1/k^2$. We can gain some idea of the practical meaning of the requirement of large Mk by comparing Eqs. (6-276a)-(6-276d) with the tables of Garrick and Rubinow. At Mach number 2.5, for example, they appear sufficiently accurate for engineering purposes (all nonzero coefficients are well within 5 percent of exact) when the reduced frequency exceeds 2. Incidentally, the "piston theory" of Lighthill permits the introduction of thickness effects into flutter calculations. When second-order terms in w_a are retained, very significant reductions in predicted flutter speeds at high supersonic Mach numbers are discovered in this way.

At the lower end of the reduced-frequency range, when

$$\bar{\omega} \ll 1 \tag{6-277}$$

for any Mach number not too close to unity, another even more important simplification of the supersonic theory is possible. It is founded on the observation that the argument of J_0 in Eq. (6-256) is always less than $\bar{\omega}$, and J_0 is within one percent of its limiting value unity whenever its argument is less than 0.2. Taking advantage of the series expansion for the exponential, we can write

$$f_\lambda(M,\bar{\omega}) \cong \int_0^1 u^\lambda [1 - i\bar{\omega}u + \cdots] \cdot 1 \cdot du$$

$$= \frac{1}{\lambda + 1} - \frac{i\bar{\omega}}{\lambda + 2} - \frac{\bar{\omega}^2}{2(\lambda + 3)} + \cdots . \tag{6-278}$$

The first three terms are all that are needed for calculations. Equation (6-278) reduces f_λ to a function of the single variable $\bar{\omega}$ and facilitates tabulating the aerodynamic coefficients enormously within its range of validity. Since a fair proportion of cases of supersonic flutter occur at reduced frequencies below 0.1, formulas based on Eq. (6-278) will become increasingly useful as flutter Mach numbers grow larger and cause corresponding reductions in $\bar{\omega}$.

A final application of the general theory of the oscillating supersonic airfoil is presented because of its significance for the gust problem. We consider the loading due to a sinusoidal gust, which, as discussed in Section 5-7, requires a normal velocity distribution:

$$w_a(x^*,t) = -\bar{w}_G e^{-ikx^*} e^{i\omega t}. \tag{5-375}$$

When Eq. (5-375) is inserted into Eq. (6-241), we obtain

$$\phi_G'(x^*,0^+,t) = \frac{b\bar{w}_G e^{i\omega t}}{\pi\sqrt{M^2 - 1}} \int_{-1}^{x^*} e^{-ik\xi^*} e^{-i(\bar{\omega}/2)(x^* - \xi^*)} J_0\left(\frac{\bar{\omega}}{2M}(x^* - \xi^*)\right) d\xi^*$$

$$= \frac{2b\bar{w}_G e^{i\omega t}}{\pi\sqrt{M^2 - 1}} e^{-ikx^*} \int_0^{(x^*+1)/2} e^{-i(\bar{\omega}u/M^2)} J_0\left(\frac{\bar{\omega}}{M}u\right) du, \tag{6-279}$$

where u is defined by Eq. (6–246). The lift and moment are found by substituting Eq. (6–279), first into Eq. (6–243), and then into Eqs. (5–106) and (5–107). The detailed steps in calculating running lift are as follows:

$$
\begin{aligned}
L_G(t) &= 2\rho_\infty U \phi_G'(1,0^+,t) + 2\rho_\infty b \frac{\partial}{\partial t} \int_{-1}^{1} \phi_G' dx^* \\
&= \frac{4\rho_\infty U b \bar{w}_G e^{i\omega t}}{\sqrt{M^2 - 1}} \left\{ e^{-ik} \int_0^1 e^{-i(\bar{\omega}/M^2)u} J_0\left(\frac{\bar{\omega}}{M} u\right) du \right. \\
&\qquad \left. + 2ik \int_0^1 e^{ik} e^{-2iku'} \int_0^{u'} e^{-i(\bar{\omega}/M^2)u} J_0\left(\frac{\bar{\omega}}{M} u\right) du\, du' \right\} \\
&= \frac{4\rho_\infty U b \bar{w}_G e^{i\omega t}}{\sqrt{M^2 - 1}} \left\{ e^{-ik} \int_0^1 e^{-i(\bar{\omega}/M^2)u} J_0\left(\frac{\bar{\omega}}{M} u\right) du \right. \\
&\qquad \left. + e^{ik} \int_0^1 e^{-i(\bar{\omega}/M^2)u} J_0\left(\frac{\bar{\omega}}{M} u\right) [e^{-2iku} - e^{-2ik}] du \right\} \\
&= \frac{4\rho_\infty U b \bar{w}_G e^{i\omega t}}{\sqrt{M^2 - 1}} e^{ik} \int_0^1 e^{-i\bar{\omega}u} J_0\left(\frac{\bar{\omega}}{M} u\right) du \\
&= \frac{4\rho_\infty U b \bar{w}_G e^{i\omega t}}{\sqrt{M^2 - 1}} e^{ik} f_0(M,\bar{\omega}). \qquad (6\text{–}280)
\end{aligned}
$$

Here, after introducing $u' = (x^* + 1)/2$ and interchanging order of integration in the second integral, we have recognized that the entire first term inside the braces is cancelled by the e^{-2ik} part of the second term.

Another calculation along the same lines as Eq. (6–280) yields for the pitching moment per unit span about an axis at $x = ba$,

$$
M_{yG}(t) = \frac{2\rho_\infty U b^2 \bar{w}_G e^{i\omega t}}{\sqrt{M^2 - 1}} e^{ik} [(2a + 1) f_0(M,\bar{\omega}) - f_1(M,\bar{\omega})]. \qquad (6\text{–}281)
$$

The first term in Eq. (6–281) can be thought of as the effect of the lift acting at the quarter-chord point $a = -\frac{1}{2}$. However, the presence of f_1 shows that the gust produces no fixed center of pressure, independent of frequency, such as occurs in the incompressible case.

It is instructive in closing this section on airfoil oscillations at supersonic speeds to review a remarkable simplification of the analysis which can be achieved by means of the Laplace transform. We start out from the assumption of simple harmonic motion

$$
\phi'(x,z,t) = \bar{\phi}'(x,z) e^{i\omega t}, \qquad (6\text{–}118)
$$

which enables us to rewrite differential equation (6-1) as

$$\frac{\partial^2 \bar{\phi}'}{\partial x^2} + \frac{\partial^2 \bar{\phi}'}{\partial z^2} = M^2 \frac{\partial^2 \bar{\phi}'}{\partial x^2} + \frac{2Mi\omega}{a_\infty} \frac{\partial \bar{\phi}'}{\partial x} - \frac{\omega^2}{a_\infty^2} \bar{\phi}'. \qquad (6\text{-}282)$$

The principal boundary condition (5-225) is

$$\frac{\partial \bar{\phi}'}{\partial z} = \bar{w}_a(x); \qquad \text{for } z = 0, \, 0 \leq x \leq 2b, \qquad (6\text{-}283)$$

where we have placed the leading edge of the airfoil at the origin to facilitate Laplace transformation.

As pointed out by Stewartson (Ref. 6-48), Eq. (6-282) is reduced to a differential equation for exponentials by applying Eq. (5-401) to its functional dependence on x. Thus, we define

$$\bar{\phi}^*(p,z) = \mathcal{L}\{\bar{\phi}'(x,z)\} = \int_0^\infty e^{-px}\bar{\phi}'(x,z)dx. \qquad (6\text{-}284)$$

This has the well-known consequences, when $\bar{\phi}' = \partial\bar{\phi}'/\partial x = 0$ ahead of $x = 0$,

$$\mathcal{L}\left\{\frac{\partial^2 \bar{\phi}'}{\partial x^2}\right\} = p^2 \bar{\phi}^*, \qquad (6\text{-}285)$$

$$\mathcal{L}\left\{\frac{\partial \bar{\phi}'}{\partial x}\right\} = p\bar{\phi}^*. \qquad (6\text{-}286)$$

Equations (6-284) through (6-286) transform Eq. (6-282) into

$$\frac{\partial^2 \bar{\phi}^*}{\partial z^2} = \left[p^2(M^2-1) + 2p\frac{Mi\omega}{a_\infty} - \frac{\omega^2}{a_\infty^2}\right]\bar{\phi}^* = \mu^2\bar{\phi}^*, \qquad (6\text{-}287)$$

where

$$\mu = \sqrt{p^2(M^2-1) + 2p\frac{Mi\omega}{a_\infty} - \frac{\omega^2}{a_\infty^2}}. \qquad (6\text{-}288)$$

At the same time, boundary condition (6-283) becomes

$$\frac{\partial \bar{\phi}^*}{\partial z} = \bar{w}_a^*(p); \qquad \text{for } z = 0. \qquad (6\text{-}289)$$

In calculating the Laplace transform $\bar{w}_a^*(p)$ of the vertical velocity distribution, it makes no difference what values we assign $\bar{w}_a(x)$ behind the trailing edge $x = 2b$. In any correct solution, the flow there cannot affect the aerodynamic loading of the wing.

We see how well suited this problem is to Laplace transformation from two properties it has in common with time-dependent transient problems in linear, lumped-parameter systems. First, $\bar{\phi}'$ vanishes when $x < 0$, just as transients are assumed to start at $t = 0$. Second, values of $\bar{w}_a(x)$ behind a certain point are unable to influence $\bar{\phi}'(x,z)$ ahead of that point, which is analogous to the fact that a future input cannot possibly affect the response of a system at times before it is applied.

The general solution of Eq. (6-287) is (the real part of μ is always positive)

$$\bar{\phi}^* = Ae^{\mu z} + Be^{-\mu z}. \tag{6-290}$$

If we restrict ourselves to points above the airfoil, $z \geq 0^+$, we must omit the term containing A so as to avoid an answer which grows infinitely large at infinity. The converse is true in the lower half space, and we can write

$$\bar{\phi}^* = \begin{cases} Ae^{-\mu z}, & z \geq 0^+, \\ Be^{\mu z}, & z \leq 0^-. \end{cases} \tag{6-291}$$

In the first instance,

$$\frac{\partial \bar{\phi}^*}{\partial z} = -\mu A e^{-\mu z}, \tag{6-292}$$

so that Eq. (6-289) gives

$$A = -\frac{\bar{w}_a^*(p)}{\mu} \tag{6-293}$$

and

$$\bar{\phi}^*(p,z) = -\frac{\bar{w}_a^*(p)}{\mu} e^{-\mu z}; \quad \text{for } z \geq 0^+. \tag{6-294}$$

Inverse transformation of Eq. (6-294) is difficult for arbitrary values of z. When we are interested only in points along the airfoil surface, however,

$$\bar{\phi}^*(p,0^+) = -\frac{\bar{w}_a^*(p)}{\mu} \tag{6-295}$$

is the product of two simple functions of p. It can be inverted by means of the so-called convolution theorem to yield

$$\bar{\phi}'(x,0^+) = -\int_0^x \bar{w}_a(\xi) \mathcal{L}^{-1} \left\{ \frac{1}{\mu} \right\} d\xi. \tag{6-296}$$

Here the inverse of $1/\mu$ must be taken as a function of $(x - \xi)$. Since

$$\frac{1}{\mu} = \frac{1}{\sqrt{M^2 - 1} \sqrt{\left[p + \frac{iM\omega}{a_\infty(M^2 - 1)} \right]^2 + \frac{\omega^2}{a_\infty^2(M^2 - 1)^2}}}, \tag{6-297}$$

we can use two formulas from the tables at the end of Ref. 6–49:

$$\mathcal{L}^{-1}\left\{\frac{1}{\sqrt{p^2 + \alpha^2}}\right\} = J_0(\alpha x) \tag{6–298}$$

and

$$\mathcal{L}^{-1}\{F(p + a)\} = e^{-ax}f(x), \tag{6–299a}$$

where

$$F = \mathcal{L}\{f\} \tag{6–299b}$$

to derive

$$\mathcal{L}^{-1}\left\{\frac{1}{\mu}\right\} = \frac{e^{-i\frac{M\omega x}{a_\infty(M^2-1)}}}{\sqrt{M^2 - 1}} J_0\left(\frac{\omega x}{a_\infty(M^2 - 1)}\right). \tag{6–300}$$

Substituting Eq. (6–300), $\bar{\omega}$, and the dimensionless x variables into Eq. (6–296) yields

$$\bar{\phi}'(x^*,0^+) = -\frac{b}{\sqrt{M^2 - 1}} \int_0^{x^*} \bar{w}_a(\xi^*)e^{-i(\bar{\omega}/2)(x^*-\xi^*)} J_0\left(\frac{\bar{\omega}}{2M}(x^* - \xi^*)\right) d\xi^*. \tag{6–301}$$

Except for a shift in the origin of coordinates, Eq. (6–301) is evidently identical with Eq. (6–241), but the latter is derived by the source-pulse method at much greater expense of space and effort.

6–7 Indicial airfoil motions in supersonic flow. Techniques for determining the aerodynamic loads caused by general unsteady motion of a chordwise-rigid airfoil were explained in the opening paragraphs of Section 6–5. Having formulas like (6–151) and (6–152) already at our disposal, we need only to derive new expressions for each indicial function to extend their usefulness into the supersonic range. This proves to be a much easier problem than for subsonic flow, because the necessary integrations of Eq. (6–236) can be carried out in terms of fully tabulated functions.

For consistency with what we have done previously, we retain the definitions of $\phi_c(s)$, $\phi_{cM}(s)$, $\phi_{cq}(s)$, and $\phi_{cMq}(s)$ given in Eqs. (6–147)–(6–150). Our reasoning with regard to the impulsive pressures at the instant of starting is not restricted to subsonic speeds, so Eqs. (6–175)–(6–178) furnish initial values for the supersonic functions. Asymptotic lifts and moments are calculated from the properties of supersonic airfoils in steady motion, developed in Section 6–3. Thus, Eqs. (6–59) and (6–60) yield, for lift and moment coefficients due to an angle of attack $\dot{h}_0/U$,

$$c_l = \frac{4}{\sqrt{M^2 - 1}} \frac{\dot{h}_0}{U}, \tag{6–302}$$

$$c_m = \frac{2a}{\sqrt{M^2 - 1}} \frac{\dot{h}_0}{U}. \tag{6-303}$$

It follows, by comparison with Eqs. (6–147) and (6–148), that

$$\phi_c(\infty) = \frac{2}{\pi\sqrt{M^2 - 1}}, \tag{6-304}$$

$$\phi_{cM}(\infty) = \frac{-1}{\pi\sqrt{M^2 - 1}}. \tag{6-305}$$

A similar derivation, based on substituting

$$z_a = \frac{-q_0(x + b)^2}{2U} \tag{6-306}$$

into Eqs. (6–56) and (6–57), leads to

$$\phi_{cq}(\infty) = \frac{1}{\pi\sqrt{M^2 - 1}}, \tag{6-307}$$

$$\phi_{cMq}(\infty) = \frac{-2}{3\pi\sqrt{M^2 - 1}}. \tag{6-308}$$

Equations (6–304), (6–305), (6–307), and (6–308) are not really asymptotic formulas. They give the lift and moment for all times after the last disturbance created during the unsteady motion at $t = 0$ is swept past the trailing edge by the supersonic stream. Since this disturbance is at the foremost point of the expanding cylinder that starts from the leading edge $x = -b$ at $t = 0$, it moves downstream with velocity $(U - a_\infty)$ relative to the airfoil, taking an interval

$$\Delta t = \frac{2b}{U - a_\infty} \tag{6-309a}$$

to traverse the chord. The corresponding value of s,

$$s = \frac{U\Delta t}{b} = \frac{2M}{M - 1}, \tag{6-309b}$$

marks the limit beyond which all indicial functions assume their final values.

As an illustration of the mathematical steps associated with supersonic indicial problems, let us consider the case of plunging motion, for which

$$w_a(x,t) = \begin{cases} 0, & t < 0, \\ -\dot{h}_0, & t \geq 0. \end{cases} \tag{6-310}$$

For insertion into Eq. (6–236), we must rewrite this

$$w_a(\xi, t - \tau) = \begin{cases} 0, & t < \tau, \\ -\dot{h}_0, & t \geq \tau. \end{cases} \tag{6–311}$$

Three different ranges of t must be considered when evaluating the inner integral I of Eq. (6–236). Using the variable θ in Eq. (6–232) to assist our integrations, and setting $z = 0$, we find

(1) for $t < \tau_1 = \dfrac{x - \xi}{a_\infty(M + 1)}$,

$$I_1 = \int_{\tau_1}^{\tau_2} \frac{w_a(\xi, t - \tau)d\tau}{\sqrt{(\tau - \tau_1)(\tau_2 - \tau)}} = 0; \tag{6–312a}$$

(2) for $\dfrac{x - \xi}{a_\infty(M + 1)} = \tau_1 \leq t \leq \tau_2 = \dfrac{x - \xi}{a_\infty(M - 1)}$,

$$I_2 = \int_{\tau_1}^{\tau_2} \frac{w_a(\xi, t - \tau)d\tau}{\sqrt{(\tau - \tau_1)(\tau_2 - \tau)}} = -\dot{h}_0 \int_{\tau_1}^{t} \frac{d\tau}{\sqrt{(\tau - \tau_1)(\tau_2 - \tau)}}$$

$$= -\dot{h}_0 \int_0^{\cos^{-1}\left(\frac{\tau_2 + \tau_1 - 2t}{\tau_2 - \tau_1}\right)} d\theta = -\dot{h}_0 \cos^{-1}\left[M - \frac{a_\infty t(M^2 - 1)}{(x - \xi)} \right]; \tag{6–312b}$$

and (3) for $\dfrac{x - \xi}{a_\infty(M - 1)} = \tau_2 \leq t$,

$$I_3 = \int_{\tau_1}^{\tau_2} \frac{w_a(\xi, t - \tau)d\tau}{\sqrt{(\tau - \tau_1)(\tau_2 - \tau)}} = -\dot{h}_0 \int_0^{\pi} d\theta = -\dot{h}_0\pi. \tag{6–312c}$$

We summarize the value of I, which we now look at as a function of ξ:

$$I = \begin{cases} 0, & \text{for } \xi \leq x - a_\infty t(M+1), \\ -\dot{h}_0 \cos^{-1}\left[M - \dfrac{a_\infty t(M^2 - 1)}{(x - \xi)} \right], & \text{for } x - a_\infty t(M+1) \leq \xi \leq x - a_\infty t(M-1), \\ -\dot{h}_0\pi, & \text{for } \xi \geq x - a_\infty t(M-1). \end{cases} \tag{6–312d}$$

Equation (6–312d) demonstrates that we must account for four possibilities when calculating the outer integral that gives us the disturbance potential. In general,

$$\phi'(x, 0^+, t) = -\frac{1}{\pi\sqrt{M^2 - 1}} \int_{-b}^{x} I(\xi, x, t)d\xi. \tag{6–313}$$

Therefore,

(1) for $x \leq x - a_\infty t(M + 1)$ (that is, for $t \leq 0$),

$$\phi_1'(x,0^+,t) = 0; \tag{6-314a}$$

(2) for $-b \leq x - a_\infty t(M + 1)$ $\left(\text{that is, for } 0 \leq t \leq \dfrac{x + b}{a_\infty(M + 1)}\right)$,

$$\phi_2'(x,0^+,t) = \frac{\dot{h}_0}{\pi\sqrt{M^2 - 1}} \int_{x - a_\infty t(M+1)}^{x - a_\infty t(M-1)} \cos^{-1}\left[M - \frac{a_\infty t(M^2 - 1)}{(x - \xi)}\right] d\xi$$

$$+ \frac{\dot{h}_0}{\sqrt{M^2 - 1}} \int_{x - a_\infty t(M-1)}^{x} d\xi; \tag{6-314b}$$

(3) for $-b \geq x - a_\infty t(M + 1)$ but $-b \leq x - a_\infty t(M - 1)$ $\left(\text{that is,}\right.$

for $\left.\dfrac{x + b}{a_\infty(M + 1)} \leq t \leq \dfrac{x + b}{a_\infty(M - 1)}\right)$,

$$\phi_3'(x,0^+,t) = \frac{\dot{h}_0}{\pi\sqrt{M^2 - 1}} \int_{-b}^{x - a_\infty t(M-1)} \cos^{-1}\left[M - \frac{a_\infty t(M^2 - 1)}{(x - \xi)}\right] d\xi$$

$$+ \frac{\dot{h}_0}{\sqrt{M^2 - 1}} \int_{x - a_\infty t(M-1)}^{x} d\xi; \tag{6-314c}$$

and (4) for $-b \geq x - a_\infty t(M - 1)$ $\left(\text{that is, for } t \geq \dfrac{x + b}{a_\infty(M - 1)}\right)$,

$$\phi_4'(x,0^+,t) = \frac{\dot{h}_0}{\sqrt{M^2 - 1}} \int_{-b}^{x} d\xi = \frac{\dot{h}_0 x}{\sqrt{M^2 - 1}}. \tag{6-314d}$$

We do not present here the detailed results of integrating all terms in Eqs. (6–314). When $t \geq 0$, the pressure distributions according to Eq. (6–243) turn out to be

$$\frac{(p_U - p_L)_2}{(\frac{1}{2}\rho_\infty)U^2} = -\frac{4}{M}\frac{\dot{h}_0}{U}, \tag{6-315a}$$

$$\frac{(p_U - p_L)_3}{(\frac{1}{2}\rho_\infty)U^2} = \frac{-4\dot{h}_0/U}{\sqrt{M^2 - 1}} \left\{\frac{1}{\pi}\cos^{-1}\left[\frac{Mx + a_\infty t}{x + Ut}\right]\right.$$

$$+ \left.\frac{\sqrt{M^2 - 1}}{\pi M}\cos^{-1}\left[\frac{-x}{a_\infty t}\right]\right\}, \tag{6-315b}$$

$$\frac{(p_U - p_L)_4}{(\frac{1}{2}\rho_\infty)U^2} = -\frac{4}{\sqrt{M^2 - 1}}\frac{\dot{h}_0}{U}. \tag{6-315c}$$

We recognize in Eq. (6–315c) the steady-state pressure. Moreover, as in the subsonic case, the pressure throughout region 2 is the same as that associated with the impulsive starting of a piston at velocity $\dot{h}_0$.

The final lift and moment expressions are found, as usual, by substituting Eqs. (6–315) into Eqs. (5–106) and (5–107). When put into the form of the dimensionless indicial functions defined by Eqs. (6–147) and (6–148), they read

$$\phi_c(s) = \begin{cases} \dfrac{2}{\pi M}, \quad \text{for } 0 \le s \le \dfrac{2M}{M+1}, \\[2ex] \dfrac{2}{\pi^2}\left\{ \dfrac{1}{M}\cos^{-1}\left[M - \dfrac{2M}{s}\right] + \dfrac{\cos^{-1}\left[\dfrac{s}{2M} + M - \dfrac{sM}{2}\right]}{\sqrt{M^2-1}} \right. \\[3ex] \left. + \dfrac{1}{M}\sqrt{\dfrac{s^2}{4M^2} - \left(1 - \dfrac{s}{2}\right)^2}\right\}, \quad \text{for } \dfrac{2M}{M+1} \le s \le \dfrac{2M}{M-1}, \\[3ex] \dfrac{2}{\pi\sqrt{M^2-1}}, \quad \text{for } \dfrac{2M}{M-1} \le s. \end{cases} \tag{6–316}$$

$$\phi_{cM}(s) = \begin{cases} -\dfrac{1}{2\pi M}\left(2 - \dfrac{s^2}{4M^2}\right), \quad \text{for } 0 \le s \le \dfrac{2M}{M+1}, \\[2ex] -\dfrac{1}{\pi^2}\left\{ \dfrac{1}{M}\left(1 - \dfrac{s^2}{8M^2}\right)\cos^{-1}\left[M - \dfrac{2M}{s}\right] \right. \\[2ex] + \dfrac{1}{\sqrt{M^2-1}}\cos^{-1}\left[\dfrac{s}{2M} + M - \dfrac{sM}{2}\right] \\[2ex] \left. + \dfrac{1}{4M}(2+s)\sqrt{\dfrac{s^2}{4M^2} - \left(1 - \dfrac{s}{2}\right)^2}\right\}, \\[2ex] \qquad\qquad \text{for } \dfrac{2M}{M+1} \le s \le \dfrac{2M}{M-1}, \\[3ex] -\dfrac{1}{\pi\sqrt{M^2-1}}, \quad \text{for } \dfrac{2M}{M-1} \le s. \end{cases} \tag{6–317}$$

Lomax *et al.* (Ref. 6–37) have arrived at Eqs. (6–316) and (6–317) by means of the method of supersonic analogy (cf. Section 6–5), which is especially convenient for supersonic indicial functions. By a series of computations which parallel those outlined above, they derive the following formulas for pitching motion about the leading edge of the airfoil.

$$\phi_{cq}(s) = \begin{cases} \dfrac{1}{\pi M}\left[1 + \dfrac{s^2}{8M^2}\right], & \text{for } 0 \le s \le \dfrac{2M}{M+1}, \\[2ex] \dfrac{1}{\pi^2}\left\{\dfrac{1}{M}\left(1 + \dfrac{s^2}{8M^2}\right)\cos^{-1}\left[M - \dfrac{2M}{s}\right]\right. \\[2ex] \quad + \dfrac{1}{\sqrt{M^2-1}}\cos^{-1}\left[\dfrac{s}{2M} + M - \dfrac{sM}{2}\right] \\[2ex] \quad \left. + \dfrac{(6-s)}{4M}\sqrt{\dfrac{s^2}{4M^2} - \left(1 - \dfrac{s}{2}\right)^2}\right\}, \\[2ex] \qquad\qquad \text{for } \dfrac{2M}{M+1} \le s \le \dfrac{2M}{M-1}, \\[2ex] \dfrac{1}{\pi\sqrt{M^2-1}}, & \text{for } \dfrac{2M}{M-1} \le s. \end{cases} \qquad (6\text{-}318)$$

$$\phi_{cMq}(s) = \begin{cases} -\dfrac{2}{3\pi M}\left[1 + \dfrac{s^3}{16M^2}\right], & \text{for } 0 \le s \le \dfrac{2M}{M+1}, \\[2ex] -\dfrac{2}{3\pi^2}\left\{\dfrac{1}{M}\left[1 + \dfrac{s^3}{16M^2}\right]\cos^{-1}\left[M - \dfrac{2M}{s}\right]\right. \\[2ex] \quad + \dfrac{1}{\sqrt{M^2-1}}\cos^{-1}\left[\dfrac{s}{2M} + M - \dfrac{sM}{2}\right] \\[2ex] \quad \left. + \dfrac{\left[8 - \dfrac{s}{2} - \dfrac{s^2}{2M^2} - \dfrac{s^2}{4}\right]}{6M}\sqrt{\dfrac{s^2}{4M^2} - \left(1 - \dfrac{s}{2}\right)^2}\right\}, \\[2ex] \qquad\qquad \text{for } \dfrac{2M}{M+1} \le s \le \dfrac{2M}{M-1}, \\[2ex] -\dfrac{2}{3\pi\sqrt{M^2-1}}, & \text{for } \dfrac{2M}{M-1} \le s. \end{cases} \qquad (6\text{-}319)$$

The reader is reminded that the pitching moments in Eqs. (6–317) and (6–319) are assumed to act about the leading edge of the airfoil and are positive in a sense to depress the trailing edge. Following Ref. 6–37, we show in Fig. 6–15 how the indicial functions vary with s for different values of M. Clearly, the transient effects tend to die out very rapidly at the higher Mach numbers. Not only is this true because the unsteady regime below $s = 2M/(M-1)$ narrows down to a single chord length, but also

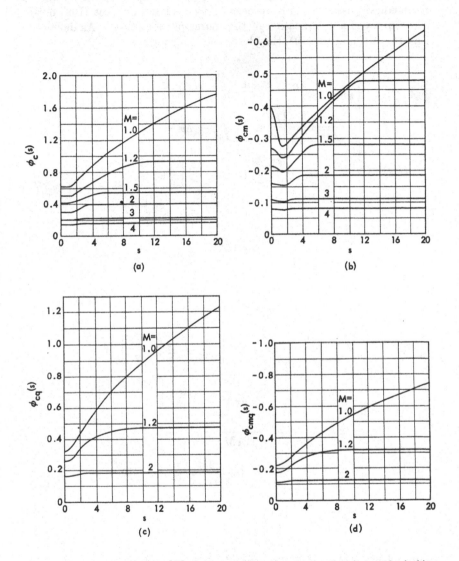

Fig. 6–15. Indicial lift and moment functions from plunging and pitching of an airfoil in sonic and supersonic flow. All quantities are defined in the text.

because the initial and final magnitudes of the various functions approach one another. As we can predict from Lighthill's simplification, $\phi_c(s)$, for instance, must equal $2/\pi M$ throughout the entire maneuver when M is large enough.

The indicial functions $\psi_c(s)$ and $\psi_{cM}(s)$ for supersonic entry into a sharp-edged gust seem to have been first calculated by Biot (Ref. 6–50), who employed a modification of the source-pulse method. As defined in Eqs. (6–218) and (6–219), they have the following forms:

$$
\psi_c(s) = \begin{cases}
\dfrac{s}{\pi M}, \quad \text{for } 0 \leq s \leq \dfrac{2M}{M+1}, \\[2ex]
\dfrac{2}{\pi\sqrt{M^2-1}}\left\{\dfrac{1}{\pi}\cos^{-1}\left[\dfrac{s}{2M}+M-\dfrac{sM}{2}\right]\right. \\[2ex]
\left. \qquad + \dfrac{s\sqrt{M^2-1}}{2\pi M}\cos^{-1}\left[M-\dfrac{2M}{s}\right]\right\}, \\[2ex]
\qquad\qquad \text{for } \dfrac{2M}{M+1} \leq s \leq \dfrac{2M}{M-1}, \\[2ex]
\dfrac{2}{\pi\sqrt{M^2-1}}, \quad \text{for } \dfrac{2M}{M-1} \leq s.
\end{cases}
\tag{6–320}
$$

$$
\psi_{cM}(s) = \begin{cases}
-\dfrac{s^2}{4\pi M}, \quad \text{for } 0 \leq s \leq \dfrac{2M}{M+1}, \\[2ex]
-\dfrac{1}{\pi\sqrt{M^2-1}}\left\{\dfrac{1}{\pi}\cos^{-1}\left[\dfrac{s}{2M}+M-\dfrac{sM}{2}\right]\right. \\[2ex]
\qquad + \dfrac{s^2\sqrt{M^2-1}}{4\pi M}\cos^{-1}\left[M-\dfrac{2M}{s}\right] \\[2ex]
\left. \qquad + \dfrac{s\sqrt{M^2-1}}{2\pi M}\sqrt{\dfrac{s^2}{4M^2}-\left(1-\dfrac{s}{2}\right)^2}\right\}, \\[2ex]
\qquad\qquad \text{for } \dfrac{2M}{M+1} \leq s \leq \dfrac{2M}{M-1}, \\[2ex]
-\dfrac{1}{\pi\sqrt{M^2-1}}, \quad \text{for } \dfrac{2M}{M-1} \leq s.
\end{cases}
\tag{6–321}
$$

Equations (6–320) and (6–321) are adapted from the work of Chang (Ref. 6–51). His paper contains a very complete discussion of the transient problem at supersonic speeds. It includes the calculation of indicial functions for a trailing-edge flap, which are related to those of the complete airfoil in the same way as the coefficients for simple harmonic motion.

Figure 6–16 illustrates curves of $\psi_c(s)$ and $\psi_{cM}(s)$ for several Mach numbers.

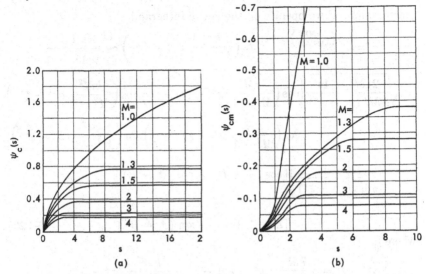

Fig. 6–16.   Indicial lift function for entry into a sharp-edged gust in sonic and supersonic flow.   Symbols are defined in the text.

6–8 Unsteady motion of airfoils at Mach number one.   A very interesting possibility from the aeroelastician's standpoint is that, whereas steady-flow airfoil theory is essentially nonlinear in the transonic range, the linearized results for unsteady motion appear to be meaningful whenever the maneuver exceeds a certain degree of unsteadiness.   We examine first the case of simple harmonic oscillation, for which the reduced frequency k furnishes a clear-cut measure of remoteness from the steady state.   This problem was first studied by Rott (Ref. 6–52).   He showed that the differential equation (6–282) possesses reasonable solutions when M is set identically equal to one.   We can obtain his solution in several ways, such as by the Laplace-transform procedure described at the end of Section 6–6.

An equally straightforward, although less rigorous, scheme consists of modifying our source-pulse solution, Eq. (6–241), by approaching the limit of sonic Mach number.   There can be no objection to this process on physical grounds, because within the linearizing assumptions, the upper and lower airfoil surfaces remain independent of one another.   Moreover, we are not forced to apply Kutta's hypothesis so long as local flow speeds do not deviate greatly from a_∞.   As M nears unity from above, $\bar\omega$ becomes indefinitely large and thus causes the arguments of the Bessel function and the exponential in Eq. (6–241) to do likewise.   Over all but an infinitesimal portion of the range of integration, J_0 can be replaced by its asymptotic representation

$$J_0\left(\frac{\bar\omega}{2M}(x^*-\xi^*)\right) \sim \sqrt{\frac{2(M^2-1)}{\pi Mk(x^*-\xi^*)}}\cos\left(\frac{kM}{M^2-1}(x^*-\xi^*)-\frac{\pi}{4}\right). \quad (6\text{–}322)$$

Furthermore, if we operate on the entire integrand,

$$\frac{1}{\sqrt{M^2 - 1}} e^{-i\left(\frac{kM^2}{(M^2-1)}\right)(x^*-\xi^*)} J_0\left(\frac{kM}{M^2 - 1}(x^* - \xi^*)\right) \sim \frac{1}{\sqrt{M^2 - 1}}$$

$$\times \sqrt{\frac{2(M^2 - 1)}{\pi M k (x^* - \xi^*)}} \left\{ \cos\left(\frac{kM}{M^2 - 1}(x^* - \xi^*) - \frac{\pi}{4}\right) \cos\left(\frac{kM^2}{M^2 - 1}(x^* - \xi^*)\right) \right.$$

$$\left. - i \cos\left(\frac{kM}{M^2 - 1}(x^* - \xi^*) - \frac{\pi}{4}\right) \sin\left(\frac{kM^2}{M^2 - 1}(x^* - \xi^*)\right) \right\}$$

$$= \frac{1}{2\sqrt{\frac{\pi M k (x^* - \xi^*)}{2}}} \left\{ [\cos(A + B) + \cos(A - B)] \right.$$

$$\left. - i[\sin(A + B) - \sin(A - B)] \right\}, \tag{6-323}$$

where

$$A + B = \frac{kM}{M^2 - 1}(x^* - \xi^*) - \frac{\pi}{4} + \frac{kM^2}{M^2 - 1}(x^* - \xi^*)$$

$$= \frac{kM}{M - 1}(x^* - \xi^*) - \frac{\pi}{4} \tag{6-324a}$$

and

$$A - B = -\frac{kM}{M + 1}(x^* - \xi^*) - \frac{\pi}{4}. \tag{6-324b}$$

The quantity $(A + B)$ becomes infinite as M approaches 1, so that the areas under those portions of the integrand containing $\cos(A + B)$ and $\sin(A + B)$ tend more and more to cancel one another. In the limit, we can show by using $(x^* - \xi^*)/(M - 1)$ as a new variable of integration that their contributions to Eq. (6–241) vanish. In the remaining terms M can be equated to one immediately, giving

$$\frac{1}{2\sqrt{\frac{\pi M k (x^* - \xi^*)}{2}}} [\cos(A - B) + i \sin(A - B)] = \frac{e^{-i(k/2)(x^*-\xi^*)}}{e^{i(\pi/4)}\sqrt{2\pi k (x^* - \xi^*)}}.$$

$$\tag{6-325}$$

Since the exponential in the denominator is just $\sqrt{i}$, the limiting form of Eq. (6–241) is

$$\bar{\phi}'(x^*, 0^+, t) = -b \int_{-1}^{x^*} w_a(\xi^*, t) \frac{e^{-i(k/2)(x^*-\xi^*)}}{\sqrt{2\pi i k (x^* - \xi^*)}} d\xi^*. \tag{6-326}$$

None of the remaining steps differs from the supersonic case, so we can regard Eq. (6–326) as the complete solution for oscillatory motion.

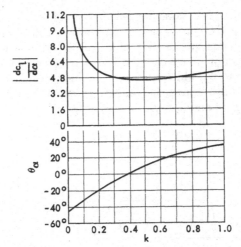

Fig. 6–17.   Amplitude and phase angle of lift due to pitching of an airfoil about its leading edge at $M = 1$.

The various aerodynamic coefficients at $M = 1$ have been computed and tabulated by Nelson and Berman (Ref. 6–53). Their notation is exactly that of Garrick and Rubinow, which we defined in Eqs. (6–262)–(6–264). An increase in simplicity over the supersonic calculations results from the fact that all integrals can be identified as Fresnel integrals (cf. Ref. 6–27).

The magnitudes of the sonic aerodynamic loads become infinite as k approaches zero. To illustrate their behavior, we plot in Fig. 6–17 the amplitude and phase angle of the dimensionless lift due to pitching motion about the leading edge:

$$4k^2[L_3{}' + iL_4{}'] = \left| \frac{dc_l}{d\alpha} \right| e^{i\theta_\alpha}. \tag{6–327}$$

It is difficult to make an exact estimate of the lowest value of reduced frequency for which these results are valid. We can state, however, that the phase-angle curve of Fig. 6–17 represents a smooth transition from subsonic through sonic to supersonic speeds when k exceeds roughly 0.15. Also the variation of predicted dimensionless flutter speed *vs.* Mach number for representative wings seems to be reasonable through the transonic range.

Turning to indicial functions for $M = 1$, we are faced with the evident difficulty that their asymptotic (steady-state) values are all infinite. This does not preclude using them within the range of the variable s for which they show reasonable agreement with their subsonic and supersonic counterparts, however, as was first observed by Heaslet, Lomax, and Spreiter (Ref. 6–5). In every case, we can carry out a continuous limiting process to sonic Mach number from the simple expressions we have already derived for $M > 1$. The same formulas may, of course, be derived by source-pulse methods, but the former alternative is far more attractive.

On examining Eqs. (6–316) through (6–321), we discover no difficulty in going to the limit except for the terms containing the arc cosine of $[s/2M + M - sM/2]$, which have $M^2 - 1$ in their denominators. Since the argument always approaches unity, we can expand in power series about the value zero and obtain

$$\lim_{M \to 1} \left\{ \frac{\cos^{-1}\left[\dfrac{s}{2M} + M - \dfrac{sM}{2}\right]}{\sqrt{M^2 - 1}} \right\}$$

$$= \lim_{M \to 1} \left\{ \frac{\sin^{-1}\sqrt{1 - \left[M - \dfrac{s}{2}\left(\dfrac{M^2 - 1}{M}\right)\right]^2}}{\sqrt{M^2 - 1}} \right\}$$

$$= \lim_{M \to 1} \left\{ \frac{\sqrt{(1 - M^2) + s(M^2 - 1) - \dfrac{s^2}{4M^2}(M^2 - 1)^2}}{\sqrt{M^2 - 1}} \right\} = \sqrt{s - 1}. \quad (6\text{–}328)$$

The ranges of the variable s in Eqs. (6–316)–(6–321) are reduced to two in number, because the upper bound $2M/(M - 1)$ of the middle range becomes infinite. The steady-state values of the various functions are therefore supplanted, as we have anticipated, by a gradual or rapid rise to an infinite limit. Being proportional to $\sqrt{s}$ as s becomes large, $\phi_c(s)$, for example, takes on the appearance of a parabola with a horizontal axis of symmetry.

The limiting expressions we obtain for $\phi_c(s)$, $\phi_{cM}(s)$, etc., are as follows:

$$\phi_c(s) = \begin{cases} \dfrac{2}{\pi}, & \text{for } 0 \le s \le 1, \\[2mm] \dfrac{2}{\pi^2}\left\{\cos^{-1}\left[1 - \dfrac{2}{s}\right] + 2\sqrt{s - 1}\right\}, & \text{for } 1 \le s. \end{cases} \quad (6\text{–}329)$$

$$\phi_{cM}(s) = \begin{cases} -\dfrac{1}{2\pi}\left[2 - \dfrac{s^2}{4}\right], & \text{for } 0 \le s \le 1, \\[3mm] -\dfrac{1}{\pi^2}\left\{\left(1 - \dfrac{s^2}{8}\right)\cos^{-1}\left[1 - \dfrac{2}{s}\right] \right. \\[2mm] \left. \quad + \left(\dfrac{3}{2} + \dfrac{s}{4}\right)\sqrt{s - 1}\right\}, & \text{for } 1 \le s. \end{cases} \quad (6\text{–}330)$$

$$\phi_{cq}(s) = \begin{cases} \dfrac{1}{\pi}\left[1 + \dfrac{s^2}{8}\right], & \text{for } 0 \le s \le 1, \\[2ex] \dfrac{1}{\pi^2}\left\{\left(1 + \dfrac{s^2}{8}\right)\cos^{-1}\left[1 - \dfrac{2}{s}\right] \right. \\[2ex] \left. + \left(\dfrac{5}{2} - \dfrac{s}{4}\right)\sqrt{s-1}\right\}, & \text{for } 1 \le s. \end{cases} \tag{6–331}$$

$$\phi_{cMq}(s) = \begin{cases} -\dfrac{2}{3\pi}\left[1 + \dfrac{s^3}{16}\right], & \text{for } 0 \le s \le 1, \\[2ex] -\dfrac{2}{3\pi^2}\left\{\left(1 + \dfrac{s^3}{16}\right)\cos^{-1}\left[1 - \dfrac{2}{s}\right] \right. \\[2ex] \left. + \left(\dfrac{7}{3} - \dfrac{s}{12} - \dfrac{s^2}{8}\right)\sqrt{s-1}\right\}, & \text{for } 1 \le s. \end{cases} \tag{6–332}$$

$$\psi_{c}(s) = \begin{cases} \dfrac{s}{\pi}, & \text{for } 0 \le s \le 1, \\[2ex] \dfrac{2}{\pi^2}\left\{\dfrac{s}{2}\cos^{-1}\left[1 - \dfrac{2}{s}\right] + \sqrt{s-1}\right\}, & \text{for } 1 \le s. \end{cases} \tag{6–333}$$

$$\psi_{cM}(s) = \begin{cases} -\dfrac{s^2}{4\pi}, & \text{for } 0 \le s \le 1, \\[2ex] -\dfrac{1}{\pi^2}\left\{\dfrac{s^2}{4}\cos^{-1}\left[1 - \dfrac{2}{s}\right] \right. \\[2ex] \left. + \left(1 + \dfrac{s}{2}\right)\sqrt{s-1}\right\}, & \text{for } 1 \le s. \end{cases} \tag{6–334}$$

We find uniform agreement between Eqs. (6–329)–(6–334) and the corresponding subsonic indicial functions, which are presented in Section 6–5 for values of s up to $2M/(M+1)$. It therefore appears that, during the first semichord length of motion, we have achieved perfectly smooth transition through transonic speeds. These formulas remain completely without experimental confirmation, of course, as do nearly all other indicial functions. They are important for aeroelastic applications just because of this lack of any other reliable information, for they do provide a tool for the analysis of many interesting maneuvers. Inasmuch as the maximum loads and stresses caused by a gust encounter or sudden pull-up often occur very shortly after the motion begins, the unrealistic size of the $M = 1$ forces and moments at large values of s is not as much of a handicap as one might suppose.

To complete the graphical presentation of indicial functions for the entire range of Mach numbers, curves computed from Eqs. (6–329)–(6–334) appear on Figs. 6–15 and 6–16.

CHAPTER 7

WINGS AND BODIES IN THREE-DIMENSIONAL
UNSTEADY FLOW

7-1 Introduction. A majority of practical aeroelastic computations, particularly those involving unsteady motions, are carried out using two-dimensional airfoil theory. It is a fortunate circumstance that this simplification is generally successful, for two reasons. First, the two-dimensional aerodynamic theories are in a fairly advanced state of development and tabulation; for instance, there exist complete tables of flutter coefficients for all important Mach numbers except in the low transonic range, where the presence of oscillating shocks on the airfoil surfaces makes experimental determination of the airloads almost mandatory. Second, the unsteady theory for three-dimensional wings at any Mach number is a formidable and unwieldy thing indeed.

As every aeroelastician knows, the proportion of problems for which the two-dimensional assumption is unsuitable grows day by day. We can cite the bending-elevon flutter of delta wings and the dynamic loads due to launching irregularities of missiles with low aspect-ratio surfaces as being typical of the many significant examples where, if any meaningful analysis is to be attempted, the predominant influence of finite span must be included from the outset. Because of the implications of the Prandtl-Glauert transformation, it is questionable whether strip methods can be employed on any wing at transonic speeds.

In order for existing theories and those under development to have a reasonable chance of achieving full acceptance and usefulness, we believe that there are certain principles which must be followed. In succinct terms, these require that the maximum amount of systematization and breadth of applicability must be sought at all costs. The total number of numerical operations per solution involved in a theory is no longer of great practical concern in an era when nearly every aircraft manufacturer and aeronautical research laboratory has at its disposal a battery of high-speed digital computing machines. What is vitally important, however, is the possibility of these computations being set up by a relatively unskilled operator, whose knowledge of aeroelasticity may be limited to an acquaintance with certain tables, matrix manipulations, and the algebra of complex numbers. We must work toward the situation where each new problem does not demand extensively the services of experienced aerodynamicists.

The difficulty with regard to narrowness of application of aerodynamic theories for three-dimensional wings is most evident at supersonic speeds.

It is not an unjust statement to say that every new geometrical shape of planform calls for reorganization of the theory. If universal techniques are to be discovered, even ones having a scope comparable with the subsonic lifting line equation, we anticipate lines of attack markedly different from those customarily pursued in the past. A considerable sacrifice of mathematical elegance may well be the price for meeting the principles stated above.

7-2 Oscillating finite wings in incompressible flow. Except in certain rather special cases, such as those described in Refs. 7-1 and 7-2, it seems unlikely that the problem of simple harmonic motion of lifting surfaces in three-dimensional flow of a liquid will ever yield to exact solution. This fact has led to a situation where there exist some twenty different treatments, no two of which produce exactly the same results. Estimates of the relative merits of the physical approximations from which these various theories set out are hampered by the paucity of experimental data on oscillating finite wings. The choice of what to use in a particular application is therefore usually based on a combination of simplicity, systematization, and subjective preference.

Most of the better established theory deals with wings of moderate to large aspect ratio, the effort being to reach a formulation which becomes exact in the limit as the span approaches infinity, and which reduces to something resembling the lifting-line equation for steady flow. The first publication along these lines was Cicala's (Ref. 7-3). From among the numerous others that have followed, we describe in this section Reissner's integral equation (Ref. 7-4), for which satisfactory experimental confirmation is known down to aspect ratios as low as two. Application of this technique is quite systematic, and tables are available to assist routine computations.

We consider oscillations of the almost plane wing pictured in Fig. 7-1, restricting ourselves to the lift-producing motion. This leads to the linearized boundary condition [cf. Eq. (5-316)]

$$\frac{\partial \phi'(x,y,0,t)}{\partial z} = \bar{w}_a(x,y)e^{i\omega t} = \left[i\omega \bar{z}_a(x,y) + U \frac{\partial \bar{z}_a}{\partial x} \right] e^{i\omega t};$$

$$\text{for } z = 0, \ (x,y) \text{ in } R_a, \quad (7\text{-}1)$$

where

$$z_a(x,y,t) = \bar{z}_a(x,y)e^{i\omega t} \qquad (7\text{-}2)$$

is the instantaneous vertical coordinate of the vibrating mean surface.

As in analogous steady-flow problems, we represent the loaded surface and its wake by a vortex sheet distributed over regions R_a and R_w of the xy-plane. Since the unsteady motion is governed by Laplace's equation we can determine the instantaneous velocities induced by this sheet from

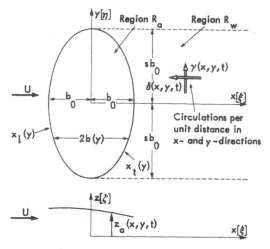

Fig. 7-1. Generalized lifting surface seen in plan and side views.

the Biot-Savart law. In fact, the only disparity between the present situation and that described by integral equation (5–128) is that here the wing circulation is continually changing, and all dependent variables in the problem are functions of time. For simple harmonic oscillation, we are justified by linearity in writing

$$\gamma_a (\xi,\eta,t) = \bar{\gamma}_a(\xi,\eta)e^{i\omega t}, \qquad \delta_a(\xi,\eta,t) = \bar{\delta}_a(\xi,\eta)e^{i\omega t}, \qquad (7\text{-}3\text{a,b})$$

$$\gamma_w(\xi,\eta,t) = \bar{\gamma}_w(\xi,\eta)e^{i\omega t}, \qquad \delta_w(\xi,\eta,t) = \bar{\delta}_w(\xi,\eta)e^{i\omega t}. \qquad (7\text{-}3\text{c,d})$$

With these substitutions, and a cancellation of the factor $e^{i\omega t}$, Eq. (5–128) becomes

$$\bar{w}_a(x,y) = -\frac{1}{4\pi} \oint\!\!\oint_{R_a} \frac{\bar{\gamma}_a(\xi,\eta)[x - \xi] + \bar{\delta}_a(\xi,\eta)[y - \eta]}{[(x - \xi)^2 + (y - \eta)^2]^{\frac{3}{2}}}\, d\xi d\eta$$

$$-\frac{1}{4\pi} \iint_{R_w} \frac{\bar{\gamma}_w(\xi,\eta)[x - \xi] + \bar{\delta}_w(\xi,\eta)[y - \eta]}{[(x - \xi)^2 + (y - \eta)^2]^{\frac{3}{2}}}\, d\xi d\eta \qquad (7\text{-}4)$$

for (x,y) in the wing-projection region R_a.

We can obtain a relationship between wing and wake circulations by the same reasoning that led to Eqs. (5–320) in the two-dimensional case. For a particular spanwise station η, the requirement that any change in total wing circulation must be matched by an equal and opposite shed vortex, which floats downstream from the trailing edge with velocity U, leads to

$$\bar{\gamma}_w(\xi,\eta) = -ik_0\bar{\Omega}(\eta)e^{-i(\omega\xi/U)}, \qquad (7\text{-}5)$$

where
$$k_0 = \frac{\omega b_0}{U} \tag{7-6}$$

and
$$\bar{\Omega}(\eta) = \frac{\bar{\Gamma}(\eta)}{b_0} e^{i(\omega x_t/U)} = \frac{e^{i(\omega x_t/U)}}{b_0} \int_{x_l(\eta)}^{x_t(\eta)} \bar{\gamma}_a(\xi,\eta) d\xi \tag{7-7}$$

is a convenient reduced-circulation function. The equation of continuity of vorticity helps us to calculate the streamwise component of wake circulation from Eq. (7-5). Thus, since $\bar{\gamma}$ vanishes ahead of the leading edge,
$$\frac{\partial \bar{\delta}_w}{\partial \xi} = \frac{\partial \bar{\gamma}_w}{\partial \eta} \tag{7-8}$$

implies

$$\bar{\delta}_w(\xi,\eta) = \int_{-\infty}^{\xi} \frac{\partial \bar{\gamma}(\xi',\eta)}{\partial \eta} d\xi' = \frac{\partial}{\partial \eta} \int_{-\infty}^{\xi} \bar{\gamma}(\xi',\eta) d\xi'$$

$$= \frac{\partial}{\partial \eta} \int_{x_l(\eta)}^{x_t(\eta)} \bar{\gamma}_a(\xi',\eta) d\xi' + \frac{\partial}{\partial \eta}\left[-ik_0 \bar{\Omega}(\eta) \int_{x_t(\eta)}^{\xi} e^{-i(\omega \xi'/U)} d\xi' \right]$$

$$= \frac{\partial}{\partial \eta}\left[b_0 \bar{\Omega}(\eta) e^{-i(\omega x_t/U)} \right]$$

$$+ \frac{\partial}{\partial \eta}\left[b_0 \bar{\Omega}(\eta) e^{-i(\omega \xi/U)} b_0 \bar{\Omega}(\eta) e^{-i(\omega x_t/U)} \right]$$

$$= b_0 \frac{d\bar{\Omega}}{d\eta} e^{-i(\omega \xi/U)}. \tag{7-9}$$

Here the integration is assumed to start slightly ahead of the leading edge $x_l(\eta)$, and account has been taken of the continuity of $\bar{\gamma}(\xi,\eta)$ at the trailing edge when interchanging orders of integration and differentiation.

If we insert Eqs. (7-5) and (7-9) into Eq. (7-4), and put in the correct integration limits, we get

$$\bar{w}_a(x,y) = -\frac{1}{4\pi} \oint_{-sb_0}^{sb_0} \oint_{x_l(\eta)}^{x_t(\eta)} \frac{\bar{\gamma}_a(\xi,\eta)[x-\xi] + \bar{\delta}_a(\xi,\eta)[y-\eta]}{[(x-\xi)^2 + (y-\eta)^2]^{\frac{3}{2}}} d\xi d\eta$$

$$- \frac{1}{4\pi} \int_{-sb_0}^{sb_0} \int_{x_t(\eta)}^{\infty} e^{-i(\omega \xi/U)} \frac{\left\{ -ik_0 \bar{\Omega}(\eta)[x-\xi] + b_0 \frac{d\bar{\Omega}}{d\eta}[y-\eta] \right\}}{[(x-\xi)^2 + (y-\eta)^2]^{\frac{3}{2}}} d\xi d\eta. \tag{7-10}$$

Since $\bar{\gamma}_a$ and $\bar{\delta}_a$ are also connected through Eq. (7-8), Eq. (7-10) may be regarded as a double integral equation between the known $\bar{w}_a$ and the unknown $\bar{\gamma}_a$. It cannot be solved exactly for arbitrary wing planforms and modes of motion, but nearly all of the single integral equations of aforementioned approximate theories are derivable from it by various modifications of the physical model or the kernel functions.

The mathematical objective of Reissner's approximations is to produce an integral containing $\bar{\gamma}_a$ that can be inverted by the Söhngen formulas (5–103) and (5–104). There are actually three distinct simplifying steps. We can state them in words and summarize their effects on the terms in Eq. (7–10), as follows:

(1) As far as their contributions to $\bar{w}_a(x,y)$ are concerned, the bound vortices can be treated as if the wing were replaced with an airfoil in two-dimensional flow, having the same chordwise load distribution as the section at station y. This assumption is equivalent to

$$\bar{\delta}_a(\xi,\eta) \cong 0, \tag{7–11}$$

$$\bar{\gamma}_a(\xi,\eta) \cong \bar{\gamma}_a(\xi,y), \tag{7–12}$$

and the first integral on the right of Eq. (7–10) is replaced by

$$I_1 \cong -\frac{1}{2\pi} \oint_{x_l(y)}^{x_t(y)} \frac{\bar{\gamma}_a(\xi,y)}{x - \xi}\, d\xi. \tag{7–13}$$

An identical physical approximation is involved in lifting-line theory for steady flow. Its areas of weakness occur where $\bar{\gamma}_a$ is changing rapidly in the spanwise direction and thus generating large chordwise circulation components, a state of affairs which we observe only near the tips on continuous, large aspect-ratio wings.

(2) The trailing vortex pattern in the wake can be projected forward from the trailing edge $x_t(\eta)$ to a spanwise line passing through the point (x,y) where the vertical velocity is to be determined. The second term in the second integral on the right of Eq. (7–10) is thereby altered to read

$$I_3 \cong -\frac{b_0}{4\pi} \oint_{-sb_0}^{sb_0} \oint_x^\infty \frac{(d\bar{\Omega}/d\eta)[y - \eta]e^{-i(\omega\xi/U)}}{[(x - \xi)^2 + (y - \eta)^2]^{\frac{3}{2}}}\, d\xi d\eta$$

$$= \frac{-b_0 e^{-i(\omega x/U)}}{4\pi} \oint_{-sb_0}^{sb_0} \frac{d\bar{\Omega}}{d\eta} \int_0^\infty \frac{e^{-i(\omega\lambda/U)}[y - \eta]d\lambda}{[\lambda^2 + (y - \eta)^2]^{\frac{3}{2}}}\, d\eta, \tag{7–14}$$

where

$$\lambda = \xi - x. \tag{7–15}$$

This is the second approximation underlying lifting-line theory. In that theory we thus obtain constant downwash along the wing chord. The approximation is obviously poorest near the leading edge of a wing with large chordwise dimensions.

(3) In the pattern of shed γ_w-vortices, those portions which represent deviations from two-dimensional flow at any spanwise station can be projected up to a line through point (x,y), just as was done with the δ_w-

vortices. This step requires separating the first term in the second integral of Eq. (7–10) into a part

$$I_2^{(2)} = \frac{ik_0\bar{\Omega}(y)}{2\pi} \int_{x_l(y)}^{\infty} \frac{e^{-i(\omega\xi/U)}}{x - \xi} d\xi \qquad (7\text{–}16)$$

which shows how it would appear if the wing had infinite span and were everywhere loaded as at station y, plus a correction term for three-dimensional effects. After the latter has been appropriately modified, there results

$$I_2 \cong \frac{ik_0\bar{\Omega}(y)}{2\pi} \int_{x_l(y)}^{\infty} \frac{e^{-i(\omega\xi/U)}}{x - \xi} d\xi + \frac{ik_0}{4\pi} e^{-i(\omega x/U)} \oint_{-sb_0}^{sb_0} \frac{d\bar{\Omega}}{d\eta}$$

$$\times \left\{ \int_0^{\infty} \frac{1}{\lambda} \left[\frac{y - \eta}{\sqrt{\lambda^2 + (y - \eta)^2}} - \frac{|y - \eta|}{(y - \eta)} \right] e^{-i(\omega\lambda/U)} d\lambda \right\} d\eta. \qquad (7\text{–}17)$$

Although it is more difficult to estimate the nature of the errors caused by Eq. (7–17), it seems to involve less severe approximations than steps (1) and (2) for those aspect ratios to which the theory is adapted.

When Eqs. (7–13), (7–14), and (7–17) are substituted into integral equation (7–10), the simplified result reads

$$\bar{w}_a(x,y) = -\frac{1}{2\pi} \oint_{x_l(y)}^{x_t(y)} \frac{\bar{\gamma}_a(\xi,y)}{x - \xi} d\xi$$

$$+ \frac{ik_0\bar{\Omega}(y)}{2\pi} \oint_{x_l(y)}^{\infty} \frac{e^{-i(\omega\xi/U)}}{x - \xi} d\xi$$

$$- \frac{e^{-i(\omega x/U)}}{4\pi} \oint_{-sb_0}^{sb_0} \frac{d\bar{\Omega}}{d\eta} K\left(\frac{\omega}{U}(y - \eta)\right) d\eta. \qquad (7\text{–}18)$$

The kernel function of the third term in Eq. (7–18) (with its complete argument $\omega(y - \eta)/U$ replaced by q for brevity) is

$$K(q) = \frac{k_0}{q} - \frac{ik_0}{q} \oint_0^{\infty} e^{-i\lambda} \left[1 + \frac{|q| - \sqrt{\lambda^2 + q^2}}{\lambda} \right] d\lambda. \qquad (7\text{–}19)$$

The integral here is known as Cicala's function, because of its occurrence in the earliest theory of the oscillating finite wing. It is tabulated in Refs. 7–3 and 7–4.

Equation (7–18) is put into a form suitable for inversion by the transformation

$$x^* = \frac{2x - x_t - x_l}{2b}, \qquad y^* = \frac{y}{b_0}, \qquad (5\text{–}135\text{a,b})$$

$$k_m = \frac{k_0}{2b}[x_t + x_l]. \qquad (7\text{–}20)$$

The parameter k_m is a measure of sweep, being zero for a planform with a straight midchord line. These definitions lead to

$$\bar{w}_a(x^*,y^*) = -\frac{1}{2\pi} \oint_{-1}^{1} \frac{\bar{\gamma}_a(\xi^*,y^*)}{x^* - \xi^*} d\xi^*$$

$$+ \frac{ik_0 e^{-ik_m}}{2\pi} \bar{\Omega}(y^*) \int_{1}^{\infty} \frac{e^{-ik\xi^*}}{x^* - \xi^*} d\xi^*$$

$$- \frac{e^{-ikx^*}e^{-ik_m}}{4\pi} \oint_{-s}^{s} \frac{d\bar{\Omega}}{d\eta^*} K(k_0(y^* - \eta^*)) d\eta^*. \quad (7\text{-}21)$$

In terms of these new symbols, incidentally, the reduced circulation function from Eq. (7–7) becomes

$$\bar{\Omega}(y^*) = \frac{b}{b_0} e^{i(k+k_m)} \int_{-1}^{1} \bar{\gamma}_a(\xi^*,y^*) d\xi^*. \quad (7\text{-}22)$$

Equation (5–335) is the appropriate formula for calculating the distribution of pressure difference over the planform, since the linearized Bernoulli equation is unchanged by going from two- to three-dimensional flow. In the modified notation, it reads

$$\frac{\Delta \bar{p}_a(x^*,y^*)}{\rho U} = -\bar{\gamma}_a(x^*,y^*) - ik \int_{-1}^{x^*} \bar{\gamma}_a(\xi^*,y^*) d\xi^*. \quad (7\text{-}23)$$

By reference to Eqs. (5–103) and (5–104), we see that Eq. (7–21) is solved thereby for the unknown $\bar{\gamma}_a$, if we take $(-\bar{w}_a)$ plus the two integrals on the right as the function $g(x^*)$. The variable y^* and the various functions which depend on it, like $\bar{\Omega}$ and k_m, are carried through the process as parameters. The result of inverting can be reduced by an interchange of order of integration to the following form (see Ref. 7–4 for details):

$$\bar{\gamma}_a(x^*,y^*) = \frac{2}{\pi} \sqrt{\frac{1-x^*}{1+x^*}} \left\{ \oint_{-1}^{1} \sqrt{\frac{1+\xi^*}{1-\xi^*}} \frac{\bar{w}_a(\xi^*,y^*)}{(x^* - \xi^*)} d\xi^* \right.$$

$$+ \frac{ik_0 e^{-ik_m}}{2} \bar{\Omega}(y^*) \int_{1}^{\infty} \sqrt{\frac{\lambda+1}{\lambda-1}} \frac{e^{-ik\lambda}}{(x^* - \lambda)} d\lambda$$

$$+ \frac{e^{-ik_m}}{4\pi} \left[\oint_{-s}^{s} \frac{d\bar{\Omega}}{d\eta^*} K(k_0(y^* - \eta^*)) d\eta^* \right]$$

$$\left. \times \left[\oint_{-1}^{1} \sqrt{\frac{1+\xi^*}{1-\xi^*}} \frac{e^{-ik\xi^*}}{(x^* - \xi^*)} d\xi^* \right] \right\}. \quad (7\text{-}24)$$

Equation (7–24) is the first important conclusion to be drawn from Reissner's approximations. It reduces directly to Eq. (5–330) of the two-dimensional theory when the wingspan becomes infinite and all reference to the variables y^* and η^* is removed, and it yields Eq. (5–105) when the flow is steady. The process by which aerodynamic loading is calculated follows closely that outlined in the latter portions of Section 5–6. Two steps must be taken: first, a relation is derived which gives the circulation $\bar{\Omega}(y^*)$; second, Eq. (7–23) is solved for the pressure distribution and integrated to find the spanwise variation of lift and moment.

A circulation formula follows from integrating Eq. (7–24) with respect to x^* between -1 and 1, eliminating $\bar{\gamma}_a$ through Eq. (7–7), and identifying certain integrals as Bessel functions. The result of these manipulations is an integrodifferential equation of the lifting-line type:

$$\bar{\Omega}(y^*) + \mu(k)\, \frac{b}{b_0} \oint_{-s}^{s} \frac{d\bar{\Omega}}{d\eta^*}\, K(k_0(y^* - \eta^*))d\eta^* = \bar{\Omega}^{(2)}(y^*). \quad (7\text{–}25)$$

In Eq. (7–25), $\mu(k)$ denotes the combination

$$\mu(k) = \frac{-[J_0(k) - iJ_1(k)]}{\pi ik[H_1^{(2)}(k) + iH_0^{(2)}(k)]}. \quad (7\text{–}26)$$

It is tabulated in Ref. 7–4. $\bar{\Omega}^{(2)}(y^*)$ is the quasi two-dimensional reduced circulation which would be generated by the motion at station y^* in the absence of spanwise induction effects:

$$\bar{\Omega}^{(2)}(y^*) = 4\, \frac{b}{b_0}\, e^{ikm} \frac{\displaystyle\int_{-1}^{1} \sqrt{(1 + \xi^*)/(1 - \xi^*)}\, \bar{w}_a(\xi^*,y^*)d\xi^*}{\pi ik[H_1^{(2)}(k) + iH_0^{(2)}(k)]}. \quad (7\text{–}27)$$

This may be checked against Eq. (5–334) by letting $b = b_0$ and equating sweep parameter k_m to zero.

Using the fact that $\mu(k)$ is $\frac{1}{2}$ when $k = 0$, it is also an easy matter to prove that Eq. (7–25) becomes identical with the lifting-line Eq. (5–147) in steady flow. The lift-curve slope a_0 must be assigned its theoretical value 2π.

Perhaps the most significant simplification achieved by Reissner's theory is found when Eq. (7–24) is substituted into Eq. (7–23). After considerable manipulation, the expression for pressure turns out to be

$$\frac{\Delta \bar{p}_a(x^*,y^*)}{\rho U}$$

$$= -\frac{2}{\pi} \oint_{-1}^{1} \left[\sqrt{\frac{1 - x^*}{1 + x^*}}\, \sqrt{\frac{1 + \xi^*}{1 - \xi^*}}\, \frac{1}{(x^* - \xi^*)} - ik\Lambda_1 \right] \bar{w}_a(\xi^*,y^*)d\xi^*$$

$$+ \frac{2}{\pi} \sqrt{\frac{1 - x^*}{1 + x^*}} \int_{-1}^{1} \sqrt{\frac{1 + \xi^*}{1 - \xi^*}}\, \bar{w}_a(\xi^*,y^*)d\xi^* \left\{ [C(k) - 1] \right.$$

$$\left. + \left[\frac{\bar{\Omega}(y^*)}{\bar{\Omega}^{(2)}(y^*)} - 1 \right] \left[C(k) + \frac{iJ_1(k)}{J_0(k) - iJ_1(k)} \right] \right\}, \quad (7\text{–}28)$$

where Λ_1 and $C(k)$ are functions associated with the oscillating airfoil:

$$\Lambda_1(x^*,\xi^*) = \tfrac{1}{2}\ln\left[\frac{1 - x^*\xi^* + \sqrt{1 - x^{*2}}\sqrt{1 - \xi^{*2}}}{1 - x^*\xi^* - \sqrt{1 - x^{*2}}\sqrt{1 - \xi^{*2}}}\right], \quad (5\text{-}336b)$$

$$C(k) = \frac{H_1^{(2)}(k)}{H_1^{(2)}(k) + iH_0^{(2)}(k)}. \quad (5\text{-}309)$$

All terms in Eq. (7–28) except the one containing

$$\sigma(y^*) = \left[\frac{\bar{\Omega}(y^*)}{\bar{\Omega}^{(2)}(y^*)} - 1\right]\left[C(k) + \frac{iJ_1(k)}{J_0(k) - iJ_1(k)}\right] \quad (7\text{-}29)$$

coincide with the results of the two-dimensional theory. $\sigma(y^*)$ vanishes when induction effects are neglected, since the ratio of reduced circulations in its first factor then equals unity. But more interesting is the fact that, when $\sigma(y^*)$ is retained, it depends only on the spanwise coordinate and the mode of oscillation of the wing. It has no disturbing influence on the chordwise integrations of $\Delta\bar{p}_a$ which yield running lift and moment, being carried through these as an additive correction to $C(k)$. We discover that no complete rederivation of two-dimensional formulas like those for L, M_y, and M_β is necessary; all we have to do is replace $C(k)$, wherever it appears, by $C(k) + \sigma$, the latter being fixed by the location of station y^* and [through Eqs. (7–25)–(7–27)] by the particular type of motion producing the desired force or moment.

The order of operations in solving a problem of spanwise load distribution by Reissner's theory can be summarized as follows:

(1) The vibration mode is completely specified. For instance, it might be pure bending,

$$h(y^*,t) = \bar{h}_R e^{i\omega t}f_h(y^*), \quad (7\text{-}30)$$

where $\bar{h}_R$ is the complex amplitude at some reference station where the amplitude-distribution function $f_h(y)$ equals unity; or rotation about a known axis

$$\alpha(y^*,t) = \bar{\alpha}_R e^{i\omega t}f_\alpha(y^*). \quad (7\text{-}31)$$

(2) The function $\bar{\Omega}^{(2)}(y^*)$ is found, and the integrodifferential equation (7–25) is solved for $\bar{\Omega}(y^*)$. Since this is the most time-consuming step, we discuss it in more detail below.

(3) The function $\sigma(y^*)$ is calculated from Eq. (7–29). We distinguish the functions σ corresponding to different wing motions by appropriate subscripts. Thus a particular rotational vibration would lead to

$$\sigma_\alpha(y^*) = \left[\frac{\bar{\Omega}_\alpha(y^*)}{\bar{\Omega}_\alpha^{(2)}(y^*)} - 1\right]H(k), \quad (7\text{-}32)$$

where
$$H(k) = C(k) + \frac{iJ_1(k)}{J_0(k) - iJ_1(k)}.\qquad(7\text{-}33)$$

(4) The two-dimensional portions of $L(y^*,t)$, $M_y(y^*,t)$, and $M_\beta(y^*,t)$ are taken from equations like (5–300)–(5–302), with k, a, and the dimensionless location of the flap leading edge assigned their values at station y^*. Existing tabulations of two-dimensional aerodynamic coefficients can be used in this process.

(5) The corrections $\Delta L(y^*,t)$, $\Delta M_y(y^*,t)$, and $\Delta M_\beta(y^*,t)$ for the influence of finite span are determined from the $\sigma(y^*)$ found in step (3) and added to the loads from step (4). Each of these corrections is derived by identifying all terms containing $C(k)$ in the expression for the two-dimensional running load due to the particular type of motion considered, and in these terms replacing $C(k)$ by the value of $\sigma(y^*)$ at the same spanwise station. For example, we can see from Eq. (5–311) that the lift contributions due to vertical translation $h(y^*,t)$ and rotation $\alpha(y^*,t)$ are

$$\Delta L_h(y^*,t) = 2\pi\rho U^2 b_0 [ik\sigma_h(y^*)]\frac{h(y^*,t)}{b_0}\qquad(7\text{-}34)$$

and

$$\Delta L_\alpha(y^*,t) = 2\pi\rho U^2 b[1 + ik(\tfrac{1}{2} - a)]\sigma_\alpha(y^*)\alpha(y^*,t).\qquad(7\text{-}35)$$

Steps (2), (3), and (5) are greatly simplified by using tables prepared by Reissner and Stevens (Ref. 7–5). Their paper presents complete expressions for three-dimensional lift, pitching moment, flap-hinge moment, and tab-hinge moment, and it discusses the systematic modification of flutter calculations when the results of Ref. 7–4 are used to refine a strip-theory analysis. We illustrate the procedure for a case of bending-torsion flutter in Chapter 9.

The terms containing $C(k)$ in the two-dimensional theory are entirely circulatory, and Eq. (7–28) shows that only these terms are altered by the influence of finite span. Hence we are forced to conclude that, within Reissner's approximations, the noncirculatory component of the flow remains unchanged.† Since this part of the liquid motion produces no wake, aerodynamic induction effects associated with it appear to be relatively small. Certainly this is true except in the immediate tip region. Other theories do show a change in the noncirculatory flow, however, so we must rely on comparisons with experiment to establish the validity of Reissner's conclusion. The most extensive measurements of airloads on finite wings performing known oscillatory motions are those of Ashley, Zartarian, and Neilson (Ref. 7–6). None of these indicates any important

---

† This fact can also be seen from the form of Eq. (7–13), which is the only term in the integral equation when $\bar\Omega = 0$.

inaccuracies in the Reissner method on rectangular planforms with aspect ratios down to two.

An important result of finite-wing theory concerns the relative unimportance of induction effects and the increasing accuracy of airload predictions by two-dimensional strip formulas as the reduced frequency becomes large. This phenomenon is brought about by the predominance of the noncirculatory flow due to rapid vibrations, since the virtual-mass force and moment contributions are proportional to k^2. It has not been confirmed experimentally, because of the difficulty in simultaneously achieving high reduced frequencies and Reynolds numbers adequate to suppress undesirable viscous effects. On physical grounds, however, there is every reason to believe its correctness. On the basis of a large number of sample calculations, Reissner and Stevens point out that finite-span effects are negligible on wings of aspect ratio 6 when k exceeds 1.0, or on wings of aspect ratio 3 when k exceeds 2.0. On the other hand, there are significant deviations from two-dimensional theory when k is less than 0.5 for aspect ratio 6, and when k is less than 1.0 for aspect ratio 3.

Inasmuch as we have emphasized the desirability of systematizing the sort of extensive computations that are required for oscillating wings in three dimensions, we point out that the solution of Reissner's circulation equation (7–25) is a very systematic process. All necessary tables are given in Ref. 7–5. Following the procedure for lifting-line theory, the unknown $\bar{\Omega}$ is replaced by a Fourier sine series

$$\bar{\Omega}_j(y^*) = \sum_{n=1}^{N} K_{nj} \frac{\sin n\phi}{n}, \qquad (7\text{–}36)$$

where

$$y^* = s \cos \phi, \qquad \eta^* = s \cos \theta, \qquad (7\text{–}37)$$

and the subscript j is replaced by h, α, β, etc., depending on the type of wing motion involved. This substitution ensures that the circulation drops off to zero at the wing tips, although it naturally cannot make the pressure discontinuity vanish along their entire chordwise extent.

When Eq. (7–36) is put into Eq. (7–25), an algebraic equation in the Fourier coefficients K_{nj} is obtained.

$$\sum_{n=1}^{N} K_{nj} \left\{ \frac{\sin n\phi}{n} + \frac{b}{b_0} \frac{\pi}{s} \mu(k) \left[\frac{\sin n\phi}{\sin \phi} \right. \right.$$
$$\left. \left. + \frac{ik_0 s}{\pi} \int_0^\pi \frac{\cos \phi - \cos \theta}{|\cos \phi - \cos \theta|} F(k_0 s |\cos \phi - \cos \theta|) \cos n\theta d\theta \right] \right\} = \bar{\Omega}_j^{(2)}(s \cos \phi).$$

$$(7\text{–}38)$$

Here F denotes Cicala's function

$$F(q) = \int_0^\infty e^{-i\lambda} \left[\frac{1}{q} + \frac{1}{\lambda} - \frac{\sqrt{\lambda^2 + q^2}}{\lambda q} \right] d\lambda, \qquad \text{for } q \geq 0. \quad (7\text{–}39)$$

The quasi two-dimensional circulation function introduces the boundary conditions into Eq. (7–38). For vertical translation and rotation about an axis at $x^* = a$, it reads†

$$\frac{\bar{\Omega}_h^{(2)}}{U} = \frac{4iC(k)}{kH_1^{(2)}(k)} e^{ik_m} \frac{h(y^*,t)}{b_0 e^{i\omega t}} \qquad (7\text{–}40)$$

and

$$\frac{\bar{\Omega}_\alpha^{(2)}}{U} = \frac{4iC(k)}{k_0 H_1^{(2)}(k)} e^{ik_m}[1 + ik(\tfrac{1}{2} - a)]\frac{\alpha(y^*,t)}{e^{i\omega t}}, \qquad (7\text{–}41)$$

as can be deduced from Eq. (7–27).

The routine solution of Eq. (7–38) is accomplished through tables, given in Ref. 7–5, of the function

$$S_n(k_0 s, \phi) = \frac{\sin n\phi}{n} + i\frac{k_0 s}{\pi}\int_0^\pi \frac{\cos\phi - \cos\theta}{|\cos\phi - \cos\theta|} F(k_0 s|\cos\phi - \cos\theta|)\cos n\theta\, d\theta,$$
$$(7\text{–}42)$$

which shortens Eq. (7–38) to the form

$$\sum_{n=1}^N K_{nj}\left\{\frac{\sin n\phi}{n} + \frac{\pi}{s}\frac{b}{b_0}\mu(k)S_n(k_0 s,\phi)\right\} = \bar{\Omega}_j^{(2)}(s\cos\phi). \qquad (7\text{–}43)$$

Using the symmetry (n odd) or the antisymmetry (n even) of the spanwise load distribution to reduce the problem to one along the half-span $0 \le \phi \le \pi/2$, we can write Eq. (7–43) for a number of stations equal to the desired number of constants K_{nj}:

$$\left(\left[\frac{\sin n\phi}{n}\right] + \frac{\pi}{s}\left[\frac{b}{b_0}\mu(k)\right][S_n]\right)\{K_{nj}\} = \{\bar{\Omega}_j^{(2)}\}. \qquad (7\text{–}44)$$

Let $\phi_1, \phi_2, \cdots, \phi_N$ represent the values of ϕ at the points where Eq. (7–44) is to be satisfied. The diagonal matrix contains the values of $b\mu(k)/b_0$ at these points, and the two square matrices read, for symmetrical loading,

$$\left[\frac{\sin n\phi}{n}\right] = \begin{bmatrix} \sin\phi_1 & \dfrac{\sin 3\phi_1}{3} & \cdots & \dfrac{\sin(2N-1)\phi_1}{(2N-1)} \\[2ex] \sin\phi_2 & \dfrac{\sin 3\phi_2}{3} & & \dfrac{\sin(2N-1)\phi_3}{(2N-1)} \\[1ex] \cdot & & & \cdot \\ \cdot & & & \cdot \\ \cdot & & & \cdot \\[1ex] \sin\phi_N & \dfrac{\sin 3\phi_N}{3} & \cdots & \dfrac{\sin(2N-1)\phi_N}{(2N-1)} \end{bmatrix}, \qquad (7\text{–}45)$$

---

† Special care must be observed in the neighborhood of a node line of the vibration mode, where the quasi two-dimensional circulation vanishes (see Ref. 7–5).

$$[S_n] = \begin{bmatrix} S_1(k_0 s, \phi_1) & S_3(k_0 s, \phi_1) & \cdots & S_{2N-1}(k_0 s, \phi_1) \\ S_1(k_0 s, \phi_2) & S_3(k_0 s, \phi_2) & & S_{2N-1}(k_0 s, \phi_2) \\ \cdot & & & \cdot \\ \cdot & & & \cdot \\ \cdot & & & \cdot \\ S_1(k_0 s, \phi_N) & S_3(k_0 s, \phi_N) & \cdots & S_{2N-1}(k_0 s, \phi_N) \end{bmatrix}. \quad (7\text{-}46)$$

Equation (7–44) is a set of N simultaneous equations, with complex coefficients, for the N complex numbers K_{nj}. The relationship between the incremental three-dimensional loads ΔL, ΔM_y, ΔM_β at a series of stations and the K_{nj} can also be expressed by a matrix equation. It is possible to eliminate the column of K_{nj} between such a formula and Eq. (7–44), thus arriving at the unsteady form of the matrix analogous to the $[A]$ in Eq. (5–157) of lifting-line theory.

When the chord of the wing is constant, k is independent of y^*. Therefore, for a particular frequency and mode of motion, $\bar{\Omega}_j^{(2)}(y^*)$ is directly proportional to the spanwise distribution of amplitude. For instance, we might substitute Eq. (7–30) into Eq. (7–40) and get (assuming a straight midchord line)

$$\frac{\bar{\Omega}_h^{(2)}}{U} = \frac{4iC(k_0)}{k_0 H_1^{(2)}(k_0)} \frac{\bar{h}_R}{b_0} f_h(y^*). \quad (7\text{-}47)$$

This is a considerable simplification, for it enables us to replace the unknown K_{nj} in Eq. (7–43) with another set of constants which reduces the right side to a set of real numbers. In the case represented by Eq. (7–47), we use

$$K_{nh}' = \frac{K_{nh}}{\dfrac{4iUC(k_0)}{k_0 H_1^{(2)}(k_0)} \dfrac{\bar{h}_R}{b_0}}. \quad (7\text{-}48)$$

Substituting Eq. (7–48) into Eq. (7–43), we find

$$\sum_{n=1}^{N} K_{nh}' \left\{ \frac{\sin n\phi}{n} + \frac{\pi}{s} \mu(k_0) S_n(k_0 s, \phi) \right\} = f_h(s \cos \phi). \quad (7\text{-}49)$$

The three-dimensional circulation is

$$\bar{\Omega}_h = \frac{4iUC(k_0)\bar{h}_R}{k_0 H_1^{(2)}(k_0) b_0} \sum_{n=1}^{N} K_{nh}' \frac{\sin n\phi}{n}, \quad (7\text{-}50)$$

so that when we come to compute the ratio of this to $\bar{\Omega}_h^{(2)}$, which is all we need to obtain σ_h and the load distribution, we can employ the elementary formula

$$\frac{\bar{\Omega}_h}{\bar{\Omega}_h^{(2)}} = \frac{1}{f_h(s \cos \phi)} \sum_{n=1}^{N} K_{nh}' \frac{\sin n\phi}{n}. \quad (7\text{-}51)$$

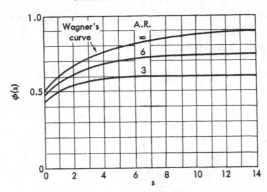

Fig. 7–2.   Development of lift following a sudden change in the angle of attack of an elliptic wing of aspect ratio AR.   The ratio of the lift to its final steady-state value is plotted *vs.* distance traveled in root semichords from the starting point.

Closely related to the problem of simple harmonic oscillation of three-dimensional wings is that of arbitrary unsteady motion. In general, we can state that it does not suffice to derive indicial functions for rigid-body plunging and pitching, since these cannot account for antisymmetric maneuvers of the aircraft or for loads due to structural deformations. A complete solution calls for calculating the pressure distribution due to sudden displacements of all important degrees of freedom, including whatever vibration modes are needed to describe aeroelastic effects. In the light of existing theory, this is a practically impossible task. It is easy to understand why such loadings have been determined almost exclusively by two-dimensional strip methods.

Once a suitable theory is established for the simple harmonic case, Fourier methods can, in principle, be adapted to the computation of any indicial function. This process is awkward and time-consuming, however, and we know only one example (Chapter V of Ref. 7–7) of where it has been successfully applied to a practical airplane. More direct theories of unsteady motion in incompressible flow have been proposed by Jones (Ref. 7–8) and Shen (Ref. 7–9), among others. The indicial functions worked out in Ref. 7–8 for sudden angle-of-attack change and gust entry of elliptical wings provide the best source of information on three-dimensional effects. We reproduce some of Jones' curves in Fig. 7–2. They show that the starting circulatory lift due to angle-of-attack or plunging motion exceeds one-half its final value, with the excess getting larger as we progress from the two-dimensional airfoil down to lower aspect ratios. This behavior is attributable to the fact that the influence of finite span is smaller on the highly unsteady flow during the starting transient than it is on the steady-state condition which is ultimately established.

Convenient exponential approximations for the two kinds of indicial functions at aspect ratios 3 and 6 are listed in Ref. 7-8. We employ the notation of Chapters 5 and 6 for total lift:

$$L(s) = 2\pi \frac{\rho}{2} U^2 S \alpha_0 \phi(s), \tag{7-52}$$

$$L_g(s) = 2\pi \frac{\rho}{2} U^2 S \frac{w_0}{U} \psi(s), \tag{7-53}$$

where s is the distance from starting, measured in midspan semichords. In these symbols, Jones' approximations read

$$\left.\begin{array}{c}\phi(s)\\\psi(s)\end{array}\right\} = b_0 + b_1 e^{-\beta_1 s} + b_2 e^{-\beta_2 s} + b_3 e^{-\beta_3 s}. \tag{7-54}$$

The various coefficients are tabulated in Table 7-1.

TABLE 7-1

COEFFICIENTS IN EXPONENTIAL APPROXIMATIONS TO THE INDICIAL FUNCTIONS
FOR PLUNGING MOTION AND GUST ENTRY BY ELLIPTICAL WINGS

Indicial function	Aspect ratio	b_0	b_1	b_2	b_3	β_1	β_2	β_3
$\phi(s)$	∞	1.000	−0.165	−0.335	0	0.0455	0.300	—
	6	0.750	−0.217	0	0	0.381	—	—
	3	0.600	−0.170	0	0	0.540	—	—
$\psi(s)$	∞	1.000	−0.236	−0.513	−0.171	0.058	0.364	2.42
	6	0.750	−0.336	−0.204	−0.145	0.290	0.725	3.00
	3	0.600	−0.407	−0.136	0	0.558	3.20	—

7-3 The influence of sweep. No one can question the importance of achieving a complete understanding of the unsteady aerodynamics of swept wings. Aeroelasticians employed by aircraft companies are repeatedly faced with the problem of predicting flutter on sharply swept-back surfaces with aspect ratios as low as two, often in the high subsonic and transonic speed ranges. No three-dimensional theory exists which is not based either on very questionable physical assumptions or on elaborate, poorly systematized computational procedures. Unfortunately, the oscillating swept wing presents a peculiar challenge to the aerodynamicist even when finite-span effects are neglected. We therefore devote the present section to discussing the use of strip theory in the presence of sweep.

In connection with the three-dimensional problem, it should be mentioned that several approximate theories appear in the literature. One of the more acceptable of these is presented by Turner (Ref. 7–10) as the culmination of a series of investigations by himself, Reissner, and Hildebrand (e.g., Ref. 7–11). The functions required for application of Turner's method have not been fully tabulated. Other approaches to the subject are suggested by Dengler and Goland (Ref. 7–12), W. P. Jones (Ref. 7–13), and Ashley, Zartarian, and Neilson (Ref. 7–6, pp. 24–39). The formulas of Ref. 7–6, although no more difficult to use than Reissner's for the straight wing, rest on such an extreme distortion of the vortex sheet simulating the wing motion that their validity is dubious for large sweep angles and moderate to small aspect ratios.

When we undertake to apply aerodynamic strip theory to a swept planform, we can make only one statement unequivocally: the aerodynamic loads due to simple harmonic motion of an infinite swept wing of constant chord are derivable directly from those on an unswept airfoil, when the motion is such that the chordwise distribution w_a of particle velocities is the same at every station. For example, consider a wing of sweep angle Λ and streamwise chord $2b$, which oscillates in vertical translation

$$h = h_0 e^{i\omega t} \tag{7-55}$$

and pitches according to

$$\alpha = \alpha_0 e^{i(\omega t + \phi)}, \tag{7-56}$$

when α is measured in a cross section parallel to U. Pitching occurs about an axis at a fixed distance ba (in the flight direction) aft of local midchord. These ensure the same motion at all stations. The angle of pitch $\bar{\alpha}$, observed in a section perpendicular to the span, is given by

$$\bar{\alpha} = \frac{\alpha}{\cos \Lambda}. \tag{7-57}$$

The lift (positive upward) and pitching moment (about an axis perpendicular to the flight direction through the point ba), both per unit distance normal to U, are

$$L(t) = \pi\rho b^2 \cos \Lambda [\ddot{h} + U\dot{\alpha} - ba\ddot{\alpha}] \\ + 2\pi\rho Ub \cos \Lambda C(k)[\dot{h} + U\alpha + b(\tfrac{1}{2} - a)\dot{\alpha}], \tag{7-58}$$

$$M_y(t) = \pi\rho b^3 \cos \Lambda [a\ddot{h} - (\tfrac{1}{2} - a)U\dot{\alpha} - b(\tfrac{1}{8} + a^2)\ddot{\alpha}] \\ + 2\pi\rho Ub^2(\tfrac{1}{2} + a) \cos \Lambda C(k)[\dot{h} + U\alpha + b(\tfrac{1}{2} - a)\dot{\alpha}], \tag{7-59}$$

where

$$k = \frac{\omega b \cos \Lambda}{U \cos \Lambda} = \frac{\omega b}{U}. \tag{7-60}$$

We derive Eqs. (7–58) and (7–59) from Eqs. (5–311) and (5–312) by the same physical argument employed for steady motion at the end of Section 5–5. They reduce to steady-state results like Eq. (5–222a) when the frequency is zero.

This same line of reasoning is valid for compressible flow. We simply multiply all aerodynamic loads by the cosine of the sweep angle and measure everything in the flight direction, with the important exception of the Mach number. Since compressibility effects are determined by the cross-flow component, we must always use for the latter

$$\frac{U \cos \Lambda}{a_\infty} = M \cos \Lambda. \qquad (7\text{-}61)$$

The lift $L(t)$ is related to the force per unit distance along the swept span by

$$\overline{L}(t) = L(t) \cos \Lambda, \qquad (7\text{-}62)$$

since unit spanwise length corresponds to $\cos \Lambda$ normal to the flight direction. $M_y(t)$, whose definition is perhaps convenient for airplane dynamic stability problems but is awkward from the structural standpoint, is related to the moment, per unit spanwise distance, about a spanwise axis located $ba \cos \Lambda$ aft of midchord by†

$$\overline{M}_y(t) = M_y(t) \cos^2 \Lambda. \qquad (7\text{-}63)$$

We can derive Eqs. (7-58) and (7-59) without resorting to the physical picture by means of the two-dimensional integral equation given in Ref. 7-10. In a nonorthogonal system of coordinates with $x = bx^*$ parallel to U and y along the swept midchord line, this reads

$$\overline{w}_a(x^*) = -\frac{1}{2\pi \cos \Lambda} \left\{ \oint_{-1}^{1} \frac{\overline{\gamma}_a(\xi^*)}{x^* - \xi^*} d\xi^* - \frac{ik\overline{\Gamma}}{b} \int_{1}^{\infty} \frac{e^{-ik\xi^*}}{x^* - \xi^*} d\xi^* \right\}, \qquad (7\text{-}64)$$

where $\overline{\gamma}_a$ is the complex amplitude of the component of circulation (per unit x-distance) normal to U. The resultant vorticity vector in this two-dimensional problem, of course, parallels the span. Equation (7-64) can be inverted by Söhngen's formulas (5-103) and (5-104). After the answer is substituted into the appropriate form of Bernoulli's equation,

$$-\Delta\overline{p}_a(x^*) = \rho U[\overline{\gamma}_a(x^*) + ik \int_{-1}^{x^*} \overline{\gamma}_a(\xi^*)d\xi^*], \qquad (5\text{-}336)$$

the only difference from the straight-wing solution is the appearance of a factor $\cos \Lambda$. We arrive by appropriate chordwise integrations at Eqs. (7-58) and (7-59).

When the vertical displacement and twist of a finite swept wing vary along the span, as they do in flutter motion, it is not as clear as for straight wings how strip theory should be applied. Two general methods of

---

† Since M_y is the moment about an axis normal to U, the only clear physical picture we can give of it is this: $M_y dy$ is the integral of the pressure difference, weighted with its arm about this axis, along a strip of width dy oriented in the flight direction. $\overline{M}_y$, on the other hand, is the actual, measurable moment about the pitching axis per unit distance along the swept span. For a further discussion, see Section 8-4.

approach have been tried. The first is essentially the use of Eqs. (7–58) and (7–59), with h and α replaced by the vertical displacement and rotation due to the assumed mode at each streamwise section of the wing. This scheme was porposed by Fettis (Ref. 7–14) and tried out by Spielberg, Fettis, and Toney in flutter calculations (Ref. 7–15). The various formulas for running lift and moment in these two references must be multiplied by $\cos \Lambda$ to bring them into accord with Eqs. (7–58) and (7–59).

In connection with the practical application of this streamwise-strip theory, we mention that some swept-wing structures have their ribs oriented in the flight direction, while others are nearly perpendicular to the midchord line or elastic axis (if any). In the former cases it is logical to assume that streamwise sections remain undeformed during bending and torsion, so that the simple strip method is peculiarly well suited. The latter structures twist more like a surface developed out of straight lines normal to the elastic axis, since perpendicular sections are the ones which do not bend. Therefore there are simple harmonic camber changes in planes parallel to U, the amount of camber being proportional to the product $\sin \Lambda [\partial \bar{\alpha}/\partial \bar{y}]$. It seems reasonable to add into Eqs. (7–58) and (7–59) contributions from parabolic chordwise bending oscillations, such as those tabulated in Ref. 7–16. As an alternative, the strip theory based on velocity components, which we outline below, may be more suitable.

If we concentrate on sections perpendicular to some spanwise reference axis, we are led to the second scheme for applying two-dimensional aerodynamics to a finite swept wing. This is another of the proposals in Ref. 7–15. Broadbent describes it briefly (Ref. 7–17), attributing the idea to Minhinnick, and a thorough derivation (in a modified form) appears in Ref. 7–18.

We assume a rectilinear reference axis oriented along the wingspan and swept at the angle Λ. Sections normal to this axis are rigid. Its bending deformation is $h(\bar{y},t)$ (positive downward) and the twist about it is $\bar{\alpha}(\bar{y},t)$ (positive nose up), where $\bar{y}$ is a coordinate directed along the axis. If $z_a(\bar{x},\bar{y},t)$ denotes the upward vertical displacement of points on the mean line, we get

$$z_a(\bar{x},\bar{y}.t) = -h - \bar{\alpha}\bar{x}; \qquad \text{for } (\bar{x},\bar{y}) \text{ in } R_a, \qquad (7\text{–}65)$$

$\bar{x}$ being measured positive aft and normal to the $\bar{y}$-direction. The vertical velocity of fluid particles forced by the wing motion is

$$w_a(\bar{x},\bar{y},t) = \frac{\partial z_a}{\partial t} + U \frac{\partial z_a}{\partial x}$$

$$= \frac{\partial z_a}{\partial t} + \frac{\partial z_a}{\partial \bar{x}} U \cos \Lambda + \frac{\partial z_a}{\partial \bar{y}} U \sin \Lambda$$

$$= -\dot{h} - U\bar{\alpha} \cos \Lambda - \dot{\bar{\alpha}}\bar{x} - U(\sigma + \tau\bar{x}) \sin \Lambda;$$

$$\text{for } z = 0, \ (x,y) \text{ in } R_a, \qquad (7\text{–}66)$$

where
$$\sigma = \frac{\partial h}{\partial \bar{y}}, \qquad \tau = \frac{\partial \bar{\alpha}}{\partial \bar{y}} \tag{7-67}$$

are the spanwise rates of change of bending and twist.

A wing with an *unswept* reference axis flying at velocity U would give rise to

$$w_a(x,y,t) = -\dot{h} - U\alpha - \dot{\alpha}x. \tag{7-68}$$

If we neglect the influence of the spanwise component $U \sin \Lambda$ on the differential equation and the geometry of the vortex sheet, introducing it only through the boundary condition (7-66), we can employ the formulas of Section 5-6 directly to write down the lift and moment on the swept wing. Terms from the $-(\dot{h} + U\alpha)$ portion of Eq. (7-68) must be replaced by similar terms containing the combination $-(\dot{h} + U\bar{\alpha} \cos \Lambda + U\sigma \sin \Lambda)$. Terms due to the $-(\dot{\alpha}x)$ in Eq. (7-68) now involve $-(\dot{\alpha} + U\tau \sin \Lambda)\bar{x}$. The speed U, chord $2b$, and axis location ba aft of midchord transform into the corresponding properties $U \cos \Lambda$, $2\bar{b}$, and $\bar{b}a$ of the perpendicular cross section. In dealing with terms containing α, we must be careful to separate the contributions of the two parts of Eq. (7-68), since the $\partial\phi/\partial t$ in Bernoulli's equation gives rise to an $\dot{\alpha}$ from the $-(U\alpha)$ in the boundary condition. As we have mentioned previously, all effects of the $-(U\alpha)$ are accompanied by identical forms containing $-(\dot{h})$. They are not replaced by $-(\dot{\alpha} + U\tau \sin \Lambda)$, as are the other $\dot{\alpha}$ terms and all the $\ddot{\alpha}$ terms.

The results of applying these similarity laws to Eqs. (5-311) and (5-312) are as follows:

$$\begin{aligned}
\bar{L}(\bar{y},t) = \ &\pi\rho\bar{b}^2\{\ddot{h} + U\dot{\alpha} \cos \Lambda + U\dot{\sigma} \sin \Lambda \\
&- \bar{b}a\ddot{\alpha} - \bar{b}aU\dot{\tau} \sin \Lambda\} \\
&+ 2\pi\rho\bar{b}U \cos \Lambda C(k)\{\dot{h} + U\bar{\alpha} \cos \Lambda + U\sigma \sin \Lambda \\
&+ \bar{b}(\tfrac{1}{2} - a)[\dot{\alpha} + U\tau \sin \Lambda]\},
\end{aligned} \tag{7-69}$$

$$\begin{aligned}
\bar{M}_y(\bar{y},t) = \ &-\pi\rho\bar{b}^3\{U(\tfrac{1}{2} - a)\dot{\alpha} \cos \Lambda + \tfrac{1}{2}U^2\tau \cos \Lambda \sin \Lambda \\
&- a\ddot{h} - Ua\dot{\sigma} \sin \Lambda + \bar{b}(\tfrac{1}{8} + a^2)[\ddot{\alpha} + U\dot{\tau} \sin \Lambda]\} \\
&+ 2\pi\rho\bar{b}^2U \cos \Lambda(\tfrac{1}{2} + a)C(k)\{\dot{h} + U\bar{\alpha} \cos \Lambda \\
&+ U\sigma \sin \Lambda + \bar{b}(\tfrac{1}{2} - a)[\dot{\alpha} + U\tau \sin \Lambda]\}.
\end{aligned} \tag{7-70}$$

Incidentally, these formulas yield Eqs. (7-58) and (7-59) when σ and τ are zero. To facilitate the use of existing tables, it is shown in Ref. 7-15 that Eqs. (7-69) and (7-70) can be written

$$\bar{L}(\bar{y},t) = -\pi\rho\omega^2\bar{b}^3 \left\{ \frac{h}{\bar{b}} L_{hh} + \sigma L_{hh'} + \bar{\alpha}L_{h\alpha} + \bar{b}\tau L_{h\alpha'} \right\}, \tag{7-71}$$

$$\bar{M}_y(\bar{y},t) = \pi\rho\omega^2\bar{b}^4 \left\{ \frac{h}{\bar{b}} M_{\alpha h} + \sigma M_{\alpha h'} + \bar{\alpha}M_{\alpha\alpha} + \bar{b}\tau M_{\alpha\alpha'} \right\} . \tag{7-72}$$

The eight dimensionless coefficients are calculated from the functions of k only, defined in Eqs. (5–350) and (5–351), through the relations

$$L_{hh} = L_h, \tag{7-73a}$$

$$L_{hh}' = -i\,\frac{\tan \Lambda}{k}\,L_h, \tag{7-73b}$$

$$L_{h\alpha} = L_\alpha - L_h(\tfrac{1}{2} + a), \tag{7-73c}$$

$$L_{h\alpha'} = -i\,\frac{\tan \Lambda}{k}\,[-\tfrac{1}{2} + L_h(\tfrac{1}{2} - a)], \tag{7-73d}$$

$$M_{\alpha h} = M_h - L_h(\tfrac{1}{2} + a), \tag{7-73e}$$

$$M_{\alpha h'} = -i\,\frac{\tan \Lambda}{k}\,[M_h - L_h(\tfrac{1}{2} + a)], \tag{7-73f}$$

$$M_{\alpha\alpha} = M_\alpha - [M_h + L_\alpha](\tfrac{1}{2} + a) + L_h(\tfrac{1}{2} + a)^2, \tag{7-73g}$$

$$M_{\alpha\alpha'} = -i\,\frac{\tan \Lambda}{k}\left[\frac{3}{8} - \frac{i}{2k} - L_h(\tfrac{1}{4} - a^2)\right]. \tag{7-73h}$$

It should be clear from the foregoing how the velocity component method can be carried over to other modes of motion in perpendicular cross sections of swept wings. For example, oscillation of a trailing-edge flap can be introduced by replacing β in the formulas for the two-dimensional airfoil by

$$\bar\beta = \frac{\beta}{\cos \Lambda}, \tag{7-74}$$

which is practically equal to the angle of rotation of the flap about its swept hinge line. As above, U, b and the hinge location be become $U \cos \Lambda$, $\bar b$, and $\bar b e$. The extension of these ideas to other types of unsteady motion is also a straightforward matter.

We close this section by calling attention to certain weaknesses of velocity-component strip theory, which are implied in the work of Barmby, Cunningham, and Garrick (Ref. 7–18). There is an important difference between the wake vortex sheet on which Eqs. (7–69)–(7–70) are based and the one actually shed from a finite swept wing, even when we overlook tip effects. The vortices do not move downstream with velocity $U \cos \Lambda$ in a direction normal to the swept span. They follow much more closely the resultant stream U. Only when the motion is exactly the same in all cross sections can we say with certainty that the velocity patterns induced by the two configurations coincide.

A second shortcoming of the theory has to do with the way in which we apply Bernoulli's equation. There are actually two principal steps in calculating the aerodynamic loads, the first being to find the disturbance velocity potential, the second to deduce therefrom the pressure distribution. In the latter step, we have assumed in effect that

$$p_U - p_L = -2\rho \left[\frac{\partial \phi'}{\partial t} + U \cos \Lambda \frac{\partial \phi'}{\partial \bar{x}} \right]. \tag{7-75}$$

However, the correct form is

$$p_U - p_L = -2\rho \left[\frac{\partial \phi'}{\partial t} + U \frac{\partial \phi'}{\partial x} \right]$$

$$= -2\rho \left[\frac{\partial \phi'}{\partial t} + U \cos \Lambda \frac{\partial \phi'}{\partial \bar{x}} + U \sin \Lambda \frac{\partial \phi'}{\partial \bar{y}} \right]. \tag{7-76}$$

The last term in Eq. (7–76) is not zero, regardless of how we find ϕ', because $\bar{y}$ enters the problem through every element in Eq. (7–66). It is no easy matter to compute $\partial \phi'/\partial \bar{y}$; its circulatory part contains $C(k)$ and k is a function of $\bar{y}$ whenever the wing is tapered. In Ref. 7–18, this difficulty is avoided by the admittedly inconsistent procedure of retaining only the noncirculatory portions of $\partial \phi'/\partial \bar{y}$. A somewhat more accurate scheme, which would introduce no more complexity, might be to include the circulatory effects but overlook the presence of taper, so that

$$\frac{dk}{d\bar{y}} \cong 0. \tag{7-77}$$

Inasmuch as neither of these alternatives is very satisfactory, it is fortunate that little change in the values of $\bar{L}$ and $\bar{M}_y$ results from leaving out $\partial \phi'/\partial \bar{y}$ entirely for wings and vibration modes of practical interest.

7–4 Wings of very low aspect ratio in unsteady motion. When dealing with the airloads on wings of moderate or large aspect ratio, it is natural to concentrate on the spanwise variation of lift and moment. For this reason lifting-line types of aerodynamic theory maintained the greatest prominence over a long period, and only recently have highly swept and other low aspect-ratio planforms created a need for changed emphasis. Wings with nearly equal spanwise and chordwise dimensions require predictions of load distribution over an area rather than along a line, and in the limit one can conceive of planforms so narrow that the chordwise distribution of force is the most important information.

With regard to unsteady lifting-surface theory in the aspect ratio range around unity, the state of current knowledge is too incomplete to justify an extensive treatment. However, we call attention to the promising methods of Lawrence-Gerber (Ref. 7–19) and Laidlaw (Ref. 7–20) for incompressible flow. The subject of low aspect-ratio wings in subsonic and supersonic flight is briefly discussed in Section 7–5.

The prediction of chordwise loading is a relatively simple task when the aspect ratio has a small fractional value. The important contribu-

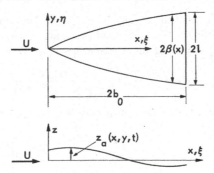

Fig. 7–3. Plan and side views of a wing with fractional aspect ratio.

tions of R. T. Jones on the steady-flow problem have already been described in Section 5–5(c). His momentum method can also be employed for unsteady motion, but a somewhat more general solution is obtained by introducing simplifying assumptions into the lifting-surface integral equation. This procedure was proposed by Lomax and Sluder (Ref. 7–21), and the development presented here follows that of Voss and Hassig (Ref. 7–22). Arbitrary small chordwise and spanwise deformations with any time dependence whatever are included; the planform must be slender and pointed with a straight trailing edge, but the only restriction on leading-edge shape is that it can nowhere form too large an angle with the flight direction.

We consider the wing illustrated in Fig. 7–3. As can be deduced from Eq. (5–128), within the limitations of small-perturbation theory, the disturbance velocity potential describing the incompressible flow generated by the motion $z_a(x,y,t)$ has the exact form

$$\phi'(x,y,z,t)$$
$$= \frac{z}{4\pi} \int_0^\infty \int_{-\beta(\xi)}^{\beta(\xi)} \frac{\gamma(\xi,\eta,t)}{[(y-\eta)^2+z^2]} \left\{ 1 + \frac{x-\xi}{\sqrt{(x-\xi)^2+(y-\eta)^2+z^2}} \right\} d\eta d\xi. \quad (7\text{–}78)$$

Here γ is the circulation per unit chordwise distance, as in the various applications of Chapter 5.

The same assumption adopted for steady flow in Section 5–5(c) is now introduced here. We observe that, over nearly the entire area of integration in Eq. (7–78), spanwise distances $|y - \eta|$ have small magnitudes compared with chordwise distances $|x - \xi|$. If this is true, and if we additionally note that we are interested in the solution close to the wing surface so that z is a small quantity, the bracket in Eq. (7–78) can be approximated as follows:

$$\left[1 + \frac{x-\xi}{\sqrt{(x-\xi)^2+(y-\eta)^2+z^2}} \right] \cong \left[1 + \frac{(x-\xi)}{|x-\xi|} \right] = \begin{cases} 2, & \text{for } \xi \leq x, \\ 0, & \text{for } \xi > x. \end{cases} \quad (7\text{–}79)$$

When substituted into Eq. (7–78), Eq. (7–79) leads to the same conclusion as in steady flow. That is, conditions at a given point on the wing are influenced only by what takes place upstream from that point.

$$\phi'(x,y,z,t) = \frac{z}{2\pi} \int_0^x \int_{-\beta(\xi)}^{\beta(\xi)} \frac{\gamma(\xi,\eta,t)}{\left[(y-\eta)^2 + z^2\right]} \, d\eta d\xi. \tag{7–80}$$

In terms of the frequently used dimensionless independent variables

$$x^* = \frac{x}{b_0}, \qquad y^* = \frac{y}{b_0}, \qquad z^* = \frac{z}{b_0}, \tag{7–81}$$

Eq. (7–80) reads

$$\phi'(x^*,y^*,z^*,t) = \frac{z^* b_0}{2\pi} \int_0^{x^*} \int_{-\beta/b_0}^{\beta/b_0} \frac{\gamma(\xi^*,\eta^*,t)}{\left[(y^* - \eta^*)^2 + z^{*2}\right]} \, d\eta^* d\xi^*. \tag{7–82}$$

To facilitate solution by reducing the order of mathematical singularities, we define a partial spanwise lift distribution [cf. Eq. (5–199)]:

$$\Delta\phi'(x^*,y^*,t) = b_0 \int_{x_l(y^*)}^{x^*} \gamma(\xi^*,y^*,t) d\xi^*. \tag{7–83}$$

Substituting $\Delta\phi'$ into Eq. (7–82), with the order of integration interchanged, we get

$$\phi'(x^*,y^*,z^*,t) = \frac{z^*}{2\pi} \int_{-\beta/b_0}^{\beta/b_0} \frac{\Delta\phi'(x^*,\eta^*,t)}{\left[(y^* - \eta^*)^2 + z^{*2}\right]} \, d\eta^*$$

$$= \frac{1}{2\pi} \int_{-\beta/b_0}^{\beta/b_0} \frac{\partial\Delta\phi'}{\partial\eta^*} \tan^{-1}\left(\frac{y^* - \eta^*}{z^*}\right) d\eta^*. \tag{7–84}$$

In the integration by parts leading to the last line of Eq. (7–84), the integrated portion vanishes because $\Delta\phi'$ itself is zero for points on the leading edge. This is so in spite of the integrable singularity γ is known to exhibit there.

The boundary condition to be satisfied by ϕ' on the planform region R_a is the familiar one that the vertical velocity of fluid particles is forced by the wing's motion:

$$\frac{\partial\phi'}{\partial z} = \frac{1}{b_0}\frac{\partial\phi'}{\partial z^*} = w_a(x^*,y^*,t)$$

$$= \frac{\partial z_a}{\partial t} + U\frac{\partial z_a}{\partial x}; \qquad \text{for } z^* = 0, \ (x^*,y^*) \text{ in } R_a. \tag{7–85}$$

Regarding w_a as a known function, we substitute Eq. (7–84) into Eq. (7–85), set $z^* = 0$, and deduce the integral equation

$$w_a(x^*,y^*,t) = -\frac{1}{2\pi b_0} \oint_{-\beta(x^*)/b_0}^{\beta(x^*)/b_0} \frac{\partial\Delta\phi'}{\partial\eta^*} \frac{d\eta^*}{y^* - \eta^*}. \tag{7–86}$$

Equation (7–86) is solved directly by the appropriate pair of Söhngen's inversion formulas from Section 5–5(c). These state that if

$$g(y^*) = \fint_{-1}^{1} \frac{f(\eta^*)}{y^* - \eta^*} \, d\eta^*, \tag{5–204}$$

with the auxiliary condition

$$\int_{-1}^{1} f(\eta^*) d\eta^* = 0, \tag{5–205}$$

then the unknown function f is given by

$$f(y^*) = -\frac{2}{\pi \sqrt{1 - y^{*2}}} \fint_{-1}^{1} g(\eta^*) \frac{\sqrt{1 - \eta^{*2}}}{(y^* - \eta^*)} \, d\eta^*. \tag{5–206}$$

When applied to Eq. (7–86), these formulas yield

$$
\frac{\partial \Delta\phi'(x^*, y^*, t)}{\partial y^*}
$$
$$
= \frac{-2b_0}{\pi \sqrt{\beta^2(x^*) - b_0^2 y^{*2}}} \fint_{-\beta(x^*)/b_0}^{\beta(x^*)/b_0} w_a(x^*, \eta^*, t) \frac{\sqrt{\beta^2(x^*) - b_0^2 \eta^{*2}}}{(y^* - \eta^*)} \, d\eta^*. \tag{7–87}
$$

The complete load distribution over the wing is calculated from Eq. (7–87) by straightforward integrations and differentiations. Thus, from Eq. (5–263), we have

$$
\Delta p_a = p_U - p_L = -\rho \left[\frac{\partial \Delta\phi'}{\partial t} + U \frac{\partial \Delta\phi'}{\partial x} \right]
$$
$$
= -\rho \left[\frac{\partial \Delta\phi'}{\partial t} + \frac{U}{b_0} \frac{\partial \Delta\phi'}{\partial x^*} \right], \tag{7–88}
$$

where

$$
\Delta\phi'(x^*, y^*, t) = \int_{-\beta(x^*)/b_0}^{y^*} \frac{\partial \Delta\phi'}{\partial \eta^*} \, d\eta^*. \tag{7–89}
$$

The time dependence of the motion is arbitrary, but if we restrict it to simple harmonic oscillation, Eq. (7–88) becomes

$$
\Delta\bar{p}_a e^{i\omega t} = -\rho \left[i\omega \Delta\bar{\phi}' + \frac{U}{b_0} \frac{\partial \Delta\bar{\phi}'}{\partial x^*} \right] e^{i\omega t}. \tag{7–90}
$$

Reference 7–22 shows how a Fourier series representation of the spanwise mode of wing deformation makes it possible to express the entire solution algebraically. If we define

$$
\theta = \cos^{-1} \left[\frac{y}{\beta(x)} \right] = \cos^{-1} \left[\frac{b_0 y^*}{\beta(x^*)} \right] \tag{7–91}
$$

and assume

$$w_a(x^*,y^*,t) = f(x^*,t) \sum_{n=0}^{N} a_n \cos n\theta, \qquad (7\text{-}92)$$

Eq. (7-87) can be integrated to yield

$$\frac{\partial \Delta \phi'}{\partial y^*} = -b_0 f(x^*,t) \left\{ 2a_0 \cot \theta + a_1 \frac{\cos 2\theta}{\sin \theta} -2 \sum_{n=2}^{N} a_n \sin n\theta \right\}. \qquad (7\text{-}93)$$

This form also permits us to evaluate $\Delta \phi'$ through Eq. (7-89):

$$\Delta \phi'(x^*,y^*,t) = f(x^*,t)\beta(x^*) \left\{ 2a_0 \sin \theta + a_1 \sin \theta \cos \theta \right.$$

$$\left. + 2 \sum_{n=2}^{N} \frac{a_n}{n^2 - 1} \left[n \sin \theta \cos n\theta - \cos \theta \sin n\theta \right] \right\}. \qquad (7\text{-}94)$$

When computing pressure distribution by means of Eqs. (7-88) and (7-94), one must be careful to note that the variable θ is itself a function of x^*.

If the wing motion is steady, the foregoing solution is still valid, independent of the Mach number in compressible flow (see Ref. 7-22, for example). It is also an accurate approximation for simple harmonic motion at low reduced frequencies. When the motion becomes more rapidly oscillatory, however, the governing differential equation is the wave equation, and the problem is substantially more difficult (e.g., Merbt and Landahl, Ref. 7-23). The momentum approach analogous to the above solution has been extended to wing-body combinations by Bryson (Ref. 7-24), among others, and to supersonic unsteady flow by Miles (Ref. 7-25).

There is considerable doubt about the size of the aspect ratio range within which slender-wing theory is valid. It is known to overestimate both the steady and unsteady airloads on a 60°-delta wing, but the aspect ratio in this case exceeds 2. Applications to flutter analyses, such as the one in Ref. 7-22, have met with mixed success. The principal difficulty seems to be that Kutta's hypothesis of zero pressure discontinuity at the trailing edge is not, in general, fulfilled, whereas experiments even on the narrowest planforms show a strong tendency for Δp_a to vanish there. To overcome this weakness and produce much better agreement with his own measurements, Laidlaw (Ref. 7-20) suggests that an empirical correction factor be applied to all chordwise pressure distributions. This factor takes the form of a semi-ellipse and can be written, in the coordinates of Fig. 7-3,

$$F(x) = \sqrt{1 - \frac{x^2}{(2b_0)^2}} = \sqrt{1 - \tfrac{1}{4}x^{*2}}. \qquad (7\text{-}95)$$

Such a correction has resulted in a remarkable improvement of the accuracy of at least one calculated flutter speed (Ref. 7-26).

7–5 The influence of compressibility on oscillating wings of finite span.

A few remarks were made in the Introduction about unsteady, three-dimensional, compressible flow theory. Important theoretical advances have lately been achieved in the subsonic, transonic, and supersonic speed ranges, but the requisite degrees of systematization and general applicability are far from attainment in any of these fields. In this necessarily brief section, we review without detailed derivation certain of the more fruitful lines of approach.

(a) *Subsonic case.* Subsonic flow presents the greatest obstacles. There are no independent upper and lower wing surfaces to permit the employment of a simple, source-pulse type of elementary solution, and the fact that the governing differential equation is hyperbolic almost rules out the concept of compressibility correction, so powerful for steady flight. Miles (Ref. 7–27) has proposed such a correction, but it is valid only up to terms of the first power in the reduced frequency and may be used for dynamic stability and low-frequency flutter prediction only. Reissner (Ref. 7–28, among others) published a more complete theory for wings of relatively large aspect ratio. The tabulations needed to apply Reissner's method are so extensive, however, that its acceptance for practical aeroelastic calculations seems unlikely.

A more systematic and therefore more promising avenue of attack for subsonic flow is based on numerical solution of the integral equation relating the vertical velocity and pressure distributions over the planform. As calculated directly from an appropriate representation of the acceleration potential or velocity potential, this equation has the form

$$\bar{w}_a(x^*,y^*) = \frac{e^{i\bar{\omega}x^*}}{4\pi\rho_\infty U} \lim_{z^*\to 0} \frac{\partial}{\partial z^*} \iint_{R_a} \Delta\bar{p}_a(\xi^*,\eta^*)e^{ik\xi^*}$$

$$\times \int_{\xi^*}^{\infty} e^{-i(k\zeta^*/\beta^2)} \frac{\partial}{\partial z^*}\left(\frac{e^{-i(\bar{\omega}/M)r}}{r}\right) d\zeta^* d\xi^* d\eta^*. \quad (7\text{--}96)$$

Here

$$r^2 = (x^* - \zeta^*)^2 + \beta^2(y^* - \eta^*)^2 + \beta^2 z^{*2}, \quad (7\text{--}97)$$

and the remaining notation, such as

$$\beta = \sqrt{1 - M^2} \quad (7\text{--}98)$$

and

$$\bar{\omega} = \frac{kM^2}{\beta^2} = \frac{\omega b M^2}{U(1 - M^2)}, \quad (7\text{--}99)$$

we have already introduced in connection with two-dimensional theory in Chapter 6.

Symbolically, Eq. (7–96) can be written

$$4\pi\rho_\infty U\bar{w}_a(x^*,y^*) = \iint_{R_a} \Delta\bar{p}_a(\xi^*,\eta^*)K(x^*,y^*,\xi^*,\eta^*;k,M)d\xi^* d\eta^*, \quad (7\text{--}100)$$

where K is known as the kernel function of unsteady, three-dimensional wing theory. (The same representation can be employed, incidentally, for any flight speed range wherein the relationship between pressure and vertical velocity is linear.) Unfortunately, the singularities of K are so strong that Eq. (7–100) is not suitable for numerical solution. An integration of the most singular term by parts with respect to the spanwise coordinate η^* enables us to set z^* equal to zero in Eq. (7–96) and rewrite it

$$4\pi\rho_\infty U\bar{w}_a(x^*,y^*)$$

$$= \oint\oint_{R_a} \frac{\partial\Delta\bar{p}_a}{\partial\eta^*} D(x^* - \xi^*, y^* - \eta^*)d\xi^*d\eta^*$$

$$+ \oint\oint_{R_a} \Delta\bar{p}_a(\xi^*,\eta^*)[F(x^* - \xi^*, y^* - \eta^*) + H(x^* - \xi^*, y^* - \eta^*)]d\xi^*d\eta^*.$$

$$(7\text{–}101)$$

In Eq. (7–101), the two singular portions of the modified kernel function are

$$D(\alpha,\delta) = \frac{e^{-ik\alpha}}{|\delta|}\left[1 + \frac{\sqrt{\alpha^2 + \beta^2\delta^2}}{\alpha}\right] \qquad (7\text{–}102)$$

and

$$F(\alpha,\delta) = ike^{-ik\alpha}\left[\frac{1}{\sqrt{\alpha^2 + \beta^2\delta^2}} + \frac{ik}{2}\ln\left(\sqrt{\alpha^2 + \beta^2\delta^2} - \alpha\right)\right]. \qquad (7\text{–}103)$$

The nonsingular increment H, which vanishes as k^2 in steady flow, is†

$$H(\alpha,\delta) = e^{-ik\alpha}\left\{-\frac{k}{|\delta|}K_1(k|\delta|) + \frac{1}{\delta^2} + \frac{i\pi k}{2|\delta|}[L_1(k|\delta|) - I_1(k|\delta|)]\right.$$

$$+ \frac{ik}{\beta|\delta|}e^{-i(kM|\delta|/\beta)} + \frac{1}{M\delta^2}e^{-i(kM|\delta|/\beta)} - k^2\int_0^{M/\beta}\sqrt{\zeta^2 + 1}\,e^{-ik|\delta|\zeta}d\zeta$$

$$- \frac{\alpha[e^{i(k/\beta^2)[\alpha - M\sqrt{\alpha^2+\beta^2\delta^2}]} - 1]}{\delta^2\sqrt{\alpha^2 + \beta^2\delta^2}} - \frac{ik}{\sqrt{\alpha^2 + \beta^2\delta^2}}$$

$$+ \frac{k^2}{2}\ln\left(\sqrt{\alpha^2 + \beta^2\delta^2} - \alpha\right) - \frac{1}{M\delta^2}e^{i(k/\beta^2)[\alpha - M\sqrt{\alpha^2+\beta^2\delta^2}]}$$

$$+ \frac{ik}{M\delta^2}\int_0^\alpha e^{i(k/\beta^2)[\lambda - M\sqrt{\lambda^2+\beta^2\delta^2}]}d\lambda\right\}. \qquad (7\text{–}104)$$

The routine calculation of aerodynamic loads through Eq. (7–101) will rely heavily on the availability of adequate tables of the function H. Two procedures have been tried out successfully. The first of these is the

---

† I_1, K_1, and L_1 are modified Bessel and Struve functions of the first order.

development of Dynamic Loads Division, Langley Aircraft Laboratory, N.A.C.A., and stems from the initial publication describing the properties of the integral equation (Watkins, Runyan, and Woolston, Ref. 7–29). The scheme is based upon a spanwise integration of the kernel function to produce what is effectively an oscillating horseshoe vortex in compressible flow. Vortices are suitably distributed over the area of the wing, and their strengths are determined by requiring the vertical velocity to satisfy the boundary condition at a number of control points equal to the number of unknowns. As so far employed, the N.A.C.A. method emphasizes spanwise load distribution and is therefore more suitable for relatively large aspect ratios. This is not a necessary limitation, however. Results of some typical calculations are described by Garrick (Ref. 7–30).

A second procedure for solving Eq. (7–101) has been suggested by Voss (Ref. 7–31). It consists of direct numerical integration of each of the three elements of the right side in terms of the values of $\Delta \bar{p}_a$ at the centers of a set of rectangular boxes distributed over the planform. A system of algebraic equations is thus generated with which the $\Delta \bar{p}_a$ can be computed from the known $\bar{w}_a$ at the same control stations. The method requires more mathematical steps than that employed by N.A.C.A., but it is highly organized and has already been proved well suited to digital machine computation. It is especially designed for low aspect-ratio wings whose structural properties or vibration mode shapes can be described by sets of deflections at the centers of the aerodynamic boxes. Sample calculations for steady flow are given by Voss, Zartarian, and Hsu (Ref. 7–32).

(b) *Supersonic case.* As observed in Section 7–1, calculation of both steady and unsteady loadings of supersonic finite wings is hampered by the limitation of most methods to particular classes of planforms. It is therefore fortunate that, for purposes of flutter analysis by the Rayleigh-Ritz method, aerodynamic strip theory is much more accurate and widely applicable than might be anticipated from experience with subsonic wings. This fact is discussed by Walsh, Voss, and Zartarian (Ref. 7–33), who study the form of the generalized forces entering flutter equations like those we derive in Section 9–5. For a wing with all supersonic leading edges and a straight trailing edge, they prove among other things that strip theory is exact for computing any generalized force acting on a deformation mode which is arbitrary chordwise but contains no higher than the first power of the spanwise variable y.

When three-dimensional effects must be accounted for in unsteady supersonic flow, two theoretical approaches have proved fruitful. These are superposition of elementary source or doublet solutions of the partial differential equation, and application of Laplace and Fourier transformations analogous to those often used in the study of light and sound wave propagation. We shall give a brief account of the first technique and furnish references to the second.

The governing differential equation for all cases was derived in Section 5–1.

$$\nabla^2 \phi' - \frac{1}{a_\infty^2} \left[\frac{\partial^2 \phi'}{\partial t^2} + 2U \frac{\partial^2 \phi'}{\partial x \partial t} + U^2 \frac{\partial^2 \phi'}{\partial x^2} \right] = 0. \tag{5–44}$$

The principal boundary condition for the lifting problem is [cf. Eq. (5–316) or (7–85)]

$$\frac{\partial \phi'}{\partial z} = w_a(x,y,t) = \frac{\partial z_a}{\partial t} + U \frac{\partial z_a}{\partial x} ; \qquad \text{for } z = 0, (x,y) \text{ in } R_a. \tag{7–105}$$

In addition, it is specified that ϕ' vanishes ahead of the wing and ahead of Mach cones from the wing tips. All disturbances must be radiating outward at infinity. Most planforms have supersonic trailing edges, and for them no reference need be made to Kutta's hypothesis or conditions in the wake, since signals cannot be propagated upstream to the wing.

In analogy with the two-dimensional Eq. (6–226), Garrick and Rubinow (Ref. 7–34) employ the following three-dimensional source pulse solution of Eq. (5–44):

$$\phi_{sp}' = \frac{A(\xi,\eta,\zeta)}{r} \left[f(t - \tau_2) + f(t - \tau_1) \right]. \tag{7–106}$$

Here

$$r = \frac{1}{\beta^2} \sqrt{(x - \xi)^2 - \beta^2 [(y - \eta)^2 + (z - \zeta)^2]} \tag{7–107}$$

and

$$\tau_{2,1} = \frac{M(x - \xi)}{a_\infty (M^2 - 1)} \pm \frac{r}{a_\infty}. \tag{7–108}$$

ϕ_{sp}' differs from the two-dimensional source pulse in that it represents the continuous generation of disturbances at point (ξ,η,ζ). Their time-dependent strength is given by $A(\xi,\eta,\zeta) f(t)$. The appearance of $f(t - \tau_1)$ and $f(t - \tau_2)$ in Eq. (7–106) reflects the fact that the disturbance released from (ξ,η,ζ) at any instant is felt at (x,y,z) just twice: at an interval τ_1 later and at an interval τ_2 later. τ_1 equals the time required by the expanding spherical wave first to reach (x,y,z), and τ_2 is the time until this wave is again swept by that point in its passage downstream. Because of this behavior, ϕ_{sp}' is sometimes called a retarded potential. Time delays τ_1 and τ_2 equal each other on the Mach cone $r = 0$ aft from (ξ,η,ζ). Outside this cone, ϕ_{sp}' is interpreted on physical grounds to be zero.

Source pulses, like their two-dimensional counterparts, are point functions with respect to the normal velocity component when distributed in sheets. Thus, we spread pulses over an area in the $\xi\eta$-plane:

$$\phi'(x,y,z,t) = \iint_S A(\xi,\eta) \frac{f(t - \tau_2) + f(t - \tau_1)}{r} d\xi d\eta, \tag{7–109}$$

where S is that portion of the sheet intercepted by the Mach forecone from (x,y,z). It can be shown that

$$\frac{\partial \phi'(x,y,0,t)}{\partial z} = -2\pi\beta^2 A(x,y)f(t). \tag{7–110}$$

In view of boundary condition (7–105), we can therefore write

$$\phi'(x,y,z,t) = -\frac{1}{2\pi} \iint_S \frac{w_a(\xi,\eta,t-\tau_2) + w_a(\xi,\eta,t-\tau_1)}{\sqrt{(x-\xi)^2 - \beta^2[(y-\eta)^2 + z^2]}}\, d\xi d\eta, \tag{7–111}$$

providing we assume that w_a is known over the entire integration area.

For simple harmonic motion,

$$w_a(\xi,\eta,t-\tau_2) + w_a(\xi,\eta,t-\tau_1) = \bar{w}_a(\xi,\eta)e^{i\omega t}\left[e^{-i\omega\tau_2} + e^{-i\omega\tau_1}\right]$$

$$= \bar{w}_a(\xi,\eta)e^{i\omega t}\left[2e^{-i\omega\left(\frac{\tau_2+\tau_1}{2}\right)}\cos\left(\omega\frac{\tau_2-\tau_1}{2}\right)\right]. \tag{7–112}$$

For $z = 0^+$, the definitions of τ_1 and τ_2 then reduce Eq. (7–111) to

$$\phi'(x,y,0^+,t) = -\frac{e^{i\omega t}}{\pi} \iint_S \bar{w}_a(\xi,\eta)\, \frac{e^{-i\bar{w}/2\left(\frac{x-\xi}{b}\right)}\cos\left(\frac{\bar{w}/2}{M}\frac{r_0}{b}\right)}{r_0}\, d\xi d\eta, \tag{7–113}$$

where

$$r_0 = \sqrt{(x-\xi)^2 - \beta^2(y-\eta)^2} \tag{7–114}$$

and $\bar{w}$ is the reduced frequency parameter for compressible flow. On the lower surface $z = 0^-$, only the sign of ϕ' must be changed.

The solution (7–113) can be used directly to represent the motion of wings with completely independent upper and lower surfaces. These are called "simple planforms" by Miles. Examples are the wide triangular or delta wing (leading edges outside the Mach lines from the vertex), and various modifications of the delta such as crescent and arrowhead planforms. Nelson (Ref. 7–35) has used the source-pulse method to find forces and moments on rigid wings of these types, oscillating in pitch and vertical translation. Some integrals arising in the process cannot be identified with tabulated functions. To avoid this difficulty, Nelson expands the kernel in powers of the parameter $\bar{w}$ and obtains results correct up to the third order. Other investigations of this type have been published by Miles (Ref. 7–36) and Froehlich (Ref. 7–37).

Further examples of the application of source pulses are the work of Stewart and Li (Refs. 7–38 through 7–40) on rectangular planforms, and Chang (Ref. 7–41) on swept planforms. These authors assume the validity of Evvard's theorem about equivalent integration areas on and off the wing (Section 6–3), so that their formulas are limited to small values of the reduced frequency.

Modified use of the source-pulse method has been made by Watkins (Ref. 7–42) in dealing with the rectangular wing. In this case there is interaction between the upper and lower surfaces in regions bounded by the inward Mach lines from the tips of the leading edges. Over such a region Watkins distributes a sheet of doublets, obtained by taking the normal derivative of the source sheet. He relates the solution of the oscillatory problem to that of the corresponding steady-state problem. His development involves expansion in $\bar{\omega}$ up to the third power, but the same process has been carried up to the seventh power by Nelson, Rainey, and Watkins (Ref. 7–43). Doublets have also been successfully used to represent the oscillation of delta wings with subsonic edges (Refs. 7–44 and 7–45). The same problem was solved by Haskind and Falkovich (Ref. 7–46) by transformation of the differential equation and use of complex variable techniques.

No account of the use of superposition methods for oscillating supersonic wings is complete without reference to the extensive work of Krasilschikova (Ref. 7–47) and Frankl-Karpovich (Ref. 7–48). We also mention Rott's solution (Ref. 7–49) for the rectangular wing. A host of related problems of indicial motion have been studied by Lomax and collaborators (Ref. 7–50), using the concept of the acoustic planform.

Since each of the aforementioned references involves a rather narrow class of wing shapes, we finish this part of this discussion by calling attention to a scheme for applying Eq. (7–113), proposed by Pines' group at Republic Aviation Corporation (Refs. 7–51 and 7–52). This technique shows promise of being able to handle, with necessary numerical approximations, the oscillatory motion of arbitrary planforms with subsonic or supersonic edges. In some convenient way, the wing surface and adjacent disturbed portion of the xy-plane are divided into elementary areas. It is assumed that $\bar{w}_a$ either is constant over each of these areas or varies in a known way related to its value at the center. For a particular point (x,y), the disturbance velocity potential (or the pressure, which is simply related to it) is then calculated by numerical evaluation of the double integral in terms of the known central values of $\bar{w}_a$ in all those areas inside the region which can influence (x,y). Thus one is led to a matrix equation of the form

$$\{\bar{\phi}'\} = [A_\phi]\{\bar{w}_a\} \tag{7–115}$$

or

$$\{\Delta\bar{p}_a\} = [A_p]\{\bar{w}_a\}, \tag{7–116}$$

where the column matrices on the left contain values of $\bar{\phi}'$ or pressure amplitude $\Delta\bar{p}_a$ at the centers of all elementary areas of interest. The complex numbers making up $[A_\phi]$ or $[A_p]$ are known as aerodynamic influence coefficients. Each one is a function of Mach number, reduced frequency based on the dimensions of the elementary area, and two coordinates giving the x- and y-distances from the point whose $\bar{w}_a$ is involved to the point at which $\bar{\phi}'$ or $\Delta\bar{p}_a$ is being computed. The details of how these influence coefficients are found will vary from case to case. For low reduced frequencies and widely separated areas, they can be approximated by the value of the kernel function,

$$\frac{e^{-i\bar{w}/2\left(\frac{x-\xi}{b}\right)}\cos\left(\frac{\bar{w}/2}{M}\frac{r_0}{b}\right)}{r_0},$$

at the center (ξ,η) of the associated area.

For wings with all supersonic edges, every $\bar{w}_a$ in the column matrices of Eqs. (7–115) and (7–116) is given. Computing the pressure distribution is then just a matter of numerical integration, and Ref. 7–51 shows that the process is quite accurate and systematic when k is small. If disturbed regions off the planform are encountered, $\bar{w}_a$ is not known there, but it can be calculated approximately by making use of the condition in Eq. (7–116) that $\Delta\bar{p}_a$ must vanish at the center of each elementary area in such a disturbed region. Incidentally, Pines' method appears to be a useful one for steady supersonic flow over wings with arbitrary deformation shapes. It is especially adapted to problems like spanwise and chordwise divergence, if structural influence coefficients are known for the same set of points associated with the aerodynamic influence coefficients.

The use of Laplace transformation on the streamwise coordinate has been illustrated (Section 6–6) for the two-dimensional airfoil oscillating in supersonic flow. This same operation, combined with Fourier transformation of the y-coordinate, has proved very efficient in dealing with finite wings having straight leading edges and with other planforms that can be conveniently obtained from them by rotations of the edges. Miles has published several papers on the transformation technique (e.g., Refs. 7–53 and 7–54). Other applications of this method are presented by Stewartson in Ref. 7–55. Temple's section of *Modern Developments in Fluid Dynamics — High Speed Flow* contains a valuable general discussion of the subject. Limitations of space prevent our reproducing any of their results here.

(c) *The reverse flow theorem.* Although relatively little practical use has been made of them in the past, there exist certain reciprocity relations for unsteady motion of lifting surfaces through a compressible gas which will unquestionably play a part in simplifying the aerodynamic treatment of aeroelastic problems. Most of these assert the equality between certain weighted integrals of the loadings developed by a wing moving in a certain direction and by the same wing moving in the opposite direction. In a few cases another property of the flow, like the circulation, is involved. Reciprocity relations are derived by applying Green's theorem to differential

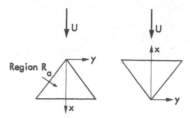

Fig. 7–4.   The same delta-wing planform in forward and reverse flow.

equation (5–44) and suitable boundary conditions. Accordingly, their validity rests on the linearizing assumption of small disturbances. A rather complete discussion, with interesting examples, is presented by Heaslet and Spreiter (Ref. 7–57).

For the flutter analyst the most useful of the reciprocity relations is the so-called reverse flow theorem of simple harmonic motion. As given by Flax (Ref. 7–58), it takes the following form:

$$\iint_{R_a} \Delta \bar{p}_{aF}(x,y)\,\bar{w}_{aR}(x,y)\,dxdy = \iint_{R_a} \Delta \bar{p}_{aR}(x,y)\,\bar{w}_{aF}(x,y)\,dxdy. \quad (7\text{--}117)$$

Here R_a denotes, as above, the portion of the xy-plane occupied by the projection of the nearly plane wing. $\Delta \bar{p}_{aF} e^{i\omega t}$ is the distribution of pressure difference produced by the wing oscillation associated with the vertical velocity distribution $\bar{w}_{aF} e^{i\omega t}$, when the undisturbed flow is moving with speed U in the positive x-direction (Fig. 7–4). $\Delta \bar{p}_{aR} e^{i\omega t}$ is the loading caused by the motion $\bar{w}_{aR} e^{i\omega t}$ when the undisturbed flow is moving with the same speed in the negative x-direction, that is, when the flight direction is reversed. This equality holds throughout any speed range — incompressible, subsonic, or supersonic — where Eq. (5–44) or some reduced form thereof governs the problem.

One elementary conclusion we can quickly draw from Eq. (7–117) is that, in steady flow ($\omega = 0$), the additional lift or lift due to angle-of-attack changes is invariant with respect to the way the wing is going. To prove this fact, we set

$$w_{aR}(x,y) = w_{aF}(x,y) = U\alpha \quad (7\text{--}118)$$

and get, after cancelling the constant factor $U\alpha$,

$$L_F = -\iint_{R_a} \Delta p_{aF}(x,y)\,dxdy = -\iint_{R_a} \Delta p_{aR}(x,y)\,dxdy = L_R. \quad (7\text{--}119)$$

Since the wing areas and flight dynamic pressures are the same, Eq. (7–119) shows the forward and reversed lift-curve slopes to be equal. Similar results for pitching moment, rolling moment, etc., are readily obtained.

There are two general ways in which Eq. (7–117) can be helpful to the aeroelastician dealing with equations of flutter motion. One of these is described in Section 9–9 and has to do with replacing the aerodynamic term in the flutter equation of a low aspect-ratio wing with an integral containing the mode shape $\bar{z}_a(x,y)$, multiplied by a known pressure distribution (called by Voss an aeroelastic influence function) related to the load-deflection influence function of the wing structure.

A second, more straightforward application of Eq. (7–117) is for simplifying the calculation of the generalized forces appearing in Rayleigh-Ritz flutter equations, such as those derived in Section 9–5. If we interpret $\bar{w}_{aR}(x,y)$ as a known mode shape, each generalized force is in exactly the form of the left side of Eq. (7–117). Should the aerodynamic theory be easier to handle for reversed than for forward flight of the planform being analyzed, the reverse flow theorem shows that the required generalized force may be equally well evaluated from the pressure distribution developed in the new direction by a known oscillation. A good example of the utility of this result is for supersonic flight of the delta wing shown in Fig. 7–4. If all the edges are supersonic, the reversed planform has a straight leading edge and a supersonic trailing edge. The integrations that must be performed to find the pressures are then little more difficult than for purely two-dimensional motion and much easier than those associated with the swept-back leading edge of the original planform. Walsh, Voss, and Zartarian (Ref. 7–33) employ this fact in their investigation of the accuracy of aerodynamic strip theory for supersonic flutter computations.

(d) *Reduction of the differential equation and boundary conditions in special cases.* Lin, Reissner, and Tsien (Ref. 7–59) opened up a new field of research in unsteady aerodynamic theory by systematically examining the circumstances in which various terms become negligibly small and can be dropped from the differential equation and boundary conditions of small-disturbance potential flow. Their original work on two-dimensional airfoils has been extended to wings of finite span by Miles (Ref. 7–60). There are four dimensionless parameters whose relative magnitudes establish the ranges in which the problem can be simplified: the Mach number, reduced frequency (or time constant of unsteady motion), wing aspect ratio, and thickness ratio or dimensionless amplitude of oscillation.

As might be expected, it is easy to identify well-known special cases like incompressible flow ($M \ll 1$) and steady flow ($k \ll 1$). There are, however, several much less familiar situations where the differential equation (5–44) can be replaced by more mathematically tractable forms. One of these, the case of the very low aspect-ratio wing executing rapid oscillations, is exploited by Merbt and Landahl in the aforementioned Ref. 7–23. "Piston theory," which we have described in Chapter 6, represents a very strong simplification, according to which the differential equation becomes the one-dimensional wave equation at very large reduced frequencies.

7-6 Unsteady motion of nonlifting bodies. As in other applications of aerodynamic theory to aircraft design, there often occur aeroelastic problems where the motion of a lifting surface is coupled with that of a large, nonlifting body. The airloads on such bodies have frequently been omitted in the past when studying flutter, divergence, and the like. However, the sizes of jet pods, external fuel tanks, and similar stores are becoming so large that such an assumption is not always justified; also degrees of freedom like fuselage pitch may contribute substantial forces and moments in the flutter equations. The simplest way of including the effect of such a body is to add onto the wing and tail loads whatever loading the body would develop when executing the same motion in the absence of the rest of the aircraft. This procedure neglects aerodynamic interference — the additional flow induced over the body by the wing or tail and vice versa. If we wish to take interference into account in unsteady motion, we are faced with much more formidable theoretical problems, which have been overcome only in a few special cases (e.g., Bryson, Ref. 7–24). A discussion of approximate interference theory is given by Ashley, Zartarian, and Neilson (Ref. 7–6).

Turning to the unsteady flow over an isolated streamlined body, we can identify three general methods that have been applied to various examples:

(1) Exact satisfaction of the differential equation and boundary conditions, such as is done for arbitrary ellipsoids in incompressible liquid by Lamb (Ref. 7–61, pp. 139–156).

(3) Linearization of the problem, under the assumption that the lateral dimensions are small compared with the length in the flight direction (cf. Miles, Ref. 7–62).

(3) Neglect of the three-dimensional details of the flow pattern on physical grounds, and use of the momentum method in planes normal to the flight direction. This process was first applied to nonlifting bodies by Munk.

We shall give a brief account of the first and third methods in this section. With regard to the results of linearized theory, many of them are equally well predicted by momentum considerations. However, one important conclusion of the more rigorous theory should be stated: as long as the lateral flow velocities are small compared with the flight speed, the first-order forces and moments are independent of Mach number in compressible flow.

Both calculation and experiment show that the linearized solutions are not accurate for nacelle-like bodies with fineness ratios less than approximately 7. Exact three-dimensional theory is useless for arbitrarily shaped objects, however, so we often replace a given external store with an equivalent ellipsoid. This procedure seems to yield satisfactory airload predictions for unsteady motion at subcritical Mach numbers, so long as the fineness ratio is not so small that a large, separated wake is produced behind the body.

We have briefly discussed the prolate ellipsoid of revolution in Section 5–2. Turning to the general ellipsoid with three unequal principal axes $2a > 2b > 2c$, we find from Ref. 7–61 that the total kinetic energy produced by its motion through a mass of liquid stationary at infinity is

$$T = \frac{\rho}{2} \, (\text{volume}) [k_1 U_1{}^2 + k_2 U_2{}^2 + k_3 U_3{}^2]$$

$$+ \frac{\rho}{2} [k_1' J_1 \Omega_1{}^2 + k_2' J_2 \Omega_2{}^2 + k_3' J_3 \Omega_3{}^2], \qquad (7\text{–}120)$$

where U_1, U_2, U_3, Ω_1, Ω_2, and Ω_3 are the absolute linear velocities along, and angular velocities about, the principal axes through the centroid. J_1, J_2, and J_3 are the principal volume moments of inertia of the ellipsoid.

$$(\text{Volume}) = \tfrac{4}{3}\pi abc, \qquad (7\text{–}121\text{a})$$

$$J_1 = \tfrac{1}{5}(b^2 + c^2)(\text{volume}). \qquad (7\text{–}121\text{b})$$

The expressions for the six inertial coefficients are

$$k_1 = \frac{\alpha_0}{2 - \alpha_0}, \qquad k_2 = \frac{\beta_0}{2 - \beta_0}, \qquad k_3 = \frac{\gamma_0}{2 - \gamma_0}, \qquad (7\text{–}122\text{a,b,c})$$

$$k_1' = \frac{(b^2 - c^2)(\gamma_0 - \beta_0)}{(b^2 + c^2)[2(b^2 - c^2) + (b^2 + c^2)(\beta_0 - \gamma_0)]}. \qquad (7\text{–}122\text{d})$$

k_2' and k_3' follow from Eq. (7–122d) by cyclic permutation. The constants α_0, β_0, γ_0 appearing in these expressions are certain definite integrals which can be identified with elliptic integrals:

$$\alpha_0 = abc \int_0^\infty \frac{d\lambda}{(a^2 + \lambda)\sqrt{(a^2 + \lambda)(b^2 + \lambda)(c^2 + \lambda)}}, \qquad (7\text{–}123\text{a})$$

$$\beta_0 = abc \int_0^\infty \frac{d\lambda}{(b^2 + \lambda)\sqrt{(a^2 + \lambda)(b^2 + \lambda)(c^2 + \lambda)}}, \qquad (7\text{–}123\text{b})$$

$$\gamma_0 = abc \int_0^\infty \frac{d\lambda}{(c^2 + \lambda)\sqrt{(a^2 + \lambda)(b^2 + \lambda)(c^2 + \lambda)}}. \qquad (7\text{–}123\text{c})$$

For example, γ_0 can be written

$$\gamma_0 = \frac{2abc}{(b^2 - c^2)\sqrt{a^2 - c^2}} \left[\frac{b}{ac} \sqrt{a^2 - c^2} - E\left(\phi, \sqrt{\frac{a^2 - b^2}{a^2 - c^2}}\right) \right], \qquad (7\text{–}124\text{a})$$

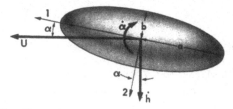

Fig. 7–5. Ellipsoid moving with constant forward speed and variable vertical velocity $\dot{h}(t)$ and angular velocity $\dot{\alpha}(t)$.

where
$$\phi = \tan^{-1} \sqrt{\frac{a^2}{c^2} - 1}. \tag{7–124b}$$

Corresponding to Eq. (7–120), we must have the following components of instantaneous linear and angular momentum of the liquid due to the moving ellipsoid:

$$\begin{array}{llll}
\text{(a)} & I_1 = k_1 \rho U_1 \times (\text{volume}), & \text{(d)} & H_1 = k_1{}' \rho \Omega_1 J_1, \\[4pt]
\text{(b)} & I_2 = k_2 \rho U_2 \times (\text{volume}), & \text{(e)} & H_2 = k_2{}' \rho \Omega_2 J_2, \quad (7\text{–}125) \\[4pt]
\text{(c)} & I_3 = k_3 \rho U_3 \times (\text{volume}), & \text{(f)} & H_3 = k_3{}' \rho \Omega_3 J_3.
\end{array}$$

Employing the notation previously used in connection with unsteady motion of thin wings, we assume that the ellipsoid is moving forward with constant speed U and vertically downward with velocity $\dot{h}(t)$, while the long principal axis 1 is inclined at a variable small angle $\alpha(t)$ to the direction of U (Fig. 7–5). The total linear momentum of the liquid, in vector form, is

$$\begin{aligned}
\mathbf{I}_{\text{tot}} &= I_1 \mathbf{i}_1 + I_2 \mathbf{i}_2 \\
&= \mathbf{i}_1 k_1 \rho \times (\text{volume}) \times [U \cos \alpha - \dot{h} \sin \alpha] \\
&\quad + \mathbf{i}_2 k_2 \rho \times (\text{volume}) \times [U \sin \alpha + \dot{h} \cos \alpha], \qquad (7\text{–}126)
\end{aligned}$$

the $\mathbf{i}$ representing a unit vector in the direction defined by its subscript. Equation (7–126) can also be written with components in the fixed directions of U and $\dot{h}$.

$$\mathbf{I}_{\text{tot}} = I_U \mathbf{i}_U + I_h \mathbf{i}_h, \tag{7–127}$$

where

$$\begin{aligned}
I_U &= I_1 \cos \alpha + I_2 \sin \alpha \\
&= \rho \times (\text{volume}) \times \{ k_1 [U \cos^2 \alpha - \dot{h} \sin \alpha \cos \alpha] \\
&\qquad\qquad\qquad\quad + k_2 [U \sin^2 \alpha + \dot{h} \sin \alpha \cos \alpha] \} \quad (7\text{–}128)
\end{aligned}$$

and

$$\begin{aligned}
I_h &= \rho \times (\text{volume}) \times \{ -k_1 [U \cos \alpha \sin \alpha - \dot{h} \sin^2 \alpha] \\
&\qquad\qquad\qquad\quad + k_2 [U \sin \alpha \cos \alpha + \dot{h} \cos^2 \alpha] \}. \quad (7\text{–}129)
\end{aligned}$$

The horizontal and vertical forces exerted by the liquid on the ellipsoid are the reactions to the substantial rates of change of the components I_U and I_h.

$$D = \frac{DI_U}{Dt} = (k_2 - k_1) \times (\text{volume}) \times \rho U \dot\alpha \sin 2\alpha$$

$$+ \tfrac{1}{2}(k_2 - k_1) \times (\text{volume}) \times \rho[\ddot h \sin 2\alpha + 2\dot h \dot\alpha \cos 2\alpha], \tag{7-130}$$

$$L = \frac{DI_h}{Dt} = (k_2 - k_1) \times (\text{volume}) \times \rho[U\dot\alpha \cos 2\alpha - \dot h \dot\alpha \sin 2\alpha]$$

$$+ (k_2 \cos^2\alpha + k_1 \sin^2\alpha) \times (\text{volume}) \times \rho \ddot h. \tag{7-131}$$

The motion described here is symmetrical (longitudinal in the aeronautical sense), so that the resultant angular momentum is always a vector in the 3-direction. The total moment exerted by the liquid on the body is therefore a pitching moment and equals the negative of the absolute rate of change of angular momentum, measured with respect to an axis fixed in space and momentarily coincident with 3. If M_3 is positive nose-up,

$$M_3 = \frac{DH_3}{Dt} = \frac{\partial H_3}{\partial t} + I_h U - I_U \dot h$$

$$= -\frac{\partial}{\partial t}[k_3{}' \rho J_3 \dot\alpha] + I_h U - I_U \dot h$$

$$= -k_3{}' \rho J_3 \ddot\alpha + \tfrac{1}{2}(k_2 - k_1) \times (\text{volume}) \times \rho(U^2 - \dot h^2) \sin 2\alpha$$
$$+ \tfrac{1}{2}(k_2 - k_1) \times (\text{volume}) \times \rho U \dot h \cos 2\alpha. \tag{7-132}$$

Equations (7–130)–(7–132) contain nonlinear terms in the small quantities h and α. They are exactly correct as they stand, but it is more convenient to have them linearized for aeroelastic calculations.

$$D = 0. \tag{7-133}$$

$$L = (k_2 - k_1) \times (\text{volume}) \times \rho U \dot\alpha + k_2 \times (\text{volume}) \times \rho \ddot h. \tag{7-134}$$

$$M_3 = -k_3{}' \rho J_3 \ddot\alpha + (k_2 - k_1) \times (\text{volume}) \times \rho[U^2\alpha + U\dot h]. \tag{7-135}$$

The second term in Eq. (7–135) may be compared with Eq. (5–58); it is the destabilizing moment experienced by a body translating at a small angle of attack $\alpha + \dot h/U$. This is the only resultant airload in steady flight, except for a very small lift which is due to separation of the boundary layer. For use in aeroelastic applications of Eqs. (7–134) and (7–135), Ref. 7–6 provides a table of the necessary inertia coefficients.

In connection with the foregoing discussion of liquid motion external to an ellipsoid, we mention that Lamb's theory also covers the ellipsoidal shell filled with incompressible fluid. The results can therefore be used to calculate the loads due to fuel motion inside a large tank whose shape can be adequately approximated by an ellipsoid. This phenomenon may have an important influence on certain wing flutter modes. Partially filled tanks and tanks provided with internal baffles cannot be treated in this way but present a more difficult theoretical problem.

When the streamlined body or fuselage is slender enough, we can employ the momentum method based on the assumption of Munk and R. T. Jones: the disturbed flow is two-dimensional in planes normal to the direction of flight. This scheme has been described briefly in Subsection 5–5(c). In order to include the possibility of bending deformations, we assume that the uniform flow U is parallel to the x-axis and that the lateral motion of the body's longitudinal centerline is defined by an arbitrary function $z_a(x,t)$. This reduces to the special case of vertical translation and pitching of a rigid body when

$$z_a = -h(t) - \alpha(t)[x - x_B], \tag{7–136}$$

where x_B is the x-coordinate of the axis of pitch.

If the local incidence of the axis is everywhere small, the z-velocity of any cross section relative to the fluid at rest is

$$w_a(x,t) = U\frac{\partial z_a}{\partial x} + \frac{\partial z_a}{\partial t}. \tag{7–137}$$

The corresponding component of momentum of the fluid contained between two yz-planes a distance dx apart is

$$dI_z = \rho_\infty S dx \left[U\frac{\partial z_a}{\partial x} + \frac{\partial z_a}{\partial t} \right]. \tag{7–138}$$

Here $\rho_\infty S(x)$ represents the virtual mass per unit length of a cylinder having the same sectional area and shape as the body. That is, for a circular section of radius R,

$$\rho_\infty S = \rho_\infty \pi R^2. \tag{7–139}$$

The z-force acting per unit length of the body at station x is the reaction to the substantial rate of change of dI_z/dx, that is, the rate at which dI_z/dx of a particular slab of fluid is increasing:

$$\frac{dL}{dx} = -\frac{D}{Dt}\left[\frac{dI_z}{dx}\right]$$

$$= -\rho_\infty \frac{DS}{Dt}\left[U\frac{\partial z_a}{\partial x} + \frac{\partial z_a}{\partial t} \right] - \rho_\infty S\left[U\frac{D}{Dt}\left(\frac{\partial z_a}{\partial x}\right) + \frac{D}{Dt}\left(\frac{\partial z_a}{\partial t}\right) \right]. \tag{7–140}$$

We use the facts that

$$\frac{D}{Dt} = U \frac{\partial}{\partial x} + \frac{\partial}{\partial t} \tag{7-141}$$

and

$$\frac{DS}{Dt} = \frac{dS}{dx} \frac{Dx}{Dt} = \frac{dS}{dx} U \tag{7-142}$$

and obtain from Eq. (7-140)

$$\frac{dL}{dx} = -\rho_\infty \frac{dS}{dx} \left[U^2 \frac{\partial z_a}{\partial x} + U \frac{\partial z_a}{\partial t} \right] - \rho_\infty S \left[U^2 \frac{\partial^2 z_a}{\partial x^2} + 2U \frac{\partial^2 z_a}{\partial x \partial t} + \frac{\partial^2 z_a}{\partial t^2} \right]. \tag{7-143}$$

No particular simplification occurs when Eq. (7-143) is integrated over x to get the total lateral force and pitching moment on the body. Certain terms vanish as a result of the fact that for bodies closed at both ends

$$\int_{\text{body}} \frac{dS}{dx} \, dx = 0. \tag{7-144}$$

When the motion is described by Eq. (7-136), the linearized resultant airloads are

$$L = \rho_\infty K_V \times \text{(volume)} \times (U\dot{\alpha} + \ddot{h}) + \rho_\infty K_M \bar{M} \ddot{\alpha}, \tag{7-145}$$

$$M_y = \rho_\infty U K_V \times \text{(volume)} \times (U\alpha + \dot{h}) - \rho_\infty K_M \bar{M} \ddot{h} - \rho_\infty K_J J_3 \ddot{\alpha}. \tag{7-146}$$

Equations (7-145) and (7-146) contain certain inertia coefficients adjusting the volume, volume static unbalance $\bar{M}$, and volume moment of inertia J_3 in accordance with the formulas

$$K_V \times \text{(volume)} = \int_{\text{body}} S \, dx, \tag{7-147a}$$

$$K_M \bar{M} = \int_{\text{body}} S[x - x_B] \, dx, \tag{7-147b}$$

$$K_J J_3 = \int_{\text{body}} S[x - x_B]^2 \, dx. \tag{7-147c}$$

K_V and K_J are unity for any body of revolution or other shape with circular cross sections, in view of Eq. (7-139).

Because the virtual mass concept is limited to incompressible flow, Eqs. (7-144)–(7-146) are valid only when the fluid motion in yz-planes is effectively incompressible. This is the case even up into the low supersonic range of flight speeds, so long as the body is slender enough and the unsteady motion is not too rapid.

Much of the limited available information shows only fair agreement between experimental measurements and potential-theory calculations for elongated bodies (e.g., Zahm, Ref. 7–63). All of these data were taken in steady flow, however, and the extension of empirical corrections derived from them to the unsteady problem is hard to justify. There is considerable evidence suggesting that unsteady motion hinders the development of viscous separation in a flow and brings it closer to its ideal counterpart. This conclusion is supported by Ref. 7–6, where resultant lifts and pitching moments are reported on an oscillating ellipsoid of revolution with fineness ratio 10. It is concluded that Lamb's theory is verified for all quantities except the destabilizing couple due to angle of attack, which must be multiplied by a correction factor that rises from 0.85 in steady flow to 1.0 at high frequencies of vibration. Reference 7–6 shows the theory to be much poorer for a blunt body of fineness ratio 5. Both these bodies were tested alone and with wings attached. It has been suggested that when a body is located adjacent to a wing, the latter tends to align the flow along the former and thus reduce the effects of viscosity. No satisfactory verification of this hypothesis has been obtained in unsteady flow.

CHAPTER 8

STATIC AEROELASTIC PHENOMENA

8–1 Introduction. Static aeroelastic problems are characterized by several simplifying features. By definition, time does not appear as an independent variable and, as a result, vibratory inertial forces are eliminated from the equilibrium equations. Aerodynamic forces can be based upon well-known steady flow results rather than the more complex unsteady flow theories.

We have seen in Chapter 1 that the majority of static aeroelastic problems divide into two main classes. The first includes the problems of load distribution and divergence. These are of primary interest to the airplane structural designer. The second includes the problems of control effectiveness and reversal, and aeroelastic effects on airplane stability. The latter are chiefly the concern of control system designers and stability analysts, but are also important to the structural designer in determining rigidity requirements.

The methods of analysis developed in this chapter are, in general, intended to apply to all speed ranges; the only requirement is that the aerodynamic theory be linear.

8–2 Twisting of simple two-dimensional wing with aileron. In order to comprehend the basic phenomena of static aeroelasticity, it is informative to consider the behavior of a rigid, elastically restrained wing segment in two-dimensional flow prior to taking up the more complicated cases involving finite span. Some of the early work on aeroelasticity was, in fact, based upon such simplified two-dimensional representations of true lifting surfaces. Theodorsen (Ref. 8–1) used a model of this kind in analyzing the flutter problem, designating it the "typical section" and choosing its properties to match a cross section 70–75% of the distance from root to tip of the actual wing. Elsewhere it has been called the "semirigid wing," but this term is not employed here because other authors have applied it to a finite wing constrained so that its mode of spanwise twist distribution is independent of the manner of loading.

Consider Fig. 8–1, which pictures a rigid segment of airfoil with a control surface attached. The wing has constant chord c and area S. Its twisting is opposed by a linear coiled spring attached at the elastic axis, which is located at a distance e behind the aerodynamic center. All angular deflections of the wing are assumed small. The total angle of attack, measured from the zero-lift attitude, is taken as the sum of an initial angle α^r and an elastic twist θ:

$$\alpha = \alpha^r + \theta. \tag{8-1}$$

(a) *Wing phenomena.* We consider first the behavior of the wing when the aileron displacement $\beta = 0$. The elastic twist is proportional to the torque about the elastic axis,

$$\theta = C^{\theta\theta}T, \tag{8-2}$$

where $C^{\theta\theta}$ is the flexibility influence coefficient of the coiled spring.

The total aerodynamic torque about the elastic axis is

$$T = (C_L e + C_{MAC}c)qS, \tag{8-3}$$

where

C_L = wing lift coefficient,
C_{MAC} = wing pitching moment coefficient about the aerodynamic center,
q = dynamic pressure,
S = area of rigid wing segment.

The total lift coefficient is related to the angle of attack measured from zero lift by

$$C_L = \frac{\partial C_L}{\partial \alpha}(\alpha^r + \theta) \tag{8-4a}$$

where $\partial C_L/\partial \alpha$ is the slope of the lift curve. The elastic twist of the wing can be computed by substituting Eq. (8-4a) into Eq. (8-3) and combining the result with Eq. (8-2):

$$\theta = \frac{C^{\theta\theta}[(\partial C_L/\partial \alpha)e\alpha^r + C_{MAC}c]qS}{1 - C^{\theta\theta}(\partial C_L/\partial \alpha)qSe}. \tag{8-4b}$$

We observe in Eq. (8-4b) that it is possible for the denominator to become zero for a particular value of q. Thus, an infinite twist angle θ is produced. This is the condition known as torsional divergence. Putting the denominator equal to zero yields

$$1 - C^{\theta\theta}\frac{\partial C_L}{\partial \alpha}qSe = 0, \tag{8-5}$$

which defines a divergence dynamic pressure

$$q_D = \frac{1}{C^{\theta\theta}(\partial C_L/\partial \alpha)Se} \tag{8-6}$$

and a divergence speed

$$U_D = \sqrt{\frac{1}{C^{\theta\theta}(\partial C_L/\partial \alpha)(\rho/2)Se}}. \tag{8-7}$$

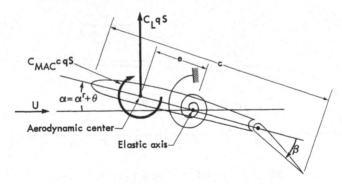

Fig. 8–1.   Elastically supported rigid airfoil segment.

As shown by Fig. 8–1, positive values of e locate the aerodynamic center ahead of the elastic axis. Thus, it is seen from Eq. (8–7) that the divergence speed depends markedly on the value of e, and when the aerodynamic center falls on the elastic axis, or is aft of it, the wing is stable at all speeds.

Examination of Eq. (8–4b) shows that the angle of twist due to aerodynamic loads is a function of the initial angle of attack, α^r. However, Eq. (8–7) indicates that the divergence speed is independent of the initial angle of attack and the airfoil camber as it affects C_{MAC}. It is apparent therefore, that wing divergence corresponds to a physical condition where the increase in aerodynamic moment about the elastic axis due to an arbitrary change in angle of attack is equal to the corresponding increase in elastic restoring moment. For an arbitrary finite change $\Delta\alpha$ in angle of attack, the change in elastic restoring moment is

$$\Delta M_e = \frac{\Delta\alpha}{C^{\theta\theta}}, \qquad (8\text{–}8)$$

and the change in aerodynamic moment is

$$\Delta M_a = \frac{\partial C_L}{\partial\alpha}\,\Delta\alpha\, qSe. \qquad (8\text{–}9)$$

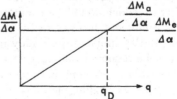

Fig. 8–2.   Change in aerodynamic moment and elastic restoring moment per unit change in angle of attack.

If these changes in moment are plotted against q, as shown in Fig. 8–2, the elastic restoring moment remains constant with q, whereas the aerodynamic moment increases linearly until an intersection is reached. This intersection corresponds to the divergence condition.

The problems of elastic wing twist and torsional divergence are analogous to the problem of lateral deflection of a column with initial eccentricity (Ref. 8–2). We may form the expression for the elastic twist, in

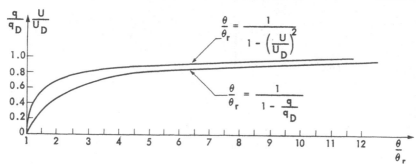

Fig. 8-3.   Variation in wing twist with speed.

the case where the twist introduces no additional aerodynamic load, by putting $\alpha = \alpha^r$ and solving for the angle of twist.

$$\theta_r = C^{\theta\theta}\left(\frac{\partial C_L}{\partial \alpha}e\alpha^r + C_{MAC}c\right)qS. \tag{8-10}$$

Dividing Eq. (8-4b) by Eq. (8-10) yields

$$\frac{\theta}{\theta_r} = \frac{1}{1 - C^{\theta\theta}(\partial C_L/\partial \alpha)qSe}. \tag{8-11}$$

Introducing Eq. (8-6),

$$\frac{\theta}{\theta_r} = \frac{1}{1 - (q/q_D)} = \frac{1}{1 - (U/U_D)^2}. \tag{8-12}$$

Equation (8-12) expresses a ratio of the actual angle of twist to the angle of twist which would be computed by neglecting the additional aerodynamic moment introduced by elastic twist. It should be observed that Eq. (8-12) applies strictly only in the incompressible speed range, where it can be assumed that lift-curve slope is independent of Mach number. Since wing stress is proportional to elastic twist, Eq. (8-12) indicates the importance of taking account of wing twist in computing wing stresses. A graph of Eq. (8-12) is shown by Fig. 8-3. It should be observed from this graph that wing deflections increase rapidly with speed as the divergence speed is approached.

(b) *Wing-aileron phenomena.* Consider next the aeroelastic phenomena which result when a control surface is placed on a wing. Referring to Fig. 8-1, let us assume that the aileron is given a downward displacement β. If the wing segment is rigidly supported, a downward aileron displacement is accompanied by an increase in lift, and an upward displacement by a decrease. However, we observe that when the wing segment is elastically supported, a downward aileron displacement twists the wing segment nose down and an upward aileron deflection twists it nose up. As the airspeed over the wing segment increases, the aerodynamic twisting moments increase with the square of the airspeed, whereas the elastic restoring moments remain constant. As a result, the effectiveness of the aileron in producing lift is decreased with airspeed until a velocity is reached at which the aileron is completely ineffective. This velocity is known as the aileron reversal speed. In the following analysis we shall consider changes in deflections, forces, and moments over and above those which exist under steady conditions prior to the aileron displacement. The changes in lift and moment coefficients produced on an elastically supported rigid wing segment by an aileron displacement β are, respectively,

$$C_L = \left(\frac{\partial C_L}{\partial \beta} \beta + \frac{\partial C_L}{\partial \alpha} \theta \right), \tag{8-13a}$$

$$C_{MAC} = \frac{\partial C_{MAC}}{\partial \beta} \beta, \tag{8-13b}$$

where
$\partial C_L/\partial \beta$ = change in wing lift coefficient per unit aileron deflection,
$\partial C_{MAC}/\partial \beta$ = change in wing moment coefficient about the aerodynamic center per unit aileron deflection.

The change in torque about the elastic axis resulting from an aileron displacement β follows from Eqs. (8-3) and (8-13):

$$T = qS \left\{ e \left(\frac{\partial C_L}{\partial \beta} \beta + \frac{\partial C_L}{\partial \alpha} \theta \right) + c \frac{\partial C_{MAC}}{\partial \beta} \beta \right\}. \tag{8-14}$$

Substituting Eq. (8-14) into Eq. (8-2) gives the following equilibrium condition:

$$\theta = C^{\theta\theta}T = C^{\theta\theta}qS \left\{ e \left(\frac{\partial C_L}{\partial \beta} \beta + \frac{\partial C_L}{\partial \alpha} \theta \right) + c \frac{\partial C_{MAC}}{\partial \beta} \beta \right\}. \tag{8-15}$$

This condition may also be expressed in terms of the ratio of elastic twist to aileron displacement, as follows,

$$\frac{\theta}{\beta} = \frac{e(\partial C_L/\partial \beta) + c(\partial C_{MAC}/\partial \beta)}{(1/C^{\theta\theta}qS) - e(\partial C_L/\partial \alpha)}. \tag{8-16}$$

Combining Eqs. (8-13a) and (8-16), we obtain the total lift coefficient of the wing segment as a linear function of the aileron displacement.

$$C_L = \frac{\left(\dfrac{\partial C_L}{\partial \beta} \dfrac{1}{(\partial C_L/\partial \alpha)C^{\theta\theta}qSe} + \dfrac{c}{e} \dfrac{\partial C_{MAC}}{\partial \beta} \right) \beta}{\dfrac{1}{(\partial C_L/\partial \alpha)C^{\theta\theta}qSe} - 1}. \tag{8-17}$$

The aileron is completely ineffective when $C_L = 0$, and if we put the numerator in Eq. (8-17) equal to zero, we obtain the following condition for aileron reversal:

$$\frac{\partial C_L}{\partial \beta} \frac{1}{(\partial C_L/\partial \alpha)C^{\theta\theta}q_R S} + c\frac{\partial C_{MAC}}{\partial \beta} = 0, \qquad (8\text{--}18)$$

where q_R is the dynamic pressure corresponding to the aileron reversal condition. Solving Eq. (8–18) for the aileron reversal speed yields

$$U_R = \sqrt{\frac{-\dfrac{\partial C_L}{\partial \beta}}{\dfrac{\partial C_L}{\partial \alpha}\dfrac{\partial C_{MAC}}{\partial \beta}C^{\theta\theta}\dfrac{\rho}{2}Sc}}. \qquad (8\text{--}19)$$

An aileron effectiveness of the two-dimensional wing, at speeds below the reversal speed, can be defined by

$$\frac{C_L}{C_L{}^r} = \frac{\dfrac{\partial C_L}{\partial \beta}\dfrac{1}{(\partial C_L/\partial \alpha)C^{\theta\theta}qSe} + \dfrac{c}{e}\dfrac{\partial C_{MAC}}{\partial \beta}}{\dfrac{\partial C_L}{\partial \beta}\left(\dfrac{1}{(\partial C_L/\partial \alpha)C^{\theta\theta}qSe} - 1\right)} = 1 - \frac{U^2}{U_R{}^2}\left(\frac{U_D{}^2 - U_R{}^2}{U_D{}^2 - U^2}\right), \qquad (8\text{--}20)$$

where $C_L{}^r$ is the lift coefficient of the rigid wing, U_R is the reversal speed, and U_D is the divergence speed given by Eq. (8–7). The final result given by Eq. (8–20) applies only in the incompressible speed range, since the lift-curve slope has been assumed independent of speed. The aileron effectiveness and reversal speeds are seen to be dependent upon the aerodynamic quantities $\partial C_L/\partial \alpha$, $\partial C_L/\partial \beta$, and $\partial C_{MAC}/\partial \beta$. In addition, the reversal speed is independent of the elastic axis offset e. This is an interesting result which is due to the fact that the aerodynamic moment producing twist at the reversal speed is a pure couple, and is independent of the elastic axis position.

In comparing the formulas for U_R and U_D given above, we see that in the possible circumstance where

$$\frac{\partial C_L}{\partial \beta} = -\frac{c}{e}\frac{\partial C_{MAC}}{\partial \beta}, \qquad (8\text{--}21)$$

the reversal speed and the divergence speed are the same. In the case when $U_R = U_D$, Eq. (8–20) indicates an anomalous situation in which the aileron retains its full effectiveness at all speeds. This comes about because the nose-down pitching moment due to the deflected aileron is cancelled by the nose-up pitching moment due to the lift on the wing, a condition which is fulfilled when Eq. (8–21) is satisfied. Figure 8–4 illustrates how $C_L/C_L{}^r$, as given by Eq. (8–20), varies with the ratio U_R/U_D. The rapid rate at which aileron effectiveness reduces with speed for low ratios of U_R/U_D is evident from this figure.

The elastic twist resulting from an aileron displacement β follows from

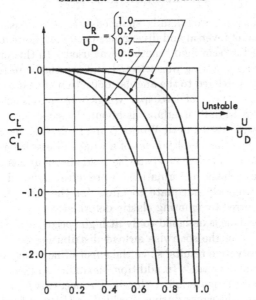

Fig. 8-4.　Variation of aileron effectiveness with speed.

Eq. (8-15):

$$\theta = \frac{C^{\theta\theta}qS[e(\partial C_L/\partial\beta) + c(\partial C_{MAC}/\partial\beta)]\beta}{1 - (\partial C_L/\partial\alpha)C^{\theta\theta}qSe}. \qquad (8\text{-}22)$$

The results of simplified theories such as described above are often employed to obtain approximate answers in practical problems, as described in Ref. 8-1. In these applications the theory is applied to a representative cross section of the wing at an arbitrary spanwise station, and the wing properties at this station are introduced into the formulas.

8-3 Slender straight wings. The simplest static aeroelastic problems of practical interest are posed by slender wings with straight elastic axes, essentially perpendicular to the fuselage center line. The problem of computing the elastic twist and wing divergence speed of straight wings was first analyzed by Hans Reissner (Ref. 8-3) in 1926. An intensive investigation of aeroelastic phenomena of straight wings followed six years later in England when Cox and Pugsley studied the problems of aileron reversal, aileron effectiveness, and wing divergence (Refs. 8-4, 8-5, and 8-6). Cox and Pugsley considered both the case of a monoplane cantilever wing which resists twisting by differential bending of two parallel spars, and the more complicated case in which twisting is resisted by both differential bending and torsional stiffness. In both cases, the stiffness of the spars was considered to vary along the span. In the treatment of aerodynamic forces, aerodynamic span effects were taken into account. In their solution of reversal and divergence problems, however, a simplified treatment called a "semirigid" method was used. In this method a reference section along the wing was selected and the elastic restoring moment was assumed to be related to the angular deflection at that section. Similar analyses of elastic twist of two spar wings were published by Datwyler (Refs. 8–7 and 8–8) in Switzerland at about the same time.

From the early 1930's until World War II, airplane designers concentrated primarily on the establishment of a wing stiffness criterion to prevent flutter. In the calculation of spanwise load distributions for structural design purposes, elastic deformations were often ignored. Margins of safety were fortunately adequate to allow for the error in spanwise lift distribution incurred by ignoring elastic deformations.

When an airplane is designed to fly at high speeds, neglect of flexibility in the calculation of the spanwise airload distribution on a straight wing leads to unconservative results, since the effect of flexibility is to move the center of pressure outward. In addition, flexibility causes the lateral control effectiveness to be materially reduced. The necessity for refined methods of analysis of straight wings during World War II led to further expansion of the original work of Reissner, Cox, and Pugsley. Flax (Ref. 8–9) formulated the aeroelastic problem of a straight wing in terms of the Rayleigh-Ritz method by expressing the twist distribution as a superposition of assumed modes with undetermined coefficients. The unknown coefficients are found by solution of a system of linear algebraic equations formed by applying the principle of virtual work. In this formulation of the problem, aerodynamic induction effects can be accounted for. Hildebrand and E. Reissner (Ref. 8–10) later developed a procedure which takes account of aerodynamic span effects in the calculation of divergence speeds of straight wings. Explicit solutions were obtained which indicate the importance of considering these effects in divergence calculations.

More recent developments in the straight wing aeroelastic problem relate to techniques for casting the equations of equilibrium in matrix form. Such techniques are useful in the treatment of wings with nonuniform properties and have found wide application in airplane design. Contributions by Lawrence and Sears (Ref. 8–11) and Pines (Ref. 8–12) have provided the basis for application of matrices to straight-wing aeroelastic problems.

(a) *Equilibrium equations.* In the following discussion, it will be assumed that straight wings are characterized essentially by an elastic axis which is nearly perpendicular to the plane of symmetry of the airplane, and that chordwise segments of the wing remain rigid, that is, camber bending is not considered. The differential equation of torsional aeroelastic equilibrium of a straight wing about its elastic axis is obtained by relating rate of twist to applied torque by the St. Venant torsion theory

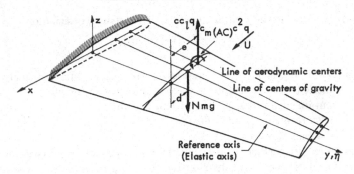

Fig. 8–5.   Slender straight wing.

[cf. Eq. (2–118)]:

$$\frac{d}{dy}\left(GJ\frac{d\theta}{dy}\right) = -t(y),\qquad(8\text{–}23a)$$

where $\theta\ (y)$ is the elastic twist distribution. From Fig. 8–5 it can be seen that the applied torque per unit span, $t(y)$, is

$$t(y) = (ecc_l + c^2 c_{mAC})q - Nmgd,\qquad(8\text{–}23b)$$

where

c_l = local lift coefficient,
c_{mAC} = local moment coefficient about the aerodynamic center,
N = load factor normal to the wing surface ($N = 1$ for level flight),
mg = weight per unit span.

The differential equation of equilibrium is obtained by substituting Eq. (8–23b) into Eq. (8–23a):

$$\frac{d}{dy}\left(GJ\frac{d\theta}{dy}\right) + qecc_l = -qc^2 c_{mAC} + Nmgd.\qquad(8\text{–}23c)$$

The boundary conditions are $\theta\ (0) = \theta'(l) = 0$.

The integral equation of torsional static equilibrium is obtained by introducing expression (8–23b) into Eq. (2–69):

$$\theta(y) = \int_0^l C^{\theta\theta}(y,\eta)[(ecc_l + c^2 c_{mAC})q - Nmgd]d\eta.\qquad(8\text{–}24)$$

Boundary conditions on θ are contained implicitly in the influence function $C^{\theta\theta}(y,\eta)$, and need not be stated separately.

Following the method used in the simplified development of Section 8–2, we can regard the total angle of attack as a superposition of a rigid angle and an elastic twist,

$$\alpha(y) = \alpha^r(y) + \theta(y),\qquad(8\text{–}25)$$

and correspondingly the local lift coefficient as a superposition of

$$c_l(y) = c_l{}^r(y) + c_l{}^e(y), \qquad (8\text{--}26)$$

where

$\alpha^r(y)$ = Local angle of attack measured from zero lift, not including elastic twist and spanwise aerodynamic induction effects. It may contain such terms as wing attitude, geometric twist, aerodynamic twist resulting from deflected control surfaces, or induced angle of attack due to a gust.

$c_l{}^r(y)$ = Local lift coefficient distribution resulting from rigid twist, $\alpha^r(y)$.

$c_l{}^e(y)$ = Local lift coefficient distribution resulting from elastic twist, $\theta(y)$.

Inserting Eq. (8–26) into Eqs. (8–23c) and (8–24) produces the following alternative differential and integral equation forms:

$$\frac{d}{dy}\left(GJ\,\frac{d\theta}{dy}\right) + qecc_l{}^e = -qecc_l{}^r - qc^2 c_{mAC} + Nmgd \qquad (8\text{--}27)$$

and

$$\theta(y) = q\int_0^l C^{\theta\theta}(y,\eta) ecc_l{}^e d\eta + f(y), \qquad (8\text{--}28)$$

where

$$f(y) = \int_0^l C^{\theta\theta}(y,\eta)(qecc_l{}^r + qc^2 c_{mAC} - Nmgd)d\eta. \qquad (8\text{--}28a)$$

The differential equation finds widest use in obtaining exact solutions of certain simplified problems. The integral equation form serves as a convenient basis for numerical solutions of complex practical problems. In both equations, $\theta(y)$ and $c_l{}^e(y)$ are regarded as unknown functions, and all other terms are assumed specified. The problem becomes mathematically determinate as soon as a second relation between the two unknowns is stated; this is supplied by some appropriate choice of aerodynamic theory. The aerodynamic theory is usually assumed to involve a linear relation between incidence and lift distribution which can be represented symbolically by

$$\alpha(y) = \mathfrak{a}[cc_l], \qquad (8\text{--}29)$$

where $\mathfrak{a}$ is a linear operator which operates on the lift distribution $cc_l(y)$ to produce the required incidence distribution $\alpha(y)$. For example, in the case of strip theory, $\mathfrak{a}$ is simply

$$\mathfrak{a} = \frac{1}{a_0 c}, \qquad (8\text{--}30)$$

where a_0 is the local two-dimensional slope of the lift coefficient curve. Equation (8–27) together with Eq. (8–29), or alternatively Eq. (8–28) together with Eq. (8–29), forms the basis for the prediction of elastic twist and lift distribution of a slender unswept wing with straight elastic axis. Because of the linearity of these equations, and because of the symmetry of wings about the airplane center line, a condition of unsymmetrical loading can be handled as a superposition of a symmetrical loading and an anti-symmetrical loading.

(b) *Torsional divergence.* The torsional divergence speed of a three-dimensional wing is determined from the lowest eigenvalue of dynamic pressure q obtained from the homogeneous differential or integral equations of equilibrium. It thus represents that speed at which the wing, arranged so that in the untwisted condition it experiences no aerodynamic moments whatever, is theoretically capable of assuming an arbitrary amount of twist and remaining in neutral equilibrium there under the airloads due to the twist alone. Since the solution to a nonhomogeneous equation becomes infinite for the eigenvalues of the corresponding homogeneous equation, we may conclude that an actual wing (which never can be adjusted so that the rigid airloads are exactly zero) would twist off and be destroyed at its divergence speed. The homogeneous forms of Eqs. (8–27) and (8–28) are

$$\frac{d}{dy}\left(GJ\frac{d\theta}{dy}\right) + qecc_l^e = 0, \tag{8-31}$$

$$\theta(y) = q\int_0^l C^{\theta\theta}(y,\eta)ecc_l^e d\eta. \tag{8-32}$$

Equation (8–31) or Eq. (8–32) can be alternatively used together with Eq. (8–29) to compute the divergence speed. They are both satisfied by the same infinite set of eigenvalues and eigenfunctions. The lowest eigenvalue is the dynamic pressure, q_D, corresponding to torsional divergence. The corresponding eigenfunction $\Theta_D(y)$, is the spanwise twist distribution at the divergence speed.

(1) *Closed-form solutions based on strip theory.* Solutions to the differential equation (Eq. 8–31) for the case of strip theory can be obtained in closed form in a few special cases. The simplest of these is that of a cantilever wing of uniform chord and stiffness, for which the homogeneous differential equation has the form

$$\frac{d^2\theta}{dy^2} + \lambda^2\theta = 0, \tag{8-33}$$

where $\lambda^2 = qcea_0/GJ$. The general solution of Eq. (8–33) is

$$\theta(y) = A\sin\lambda y + B\cos\lambda y, \tag{8-34}$$

with the boundary conditions

$$\theta(0) = 0, \tag{8-35}$$

$$\theta'(l) = 0. \tag{8-36}$$

Introducing Eq. (8–34) into the boundary conditions, we obtain

$$B = 0, \tag{8-37}$$

$$\cos \lambda l = 0. \tag{8-38}$$

Equation (8–38) indicates that a solution exists only when

$$\lambda l = (2n + 1)\frac{\pi}{2}, \qquad (n = 0, 1, 2, \cdots, \infty). \tag{8-39}$$

The smallest of these values, $\lambda = \pi/2l$, corresponds to the torsional divergence speed

$$q_D = \frac{GJ\pi^2}{4cea_0 l^2} \tag{8-40}$$

and

$$U_D = \frac{\pi}{2l}\sqrt{\frac{GJ}{cea_0(\rho/2)}}. \tag{8-41}$$

The associated deflection mode at divergence is

$$\Theta_D(y) = A \sin\frac{\pi y}{2l}. \tag{8-42}$$

It can easily be shown that the results given by Eqs. (8–41) and (8–42) can also be derived from the integral equation (8–32) for the special case of uniform chord and stiffness.

It should be observed that Eq. (8–7), derived from the case of a rigid wing segment, is quite similar to Eq. (8–41). In fact, Eq. (8–7) could be used to compute the divergence speed of a uniform, three-dimensional wing providing $C^{\theta\theta}$ were appropriately defined.

Other explicit solutions are available in which the eigenvalues of the homogeneous differential equation are given in closed form (Ref. 8–10). The most useful of these is constructed on the assumptions that

$$\frac{c(y)}{c_R} = \frac{e(y)}{e_R} = \left(1 - \beta\frac{y}{l}\right)^{\gamma_1} = y_1{}^{\gamma_1}, \tag{8-43}$$

$$\frac{GJ}{GJ_R} = \left(1 - \beta\frac{y}{l}\right)^{\gamma_2} = y_1{}^{\gamma_2}, \tag{8-44}$$

where the subscript R denotes properties of the wing-root section, and y_1 is a dimensionless spanwise coordinate replacing y. When Eqs. (8–43) and (8–44) are substituted into Eq. (8–31) and the assumption of strip theory is applied, one obtains

$$\frac{d}{dy_1}\left(y_1{}^{\gamma_2}\frac{d\theta}{dy_1}\right) + \left(\frac{\lambda_R l}{\beta}\right)^2 y_1{}^{2\gamma_1}\theta = 0, \tag{8-45}$$

where
$$\lambda_R{}^2 = \frac{q c_R e_R a_0}{G J_R} \tag{8-46}$$

and a_0 is assumed constant.

Referring to Kármán and Biot (Ref. 8–13), Eq. (8–45) is solved by

$$\theta = y_1{}^{\nu/\delta} \left[A J_\nu \left(y_1{}^{1/\delta} \frac{\lambda_R l \delta}{\beta} \right) + B Y_\nu \left(y_1{}^{1/\delta} \frac{\lambda_R l \delta}{\beta} \right) \right], \tag{8-47}$$

with

$$\frac{1}{\delta} = \frac{2\gamma_1 - \gamma_2 + 2}{2}, \tag{8-48}$$

$$\frac{\nu}{\delta} = \frac{1 - \gamma_2}{2}, \tag{8-49}$$

where J_ν and Y_ν are Bessel functions of the first and second kinds and order ν; except when ν is an integer, $Y_\nu = J_{-\nu}$. The eigenvalues and eigenfunctions for any particular choice of γ_1, γ_2, and β are established by applying boundary conditions (8–35) and (8–36) to Eq. (8–47). Generally, the constants A and B both remain finite, and the eigenvalues λ_R are roots of a transcendental equation.

A particularly important subcase of Eq. (8–45) occurs when $\gamma_2 = 2(\gamma_1 + 1)$. For example, $\gamma_1 = 1$ and $\gamma_2 = 4$ describe a straight-tapered wing, having taper ratio $(1 - \beta)$ and cross sections geometrically similar to one another but with all linear dimensions reduced in direct proportion to the local chord. Since Eqs. (8–48) and (8–49) now yield infinite values of δ and ν, we must return for solution to the basic differential equation. As shown in Ref. 8–10, the change of variable $z_1 = \ln y_1$ in Eq. (8–45) leads to the general result

$$\theta = y_1{}^{-\left(\frac{1+2\gamma_1}{2}\right)} [A \sin (\mu \ln y_1) + B \cos (\mu \ln y_1)], \tag{8-50}$$

where

$$\mu = \sqrt{\left(\frac{\lambda_R l}{\beta}\right)^2 - \left(\frac{2\gamma_1 + 1}{2}\right)^2}. \tag{8-51}$$

Boundary condition (8–35) shows that $B = 0$, and Eq. (8–50) yields, upon substitution into Eq. (8–36),

$$A y_1{}^{-\left(\frac{1+2\gamma_1}{2}\right)} \left[-\left(\frac{1 + 2\gamma_1}{2}\right) \frac{\sin (\mu \ln y_1)}{y_1} + \frac{\mu}{y_1} \cos (\mu \ln y_1) \right] \Big|_{y_1 = 1-\beta} = 0. \tag{8-52}$$

Reorganization of Eq. (8–52) leads to

$$\tan [\mu \ln (1 - \beta)] = \frac{2\mu}{1 + 2\gamma_1}, \tag{8-53}$$

from which the lowest eigenvalue μ and the corresponding divergence speed U_D are found by a simple graphical construction. The eigenfunction $\Theta_D(y)$ follows from equation (8–50) with B set equal to zero. It can easily be seen, for example, that for a straight-tapered wing with a taper ratio of one-half ($\gamma_1 = 1$, $\gamma_2 = 4$, $\beta = 0.5$), the lowest eigenvalue becomes $\lambda_R = 1.653/l$, which corresponds to a divergence speed of

$$U_D = \frac{1.653}{l} \sqrt{\frac{GJ_R}{c_R e_R a_0 (\rho/2)}}.$$ (8–53a)

The latter is slightly higher than the corresponding value for the straight wing given by Eq. (8–41).

Although mathematical solutions like those implied in Eqs. (8–39) and (8–53) rarely apply with engineering precision to an actual wing or tail surface, they are frequently valuable for estimating the order of magnitude of U_D. They lead to distinctly conservative results when the two-dimensional value of a_0 is inserted. However, it has been shown that when a_0 is replaced by the three-dimensional lift-curve slope of the finite wing, values of U_D calculated from the simplified equations often approximate closely those found from much more elaborate statements of the problem, including more rigorous aerodynamic theories.

(2) *Solution by matrices using strip theory.* When treating practical airplane problems where assured accuracy is required, it is usually necessary to employ numerical solutions. In such cases, the integral equation (8–32), together with matrix methods, possesses marked advantages. Applying strip theory, Eq. (8–32) has the form

$$\frac{c_l^e}{dC_L/d\alpha} = q \int_0^l C^{\theta\theta}(y,\eta) e c c_l^e d\eta.$$ (8–54)

For the reasons discussed above, $a_0(y)$ in Eq. (8–54) has been replaced by a constant, $dC_L/d\alpha$, the effective lift-curve slope corrected for aspect ratio. The matrix form of Eq. (8–54) can be written as follows:

$$[A]\{cc_l^e\} = q[E]\{cc_l^e\},$$ (8–55)

where

$$[A] = \frac{1}{dC_L/d\alpha}\lceil 1/c \rfloor, \qquad [E] = [C^{\theta\theta}]\lceil e \rfloor\lceil \overline{W} \rfloor.$$ (8–55a)

Equation (8–55) has the same form as the characteristic value simultaneous equations derived in Chapter 4 for free-vibration problems. The basic requirement for a nonzero solution is the vanishing of the determinant of the coefficients of cc_l^e.

$$|[A] - q[E]| = 0.$$ (8–56)

This determinant, when expanded, yields a polynomial in q, the smallest root of which corresponds to the divergence condition. The various approximate solutions outlined in Section 4–5 are applicable here; in particular, matrix iteration has been found useful, since it converges automatically upon the lowest eigenvalue.

(3) *Solution by matrices accounting for aerodynamic span effects.* When aerodynamic span effects are taken into consideration, the divergence speed can be found from the lowest eigenvalue of the homogeneous integral equation formed from Eqs. (8–32) and (8–29):

$$\mathfrak{A}[cc_l{}^e(y)] = q \int_0^l C^{\theta\theta}(y,\eta)ecc_l{}^e d\eta. \tag{8–57}$$

When induction effects are taken into account according to Prandtl's lifting-line theory, the functional relation $\mathfrak{A}[cc_l{}^e(y)]$ can be derived from Eq. (5–147):

$$\Gamma(y) = a_0 U \frac{c}{2}\left[\alpha(y) - \frac{1}{4\pi U}\oint_{-l}^l \frac{d\Gamma}{d\eta}\frac{d\eta}{y-\eta}\right]. \tag{5–147}$$

Introducing $cc_l = 2\Gamma/U$ and transposing yields

$$\alpha(y) = \mathfrak{A}[cc_l(y)] = \frac{cc_l(y)}{a_0 c} + \frac{1}{8\pi}\oint_{-l}^l \frac{d}{d\eta}(cc_l)\frac{d\eta}{y-\eta}. \tag{8–58}$$

Combining Eqs. (8–57) and (8–58),

$$\frac{cc_l{}^e}{a_0 c} + \frac{1}{8\pi}\oint_{-l}^l \frac{d}{d\eta}(cc_l{}^e)\frac{d\eta}{y-\eta} = q\int_0^l C^{\theta\theta}(y,\eta)ecc_l{}^e d\eta, \tag{8–59}$$

where the local lift coefficient slope a_0 has been assumed constant.

Equation (8–59) is satisfied by an infinite set of eigenvalues q_j and eigenfunctions $(cc_l{}^e)_j$. They correspond to the solutions of Eq. (8–54) except that, in this case, the influence of finite span on aerodynamic forces is included. The lowest eigenvalue, q_D, is the dynamic pressure corresponding to torsional divergence. The eigenfunctions of Eq. (8–59) may be either even or odd in y, depending on whether the associated distribution of wing twist is symmetrical or antisymmetrical; these functions correspond, in general, to different divergence speeds. The process of solving Eq. (8–59) requires considering the symmetrical and antisymmetrical solutions separately. The former is designated by $(cc_l{}^{es})$ and the latter by $(cc_l{}^{ea})$. Since the most general $(cc_l{}^e)$ can be represented as the sum of $(cc_l{}^{es})$ and $(cc_l{}^{ea})$, this sum can be substituted into Eq. (8–59) and the result separated into two independent parts. When approximation formulas are used to evaluate the integrals, one obtains the following matrix equations:

$$[A^s]\{cc_l{}^{es}\} = q[E]\{cc_l{}^{es}\}, \tag{8–60}$$

$$[A^a]\{cc_l{}^{ea}\} = q[E]\{cc_l{}^{ea}\}, \tag{8–61}$$

where

$$[E] = [C^{\theta\theta}][e][\bar{W}].$$

The aerodynamic matrices $[A^s]$ and $[A^a]$, which represent, respectively, the symmetrical and antisymmetrical relations between $cc_l(y)$ and $\alpha(y)$, are constructed from Eq. (8–58). An explicit expression for $[A^s]$ is given by Eq. (5–157).

When $[A^s]$ is substituted into Eq. (8–60), multiplication by $[A^s]^{-1}$ produces a form suitable for matrix iteration to determine the symmetric divergence modes. A similar iteration of Eq. (8–61) results in the anti-symmetric modes. In each case, the mode corresponding to the lowest value of dynamic pressure q, which is the one obtained without sweeping, is the one of greatest practical interest.

EXAMPLE 8–1. To calculate the divergence speed of the wing of the jet transport introduced in Example 2–1. Compare the symmetrical and antisymmetrical divergence speeds, computed by applying lifting-line theory with the results of strip theory.

Solution. (a) *Strip theory.* The divergence speed for the case of strip theory is computed from

$$[A]\{cc_l{}^e\} = q[E]\{cc_l{}^e\}, \qquad (8\text{–}55)$$

where

$$[A] = \frac{1}{dC_L/d\alpha}\,[1/c], \qquad [E] = [C^{\theta\theta}][e][\overline{W}]. \qquad (8\text{–}55a)$$

The wing is divided into four stations over the semispan, as illustrated by Fig. 8–6. These particular stations, called Multhopp's stations, (cf. Appendix B), are selected for convenience in computing the aerodynamic matrices.

The elastic properties of the wing are given in Example 2–1. Assuming a straight elastic axis, the matrix of torsional influence coefficients associated with the stations shown in Fig. 8–6 can be computed from Eqs. (2–120) and (2–121).

$$[C^{\theta\theta}] = \begin{bmatrix} 424.3 & 186.6 & 78.45 & 0 \\ 186.6 & 186.6 & 78.45 & 0 \\ 78.45 & 78.45 & 78.45 & 0 \\ 0 & 0 & 0 & 0 \end{bmatrix} \times 10^{-10}\,\text{rad/in}\cdot\text{lb}. \qquad (a)$$

From the geometric properties of the wing shown in Fig. 8–6, the $[e]$ and $[c]$ matrices have the following numerical values:

$$[e] = \begin{bmatrix} 10.95 & & & \\ & 13.66 & & \\ & & 17.72 & \\ & & & 22.50 \end{bmatrix} \text{inches}, \qquad (b)$$

$$[c] = \begin{bmatrix} 109.515 & & & \\ & 136.612 & & \\ & & 177.165 & \\ & & & 225 \end{bmatrix} \text{inches}. \qquad (c)$$

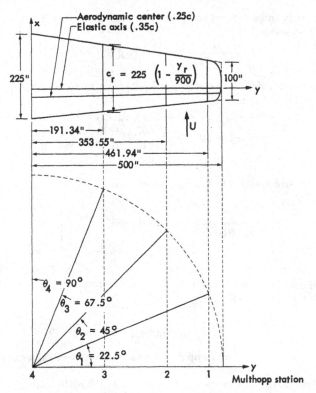

Fig. 8–6.   Multhopp stations on jet transport wing.

In the present example we shall obtain the weighting matrix $\lceil\overline{W}\rceil$ by applying Multhopp's quadrature formula.   Application of this formula is discussed in detail in Appendix B, where it is shown that for this particular four-station configuration, the weighting matrix has the form

$$\lceil\overline{W}\rceil = \frac{\pi l}{8}\begin{bmatrix} \sin \pi/8 & & & \\ & \sin \pi/4 & & \\ & & \sin 3\pi/8 & \\ & & & \tfrac{1}{2}\sin \pi/2 \end{bmatrix} \text{inches,} \qquad (d)$$

where $l = 500$ in. (semispan) in the present example.

The matrix $[E]$ is obtained by combining matrices (a), (b), and (d) as follows:

$$[E] = [C^{\theta\theta}]\lceil e\rceil\lceil\overline{W}\rceil = \begin{bmatrix} 34.9105 & 35.3897 & 25.2175 & 0 \\ 15.3531 & 35.3897 & 25.2175 & 0 \\ 6.4547 & 14.8785 & 25.2175 & 0 \\ 0 & 0 & 0 & 0 \end{bmatrix} \times 10^{-6}\frac{\text{rad}\cdot\text{in}}{\text{lb}}. \qquad (e)$$

The matrix to be iterated to obtain the strip-theory result derives from Eq. (8–55), as follows:

$$\frac{1}{q(dC_L/d\alpha)}\{cc_l^e\} = [c][E]\{cc_l^e\}. \tag{f}$$

Substituting numerical values of $[c]$ and $[E]$ from matrices (c) and (e), respectively, into Eq. (f) yields

$$\frac{1}{q(dC_L/d\alpha)}\{cc_l^e\} = 10^{-2} \times \begin{bmatrix} .3823 & .3875 & .2762 & 0 \\ .2097 & .4835 & .3445 & 0 \\ .1143 & .2636 & .4467 & 0 \\ 0 & 0 & 0 & 0 \end{bmatrix} \{cc_l^e\}. \tag{g}$$

Applying matrix iteration to Eq. (g), we obtain

$$\frac{1}{q_D(dC_L/d\alpha)}\{cc_l^e\} = 0.00957 \begin{bmatrix} 1 \\ .9684 \\ .7237 \\ 0 \end{bmatrix}. \tag{h}$$

The column matrix in Eq. (h) represents the normalized torsional divergence mode shape for the case of strip theory. The divergence speed is obtained from the eigenvalue, as follows:

$$U_D = \sqrt{\frac{1}{(\rho/2)(.00957)(dC_L/d\alpha)}}. \tag{i}$$

Numerical values of U_D are computed for various assumed values of $dC_L/d\alpha$ at the conclusion of the example.

(b) *Symmetrical divergence speed according to lifting-line theory.* The symmetrical divergence speed according to lifting-line theory is computed from

$$[A^s]\{cc_l^{es}\} = q[E]\{cc_l^{es}\}. \tag{8–60}$$

The $[E]$ matrix in Eq. (8–60) has been computed above, and is given by Eq. (e). The $[A^s]$ matrix derives from Eq. (5–157) and is given by

$$[A^s] = \begin{bmatrix} 4.2733 & -.95711 & 0 & -.07322 \\ -.51798 & 2.7451 & -.59724 & 0 \\ 0 & -.45711 & 2.10866 & -.42677 \\ -.05604 & 0 & -.78858 & 1.80808 \end{bmatrix} \times 10^{-3} \text{ rad/in.} \tag{j}$$

The matrix to be iterated derives from Eq. (8–60), as follows:

$$\frac{1}{q}\{cc_l^{es}\} = [A^s]^{-1}[E]\{cc_l^{es}\}. \tag{k}$$

Inverting Eq. (j), we obtain

$$[A^s]^{-1} = \begin{bmatrix} .24525 & .09089 & .03231 & .01756 \\ .04919 & .40238 & .12582 & .03169 \\ .01338 & .09629 & .55029 & .13043 \\ .01344 & .04482 & .24100 & .61050 \end{bmatrix} \times 10^3 \text{ in/rad.} \tag{l}$$

Substituting matrices (e) and (l) into Eq. (k) gives

$$\frac{1}{q}\{cc_l{}^{es}\} = 10^{-3} \times \begin{bmatrix} 10.1660 & 12.3768 & 9.2915 & 0 \\ 8.7072 & 17.8530 & 14.5604 & 0 \\ 5.4976 & 12.0690 & 16.6427 & 0 \\ 2.7128 & 5.6474 & 7.5466 & 0 \end{bmatrix}\{cc_l{}^e\}. \tag{m}$$

Applying matrix iteration to Eq. (m) gives

$$\frac{1}{q_D}\{cc_l{}^{es}\} = .036439 \begin{bmatrix} .7615 \\ 1 \\ .8211 \\ .3817 \\ 0 \end{bmatrix}. \tag{n}$$

Ordinates to the symmetrical mode shape are given by the column matrix, and the divergence speed is obtained from

$$U_D = \sqrt{\frac{1}{(\rho/2)(.036439)}}. \tag{o}$$

(c) *Antisymmetrical divergence speed according to lifting-line theory.* The antisymmetrical divergence speed according to lifting-line theory is obtained from

$$[A^a]\{cc_l{}^{ea}\} = q[E]\{cc_l{}^{ea}\}, \tag{8-61}$$

where $[E]$ is given by the matrix (e) above. The matrix $[A^a]$ is obtained from Eq. (5–157) by using even rather than odd values of r, and putting the midspan station terms equal to zero. Thus we obtain for the present example,

$$[A^a] = \begin{bmatrix} 4.2733 & -.92387 & 0 & 0 \\ -.5000 & 2.7451 & -.5000 & 0 \\ 0 & -.38268 & 2.10865 & 0 \\ 0 & 0 & 0 & .80808 \end{bmatrix} \times 10^{-3}\,\text{rad/in.} \tag{p}$$

The matrix to be iterated is obtained from Eq. (8–61), as follows:

$$\frac{1}{q}\{cc_l{}^{ea}\} = [A^a]^{-1}[E]\{cc_l{}^{ea}\}. \tag{q}$$

Inverting Eq. (p) gives

$$[A^a]^{-1} = \begin{bmatrix} .24394 & .08491 & .02013 & 0 \\ .04595 & .39273 & .09312 & 0 \\ .00834 & .07127 & .49113 & 0 \\ 0 & 0 & 0 & 0 \end{bmatrix} \times 10^3\,\text{in/rad.} \tag{r}$$

Substituting matrices (e) and (r) into Eq. (q),

$$\frac{1}{q}\{cc_l{}^{ea}\} = 10^{-3} \times \begin{bmatrix} 9.9497 & 11.9374 & 8.8004 & 0 \\ 8.2348 & 16.9103 & 13.4107 & 0 \\ 4.5555 & 10.1248 & 14.3928 & 0 \\ 0 & 0 & 0 & 0 \end{bmatrix}\{cc_l{}^{ea}\}. \tag{s}$$

Iterating Eq. (s), we obtain

$$\frac{1}{q_D} \{cc_l{}^{ea}\} = .033198 \begin{bmatrix} .78966 \\ 1 \\ .72968 \\ 0 \end{bmatrix}. \tag{t}$$

Ordinates to the antisymmetrical mode shape are given by the column matrix, and the divergence speed is obtained from

$$U_D = \sqrt{\frac{1}{(\rho/2)(.033198)}}. \tag{u}$$

(d) *Summary.* Results of the above calculations are summarized by the table given below for the case of flight at sea level.

Method of analysis	Divergence speed, U_D (ft/sec)
Strip theory (no finite span correction): $\dfrac{dC_L}{d\alpha} = a_0 = 5.5$	1516.6
Strip theory (corrected by Eq. 5–166): $\dfrac{dC_L}{d\alpha} = a_0 \dfrac{AR}{AR + 2} = 4.1497$	1745.9
Strip theory (corrected by Eq. 5–219): $\dfrac{dC_L}{d\alpha} = a_0 \dfrac{AR}{AR + 4} = 3.3325$	1948.2
Lifting-line theory: Symmetric	1823.1
Lifting-line theory: Antisymmetric	1910.0

Although the divergence speeds tabulated above have no real physical meaning, since incompressible aerodynamic theory has been used, we do not attempt to correct them for Mach number in the present example. It may be concluded, however, that the divergence speed is obviously above the speed range of the jet transport airplane under discussion.

In comparing the tabulated data, we see that when lifting-line theory results are compared, the antisymmetric divergence speed is higher than the symmetric. This is generally true for straight wings and, consequently, there is usually no practical need for computing antisymmetric divergence speeds. The strip-theory result obtained by putting $dC_L/d\alpha = a_0$ is lower than the symmetric divergence speed using lifting-line theory. This has also been found to be generally true

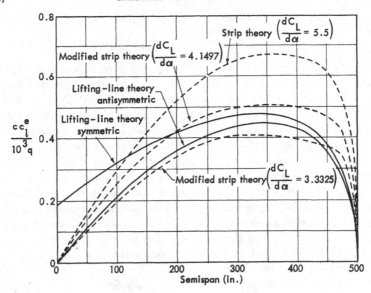

Fig. 8–7. Spanwise lift distribution for assigned twisting mode, $\theta = \sin(\pi y/2l)$.

for straight wings by Hildebrand and E. Reissner (Ref. 8–10). Results obtained by modifying the strip theory by means of a finite span correction on the lift-curve slope are seen to bring the strip-theory divergence speed closer to the symmetrical divergence speed obtained by lifting-line theory. The result obtained by applying $dC_L/d\alpha = a_0[\mathrm{AR}/(\mathrm{AR}+2)]$ gives a closer, but conservative, approximation to the lifting-line theory in this particular example.*

These conclusions can also be deduced by reasoning physically in the following way. Assume that the wing has approximately the same spanwise mode of twist in all methods of analysis. If this twisting mode is given a finite displacement from its zero position, the resulting aerodynamic forces on the wing produce aerodynamic twisting moments. The magnitudes of these aerodynamic twisting moments at the various wing stations are, of course, dependent upon the spanwise distribution of aerodynamic forces. It would be expected, then, that the twisting moment at a given arbitrary spanwise station would be largest and the divergence speed smallest for lift distributions which tend to have large values near the tip. Figure 8–7 compares the various lift distributions for a given assigned spanwise twisting mode, $\Theta_D = \sin(\pi y/2l)$. In comparing the strip-theory and lifting-line distributions in Fig. 8–7, we would expect the strip-theory divergence speed to be the lowest, since the strip-theory distribution tends to load the tip to a higher degree. Similarly, in comparing symmetric and antisymmetric lifting-line distributions, it would be concluded that the symmetric divergence mode has the lower divergence speed. Similar deductions from Fig.

---

* In Chapter 5, Section 5–5, we have seen that the formula $dC_L/d\alpha = a_0[(\mathrm{AR}/(\mathrm{AR}+2)]$ derives from lifting-line theory, whereas the formula $dC_L/d\alpha = a_0[(\mathrm{AR}/(\mathrm{AR}+4)]$ has an empirical basis.

8-7 can also be made on this basis concerning the relative magnitude of the strip theory and the modified strip theory results.

---

(4) *Energy solutions.* It is sometimes useful to consider the divergence of a slender straight wing in terms of an elementary energy solution. Since the eigenvalue problem considered here is identical to that of a vibrating beam or of an Euler column, it would be expected that the Rayleigh method, so useful in these problems, would apply equally well to torsional divergence. If the wing is given an arbitrary increase in angular distortion while operating at a specified speed, and if this increase is accompanied by a greater change in elastic strain energy than the work done by the aerodynamic forces, the wing is statically stable. However, if the work done by the aerodynamic forces exceeds the change in strain energy, the wing is statically unstable. The borderline case of neutral stability is thus defined by the condition that the change in strain energy be equal to the work done by the aerodynamic forces during an arbitrary finite increase in angular distortion.

Suppose that the wing is given an arbitrary finite change in angular distortion defined by a function $\Theta(y)$. Equating the change in strain energy to the work done by the aerodynamic forces, we obtain

$$\int_0^l \frac{GJ}{2} \left(\frac{d\Theta}{dy}\right)^2 dy = \tfrac{1}{2} q_D \int_0^l c c_l{}^e e \Theta dy, \qquad (8\text{--}62)$$

where q_D is the divergence dynamic pressure. The latter becomes simply

$$q_D = \frac{\displaystyle\int_0^l GJ (d\Theta/dy)^2 dy}{\displaystyle\int_0^l c c_l{}^e e \Theta dy}. \qquad (8\text{--}63)$$

The form of Eq. (8–63) is similar to the well-known Rayleigh formula for natural frequencies of vibrating beams (Eq. 4–17). The divergence dynamic pressure can be calculated by introducing the divergence mode shape $\Theta(y)$. The exact divergence mode shape produces a minimum value of q_D, and any other mode shape produces a greater value. Equation (8–63) is useful in that any reasonable function which satisfies the boundary conditions $\Theta(0) = \Theta'(l) = 0$ will yield a close approximation to q_D. For example, nonuniform wings can be handled with ease and considerable accuracy by introducing the fundamental divergence mode of a uniform cantilever wing given by Eq. (8–42) and evaluating the integrals numerically. Spanwise induction effects can be taken into account by computing the $c c_l{}^e$ distribution corresponding to the assumed $\Theta(y)$ distribution. This is a separate problem and requires the solution of

$$\Theta(y) = \alpha[c c_l{}^e(y)]. \qquad (8\text{--}64)$$

The $cc_l{}^e(y)$ distribution and the assumed $\Theta(y)$ distribution can be substituted into Eq. (8–63) to obtain an approximate value for q_D. When strip theory is employed, Eq. (8–63) reduces to the following simplified form useful for quick estimates of the divergence speed of a straight wing:

$$q_D = \frac{\displaystyle\int_0^l GJ\,(d\Theta/dy)^2\,dy}{\displaystyle\int_0^l a_0\Theta^2 ce\,dy}. \tag{8–65}$$

EXAMPLE 8–2. Using the approximate deflection mode shape $\Theta = 2(y/l) - (y/l)^2$, compute the torsional divergence speed of a uniform wing by use of Rayleigh's method and strip theory, and compare with the exact solution.

  Solution. The assumed function satisfies the boundary conditions and represents approximately the exact mode shape. From Eq. (8–65),

$$q_D = \frac{GJ \displaystyle\int_0^l \left(\frac{2}{l} - 2\frac{y}{l^2}\right)^2 dy}{ce a_0 \displaystyle\int_0^l \left(2\frac{y}{l} - \left(\frac{y}{l}\right)^2\right)^2 dy} = \frac{15}{6}\frac{GJ}{ce a_0 l^2}.$$

Comparing the result with the exact result given by Eq. (8–40), the error is seen to be 1.4%.

  (c) *Symmetrical lift distribution.* Aeroelastic effects on the spanwise lift distribution of straight wings during symmetrical flight conditions are of importance to structural designers because of an outboard shift in spanwise center of pressure due to wing twist and a resulting increase in spanwise bending moments.

  (1) *Closed form solutions of uniform wings based on strip theory.* A closed solution of the spanwise lift distribution on a uniform wing in terms of strip theory can be easily obtained by solving the differential equation of equilibrium. Equation (8–27) reduces in the case of a uniform wing and strip theory to

$$\frac{d^2\theta}{dy^2} + \lambda^2\theta = K^r, \tag{8–66}$$

where

$$K^r = -\frac{1}{GJ}\,(qcec_l{}^r + qc_{mAC}c^2 - mNgd), \qquad \lambda^2 = \frac{qcea_0}{GJ}.$$

A general solution to Eq. (8–66) is given by

$$\theta(y) = A \sin \lambda y + B \cos \lambda y + (K^r/\lambda^2), \tag{8–67}$$

where the constants A and B are to be computed from the boundary conditions. In introducing the boundary conditions, two types of practical problems must be recognized.

Attitude of rigid wing specified. When the attitude of the rigid wing is specified, the total lift as well as the lift distribution remains to be computed. In a practical problem, this would be equivalent to specifying the attitude of the fuselage of an airplane in flight, or the attitude of the support of a wind-tunnel model. Let us consider a solution of Eq. (8–66) for the case of an airplane in flight with a uniform wing at a specified attitude of the wing root, $\alpha(0)$. The following conditions must be satisfied:

$$c_l{}^r = a_0\alpha(0), \qquad \theta(0) = 0, \qquad \theta'(l) = 0. \tag{8–68}$$

Evaluating the constants A and B subject to conditions (8–68) yields

$$\theta(y) = \frac{K^r}{\lambda^2} (1 - \tan \lambda l \sin \lambda y - \cos \lambda y), \tag{8–69}$$

where

$$K^r = -\frac{1}{GJ} (qcea_0\alpha(0) + qc_{mAC}c^2 - mNgd). \tag{8–69a}$$

The total lift coefficient is a superposition of the rigid lift coefficient and the incremental lift coefficient due to elastic twist, as follows:

$$c_l = a_0\alpha(0) + a_0\theta(y). \tag{8–70}$$

The lift distribution is not known explicitly, however, until a numerical value of the load factor N has been obtained. This value is obtained by substituting Eq. (8–70) into the following condition:

$$N = \frac{2q}{W} \int_0^l cc_l dy, \tag{8–70a}$$

where W is the airplane weight. This yields

$$N = \frac{a_0\alpha(0) \dfrac{\tan \lambda l}{\lambda l} + \dfrac{c}{e} c_{mAC} \left(\dfrac{\tan \lambda l}{\lambda l} - 1 \right)}{\dfrac{1}{qc} \left\{ \dfrac{W}{2l} - \dfrac{mgd}{e} \left(1 - \dfrac{\tan \lambda l}{\lambda l} \right) \right\}}. \tag{8–70b}$$

The value of N given above is the value required for equilibrium of the airplane when the wing root angle of attack is $\alpha(0)$. Thus the lift coefficient distribution resulting from elastic twist on a uniform wing of an airplane in flight at a specified attitude is given by

$$c_l{}^e(y) = a_0 \frac{K^r}{\lambda^2} (1 - \tan \lambda l \sin \lambda y - \cos \lambda y), \tag{8–70c}$$

where K^r is defined by Eq. (8–69a) and N is given by Eq. (8–70b).

A particularly simple result is obtained if we neglect inertial twisting moments and assume a symmetrical airfoil. This is obtained by putting

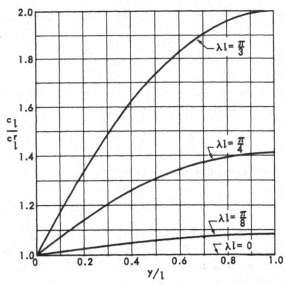

Fig. 8–8. Variation of list distribution on a uniform wing with the parameter $\lambda l = l \sqrt{\dfrac{qcea_0}{GJ}}$ (attitude specified).

$c_{mAC} = d = 0$ in the previous results. The rigid lift coefficient distribution due to the specified attitude is the same as before:

$$c_l{}^r = a_0\alpha(0);\qquad (8–71a)$$

however, the lift coefficient distribution resulting from elastic twist reduces to the simpler form

$$c_l{}^e(y) = a_0\alpha(0)[\cos \lambda y + \tan \lambda l \sin \lambda y - 1].\qquad (8–71b)$$

It is instructive to consider the ratio $c_l/c_l{}^r$ as follows:

$$\frac{c_l}{c_l{}^r} = \frac{c_l{}^e + c_l{}^r}{c_l{}^r} = \cos \lambda y + \tan \lambda l \sin \lambda y.\qquad (8–72)$$

Figure 8–8 illustrates a plot of this ratio at various values of λl less than $\pi/2$. The curves illustrate the profound effect of elastic twist on spanwise lift distribution with increasing values of λl. Although these curves have been plotted only up to a value of $\lambda l = \pi/3$, we have already seen by Eq. (8–39) that when $\lambda l = \pi/2$ the wing diverges and the ratio $c_l/c_l{}^r$ approaches infinity.

Total lift on wing specified. In the case of structural design conditions, load factors and speeds are frequently specified as points on a V-n diagram. Thus the total lift on the wing is given and the attitude and lift distribution must both be computed. In this case, the following additional condition applies:

$$q \int_0^l cc_l^r dy + q \int_0^l cc_l^e dy = \frac{WN}{2}, \qquad (8\text{-}73)$$

where N is a specified load factor. It is convenient to compute the rigid lift distribution according to the assumption that

$$q \int_0^l cc_l^r dy = \frac{WN}{2}, \qquad (8\text{-}74)$$

and the corrective elastic distribution according to

$$\int_0^l cc_l^e dy = 0. \qquad (8\text{-}75)$$

With these assumptions, the lift distribution on the rigid wing is obtained in the usual way, and a corrective lift distribution is superposed which produces no additional total lift. The rigid lift coefficient distribution is given immediately by

$$c_l^r = \frac{WN}{2qcl}, \qquad (8\text{-}76a)$$

where N is a specified load factor. The wing twist measured relative to the root section is obtained from Eq. (8-67) by introducing Eq. (8-76a) and the boundary conditions

$$\theta(0) = 0, \qquad \theta'(l) = 0, \qquad (8\text{-}76b)$$

which results in

$$\theta(y) = \frac{K^r}{\lambda^2} (1 - \tan \lambda l \sin \lambda y - \cos \lambda y), \qquad (8\text{-}76c)$$

where

$$K^r = -\frac{1}{GJ} \left(\frac{WNe}{2l} + qc_{mAC}c^2 - mNgd \right).$$

The corrective angle of attack distribution $\alpha^e(y)$, corresponding to the corrective lift distribution, is obtained by adding a constant value $\alpha^e(0)$ to the elastic twist $\theta(y)$:

$$\alpha^e(y) = \alpha^e(0) + \frac{K^r}{\lambda^2} (1 - \tan \lambda l \sin \lambda y - \cos \lambda y). \qquad (8\text{-}77)$$

The quantity $\alpha^e(0)$ can be interpreted physically as the nose-down pitching attitude that must be assumed by the wing root in order that the total lift resulting from elastic twist be zero. It is obtained by multiplying Eq. (8-77) by a_0, which yields c_l^e, and substituting the results into Eq. (8-75):

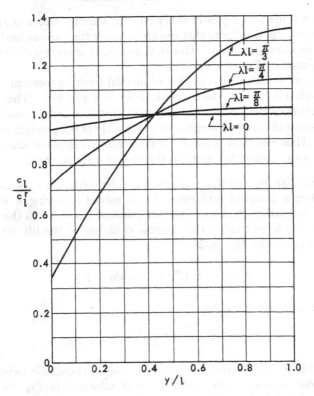

Fig. 8–9. Variation of lift distribution on a uniform wing with the parameter $\lambda l = l \sqrt{\dfrac{qcea_0}{GJ}}$ (total lift specified).

$$\alpha^e(0) = \frac{K^r}{\lambda^2}\left(\frac{\tan \lambda l}{\lambda l} - 1\right). \tag{8-78}$$

Substituting Eq. (8–78) into Eq. (8–77) and multiplying by a_0, we obtain an expression for the corrective lift coefficient distribution, as follows:

$$c_l^{\,e} = \frac{a_0 K^r}{\lambda^2}\left[1 - \lambda l \sin \lambda y - \frac{\lambda l \cos \lambda y}{\tan \lambda l}\right]\frac{\tan \lambda l}{\lambda l}. \tag{8-79}$$

The total lift coefficient distribution is a superposition of $c_l^{\,r}$ given by Eq. (8–76a) and $c_l^{\,e}$ given by Eq. (8–79). The effect on lift distribution due to elasticity, when the total load factor is specified, can be simply illustrated if we again assume $c_{mAC} = d = 0$. In this case, the ratio $c_l/c_l^{\,r}$ reduces to the simple result

$$\frac{c_l}{c_l^{\,r}} = \frac{c_l^{\,r} + c_l^{\,e}}{c_l^{\,r}} = 1 - \frac{\tan \lambda l}{\lambda l} + \sin \lambda y \tan \lambda l + \cos \lambda y. \tag{8-80}$$

Figure 8–9 illustrates a plot of this result for various values of λl. It can be observed from this figure that the areas under the various load distribution curves remain the same. This is, of course, consistent with the requirement imposed by Eq. (8–73).

It can be remarked that neglect of inertial twisting moments probably has a small influence on the curves in Figs. 8–8 and 8–9. This is due to the fact that the ratio of inertial twisting moments to aerodynamic twisting moments is of the order of magnitude of the ratio of wing weight to airplane weight. Thus the ratio is small for conventional airplane configurations but, of course, approaches unity in the case of a flying wing.

(2) *Solution by matrices.* Let us consider next the approach to be taken when a practical problem is to be solved involving a wing with spanwise properties which vary in a complicated manner. In this case, we find it desirable to employ the integral equation of the lift distribution problem as given by Eq. (8–28):

$$\theta(y) = q \int_0^l C^{\theta\theta}(y,\eta) e c c_l{}^e d\eta + f(y). \qquad (8\text{–}28)$$

Introducing Eq. (8–29) gives

$$\mathcal{C}[cc_l{}^e] = q \int_0^l C^{\theta\theta}(y,\eta) e c c_l{}^e d\eta + f(y), \qquad (8\text{–}81)$$

where $cc_l{}^e$ is the lift distribution corresponding to the elastic twist $\theta(y)$.

In order to obtain practical numerical solutions to Eq. (8–81), we transform it to the following matrix form:

$$[A^s]\{cc_l{}^e\} = q[E]\{cc_l{}^e\} + \{f\}, \qquad (8\text{–}82)$$

where

$[A^s]$ = matrix of aerodynamic influence coefficients for symmetrical loading (Eq. 5–157),
$\{f\} = q[E]\{cc_l{}^r\} + q[F]\{c_{mAC}\} - N[G]\{mg\}$,
$[E] = [C^{\theta\theta}][e][\overline{W}]$,
$[F] = [C^{\theta\theta}][c^2][\overline{W}]$,
$[G] = [C^{\theta\theta}][d][\overline{W}]$.

It should be observed that the first component of the $\{f\}$ matrix, $q[E]\{cc_l{}^r\}$ represents merely a column matrix of wing twist due to the rigid lift distribution. Similarly, the second and third components represent, respectively, wing twist due to aerodynamic moments about the aerodynamic center and wing twist due to inertial forces. An alternative form of Eq. (8–82) can be derived by putting $\{cc_l{}^e\} = \{cc_l\} - \{cc_l{}^r\}$ into Eq. (8–82), which yields

$$[A^s]\{cc_l\} = q[E]\{cc_l\} + \{\alpha^r\} + q[F]\{c_{mAC}\} - N[G]\{mg\}, \quad (8\text{–}83)$$

where $$\{\alpha^r\} = [A^s]\{cc_l{}^r\}.$$

In Eq. (8-82) the unknown quantity is the column matrix $\{cc_l{}^e\}$, and in Eq. (8-83) the unknown quantity is $\{cc_l\}$. All other quantities are assumed to be specified by the properties of the wing and the condition of the problem. The particular flight condition under consideration serves to define $\{\alpha^r\}$, $\{cc_l{}^r\}$, and N. We have seen that these flight conditions fall generally into two categories.

Attitude of rigid wing specified. When the attitude of an airplane in accelerated flight is specified, the matrix $\{\alpha^r\}$ is given, and the total lift or load factor, and the lift distribution, remain to be computed. The load factor N is computed from the additional condition that the total lift on the airplane is balanced by gravity and inertial forces. In matrix form, this condition becomes

$$q\lfloor 1 \rfloor [\overline{W}]\{cc_l\} = \tfrac{1}{2}WN, \tag{8-84}$$

where $\lfloor 1 \rfloor$ is a row matrix, each element of which is one. Hence, by using the specified matrix $\{\alpha^r\}$, together with the condition given by Eq. (8-84), the total lift distribution can be computed from Eq. (8-83). Solving Eq. (8-84) for N and putting the result into Eq. (8-83) yields

$$[A^s]\{cc_l\}$$
$$= q\left([E] - \frac{2}{W}[G]\{mg\}\lfloor 1 \rfloor [\overline{W}]\right)\{cc_l\} + \{\alpha^r\} + q[F]\{c_{mAC}\}. \tag{8-85}$$

Equations (8-85) are a set of simultaneous equations in which the unknowns are ordinates to the total lift distribution curves. When the attitude $\{\alpha^r\}$ is specified, Eq. (8-85) can be solved explicitly for the column matrix $\{cc_l\}$. In some problems it may be desired to compute the *elastic* lift distribution directly for the case of specified attitude. In order to accomplish this, we rewrite Eq. (8-84) in the form

$$N = N^r + \frac{2q}{W}\lfloor 1 \rfloor [\overline{W}]\{cc_l{}^e\}, \tag{8-86}$$

where N^r is the load factor corresponding to the rigid angle of attack matrix $\{\alpha^r\}$. If we now substitute the value of N given by Eq. (8-86) into Eq. (8-82), we obtain a set of simultaneous equations that can be solved directly for the *elastic* lift distribution:

$$[A^s]\{cc_l{}^e\} = q\left([E] - \frac{2}{W}[G]\{mg\}\lfloor 1 \rfloor [\overline{W}]\right)\{cc_l{}^e\}$$

$$+ q[E]\{cc_l{}^r\} + q[F]\{c_{mAC}\} - N^r[G]\{mg\}. \tag{8-87}$$

Equations (8-87) can be solved explicitly for the column matrix $\{cc_l{}^e\}$ when the attitude $\{\alpha^r\}$ and the load factor N^r are specified.

Total lift on wing specified. We consider now the integral equation and matrix forms to be applied when the total lift or load factor N is specified. We have seen in connection with closed-form solutions of the differential equation that it is convenient to assume that N is produced entirely by the rigid lift distribution and that the integral of the corrective elastic lift distribution over the span is zero. The angle of attack distribution $\alpha^e(y)$, corresponding to the corrective elastic lift distribution, is

$$\alpha^e(y) = \alpha^e(0) + \theta(y), \tag{8-88}$$

where $\alpha^e(0)$ is a constant determined by the condition that the integral of the corrective elastic lift distribution over the span is zero. Substituting Eq. (8-88) into Eq. (8-28) gives

$$\alpha^e(y) = q \int_0^l C^{\theta\theta}(y,\eta) ecc_l{}^e d\eta + f(y) + \alpha^e(0), \tag{8-89a}$$

where $cc_l{}^e$ now corresponds to the incidence distribution $\alpha^e(y)$. Introducing Eq. (8-29) gives

$$\mathcal{C}[cc_l{}^e] = q \int_0^l C^{\theta\theta}(y,\eta) ecc_l{}^e d\eta + f(y) + \alpha^e(0), \tag{8-89b}$$

which reduces to the matrix form

$$[A^s]\{cc_l{}^e\} = q[E]\{cc_l{}^e\} + \{f\} + \alpha^e(0)\{1\}. \tag{8-89c}$$

The condition that the total lift due to the corrective elastic twist is zero can be expressed in matrix form as

$$\lfloor 1 \rfloor [\overline{W}]\{cc_l{}^e\} = 0. \tag{8-89d}$$

Equations (8-89c) and (8-89d) can be solved simultaneously to obtain the corrective lift distribution due to elastic twist and the constant $\alpha^e(0)$:

$$\begin{bmatrix} [A^s] - q[E] & \{-1\} \\ \lfloor 1 \rfloor [\overline{W}] & 0 \end{bmatrix} \begin{bmatrix} \{cc_l{}^e\} \\ \alpha^e(0) \end{bmatrix} = \begin{bmatrix} \{f\} \\ 0 \end{bmatrix}. \tag{8-89e}$$

Thus, when a total load factor N is specified, the elastic lift distribution $\{cc_l{}^e\}$ and the constant $\alpha^e(0)$ can be computed directly by solving the simultaneous equations given by matrix equation (8-89e).

Matrix equations (8-85), (8-87), and (8-89e) can be solved formally by matrix inversion. If the wingspan is divided into n subintervals, inversion of a square matrix of order n is required in the case of Eqs. (8-85) and (8-87), and inversion of an $n + 1$ order matrix is required in the case of Eq. (8-89e). Matrix iteration can also be applied as a method of solution. The iteration process converges rapidly providing $q \ll q_D$.

Rigid-wing lift distributions can be obtained either experimentally or analytically. Various analytical methods are applicable to clean wings without discontinuities. Of these, the Prandtl lifting-line theory described in Chapter 5 forms the basis for most of the methods applied to straight wings. Wind-tunnel tests of rigid models have been extensively used as a means of obtaining rigid-wing lift distributions. Such tests are often necessary when the airplane configuration involves complicated combinations of nacelles and external fuel tanks.

EXAMPLE 8-3. To calculate the effect of elastic twist on the span load distribution of the jet transport introduced in Example 2-1. Assume that the airplane is executing a symmetrical pull-up at a load factor of $N = 4$ at 450 mph.

Solution. A sketch of the wing plan form is given by Fig. 8-6. Basic data needed for the computations in this example will be taken from Examples 2-1 and 8-1. In addition, the following data are required:

Gross weight, $W = 83,838$ lb.

Dynamic pressure at 450 mph, $q = 3.596$ psi.

Lift-curve slope, $a_0 = 5.5$ per rad.

Moment coefficient of all sections about the aerodynamic center, $c_{mAC} = -.015$. The same Multhopp stations are used in the present example as were used in Example 8-1, and as are shown by Fig. 8-6. Since the load factor is specified in the present example, we apply Eq. (8-89e) to compute the lift distribution resulting from elastic twist:

$$\begin{bmatrix} [A^s] - q[E] & \{-1\} \\ \lfloor 1 \rfloor \lceil \overline{W} \rceil & 0 \end{bmatrix} \begin{bmatrix} \{cc_l^e\} \\ \alpha^e(0) \end{bmatrix} = \begin{bmatrix} \{f\} \\ 0 \end{bmatrix}. \tag{8-89e}$$

The matrices $[A^s]$, $[E]$, and $\lceil \overline{W} \rceil$ which appear in Eq. (8-89e) have been previously tabulated in Example 8-1. In forming Eq. (8-89e) the remaining task is to compute the column matrix $\{f\}$ as defined by Eq. (8-82). In order to compute $\{f\}$, it is necessary to establish the lift distribution on the rigid wing for $N = 4$. If it is assumed that the wing has no geometric or aerodynamic twist, we can write for the case of the rigid wing,

$$[A^s]\{cc_l^r\} = \alpha^r\{1\}, \tag{a}$$

where α^r is the rigid angle of attack. Since the integral of the rigid lift distribution over the span must correspond to a load factor of 4, we have the additional condition that

$$q\lfloor 1 \rfloor \lceil \overline{W} \rceil \{cc_l^r\} = \tfrac{1}{2}WN. \tag{b}$$

Combining Eqs. (a) and (b), the rigid-wing angle of attack is

$$\alpha^r = \frac{WN}{2q} \frac{1}{\lfloor 1 \rfloor \lceil \overline{W} \rceil [A^s]^{-1}\{1\}}. \tag{c}$$

Substituting numerical data and matrices given above and in Example (8-1), we obtain for the rigid angle of attack

$$\alpha^r = 0.1346 \text{ rad.} \tag{d}$$

The rigid span load distribution $\{cc_l{}^r\}$ becomes

$$\{cc_l{}^r\} = 0.1346[A^s]^{-1}\{1\} = \begin{bmatrix} 51.9704 \\ 82.0027 \\ 106.4141 \\ 122.4845 \end{bmatrix} \text{inches.} \tag{e}$$

The first component of the $\{f\}$ matrix, $q\,[E]\{cc_l{}^r\}$, is a column matrix of wing twist due to the rigid lift distribution:

$$q[E]\{cc_l{}^r\} = \{\theta\}_{\text{rigid lift distribution}} = \begin{bmatrix} .026615 \\ .022959 \\ .015246 \\ 0 \end{bmatrix} \text{rad.} \tag{f}$$

The second component of the $\{f\}$ matrix, $q[F]\{c_{mAC}\}$, is a column matrix of wing twist due to the aerodynamic moment about the aerodynamic center. This is obtained by first calculating the $[F]$ matrix:

$$[F] = [C^{\theta\theta}][c^2][\overline{W}] = \begin{bmatrix} .038237 & .048351 & .044667 & 0 \\ .016816 & .048351 & .044667 & 0 \\ .007069 & .020327 & .044667 & 0 \\ 0 & 0 & 0 & 0 \end{bmatrix} \frac{\text{rad}\cdot\text{in}^2}{\text{lb}}, \tag{g}$$

and substituting the result into

$$q[F]\{c_{mAC}\} = -.015q[F] = \{\theta\}_{\text{aerodynamic moment}}$$

$$= \begin{bmatrix} -.00708 \\ -.00592 \\ -.00388 \\ 0 \end{bmatrix} \text{rad.} \tag{h}$$

The third component of the $\{f\}$ matrix, $N[G]\{mg\}$, is a column matrix of wing twist due to the inertial loads on the wing. This can be obtained directly from the mass unbalance data given in Example 4–7. These calculations will not be reproduced here; however, they involve merely a transformation of angles of twist computed at the lumped mass wing stations of Example 4–7 to the wing stations employed in the present analysis.

$$N[G]\{mg\} = 4[G]\{mg\} = \{\theta\}_{\text{inertial forces}}$$

$$= \begin{bmatrix} -.003894 \\ -.003943 \\ -.004354 \\ 0 \end{bmatrix} \text{rad.} \tag{i}$$

The matrix $\{f\}$ is calculated by superposition of matrices (f), (h), and (i) given above.

$$\{f\} = \begin{bmatrix} .026615 \\ .022959 \\ .015246 \\ 0 \end{bmatrix} - \begin{bmatrix} .00708 \\ .00592 \\ .00388 \\ 0 \end{bmatrix} - \begin{bmatrix} .003894 \\ .003943 \\ .004354 \\ 0 \end{bmatrix} = \begin{bmatrix} .015639 \\ .013090 \\ .00700 \\ 0 \end{bmatrix} \text{rad.} \tag{j}$$

Inserting this value of $\{f\}$, together with previously tabulated data, into Eq. (8-89e), we obtain

$$
\begin{bmatrix}
.0041477 & -.001084 & -.0000907 & -.0000732 & -1 \\
-.0005732 & .0026178 & -.0006879 & 0 & -1 \\
-.0000232 & -.0005106 & .002018 & -.000427 & -1 \\
-.0000560 & 0 & -.0007885 & .001808 & -1 \\
\\
75.139 & 13.840 & 181.43 & 98.173 & 0
\end{bmatrix}
\begin{bmatrix}
(cc_l{}^e)_1 \\
(cc_l{}^e)_2 \\
(cc_l{}^e)_3 \\
(cc_l{}^e)_4 \\
\\
\alpha^e(0)
\end{bmatrix}
$$

$$
=
\begin{bmatrix}
.015639 \\
.013090 \\
.00700 \\
0 \\
\\
0
\end{bmatrix}. \qquad \text{(k)}
$$

Solving Eq. (k) by inversion,

$$
\begin{bmatrix}
\{cc_l{}^e\} \\
\\
\alpha^e(0)
\end{bmatrix}
=
\begin{bmatrix}
2.4966 \\
2.4899 \\
-.56894 \\
-4.3808 \\
\\
-.007612
\end{bmatrix}
\begin{bmatrix}
\text{inches} \\
\\
\text{radians}
\end{bmatrix}. \qquad \text{(l)}
$$

The total lift distribution is obtained by adding $\{cc_l{}^e\}$ and $\{cc_l{}^r\}$ and multiplying the result by the dynamic pressure q.

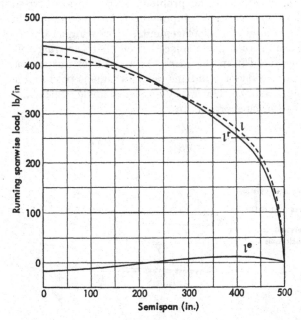

Fig. 8-10. Lift distribution on jet transport wing.

$$q\{cc_l\} = q\{cc_l{}^r\} + q\{cc_l{}^e\} = \begin{bmatrix} 186.9234 \\ 294.9414 \\ 382.7422 \\ 440.5431 \end{bmatrix} + \begin{bmatrix} 8.9796 \\ 8.9555 \\ -2.0463 \\ -15.7565 \end{bmatrix} = \begin{bmatrix} 195.9030 \\ 303.8969 \\ 380.6959 \\ 424.7866 \end{bmatrix} \frac{lb}{in}. \quad (m)$$

The results are plotted in Fig. 8–10, which illustrates the elastic, rigid, and total lift distributions. The outboard shift of the spanwise center of pressure resulting from wing elasticity is evident by comparing the total lift distribution l and the rigid lift distribution l_r. It is apparent from these results that the effect of wing elasticity on span load distribution is small in this airplane. The primary reason for this small effect is the small offset distance between aerodynamic center and elastic axis in this particular wing ($e/c = 0.1$). The role played by the aerodynamic moments about the aerodynamic center and by the inertial forces should also be observed. The moments about the elastic axis due both to aerodynamic moments about the aerodynamic center and to inertial forces are such as to twist the wing nose down. Thus they tend to reduce wing twist and hence reduce the effect of elasticity on span load distribution.

The angle $\alpha^e(0)$ represents physically a nose-down pitching of the fuselage of $-.007612$ radian or $-.436$ degree.

(d) *Antisymmetrical lift distribution.* In Section 8–3(a), equilibrium equations of a straight wing were derived which are applicable to general unsymmetrical loading conditions. It was stated that such conditions, because of the linearity of the problem, could be synthesized by symmetrical and antisymmetrical components. Antisymmetrical loadings result primarily from aileron displacements. Referring to Fig. 8–11, let us assume that the ailerons are displaced such that the wing rolls counterclockwise when faced from the rear. As a result of this aileron displacement, a rolling velocity of p radians per unit time is produced and a twist distribution $\theta(y)$ results. The antisymmetrical angle of attack distribution along the wing can be expressed by

$$\alpha(y) = \frac{\partial \alpha^r}{\partial \beta} \beta - \frac{py}{U} + \theta(y), \quad (8\text{–}90)$$

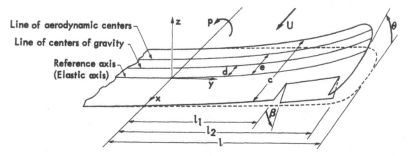

Fig. 8–11. Elastic straight wing with displaced aileron.

where

$\partial\alpha^r/\partial\beta$ = aerodynamic twist of rigid wing resulting from a unit aileron
displacement β,

py/U = induced angle of attack due to rolling velocity p.

(1) *Equations of equilibrium.* Referring to the general differential and integral equation forms given by Eqs. (8–27) and (8–28), we see that the following quantities can be introduced:

$$c_{mAC}(y) = \frac{\partial c_{mAC}}{\partial\beta}\beta, \tag{8-91a}$$

$$c_l^r(y) = \frac{\partial c_l^r}{\partial\beta}\beta + \frac{\partial c_l^r}{\partial(pl/U)}\left(\frac{pl}{U}\right), \tag{8-91b}$$

$$N(y) = \frac{y\dot{p}}{g}. \tag{8-91c}$$

Substituting these quantities into Eq. (8–27), we obtain the following differential equation applicable to the problem of antisymmetric loading:

$$\frac{d}{dy}\left[GJ\frac{d\theta}{dy}\right] + qecc_l^e$$

$$= -q\left(ec\frac{\partial c_l^r}{\partial\beta} + \frac{\partial c_{mAC}}{\partial\beta}c^2\right)\beta - qec\frac{\partial c_l^r}{\partial(pl/U)}\left(\frac{pl}{U}\right) + m\dot{p}yd. \tag{8-92}$$

Similarly, by substituting Eqs. (8–91) into Eq. (8–28), we get the appropriate integral equation, as follows:

$$\theta(y) = q\int_0^l C^{\theta\theta}(y,\eta)ecc_l^e d\eta + f_a(y), \tag{8-93}$$

where

$$f_a(y) = \int_0^l C^{\theta\theta}(y,\eta)\left[qec\left(\frac{\partial c_l^r}{\partial\beta}\beta + \frac{\partial c_l^r}{\partial(pl/U)}\left(\frac{pl}{U}\right)\right) + qc^2\frac{\partial c_{mAC}}{\partial\beta}\beta - mn\dot{p}d\right]d\eta. \tag{8-93a}$$

In the differential and integral equations given above, all terms except the twist distribution $\theta(y)$ and the elastic lift distribution cc_l^e are regarded as known quantities defined by the physical properties of the wing and the conditions of the problem. Either the differential or the integral equation can be combined with an appropriate linear relation between incidence and lift distribution to compute the antisymmetrical lift distribution.

(2) *Symbolic form of solution.* The antisymmetrical twist distribution which results from solution of either Eq. (8–92) or Eq. (8–93) is a linear function of the aileron angle β, the wing tip helix angle pl/U, and the

rolling acceleration $\dot{p}$. Let us assume for purpose of discussion that the antisymmetrical twist distribution can be expressed by

$$\theta(y) = C_1(y)\beta + C_2(y)\frac{pl}{U} + C_3(y)\dot{p}, \qquad (8\text{-}94)$$

where the functions $C_1(y)$, $C_2(y)$, and $C_3(y)$ represent twist distributions per unit values of β, pl/U, and $\dot{p}$, respectively. Each of these twist functions satisfies the boundary condition $\theta(0) = \theta'(l) = 0$. The corrective lift distribution resulting from elastic twist derives from Eq. (8-94). It can can be expressed for the purpose of discussion in the following symbolic form:*

$$cc_l{}^e = \mathcal{P}[\theta(y)] = \mathcal{P}[C_1(y)]\beta + \mathcal{P}[C_2(y)]\frac{pl}{U} + \mathcal{P}[C_3(y)]\dot{p}. \quad (8\text{-}95)$$

(3) *Lift distribution due to sudden aileron displacement.* A common condition arising in aircraft structural load analysis is one in which the ailerons are given a sudden displacement. The object is to compute the rolling acceleration and the antisymmetrical lift distribution at the outset of the motion when the rolling velocity p is small and can be assumed zero. The equation of equilibrium of rolling moments, assuming $p = 0$, is

$$I_x\dot{p} = 2q\int_0^l cc_l y\,dy = 2q\beta\int_0^l c\frac{\partial c_l{}^r}{\partial\beta}\,y\,dy + 2q\int_0^l cc_l{}^e y\,dy. \quad (8\text{-}97)$$

Substituting Eq. (8-95) and solving for $\dot{p}$ yields

$$\dot{p} = \beta\frac{\displaystyle\int_0^l \left(c\frac{\partial c_l{}^r}{\partial\beta} + \mathcal{P}[C_1(y)]\right)y\,dy}{\displaystyle\frac{I_x}{2q} - \int_0^l \mathcal{P}[C_3(y)]y\,dy}. \qquad (8\text{-}98)$$

If the value of $\dot{p}$ thus computed is substituted into Eq. (8-95), together with $p = 0$ and the prescribed value of β, the corrective lift distribution is obtained. This, together with the rigid lift distribution produced by the aileron, gives the total antisymmetrical lift distribution. An analysis such as this does not, of course, account for transient vibrations of the wing or yawing motions of the airplane, which may have secondary effects upon the results.

(4) *Lift distribution during steady roll.* Span load distributions while the airplane is in a steady roll at a prescribed value of pl/U are often desired. The equilibrium condition for steady rolling motion is a condition

---

* The symbol $\mathcal{P}$ is employed to denote a linear aerodynamic operator which can be applied to the incidence distribution to produce the corresponding lift distribution as follows:

$$cc_l = \mathcal{P}[\alpha(y)] \qquad (8\text{-}96)$$

The operation $\mathcal{P}$ may be regarded as reciprocal to the operation $\mathcal{Q}$ defined by Eq. (8-29).

in which the rolling moments due to aileron displacement, wing twist, and damping are in equilibrium:

$$\int_0^l cc_l y\,dy = \int_0^l cc_l{}^r y\,dy + \int_0^l cc_l{}^o y\,dy = 0. \tag{8-99}$$

Substituting Eqs. (8-91b) and (8-95) into Eq. (8-99), putting $\dot{p} = 0$, and solving for β gives

$$\beta = -\frac{pl}{U}\frac{\displaystyle\int_0^l \left(c\frac{\partial c_l{}^r}{\partial(pl/U)} + \mathscr{O}[C_2(y)]\right)y\,dy}{\displaystyle\int_0^l \left(c\frac{\partial c_l{}^r}{\partial\beta} + \mathscr{O}[C_1(y)]\right)y\,dy}. \tag{8-100}$$

If pl/U is prescribed, β can be computed by Eq. (8-100). Thus the corrective lift distribution can be obtained from Eq. (8-95) by introducing the prescribed value of pl/U, the value of β given by Eq. (8-100), and $\dot{p} = 0$.

(5) *Aileron effectiveness.* The effectiveness of lateral control devices is often measured in terms of wing-tip helix angle per unit aileron displacement. This quantity, called the aileron effectiveness, is obtained by taking the reciprocal of Eq. (8-100):

$$\frac{\partial\left(\dfrac{pl}{U}\right)}{\partial\beta} = -\frac{\displaystyle\int_0^l \left(c\frac{\partial c_l{}^r}{\partial\beta} + \mathscr{O}[C_1(y)]\right)y\,dy}{\displaystyle\int_0^l \left(c\frac{\partial c_l{}^r}{\partial(pl/U)} + \mathscr{O}[C_2(y)]\right)y\,dy}. \tag{8-101}$$

The aileron effectiveness is profoundly affected by wing twist and hence by airplane speed. The speed at which the wing twist causes the aileron effectiveness to be zero, called the aileron reversal speed, can be obtained by putting the numerator of Eq. (8-101) equal to zero. It is sometimes convenient to express Eq. (8-101) in terms of rolling moment coefficients of the rigid airplane, as follows:

$$\frac{\partial\left(\dfrac{pl}{U}\right)}{\partial\beta} = -\frac{C_{l_\beta} + \dfrac{1}{Sl}\displaystyle\int_0^l \mathscr{O}[C_1(y)]\,y\,dy}{C_{l_p} + \dfrac{1}{Sl}\displaystyle\int_0^l \mathscr{O}[C_2(y)]\,y\,dy}, \tag{8-101a}$$

where

$C_{l_\beta} = \dfrac{1}{Sl}\displaystyle\int_0^l c\frac{\partial c_l{}^r}{\partial\beta}y\,dy$ = derivative of rolling moment coefficient of the rigid airplane with respect to aileron deflection β,

$C_{l_p} = \dfrac{1}{Sl}\displaystyle\int_0^l c\frac{\partial c_l{}^r}{\partial(pl/U)}y\,dy$ = derivative of rolling moment coefficient of the rigid airplane with respect to wing tip helix angle pl/U.

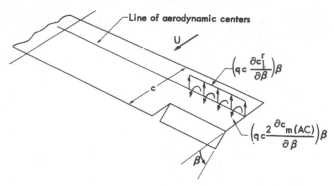

Fig. 8–12.　Forces and moments due to displaced aileron.

(6) *Closed-form solutions of uniform wing based on strip theory.* Prior to taking up more complex cases, it is instructive to consider first the simple behavior of a uniform wing where closed-form solutions can be obtained. When the aileron is displaced, forces and moments on the rigid wing are assumed to act over the portion of the span covered by the aileron, as illustrated by Fig. 8–12. As a result of these forces and moments, the wing commences to roll and twist. Figure 8–13 illustrates the assumed distribution of aerodynamic damping forces and inertial forces. The elastic twist is obtained by solving the following differential equation derived from Eq. (8–92):

$$\frac{d^2\theta}{dy^2} + \lambda^2\theta = \left(\lambda^2 \frac{p}{U} + k_1 \dot{p}\right) y - k_2 \lambda^2 \beta 1_a(y), \qquad (8\text{–}102)$$

where

$$k_1 = \frac{md}{GJ}, \qquad k_2 = \frac{1}{ea_0}\left(e \frac{\partial c_l^r}{\partial \beta} + c \frac{\partial c_{mAC}}{\partial \beta}\right), \qquad \lambda^2 = \frac{qcea_0}{GJ}.$$

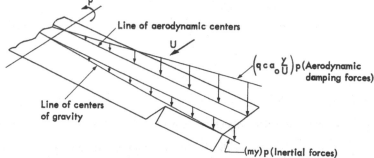

Fig. 8–13.　Damping and inertial forces on rolling wing.

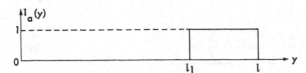

Fig. 8–14. Step function used in representing aileron forces and moments.

The function $1_a(y)$ has a value of unity over the aileron span, and is zero over the remainder of the wing span, as shown by Fig. 8–14.

It can be easily verified that a closed solution to Eq. (8–102) satisfying the boundary conditions $\theta(0) = \theta'(l) = 0$ is

$$\theta(y) = \frac{1}{\lambda^2}\left(\lambda^2\frac{p}{U} + k_1\dot{p}\right)\left(y - \frac{\sin\lambda y}{\lambda\cos\lambda l}\right)$$

$$- k_2\left[1_a(y)(1 - \cos\lambda(y - l_1)) - \frac{\sin\lambda(l - l_1)}{\cos\lambda l}\sin\lambda y\right]\beta, \quad (8\text{–}103)$$

where $\cos\lambda(y - l_1) = 0$, when $y < l_1$. Hence, in the notation of Eq. (8–94), we have for a uniform wing

$$C_1(y) = -k_2\left[1_a(y)(1 - \cos\lambda(y - l_1)) - \frac{\sin\lambda(l - l_1)}{\cos\lambda l}\sin\lambda y\right], \quad (8\text{–}104)$$

$$C_2(y) = \left(\frac{y}{l} - \frac{\sin\lambda y}{\lambda l\cos\lambda l}\right), \qquad (8\text{–}105)$$

$$C_3(y) = \frac{k_1}{\lambda^2}\left(y - \frac{\sin\lambda y}{\lambda\cos\lambda l}\right). \qquad (8\text{–}106)$$

The antisymmetric lift distribution due to elastic twist follows from Eq. (8–95). For the case of strip theory applied to a uniform wing, Eq. (8–95) becomes simply

$$cc_l{}^e(y) = a_0cC_1(y)\beta + a_0cC_2(y)\frac{pl}{U} + a_0cC_3(y)\dot{p}, \qquad (8\text{–}107)$$

where $C_1(y)$, $C_2(y)$, and $C_3(y)$ are given by Eqs. (8–104), (8–105), and (8–106), respectively. Equation (8–107) can be used to calculate the span load distribution due to elastic twist for any prescribed values of β, pl/U, and $\dot{p}$.

Sudden aileron displacement. Suppose, for example, that we wish to compute the lift distribution on a uniform wing as a result of a sudden aileron deflection β. The resulting acceleration at the outset of the motion before appreciable rolling velocity has been developed is obtained from

Eq. (8–98):

$$\dot{p} = \beta \frac{\int_0^l \left(\frac{\partial c_l^r}{\partial \beta} + a_0 C_1(y)\right) cy\,dy}{\frac{I_x}{2q} - a_0 \int_0^l C_3(y) cy\,dy}$$

$$= \beta \frac{\dfrac{\partial c_l^r}{\partial \beta}\left(\dfrac{\cos \lambda l_1}{\cos \lambda l} - 1\right) + \dfrac{c}{e}\dfrac{\partial c_{mAC}}{\partial \beta}\left(\dfrac{\cos \lambda l_1}{\cos \lambda l} - 1 - \lambda^2\dfrac{l^2 - l_1{}^2}{2}\right)}{\dfrac{ea_0}{2GJ}\left[I_x - 2m\dfrac{d}{e}\left(\dfrac{l^3}{3} + \dfrac{l}{\lambda^2} - \dfrac{\tan \lambda l}{\lambda^3}\right)\right]}. \qquad (8\text{–}108)$$

Inserting the value of $\dot{p}$ given above, together with the prescribed value of β, into Eq. (8–107), and putting $p = 0$ gives the lift distribution due to elastic twist. The total lift distribution is obtained by superposing this result and the rigid lift distribution, as follows:

$$cc_l(y) = cc_l^r(y) + cc_l^e(y) = c\left(\frac{\partial c_l^r}{\partial \beta} 1_a(y)\right)\beta + cc_l^e(y). \qquad (8\text{–}109)$$

Aileron effectiveness. The aileron effectiveness is obtained by substituting Eqs. (8–104) and (8–105) into Eq. (8–101) and applying strip theory:

$$\frac{\partial \left(\dfrac{pl}{U}\right)}{\partial \beta} = \frac{\int_0^l \left(\dfrac{\partial c_l^r}{\partial \beta} + a_0 C_1(y)\right) cy\,dy}{a_0 \int_0^l \left(\dfrac{y}{l} - C_2(y)\right) cy\,dy}$$

$$= \frac{\left(\dfrac{\cos \lambda l_1}{\cos \lambda l} - 1\right)\dfrac{1}{a_0}\dfrac{\partial c_l^r}{\partial \beta} + \left(\dfrac{\cos \lambda l_1}{\cos \lambda l} - 1 - \lambda^2\dfrac{l^2 - l_1{}^2}{2}\right)\dfrac{c}{ea_0}\dfrac{\partial c_{mAC}}{\partial \beta}}{\left(\dfrac{\tan \lambda l}{\lambda l} - 1\right)}.$$

$$(8\text{–}110)$$

The aileron effectiveness as defined by Eq. (8–110) is a function of the parameters λl, $(c/ea_0)(\partial c_{mAC}/\partial \beta)$, $(1/a_0)(\partial c_l^r/\partial \beta)$, and l_1/l. Figure 8–15 is a plot of Eq. (8–110), which illustrates the dependence of aileron effectiveness on λl when numerical values are assigned to the latter three parameters. The aileron effectiveness at $\lambda l = 0$ is the maximum that can be obtained, and it corresponds to a rigid wing:

$$\lim_{\lambda l \to 0} \frac{\partial (pl/U)}{\partial \beta} = \frac{3}{2}\left[1 - \frac{l_1{}^2}{l^2}\right]\frac{1}{a_0}\frac{\partial c_l^r}{\partial \beta}. \qquad (8\text{–}111)$$

As λl increases, the absolute value of the negative rolling moment due to

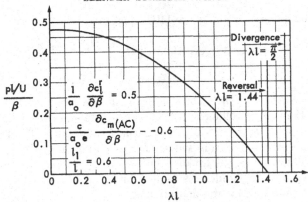

Fig. 8–15.   Aileron effectiveness of uniform wing as a function of $\lambda l = l \sqrt{\dfrac{qcea_0}{GJ}}$.

elastic twist approaches the absolute value of the positive rolling moment due to aileron deflection until the reversal speed is reached, where they cancel each other entirely.   Figure 8–16 indicates the manner in which the three rolling moments due to aileron deflection, elastic twist, and aerodynamic damping vary with λl up to the reversal speed.   The value of λl at the reversal speed, $\lambda l = 1.44$, for the particular uniform wing under discussion, is indicated by Figs. 8–15 and 8–16.   This value can be computed independently by solving the transcendental equation obtained by

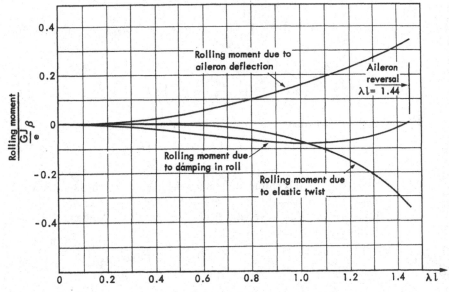

Fig. 8–16.   Synthesis of rolling moments on uniform wing.

putting the numerator in Eq. (8–110) equal to zero:

$$\left(\frac{\partial c_l{}^r}{\partial \beta} + \frac{c}{e}\frac{\partial c_{mAC}}{\partial \beta}\right)(\cos \lambda l - \cos \lambda l_1) + \left(\lambda^2 \frac{l^2 - l_1{}^2}{2}\frac{c}{e}\frac{\partial c_{mAC}}{\partial \beta}\right)\cos \lambda l = 0.$$

(8–112)

Steady rolling motion. The corrective lift distribution due to elastic twist for the steady rolling condition can be obtained from Eq. (8–107) by putting $\dot{p} = 0$ and introducing a specified value of β. The helix angle, pl/U, corresponding to this value of β is obtained from Eq. (8–110). Figure 8–17 illustrates the lift distribution due to elastic twist on the particular uniform wing under discussion during a steady rolling condition resulting from a unit aileron deflection β.

It should be observed that this analysis assumes the reversal speed to be less than the divergence speed, a condition usually obtained with straight wings. It can be seen from Eq. (8–112) that the two speeds are identical when $\partial c_l{}^r/\partial \beta = -(c/e)(\partial c_{mAC}/\partial \beta)$. This condition is, of course, similar to the condition previously stated by Eq. (8–21) for the case of a two-dimensional rigid wing segment.

(7) *Solution by matrices.* Let us consider next the analytical processes to be applied to antisymmetrical lift distribution problems involving wings with irregular spanwise properties. As in the case of symmetrical loading, we proceed by using the integral equation as a basis. The appropriate integral equation is given by Eq. (8–93). Introducing the aerodynamic operator of Eq. (8–29) into Eq. (8–93), we obtain

$$\mathfrak{A}[cc_l{}^e] = q\int_0^l C^{\theta\theta}(y,\eta)ecc_l{}^e d\eta + f_a(y),$$

(8–113)

where

$$f_a(y) = \int_0^l C^{\theta\theta}(y,\eta)\left[qec\left(\frac{\partial c_l{}^r}{\partial \beta}\beta + \frac{\partial c_l{}^r}{\partial(pl/U)}\left(\frac{pl}{U}\right)\right) + qc^2\frac{\partial c_{mAC}}{\partial \beta}\beta - m\eta\dot{p}d\right]d\eta.$$

Equation (8–113) transforms to the following matrix form:

$$[A^a]\{cc_l{}^e\} = q[E]\{cc_l{}^e\} + \{f_a\},$$

(8–114)

where

$[A^a]$ = matrix of aerodynamic influence coefficients for antisymmetrical loading on a straight wing (derivable from Eq. (5–157) by using even rather than odd values of r, and putting midspan station terms equal to zero),

$$\{f_a\} = q[E]\left(\left\{c\frac{\partial c_l{}^r}{\partial \beta}\right\}\beta + \left\{c\frac{\partial c_l{}^r}{\partial(pl/U)}\right\}\frac{pl}{U}\right) + q[F]\left\{\frac{\partial c_{mAC}}{\partial \beta}\right\}\beta - [G]\lceil y\rfloor\{m\}\dot{p},$$

$$[E] = [C^{\theta\theta}]\lceil e\rfloor\lceil \overline{W}\rfloor,$$

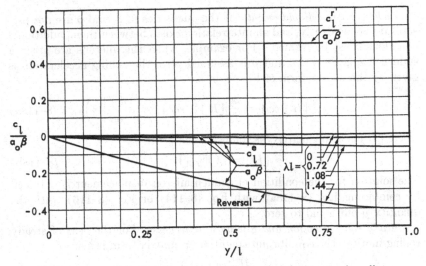

Fig. 8–17.　Lift distribution on uniform wing during steady roll.

$$[F] = [C^{\theta\theta}][c^2][\overline{W}],$$

$$[G] = [C^{\theta\theta}][d][\overline{W}].$$

An alternative form of Eq. (8–114), in terms of the total antisymmetrical lift distribution, can be derived by substituting

$$\{cc_l{}^e\} = \{cc_l\} - \{cc_l{}^r\} = \{cc_l\} - \left\{c\frac{\partial c_l{}^r}{\partial\beta}\right\}\beta - \left\{c\frac{\partial c_l{}^r}{\partial(pl/U)}\right\}\frac{pl}{U} \quad (8\text{–}115)$$

into Eq. (8–114). This substitution gives the result

$$[A^a]\{cc_l\} = q[E]\{cc_l\} + \{f_a{}'\}, \qquad (8\text{–}116)$$

where

$$\{f_a{}'\} = \left\{\frac{\partial\alpha^r}{\partial\beta}\right\}\beta + \left\{\frac{y}{l}\right\}\frac{pl}{U} + q[F]\left\{\frac{\partial c_{mAC}}{\partial\beta}\right\}\beta - [G][y]\{m\}\dot{p},$$

$$\left\{\frac{\partial\alpha^r}{\partial\beta}\right\} = [A^a]\left\{c\frac{\partial c_l{}^r}{\partial\beta}\right\}, \qquad \left\{\frac{y}{l}\right\} = [A^a]\left\{c\frac{\partial c_l{}^r}{\partial(pl/U)}\right\}.$$

The quantities β, p, and $\dot{p}$ are regarded as known quantities and must be specified by the conditions of the problem. Equations (8–114) and (8–116) can be solved by matrix inversion or iteration. For example, by the former method, the column matrix of corrective lift coefficients from Eq. (8–114) becomes

$$\{cc_l{}^e\} = ([A^a] - q[E])^{-1}\{f_a\}. \qquad (8\text{–}117)$$

Sudden aileron displacement. If the quantities β, p, and $\dot{p}$ are not prescribed independently, and an interrelation exists between them, additional equations must be formed. For example, when the ailerons are given a sudden displacement β, and p is assumed zero, the rolling acceleration is given by the matrix form of Eq. (8–97):

$$\dot{p} = \frac{2q}{I_x} \lfloor H \rfloor \{cc_l\} = \frac{2q}{I_x} \lfloor H \rfloor \left(\{cc_l{}^e\} + \left\{ c\, \frac{\partial c_l{}^r}{\partial \beta} \right\} \beta \right), \qquad (8\text{–}118\text{a})$$

where $\lfloor H \rfloor$ is a row matrix, as follows:

$$\lfloor H \rfloor = \lfloor 1 \rfloor [y] [\overline{W}]. \qquad (8\text{–}118\text{b})$$

The solution for the condition of sudden aileron displacement is obtained by combining Eq. (8–118a) with Eq. (8–114) or Eq. (8–116), with the quantity p put equal to zero.

Steady rolling motion and aileron effectiveness. For the case of steady rolling motion, the equilibrium condition in matrix form is simply

$$\lfloor H \rfloor \{cc_l\} = 0. \qquad (8\text{–}119)$$

Combining this result with Eq. (8–116), there is obtained the following result for aileron effectiveness:

$$\frac{\partial (pl/U)}{\partial \beta} = - \frac{\lfloor H \rfloor ([A^a] - q[E])^{-1}(\{\partial \alpha^r/\partial \beta\} + q[F]\{\partial c_{mAC}/\partial \beta\})}{\lfloor H \rfloor ([A^a] - q[E])^{-1}\{y/l\}}. \qquad (8\text{–}120)$$

The lift distribution under steady rolling conditions can be computed from Eqs. (8–114) or (8–116) by putting $\dot{p} = 0$. pl/U and β cannot both be specified independently, since they are related through Eq. (8–120). It is common practice in design specifications to specify pl/U. In such cases, the appropriate value of β can be computed by Eq. (8–120).

Aileron reversal. Aileron reversal occurs at the speed where the aileron effectiveness is zero. This condition is obtained by allowing the numerator in Eq. (8–120) to vanish. Thus the aileron reversal condition can be put in the form

$$q_R = - \frac{\lfloor H \rfloor ([A^a] - q_R[E])^{-1}\{\partial \alpha^r/\partial \beta\}}{\lfloor H \rfloor ([A^a] - q_R[E])^{-1}[F]\{\partial c_{mAC}/\partial \beta\}}. \qquad (8\text{–}120\text{a})$$

To solve Eq. (8–120a) for the reversal dynamic pressure, a useful iteration process has been developed by Lawrence and Sears (Ref. 8–11). A value of q is assumed and the right side of Eq. (8–120a) is evaluated. The result is a new value for q. The new value is inserted into the right side and the process is continued until agreement between two successive values of q is obtained. It can be observed that the matrix multiplication in the numerator should be carried out from left to right, so that each step involves multiplication by a row matrix. The resulting row matrix, before multiplication by $\{\partial \alpha^r/\partial \beta\}$, can be used to evaluate the denominator as well. Thus the numerical calculations are not excessively tedious.

Alternative forms of aileron effectiveness and reversal expressions. Equations (8–120) and (8–120a) can be expressed in alternative forms in which the rolling moment coefficients of the rigid airplane appear explicitly. These forms can be obtained by a transformation of Eqs. (8–120) and (8–120a), or more simply by writing Eq. (8–119) as follows:

$$\lfloor H \rfloor \{cc_l{}^e\} + \lfloor H \rfloor \{cc_l{}^r\} = 0, \tag{8-121}$$

and substituting appropriate results for the elastic and rigid lift distributions during steady rolling motion. We obtain by this means the following expressions for aileron effectiveness and reversal:

$$\frac{\partial (pl/U)}{\partial \beta} = -\frac{C_{l_\beta} + \dfrac{q}{Sl} \lfloor H \rfloor ([A^a] - q[E])^{-1} \left([E] \left\{ c\, \dfrac{\partial c_l{}^r}{\partial \beta} \right\} + [F] \left\{ \dfrac{\partial c_{mAC}}{\partial \beta} \right\} \right)}{C_{l_p} + \dfrac{q}{Sl} \lfloor H \rfloor ([A^a] - q[E])^{-1} [E] \left\{ c\, \dfrac{\partial c_l{}^r}{\partial (pl/U)} \right\}}, \tag{8-122}$$

$$q_R = -\frac{C_{l_\beta}}{\dfrac{1}{Sl} \lfloor H \rfloor ([A^a] - q_R[E])^{-1} \left([E] \left\{ c\, \dfrac{\partial c_l{}^r}{\partial \beta} \right\} + [F] \left\{ \dfrac{\partial c_{mAC}}{\partial \beta} \right\} \right)}, \tag{8-122a}$$

where

$$C_{l_\beta} = \frac{1}{Sl} \lfloor H \rfloor \left\{ c\, \frac{\partial c_l{}^r}{\partial \beta} \right\},$$

$$C_{l_p} = \frac{1}{Sl} \lfloor H \rfloor \left\{ c\, \frac{\partial c_l{}^r}{\partial (pl/U)} \right\}.$$

The reader should observe that the quantities $\left([E] \left\{ c\, \dfrac{\partial c_l{}^r}{\partial \beta} \right\} + [F] \left\{ \dfrac{\partial c_{mAC}}{\partial \beta} \right\} \right)$ and $[E] \left\{ c\, \dfrac{\partial c_l{}^r}{\partial (pl/U)} \right\}$ are merely column matrices of wing twist resulting from the aerodynamic forces produced by unit values of β and pl/U, respectively.

When assured numerical results are desired in antisymmetrical lift distribution problems, it has been found necessary to include spanwise aerodynamic effects. Some evidence indicates that strip theory can be used to obtain the corrective lift distribution due to elastic twist, providing the basic rigid lift distribution due to aileron displacement is obtained by taking proper account of spanwise aerodynamic effects. The quantities $\partial c_l{}^r/\partial \beta$ or $\partial \alpha^r/\partial \beta$ and $\partial c_{mAC}/\partial \beta$ have a most important effect on the antisymmetrical lift distribution. In general, they will vary primarily with flap-chord ratio and Mach number and secondarily with other variables such as nose shape, gap, aileron section, and angle of attack. For accurate data on these aerodynamic coefficients and their spanwise distribution, it is necessary to employ wind-tunnel tests on the specific configuration being investigated.

(8) *Approximate solutions of the integral equation.* Solutions to the integral equation (8–93) based upon simplifying assumptions have found wide application in computing aileron effectiveness and reversal characteristics of straight wings. For the case of steady roll, Eq. (8–93) can be written in the following form:

$$\theta(y) = q \int_0^l C^{\theta\theta}(y,\eta) e c c_i{}^e d\eta + q\beta \int_0^l C^{\theta\theta}(y,\eta) \left(e \frac{\partial c_l{}^r}{\partial \beta} + c \frac{\partial c_{mAC}}{\partial \beta} \right) c d\eta$$

$$+ q\left(\frac{pl}{U}\right) \int_0^l C^{\theta\theta}(y,\eta) \frac{\partial c_l{}^r}{\partial (pl/U)} c e d\eta. \qquad (8\text{–}123)$$

A simplifying assumption often made in the solution of Eq. (8–123) to obtain rolling effectiveness is that only the twist due to aileron deflection need be considered (Refs. 8–14 and 8–15). This is the same as assuming a coincidence of the line of aerodynamic centers and the elastic axis. In Eq. (8–123), it is equivalent to putting $e = 0$. Thus Eq. (8–123) reduces immediately to an explicit result for wing twist per unit aileron deflection.

$$\frac{\theta(y)}{\beta} = q \int_0^l C^{\theta\theta}(y,\eta) \frac{\partial c_{mAC}}{\partial \beta} c^2 d\eta. \qquad (8\text{–}124)$$

In terms of the notation introduced in Eq. (8–94), we have

$$C_1(y) = q \int_0^l C^{\theta\theta}(y,\eta) \frac{\partial c_{mAC}}{\partial \beta} c^2 d\eta, \qquad C_2(y) = 0. \qquad (8\text{–}125)$$

Putting the results of Eqs. (8–125) into Eq. (8–101a), we obtain the following approximate formula for aileron effectiveness:

$$\frac{\partial \left(\dfrac{pl}{U}\right)}{\partial \beta} = -\frac{C_{l_\beta} + \dfrac{q}{Sl} \displaystyle\int_0^l \mathcal{P}\left[\int_0^l C^{\theta\theta}(y,\eta) \frac{\partial c_{mAC}}{\partial \beta} c^2 d\eta \right] y\, dy}{C_{l_p}}. \qquad (8\text{–}126)$$

The dynamic pressure corresponding to aileron reversal can be computed by putting the numerator in Eq. (8–126) equal to zero.

$$q_R = -\frac{C_{l_\beta}}{\dfrac{1}{Sl} \displaystyle\int_0^l \mathcal{P}\left[\int_0^l C^{\theta\theta}(y,\eta) \frac{\partial c_{mAC}}{\partial \beta} c^2 d\eta \right] y\, dy}. \qquad (8\text{–}127)$$

In Eqs. (8–126) and (8–127) all of the quantities are affected by spanwise induction effects and by Mach number. For precise results on a given configuration, the quantity C_{l_β} can be obtained from high-speed wind-tunnel tests. The quantity C_{l_p} can be obtained from computation using lifting-line theory (Ref. 8–14) or from low-speed wind-tunnel tests. The integral quantity in Eqs. (8–126) and (8–127) must usually be obtained by computation. Reference 8–14 gives curves and charts useful in estimating values of the aerodynamic quantities which appear in Eqs. (8–126) and (8–127).

If strip theory is applied, and if the coefficients $\partial c_l{}^r/\partial\beta$ and $\partial c_{mAC}/\partial\beta$ are assumed finite only over the portion of the span covered by the aileron, Eqs. (8–126) and (8–127) reduce to

$$\frac{\partial\left(\dfrac{pl}{U}\right)}{\partial\beta} = -\frac{C_{l_\beta} + \dfrac{q}{Sl}\displaystyle\int_0^l a_0 \int_{l_1}^{l_2} C^{\theta\theta}(y,\eta)\,\frac{\partial c_{mAC}}{\partial\beta}\,c^2 d\eta c y dy}{C_{l_p}} \qquad (8\text{–}128)$$

and

$$q_R = -\frac{C_{l_\beta}}{\dfrac{1}{Sl}\displaystyle\int_0^l a_0 \int_{l_1}^{l_2} C^{\theta\theta}(y,\eta)\,\frac{\partial c_{mAC}}{\partial\beta}\,c^2 d\eta c y dy}, \qquad (8\text{–}129)$$

where

$$C_{l_\beta} = \frac{1}{Sl}\int_{l_1}^{l_2} a_0\,\frac{\partial\alpha^r}{\partial\beta}\,c y dy,$$

$$C_{l_p} = -\frac{1}{Sl}\int_0^l a_0\,\frac{y^2}{l}\,c dy.$$

Equations (8–128) and (8–129) can be combined to yield the following simple form applicable in the incompressible range:

$$\frac{\partial(pl/U)}{\partial\beta} = \left(\frac{\partial(pl/U)}{\partial\beta}\right)_{\text{rigid}}\left(1 - \frac{q}{q_R}\right), \qquad (8\text{–}130)$$

where

$$\left(\frac{\partial(pl/U)}{\partial\beta}\right)_{\text{rigid}} = -\frac{C_{l_\beta}}{C_{l_p}}.$$

In Eq. (8–130), the quantity $[1 - (q/q_R)]$ acts as an attenuation factor on the rolling effectiveness. The magnitude of this factor depends upon the reversal speed. The latter therefore plays a useful role as a reference value in estimating rolling performance.

EXAMPLE 8–4. To calculate the aileron effectiveness of the jet transport introduced in Example 2–1 as a function of Mach number. In particular, to obtain its aileron effectiveness at the limit dive speed, and its aileron reversal speed.

Solution. The following data are required:

Limit dive speed = 553 mi/hr at sea level.
Mach number = 0.728.
Dynamic pressure at limit dive speed = 782.8 lb/ft².
$\partial\alpha^r/\partial\beta = 0.40$, $\partial c_{mAC}/\partial\beta = -0.45$.
Wing semispan = 41.66 ft.
Aspect ratio = 6.15.
Distance to inner end of aileron, $l_1 = 30.83$ ft.
Distance to outer end of aileron, $l_2 = 40.58$ ft.

Equation (8–128) is applied for calculating the aileron effectiveness. If a Mach number correction according to $1/\sqrt{1 - M^2}$ is applied to each aerodynamic derivative, the form of Eq. (8–128) corrected for Mach number reduces to

$$\frac{\partial (pl/U)}{\partial \beta} = - \frac{C_{l_\beta} + \dfrac{q}{Sl\sqrt{1 - M^2}} \displaystyle\int_0^l a_0 \int_{l_1}^{l_2} C^{\theta\theta}(y,\eta) \frac{\partial c_{mAC}}{\partial \beta} c^2 d\eta cydy}{C_{l_p}}. \tag{a}$$

Using the planform geometry illustrated in Fig. 8–18(a), the following integrals can be easily evaluated:

$$C_{l_\beta} = \frac{a_0}{Sl} \frac{\partial \alpha^r}{\partial \beta} \int_{l_1}^{l_2} cydy = 1328.04 \left(\frac{a_0}{Sl}\right), \tag{b}$$

$$C_{l_p} = - \frac{a_0}{Sl} \int_0^l \frac{y^2}{l} cdy = -6655.4 \left(\frac{a_0}{Sl}\right). \tag{c}$$

The double integral in Eq. (a) can be evaluated in terms of weighting matrices. The wing is divided into five stations as illustrated by Fig. 8–18(a), and the trapezoidal rule is employed.

$$\left(\frac{a_0}{Sl}\right)\left(\frac{\partial c_{mAC}}{\partial \beta}\right) \int_0^l \int_{l_1}^{l_2} C^{\theta\theta}(y,\eta) c^2 d\eta cydy = \left(\frac{a_0}{Sl}\right)\left(\frac{\partial c_{mAC}}{\partial \beta}\right) \lfloor f \rfloor [\overline{W}]\{cy\}$$

$$= -.80913 \left(\frac{a_0}{Sl}\right), \tag{d}$$

where

$$\{f\} = [C^{\theta\theta}][\overline{W}]\{c^2\} = \begin{bmatrix} 0.5255 \\ 1.0834 \\ 2.2075 \\ 3.0178 \\ 3.2605 \end{bmatrix} \times 10^{-4}. \tag{e}$$

The aileron effectiveness as a function of dynamic pressure and Mach number is obtained by substituting Eqs. (b), (c), and (d) into Eq. (a).

$$\frac{\partial (pl/U)}{\partial \beta} = \frac{1328.04 - (q/\sqrt{1 - M^2})(0.80913)}{6655.4}. \tag{f}$$

Equation (f) is plotted in Fig. 8–18(b), where it can be observed that the aileron effectiveness at limit dive speed is approximately 0.061, as compared with a fully effective value of 0.20. Thus at limit dive speed, the ailerons have lost approximately 70% of their effectiveness due to aeroelastic effects. The aileron reversal speed is obtained by putting the numerator of Eq. (a) equal to zero, as follows:

$$\frac{q_R}{\sqrt{1 - M^2}} = - \frac{C_{l_\beta}}{\dfrac{a_0}{Sl} \dfrac{\partial c_{mAC}}{\partial \beta} \displaystyle\int_0^l \int_{l_1}^{l_2} C^{\theta\theta}(y,\eta) c^2 d\eta cydy} = 1641.3 \frac{\text{lb}}{\text{ft}^2}. \tag{g}$$

Solving the equation

$$\frac{q_R}{\sqrt{1 - M_R^2}} = 1641.3 \tag{h}$$

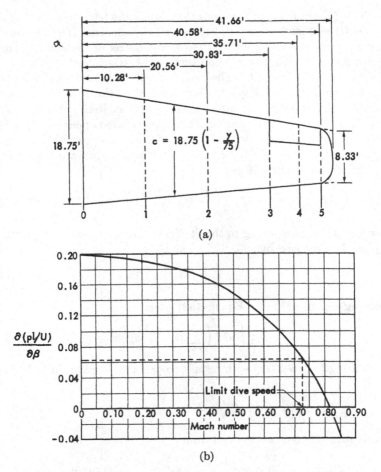

Fig. 8–18.   (a) Planform geometry of jet transport wing.   (b) Aileron effectiveness as a function of Mach number for jet transport.

for airspeed, we obtain after converting to mph units, a reversal speed of $U_R = 615$ mph at sea level.

(e) *Solution of slender unswept wing problems in terms of generalized coordinates.* We have seen in Chapters 2, 3, and 4, in connection with approximate solutions to static and dynamic problems, that the concept of generalized coordinates plays a useful role. Let us consider similar applications to static aeroelastic problems of straight wings. We assume first that the spanwise twist distribution can be expanded in the series

$$\theta(y) = \sum_{j=1}^{n} \Theta_j(y)q_j, \tag{8–131}$$

where the $\Theta_j(y)$ are assumed functions that satisfy the boundary conditions $\Theta(0) = \Theta'(l) = 0$, and the q_j are generalized coordinates. The functions $\Theta_j(y)$ are explicit functions of y inserted at the outset of the solution, and the generalized coordinates q_j are constants to be determined.

(1) *Rayleigh-Ritz method.* The technique of obtaining approximate solutions by applying the Rayleigh-Ritz method to the static aeroelastic problem of straight wings was introduced by Flax in Ref. 8–9. We have seen in Chapter 2 that the Rayleigh-Ritz method is based upon the principle of virtual work. When the deformation of the structure can be described in terms of generalized coordinates, the principle of virtual work assumes the following simple form [cf. Section 2–11(b)]:

$$\frac{\partial U}{\partial q_i} = Q_i. \tag{2-97}$$

For a straight wing, according to the St. Venant torsion theory, the strain energy can be expressed by

$$U = \tfrac{1}{2} \int_0^l GJ(\theta')^2 dy. \tag{8-132}$$

Introducing Eq. (8–131) into Eq. (8–132) gives

$$U = \tfrac{1}{2} \int_0^l GJ \left(\sum_{j=1}^n \Theta_j' q_j \right)^2 dy. \tag{8-132a}$$

Differentiating Eq. (8–132a) with respect to q_i, as required by Eq. (2–97), yields

$$\frac{\partial U}{\partial q_i} = \sum_{j=1}^n \left\{ \int_0^l GJ\Theta_i'\Theta_j' dy \right\} q_j, \qquad (i = 1, 2, \cdots, n). \tag{8-133}$$

The generalized force Q_i is obtained from

$$Q_i = \int_0^l t(y)\Theta_i(y)dy = \int_0^l (qecc_l{}^e + t^*(y))\Theta_i(y)dy,$$
$$(i = 1, 2, \cdots, n), \tag{8-134}$$

where

$$t^*(y) = qecc_l{}^r + qc_{mAC}c^2 - mNgd \tag{8-134a}$$

for symmetrical loading, and

$$t^*(y) = qec \left(\frac{\partial c_l{}^r}{\partial \beta} \beta + \frac{\partial c_l{}^r}{\partial (pl/U)} \left(\frac{pl}{U} \right) \right) + qc^2 \frac{\partial c_{mAC}}{\partial \beta} \beta - m\dot{p}yd \tag{8-134b}$$

for antisymmetrical loading.

It is convenient for purposes of including spanwise aerodynamic effects to put

$$c_l{}^e(y) = \sum_{j=1}^n c_{l_j}(y)q_j, \tag{8-135}$$

where $c_{l_j}(y)$ is the spanwise lift coefficient distribution which would result if the wing were twisted in the spanwise twisting mode $\Theta_j(y)$. Putting Eq. (8–135) into Eq. (8–134) and substituting the result together with Eq. (8–133) into Eq. (2–97) yields

$$\sum_{j=1}^{n} \left\{ \int_0^l GJ\Theta_i'\Theta_j' \, dy - q \int_0^l c_{l_j}\Theta_i ce \, dy \right\} q_j = \int_0^l t^*\Theta_i \, dy,$$

$$(i = 1, 2, \cdots, n). \quad (8\text{–}136)$$

Allowing i and j to take on n values, we obtain from Eq. (8–136) n simultaneous equations in the unknowns $q_1 \cdots q_n$.

$$A_{11}q_1 + A_{12}q_2 + \cdots + A_{1n}q_n = B_1,$$

$$A_{21}q_1 + A_{22}q_2 + \cdots + A_{2n}q_n = B_2,$$

$$\cdot \qquad\qquad (8\text{–}136a)$$

$$\cdot$$

$$\cdot$$

$$A_{n1}q_1 + A_{n2}q_2 + \cdots + A_{nn}q_n = B_n,$$

where

$$A_{ij} = \int_0^l (GJ\Theta_i'\Theta_j' - qc_{l_j}\Theta_i ce) \, dy, \qquad B_i = \int_0^l t^*\Theta_i \, dy.$$

An alternative form of the quantities A_{ij} can be obtained by an integration by parts, as follows:

$$\int_0^l GJ\Theta_i'\Theta_j' \, dy = [GJ\Theta_j'\Theta_i]_0^l - \int_0^l [GJ\Theta_j']'\Theta_i \, dy = -\int_0^l [GJ\Theta_j']'\Theta_i \, dy,$$

$$(8\text{–}137)$$

where we have applied the boundary conditions $\Theta(0) = \Theta'(l) = 0$. Thus A_{ij} becomes

$$A_{ij} = -\int_0^l \{ [GJ\Theta_j']' + qc_{l_j}ce \} \Theta_i \, dy. \qquad (8\text{–}138)$$

A solution of Eqs. (8–136a) yields numerical values of $q_1, \cdots, q_n$ which can be substituted into Eq. (8–131) to obtain the elastic twist distribution, or into Eq. (8–135) to obtain the lift distribution due to elastic twist. It can be seen that the coefficients A_{ij} include the dynamic pressure, which has a specified value in a lift distribution problem. Equations (8–136a) can also be used to calculate the divergence speed by putting $B_i = 0$ and equating the determinant formed by the A_{ij} quantities equal to zero. The dynamic pressure q must, of course, be left as a variable quantity in this case, and a polynomial of the nth degree in q, similar to Eq. (8–56), must be factored to obtain the divergence dynamic pressure.

It should be observed that three-dimensional aerodynamics can be conveniently included in the procedure described above. Computation of the $c_{l_j}(y)$ terms in Eq. (8–135) can be carried out as a separate problem by taking account of spanwise induction effects in any desired manner.

(2) *Galerkin's method.* When Galerkin's method is applied to the differential equation of torsional equilibrium of a straight elastic wing, the result obtained is identical to that derived by the Rayleigh-Ritz method in the previous section. However, if we apply Galerkin's method to the integral equation of equilibrium, we obtain a different result which is somewhat more convenient to apply to practical problems, and which has a more rapid convergence. The advantages of applying Galerkin's method to the integral equation rather than the differential equation, in the present problem, are analogous to the advantages of the modified Rayleigh-Ritz method over the Rayleigh-Ritz method in the vibration problem, as discussed in Chapter 4.

From Eqs. (8–28) and (8–93) we have seen that the integral equation applicable to slender straight wings is

$$\theta(y) = q \int_0^l C^{\theta\theta}(y,\eta)ecc_i^e d\eta + f(y), \qquad (8\text{–}139)$$

where $f(y)$ is given by Eq. (8–28a) for symmetrical loading and by Eq. (8–93a) for antisymmetrical loading. Let us introduce Eqs. (8–131) and (8–135) into Eq. (8–139). This results in an error function $\mathcal{E}(y)$ such that

$$\mathcal{E}(y) = \sum_{j=1}^n \left\{ \theta_j - q \int_0^l C^{\theta\theta}(y,\eta)c_{l_j}ced\eta \right\} q_j - f(y). \qquad (8\text{–}140)$$

According to the Galerkin process, we require $\mathcal{E}(y)$ to be orthogonal to $\Theta_i ce$, $(i = 1, 2, \cdots, n)$, that is, the average value with respect to the weighting function $\Theta_i ce$ is zero.

$$\int_0^l \mathcal{E}(y)\Theta_i cedy = 0, \qquad (i = 1, 2, \cdots, n). \qquad (8\text{–}140a)$$

Thus we obtain from Eq. (8–140)

$$\sum_{j=1}^n \left\{ \int_0^l \Theta_i \Theta_j cedy - q \int_0^l \Theta_i ce \int_0^l C^{\theta\theta}(y,\eta)c_{l_j}ced\eta dy \right\} q_j = \int_0^l f(y)\Theta_i cedy,$$
$$(i = 1, 2, \cdots, n). \qquad (8\text{–}140b)$$

Allowing i and j to take on n values, we obtain n simultaneous equations having a form identical to those of Eq. (8–136a), except that now the coefficients A_{ij} and B_i are defined by

$$A_{ij} = \int_0^l \Theta_i \Theta_j cedy - q \int_0^l \Theta_i ce \int_0^l C^{\theta\theta}(y,\eta)c_{l_j}ced\eta dy, \qquad (8\text{–}141a)$$

$$B_i = \int_0^l f(y)\Theta_i cedy. \qquad (8\text{–}141b)$$

When Eqs. (8–136a) are solved with the coefficients given by Eqs. (8–141a) and (8–141b), values of the generalized coordinates $q_1, \cdots, q_n$ will be obtained, which, when substituted into Eq. (8–131) give the twist distribution, and when substituted into Eq. (8–135) give the lift coefficient distribution.

A particularly simple result can be obtained if only one term of the series in Eq. (8–131) is taken, as follows:

$$\theta(y) = \Theta_D(y)q_1, \qquad (8\text{–}142)$$

where $\Theta_D(y)$ is the twist distribution at the divergence speed. It is defined by a solution to

$$\Theta_D(y) = q_D \int_0^l C^{\theta\theta}(y,\eta)c_{l_D}ce\,d\eta, \qquad (8\text{–}142\text{a})$$

where c_{l_D} is the lift coefficient distribution corresponding to the twist distribution $\Theta_D(y)$. Combining Eqs. (8–141a) and (8–142a), we obtain the following single value of A_{ij},

$$A = \left(1 - \frac{q}{q_D}\right)\int_0^l \Theta_D{}^2ce\,dy. \qquad (8\text{–}142\text{b})$$

A single value of B_i is obtained when we put $\Theta_i = \Theta_D$ in Eq. (8–141b), as follows:

$$B = \int_0^l f(y)\Theta_D ce\,dy. \qquad (8\text{–}142\text{c})$$

Inserting these values of A and B into Eq. (8–136a) and solving for q_1, the twist distribution becomes

$$\theta(y) = \Theta_D(y)q_1 = \Theta_D(y)\frac{\displaystyle\int_0^l f(y)\Theta_D ce\,dy}{[1 - (q/q_D)]\displaystyle\int_0^l \Theta_D{}^2ce\,dy}. \qquad (8\text{–}142\text{d})$$

Equation (8–142d) can be written in the equivalent form

$$\theta(y) = \Theta_r(y)\frac{1}{1 - (q/q_D)}, \qquad (8\text{–}142\text{e})$$

where

$$\Theta_r(y) = \frac{\Theta_D(y)\displaystyle\int_0^l f(y)\Theta_D(y)ce\,dy}{\displaystyle\int_0^l \Theta_D{}^2ce\,dy} \qquad (8\text{–}142\text{f})$$

is an approximate expression for the elastic twist computed on the assumption that the twist introduces no additional aerodynamic forces. Equation (8-142e) is identical to the result given by Eq. (8-12), which was derived for the case of an elastically suspended rigid wing segment. It should be observed that Eq. (8-142e) is rigorously valid only in the incompressible speed range, where it can be assumed that lift-curve slope is independent of speed. The ratio $1/[1 - (q/q_D)]$ plays the role of an amplification factor, and it is analogous to the concept of that name in vibration theory. The magnitude of the amplification factor depends upon the ratio of the airspeed to the divergence speed. The divergence speed thus plays a useful role as a reference value even though it may be well beyond the speed capabilities of the airplane.

8-4 Swept wings. In the previous section, a straight wing is characterized essentially by the existence of an elastic axis, and approximate perpendicularity between the elastic axis and the airplane plane of symmetry. In such wings, aeroelastic phenomena are dependent entirely upon wing twist about the elastic axis, and wing bending is not a factor. We turn our attention next to swept wings where spanwise axes are no longer perpendicular to the airplane center line. Here, wing bending has an important and complicating effect upon aeroelastic behavior. Consider, for example, Fig. 8-19, where it can be seen that when a swept-back wing bends, its angle of attack in the streamwise direction is reduced. If an upward force is applied to the reference axis, the points A and B will deflect through approximately the same distance, which will be clearly greater than the deflection of point C. The streamwise segment BC will therefore assume a lesser angle of attack, and a negative increment of lift will be developed. This negative lift has a stabilizing influence, since it opposes the lift resulting from nose-up twist of the segment BC. As a result of this action, bending of a swept-back wing tends to shift the center of pressure of the aerodynamic loads inboard, and bending of a swept-forward wing shifts the center of pressure outboard. The possibility of wing di-

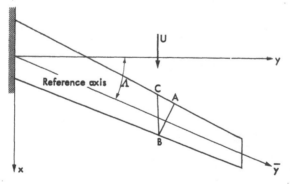

Fig. 8-19. Streamwise and chordwise segments of swept wing.

vergence for a swept-back wing is reduced, while the possibility of control reversal is increased. The converse of this holds for swept-forward wings and, in fact, swept-forward wings are practically ruled out by considerations of divergence. In addition, elastic deflections of swept wings produce large fore-and-aft center of pressure shifts which have a marked effect upon static longitudinal stability.

Aeroelastic phenomena of swept wings did not receive serious attention until the post World War II period. Collar (Ref. 8–16) in a discussion of aeroelastic problems at high speed in 1947, described qualitatively for the first time some of the aeroelastic phenomena associated with sweptback wings. Diederich and Budiansky (Ref. 8–17) in 1948 gave the first comprehensive treatment of the divergence phenomenon of slender swept wings. Their results indicated that the divergence speed drops rapidly as sweep forward increases but that wings with moderate or large sweep back cannot diverge. Subsequent investigations published by Pai and Sears (Ref. 8–18) and Miles (Ref. 8–19) described methods of calculating symmetric and antisymmetric span loadings on elastic swept wings. Pai and Sears demonstrated how the integral equation of equilibrium can be reduced to matrix form. The latter form is suitable for application to nonuniform wings. Miles applied assumed deflection functions, together with Galerkin's method, to obtain approximate solutions for slender swept wings with straight elastic axes. A subsequent report by Pian and Lin (Ref. 8–20) contains an excellent summary of swept-wing static aeroelastic problems, in addition to a formulation of the problem applicable to the various conditions required for structural design.

In the discussion which follows we shall apply both differential and integral equations of equilibrium to the solution of swept-wing aeroelastic problems. In general, the differential equation applications are limited to slender swept wings where an elastic axis can be assumed. In such cases, the reference axis $\bar{y}$ in Fig. 8–19 can be located along the elastic axis, and the wing deformation can be described in terms of the elementary engineering theories of bending and twisting along and about the elastic axis. The integral equation formulation of the problem is not limited by this assumption and is applicable to wings of arbitrary planform and stiffness, providing streamwise segments of the wing can be assumed rigid.

(a) *Differential equations of equilibrium of slender swept wings (streamwise segments).* Let us consider a slender swept wing with elastic properties such that the $\bar{y}$-axis in Fig. 8–20 can be regarded as an elastic axis. That is, the wing is assumed to behave elastically in the same manner as a slender beam with a skewed effective root. Figure 8–20 also illustrates a streamwise wing segment with a unit dimension along the y-axis, and a chord c in the streamwise direction. The total force on this segment is

$$Z(y) = qcc_l - mNg, \qquad (8-143)$$

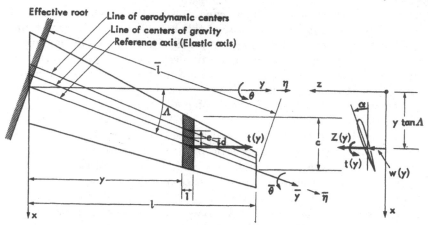

Fig. 8–20. Slender swept wing with streamwise segments.

where

c_l = local lift coefficient on streamwise segment,

m = mass per unit length along the y-axis.

The moment of the segment about an axis parallel to the y-axis and passing through the point of intersection of the elastic axis with the streamwise segment, that is, a chordwise location $x = y \tan \Lambda$, as shown by Fig. 8–20, is

$$t(y) = qecc_l + qc^2 c_{mAC} - mNgd, \qquad (8\text{–}144)$$

where the quantities e and d are measured in the streamwise direction, positive forward. In forming differential equations of equilibrium along the $\bar{y}$-axis, it is necessary to transform the quantities given by Eqs. (8–143) and (8–144) into running forces and torques per unit length along the $\bar{y}$-axis. This is accomplished by a multiplication by $\cos \Lambda$, as follows:

$$Z(\bar{y}) = Z(y) \cos \Lambda = (qcc_l - mNg) \cos \Lambda,$$

$$t(\bar{y}) = t(y) \cos \Lambda = (qecc_l + qc^2 c_{mAC} - mNgd) \cos \Lambda. \qquad (8\text{–}145)$$

All quantities referred to chordwise wing segments, such as segment AB in Fig. 8–19, will be written with bars. For example, $\bar{c}$, $\bar{c}_l$, and $\bar{c}_{mAC}$ represent respectively the chord, lift coefficient, and moment coefficient on a chordwise segment. All quantities referred to streamwise segments, such as BC in Fig. 8–19, will be written without bars. For example, c, c_l, and c_{mAC} represent similar quantities referred to streamwise segments. Relations between these quantities exist in all cases. Relations based on geometry alone are evident in the discussion. Relations based on aerodynamic considerations are obtained from Chapter 5. The running torque $t(\bar{y})$ can be resolved into a running torque about the elastic axis

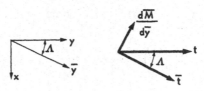

Fig. 8–21. Vector diagram of running torque and bending moment on swept wing.

and a running bending moment about a line perpendicular to the elastic axis, as illustrated by the vector diagram in Fig. 8–21. The running torque about the elastic axis is

$$\bar{l}(\bar{y}) = t(\bar{y}) \cos \Lambda, \tag{8–146}$$

and the running bending moment is

$$\bar{m}(\bar{y}) = t(\bar{y}) \sin \Lambda. \tag{8–147}$$

The total bending moment is the sum of the accumulated bending moments due to the running side load, $Z(\bar{y})$, and the running bending moment $\bar{m}(\bar{y})$. Thus

$$EI\frac{d^2w(\bar{y})}{d\bar{y}^2} = M(\bar{y}) = \int_{\bar{y}}^{l} Z(\bar{\eta})(\bar{\eta} - \bar{y})d\bar{\eta} - \int_{\bar{y}}^{l} \bar{m}(\bar{\eta})d\bar{\eta}. \tag{8–148}$$

Differentiating Eq. (8–148) twice with respect to $\bar{y}$, and substituting Eq. (8–147), we obtain a differential equation which describes the bending action of the wing, as follows:

$$\frac{d^2}{d\bar{y}^2}\left(EI\frac{d^2w}{d\bar{y}^2}\right) = Z(\bar{y}) + \frac{dt(\bar{y})}{d\bar{y}}\sin \Lambda. \tag{8–149a}$$

The accompanying torsional differential equation derives from the St. Venant torsion theory.

$$\frac{d}{d\bar{y}}\left(GJ\frac{d\bar{\theta}}{d\bar{y}}\right) = -\bar{l}(\bar{y}) = -t(\bar{y})\cos \Lambda, \tag{8–149b}$$

where $\bar{\theta}$ is the wing twist about the $\bar{y}$-axis. Introducing Eq. (8–145) into Eqs. (8–149a) and (8–149b), we obtain the following simultaneous differential equations:

$$\frac{1}{\cos \Lambda}\frac{d^2}{d\bar{y}^2}\left(EI\frac{d^2w}{d\bar{y}^2}\right) = qcc_l - mNg + \frac{d}{d\bar{y}}(qecc_l + qc^2c_{mAC} - mNgd)\sin \Lambda, \tag{8–150}$$

$$\frac{d}{d\bar{y}}\left(GJ\frac{d\bar{\theta}}{d\bar{y}}\right) = -(qecc_l + qc^2c_{mAC} - mNgd)\cos^2 \Lambda. \tag{8–151}$$

Assuming that c_l is a superposition of rigid and elastic components, Eqs. (8–150) and (8–151) reduce to the forms

$$\frac{1}{\cos \Lambda} \frac{d^2}{d\bar{y}^2} \left(EI \frac{d^2 w}{d\bar{y}^2} \right) - \frac{d}{d\bar{y}} (ecc_l{}^e) q \sin \Lambda - q c c_l{}^e$$

$$= q c c_l{}^r - mNg + \sin \Lambda \frac{d}{d\bar{y}} f(\bar{y}), \quad (8\text{–}152)$$

$$\cdot \frac{d}{d\bar{y}} \left(GJ \frac{d\bar{\theta}}{d\bar{y}} \right) + q e c c_l{}^e \cos^2 \Lambda = -f(\bar{y}) \cos^2 \Lambda, \quad (8\text{–}153)$$

where

$$f(\bar{y}) = q e c c_l{}^r + q c^2 c_{mAC} - mNgd.$$

In order to relate streamwise angle of attack to wing deformation, let us consider the vector diagram in Fig. 8–22. Since angular displacements can be treated as vectors when they are small quantities, it can be seen from Fig. (8–22) that the streamwise angle of attack resulting from elastic deformation is made up of a component of twist and a component of slope according to the relation

$$\theta = \bar{\theta} \cos \Lambda - \frac{dw}{d\bar{y}} \sin \Lambda. \quad (8\text{–}154)$$

In writing Eq. (8–154) it is, of course, assumed that streamwise segments of the wing are rigid. If we assume a symbolic linear relation between streamwise angle of attack and lift distribution, Eq. (8–154) becomes

$$\bar{\theta} \cos \Lambda - \frac{dw}{d\bar{y}} \sin \Lambda = \mathcal{Q}[c c_l{}^e], \quad (8\text{–}155)$$

where $\mathcal{Q}$ is a linear operator depending upon the aerodynamic theory that is employed. For example, in the case of strip theory, the operator $\mathcal{Q}$ is simply

$$\mathcal{Q} = \frac{1}{a_0 c \cos \Lambda}, \quad (8\text{–}156)$$

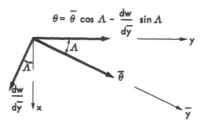

Fig. 8–22.   Vector diagram of streamwise angle of attack due to elastic deformation.

where a_0 is the two-dimensional lift-curve slope of the wing section. Since it is assumed that streamwise segments are rigid, c_{mAC} is unaffected by wing deformations.

The coupled differential equations (8–152) and (8–153), together with the auxiliary equation (8–155), define the behavior of swept wings with straight elastic axes. To obtain solutions, the boundary conditions must be stated. The slope and twist at the root are zero, which requires that

$$\frac{dw(0)}{d\bar{y}} = \bar{\theta}(0) = 0 \tag{8–157}$$

and the moment, torque, and shear at the tip are zero:

$$EI \frac{d^2w(l)}{d\bar{y}^2} = GJ \frac{d\bar{\theta}(l)}{d\bar{y}} = 0, \tag{8–158}$$

$$\frac{d}{d\bar{y}}\left(EI \frac{d^2w(l)}{d\bar{y}^2}\right) = 0. \tag{8–159}$$

(b) *Differential equations of equilibrium of slender swept wings (chordwise segments).* Differential equations of equilibrium of a slender swept wing can also be obtained by computing the aerodynamic forces and moments on segments of the wing taken perpendicular to the elastic axis, as illustrated by Fig. 8–23. It is necessary to assume an effective root and an effective tip. Although this assumption leads to some error, the method produces simple equations that give satisfactory results for large aspect ratios.

The total force on the chordwise segment of unit width along the $\bar{y}$-axis shown in Fig. 8–23 is given by

$$Z(\bar{y}) = \bar{c}\bar{c}_l q \cos^2 \Lambda - \bar{m}Ng, \tag{8–160}$$

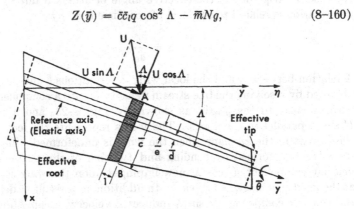

Fig. 8–23. Slender swept wing with chordwise segments.

where

$\bar{c}_l$ = local lift coefficient on the chordwise segment,
$\bar{c}$ = chord measured perpendicular to the $\bar{y}$-axis,
$\bar{m}$ = mass per unit length of wing along the $\bar{y}$-axis.

The moment of the chordwise segment about the $\bar{y}$-axis is

$$\bar{t}(\bar{y}) = \bar{c}\bar{c}_l\bar{e}q\cos^2\Lambda + \bar{c}_{mAC}\bar{c}^2q\cos^2\Lambda - \bar{m}Ng\bar{d}, \qquad (8\text{–}161)$$

where $\bar{c}_{mAC}$ is the local moment coefficient of the chordwise segment, and the quantities $\bar{e}$ and $\bar{d}$ are measured perpendicular to the elastic axis.

Forming the differential equations in bending and torsion, we obtain

$$\frac{d^2}{d\bar{y}^2}\left(EI\frac{d^2w}{d\bar{y}^2}\right) = \frac{d^2}{d\bar{y}^2}\int_y^l Z(\bar{\eta})(\bar{\eta}-\bar{y})d\bar{\eta} = \bar{c}\bar{c}_lq\cos^2\Lambda - \bar{m}Ng, \quad (8\text{–}162)$$

$$\frac{d}{d\bar{y}}\left(GJ\frac{d\bar{\theta}}{d\bar{y}}\right) = -\bar{t}(\bar{y}) = -\bar{c}\bar{c}_l\bar{e}q\cos^2\Lambda - \bar{c}_{mAC}\bar{c}^2q\cos^2\Lambda + \bar{m}Ng\bar{d}. \quad (8\text{–}163)$$

Taking $\bar{c}_l$ as a superposition of rigid and elastic components, Eqs. (8–162) and (8–163) become

$$\frac{d^2}{d\bar{y}^2}\left(EI\frac{d^2w}{d\bar{y}^2}\right) - \bar{c}\bar{c}_l{}^eq\cos^2\Lambda = \bar{c}\bar{c}_l{}^rq\cos^2\Lambda - \bar{m}Ng, \qquad (8\text{–}164)$$

$$\frac{d}{d\bar{y}}\left(GJ\frac{d\bar{\theta}}{d\bar{y}}\right) + \bar{c}\bar{c}_l{}^e\bar{e}q\cos^2\Lambda = -\bar{c}\bar{c}_l{}^r\bar{e}q\cos^2\Lambda - \bar{c}_{mAC}\bar{c}^2q\cos^2\Lambda + \bar{m}Ng\bar{d}.$$
$$(8\text{–}165)$$

Equations (8–164) and (8–165) must be supplemented by an additional statement relating $\bar{c}_l$ to the wing deformation. It is not clear what this statement is if aerodynamic induction effects are considered; however, in the case of strip theory, the effective angle of attack $\bar{\alpha}$ due to structural deformation is related to the local lift coefficient $\bar{c}_l{}^e$ by

$$\bar{\alpha} = \frac{\bar{c}_l{}^e}{a_0}. \qquad (8\text{–}166)$$

A relation between $\bar{\alpha}$ and the local wing deformations $\bar{\theta}$ and $dw/d\bar{y}$ can be obtained by resolving the free-stream velocity U into a component $U\cos\Lambda$ perpendicular to the $\bar{y}$-axis, as shown by Fig. 8–23, and a component $U\sin\Lambda$ parallel to the $\bar{y}$-axis. The $\bar{y}$-axis represents the location of the elastic axis in the xy-plane when the wing is undeformed. Suppose that the wing is deformed by bending and twisting of the elastic axis. The velocity component, $U\cos\Lambda$, now impinges upon the wing segment AB at the geometric angle of attack $\bar{\theta}$. In addition, as a result of deformation, the velocity component $U\sin\Lambda$ induces a velocity $U\sin\Lambda(dw/d\bar{y})$ perpendicular to the wing surface. Thus the total angle of attack of the seg-

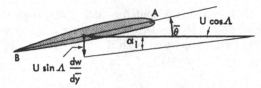

Fig. 8–24. Induced velocity component on chordwise segment of swept wing.

ment AB consists of the sum of a geometric component and an induced component, as illustrated by Fig. 8–24.

$$\bar{\alpha} = \bar{\theta} - \bar{\alpha}_i = \bar{\theta} - \frac{dw}{d\bar{y}} \tan \Lambda. \qquad (8\text{–}167)$$

Combining Eqs. (8–166) and (8–167), we obtain the required additional relation between lift coefficient and wing deformation, as follows:

$$\frac{\bar{c}_l{}^e}{a_0} = \bar{\theta} - \frac{dw}{d\bar{y}} \tan \Lambda. \qquad (8\text{–}168)$$

Differential equations (8–164) and (8–165), together with Eq. (8–168) and boundary conditions (8–157), (8–158), and (8–159), can be solved to obtain simplified solutions for large aspect-ratio swept wings. It can be seen that Eqs. (8–152) and (8–153) are similar in form to Eqs. (8–164) and (8–165), except that the latter do not contain a first-derivative term.

(c) *Integral equation of equilibrium of swept wing of arbitrary planform and stiffness.* Consider next the integral equation of equilibrium applying to a swept wing of arbitrary planform and stiffness, as illustrated by Fig. 8–25. We require the assumption that streamwise segments such as

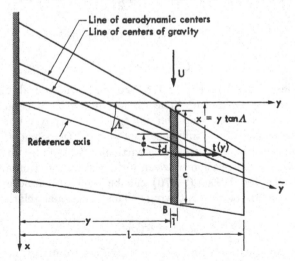

Fig. 8–25. Swept wing of arbitrary planform and stiffness.

BC in Fig. 8–25 be essentially undeformable. Although this assumption implies a certain degree of slenderness, the wing may have any planform and stiffness properties consistent with it. The equation of equilibrium is formed by applying Eq. (2–64).

$$\theta(y) = \int_0^l C^{\theta z}(y,\eta)Z(\eta)d\eta + \int_0^l C^{\theta\theta}(y,\eta)t(\eta)d\eta. \qquad (2\text{–}64)$$

In applying Eq. (2–64) we employ a $\bar{y}$ reference axis, as explained in Section 2–9. That is, the influence functions are defined such that the running torque $t(y)$ can be taken as the torque on each streamwise segment about an axis parallel to the y-axis. The reader may verify the location of the running torque $t(y)$ by referring to Fig. 8–25. Expressions for the running force $Z(y)$ and the running torque $t(y)$ which appear in Eq. (2–64) are the same as those given by Eqs. (8–143) and (8–144), respectively.

Substituting Eqs. (8–143) and (8–144) into Eq. (2–64) gives

$$\theta(y) = \int_0^l C^{\theta z}(y,\eta)(qcc_l - mNg)d\eta + \int_0^l C^{\theta\theta}(y,\eta)(qecc_l + qc^2c_{mAC} - mNgd)d\eta.$$

$$(8\text{–}169)$$

Substituting $c_l = c_l^r + c_l^e$ into Eq. (8–169) and reducing yields

$$\theta(y) = q\int_0^l \bar{C}(y,\eta)cc_l^e d\eta + \tilde{f}(y), \qquad (8\text{–}170)$$

where

$$\tilde{f}(y) = q\int_0^l \bar{C}(y,\eta)cc_l^r d\eta + q\int_0^l C^{\theta\theta}(y,\eta)c_{mAC}c^2 d\eta$$

$$- g\int_0^l [C^{\theta z}(y,\eta) + C^{\theta\theta}(y,\eta)d(\eta)]mN d\eta, \qquad (8\text{–}171)$$

$$\bar{C}(y,\eta) = C^{\theta z}(y,\eta) + e(\eta)C^{\theta\theta}(y,\eta). \qquad (8\text{–}172)$$

The quantities c_l^e and θ in Eq. (8–170) are unknown, and all other quantities are regarded as known. The known quantities depend upon the physical properties of the wing and the type of maneuver under consideration. The nature of the influence functions $C^{\theta z}(y,\eta)$, and $C^{\theta\theta}(y,\eta)$ depend upon the elastic properties of the particular wing under investigation.

When a functional relation between incidence distribution and lift distribution is introduced, Eq. (8–170) can be used to compute the static aeroelastic behavior of a swept wing. This functional relation has the form

$$\theta(y) = \mathcal{C}[cc_l^e]. \qquad (8\text{–}173)$$

The integral equation of a swept wing differs from the corresponding equation of a straight wing (Eq. 8–28) in one important respect. The influence function $\bar{C}(y,\eta)$ is not symmetrical, since one of its components is $C^{\theta z}(y,\eta)$, a nonsymmetrical function. The type of mathematical problem represented here is known as non self-adjoint, whereas the static aeroelastic problem of the straight wing and the vibration problems treated in Chapters 3 and 4 are of the self-adjoint type. It will be seen later in this chapter that non self-adjointness has an important bearing on the mathematical nature of the solution.

(d) *Divergence*. The divergence speed of a swept wing is obtained by computing the lowest characteristic value of the homogeneous simultaneous differential equations or of the homogeneous integral equation. In the case of swept wings, unlike the straight wing, bending has a profound effect upon the results. Divergence phenomena of swept wings should therefore be more correctly referred to as bending-torsion divergence, in contrast to torsional divergence of straight wings. It will become apparent to the reader in this discussion that divergence speeds of swept-back wings are so high that divergence itself is not a consideration of primary importance. Conversely, divergence speeds of swept-forward wings will be seen to be so deleteriously influenced by forward sweep that such wings are practically ruled out by divergence considerations. The reader may logically wonder, therefore, what practical importance can be attached to the divergence problem of swept wings. This question may at least be partially answered in Section 8–4(g), where a useful approximate method is presented of analyzing swept-back wing load distribution problems. The method expresses the deformation of a loaded swept-back wing in terms of the divergence mode shape of the same wing swept forward. Here the divergence speed of the swept-forward wing plays an important role as a reference value which determines the attenuation in elastic twist of the swept-back wing.

(1) *Closed form solutions of the differential equations of slender swept wings*. Solutions in closed form of the simultaneous differential equations can be obtained in a few special cases when strip theory is assumed. Perhaps the simplest case with some practical significance is that of a uniform slender swept wing with the elastic axis coincident with the line of aerodynamic centers. Thus e is zero and there can be no twist about the elastic axis. The applicable differential equation is obtained by putting $e = 0$ in the homogeneous form of Eq. (8–152):

$$\frac{d^2}{d\bar{y}^2}\left(EI\,\frac{d^2w}{d\bar{y}^2}\right) - qcc_l^e \cos \Lambda = 0. \tag{8–174}$$

Substituting Eq. (8–156) into Eq. (8–155), and putting $\bar{\theta} = 0$, we obtain for the case of strip theory

$$c_l^e = -a_0 \frac{dw}{d\bar{y}} \sin \Lambda \cos \Lambda. \tag{8–175}$$

Specializing Eq. (8–174) to a uniform wing and introducing Eq. (8–175),

$$\frac{d^4 w}{d\bar{y}^4} + \frac{a_0 qc \sin \Lambda \cos^2 \Lambda}{EI} \frac{dw}{d\bar{y}} = 0. \tag{8–176}$$

It is convenient to transform Eq. (8–176) to the following form:

$$\frac{d^3 \Gamma}{d\xi^3} - b\Gamma = 0, \tag{8–177}$$

where

$$\Gamma = \frac{dw}{d\bar{y}}, \qquad \xi = 1 - \frac{\bar{y}}{l}, \qquad b = \frac{a_0 qcl^3}{EI} \sin \Lambda \cos^2 \Lambda.$$

The boundary conditions on Eq. (8–177) follow from expressions (8–157), (8–158), and (8–159).

$$\Gamma(1) = 0, \qquad \frac{d\Gamma(0)}{d\xi} = 0, \qquad \frac{d^2 \Gamma(0)}{d\xi^2} = 0. \tag{8–178}$$

Equation (8–177) is a third-order constant coefficient linear differential equation with the following solution:

$$\Gamma = \sum_{i=1}^{3} A_i e^{r_i \xi}, \tag{8–179}$$

where r_1, r_2, and r_3 are roots of the characteristic equation

$$r^3 - b = 0. \tag{8–180}$$

Putting the solution given by Eq. (8–179) into boundary conditions (8–178) gives

$$\begin{bmatrix} e^{r_1} & e^{r_2} & e^{r_3} \\ r_1 & r_2 & r_3 \\ r_1^2 & r_2^2 & r_3^2 \end{bmatrix} \begin{bmatrix} A_1 \\ A_2 \\ A_3 \end{bmatrix} = 0. \tag{8–181}$$

In order for the A's to be finite, the determinant of their coefficients must be zero. Expansion of this determinant gives

$$1 + \frac{r_1}{r_2}\left(\frac{r_1 - r_3}{r_3 - r_2}\right) e^{r_2 - r_1} + \frac{r_1}{r_3}\left(\frac{r_2 - r_1}{r_3 - r_2}\right) e^{r_3 - r_1} = 0. \tag{8–182}$$

Let us now assume that the wing is swept forward. In this case the quantity b is negative, since the sweep angle Λ is negative. The characteristic equation is

$$r^3 + |b| = 0, \tag{8–183}$$

which has the following roots:

$$r_1 = -\sqrt[3]{|b|}, \qquad r_2 = \tfrac{1}{2}(1 + i\sqrt{3})\sqrt[3]{|b|}, \qquad r_3 = \tfrac{1}{2}(1 - i\sqrt{3})\sqrt[3]{|b|}. \tag{8–184}$$

Substituting these roots into Eq. (8–182), we obtain a transcendental equation in $|b|$, as follows:

$$e^{-\frac{3}{2}\sqrt[3]{|b|}} + 2\cos\left(\frac{\sqrt{3}}{2}\sqrt[3]{|b|}\right) = 0. \qquad (8\text{–}185)$$

A trial-and-error solution of (8–185) yields a minimum value for $|b|$ of 6.33. Thus the divergence dynamic pressure of a uniform swept-forward wing with freedom only in bending is given by

$$q_D = \frac{6.33 EI}{a_0 c \bar{l}^3 \cos^2 \Lambda |\sin \Lambda|}. \qquad (8\text{–}186)$$

This result indicates that as the degree of sweep forward is reduced, that is, as Λ becomes a smaller negative quantity, the divergence speed increases, and as Λ approaches zero the divergence speed approaches infinity. Equation (8–186) does not, of course, apply to the case of swept-back wings; however, we can reason physically that a swept-back wing with freedom only in bending has an infinite divergence speed.

An interesting and useful solution of Eqs. (8–164) and (8–165) for divergence has been obtained by Diederich and Budiansky (Ref. 8–17) for the case of slender tapered wings in which the chord is assumed to vary linearly and the bending and torsional stiffnesses are assumed to vary as the fourth power of the chord. For these assumptions we can put

$$\bar{c} = \xi \bar{c}_R, \qquad \bar{e} = \bar{e}_1 \bar{c}, \qquad GJ = GJ_R \xi^4, \qquad EI = EI_R \xi^4, \qquad (8\text{–}187)$$

where

$$\xi = 1 - (1 - \lambda)\bar{y}/\bar{l}, \qquad \bar{c}_R = \text{wing chord at the root,}$$
$$\lambda = \text{taper ratio} = \bar{c}_T/\bar{c}_R, \qquad GJ_R = \text{torsional stiffness at the root,}$$
$$\bar{c}_T = \text{wing chord at the tip,} \qquad EI_R = \text{bending stiffness at the root.}$$

Putting the right sides of Eqs. (8–164) and (8–165) equal to zero, and substituting Eq. (8–168), we obtain

$$\frac{d^2}{d\bar{y}^2}\left(EI\frac{d\Gamma}{d\bar{y}}\right) - a_0 q \bar{c} \cos^2 \Lambda (\bar{\theta} - \Gamma \tan \Lambda) = 0, \qquad (8\text{–}188)$$

$$\frac{d}{d\bar{y}}\left(GJ\frac{d\bar{\theta}}{d\bar{y}}\right) + a_0 q \bar{c}\bar{e} \cos^2 \Lambda (\bar{\theta} - \Gamma \tan \Lambda) = 0, \qquad (8\text{–}189)$$

where $\Gamma = dw/d\bar{y}$. Transforming Eqs. (8–188) and (8–189) to the independent variable ξ yields

$$\frac{d^2}{d\xi^2}\left(EI\frac{d\Gamma}{d\xi}\right) + \frac{a_0 q \bar{c} \bar{l}^3 \cos^2 \Lambda}{(1 - \lambda)^3}(\bar{\theta} - \Gamma \tan \Lambda) = 0, \qquad (8\text{–}190)$$

$$\frac{d}{d\xi}\left(GJ\frac{d\bar{\theta}}{d\xi}\right) + \frac{a_0 q \bar{e} \bar{c} \bar{l}^2 \cos^2 \Lambda}{(1 - \lambda)^2}(\bar{\theta} - \Gamma \tan \Lambda) = 0. \qquad (8\text{–}191)$$

Introducing the assumed properties of the tapered wing from (8–187), the simultaneous differential equations reduce to

$$\tan \Lambda \left[\xi^3 \frac{d^3 \Gamma}{d\xi^3} + 8\xi^2 \frac{d^2 \Gamma}{d\xi^2} + 12\xi \frac{d\Gamma}{d\xi} - b\Gamma \right] + b\bar{\theta} = 0, \quad (8\text{–}192)$$

$$\xi^2 \frac{d^2 \bar{\theta}}{d\xi^2} + 4\xi \frac{d\bar{\theta}}{d\xi} + a\bar{\theta} - a\Gamma \tan \Lambda = 0, \quad (8\text{–}193)$$

where

$$a = \frac{q a_0 \bar{e}_1 \bar{c}_R{}^2 l^2 \cos^2 \Lambda}{G J_R (1 - \lambda)^2}, \qquad b = \frac{q a_0 \bar{c}_R l^3 \cos^2 \Lambda \tan \Lambda}{E I_R (1 - \lambda)^3}.$$

These two differential equations can be combined to form a single equation by differentiating Eq. (8–193) once with respect to ξ, multiplying it by ξ, and combining the result with Eqs. (8–192) and (8–193). The result is an Euler differential equation as follows:

$$\xi^3 \frac{d^3 \bar{\alpha}}{d\xi^3} + 8\xi^2 \frac{d^2 \bar{\alpha}}{d\xi^2} + (12 + a)\xi \frac{d\bar{\alpha}}{d\xi} + (2a - b)\bar{\alpha} = 0, \quad (8\text{–}194)$$

where

$$\bar{\alpha} = \bar{\theta} - \Gamma \tan \Lambda.$$

The Euler differential equation, since it is equidimensional, may be reduced to a linear equation with constant coefficients by the change of variable $\xi = e^t$, and the resulting equation can be easily solved. This entire process is accomplished in one step (Ref. 8–21) by putting as a solution the following power series:

$$\bar{\alpha} = \sum_{i=1}^{3} B_i \xi^{S_i}, \quad (8\text{–}195)$$

where the S_i's are the roots of

$$S^3 + 5S^2 + (6 + a)S + (2a - b) = 0 \quad (8\text{–}196)$$

and the B_i's are arbitrary constants. The boundary conditions are obtained by applying Eqs. (8–157), (8–158), and (8–159). From Eq. (8–157),

$$\bar{\alpha}(1) = \bar{\theta}(1) - \Gamma(1) \tan \Lambda = 0. \quad (8\text{–}197)$$

From Eq. (8–158),

$$\frac{d\bar{\alpha}(\lambda)}{d\xi} = \frac{d}{d\xi} (\bar{\theta}(\lambda) - \Gamma(\lambda) \tan \Lambda) = 0. \quad (8\text{–}198)$$

From Eqs. (8–158), (8–159), and (8–193),

$$\lambda^2 \frac{d^2 \bar{\alpha}(\lambda)}{d\xi^2} + a\bar{\alpha}(\lambda) = 0. \quad (8\text{–}199)$$

Substituting the solution given by Eq. (8–195) into these three boundary conditions yields

$$\sum_{i=1}^{3} B_i = 0, \qquad \sum_{i=1}^{3} S_i B_i \lambda^{S_i} = 0, \qquad \sum_{i=1}^{3} [S_i(S_i - 1) + a] B_i \lambda^{S_i} = 0.$$

$$(8\text{–}200)$$

For divergence, the determinant of the coefficients of the B_i's must vanish:

$$\begin{vmatrix} 1 & 1 & 1 \\ S_1 \lambda^{S_1} & S_2 \lambda^{S_2} & S_3 \lambda^{S_3} \\ (S_1{}^2 - S_1 + a) \lambda^{S_1} & (S_2{}^2 - S_2 + a) \lambda^{S_2} & (S_3{}^2 - S_3 + a) \lambda^{S_3} \end{vmatrix} = 0. \quad (8\text{–}201)$$

Combinations of the parameters a and b which correspond to a divergence condition are those combinations which produce roots S_1, S_2, and S_3 that satisfy (8–201). The system has been solved numerically by Diederich and Budiansky in Ref. 8–17 by selecting values of the taper ratio λ, assuming values of a, and solving for critical values of b. These results are reproduced in Figs. 8–26(a) and 8–26(b) for taper ratios of 0.2, 0.5, 1.0, and 1.5. In order to apply Fig. 8–26(a), the ratio b'/a', which depends upon the physical properties of the wing, is computed from

$$a' = \frac{q a_0 \bar{e}_1 \bar{c}_R{}^2 \bar{l}^2 \cos^2 \Lambda}{G J_R}, \qquad b' = \frac{q a_0 \bar{c}_R \bar{l}^3 \cos^2 \Lambda \tan \Lambda}{E I_R}. \quad (8\text{–}202)$$

Using this value of b'/a', the critical value of a' can be read from Fig. 8–26(a). From this result, the divergence dynamic pressure can be computed from

$$q_D = \frac{a' G J_R (1 - \lambda)^2}{a_0 \bar{e}_1 c_R{}^2 \bar{l}^2 \cos^2 \Lambda}. \quad (8\text{–}203)$$

The curves in Fig. 8–26(a) are somewhat restricted in range of the variable b'/a' when applied to slender wings. Figure 8–26(b) should be used in place of Fig. 8–26(a) for large values of b'/a'. If the value of q_D computed above is negative, the wing cannot diverge, since a negative dynamic pressure does not correspond to a real speed.

The effect of wing sweep is evident from examination of Fig. 8–26(a). As the wing is swept forward, the parameter b'/a' is increased negatively. This is accompanied by a rapid decrease in the parameter a' and hence a decrease in the divergence speed, if other parameters are kept constant. Beyond a certain swept-forward angle, the divergence speed tends to increase due to the presence of the $\cos^2 \Lambda$ term in the denominator of Eq. (8–203). As the wing is swept back, the parameter b'/a' is positive and the corresponding value of a' increases very rapidly. Figure 8–26(a) indicates that beyond values of about 2 for b'/a', the wing cannot diverge at subsonic speeds.

The results obtained above can be improved by the use of an effective lift-curve slope $dC_L/d\alpha$ corrected for finite span.   Diederich and Budiansky in Ref. 8–17 suggest the following empirical aspect ratio and sweep correction:

$$\frac{dC_L}{d\alpha} = a_0 \frac{AR}{AR + 4 \cos \Lambda}$$ (8–203a)

for subsonic and subcritical Mach numbers.

EXAMPLE 8–5.   To compute the variation of divergence speed with sweep angle for a tapered wing with the properties

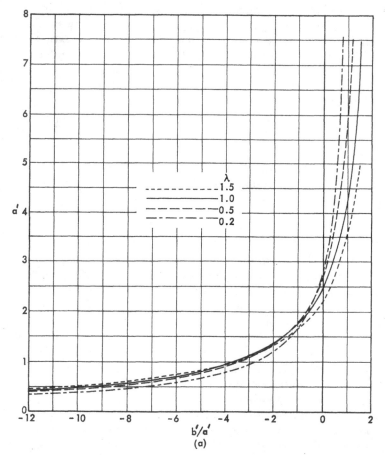

Fig. 8–26.   (a) Small values of b'/a'.   (*Reproduced from NACA Technical Note No.* 1680.)

$$\lambda = 0.5, \quad \frac{l}{\bar{e}_1 \bar{c}_R} = 70, \quad \frac{GJ_R}{EI_R} = 1,$$

neglecting variations in a_0 and $\bar{e}_1$ with speed as a first approximation.

Solution. The divergence speed can be calculated in terms of the divergence speed at zero sweep by applying Eq. (8–203):

$$q_D = q_D(\Lambda = 0)\, \frac{a'(\Lambda)}{a'(\Lambda = 0)}\, \frac{1}{\cos^2 \Lambda}. \tag{a}$$

From Fig. 8–26(a) for a taper ratio of 0.5, we obtain $a'\,(\Lambda = 0) = 2.7$. Values of $a'(\Lambda)$ are obtained from Fig. 8–26(a) by entering the appropriate values of b'/a':

$$\frac{b'}{a'} = \frac{l}{\bar{e}_1 \bar{c}_R}\, \frac{GJ_R}{EI_R}\, \tan \Lambda = 70 \tan \Lambda. \tag{b}$$

The variation in divergence speed with sweep angle is plotted in Fig. 8–26(c), where the profound effect of forward sweep in reducing the divergence speed is portrayed. At a forward angle of sweep of 45°, the divergence speed is only 5% of its value at zero sweep. On the other hand, a few degrees of rearward sweep causes the divergence speed to rise rapidly to an infinite value.

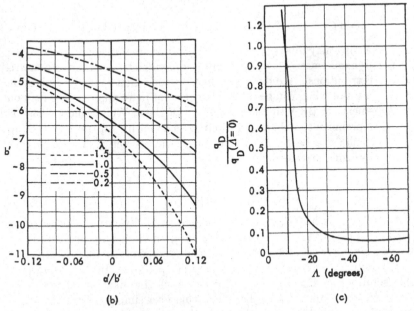

(b)　　　　　　　　　　　　　　　　(c)

Fig. 8–26. (b) Large values of b'/a'. (*Reproduced from NACA Technical Note No.* 1680.) (c) Variation in divergence speed with sweep angle.

(2) *Solutions in terms of integral equations and matrices for swept wings of arbitrary planform and stiffness.* The integral equation form applicable to the swept-wing divergence problem follows from the homogeneous form of Eq. (8–170):

$$\theta(y) = q \int_0^l \bar{C}(y,\eta) c c_l{}^e d\eta, \qquad (8\text{–}204)$$

where $\bar{C}(y,\eta)$ is defined by Eq. (8–172), and the functional relation between angle of attack and lift coefficient is represented by Eq. (8–173). Since $\bar{C}(y,\eta)$ is not a symmetrical function, it cannot be said with certainty that real characteristic values of Eq. (8–204) exist in a given case. They may be complex or may not exist at all. No complete statement is available on the conditions for existence of real characteristic values in non self-adjoint problems. When real characteristic values exist, solutions of the characteristic value problem formed by Eq. (8–204) have been shown by experience to yield results which are physically correct. In problems involving nonuniform wings, Eqs. (8–204) and (8–173) can be combined in matrix form to obtain the following characteristic equations:

$$[A]\{c c_l{}^e\} = q[\bar{E}]\{c c_l{}^e\}, \qquad (8\text{–}205)$$

where

$[A]$ = matrix of aerodynamic influence coefficients for symmetrical or antisymmetrical loading on a swept wing,

$[\bar{E}] = [\bar{C}][\bar{W}]$,

$[\bar{C}] = [C^{\theta z}] + [C^{\theta \theta}][e] =$ matrix of flexibility influence coefficients.

Since the wing may diverge in a symmetrical or an antisymmetrical mode, the aerodynamic matrix $[A]$ may represent symmetrical or antisymmetrical lift distributions. We have seen in Chapter 5 that spanwise induction effects on swept wings can be taken into account in an approximate fashion by refinements of the lifting-line theory. Of these, Weissinger's theory is the most prominent, and the $[A]$ matrix is given by Eq. (5–195). Strip theory results can be altered by an empirical correction, in which case the aerodynamic matrix is simply

$$[A] = \frac{1}{dC_L/d\alpha} [1/c], \qquad (8\text{–}206)$$

where $dC_L/d\alpha$ is an effective three-dimensional lift-curve slope such as given by Eq. (5–224). In the latter case, there can be no distinction between symmetrical and antisymmetrical divergence speeds.

(2) *Solution in terms of adjoint integral equations.* In discussing possible solutions to Eq. (8–204), mention should be made of adjoint integral equations. In order to introduce this idea simply, let us specialize Eq. (8–204) by applying strip theory to the case of a slender swept-forward wing which is free to bend without twisting. That is, we put $e = 0$ in

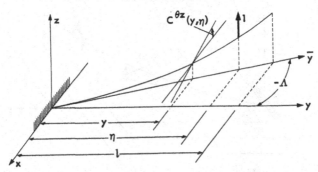

Fig. 8-27. Schematic representation of swept-wing influence function.

the expression for $\bar{C}(y,\eta)$ given by Eq. (8-172), and put $c_l^e = a_0\theta \cos \Lambda$. With these simplifying assumptions, Eq. (8-204) reduces to

$$\theta(y) = qa_0 \cos \Lambda \int_0^l C^{\theta z}(y,\eta)\theta c\,d\eta, \qquad (8\text{-}207)$$

where $C^{\theta z}(y,\eta)$ is given by Eqs. (2-130) and (2-131). The homogeneous equation

$$w(y) = qa_0 \cos \Lambda \int_0^l C^{\theta z}(\eta,y)wc\,d\eta \qquad (8\text{-}208)$$

is an integral equation which is said to be adjoint to Eq. (8-207). Notice that the kernel in Eq. (8-208) is derived from the kernel in Eq. (8-207) by interchanging the arguments y and η. There exist useful mathematical relations between the solutions of Eqs. (8-207) and (8-208). Before stating these relations, it is desirable to observe the physical nature of Eq. (8-208) and its connection with Eq. (8-207) by examining the properties of the kernel function. The function $C^{\theta z}(y,\eta)$ in Eq. (8-207) gives the streamwise angle of twist at y due to a unit lift force applied to the elastic axis at η. The definition is illustrated schematically by Fig. 8-27. If we interchange y and η in the function $C^{\theta z}(y,\eta)$, from the reciprocal properties of elastic systems, the new kernel can be expressed by

$$C^{\theta z}(\eta,y) = C^{z\theta}(y,\eta). \qquad (8\text{-}209)$$

Thus the adjoint integral equation can be written in the form

$$w(y) = qa_0 \cos \Lambda \int_0^l C^{z\theta}(y,\eta)wc\,d\eta, \qquad (8\text{-}210)$$

where the kernel $C^{z\theta}(y,\eta)$ represents the linear displacement of the elastic axis at y due to a unit streamwise torque applied to the elastic axis at η. This definition is illustrated schematically by Fig. 8-28.

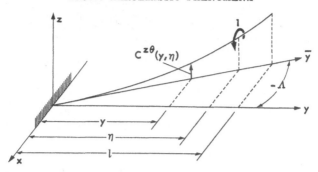

Fig. 8-28. Schematic representation of influence function of adjoint problem.

Figures 8-27 and 8-28 thus illustrate the conjugate nature of the integral equation and its adjoint. Whereas Eq. (8-207) represents the streamwise angle of twist due to a running lift proportional to the local angle of twist, the adjoint equation (8-210) represents the linear deflection of the elastic axis due to a running streamwise torque which is proportional to the local deflection of the elastic axis. It can be seen that if $C^{\theta z}(y,\eta)$ were a symmetrical function, the two problems would be identical. In this case, the system is said to be self-adjoint.

Perhaps the most important mathematical relation between Eqs. (8-207) and (8-210) concerns their eigenvalues. It can be shown by the theory of adjoint integral equations (Ref. 8-22) that if q_D is a divergence dynamic pressure computed from Eq. (8-207), the same result is obtained from the adjoint equation (8-210). This correspondence holds not only for the lowest eigenvalue, but also for all higher eigenvalues.

A second useful mathematical relation between the two equations relates to orthogonality between their eigenfunctions. Let us suppose that two separate eigenvalues λ_i and λ_j correspond to eigenfunctions Θ_i and Θ_j, respectively, in Eq. (8-207), and to W_i and W_j, respectively, in Eq. (8-210). Since $C^{\theta z}(y,\eta)$ and $C^{z\theta}(y,\eta)$ are not symmetrical functions, then it is apparent that two distinct eigenfunctions from the same equation cannot be orthogonal. That is,

$$\int_0^l \Theta_i\Theta_j c\,dy \neq 0, \qquad (i \neq j), \qquad (8\text{-}211\text{a})$$

$$\int_0^l W_i W_j c\,dy \neq 0, \qquad (i \neq j). \qquad (8\text{-}211\text{b})$$

However, it can be shown that distinct eigenfunctions from adjoint equations are orthogonal. Let us write Eqs. (8-207) and (8-208) in the following forms, where λ_i and λ_j are distinct eigenvalues:

$$\Theta_i(y) = \lambda_i \int_0^l C^{\theta z}(y,\eta)\Theta_i c\,d\eta, \qquad (8\text{–}212a)$$

$$W_j(y) = \lambda_j \int_0^l C^{z\theta}(y,\eta)W_j c\,d\eta. \qquad (8\text{–}212b)$$

Multiplying through Eq. (8–212a) by $\lambda_j W_j c$ and integrating from zero to l, multiplying through Eq. (8–212b) by $\lambda_i \Theta_i c$ and integrating from zero to l, and subtracting the second equation from the first, yields

$$(\lambda_j - \lambda_i)\int_0^l \Theta_i W_j c\,dy = \lambda_i\lambda_j \int_0^l W_j c \int_0^l C^{\theta z}(y,\eta)\Theta_i c\,d\eta\,dy$$

$$- \lambda_i\lambda_j \int_0^l \Theta_i c \int_0^l C^{z\theta}(y,\eta)W_j c\,d\eta\,dy. \quad (8\text{–}213)$$

The two integrals on the right side cancel if we interchange y and η in the second term and apply Eq. (8–209). Thus we obtain the orthogonality relation

$$\int_0^l \Theta_i W_j c\,dy = 0, \qquad (i \neq j). \qquad (8\text{–}214)$$

Eigenfunctions such as $\Theta(y)$ and $W(y)$ arising from a system and its adjoint are sometimes referred to as biorthogonal functions.

If Eqs. (8–207) and (8–210) are cast in matrix form, we have the following pair of adjoint matrix equations:

$$\{\theta\} = qa_0 \cos \Lambda [C^{\theta z}][c][\overline{W}]\{\theta\}, \qquad (8\text{–}215a)$$

$$\{w\} = qa_0 \cos \Lambda [C^{z\theta}][c][\overline{W}]\{w\}. \qquad (8\text{–}215b)$$

The elastic matrix $[C^{z\theta}]$, of Eq. (8–215b) is the transpose of the elastic matrix $[C^{\theta z}]$ in Eq. (8–215a). The eigenvalues of the two matrix equations are identical; however, their eigenfunctions have different physical meanings, as explained above.

Although the mathematical relations stated above were specialized for simplicity to the case of a slender swept wing free only in bending, the same principles can be extended to the more general case where twisting action is included.

EXAMPLE 8–6. To compute the divergence dynamic pressure of a uniform swept-forward wing in terms of strip theory by applying Eqs. (8–215a) and (8–215b).

Solution. Matrix iteration will be applied to the characteristic value simultaneous equations obtained from Eqs. (8–215a) and (8–215b). Since the chord is constant, Eq. (8–215a) has the form

$$\{\theta\} = qa_0 c \cos \Lambda [C^{\theta z}][\overline{W}]\{\theta\}. \qquad (a)$$

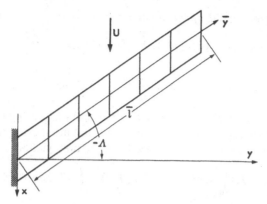

Fig. 8–29.   Uniform swept-forward wing divided in six equal segments.

The wing is divided into six equal segments, as illustrated by Fig. 8–29. The influence coefficient matrix $[C^{\theta z}]$ corresponding to the seven wing stations can be computed from Eqs. (2–130) and (2–131), from which we obtain the following result:

$$[C^{\theta z}] = \frac{\bar{l}^2}{EI} \, |\sin \Lambda| [n], \tag{b}$$

where

$$[n] = \begin{bmatrix} 0 & 0 & 0 & 0 & 0 & 0 & 0 \\ 0 & \frac{1}{72} & \frac{3}{72} & \frac{5}{72} & \frac{7}{72} & \frac{9}{72} & \frac{11}{72} \\ 0 & & \frac{1}{18} & \frac{2}{18} & \frac{3}{18} & \frac{4}{18} & \frac{5}{18} \\ 0 & & & \frac{3}{24} & \frac{5}{24} & \frac{7}{24} & \frac{9}{24} \\ 0 & & & & \frac{2}{9} & \frac{3}{9} & \frac{4}{9} \\ 0 & & & & & \frac{25}{72} & \frac{35}{72} \\ 0 & & & & & & \frac{1}{2} \end{bmatrix}. \tag{c}$$

Since we have an even number of equal segments, Simpson's rule can be applied to obtain a weighting matrix (cf. Appendix B) as follows:

$$[\overline{W}] = \frac{\bar{l} \cos \Lambda}{18} [n'], \tag{d}$$

where

$$[n'] = \begin{bmatrix} 1 & & & & & & \\ & 4 & & & & & \\ & & 2 & & & & \\ & & & 4 & & & \\ & & & & 2 & & \\ & & & & & 4 & \\ & & & & & & 1 \end{bmatrix}. \tag{e}$$

Introducing the values of $[C^{\theta z}]$ and $[\overline{W}]$ given above into Eq. (a) gives

$$\frac{18EI}{qa_0c\bar{l}^3 \cos^2 \Lambda |\sin \Lambda|} \{\theta\}$$

$$= \begin{bmatrix}
0 & 0 & 0 & 0 & 0 & 0 & 0 \\
0 & .05555 & .08333 & .2778 & .1944 & .500 & .1528 \\
0 & .05555 & .1111 & .4444 & .3333 & .8889 & .2778 \\
0 & .05555 & .1111 & .5000 & .4167 & 1.167 & .3750 \\
0 & .05555 & .1111 & .5000 & .4444 & 1.333 & .4444 \\
0 & .05555 & .1111 & .5000 & .4444 & 1.389 & .4861 \\
0 & .05555 & .1111 & .5000 & .4444 & 1.389 & .5000
\end{bmatrix} \{\theta\}. \quad (f)$$

Applying iteration to the above matrix equation, we obtain the following eigenvalue and eigenfunction:

$$\frac{18EI}{q_{Da_0}c\bar{l}^3 \cos^2 \Lambda |\sin \Lambda|} = 2.843; \quad \{\theta_1\} = \begin{bmatrix} 0 \\ .4076 \\ .6923 \\ .8689 \\ .9610 \\ .9951 \\ 1.000 \end{bmatrix}.$$

The divergence dynamic pressure thus becomes

$$q_D = \frac{6.33EI}{a_0c\bar{l}^3 \cos^2 \Lambda |\sin \Lambda|}, \quad (g)$$

which agrees with the result already given by expression (8–186).

The adjoint matrix equation for the case of constant chord derives from Eq. (8–215b), as follows:

$$\{w\} = qa_0c \cos \Lambda [C^{z\theta}][\bar{W}]\{w\}. \quad (h)$$

The matrix $[C^{z\theta}]$ in Eq. (h) can be formed by taking the transpose of the matrix $[C^{\theta z}]$ given above. The weighting matrix is the same as that given by matrix (c). Forming the adjoint matrix equation and applying matrix iteration, we obtain the following eigenvalue and eigenfunction:

$$\frac{18EI}{q_{Da_0}c\bar{l}^3 \cos^2 \Lambda |\sin \Lambda|} = 2.843; \quad \{W_1\} = \begin{bmatrix} 0 \\ .03099 \\ .1234 \\ .2753 \\ .4803 \\ .7281 \\ 1.000 \end{bmatrix}.$$

We observe that the numerical value of the eigenvalue is the same for the system and its adjoint. The eigenfunctions represented by the column matrices are, of course, different, since they have a different physical meaning.

(e) *Symmetrical lift distribution.* Aeroelastic effects on the symmetrical spanwise lift distribution of a swept wing produce an inboard and forward shift of the center of pressure if the wing is swept back, and an outboard and forward shift if the wing is swept forward. These phenomena are important factors in the longitudinal stability of swept-wing aircraft, and are of importance also to the structural designer in obtaining span-wise load distributions. The aeroelastic behavior of swept wings under symmetrical loading conditions can be obtained by solutions to the differential and integral equations derived in Sections 8–4(a), (b), and (c).

(1) *Closed-form solutions of the differential equations of slender swept wings.* A useful closed-form solution to the problem of symmetrical loading on uniform slender swept wings has been given by Diederich and Foss in Ref. 8–23. They give exact solutions of the simultaneous nonhomogeneous equations (8–164) and (8–165) for the case of a uniform wing and strip theory. If Eq. (8–168) is introduced into Eqs. (8–164) and (8–165), and it is assumed that the bending stiffness, torsional stiffness, and chord are constant, we obtain

$$\frac{d^3 \Gamma}{d\bar{y}^3} - \frac{a_0 q \bar{c} \cos^2 \Lambda}{EI}(\bar{\theta} - \Gamma \tan \Lambda) = \frac{q \bar{c} \bar{c}_l{}^r \cos^2 \Lambda - \bar{m}Ng}{EI}, \qquad (8\text{--}216)$$

$$\frac{d^2 \bar{\theta}}{d\bar{y}^2} + \frac{a_0 q \bar{c} \bar{e} \cos^2 \Lambda}{GJ}(\bar{\theta} - \Gamma \tan \Lambda) = - \frac{q \cos^2 \Lambda (\bar{c} \bar{c}_l{}^r \bar{e} + \bar{c}_{m\mathrm{AC}} \bar{c}^2) - \bar{m}Ng\bar{d}}{GJ} \cdot$$

$$(8\text{--}217)$$

Multiplying Eq. (8–216) by $\tan \Lambda$ and introducing a change in independent variable according to the transformation $\xi = 1 - (\bar{y}/\bar{l})$, the differential equations become

$$\frac{d^2 \bar{\theta}}{d\xi^2} + a(\bar{\theta} - \Gamma \tan \Lambda) = f_1(\xi), \qquad (8\text{--}218)$$

$$\frac{d^3 \Gamma}{d\xi^3} \tan \Lambda + b(\bar{\theta} - \Gamma \tan \Lambda) = f_2(\xi), \qquad (8\text{--}219)$$

where

$$a = \frac{a_0 q \bar{c} \bar{e} \bar{l}^2 \cos^2 \Lambda}{GJ}, \qquad b = \frac{a_0 q \bar{c} \bar{l}^3 \sin \Lambda \cos \Lambda}{EI},$$

$$f_1(\xi) = - \frac{q \cos^2 \Lambda (\bar{c} \bar{c}_l{}^r \bar{e} + \bar{c}_{m\mathrm{AC}} \bar{c}^2) - \bar{m}Ng\bar{d}}{GJ/\bar{l}^2},$$

$$f_2(\xi) = - \frac{q \bar{c} \bar{c}_l{}^r \sin \Lambda \cos \Lambda - \bar{m}Ng \tan \Lambda}{EI/\bar{l}^3} \cdot$$

Differentiating Eq. (8–218) once with respect to ξ and subtracting Eq. (8–219) from the result gives

$$\frac{d^3 \bar{\alpha}}{d\xi^3} + a \frac{d\bar{\alpha}}{d\xi} - b\bar{\alpha} = f_1{}'(\xi) - f_2(\xi), \qquad (8\text{--}220)$$

where
$$\bar{\alpha} = \bar{\theta} - \Gamma \tan \Lambda.$$

Boundary conditions on Eq. (8–220) are obtained from Eqs. (8–157), (8–158), and (8–159). From Eq. (8–157),

$$\bar{\alpha}(1) = 0. \tag{8–221}$$

From Eq. (8–158),

$$\frac{d\bar{\alpha}(0)}{d\xi} = 0. \tag{8–222}$$

From Eqs. (8–159) and (8–218),

$$\frac{d^2\bar{\alpha}(0)}{d\xi^2} = -\bar{\alpha}(0) + f_1(0). \tag{8–223}$$

Closed-form solutions to Eq. (8–220) can be obtained for a variety of prescribed lift distributions on the rigid wing. If, for example, the rigid wing is untwisted and the attitude is specified, the functions $f_1(\xi)$ and $f_2(\xi)$ are constants designated by f_1 and f_2, respectively, and $f_1'(\xi)$ is zero. In this case it can be verified (Ref. 8–23) that the complete solution to (8–220) satisfying the boundary conditions is

$$\bar{\alpha}(\xi) = \frac{f_2}{b}\left[1 - \frac{f_3(\xi)}{f_3(1)}\right], \tag{8–224}$$

where

$$f_3(\xi) = \left(\frac{4\beta^2}{9\beta^2 + \gamma^2}\right)e^{-2\beta\xi} + e^{\beta\xi}\left\{\left(\frac{5\beta^2 + \gamma^2}{9\beta^2 + \gamma^2}\right)\cos\gamma\xi + \left(\frac{3\beta^3 - \beta\gamma^2}{9\beta^2\gamma + \gamma^3}\right)\sin\gamma\xi\right\}, \tag{8–225}$$

and where -2β and $\beta \pm i\gamma$ are the roots of the characteristic equation

$$r^3 + ar - b = 0. \tag{8–226}$$

The nature of the result given by Eq. (8–224) can be examined simply if we neglect the inertial forces. In this case the total lift coefficient divided by the rigid lift coefficient is simply

$$\frac{\bar{c}_l}{\bar{c}_l^r} = \frac{f_3(\xi)}{f_3(1)}. \tag{8–227}$$

It is apparent that the ratio given by Eq. (8–227) will depend upon the roots of the characteristic equation, which in turn depend upon the absolute values of the constants a and b. Figure 8–30 compares plots of Eq. (8–227) for a hypothetical uniform wing with 30° sweep back and 30° sweep forward. The wing is assumed to be operating in each case at a dynamic pressure equal to one-half of the divergence dynamic pressure of the swept-forward wing. The effect of sweep back in attenuating the aerodynamic loads due to deformations and the effect of sweep forward in amplifying them is apparent.

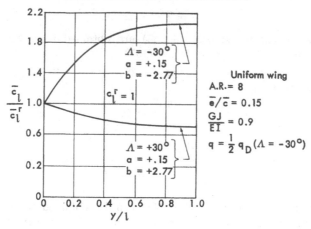

Fig. 8–30. Span load distributions on hypothetical swept-forward and swept-back wings.

(2) *Solutions in terms of matrices for swept wings of arbitrary planform and stiffness.* The symmetrical lift distribution problem can be formulated in matrix notation from the integral equation form given by Eq. (8–170). Combining Eqs. (8–170) and (8–173) gives

$$\mathfrak{a}[cc_l^e] = q \int_0^l \bar{C}(y,\eta)cc_l^e d\eta + \bar{f}(y), \qquad (8\text{--}228)$$

where cc_l^e is the lift distribution corresponding to the elastic twist $\theta(y)$, and $\mathfrak{a}$ is an appropriate linear aerodynamic operator accounting for spanwise aerodynamic effects. In matrix form, Eq. (8–228) becomes

$$[A^s]\{cc_l^e\} = q[\bar{E}]\{cc_l^e\} + \{\bar{f}\}, \qquad (8\text{--}229)$$

where

$[A^s] =$ matrix of aerodynamic influence coefficients for symmetrical loading on a swept wing,

$\{\bar{f}\} = q[\bar{E}]\{cc_l^r\} + q[\bar{F}]\{c_{mAC}\} - N[\bar{G}]\{mg\},$

$[\bar{E}] = ([C^{\theta z}] + [C^{\theta\theta}][\![e]\!])[\![\bar{W}]\!],$

$[\bar{F}] = [C^{\theta\theta}][\![c^2]\!][\![\bar{W}]\!],$

$[\bar{G}] = ([C^{\theta z}] + [C^{\theta\theta}][\![d]\!])[\![\bar{W}]\!].$

The matrix of aerodynamic influence coefficients $[A^s]$, according to Weissinger's theory, is given by Eq. (5–195).

Following the procedure used in the case of the straight wing, an alternative form of Eq. (8–229) can be derived by putting $\{cc_l^e\} = \{cc_l\} - \{cc_l^r\}$ into Eq. (8–229), which yields

$$[A^s]\{cc_l\} = q[\bar{E}]\{cc_l\} + \{\alpha^r\} + q[\bar{F}]\{c_{mAC}\} - N[\bar{G}]\{mg\}, \qquad (8\text{--}230)$$

where $$\{\alpha^r\} = [A^s]\{cc_l{}^r\}.$$

The unknown quantities in Eqs. (8–229) and (8–230) are the column matrices $\{cc_l{}^e\}$ and $\{cc_l\}$, respectively. All other quantities in each equation are assumed to be fixed by the physical properties of the wing and the conditions of the problem in exactly the same way as described for the straight wing in Section 8–3(c) (2). In fact, the entire development in Section 8–3(c) (2) applies also to the case of swept wings if the matrices $\{f\}$, $[E]$, $[F]$, and $[G]$ which appear there are replaced by the corresponding matrices $\{\tilde{f}\}$, $[\tilde{E}]$, $[\tilde{F}]$, and $[\tilde{G}]$ defined in this section. For example, when the attitude is specified, it follows from Eq. (8–85) that

$$[A^s]\{cc_l\} = q\left([\tilde{E}] - \frac{2}{W}[\tilde{G}]\{mg\}\lfloor 1 \rfloor [\overline{W}]\right)\{cc_l\} + \{\alpha^r\} + q[\tilde{F}]\{c_{mAC}\}.$$

$$(8\text{–}231)$$

Equation (8–231) can be solved by matrix inversion, and in some cases by matrix iteration, in the same manner as the straight wing equation. Experience has shown that if the sweep back is appreciable, convergence of the iteration process is poor, and the process may, in fact, diverge in some cases. In general, matrix inversion is the most satisfactory technique to apply to swept-back wings for purposes of computing span load distribution.

Matrix equation (8–231) is useful in practical application for computing the influence of the elastic swept-back wing on longitudinal static stability characteristics. Let us consider how Eq. (8–231) can be employed to compute $\partial C_L/\partial \alpha$ and $\partial C_M/\partial \alpha$. Suppose that the airplane is in flight and we impose an incremental angle of attack $\Delta\alpha$ on the wing. According to Eq. (8–231), the corresponding change in lift distribution is given by

$$\{\Delta cc_l\} = \left([A^s] - q[\tilde{E}] + \frac{2q}{W}[\tilde{G}]\{mg\}\lfloor 1 \rfloor[\overline{W}]\right)^{-1}\{\Delta\alpha\}. \quad (8\text{–}232)$$

The corresponding change in wing lift coefficient is obtained by numerical integration, as follows:

$$\Delta C_L = \frac{2}{S}\lfloor 1 \rfloor[\overline{W}]\{\Delta cc_l\}, \quad (8\text{–}233)$$

and the corresponding change in wing moment coefficient is

$$\Delta C_M = -\frac{2\tan\Lambda}{S(\text{MAC})}\lfloor y \rfloor[\overline{W}]\{\Delta cc_l\}, \quad (8\text{–}234)$$

where S is the wing area and MAC is the mean aerodynamic chord of the wing. The moment coefficient in this case is taken about the y-axis for convenience. Combining Eqs. (8–232) and (8–233), we obtain the lift-curve slope of the elastic wing, as follows:

$$\frac{\partial C_L}{\partial \alpha} = \frac{2}{S} \lfloor 1 \rfloor [\bar{W}] \left([A^s] - q[\bar{E}] + \frac{2q}{W}[\bar{G}]\{mg\}\lfloor 1 \rfloor [\bar{W}]\right)^{-1} \{1\}. \quad (8\text{-}235\text{a})$$

The moment-curve slope of the elastic wing is obtained by substituting Eq. (8-232) into Eq. (8-234):

$$\frac{\partial C_M}{\partial \alpha} = -\frac{2 \tan \Lambda}{S(\text{MAC})} \lfloor y \rfloor [\bar{W}] \left([A^s] - q[\bar{E}] + \frac{2q}{W}[\bar{G}]\{mg\}\lfloor 1 \rfloor [\bar{W}]\right)^{-1} \{1\}.$$

$$(8\text{-}235\text{b})$$

In general, the effects are to reduce $\partial C_L/\partial \alpha$ and to produce a destabilizing effect on $\partial C_M/\partial \alpha$. The latter is, of course, due to the forward movement of the center of pressure which accompanies the upward bending deflection of a swept-back wing. It can be observed by inspection of Eqs. (8-235a) and (8-235b) that the rigid values would be obtained by putting $[\bar{E}] = [\bar{G}] = 0$. The terms associated with $[\bar{E}]$ and $[\bar{G}]$ are of opposite signs, that is, the aeroelastic effects due to airloads and inertial loads are in opposition. It can be concluded from this that the inertial loads tend to reduce the aeroelastic effects on the lift- and moment-curve slopes of a swept-back wing. This phenomenon was pointed out by Pai and Sears in Ref. 8-18, where several illustrative numerical examples are given.

EXAMPLE 8-7. An airplane with a swept-back wing is subjected to a symmetrical $3g$ pull-out at a Mach number of 0.8 at 27,000 ft. Calculate the spanwise lift distribution. Apply the Weissinger method to calculate the rigid and corrective elastic lift distributions.

Solution. A sketch of the wing planform is given in Fig. 8-31. The following basic airplane data are required:

Wing span, $2l = 1392$ in.
Aspect ratio, $\text{AR} = 9.43$.
Gross weight, $W = 125,000$ lb.
Dynamic pressure corresponding to $M = 0.8$ at 27,000 ft, $q = 2.277$ lb/in².
Angle of sweep, $\Lambda = 34°$ to elastic axis and 35° to quarter-chord line.
Lift-curve slope at all sections, $a_0 = 6$ per radian.
Moment coefficient in streamwise direction at all sections, $c_{mAC} = -0.01$.

Four Multhopp stations are selected along the lift span, as illustrated by Fig. 8-31. Pertinent data at these four stations are given in the table.
Since the total load factor on the airplane is specified, we apply

$$\begin{bmatrix} [A^s] - q[\bar{E}] & \{-1\} \\ \lfloor 1 \rfloor [\bar{W}] & 0 \end{bmatrix} \begin{bmatrix} \{cc_l^e\} \\ \alpha^e(0) \end{bmatrix} = \begin{bmatrix} \{\bar{f}\} \\ 0 \end{bmatrix}, \quad \text{(a)}$$

where $\{\bar{f}\} = q[\bar{E}]\{cc_l{}^r\} + q[\bar{F}]\{c_{mAC}\} - N[\bar{G}]\{mg\},$

$$[\bar{E}] = ([C^{\theta z}] + [C^{\theta\theta}][e])[\bar{W}],$$

$$[\bar{F}] = [C^{\theta\theta}][c^2][\bar{W}],$$

$$[\bar{G}] = ([C^{\theta z}] + [C^{\theta\theta}][d])[\bar{W}].$$

The wing is sufficiently slender so that the elastic matrices can be obtained by

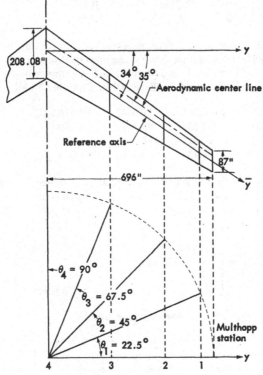

Fig. 8–31.   Multhopp stations on swept-back wing.

Multhopp station	1	2	3	4
% of spanwise location	0.9239	0.7071	0.3827	0
EI (lb · in² × 10⁹)	6.998	11.99	41.04	160.1
GJ (lb · in² × 10⁹)	6.005	10.01	27.50	100.1
mg (lb/in)	12.562	20.382	31.264	43.45
c (in)	96.24	122.5	161.8	208.1
e (in)	12.50	15.92	21.02	27.05
d (in)	0	0	0	0

applying Eqs. (2–128) through (2–131).

$$[C^{\theta z}] = \begin{bmatrix} -.4093 & -.1584 & -.02586 & 0 \\ -.3178 & -.1584 & -.02586 & 0 \\ -.1290 & -.0877 & -.02586 & 0 \\ 0 & 0 & 0 & 0 \end{bmatrix} \times 10^{-5} \, \text{rad/lb,} \qquad (b)$$

$$[C^{\theta\theta}] = \begin{bmatrix} .004111 & .001897 & .000453 & 0 \\ .001897 & .001897 & .000453 & 0 \\ .000453 & .000453 & .000453 & 0 \\ 0 & 0 & 0 & 0 \end{bmatrix} \times 10^{-5} \, \text{rad/in·lb.} \qquad (c)$$

The matrix of aerodynamic influence coefficients, according to Weissinger's theory, is obtained by substituting data for the present problem into Eq. (5–195). Since the Mach number is 0.8, a suitable Mach number correction has been applied to the Weissinger matrix, as discussed in Chapters 5 and 6.

$$[A^s] = \begin{bmatrix} 419.971 & -141.681 & 17.0941 & -16.0165 \\ -27.3168 & 250.158 & -99.4217 & 12.1300 \\ -1.1782 & -23.9116 & 185.711 & -65.5280 \\ -3.1070 & 3.4591 & -59.4864 & 171.074 \end{bmatrix} \times 10^{-5} \, \text{rad/in.} \qquad (d)$$

The weighting matrix $[\overline{W}]$ to be employed is obtained by applying Multhopp's quadrature formula. Application of this formula is discussed in Appendix B, where it is shown that for the particular four-station configuration shown by Fig. 8–32, the weighting matrix is

$$[\overline{W}] = \frac{\pi l}{8} \begin{bmatrix} \sin \pi/8 & & & \\ & \sin \pi/4 & & \\ & & \sin 3\pi/8 & \\ & & & \tfrac{1}{2} \sin \pi/2 \end{bmatrix}, \qquad (e)$$

where $l = 696$ in. (semispan) in the present example.

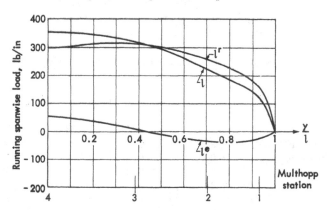

Fig. 8–32. Lift distribution on swept-back wing.

The matrices $[\bar{E}]$, $[\bar{F}]$, and $[\bar{G}]$, obtained from data given above, have the following values:

$$[\bar{E}] = \begin{bmatrix} -37.434 & -24.7764 & -4.1232 & 0 \\ -30.7608 & -24.7764 & -4.1232 & 0 \\ -12.8964 & -15.5472 & -4.1232 & 0 \\ 0 & 0 & 0 & 0 \end{bmatrix} \times 10^{-5} \frac{\text{rad·in}}{\text{lb}}, \qquad (f)$$

$$[\bar{F}] = \begin{bmatrix} 3982.32 & 5504.544 & 2995.20 & 0 \\ 1838.16 & 5504.544 & 2995.20 & 0 \\ 439.20 & 1315.152 & 2995.20 & 0 \\ 0 & 0 & 0 & 0 \end{bmatrix} \times 10^{-5} \frac{\text{rad·in}^2}{\text{lb}}, \qquad (g)$$

$$[\bar{G}] = \begin{bmatrix} -42.8100 & -30.6120 & -6.5292 & 0 \\ -33.2388 & -30.6120 & -6.5292 & 0 \\ -13.4928 & -16.9488 & -6.5292 & 0 \\ 0 & 0 & 0 & 0 \end{bmatrix} \times 10^{-5} \frac{\text{rad·in}}{\text{lb}}. \qquad (h)$$

In order to compute the matrix $\{\bar{f}\}$, it is necessary to establish the lift distribution on the rigid wing for $N = 3$. In the absence of model test data for the rigid wing, the Weissinger theory will be employed. If it is assumed that the wing does not have geometric or aerodynamic twist, we can write

$$[A^s]\{cc_l{}^r\} = \alpha^r\{1\}, \qquad (i)$$

where α^r is the rigid angle of attack. Since the integral of the rigid lift distribution must correspond to a load factor of 3, we have the additional condition that

$$q\lfloor 1 \rfloor[\bar{W}]\{cc_l{}^r\} = \frac{WN}{2}. \qquad (j)$$

Combining Eqs. (i) and (j), the wing angle of attack is

$$\alpha^r = \frac{WN}{2q} \times \frac{1}{\lfloor 1 \rfloor[\bar{W}][A^s]^{-1}\{1\}}. \qquad (k)$$

Inserting the matrices and numerical data listed above, we obtain for the rigid angle of attack

$$\alpha^r = 0.1432 \text{ radian}. \qquad (l)$$

The rigid lift distribution follows from Eq. (i):

$$\{cc_l{}^r\} = 0.1432[A^s]^{-1}\{1\} = \begin{bmatrix} 71.818 \\ 113.74 \\ 138.36 \\ 130.83 \end{bmatrix} \text{ inches}. \qquad (m)$$

The matrix $\{\bar{f}\}$ can now be calculated as follows:

$$\{\bar{f}\} = q[\bar{E}]\{cc_l{}^r\} + q[\bar{F}]\{c_{mAC}\} - N[\bar{G}]\{mg\}, \qquad (n)$$

$$\{\bar{f}\} = 2.277[\bar{E}]\begin{bmatrix} 71.818 \\ 113.74 \\ 138.36 \\ 130.83 \end{bmatrix} + 2.277[\bar{F}]\begin{bmatrix} -.01 \\ -.01 \\ -.01 \\ -.01 \end{bmatrix} - 3[\bar{G}]\begin{bmatrix} 12.562 \\ 20.382 \\ 31.264 \\ 43.450 \end{bmatrix}. \qquad (o)$$

Inserting the numerical values of $[\bar{E}]$, $[\bar{F}]$, and $[\bar{G}]$ given above and expanding gives

$$\{\bar{f}\} = \begin{bmatrix} -.13843 \\ -.12751 \\ -.07437 \\ 0 \end{bmatrix} + \begin{bmatrix} -.00284 \\ -.00235 \\ -.00108 \\ 0 \end{bmatrix} + \begin{bmatrix} .04097 \\ .03737 \\ .02157 \\ 0 \end{bmatrix} = \begin{bmatrix} -.1003 \\ -.09249 \\ -.05388 \\ 0 \end{bmatrix} \text{ radians.} \quad (p)$$

Inserting the data obtained above into Eq. (a) yields the following simultaneous equations:

$$\begin{bmatrix} .005052 & -.0008525 & .000265 & -.000160 & -1 \\ .0004275 & .003072 & -.00090 & .0001217 & -1 \\ .0002817 & .0001150 & .00195 & -.000655 & -1 \\ -.0000308 & .0000342 & -.000595 & .001708 & -1 \\ 104.59 & 193.26 & 252.50 & 136.66 & 0 \end{bmatrix} \begin{bmatrix} (cc_l{}^e)_1 \\ (cc_l{}^e)_2 \\ (cc_l{}^e)_3 \\ (cc_l{}^e)_4 \\ \alpha^e(0) \end{bmatrix}$$

$$= \begin{bmatrix} -.1003 \\ -.09249 \\ -.05388 \\ 0 \\ 0 \end{bmatrix} \cdot \quad (q)$$

Solving by matrix inversion, we obtain

$$\begin{bmatrix} \{cc_l{}^e\} \\ \alpha^e(0) \end{bmatrix} = \begin{bmatrix} -13.948 \\ -15.046 \\ 3.9444 \\ 24.664 \\ \\ .039754 \end{bmatrix} \begin{bmatrix} \text{inches} \\ \\ \text{radians} \end{bmatrix} \cdot \quad (r)$$

The total lift distribution is obtained by adding $\{cc_l{}^e\}$ and $\{cc_l{}^r\}$ and multiplying the result by the dynamic pressure q.

$$q\{cc_l\} = \begin{bmatrix} 163.58 \\ 259.09 \\ 315.17 \\ 298.02 \end{bmatrix} + \begin{bmatrix} -31.768 \\ -34.270 \\ 8.983 \\ 56.177 \end{bmatrix} = \begin{bmatrix} 131.81 \\ 224.82 \\ 324.15 \\ 354.20 \end{bmatrix} \text{ lb/in.} \quad (s)$$

Figure 8–32 illustrates a plot of the elastic, rigid, and total lift distributions. The inboard shift of the spanwise center of pressure is apparent from these results. In the present example, since only 7 Multhopp stations are used, the curve in Fig. 8–32 is subject to plotting error because of the small number of points. However, a seperate investigation using 15 stations has shown that the above results are nearly as accurate, and are satisfactory for engineering purposes. It can be observed that in the present example $\alpha^e(0)$ is a positive number, in contrast to the jet transport straight-wing problem of Example 8–3, in which $\alpha^e(0)$ is negative.

EXAMPLE 8–8. To compute the influence of an elastic swept-back wing on the slopes of the wing lift curve and wing pitching moment curve. Assume a uniform wing with strip theory and compute the ratios of

$$\frac{(\partial C_L/\partial\alpha)_{\text{elastic}}}{(\partial C_L/\partial\alpha)_{\text{rigid}}} \quad \text{and} \quad \frac{(\partial C_M/\partial\alpha)_{\text{elastic}}}{(\partial C_M/\partial\alpha)_{\text{rigid}}}$$

at values of dynamic pressure equal to one-half and twice the divergence dynamic pressure of the same wing swept forward. Assume that the wing is free to bend only, and that inertial effects are negligible.

Solution. Equation (8–235a) can be employed to compute the ratio of elastic to rigid lift-curve slope, as follows:

$$\frac{(\partial C_L/\partial\alpha)_{\text{elastic}}}{(\partial C_L/\partial\alpha)_{\text{rigid}}} = \frac{\lfloor 1 \rfloor \lceil \overline{W} \rceil ([A^s] - q[\overline{E}])^{-1}\{1\}}{\lfloor 1 \rfloor \lceil \overline{W} \rceil [A^s]^{-1}\{1\}}. \tag{a}$$

Similarly, the ratio of elastic to rigid moment-curve slope follows from Eq. (8–235b):

$$\frac{(\partial C_M/\partial\alpha)_{\text{elastic}}}{(\partial C_M/\partial\alpha)_{\text{rigid}}} = \frac{\lfloor y \rfloor \lceil \overline{W} \rceil ([A^s] - q[\overline{E}])^{-1}\{1\}}{\lfloor y \rfloor \lceil \overline{W} \rceil [A^s]^{-1}\{1\}}. \tag{b}$$

Since we are dealing with a uniform swept-back wing with strip theory, we can put

$$[A^s] - q[\overline{E}] = \frac{1}{a_0 c \cos \Lambda}[1] - q_D\left(\frac{q}{q_D}\right)[C^{\theta z}][\overline{W}], \tag{c}$$

where q_D is the divergence speed of the same wing swept forward.

If the wing is divided spanwise into six equal segments, as illustrated by Fig. 8–29, we can make use of data already computed in Example 8–6. Thus, we have for influence coefficients of a slender swept-back wing free only in bending, a matrix equal to minus the influence coefficient matrix of the same wing swept forward given by Eq. (b) in Example 8–6.

$$[C^{\theta z}] = -\frac{l^2}{EI}\,|\sin \Lambda|[n], \tag{d}$$

where the matrix of numbers $[n]$ is given by matrix (c) in Example 8–6. The matrix of weighting numbers, $[\overline{W}]$, according to Simpson's rule, is

$$[\overline{W}] = \frac{l \cos \Lambda}{18}[n'], \tag{e}$$

where the matrix $[n']$ is given by matrix (e) in Example 8–6.

Substituting Eqs. (c), (d), and (e) into Eq. (a) and eliminating q_D by means of Eq. (8–186), we obtain the following ratio of elastic to rigid lift-curve slope:

$$\frac{(\partial C_L/\partial\alpha)_{\text{elastic}}}{(\partial C_L/\partial\alpha)_{\text{rigid}}} = \frac{\lfloor 1 \rfloor \lceil n' \rceil \left([1] + \dfrac{q/q_D}{2.843}[n][n']\right)^{-1}\{1\}}{\lfloor 1 \rfloor \lceil n' \rceil [1]\{1\}}. \tag{f}$$

Similarly, the ratio of elastic to rigid moment-curve slope is obtained by substituting Eqs. (c), (d), and (e) into Eq. (b) and eliminating q_D by Eq. (8–186):

$$\frac{(\partial C_M/\partial\alpha)_{\text{elastic}}}{(\partial C_M/\partial\alpha)_{\text{rigid}}} = \frac{\lfloor y \rfloor \lceil n' \rceil \left([1] + \dfrac{q/q_D}{2.843}[n][n']\right)^{-1}\{1\}}{\lfloor y \rfloor \lceil n' \rceil [1]\{1\}}. \tag{g}$$

The matrix $\lfloor y \rfloor$ in Eq. (g) for the seven stations shown in Fig. 8–29 is

$$\lfloor y \rfloor = l\lfloor \tfrac{1}{6} \quad \tfrac{2}{6} \quad \tfrac{3}{6} \quad \tfrac{4}{6} \quad \tfrac{5}{6} \quad 1 \rfloor. \tag{h}$$

Evaluating Eqs. (f) and (g) successively for values of $q/q_D = \tfrac{1}{2}$ and 2, we obtain the values listed in the following table:

	$q/q_D = 0$	$q/q_D = \tfrac{1}{2}$	$q/q_D = 2$
$\dfrac{(\partial C_L/\partial \alpha)_{\text{elastic}}}{(\partial C_L/\partial \alpha)_{\text{rigid}}}$	1	.7355	.4671
$\dfrac{(\partial C_M/\partial \alpha)_{\text{elastic}}}{(\partial C_M/\partial \alpha)_{\text{rigid}}}$	1	.6831	.3648

The reduction effects due to aeroelasticity are apparent from the table. These effects will be reduced if inertial loads are taken into account, as discussed in the previous section.

(f) *Antisymmetrical lift distribution, aileron effectiveness, and reversal.* In Sections 8–4(a), 8–4(b), and 8–4(c), swept-wing equations applicable to a general unsymmetrical loading condition were derived. The equations can be specialized to the case of antisymmetrical loading by replacing c_l^r, c_{mAC} and N by appropriate values. Let us consider, for example, the integral equation form applicable to the case of antisymmetrical loading on swept wings of arbitrary planform and stiffness. If we introduce the following quantities into Eq. (8–169):

$$c_l^r = \frac{\partial c_l^r}{\partial \beta}\beta + \frac{\partial c_l^r}{\partial (pl/U)}\left(\frac{pl}{U}\right), \qquad c_{mAC} = \frac{\partial c_{mAC}}{\partial \beta}\beta, \qquad N = \frac{y\dot{p}}{g}, \tag{8–236}$$

we obtain

$$\theta(y) = q\int_0^l \bar{C}(y,\eta)cc_l^e d\eta + \dot{f}_a(y), \tag{8–237}$$

where

$$\dot{f}_a(y) = q\int_0^l \bar{C}(y,\eta)\left[c\frac{\partial c_l^r}{\partial \beta}\beta + c\frac{\partial c_l^r}{\partial (pl/U)}\left(\frac{pl}{U}\right)\right]d\eta$$

$$+ q\int_0^l C^{\theta\theta}(y,\eta)\frac{\partial c_{mAC}}{\partial \beta}\beta c^2 d\eta - \dot{p}\int_0^l [C^{\theta z}(y,\eta) + C^{\theta\theta}(y,\eta)d]m\eta d\eta. \tag{8–238}$$

When an appropriate relation between antisymmetric lift distribution and twist distribution is introduced, Eq. (8–237) assumes the form

$$\mathcal{a}[cc_l^e] = q\int_0^l \bar{C}(y,\eta)cc_l^e d\eta + \dot{f}_a(y), \tag{8–239}$$

where α is a linear aerodynamic operator.

We have seen that the procedure to be followed in obtaining numerical solutions is one of replacing Eq. (8–239) by its approximate matrix form, as follows:

$$[A^a]\{cc_l{}^e\} = q[\bar{E}]\{cc_l{}^e\} + \{\bar{f}_a\}, \tag{8-240}$$

where

$[A^a]$ = matrix of aerodynamic influence coefficients for antisymmetrical loading on a swept wing,

$$\{\bar{f}_a\} = q[\bar{E}]\left(\left\{c\frac{\partial c_l{}^r}{\partial \beta}\right\}\beta + \left\{c\frac{\partial c_l{}^r}{\partial (pl/U)}\right\}\left(\frac{pl}{U}\right)\right)$$

$$+ q[\bar{F}]\left\{\frac{\partial c_{mAC}}{\partial \beta}\right\}\beta - [\bar{G}][y]\{m\}\bar{p},$$

$$[\bar{E}] = ([C^{\theta z}] + [C^{\theta\theta}][e])[\bar{W}],$$
$$[\bar{F}] = [C^{\theta\theta}][c^2][\bar{W}],$$
$$[\bar{G}] = ([C^{\theta z}] + [C^{\theta\theta}][d])[\bar{W}].$$

The matrix of aerodynamic influence coefficients $[A^a]$, according to Weissinger's theory, is given by Eq. (5–195). Matrix equation (8–240) contains as its unknown quantity the column matrix $\{cc_l{}^e\}$. All other quantities are assumed to be fixed by the physical properties of the wing and the conditions of the problem. Equation (8–240) is identical in form to Eq. (8–114), which is the corresponding equation for the case of a straight wing. In fact, the entire development in Section 8–3(d) (7) applies also to the case of the swept wing. For example, the matrix equations which define the aileron effectiveness and aileron reversal conditions for a swept wing of arbitrary planform and stiffness can be written down at once by referring to Eqs. (8–122) and (8–122a). The aileron effectiveness is defined explicitly by

$$\frac{\partial\left(\frac{pl}{U}\right)}{\partial\beta} = -\frac{C_{l_\beta}+\frac{q}{Sl}\lfloor H\rfloor([A^a]-q[\bar{E}])^{-1}\left([\bar{E}]\left\{c\frac{\partial c_l{}^r}{\partial\beta}\right\}+[\bar{F}]\left\{\frac{\partial c_{mAC}}{\partial\beta}\right\}\right)}{C_{l_p}+\frac{q}{Sl}\lfloor H\rfloor([A^a]-q[\bar{E}])^{-1}[\bar{E}]\left\{c\frac{\partial c_l{}^r}{\partial(pl/U)}\right\}}, \tag{8-241}$$

and the aileron reversal dynamic pressure can be obtained by solution of

$$q_R = -\frac{C_{l_\beta}}{\frac{1}{Sl}\lfloor H\rfloor([A^a]-q_R[\bar{E}])^{-1}\left([\bar{E}]\left\{c\frac{\partial c_l{}^r}{\partial\beta}\right\}+[\bar{F}]\left\{\frac{\partial c_{mAC}}{\partial\beta}\right\}\right)}, \tag{8-241a}$$

where $\lfloor H\rfloor$ is a row matrix defined by Eq. (8–118b) and C_{l_β} and C_{l_p} are derivatives of the rolling moment coefficients of the rigid airplane as defined by Eq. (8–101a).

Perhaps the most important aspect of sweep in connection with anti-symmetrical lift distribution relates to the influence of sweep on aileron effectiveness. In general, we can say that a swept-forward wing has a high aileron reversal speed, whereas a swept-back wing has a low aileron reversal speed. Let us consider briefly the action of a swept-forward wing when an aileron is given a downward displacement. The bending of the wing under the upward load resulting from the displaced aileron produces a nose-up deflection of streamwise segments. This is in opposition to the twisting produced by the pitching moments resulting from the downward deflected aileron. Since the two effects tend to cancel each other, the wing twist resulting from the deflected aileron is small and the ailerons remain effective. Conversely, in the case of swept-back wings, streamwise angles of twist resulting from wing bending and from aileron pitching moment add together, resulting in a large amount of nose-down wing twist and hence a low aileron effectiveness.

EXAMPLE 8-9. To compute the aileron effectiveness and the aileron reversal speed at sea level of the swept-wing airplane of Example 8-7. Apply the Weissinger method to calculate the rigid and corrective elastic lift distributions.

Solution. The aileron effectiveness is obtained by applying Eq. (8-241).

Let us assume that rigid wing distributions of $c\frac{\partial c_{i}^{r}}{\partial\beta}$, $c\frac{\partial c_{i}^{r}}{\partial(pl/U)}$, and $\frac{\partial c_{mAC}}{\partial\beta}$ have been computed by methods such as those of Ref. 8-24, and are as illustrated graphically by Fig. 8-33. The matrix forms of these quantities evaluated at the Multhopp stations of Fig. 8-33 are

$$\left\{c\frac{\partial c_{i}^{r}}{\partial\beta}\right\} = \begin{bmatrix} 144.98 \\ 153.18 \\ -15.59 \end{bmatrix}, \quad \left\{c\frac{\partial c_{i}^{r}}{\partial(pl/U)}\right\} = \begin{bmatrix} -294.75 \\ -329.52 \\ -207.05 \end{bmatrix}, \quad \left\{\frac{\partial c_{mAC}}{\partial\beta}\right\} = \begin{bmatrix} -.4532 \\ -.4532 \\ 0 \end{bmatrix}.$$

In addition, application of the formulas for C_{l_β} and C_{l_p} gives for the rigid airplane

$$C_{l_\beta} = 0.16, \qquad C_{l_p} = -0.454.$$

The matrix of aerodynamic influence coefficients for antisymmetric loading, according to Weissinger's theory, is obtained by substituting data for the present problem into Eq. (5-195). Mach number corrections are omitted here in order to simplify the illustration

$$[A^a] = \begin{bmatrix} 45.679 & -13.221 & .63610 \\ -2.1228 & 28.004 & -7.3948 \\ -.63000 & -.78920 & 20.636 \end{bmatrix} \times 10^{-4}.$$

The row matrix $\lfloor H \rfloor$ in Eq. (8-241) is evaluated from Eq. (8-118b):

$$\lfloor H \rfloor = \lfloor 1 \rfloor [y][\overline{W}] = l\lfloor 96.634 \quad 131.736 \quad 96.635 \rfloor,$$

where $[\overline{W}]$ is the Multhopp weighting matrix given by Eq. (e) in Example 8-7.

Substitution of the data given above, together with numerical values of the matrices $[\overline{E}]$ and $[\overline{F}]$ given in Example 8-7, into Eq. (8-241) gives an explicit

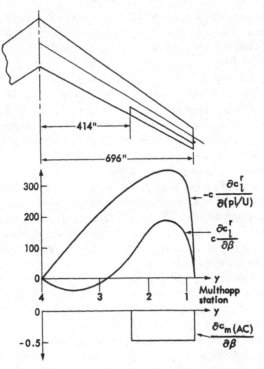

Fig. 8–33. Rigid wing distributions due to deflected aileron and steady roll.

expression for the aileron effectiveness in terms of q. Figure 8–34 illustrates a plot of the results showing the variation of aileron effectiveness with airspeed. An approximate Mach number correction based on the validity of strip theory, $U_{comp.} = U_{inc.} \sqrt[4]{1 - M^2}$ (cf. Section 6–2), can be applied to the results of Fig. 8–34.

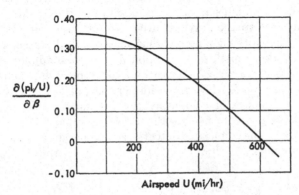

Fig. 8–34. Aileron effectiveness as a function of airspeed for swept wing.

(g) *Solutions of swept-wing problems in terms of generalized coordinates.* We have seen in the case of straight wings in Section 8–3(e) that useful methods of obtaining approximate solutions can be based upon the idea of expanding a solution in terms of assumed deformation shapes that satisfy the boundary conditions. Such methods can be applied also to the swept wing; however, their mathematical validity has not yet been as rigorously established as in the case of the straight wing. Intuitively, one does expect such methods to yield good results if the assumed deformation shapes can be superimposed to give a close approximation to the actual deformation shape under load.

(1) *Rayleigh-Ritz method.* It was pointed out in a previous section that the problem of static aeroelasticity of a swept wing is a non self-adjoint boundary value problem, whereas the comparable problem involving a straight wing is self-adjoint. It is well known that the principle of minimum potential energy is rigorously applicable only when the resulting differential or integral equation is self-adjoint (Ref. 8–25). The self-adjointness condition is satisfied by the vast majority of problems in elasticity and mechanics, and the principle of minimum potential energy is often justified in such problems purely on a basis of physical reasoning. However, from a more rigorous point of view, employment of a variational condition to derive a given differential or integral equation of equilibrium requires that the function which is to be minimized must be obtained separately for each problem or class of problems. In the case of conservative elastic systems, for example, it so happens that the function which is to be minimized is the potential energy of the system [cf. Section 2–11(a)]. In other systems, such as the swept wing, this function is not amenable to simple physical interpretation. The interested reader is referred to Ref. 8–26, in which Flax suggests a formal extension of the ordinary Rayleigh-Ritz method to the swept-wing problem.

(2) *Galerkin's method.* Galerkin's method offers a direct approach to the approximate solution of swept-wing problems. We have seen in Chapter 4 that Galerkin's method, as usually applied in Mechanics, is entirely equivalent to the Rayleigh-Ritz method, when applied to self-adjoint systems. Although no rigorous mathematical statement of the validity of Galerkin's method when applied to non self-adjoint systems is available, practical numerical applications to swept-wing problems have given satisfactory results. Let us consider first an application of Galerkin's method to the problem of computing the deformation under load of a slender swept wing free to bend without twisting, under the assumption of strip theory. The applicable integral equation can be derived from Eq. (8–169) by putting $C^{\theta\theta}(y,\eta) = 0$ and $c_l{}^e = a_0\theta \cos \Lambda$.

$$\theta(y) = qa_0 \cos \Lambda \int_0^l C^{\theta z}(y,\eta)c\theta d\eta + \bar{f}(y), \qquad (8\text{--}242)$$

where $\tilde{f}(y)$ is obtained from Eq. (8–171) for symmetrical loading, and from Eq. (8–238) for antisymmetrical loading.

Although Eq. (8–242) applies to both swept-forward and swept-back wings, the swept-back wing loading problem is of most practical interest and subsequent remarks are confined to this case. Since the influence function in Eq. (8–242) is a negative quantity for swept-back wings, we can specialize Eq. (8–242) to the case of a swept-back wing, as follows:

$$\theta(y) = -qa_0 \cos \Lambda \int_0^l |C^{\theta z}(y,\eta)|c\theta d\eta + \tilde{f}(y). \qquad (8\text{–}243)$$

Putting as a solution

$$\theta(y) = \sum_{j=1}^{n} \Theta_j q_j, \qquad (8\text{–}244)$$

where the $\Theta_j(y)$ are functions that satisfy the boundary conditions $\theta(0) = \theta'(l) = 0$, we obtain the error function

$$\mathcal{E}(y) = \sum_{j=1}^{n} \left\{ \Theta_j(y) + qa_0 \cos \Lambda \int_0^l |C^{\theta z}(y,\eta)|\Theta_j(\eta)cd\eta \right\} q_j - \tilde{f}(y). \quad (8\text{–}245)$$

Applying the Galerkin process of multiplying through by $\Theta_i(y)c(y)$, integrating from 0 to l, and requiring that $\int_0^l \mathcal{E}\Theta_i cdy = 0$, we obtain the following set of simultaneous linear algebraic equations:

$$\sum_{j=1}^{n} A_{ij}q_j = B_i, \qquad (i = 1, 2, \cdots, n), \qquad (8\text{–}246)$$

where

$$A_{ij} = \int_0^l \Theta_i \Theta_j cdy + qa_0 \cos \Lambda \int_0^l \Theta_i c \int_0^l |C^{\theta z}(y,\eta)|\Theta_j cd\eta dy,$$

$$B_i = \int_0^l \tilde{f}(y)\Theta_i cdy.$$

The streamwise twist distribution can be obtained by solving Eq. (8–246) simultaneously for $q_1, \cdots, q_n$ and substituting the results into Eq. (8–244).

If only a single mode shape, Θ_1, is employed, the generalized coordinate can be computed at once from Eq. (8–246):

$$q_1 = \frac{\displaystyle\int_0^l \tilde{f}(y)\Theta_1 cdy}{\displaystyle\int_0^l \Theta_1{}^2 cdy + qa_0 \cos \Lambda \int_0^l \Theta_1 c \int_0^l |C^{\theta z}(y,\eta)|\Theta_1 cd\eta dy}. \qquad (8\text{–}247)$$

Following a suggestion by Flax in Ref. 8–26, Eq. (8–247) can be simplified further by selecting as the single assumed mode shape $\Theta_1(y)$, the divergence mode shape of the same wing swept forward. This is given by the eigenfunction corresponding to the lowest eigenvalue of

$$\Theta_1(y) = q_D a_0 \cos \Lambda \int_0^l |C^{\theta z}(y,\eta)| \Theta_1 c d\eta, \qquad (8\text{-}248)$$

where q_D is the divergence speed of the swept-forward wing. Substituting Eq. (8-248) into Eq. (8-247), we obtain a simpler expression for the single generalized coordinate:

$$q_1 = \frac{\displaystyle\int_0^l \bar{f}(y)\Theta_1 c d y}{[1 + (q/q_D)]\displaystyle\int_0^l \Theta_1{}^2 c d y}. \qquad (8\text{-}249)$$

Thus, the twist distribution under the single-mode approximation is given by

$$\theta(y) = \Theta_1(y)q_1, \qquad (8\text{-}250)$$

where q_1 is given by Eq. (8-247) if Θ_1 is arbitrarily assumed, or is given by Eq. (8-249) if Θ_1 is the divergence mode shape of the same wing swept forward.

The simultaneous equations (8-246) can be reduced to diagonal form by expanding the solution in terms of biorthogonal eigenfunctions. Let us assume the existence of a set of biorthogonal eigenfunctions $\Theta_1, \cdots, \Theta_n$ and $W_1, \cdots, W_n$ which satisfy, respectively, the following adjoint homogeneous integral equations:

$$\Theta_j = \lambda_j a_0 \cos \Lambda \int_0^l |C^{\theta z}(y,\eta)| \Theta_j c d\eta, \qquad (8\text{-}251)$$

$$W_j = \lambda_j a_0 \cos \Lambda \int_0^l |C^{z\theta}(y,\eta)| W_j c d\eta. \qquad (8\text{-}252)$$

For a given swept-back wing, Eq. (8-251) represents physically the integral equation of the same wing swept forward. Let us put as a solution to Eq. (8-243)

$$\theta(y) = \sum_{j=1}^n \Theta_j(y)q_j, \qquad (8\text{-}253)$$

where the $\Theta_j(y)$ are eigenfunctions of Eq. (8-251). Substituting Eq. (8-253) into Eq. (8-243), we obtain the error function:

$$\mathcal{E}(y) = \sum_{j=1}^n \left\{\Theta_j + q a_0 \cos \Lambda \int_0^l |C^{\theta z}(y,\eta)| \Theta_j c d\eta\right\} q_j - \bar{f}(y). \qquad (8\text{-}254)$$

Proceeding to the next step in the Galerkin process of multiplying through Eq. (8-254) by $W_i(y)c(y)$, integrating from 0 to l, and putting the result equal to zero gives

$$\sum_{j=1}^{n} \left\{ \int_0^l \Theta_j W_i c\, dy + qa_0 \cos \Lambda \int_0^l W_i c \int_0^l |C^{\theta z}(y,\eta)| \Theta_j c\, d\eta\, dy \right\} q_j$$

$$- \int_0^l \bar{f}(y) W_i c\, dy = 0. \qquad (8\text{–}255)$$

Equation (8–255) can be simplified by applying Eq. (8–251) and the orthogonality condition between Θ_i and W_i expressed by Eq. (8–214):

$$q_i = \frac{\displaystyle\int_0^l \bar{f}(y) W_i c\, dy}{[1 + (q/\lambda_i)] \displaystyle\int_0^l \Theta_i W_i c\, dy}. \qquad (8\text{–}256)$$

Thus, an explicit result for the twist distribution of a swept-back wing can be expressed by

$$\theta(y) = \sum_{i=1}^{n} \frac{\Theta_i(y) \displaystyle\int_0^l \bar{f}(y) W_i c\, dy}{[1 + (q/\lambda_i)] \displaystyle\int_0^l \Theta_i W_i c\, dy}, \qquad (8\text{–}257)$$

where the Θ_i are eigenfunctions of the same wing swept forward defined by Eq. (8–251), the W_i are eigenfunctions of its adjoint, Eq. (8–252), and the λ_i are eigenvalues of both equations. Useful results can be obtained from Eq. (8–257) by taking only the first term in the series, in which case we obtain

$$\theta(y) = \frac{\Theta_1(y) \displaystyle\int_0^l \bar{f}(y) W_1 c\, dy}{[1 + (q/q_D)] \displaystyle\int_0^l \Theta_1 W_1 c\, dy}, \qquad (8\text{–}258)$$

where q_D is the divergence speed of the same wing swept forward. The quantity $[1 + (q/q_D)]$ acts as an attenuation factor in determining the elastic twist of a swept-back wing. Here the swept-forward divergence speed q_D plays an important role as a reference value. It can be observed that if the same analysis were carried out for a swept-forward wing, we would obtain the quantity $[1 - (q/q_D)]$ in the denominator. This quantity acts as an amplification factor.

The reader who is interested in a further and more rigorous discussion of the mathematical validity of an expansion of the solution of the swept-back wing aeroelastic problem in terms of biorthogonal eigenfunctions is referred to a paper by Seifert (Ref. 8–27).

Galerkin's method, as applied above to a slender swept-back wing free to bend, can be generalized to include the case where twisting action is also permitted. In this case, we commence the solution with Eq. (8–170):

$$\theta(y) = q \int_0^l \bar{C}(y,\eta)cc_l{}^e d\eta + \bar{f}(y), \qquad (8\text{–}170)$$

where $\bar{f}(y)$ is defined by Eq. (8–171) for symmetrical cases, and by Eq. (8–238) for antisymmetrical cases. Let us introduce the following series into Eq. (8–170):

$$\theta(y) = \sum_{j=1}^n \Theta_j(y)q_j, \qquad (8\text{–}259)$$

$$c_l{}^e(y) = \sum_{j=1}^n c_{l_j}(y)q_j, \qquad (8\text{–}260)$$

where the $\Theta_j(y)$ are arbitrarily assumed functions satisfying the boundary conditions* $\theta(0) = \theta'(l) = 0$, and the $c_{l_j}(y)$ are spanwise lift coefficient distributions corresponding to the spanwise twisting modes $\Theta_j(y)$. Putting Eqs. (8–259) and (8–260) into Eq. (8–170) and applying Galerkin's method, the following set of simultaneous algebraic equations in the unknowns $q_1, \cdots, q_n$ is obtained:

$$\sum_{j=1}^n A_{ij}q_j = B_i, \qquad (i = 1, 2, \cdots, n), \qquad (8\text{–}261)$$

where

$$A_{ij} = \int_0^l \Theta_i\Theta_j c\,dy - q \int_0^l \Theta_i c \int_0^l \bar{C}(y,\eta)cc_{l_j}d\eta\,dy, \qquad (8\text{–}262)$$

$$B_i = \int_0^l \bar{f}(y)\Theta_i c\,dy. \qquad (8\text{–}263)$$

The streamwise twist distribution $\theta(y)$ and the spanwise lift coefficient distribution $c_l{}^e(y)$ can be obtained by solving Eqs. (8–261) simultaneously for $q_1, \cdots, q_n$ and substituting the results into Eqs. (8–259) and (8–260), respectively.

Although Galerkin's method has been applied in the present section to the integral equations of equilibrium, it is, of course, applicable also to the differential equations derived in Sections 8–4(a) and (b). Miles, in Ref. 8–19, applies Galerkin's method to differential equations of the form of (8–164) and (8–165). Another variation of the technique of solving static aeroelastic problems in terms of a superposition of selected mode shapes is described by Brown, Holtby, and Martin in Ref. 8–28. Their

---

* These boundary conditions can be verified by observing that $\theta(y) = \bar{\theta}$ $\cos \Lambda - (dw/d\bar{y}) \sin \Lambda$. Since $\bar{\theta}(0) = dw(0)/d\bar{y} = 0$, we can conclude that $\theta(0) = 0$. Since $d\bar{\theta}/d\bar{y} = d^2w/d\bar{y}^2 = 0$ at $y = l$, it follows that $\theta'(l) = 0$.

application was developed principally for use in design calculations involving slender swept-back wings where simplicity and rapidity of solution are important considerations.

EXAMPLE 8–10. To compute the twist distribution of a uniform swept-back wing, free only in bending and operating at a specified constant rigid angle of attack α^r, in terms of biorthogonal functions. Assuming strip theory and neglecting inertial forces, compare the results with the exact results for dynamic pressures of one-half and twice the divergence dynamic pressure of the same wing swept forward.

Solution. The solution is obtained by applying Eq. (8–258), as follows:

$$\theta(y) = \frac{\Theta_1(y) \int_0^l \bar{f}(y) W_1 c\, dy}{[1 + (q/q_D)] \int_0^l \Theta_1 W_1 c\, dy}, \tag{8–258}$$

where

$$\bar{f}(y) = q a_0 \alpha^r c \cos \Lambda \int_0^l C^{\theta z}(y,\eta)\, d\eta. \tag{a}$$

We shall use the numerical results for Θ_1 and W_1 computed in Example 8–6. It is necessary, therefore, to put Eqs. (8–258) and (a) into matrix form. The matrix form of Eq. (8–258) is

$$\{\theta\} = \frac{\lfloor \bar{f} \rfloor \lceil \overline{W} \rceil \{W_1\}}{\lfloor \Theta_1 \rfloor \lceil \overline{W} \rceil \{W_1\}} \times \frac{\{\Theta_1\}}{[1 + (q/q_D)]}, \tag{b}$$

and of Eq. (a) is

$$\{\bar{f}\} = q a_0 \alpha^r c \cos \Lambda [C^{\theta z}] \lceil \overline{W} \rceil \{1\}. \tag{c}$$

Evaluating Eq. (c) by substituting values for $[C^{\theta z}]^*$ and $\lceil \overline{W} \rceil$ obtained from Example 8–6, there results

$$\{\bar{f}\} = -\frac{q a_0 c \bar{l}^3 \cos^2 \Lambda \,|\sin \Lambda|\, \alpha^r}{18EI} \begin{bmatrix} 0 \\ 1.26378 \\ 2.11095 \\ 2.62485 \\ 2.88875 \\ 2.98595 \\ 2.99975 \end{bmatrix}. \tag{d}$$

Introducing this result into Eq. (b) and substituting numerical values for $\lceil \overline{W} \rceil$, $\{W_1\}$, and $\{\Theta_1\}$ given in Example 8–6, we obtain, after expansion,

$$\left\{\frac{\theta}{\alpha^r}\right\} = -\frac{q a_0 c \bar{l}^3 \cos^2 \Lambda \,|\sin \Lambda|}{18EI} \times \frac{18.0398}{5.9997} \times \frac{\{\Theta_1\}}{[1 + (q/q_D)]}. \tag{e}$$

---

* Notice that the appropriate value of $[C^{\theta z}]$ for a swept-back wing is obtained by placing a minus sign in front of the $[C^{\theta z}]$ matrix given in Example 8–6.

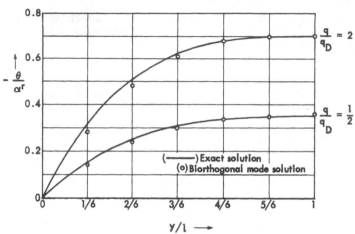

Fig. 8-35.   Comparison of biorthogonal mode solution with exact solution.

Introducing Eq. (8-186) and numerical values for $\{\Theta_1\}$, we obtain as the final solution

$$\left\{\frac{\theta}{\alpha^r}\right\} = -\frac{q/q_D}{1+(q/q_D)}\begin{bmatrix}0\\.4311\\.7322\\.9189\\1.0164\\1.0524\\1.0576\end{bmatrix}. \tag{f}$$

The results given by Eq. (f) are plotted in Fig. 8-35, where they are compared with the exact solution for ratios of $q/q_D = \frac{1}{2}$ and 2. The agreement between the approximate and exact solutions is seen to be excellent in this simple case with the exception of inboard stations for $q/q_D = 2$. The lack of agreement is probably due to the fact that higher mode contributions become important at higher speeds.

8-5 Low aspect-ratio lifting surfaces of arbitrary planform and stiffness. Previous sections of this chapter have dealt with wings in which camber bending in the streamwise direction can be neglected. A sufficient degree of slenderness is implied in these analyses such that it is possible to apply aerodynamic and structural theories that are reasonably well established. This fortuitous state of affairs is due largely to the implicit simplifying assumptions that spanwise flow components are small compared with chordwise components, and that spanwise normal stresses are large compared with chordwise values. The static aeroelastic problem is thus characterized by a single independent variable along the spanwise axis. We turn our attention next to low aspect-ratio wings in which camber bending

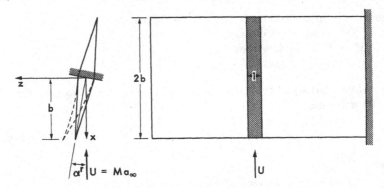

Fig. 8–36.   Deformation of chordwise strip of rectangular supersonic wing.

in the streamwise direction can no longer be ignored, and which may, in fact, have spanwise and chordwise dimensions of the same order of magnitude.   Here we are confronted with a problem in which spanwise and chordwise components of flow and normal stress assume nearly equal proportions. It is hardly necessary to remark that the complexities of the aeroelastic problem are multiplied, and that techniques available for obtaining practical solutions are thereby fewer.   It can be said, however, that this comparative state of affairs is perhaps fortunate, since aeroelastic phenomena associated with relatively slender wings are, in general, more pronounced than with low aspect-ratio lifting surfaces.   It is also generally true that the aeroelastic phenomenon of most importance in low aspect-ratio lifting surfaces occurs in the supersonic speed range.

(a) *Aeroelastic behavior of a chordwise strip at supersonic speed.*   In order to introduce the phenomenon of chordwise bending, let us consider the behavior of a unit width chordwise strip of a thin rectangular wing, as shown by Fig. 8–36.   We assume at first that the wing is incapable of bending in the spanwise direction or twisting about a spanwise axis.   With these simplifying assumptions, the chordwise segment may be treated as a cantilever beam of unit width clamped rigidly at the midchord line.   The origin of coordinates is taken at the midchord as illustrated by Fig. 8–36.

The simplified problem posed here was discussed in detail first by Biot in Ref. 8–29, and the analyses which follow are based essentially upon his development.

Differential equation of equilibrium.   Assuming that the slope dw/dx of the deflection curve is small, the local lift per unit area according to linearized supersonic wing theory is given by Eq. (6–55) as

$$\Delta p_a = \frac{4M^2}{\sqrt{M^2 - 1}} \frac{\rho_\infty a_\infty^2}{2} \frac{dw}{dx},$$ (8–264)

where Δp_a = pressure difference on the airfoil,

ρ_∞ = air density,

$M = U/a_\infty$ = Mach number,

a_∞ = speed of sound.

The differential equation of the deflection of the unit chordwise segment can be written as

$$\frac{d^2}{dx^2}\left(E_1 I(x)\frac{d^2 w}{dx^2}\right) = \frac{4M^2}{\sqrt{M^2-1}}\frac{\rho_\infty a_\infty^2}{2}\frac{dw}{dx}, \qquad (8\text{–}265)$$

where

$I(x)$ = area moment of inertia of the strip,

$E_1 = \dfrac{E}{1-\nu^2}$ = reduced modulus of elasticity for two-dimensional strain,*

E = modulus of elasticity of wing material,

ν = Poisson's ratio of wing material.

If the segment is at an initial rigid angle of attack α^r,

$$\frac{dw}{dx} = \alpha^r + \alpha^e(x), \qquad (8\text{–}266)$$

where $\alpha^e(x)$ is the local slope due to elastic deformation. Thus Eq. (8–265) becomes

$$\frac{d^2}{dx^2}\left(E_1 I\frac{d\alpha^e}{dx}\right) - \frac{4M^2}{\sqrt{M^2-1}}\frac{\rho_\infty a_\infty^2}{2}\alpha^e = \frac{4M^2}{\sqrt{M^2-1}}\frac{\rho_\infty a_\infty^2}{2}\alpha^r. \qquad (8\text{–}267)$$

Equation (8–267) is of the non self-adjoint type. Its form is similar to that of the differential equations derived previously for the bending of swept wings.

Slab of uniform thickness. A simple form of Eq. (8–267) can be derived by assuming constant thickness t. Putting $I = t^3/12$ and $\xi = 1 - x/b$, Eq. (8–267) reduces to

$$\frac{d^3\alpha^e}{d\xi^3} + k\alpha^e = -k\alpha^r, \qquad (8\text{–}268)$$

where

$$k = \frac{24M^2}{\sqrt{M^2-1}} \times \frac{\rho_\infty a_\infty^2}{E_1}\left(\frac{b}{t}\right)^3. \qquad (8\text{–}269)$$

The general solution of Eq. (8–268) is

$$\alpha^e(x) = C_1 e^{\lambda_1\xi} + C_2 e^{\lambda_2\xi} + C_3 e^{\lambda_3\xi} - \alpha^r, \qquad (8\text{–}270)$$

---

* We assume as a first approximation that the chordwise strip is in a state of plane strain (Ref. 8–30).

where the characteristic roots λ_1, λ_2, and λ_3 have the values

$$\lambda_1 = -\sqrt[3]{k}, \qquad \lambda_2 = \tfrac{1}{2}(1 + i\sqrt{3})\sqrt[3]{k}, \qquad \lambda_3 = \tfrac{1}{2}(1 - i\sqrt{3})\sqrt[3]{k}, \quad (8\text{--}271)$$

and where the constants C_1, C_2, and C_3 are obtained from the boundary conditions

$$\alpha^e(1) = \frac{d\alpha^e(0)}{d\xi} = \frac{d^2\alpha^e(0)}{d\xi^2} = 0. \tag{8--272}$$

Inserting the solution (8–270) into the boundary conditions (8–272) gives the simultaneous equations

$$C_1\lambda_1 + C_2\lambda_2 + C_3\lambda_3 = 0,$$
$$C_1\lambda_1{}^2 + C_2\lambda_2{}^2 + C_3\lambda_3{}^2 = 0,$$
$$C_1e^{\lambda_1} + C_2e^{\lambda_2} + C_3e^{\lambda_3} = \alpha^r, \tag{8--273}$$

which yield upon solution,

$$C_1 = \frac{\lambda_2\lambda_3(\lambda_3 - \lambda_2)\alpha^r}{\Delta},$$

$$C_2 = \frac{\lambda_1\lambda_3(\lambda_1 - \lambda_3)\alpha^r}{\Delta}, \tag{8--274}$$

$$C_3 = \frac{\lambda_1\lambda_2(\lambda_2 - \lambda_1)\alpha^r}{\Delta},$$

where

$$\Delta = e^{\lambda_1}\lambda_2\lambda_3(\lambda_3 - \lambda_2) + e^{\lambda_2}\lambda_1\lambda_3(\lambda_1 - \lambda_3) + e^{\lambda_3}\lambda_1\lambda_2(\lambda_2 - \lambda_1).$$

The complete solution of the differential equation is thus given by Eq. (8–270), where the λ's and C's are defined by Eqs. (8–271) and (8–274), respectively.

Biot (Ref. 8–31) has examined the solution in terms of the bending moment amplification ratio

$$\frac{M(b)}{M_r(b)} = -\frac{(E_1/12)(t^3/b)(d\alpha^e/d\xi)_{\xi=1}}{(M^2/\sqrt{M^2 - 1})\rho_\infty a_\infty{}^2 b^2 \alpha^r}, \tag{8--275}$$

where $M(b)$ is the actual bending moment at $\xi = 1$ and $M_r(b)$ is the bending moment at $\xi = 1$, assuming that the strip is rigid. Introducing Eq. (8–270) and simplifying,

$$\frac{M(b)}{M_r(b)} = -\frac{2}{k\alpha^r}(\lambda_1 C_1 e^{\lambda_1} + \lambda_2 C_2 e^{\lambda_2} + \lambda_3 C_3 e^{\lambda_3}). \tag{8--276}$$

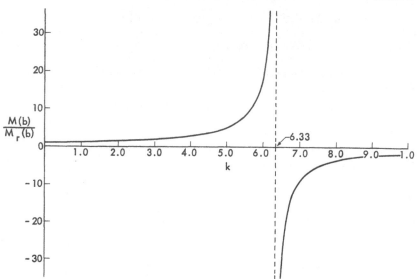

Fig. 8–37.   Bending moment amplification ratio.

Inserting the λ's and C's given by Eqs. (8–271) and (8–274) and reducing yields

$$\frac{M(b)}{M_r(b)} = \frac{2}{k^{\frac{2}{3}}} \times \frac{e^{-\frac{3}{2}\sqrt[3]{k}} + \sqrt{3}\sin\left(\frac{\sqrt{3}}{2}\sqrt[3]{k}\right) - \cos\left(\frac{\sqrt{3}}{2}\sqrt[3]{k}\right)}{e^{-\frac{3}{2}\sqrt[3]{k}} + 2\cos\left(\frac{\sqrt{3}}{2}\sqrt[3]{k}\right)}. \qquad (8\text{–}277)$$

Equation (8–277), which relates bending moment amplification ratio to the parameter k, is plotted in Fig. 8–37.   It can be seen from this plot that a critical value for k of 6.33 causes the denominator of Eq. (8–277) to be zero.   This corresponds to a critical speed of chordwise divergence of the strip.*

Integral equation of equilibrium.   When the chordwise strip is nonuniform, a solution of the differential equation is not practical except in the simplest application.   In such cases, it is preferable to obtain numerical solutions of the integral equation of equilibrium.   We refer to Fig. 8–38, which illustrates a blunt symmetric elastic slab of arbitrary thickness distribution in a supersonic flow of velocity U.   Since Biot has shown in Ref. 8–29 that a sharp leading edge introduces a singular mathematical

---

* The value 6.33 also represents the lowest eigenvalue of the homogeneous differential equation of a uniform swept-forward wing [cf. Eq. (8–185)].

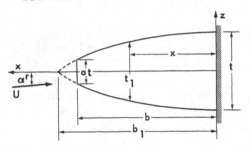

Fig. 8–38.   Blunt symmetric slab of arbitrary thickness.

behavior,* we assume that the profile is cut off at a certain distance from the leading edge such that the slab has a leading edge bluntness at. Introducing the assumptions of small thickness and deflection, the integral equation of equilibrium can be expressed as

$$\alpha^e(x) = \frac{4M^2}{\sqrt{M^2 - 1}} \frac{\rho_\infty a_\infty^2}{2} \int_0^b C(x,\xi)[\alpha^e(\xi) + \alpha^r(\xi)]d\xi, \quad (8\text{–}278)$$

where $C(x,\xi)$ is an influence function expressing the slope of the slab at x due to a unit load at ξ, and where b is the semichord.

Slab of symmetric biconvex cross section. Let us assume that the non-uniform slab has a symmetric biconvex cross section such that the top and bottom of the profile are parabolic arcs and the thickness t_1 at a point x along the chord is expressed by

$$t_1 = t\left[1 - \left(\frac{x}{b_1}\right)^2\right], \quad (8\text{–}279)$$

where t is the maximum thickness and b_1 is the distance from the origin of coordinates to a hypothetical nose, as illustrated by Fig. 8–38. For this variation in thickness, Biot (Ref. 8–33) has shown by integration of the beam bending differential equation that the following influence coefficient expressions apply for a unit strip:

$$C(\xi,\eta) = \frac{12}{b_1 E_1}\left(\frac{b_1}{t}\right)^3 \gamma(\xi,\eta), \quad (8\text{–}280)$$

---

* The assumption of an infinitely sharp leading edge, together with the employment of linearized supersonic aerodynamic theory, leads to difficulty because of the nature of some of the mathematical results. These results indicate that an infinitely sharp leading edge is unstable at all supersonic speeds.   Fung (Ref. 8–32) has also investigated the behavior of infinitely sharp leading edges. His results indicate that an infinitely sharp leading edge is unstable at all speeds, including subsonic as well as supersonic speeds.

where for $\xi < \eta$,

$$\gamma(\xi,\eta) = \frac{1}{4}\frac{\xi\eta}{(1-\xi^2)^2} + \frac{3}{8}\frac{\xi\eta}{(1-\xi^2)} + \frac{3}{16}\eta\ln\left(\frac{1+\xi}{1-\xi}\right) - \frac{1}{4}\frac{1}{(1-\xi^2)^2} + \frac{1}{4},$$

and $\gamma(\xi,\eta) = \gamma(\eta,\eta)$ for $\xi > \eta$. In these expressions the variables ξ and η denote distances from the origin of coordinates as a fraction of the length b_1.

Reducing the independent variables in Eq. (8–278) to nondimensional form and introducing (8–279) and (8–280) gives

$$\alpha^e(\xi) = k_1\int_0^{\sqrt{1-a}}\gamma(\xi,\eta)\alpha^e(\eta)d\eta + f(\xi), \qquad (8\text{–}281)$$

where

$$k_1 = 24\frac{M^2}{\sqrt{M^2-1}} \times \frac{\rho_\infty a_\infty^2}{E_1}\left(\frac{b_1}{t}\right)^3,$$

$$f(\xi) = k_1\int_0^{\sqrt{1-a}}\gamma(\xi,\eta)\alpha^r(\eta)d\eta.$$

The characteristic parameter referred to the actual semichord b is

$$k = 24\frac{M^2}{\sqrt{M^2-1}} \times \frac{\rho_\infty a_\infty^2}{E_1}\left(\frac{b}{t}\right)^3,$$

and it is related to k_1 by

$$k = k_1(\sqrt{1-a})^3. \qquad (8\text{–}282)$$

Equation (8–281) may be employed to compute the deformed shape of the chordwise strip, or if α^r is put equal to zero, the chordwise divergence speed can be obtained from the lowest characteristic value.

Biot (Ref. 8–33) has applied a collocation method employing Lagrangian interpolation functions to the computation of the divergence speed. Let us suppose that the slab is divided into two segments such that

$$\begin{aligned}\alpha^e(\xi) &= \alpha_0 = 0 &&\text{for} && \xi = 0,\\ \alpha^e(\xi) &= \alpha_1 &&\text{for} && \xi = \xi_1,\\ \alpha^e(\xi) &= \alpha_2 &&\text{for} && \xi = \xi_2.\end{aligned} \qquad (8\text{–}283)$$

Applying Lagrange's interpolation formula [Appendix B, Eqs. (5) and (9)] yields

$$\alpha^e(\xi) = \alpha_1\frac{\xi(\xi-\xi_2)}{\xi_1(\xi_1-\xi_2)} + \alpha_2\frac{\xi(\xi-\xi_1)}{\xi_2(\xi_2-\xi_1)}. \qquad (8\text{–}284)$$

If the slab is divided into equal segments such that $\xi_1 = \frac{1}{2}\sqrt{1-a}$ and $\xi_2 = \sqrt{1-a}$, expression (8–284) becomes

$$\alpha^e(\xi) = \frac{\xi}{\sqrt{1-a}}(4\alpha_1 - \alpha_2) + \frac{\xi^2}{1-a}(-4\alpha_1 + 2\alpha_2). \qquad (8\text{–}285)$$

If we introduce expression (8–285) into Eq. (8–281) and put $f(\xi) = 0$, we obtain by application of the collocation process for the case of $a = 0.5$, the following characteristic equations:

$$\frac{1}{k_1}\begin{bmatrix} \alpha_1 \\ \alpha_2 \end{bmatrix} = \begin{bmatrix} .03532 & .02907 \\ .04096 & .03460 \end{bmatrix}\begin{bmatrix} \alpha_1 \\ \alpha_2 \end{bmatrix}. \qquad (8\text{–}286)$$

Solution of Eq. (8–286) for the lowest characteristic value yields

$$\frac{1}{k_1} = 0.0664, \quad \text{and} \quad k = k_1(\sqrt{1 - a})^3 = 5.324.$$

The Mach number corresponding to chordwise divergence is obtained by solution of

$$M^4 - \left[\frac{k}{24(\rho_\infty a_\infty{}^2/E_1)(b/t)^3} \right]^2 (M^2 - 1) = 0. \qquad (8\text{–}287)$$

When a sharper profile is analyzed, higher degree polynomials should be employed.

(b) *General discussion of aeroelastic behavior of low aspect-ratio lifting surfaces.* Let us consider next a somewhat more general formulation of the problem of the static aeroelastic behavior of low aspect-ratio lifting surfaces. We have seen by Eq. (2–55) that the lateral displacement $w(x,y)$ of an elastic surface is related to the lateral load intensity per unit area $Z(x,y)$ by

$$w(x,y) = \iint_S C(x,y;\xi,\eta)Z(\xi,\eta)d\xi d\eta. \qquad (2\text{–}55)$$

Following the procedures employed in Section 8–5(a), it will be convenient for aerodynamic reasons to discuss wing deformation in terms of local streamwise angles of attack. Differentiating Eq. (2–55) with respect to x yields

$$\alpha^e(x,y) = -\iint_S C^{\alpha z}(x,y;\xi,\eta)Z(\xi,\eta)d\xi d\eta, \qquad (8\text{–}288)$$

where $\alpha^e(x,y) = -\partial w/\partial x$ is the local streamwise angle of attack resulting from wing deformations, and $C^{\alpha z}(x,y;\xi,\eta) = \partial C(x,y;\xi,\eta)/\partial x$ is an influence function that defines the local slope in the x-direction at (x,y) due to a unit load at (ξ,η). The lateral load $Z(x,y)$ results from aerodynamic pressures and inertial loads as follows:

$$Z(x,y) = \Delta p_a - \rho N g, \qquad (8\text{–}289)$$

where Δp_a is the net upward aerodynamic pressure on the wing and ρ is the mass per unit area. If Δp_a is divided into a component $\Delta p_a{}^r$ due to the original angle of attack of the rigid undeformed wing, and a component $\Delta p_a{}^e$ due to elastic deformation, Eq. (8–289) becomes

$$Z(x,y) = \Delta p_a{}^e + \Delta p_a{}^r - \rho N g. \qquad (8\text{–}290)$$

Combining Eqs. (8–288) and (8–290), we obtain the following integral equation of equilibrium:

$$-\alpha^e(x,y) = \iint_S C^{\alpha z}(x,y;\xi,\eta)\Delta p_a{}^e(\xi,\eta)d\xi d\eta + f(x,y), \qquad (8\text{–}291)$$

where

$$f(x,y) = \iint_S C^{\alpha z}(x,y;\xi,\eta)(\Delta p_a{}^r - \rho Ng)d\xi d\eta. \qquad (8\text{–}292)$$

The unknown quantities in Eq. (8–291) are $\alpha^e(x,y)$ and $\Delta p_a{}^e(x,y)$. The problem of solving Eq. (8–291) becomes mathematically determinate when we supply an additional relation based upon aerodynamic theory between the unknown quantities. This relation may be stated, for example, according to the notation of Chapter 5, in the following symbolic form:

$$\alpha(x,y) = \frac{1}{U}\,\mathfrak{a}(\gamma_a) = \frac{1}{\rho U^2}\,\mathfrak{a}(\Delta p_a), \qquad (8\text{–}293)$$

where $\Delta p_a = \rho U \gamma_a$. Equations (8–291) and (8–293) form the basis for predicting the static aeroelastic phenomena of a lifting surface. For example, the divergence speed is obtained from the lowest eigenvalue of the homogeneous equation obtained by putting $f(x,y) = 0$. Symmetric and antisymmetric loading conditions are obtained by introducing appropriate values of $\Delta p_a{}^r$ and N.

In considering possible solutions, we must be guided by the availability of the aerodynamic and elastic ingredients which are involved. We have seen in Chapters 2 and 5 that although steady state aerodynamic and elastic structural theories are available for lifting surfaces of arbitrary planform, they are so complex that we are forced to conclude that elaborate numerical methods are surely necessary.

Solution by collocation. One method of solution is that of a collocation process which requires that Eq. (8–291) be satisfied at n separate points on the surface of the wing. Thus, for the ith point, Eq. (8–291) has the form

$$-\alpha_i{}^e = \iint_S C^{\alpha z}(x_i,y_i;\xi,\eta)\Delta p_a{}^e(\xi,\eta)d\xi d\eta + f_i, \qquad (i = 1, 2, \cdots, n). \quad (8\text{–}294)$$

Applying numerical integration by means of weighting numbers, Eq. (8–294) reduces to the following summation form:

$$-\alpha_i{}^e = \sum_{j=1}^{n} C_{ij}{}^{\alpha z}\overline{W}_j \Delta p_{aj}{}^e + f_i, \qquad (i = 1, 2, \cdots, n). \quad (8\text{–}295)$$

Writing Eq. (8–295) in terms of matrices and introducing the matrix form of Eq. (8–293), we obtain

$$-\frac{1}{\rho U^2}[A]\{\Delta p_a{}^e\} = [C^{\alpha z}][\overline{W}]\{\Delta p_a{}^e\} + \{f\}, \qquad (8\text{–}296)$$

where
$$\{f\} = [C^{\alpha z}][\overline{W}](\{\Delta p_a{}^r\} - N\{\rho g\}).$$

Once the matrices of aerodynamic influence coefficients $[A]$ and structural influence coefficients $[C^{\alpha z}]$ are defined explicitly, Eq. (8–296) can be solved by the methods employed in solving the same form of matrix equation in the case of slender wings. It is in the definition of these matrices that the problem becomes one of exceptional difficulty. Matters relating to the structural and aerodynamic matrices of influence coefficients are discussed in detail in Chapters 2, 5, 6, and 7.

Solution in terms of generalized coordinates. A second method of solution is based upon the use of generalized coordinates and assumed modes in much the same manner as discussed in Sections 8–3(e) and 8–4(g). Let us assume that the streamwise slopes resulting from wing deformations can be represented by a superposition of assumed modes in the following way:

$$\alpha^e(x,y) = \sum_{k=1}^{n} \gamma_k(x,y)q_k, \qquad (8\text{–}297)$$

where the $\gamma_k(x,y)$ are assumed deformation shapes that satisfy the boundary conditions $\gamma_k(x,0) = \partial\gamma_k(x,l)/\partial y = 0$. If we now suppose that for each deformation shape we can obtain, by means of an appropriate aerodynamic theory, a corresponding pressure distribution $\Delta p_k(x,y)$, the pressure distribution $\Delta p_a{}^e$ can be represented approximately by

$$\Delta p_a{}^e = \sum_{k=1}^{n} \Delta p_k(x,y)q_k. \qquad (8\text{–}298)$$

Substituting expressions (8–297) and (8–298) into Eq. (8–291) yields

$$\sum_{k=1}^{n}\left\{\gamma_k(x,y) + \iint_S C^{\alpha z}(x,y;\xi,\eta)\Delta p_k d\xi d\eta\right\}q_k = -f(x,y). \qquad (8\text{–}299)$$

A solution to Eq. (8–299) can be obtained by Galerkin's method or by a simple collocation process in which the equation is satisfied at a network of n points on the wing surface. If, for example, Eq. (8–299) is satisfied at n stations on the wing surface, we obtain n simultaneous algebraic equations:

$$\sum_{k=1}^{n}\left\{\gamma_k(x_i,y_i) + \iint_S C^{\alpha z}(x_i,y_i;\xi,\eta)\Delta p_k d\xi d\eta\right\}q_k = -f(x_i,y_i),$$
$$(i = 1, 2, \cdots, n). \qquad (8\text{–}300)$$

After evaluating the integral by a weighting matrix, Eq. (8–300) reduces to

$$\sum_{k=1}^{n}\left\{\gamma_{ki} + \sum_{j=1}^{n} C_{ij}{}^{\alpha z}\overline{W}_j\Delta p_{kj}\right\}q_k = -f_i, \qquad (i = 1, 2, \cdots, n). \qquad (8\text{–}301)$$

The solution for the pressure distribution due to elastic deformation is obtained by putting the values of the generalized coordinates obtained by solving Eq. (8–301) into Eq. (8–298). In selecting assumed deformation functions, the analyst must be guided by the existence of corresponding pressure distributions $\Delta p_k(x,y)$. Because of this important consideration, Lin and Pian in Ref. 8–34 suggest the following form of Eq. (8–297):

$$\alpha^e(x,y) = yq_1 + xyq_2 + y^2q_3 + x^2yq_4 + xy^2q_5 + y^3q_6 + \cdots. \quad (8\text{–}302)$$

Prediction of the pressure distributions corresponding to these deformation functions has been the subject of extensive investigation. Explicit and relatively simple methods are available for computing these pressure distributions in the case of very low aspect-ratio wings and, of course, high aspect-ratio wings. At medium aspect ratios, in the vicinity of one and two, no easily applied theory is available. These and other matters relating to pressure distributions over wings of arbitrary planform are discussed in Chapters 5, 6, and 7.

CHAPTER 9

FLUTTER

9–1 Introduction. The nature of flutter. Flutter can be defined as the dynamic instability of an elastic body in an airstream. It is most commonly encountered on bodies subjected to large lateral aerodynamic loads of the lift type, such as aircraft wings, tails, and control surfaces. Perhaps more characteristically than any other, it is an aeroelastic problem. Like divergence, the only air forces necessary to produce it are those due to deflections of the elastic structure from the undeformed state.

In a large proportion of all cases we get an adequate definition of the flutter properties of a system by studying the stability of infinitesimal motions. It is then sufficient to analyze a vibration with exponential time dependence e^{pt} (p complex), since all other small motions can be built up therefrom by superposition. If slight deformations are dynamically unstable, it is an undesirable situation on any piloted or automatically controlled aircraft, regardless of the stability of bigger ones. In practice, the larger displacements are usually stable if the small ones are. In fact, they may have much greater stability, as in the case of the amplitude limitation due to stalling on flutter oscillations, which is often observed on very flexible aeroelastic models in the wind tunnel. There exists an excellent possibility of improving flutter performance by somehow installing in the structure a properly designed, rapidly responding automatic control system, actuated in closed-loop fashion by the motion to be stabilized. However, we cannot count on the human pilot to compensate for flutter, as he does for certain instabilities of the rigid airplane, because the frequencies are too high to permit his effective response.

The flutter or critical speed U_F and frequency ω_F are defined, respectively, as the *lowest* airspeed and the corresponding circular frequency at which a given structure flying at given atmospheric density and temperature will exhibit sustained, simple harmonic oscillations. Flight at U_F represents a borderline condition or neutral stability boundary, because all small motions must be stable at speeds below U_F, whereas divergent oscillations can ordinarily occur in a range of speeds (or at all speeds) above U_F. Because it is easier mathematically to describe the aerodynamic loads due to simple harmonic motion, theoretical flutter analysis often consists of assuming in advance that all dependent variables are proportional to $e^{i\omega t}$ (ω real), and finding combinations of U and ω for which this actually occurs. One is thus led to a complex or double eigenvalue problem, where two characteristic numbers determine the speed and frequency. This is in

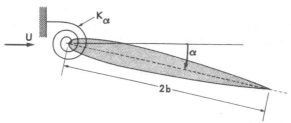

Fig. 9–1. Rigid, symmetrical airfoil restrained to rotate about its leading edge in a two-dimensional flow.

contrast to free vibration of a linear structure in vacuum, which is a real or single eigenvalue problem.

Flutter prediction has been variously done by purely theoretical means, by analog computation, by wind-tunnel or rocket experiments on scaled dynamic models, and by flight testing full-scale aircraft. The decision as to which of these is most economical in a given case depends on a multitude of factors, such as the anticipated margin of safety from flutter, the Mach number range, and the number of different mass and structural configurations to be analyzed. At times a rapid calculation based on extreme simplifying assumptions is sufficient to assure that the designer has nothing to fear from flutter. Again, an aeroelastician faced with the study of several hundred combinations of externally mounted masses on a very flexible wing once remarked that a dynamic model was the most economical and accurate analog computer he could buy. Some of these questions are discussed at more length in Chapters 12 and 13. The present chapter is limited to a review of the theoretical approach.

Since flutter is such a complicated phenomenon, as encountered in practice, it is very easy to lose sight of its physical aspects. For this reason, we preface our plunge into matters like multiple degrees of freedom and typical airplane structures by examining a very simple system which exhibits many of the properties we shall later meet in more formidable guise. Consider a two-dimensional, rigid airfoil of unit span, which is hinged at its leading edge but elastically restrained from rotating about this axis by a torsion spring with constant K_α ft·lb/rad. The airfoil is placed in a low-speed airstream so that the unstrained position of the spring corresponds to zero angle of attack α (Fig. 9–1). If the moment of inertia about the leading edge is I_α, the equation of motion for this single-degree-of-freedom system reads

$$I_\alpha \ddot{\alpha} + K_\alpha \alpha = M_y, \qquad (9\text{–}1)$$

where M_y is the aerodynamic moment due to $\alpha(t)$. For purposes of the present illustration, we assume the air density ρ to be so low and the airfoil to be so heavily weighted that the combination

$$\frac{I_\alpha}{\pi\rho b^4} = 1000. \tag{9-2}$$

(This figure is unrealistic, since for sea-level density it corresponds roughly to a 12%-thick airfoil made of solid iron, and it is some 25–50 times the values normally encountered on aircraft! However, our simple example requires it before flutter can be obtained.)

As defined in Chapter 5, the axis of rotation is located at $a = -1$. We first solve directly for the flutter condition by assuming

$$\alpha = \bar\alpha_0 e^{i\omega t}, \tag{9-3}$$

substituting Eq. (9–3) into Eq. (9–1), and dividing through by $\pi\rho b^4\omega^2\bar\alpha_0 e^{i\omega t}$:

$$\frac{I_\alpha}{\pi\rho b^4}\left[1 - \left(\frac{\omega_\alpha}{\omega}\right)^2\right] + m_y = 0. \tag{9-4}$$

Here we have used ω_α to denote the natural frequency of torsional vibration in vacuum:

$$\omega_\alpha = \sqrt{\frac{K_\alpha}{I_\alpha}}, \tag{9-5}$$

and m_y is shorthand for the dimensionless aerodynamic coefficients, which in the notation of Ref. 9–1 would be written (cf. Section 5–6)

$$m_y = \frac{M_y}{\pi\rho b^4\omega^2\bar\alpha_0 e^{i\omega t}} = M_\alpha - (L_\alpha + \tfrac{1}{2})(\tfrac{1}{2} + a) + L_h(\tfrac{1}{2} + a)^2. \tag{9-6}$$

Since $a = -1$, m_y is a function only of the reduced frequency of oscillation $k = \omega b/U$. m_y is a complex number, so that Eq. (9–4) can be split into real and imaginary parts:

$$\operatorname{Re}\{m_y\} = \frac{I_\alpha}{\pi\rho b^4}\left[\left(\frac{\omega_\alpha}{\omega}\right)^2 - 1\right] = 1000\left[\left(\frac{\omega_\alpha}{\omega}\right)^2 - 1\right], \tag{9-7a}$$

$$\operatorname{Im}\{m_y\} = 0. \tag{9-7b}$$

We see that flutter occurs at that value of the reduced frequency which just makes the out-of-phase component of the aerodynamic moment vanish, provided that the corresponding in-phase part is of such magnitude that Eq. (9–7a) yields a nonimaginary flutter frequency ω. As shown by Smilg (Ref. 9–2), the latter condition is met for this axis location so long as $I_\alpha/\pi\rho b^4$ exceeds roughly 550. On our wing, k_F turns out to be about 0.038, and $\omega_\alpha/\omega_F = 0.68$. For example, if the wing had a chord of 1 ft and a torsion spring adjusted to give a natural frequency of 1 cycle/sec, it would flutter at $U_F = 120$ ft/sec and a frequency of 1.47 cycles/sec.

To see what happens to bring about this instability, we imagine an experiment in which the airspeed U is gradually brought up from zero while we examine the airfoil's motion. At any speed, it is plausible that we can excite natural oscillations. Their frequency will rise slowly from very nearly 1 cycle/sec at $U = 0$ ft/sec to 1.47 cycles/sec at flutter. The frequency at zero U is determined by the fact that the small virtual moment of inertia $(\frac{9}{8})\pi\rho b^4$ due to the air at rest is negligible alongside of I_α, so that the wing is practically vibrating in vacuum. At any airspeed the frequency can be estimated quite satisfactorily by augmenting the torsion spring K_α by an effective aerodynamic spring, whose stiffness comes from the restoring moment $(M_y)_{k\to0}$ due to steady-state displacement of the airfoil:

$$(M_y)_{k\to0} = -\pi\rho U^2 b^2 \alpha. \tag{9-8}$$

When Eq. (9-8) is inserted into Eq. (9-1), we find for the frequency of the approximating spring-mass system

$$\omega^2 = \frac{K_\alpha}{I_\alpha} + \frac{\pi\rho U^2 b^2}{I_\alpha} = \omega_\alpha{}^2 + \frac{\pi\rho b^4}{I_\alpha} \cdot \frac{U^2}{b^2} = (2\pi)^2 + \frac{U^2}{1000b^2}. \tag{9-9}$$

Equation (9-9) estimates the frequency of free oscillation or, alternatively, the value of ω at which forced simple harmonic motion of the system will pass through a resonance peak.

The gradual rise of ω with airspeed is accompanied by a rapid fall (from ∞ at $U = 0$) of the reduced frequency k. Since we are concerned here with stability, it is the *damping* present in the actual system which interests us. This is entirely of aerodynamic origin and depends only on k. For let us imagine the airfoil artificially maintained in sinusoidal motion at the frequency given by Eq. (9-9). Then Eq. (9-6) applies, and Fig. 9-2 shows how the in-phase and out-of-phase parts of m_y behave as functions of k. We can regard the inertia, spring, and aerodynamic torques in Eq. (9-1) as rotating complex vectors with fixed length, and angular position increasing at the rate ωt. Using the position vector $\alpha = \bar{\alpha}_0 e^{i\omega t}$ as a reference, the relative locations (at $t = 0$) of the three torque vectors on the system with unit torsional frequency are plotted at five values of U (Fig. 9-3). We see for airspeeds below U_F (relatively high k), that Im $\{M_y\}$ is negative. It goes through zero at U_F and becomes positive above, where k is lower than the flutter value 0.038. This out-of-phase component Im $\{M_y\}$ is the only source of damping or instability for the system; it can easily be shown that the work per cycle done by the air on the airfoil in simple harmonic motion is just

$$W = \pi\bar{\alpha}_0 \text{ Im} \left\{\frac{M_y}{e^{i\omega t}}\right\}. \tag{9-10}$$

Thus an M_y which lags that motion (Im $\{M_y\} < 0$) removes energy from the oscillation, providing damping, and vice versa. The out-of-phase

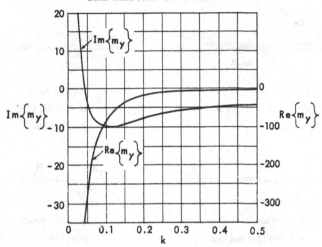

Fig. 9–2. Variation with reduced frequency k of the real and imaginary parts of the dimensionless aerodynamic moment m_y due to pitching of an airfoil about its leading edge in incompressible flow.

moments are not large compared with others in the system (cf. Fig. 9–2), but they have complete control of its stability. It is just the very small phase shift of M_y, critically dependent on the airspeed, that leads to flutter.

The function of U_F as a stability boundary, beyond which divergent oscillations will occur, is thus explained. When the airspeed becomes very large, if we neglect compressibility, we can estimate the asymptotic values of the frequency and reduced frequency from Eq. (9–9). Ultimately the term involving the spring stiffness can be neglected compared with the one containing U^2, so that

$$\omega^2 \cong \frac{U^2}{1000b^2} \qquad (9\text{–}11a)$$

or

$$k = \frac{\omega b}{U} \cong \sqrt{\frac{1}{1000}} = 0.0316. \qquad (9\text{–}11b)$$

This k is in the unstable range of Im $\{M_y\}$, so that we can expect the flutter instability to persist indefinitely. Incidentally, the moment vectors in Fig. 9–3 are not exactly in balance, except at flutter, because of the intentional omission of the artificial forcing moment needed to sustain sinusoidal motion. The main function of this other moment is to restore the energy removed, or take out the excess added, by Im $\{M_y\}$.

We mention in concluding this example that it illustrates how unsteady aerodynamic effects are of the essence in controlling flutter speed, even at

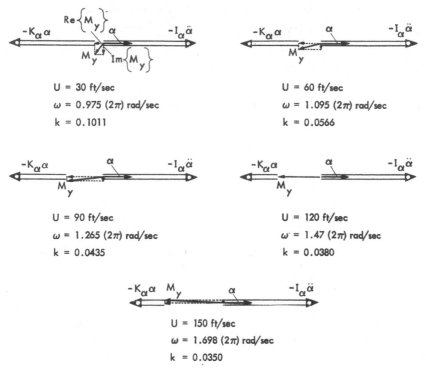

Fig. 9–3. Aerodynamic torque M_y, spring torque $-K_\alpha\alpha$, and inertia torque $-I_\alpha\ddot{\alpha}$ acting on an airfoil in simple harmonic motion at resonance frequency ω as the airspeed U is increased. Torques are shown as rotating complex vectors at the instant $t = 0$. (*Note:* The values of $\mathrm{Im}(M_y)$ have been increased ten times.)

the very low reduced frequencies involved here. Had we substituted for M_y the quasi-steady aerodynamic moment from Section 5–6,

$$M_{y\mathrm{QS}} = -\pi\rho U^2 b^2 \alpha - 3\pi\rho U b^3 \dot{\alpha} - \tfrac{9}{8}\pi\rho b^4 \ddot{\alpha}, \qquad (9\text{–}12)$$

we should have found a positive damping term proportional to airspeed and therefore an entirely stable system. Although not encountered in primary structures at low speeds, this single-degree-of-freedom torsional flutter is a very real phenomenon. It has been observed on small, relatively massive objects like yaw vanes, and it may also occur on wings in supersonic flow.

9–2 Flutter of a simple system with two degrees of freedom. Probably the most dangerous, although not the most frequently encountered, type of aircraft flutter results from coupling between the bending and torsional motions of a relatively large aspect-ratio wing and tail. A great deal of qualitative information can be obtained about the influence of

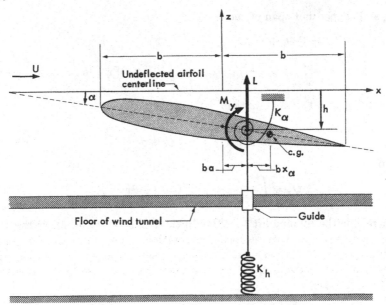

Fig. 9–4. Airfoil restrained from bending and torsional motion in an air-stream by springs K_h and K_α, acting a distance ba aft of midchord. Also shown are lift L and pitching moment M_y about the axis of twist.

various system parameters on this kind of flutter by studying the stability of the simple airfoil illustrated in Fig. 3–22. Figure 9–4 repeats the picture, showing the location of the coordinate system along with some quantities of primary aerodynamic interest. This airfoil is just the "representative section" of Theodorsen and Garrick (Refs. 9–3 and 9–4), specialized for the bending and torsion degrees of freedom. These authors suggest, and subsequent experience has confirmed, that for purposes of flutter prediction it can be made to represent fairly well a straight wing of large span by giving it the geometric and inertial properties of the station three-quarters of the way from center line to wingtip.

Section 3–8(c) demonstrates how Lagrange's equation may be used to derive the free-vibration equations for this representative section. If the thin airfoil is subjected to a distribution of pressure difference $(p_U - p_L)$ because of the air flowing past, we must include generalized external forces in the equations of motion:

$$m\ddot{h} + S_\alpha\ddot{\alpha} + m\omega_h{}^2 h = Q_h, \tag{9–13a}$$

$$S_\alpha\ddot{h} + I_\alpha\ddot{\alpha} + I_\alpha\omega_\alpha{}^2\alpha = Q_\alpha. \tag{9–13b}$$

Specialization of Eq. (3–210) for the work δW_e done by the air pressures during infinitesimal virtual displacements δh and $\delta\alpha$ yields expressions for

Q_h and Q_α per unit span of the airfoil:

$$\delta W_e = Q_h \delta h + Q_\alpha \delta \alpha$$

$$= \int_{-b}^{b} \{(p_U - p_L)dx\delta h + (p_U - p_L)[x - ba]dx\delta\alpha\}. \quad (9\text{--}14)$$

From Eq. (9–14) it follows that

$$Q_h = \int_{-b}^{b} (p_U - p_L)dx = -L \qquad (9\text{--}15a)$$

and

$$Q_\alpha = \int_{-b}^{b} (p_U - p_L)[x - ba]dx = M_y, \qquad (9\text{--}15b)$$

where L is the running lift (positive in the z-direction) and M_y is the running moment (positive for positive rotation about the spanwise y-axis along $x = ba$).* To use these directly, we agree that the section has unit length in the y-direction. We summarize the other constants appearing in Eqs. (9–13):

m = mass per unit span,

I_α = mass moment of inertia per unit span about axis $x = ba$,

$S_\alpha = mbx_\alpha$ = static mass moment per unit span about $x = ba$, positive when the center of gravity is aft,

$\omega_h = \sqrt{K_h/m}$ = uncoupled natural bending frequency,

$\omega_\alpha = \sqrt{K_\alpha/I_\alpha}$ = uncoupled natural torsional frequency.

For the prediction of flutter, we are interested only in dynamic variations of h and α from whatever steady-state values they assume. The wing would probably be flying at some positive angle of attack, and there might also be contributions to L and M_y due to camber of the airfoil. The subject of static elastic deformations associated with such loading is examined in Chapter 8. For our purposes, however, we subtract the equilibrated system of steady-state forces and moments from Eqs. (9–13), leaving only the time-dependent portions.

The standard scheme of flutter analysis resembles the one for free vibrations in that we specify simple harmonic motion in advance by setting

$$h = \bar{h}_0 e^{i\omega t}, \qquad (9\text{--}16a)$$

$$\alpha = \alpha_0 e^{i(\omega t + \phi)} = \bar{\alpha}_0 e^{i\omega t}. \qquad (9\text{--}16b)$$

---

* The reader will, of course, have no difficulty in deriving Eqs. (9–13) and (9–15) by direct application of Newton's Second Law of Motion to this simple system.

This complex representation is justified because the linearity of the equations of motion and the aerodynamic theories to be employed shows that all dependent variables in the problem contain time only as the factor $e^{i\omega t}$. We tacitly agree that the actual quantities are always found by taking the real parts of their complex counterparts, recognizing that the algebraic simplification achieved by complex notation outweighs any loss of physical clarity. Since phase shifts in the aerodynamic loads produce a phase difference between h and α, we allow for this by letting one or both of the amplitudes $\bar{h}_0$ and $\bar{\alpha}_0$ be complex numbers. If the time origin is chosen so as to make $\bar{h}_0$ real, the angle by which α leads h is defined in Eq. (9–16b) as ϕ, the argument of $\bar{\alpha}_0$.

The assumption of simple harmonic motion changes Eqs. (9–13) into

$$-\omega^2 mh - \omega^2 S_\alpha \alpha + \omega_h{}^2 mh = -L, \tag{9–17a}$$

$$-\omega^2 S_\alpha h - \omega^2 I_\alpha \alpha + \omega_\alpha{}^2 I_\alpha \alpha = M_y. \tag{9–17b}$$

Our choice of particular aerodynamic expressions to replace L and M_y depends on the Mach-number range in which flutter is expected to occur. Various forms are derived in Chapters 5 and 6, where we also refer to sources of tabulated data. We simplify the discussion in the present section by assuming low-speed flow; this permits us to put L and M_y into the notation of Ref. 9–1:

$$L = -\pi \rho b^3 \omega^2 \left\{ L_h \frac{h}{b} + [L_\alpha - L_h(\tfrac{1}{2} + a)]\alpha \right\}, \tag{9–18a}$$

$$M_y = \pi \rho b^4 \omega^2 \left\{ [M_h - L_h(\tfrac{1}{2}+a)]\frac{h}{b} + [M_\alpha - (L_\alpha + M_h)(\tfrac{1}{2}+a) + L_h(\tfrac{1}{2}+a)^2]\alpha \right\}, \tag{9–18b}$$

where L_h, L_α, and M_α are elaborate functions of the reduced frequency k only, and M_h is just $\tfrac{1}{2}$ in the incompressible case. Substituting Eqs. (9–18) into Eqs. (9–17), and dividing by $\pi \rho b^3 \omega^2 e^{i\omega t}$ and $\pi \rho b^4 \omega^2 e^{i\omega t}$, we get the dimensionless flutter equations

$$\frac{\bar{h}_0}{b}\left\{ \frac{m}{\pi \rho b^2}\left[1 - \frac{\omega_h{}^2}{\omega^2}\right] + L_h \right\} + \bar{\alpha}_0 \left\{ x_\alpha \frac{m}{\pi \rho b^2} + [L_\alpha - L_h(\tfrac{1}{2}+a)] \right\} = 0, \tag{9–19a}$$

$$\frac{\bar{h}_0}{b}\left\{ x_\alpha \frac{m}{\pi \rho b^2} + [\tfrac{1}{2} - L_h(\tfrac{1}{2}+a)] \right\} + \bar{\alpha}_0 \left\{ r_\alpha{}^2 \frac{m}{\pi \rho b^2}\left[1 - \frac{\omega_\alpha{}^2}{\omega^2}\right] \right.$$

$$\left. + M_\alpha - (L_\alpha + \tfrac{1}{2})(\tfrac{1}{2}+a) + L_h(\tfrac{1}{2}+a)^2 \right\} = 0. \tag{9–19b}$$

Since Eqs. (9–19) are homogeneous, they constitute an algebraic eigenvalue problem with finite solutions occurring at those combinations of speed and frequency for which the characteristic determinant vanishes:

$$
\begin{vmatrix}
\left\{ \dfrac{m}{\pi\rho b^2}\left[1 - \dfrac{\omega_h^2}{\omega_\alpha^2}\dfrac{\omega_\alpha^2}{\omega^2} \right] + L_h \right\} & \left\{ x_\alpha \dfrac{m}{\pi\rho b^2} + L_\alpha - L_h(\tfrac{1}{2} + a) \right\} \\[2em]
\left\{ x_\alpha \dfrac{m}{\pi\rho b^2} + \tfrac{1}{2} - L_h(\tfrac{1}{2} + a) \right\} & \left\{ r_\alpha^2 \dfrac{m}{\pi\rho b^2}\left[1 - \dfrac{\omega_\alpha^2}{\omega^2} \right] + M_\alpha \right. \\[1.5em]
 & \left. - (L_\alpha + \tfrac{1}{2})(\tfrac{1}{2} + a) + L_h(\tfrac{1}{2} + a)^2 \right\}
\end{vmatrix} = 0.
$$

$$(9\text{--}20)$$

Seen in the light of dimensional analysis (cf. Chapter 11), Eq. (9–20) shows that the flutter frequency ratio ω_α/ω and reduced frequency k (or reduced velocity $1/k = U/b\omega$) are implicitly dependent on the values of five dimensionless system parameters. These are defined as follows:

$a =$ the axis location,

$\omega_h/\omega_\alpha =$ the bending-torsion frequency ratio,

$x_\alpha = S_\alpha/mb =$ the dimensionless static unbalance,

$r_\alpha = \sqrt{I_\alpha/mb^2} =$ the dimensionless radius of gyration,

$m/\pi\rho b^2 =$ the density ratio.

All five of these are given in the case of a particular wing, but for making parametric studies it is more convenient to specify four of them and use the fifth artificially as an unknown, along with ω_α/ω, during solution of the determinant.

We comment only briefly here about the details of solving Eq. (9–20), for this topic is taken up in a subsequent section. We do point out, however, that the complex form of the aerodynamic coefficients causes the determinant to expand into real and imaginary equations, containing all the parameters except reduced frequency in powers up to the second. Because of the nature of these aerodynamic terms, both equations do include complicated transcendental functions of k. It therefore requires an unwieldy trial-and-error process to solve for k and ω_α/ω, given the physical properties of the system. If we assume k in advance and use ω_α/ω plus one other parameter as unknowns, we have at most to solve two simultaneous quadratic relations. Unless a is employed, at least one of these equations is linear in both unknowns, permitting explicit solutions to be calculated by elementary algebraic methods.

Regardless of how we find the eigenvalues U_F and ω_F that characterize flutter of the elastically suspended airfoil, we can afterward calculate the flutter mode shape from Eqs. (9–19). This is done by substituting the values of all wing parameters, ω_α/ω_F, and k_F, so that the four bracketed coefficients become complex numbers. We can then solve either equation for the complex ratio $\bar{\alpha}_0/\bar{h}_0$, which gives the relative amplitudes of the torsional and bending oscillations, and the phase angle ϕ by which torsion leads bending. As in the case of free vibrations (Chapter 3), the flutter mode is arbitrary up to the amplitude of one of the degrees of freedom. It is customary in most analyses not to bother with determining the exact mode shape, since by far the most important information is the existence and location of a stability boundary at airspeed U_F.

Our main objective in the present section is to pick out typical values of the major parameters and demonstrate the effect on bending-torsion flutter of changing each one individually. The curves we reproduce are adapted from Ref. 9–3, which provides an especially valuable source of parametric data on two-degree-of-freedom flutter (with the limitation that the range of $m/\pi\rho b^2$ is lower than that representative of most high-speed aircraft).

The various parts of Fig. 9–5 demonstrate how the dimensionless speed $U_F/b\omega_\alpha$ varies with ω_h/ω_α, $m/\pi\rho b^2$, a, and x_α for series of fixed values of the other governing quantities. In Fig. 9–5(A) we append to each set of curves the dimensionless torsional-divergence speed $U_D/b\omega_\alpha$. As shown in Chapter 8, U_D is given by the formula

$$U_D = \sqrt{\frac{K_\alpha}{2\pi\rho b^2[\frac{1}{2} + a]}}, \tag{8–7}$$

where we have replaced $1/C^{\theta\theta}$ by K_α, the lift-curve slope by its theoretical value 2π, and the distance from the aerodynamic center to the axis of rotation by $b(\frac{1}{2} + a)$. If we divide both sides of Eq. (8–7) by $b\omega_\alpha$, making use of Eq. (9–5) and the definitions of certain system parameters, we obtain

$$\frac{U_D}{b\omega_\alpha} = \frac{1}{b\omega_\alpha}\sqrt{\frac{K_\alpha b^2}{I_\alpha}}\sqrt{\frac{I_\alpha}{mb^2}\frac{m}{\pi\rho b^2[1 + 2a]}} = \sqrt{\frac{m}{\pi\rho b^2}\frac{r_\alpha^2}{[1 + 2a]}}. \tag{9–21}$$

Both Eq. (9–21) and the flutter speeds in Fig. 9–5 contain the same element of conservatism* that usually results from assuming two-dimensional aerodynamic theory, so it is fair to conclude that flutter will not be encountered in practice on a particular configuration when the predicted U_F exceeds U_D by any substantial amount.

Because of the generality achieved with dimensionless variables, we are able to draw a host of conclusions from Fig. 9–5. Rough estimates of incompressible-bending-torsion flutter can even be deduced for many practical wings. Although we cannot attempt extensive interpretation, we shall call attention to a few particularly significant facts. One of these is the marked dip near frequency ratio unity which occurs in the curves

---

* As applied by aeroelasticians to a theory, the term "conservative" means that the predicted value of some critical speed nearly always falls below the speed at which the corresponding instability actually occurs on the airplane. A theory is usually conservative if it systematically overestimates the magnitudes of aerodynamic loads. Although more accurate answers are sometimes required, theories with a known probability of conservatism, backed by many tests, can be very valuable in preliminary aeroelastic studies.

Fig. 9–5(A). The graphs on the following three pages show dimensionless flutter speed $U_F/b\omega_\alpha$ plotted against frequency ratio ω_h/ω_α for various values of dimensionless static unbalance x_α; $r_\alpha{}^2 = \frac{1}{4}$.

(a) $\dfrac{m}{\pi\rho b^2} = 2$; $a = -0.4$; $U_D/b\omega_\alpha = 1.58$ (stable for $x_\alpha = 0$ and 0.1)

(b) $\dfrac{m}{\pi\rho b^2} = 2$; $a = -0.3$; $U_D/b\omega_\alpha = 1.12$ (stable for $x_\alpha = 0$)

(c) $\dfrac{m}{\pi\rho b^2} = 2$; $a = -0.2$; $U_D/b\omega_\alpha = 0.913$

(d) $\dfrac{m}{\pi\rho b^2} = 3$; $a = -0.4$; $U_D/b\omega_\alpha = 1.94$ (stable for $x_\alpha = 0$)

(e) $\dfrac{m}{\pi\rho b^2} = 3$; $a = -0.3$; $U_D/b\omega_\alpha = 1.37$

(f) $\dfrac{m}{\pi\rho b^2} = 3$; $a = -0.2$; $U_D/b\omega_\alpha = 1.12$

(g) $\dfrac{m}{\pi\rho b^2} = 4$; $a = -0.4$; $U_D/b\omega_\alpha = 2.24$ (stable for $x_\alpha = 0$)

(h) $\dfrac{m}{\pi\rho b^2} = 4$; $a = -0.3$; $U_D/b\omega_\alpha = 1.58$

(i) $\dfrac{m}{\pi\rho b^2} = 4$; $a = -0.2$; $U_D/b\omega_\alpha = 1.29$

(j) $\dfrac{m}{\pi\rho b^2} = 5$; $a = -0.4$; $U_D/b\omega_\alpha = 2.5$

         (curve for $x_\alpha = 0$ starts at $U_F/b\omega_\alpha = 6.9$ and goes up)

(k) $\dfrac{m}{\pi\rho b^2} = 5$; $a = -0.3$; $U_D/b\dot\omega_\alpha = 1.77$

(l) $\dfrac{m}{\pi\rho b^2} = 5$; $a = -0.2$; $U_D/b\omega_\alpha = 1.45$

(m) $\dfrac{m}{\pi\rho b^2} = 10$; $a = -0.4$; $U_D/b\omega_\alpha = 3.54$

(n) $\dfrac{m}{\pi\rho b^2} = 10$; $a = -0.3$; $U_D/b\omega_\alpha = 2.5$

(o) $\dfrac{m}{\pi\rho b^2} = 10$; $a = -0.2$; $U_D/b\omega_\alpha = 2.05$

(p) $\dfrac{m}{\pi\rho b^2} = 20$; $a = -0.4$; $U_D/b\omega_\alpha = 5.0$

(q) $\dfrac{m}{\pi\rho b^2} = 20$; $a = -0.3$; $U_D/b\omega_\alpha = 3.54$

(r) $\dfrac{m}{\pi\rho b^2} = 20$; $a = -0.2$; $U_D/b\omega_\alpha = 2.88$

(s) $\dfrac{m}{\pi\rho b^2} = 5$; $a = -0.45$; $U_D/b\omega_\alpha = 3.54$

(t) $\dfrac{m}{\pi\rho b^2} = 10$; $a = -0.45$; $U_D/b\omega_\alpha = 5.0$

(u) $\dfrac{m}{\pi\rho b^2} = 20$; $a = -0.45$; $U_D/b\omega_\alpha = 7.07$

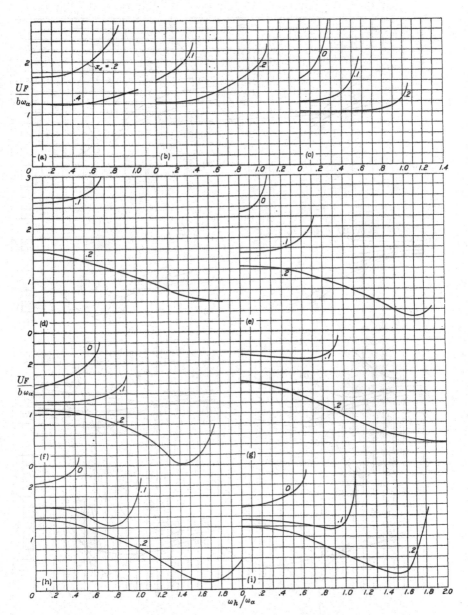

Fig. 9–5(A). See legend on page 538.

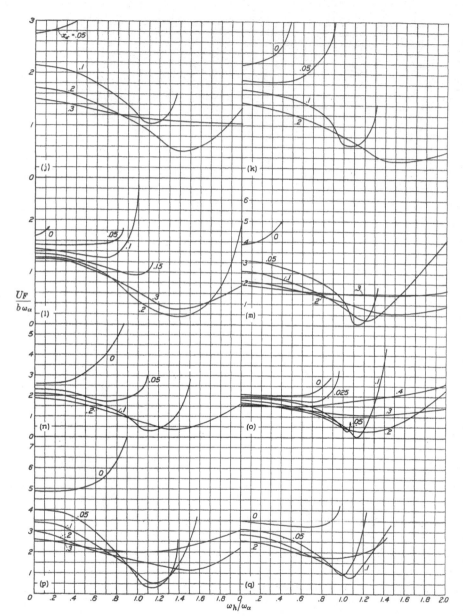

Fig. 9–5(A). (Cont'd.) See legend on page 538.

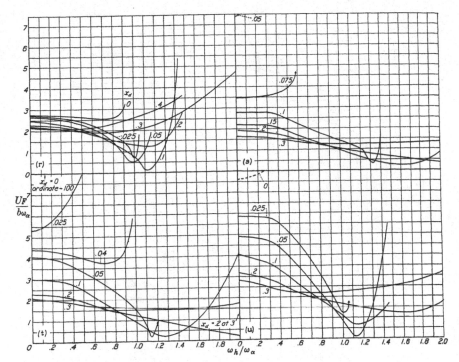

Fig. 9-5(A). (Cont'd.) See legend on page 538.

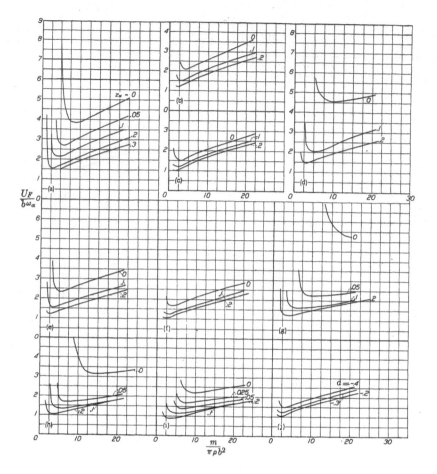

Fig. 9–5(B). Dimensionless flutter speed $U_F/b\omega_\alpha$ plotted against density ratio $m/\pi\rho b^2$ for various values of static unbalance x_α; $r_\alpha^2 = \frac{1}{4}$.

(a) $\left(\dfrac{\omega_h}{\omega_\alpha}\right)^2 = 0;$ $a = -0.4$

(b) $\left(\dfrac{\omega_h}{\omega_\alpha}\right)^2 = 0;$ $a = -0.3$

(c) $\left(\dfrac{\omega_h}{\omega_\alpha}\right)^2 = 0;$ $a = -0.2$

(d) $\left(\dfrac{\omega_h}{\omega_\alpha}\right)^2 = \frac{1}{10};$ $a = -0.4$

(e) $\left(\dfrac{\omega_h}{\omega_\alpha}\right)^2 = \frac{1}{10};$ $a = -0.3$

(f) $\left(\dfrac{\omega_h}{\omega_\alpha}\right)^2 = \frac{1}{10};$ $a = -0.2$

(g) $\left(\dfrac{\omega_h}{\omega_\alpha}\right)^2 = \frac{1}{2};$ $a = -0.4$

(h) $\left(\dfrac{\omega_h}{\omega_\alpha}\right)^2 = \frac{1}{2};$ $a = -0.3$

(i) $\left(\dfrac{\omega_h}{\omega_\alpha}\right)^2 = \frac{1}{2};$ $a = -0.2$

(j) $\left(\dfrac{\omega_h}{\omega_\alpha}\right)^2 = \frac{1}{10};$ $x_\alpha = 0.2$ (curves shown for $a = -0.4, -0.3,$ and -0.2)

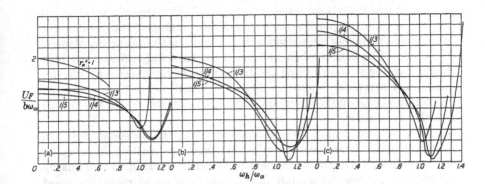

Fig. 9-5(C). Dimensionless flutter speed $U_F/b\omega_\alpha$ plotted against frequency ratio ω_h/ω_α for various values of radius of gyration r_α^2; $a = -0.2$, $x_\alpha = 0.1$.

$$(a)\quad \frac{m}{\pi\rho b^2} = 5$$

$$(b)\quad \frac{m}{\pi\rho b^2} = 10$$

$$(c)\quad \frac{m}{\pi\rho b^2} = 20$$

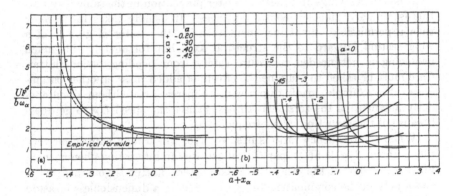

Fig. 9-5(D). Dimensionless flutter speed $U_F/b\omega_\alpha$ plotted against center-of-gravity location $a + x_\alpha$ for various values of axis location a; $r_\alpha^2 = \frac{1}{4}$.

$$(a)\quad \left(\frac{\omega_h}{\omega_\alpha}\right)^2 = 0; \qquad \frac{m}{\pi\rho b^2} = 10$$

$$(b)\quad \left(\frac{\omega_h}{\omega_\alpha}\right)^2 = \frac{1}{2}; \qquad \frac{m}{\pi\rho b^2} = 10$$

of 9–5(A) and 9–5(C) for larger values of x_α. This dip will be observed in the U_F vs. ω_h/ω_α plot for the wings of most high-performance aircraft, which have relatively large density ratios and positive static unbalances. It is only to be feared when unusually low torsional stiffness and high bending stiffness tend to bring the frequencies in resonance with one another—not a typical condition. Moreover, small amounts of structural friction can bring up the minimum value of U_F (cf. Ref. 9–3) or even wipe out the dip entirely.

Increasing the density ratio normally raises the flutter speed, as shown in Fig. 9–5(B). These curves can be interpreted as plotting flutter speed vs. altitude for a particular wing, so that the critical condition is ordinarily at sea level, unless the aircraft has a large increase of maximum speed with altitude. A few light personal airplanes and gliders fall in the range of $m/\pi\rho b^2$ where U_F decreases with altitude. However, model tests like those of Ref. 9–7 suggest that the theory may be unconservative in this region, where the aerodynamic forces dominate the inertial forces on the system.

With regard to wing mass properties, Figs. 9–5(A) and 9–5(B) show that increasing x_α and decreasing r_α both reduce $U_F/b\omega_\alpha$, although in the practical range of these parameters the speed is more sensitive to a given percentage change of the former. These facts have led to a concept of mass balancing wings and tails for flutter prevention in the same way that this is regularly done to control surfaces. It usually takes much more than a permissible weight increment to bring the center of gravity of a wing section forward to the structural axis, but fortunately this much balancing is not needed to assure safety from bending-torsion flutter. Care must be taken to examine the effects of added mass on other parameters. For example, loading up a wing to increase r_α^2 without simultaneously augmenting the torsional stiffness causes the speed U_F to go down because of the reduction in the frequency ω_α.

When the frequency ratio is near to zero, which may happen on very thin surfaces, Fig. 9–5(D), part (a), illustrates an interesting phenomenon. For fixed values of the other three parameters, flutter is governed by x_α and a only in the combination $(a + x_\alpha)$. This is a dimensionless measure of the chordwise center-of-gravity location; the farther aft it lies, the lower U_F, although the influence is not pronounced for centers of gravity behind midchord. By analogy with Eq. (9–21), Ref. 9–3 proposes an empirical formula

$$\frac{U_F}{b\omega_\alpha} \cong \sqrt{\frac{m}{\pi\rho b^2} \frac{r_\alpha^2}{\left[1 + 2(a + x_\alpha)\right]}} \qquad (9\text{–}22)$$

to be employed when ω_h/ω_α is very small and the density ratio does not exceed 10. Equation (9–22) is shown on Fig. 9–5(D) as a dashed line. The bracket in the denominator is the distance by which the center of gravity lies behind the aerodynamic center. When it is positive, the unrestrained wing would be statically unstable in flight.

Curiously enough, the flutter speed in Eq. (9–22) is imaginary when the center of gravity lies ahead of the quarter-chord point. On the other hand, the example of Section 9–1, which is effectively one of *infinite* frequency ratio, concerns a system which can flutter only when its axis of rotation is ahead of the quarter-chord point. In that case, flutter instability accompanies static stability of the unrestrained wing, thus illustrating one of the dangers of trying to generalize about the flutter problem.

In closing this section, we remark with respect to the frequency ω of bending-torsion flutter that it normally lies somewhere between ω_h and ω_α, ω_h being the smaller of the two. However, situations do arise in theory where ω exceeds ω_α.

9–3 Exact treatment of the bending-torsion flutter of a uniform cantilever wing. As we shall see in subsequent sections, most practical flutter calculations are based on the assumption that the motion of the system consists of superposition of a finite number of preassigned modes of deformation. This procedure was described, as it is used for the general study of externally excited vibrations, in Sections 3–6 through 3–10. It is almost forced on the analyst when he is confronted with a complicated structure, whose inertial and elastic properties are usually furnished at a set of points and are inconvenient to define in terms of simple mathematical functions. Therefore, it is fortunate that a few simple classes of wings exist for which the flutter problem can be solved exactly, within the limits of the linearized aerodynamic and elastic theories; these solutions may be regarded as standards of comparison to estimate the accuracy of approximate methods.

One such class includes lumped-parameter systems, like the ones discussed in Sections 9–1 and 9–2. It also comprises representative sections having flaps or tabs, and lumped-mass approximations to large aspect-ratio wings such as those used by Myklestad (Ref. 9–5). Unless this latter sort of approximation is given an inefficiently large number of degrees of freedom, it is not likely to provide a true picture of the actual structure for flutter purposes. For comparison with approximate procedures, therefore, we turn to the known exact solutions for uniform wings. These are unique in that they treat a case of distributed parameters, governed by a partial differential equation, and yield the flutter mode shape as a function of the spanwise coordinate. The first result of this kind was given by Goland (Ref. 9–6), who dealt with bending-torsion flutter of a bare restrained (cantilever) wing with constant chord and uniformly distributed mass and elastic properties.

Assume that the structure deforms like a slender beam of stiffnesses EI in bending and GJ in torsion. Any section is infinitely rigid chord-

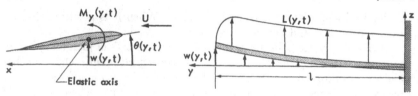

Fig. 9–6. Two views of a uniform restrained wing deforming in bending and torsion under running aerodynamic loads $L(y,t)$ and $M_y(y,t)$.

wise, and the elastic axis is straight. The complete state of strain is known once we prescribe the normal displacement $w(y,t)$ of the elastic axis (positive upward) and the twist $\theta(y,t)$ about this axis (positive nose-up). The deformed wing in the airstream loaded with running aerodynamic force $L(y,t)$ and moment $M_y(y,t)$, is illustrated in Fig. 9–6.

We can take the dynamical equations of this system from the treatment in Chapter 3 of the forced motion of restrained wings. Thus Eqs. (3–155) and (3–156) describe a nonuniform beam under generalized loading. If we substitute the aerodynamic loads into their right sides and take EI, GJ, m, etc. to be constants, we obtain

$$m\ddot{w}(y,t) - S_y\ddot{\theta}(y,t) + EI\frac{\partial^4 w(y,t)}{\partial y^4} = L(y,t), \qquad (9\text{–}23)$$

$$I_y\ddot{\theta}(y,t) - S_y\ddot{w}(y,t) - GJ\frac{\partial^2 \theta(y,t)}{\partial y^2} = M_y(y,t), \qquad (9\text{–}24)$$

where the superscript dots indicate differentiation with respect to t. We assume simple harmonic motion but do not resort to the complex representation.

$$w(y,t) = W_1(y)\cos\omega t + W_2(y)\sin\omega t, \qquad (9\text{–}25)$$

$$\theta(y,t) = \Theta_1(y)\cos\omega t + \Theta_2(y)\sin\omega t, \qquad (9\text{–}26)$$

where $W_1(y)$, $W_2(y)$ and $\Theta_1(y)$, $\Theta_2(y)$ are functions yet to be determined which describe the bending and twisting deformations of the wing, respectively.

The flight speed is presumed to be low enough so that incompressible aerodynamics is permissible, and the large aspect ratio justifies our use of two-dimensional strip theory. In these circumstances, we know from Section 5–6 that the lift and moment can be written symbolically (following Ref. 9–6):

$$L(y,t) = \omega^2 L_w w + \omega L_w'\dot{w} + \omega^2 L_\theta\theta + \omega L_\theta'\dot{\theta}, \qquad (9\text{–}27)$$

$$M_y(y,t) = \omega^2 M_w w + \omega M_w'\dot{w} + \omega^2 M_\theta\theta + \omega M_\theta'\dot{\theta}. \qquad (9\text{–}28)$$

The eight coefficients L_w, L_w', etc., are functions of the air density, elastic axis location, and wing chord $2b$; they also depend in a complicated way on the reduced frequency k. They can be identified with the tabulated dimensionless quantities L_h, L_α, M_h, and M_α by substituting complex equivalents of $w(y,t)$ and $\theta(y,t)$ for $(-h)$ and α in Eqs. (5–350) and (5–351), then equating the real parts of these two formulas to Eqs. (9–27) and (9–28).

If we insert Eqs. (9–25) through (9–28) into the equations of motion (9–23) and (9–24), we can collect each of the resulting relations into one bracket multiplied by $\cos \omega t$, plus another multiplied by $\sin \omega t$. Since the sums equal zero at all instants of time, the individual brackets must themselves vanish, and we obtain four independent equations:

$$EI \frac{d^4 W_1}{dy^4} - m\omega^2 W_1 - \omega^2 L_w W_1 - \omega^2 L_w{}' W_2$$

$$+ S_y\omega^2\Theta_1 - \omega^2 L_\theta\Theta_1 - \omega^2 L_\theta{}'\Theta_2 = 0, \qquad (9\text{–}29)$$

$$\omega^2 L_w{}' W_1 + EI \frac{d^4 W_2}{dy^4} - m\omega^2 W_2 - \omega^2 L_w W_2$$

$$+ \omega^2 L_\theta{}'\Theta_1 + S_y\omega^2\Theta_2 - \omega^2 L_\theta\Theta_2 = 0, \qquad (9\text{–}30)$$

$$-\omega^2 S_y W_1 + \omega^2 M_w W_1 + \omega^2 M_w{}' W_2 + GJ \frac{d^2\Theta_1}{dy^2}$$

$$+ I_y\omega^2\Theta_1 + \omega^2 M_\theta\Theta_1 + \omega^2 M_\theta{}'\Theta_2 = 0, \qquad (9\text{–}31)$$

$$-\omega^2 M_w{}' W_1 - \omega^2 S_y W_2 + \omega^2 M_w W_2 - \omega^2 M_\theta{}'\Theta_1$$

$$+ GJ \frac{d^2\Theta_2}{dy^2} + I_y\omega^2\Theta_2 + \omega^2 M_\theta\Theta_2 = 0. \qquad (9\text{–}32)$$

The boundary conditions accompanying Eqs. (9–29)–(9–32) prescribe no twist, bending deflection, or slope at the built-in end $y = 0$, and no torque, shear, or bending moment at the free end $y = l$. In terms of the amplitude functions, they read

$$W_1(0) = W_2(0) = \frac{dW_1(0)}{dy} = \frac{dW_2(0)}{dy} = \Theta_1(0) = \Theta_2(0) = 0, \quad (9\text{–}33)$$

$$\frac{d^2 W_1(l)}{dy^2} = \frac{d^2 W_2(l)}{dy^2} = \frac{d^3 W_1(l)}{dy^3} = \frac{d^3 W_2(l)}{dy^3} = \frac{d\Theta_1(l)}{dy} = \frac{d\Theta_2(l)}{dy} = 0. \quad (9\text{–}34)$$

This problem is solved by a procedure not essentially different from that customarily used on ordinary differential equations with constant coefficients (for instance, those describing free vibration of a uniform beam). Some computational efficiency can be gained by introducing the complex representation of simple harmonic motion and taking the Laplace transform on the variable x (cf. Refs. 9–8 and 9–9). We do not believe these mathematical refinements contribute any understanding of the physical implications in the present simple example.

As with all flutter analyses, the complicated dependence of the aerodynamic coefficients on k necessitates a trial-and-error process. This consists of two steps. First, for a particular wing configuration, frequency, and reduced frequency, the general solution of the differential equation system (9–29)–(9–32) is found by assuming

$$W_1 = A e^{\lambda y}, \tag{9-35}$$

$$W_2 = B e^{\lambda y}, \tag{9-36}$$

$$\Theta_1 = C e^{\lambda y}, \tag{9-37}$$

$$\Theta_2 = D e^{\lambda y}. \tag{9-38}$$

The factors multiplying all the unknowns in Eqs. (9–29)–(9–32) have numerical values, so there are produced four simultaneous, algebraic equations in A, B, C, and D. Since all the right sides equal zero, there can be nontrivial solutions only when the determinant of the coefficients vanishes. This fact leads to a polynomial of sixth degree in the unknown λ^2, which is the result of expanding

$$\begin{vmatrix} \left[\dfrac{EI}{\omega^2}\lambda^4 - m - L_w\right] & -L_w{}' & [S_y - L_\theta] & -L_\theta{}' \\[2ex] L_w{}' & \left[\dfrac{EI}{\omega^2}\lambda^4 - m - L_w\right] & L_\theta{}' & [S_y - L_\theta] \\[2ex] [M_w - S_y] & M_w{}' & \left[\dfrac{GJ}{\omega^2}\lambda^2 + I_y + M_\theta\right] & M_\theta{}' \\[2ex] -M_w{}' & [M_w - S_y] & -M_\theta{}' & \left[\dfrac{GJ}{\omega^2}\lambda^2 + I_y + M_\theta\right] \end{vmatrix} = 0. \tag{9-39}$$

All the coefficients of this polynomial are real, and it possesses, in the most general case, six complex roots λ^2, in three complex-conjugate pairs. Any particular $\lambda_r{}^2$ yields two complex-conjugate values of λ, which are denoted

$$\lambda_r = \xi_r + i\eta_r, \tag{9-40a}$$

$$(\lambda_r)_{\text{conj}} = \xi_r - i\eta_r. \tag{9-40b}$$

With each λ_r, we can associate a complex constant A_r in the expression for W_1, B_r in the expression for W_2, etc. However, in order to assure that the functions so obtained are real, the constant associated with $(\lambda_r)_{\text{conj}}$ must be $(A_r)_{\text{conj}}$. This makes each conjugate pair coalesce into

$$A_r e^{(\xi_r + i\eta_r)y} + (A_r)_{\text{conj}} e^{(\xi_r - i\eta_r)y} = e^{\xi_r y}[a_r \cos \eta_r y + a_r{}' \sin \eta_r y], \tag{9-41}$$

where

$$a_r = A_r + (A_r)_{\text{conj}}, \tag{9-42a}$$

$$a_r{}' = i[A_r - (A_r)_{\text{conj}}] \tag{9-42b}$$

are real constants. The most general solution is found by summing all possible terms of the type (9–41):

$$W_1(y) = \sum_{r=1}^{6} e^{\varrho_r y}[a_r \cos \eta_r y + a_r' \sin \eta_r y], \qquad (9\text{–}43)$$

$$W_2(y) = \sum_{r=1}^{6} e^{\varrho_r y}[b_r \cos \eta_r y + b_r' \sin \eta_r y], \qquad (9\text{–}44)$$

$$\Theta_1(y) = \sum_{r=1}^{6} e^{\varrho_r y}[c_r \cos \eta_r y + c_r' \sin \eta_r y], \qquad (9\text{–}45)$$

$$\Theta_2(y) = \sum_{r=1}^{6} e^{\varrho_r y}[d_r \cos \eta_r y + d_r' \sin \eta_r y]. \qquad (9\text{–}46)$$

The 36 quantities b_r, b_r', c_r, c_r', d_r, and d_r' are related to the twelve a_r and a_r', as we discover by substituting any one set of solutions (9–35)–(9–38) back into Eqs. (9–29)–(9–32) and computing the ratios B_r/A_r, C_r/A_r, and D_r/A_r. This process yields a set of relations like

$$\frac{B_r}{A_r} = \alpha_r + i\beta_r, \qquad \frac{(B_r)_{\text{conj}}}{(A_r)_{\text{conj}}} = \alpha_r - i\beta_r \qquad (9\text{–}47)$$

which, in turn, imply

$$b_r = \alpha_r a_r + \beta_r a_r', \qquad b_r' = -\beta_r a_r + \alpha_r a_r'. \qquad (9\text{–}48)$$

Accordingly, there remain twelve real, independent constants a_r and a_r' in Eqs. (9–43)–(9–46) to be fixed by the twelve boundary conditions (9–33)–(9–34).

The second step in the procedure is to specialize the general solution to the particular beam being analyzed. Here there is a difference from free-vibration problems, where the boundary conditions generate a characteristic equation for the eigenvalues of the frequency ω. The nature of the aerodynamic coefficients forces the selection of this frequency in advance, and a series of choices of ω and k must be made until one is found which permits all of Eqs. (9–33) and (9–34) to be satisfied. To summarize the details, the literal coefficients in Eqs. (9–43)–(9–46) are put in terms of the 12 a_r and a_r', and the expressions so obtained are inserted into Eqs. (9–33)–(9–34). Twelve real, homogeneous, simultaneous equations are the outcome. All the coefficients are real numbers but, in general, the determinant composed of them does not vanish. All steps are repeated until a k_F and an ω_F are reached which make the determinant zero. A critical condition is thus identified, since all the physical requirements of the problem are met by neutrally stable, simple harmonic flutter at frequency ω_F and speed

$$U_F = \frac{b\omega_F}{k_F}. \qquad (9\text{–}49)$$

Incidentally, the mode shape of bending and torsional deformation at flutter is calculated by substituting the correct set of a_r and a_r' back through Eqs. (9–43)–(9–46) into Eqs. (9–25) and (9–26). The magnitude of one of these constants remains undetermined, since the over-all amplitude of any self-sustained oscillation of a linear system can be arbitrarily set by the amount of energy put in at starting. We observe one important difference between flutter modes and characteristic vibrations in vacuum. Since the ratios W_1/W_2, Θ_1/Θ_2, and w/θ are not independent of y, the bending and torsion motions at various stations do not pass through zero simultaneously. These degrees of freedom are therefore not in phase with each other, nor is either one in phase with itself at various points along the span. Such a vibration would, in its most general form, exhibit traveling waves of deflection and twist. These are never seen when an undamped system, composed only of inertias and stiffnesses, vibrates without the action of external forces. If such waves are apparent in a flutter mode, their presence must be attributed entirely to the phase shifts between the aerodynamic loads and the motions producing them.

In Ref. 9–6, Goland applies the method described above to a typical cantilever wing with a 6-ft chord and 20-ft semispan. The other properties entering Eqs. (9–23) and (9–24) are

$$m = 0.746 \text{ slug/ft,} \qquad I_y = 1.943 \text{ slug·ft}^2/\text{ft,}$$
$$S_y = 0.447 \text{ slug·ft/ft,} \qquad EI = 23.6 \times 10^6 \text{ lb·ft}^2,$$
$$GJ = 2.39 \times 10^6 \text{ lb·ft}^2. \tag{9–50}$$

He compares the results of an exact flutter calculation with those of one based on the assumption that the bending and torsion mode shapes are identical with the fundamental modes of free vibration of the same beam without mass coupling S_y. The latter procedure is common practice in bending-torsion flutter analysis, and we describe it in Section 9–5. The flutter eigenvalues come out to be as follows:

	Exact solution	Assumed-mode solution
U_F	393 mph	385 mph
ω_F	66.2 rad/sec	67.4 rad/sec
$1/k_F$	2.90	2.80

Although the agreement shown by the two solutions is good, the most interesting comparison is between the flutter modes. The calculation based on assumed modes leads to

$$w(y,t) = W_b(y) \cos \omega t, \tag{9–51a}$$

$$\theta(y,t) = 0.17\Theta_t(y) \cos(\omega t - 56°51'), \qquad (9\text{-}51b)$$

where the two normalized cantilever shapes are (cf. Chapter 3)

$$W_b(y) = \tfrac{1}{2}\left\{ \cosh\frac{0.597\pi y}{l} - 0.734\sinh\frac{0.597\pi y}{l} - \cos\frac{0.597\pi y}{l}\right.$$

$$\left. + 0.734\sin\frac{0.597\pi y}{l}\right\}, \qquad (9\text{-}52a)$$

$$\Theta_t(y) = \sin\frac{\pi}{2}\cdot\frac{y}{l}. \qquad (9\text{-}52b)$$

These cannot display the traveling-wave phenomenon, although bending does lead torsion in phase by some 57 degrees. Goland discovered that the exact modes could be very accurately represented by

$$w(y,t) = W_E(y) \cos\omega t, \qquad (9\text{-}53a)$$

$$\theta(y,t) = \Theta_E(y) \cos(\omega t - 56°57'). \qquad (9\text{-}53b)$$

Moreover, the function $W_E(y)$ is indistinguishable from $W_b(y)$, and $\Theta_E(y)$ differs from $0.17\Theta_t(y)$ only in having a tip amplitude of about 0.15 rather than 0.17. We see that the difference between the phase angles in Eqs. (9-51b) and (9-53b) is probably less than the error of computation. The possibility of traveling waves in either bending or torsion was allowed for in arriving at Eqs. (9-53), yet such waves do not actually appear.

We are fully justified in concluding that the assumed-mode type of flutter analysis, which is a relatively simple process, is accurate for bare, straight wings of large aspect ratio. There is no reason for not extending this conclusion to tapered wings with nonuniform stiffness and mass properties, so long as there are no sharp discontinuities in the distributions. Experience shows, however, that similar accuracy is not usually obtained when the structure carries large concentrated masses like engine nacelles and external stores. The same can be said of flutter involving complicated control-surface configurations.

9-4 Aeroelastic modes. We learned in Section 9-1 how a simple, two-dimensional airfoil exhibits vibrations in an airstream at any speed, the rate of decay or divergence being determined by the unsteady aerodynamic loads produced by the motion. This is also true of a flexible airplane in flight, except that an infinite number of "aeroelastic modes" can be theoretically imagined because of the infinity of degrees of freedom. These modes are in one-to-one correspondence with the free vibration modes of the structure at rest. We can devise a wind-tunnel experiment in which the airspeed is gradually brought up from zero, while the frequency and damping of each important mode are traced from the starting point. At low speeds, the air withdraws energy from all vibrations, and

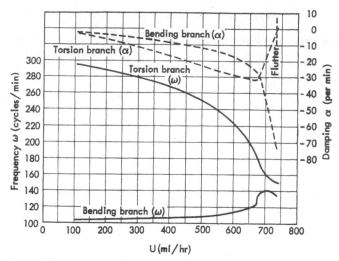

Fig. 9-7. Frequency and damping of the fundamental aeroelastic modes plotted vs. airspeed for an airplane wing with natural bending and torsional frequencies 103.5 cycles/min and 304 cycles/min.

their damping ratios or logarithmic decrements grow with increasing U. Ultimately the decay rate of one mode reverses its trend and drops to zero at the minimum speed U_F for flutter of the airplane.

On certain aircraft, even the lowest aeroelastic modes involve participation of the entire structure. Wings, tail, and fuselage move in unison, with no one element displaying relatively large amplitudes or supplying most of the energy taken out by the air. Fortunately for the flutter analyst, it is much more common for particular modes to concentrate in particular aerodynamic surfaces. When this is the case, we can study such modes while completely neglecting the presence of the rest of the aircraft or (more commonly) regarding some structural junction, such as the point where the wing enters the fuselage, as a rigid cantilever support. Whether this simplification is permissible or not, it is a general rule that the modes with the lowest frequencies are the ones which should be examined first for evidence of flutter.

Let us suppose that the structural deformations of major importance are bending and torsion of a relatively large aspect-ratio wing or tail, which we can assume to be built into an unyielding fuselage. The aeroelastic modes to be studied are then those which, at zero airspeed, merge into the lowest-frequency coupled modes of the cantilever beam. One of these still-air modes, usually the lowest in frequency, is predominantly bending, and the other involves mostly torsion. Experience has shown that either "first bending" or "first torsion" leads to the critical flutter mode.

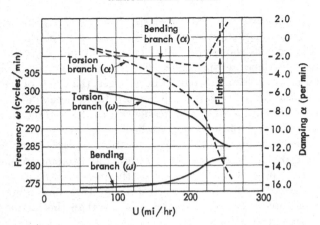

Fig. 9–8. Frequency and damping of the fundamental aeroelastic modes plotted vs. airspeed for an airplane wing with natural bending and torsional frequencies 276 cycles/min and 304 cycles/min.

Because the aerodynamic theory of transient motion is more complicated than for simple harmonic vibration, typical flutter analyses do not predict the stability of aeroelastic modes as a function of U. Such a calculation is feasible, however, for simple systems like straight cantilever wings, where the airloads can be obtained from strip theory. Goland and Luke (Ref. 9–10) make this kind of study on a large aspect-ratio wing with a tip mass. The lift and moment are determined in an approximate fashion from the indicial functions for incompressible flow, which we describe in Section 5–7. The elastic and inertial properties are introduced through preassigned deformation shapes, as in the Rayleigh-Ritz method. We reproduce, in Figs. 9–7 and 9–8, the results of Ref. 9–10 for two values of bending-to-torsion frequency ratio in vacuum.

We do not list here all the properties of the airplane, since only the behavior of the aeroelastic modes is important for our purposes. The ordinates in both figures are the real part α and the imaginary part ω of the complex time dependence

$$e^{(-\alpha + i\omega)t}$$

of the two modes; the abscissas are flight speeds in miles per hour. The degree of damping is measured closely by the magnitude of α, which is related to the damping ratio of the oscillation by

$$\text{D.R.} = \frac{\alpha}{\sqrt{\alpha^2 + \omega^2}}. \tag{9–54}$$

Figure 9–7 corresponds to a natural bending frequency of 103.5 cycles/min and a torsion frequency of 304 cycles/min. The ratio of these values is quite characteristic of all-metal wings with stressed-skin construction. In this instance, the aeroelastic mode which starts at $U = 0$ from first torsion (marked "torsion branch") proves to be the one which flutters. Both modes show quite satisfactory damping, as a function of airspeed, up to within a few percent of U_F. The damping of the bending branch is numerically less, although D.R. is larger, than that of the torsion branch, until the former begins to show very rapid decay rates close to flutter. As is always the case, the frequencies of the two modes tend to merge as the critical condition is approached.

Figure 9–8 illustrates a less typical situation, where the natural frequencies are 276 and 304 cycles/min in bending and torsion, respectively. Here the flutter speed is much lower, since the frequency resonance effect discussed in Section 9–2 has begun to assert itself. It is the bending branch which becomes unstable, and its damping is less than the torsion branch throughout the airspeed range.

An interesting observation which we make on both figures is that no clue as to the imminence of flutter can be drawn from the decay rate of either aeroelastic mode until U comes relatively close to U_F. On the one hand, this means that an aircraft may be able to fly near U_F without the appearance of undesirable marginal stability in its structural vibrations. On the other, we see how suddenly and violently bending-torsion flutter can emerge during accelerated flight. At five miles per hour above U_F, the wings in Figs. 9–7 and 9–8 will destroy themselves after two or three cycles of oscillation. These facts must be very carefully considered when planning flight tests to investigate a suspected flutter condition, particularly when there are no reliable advance estimates. The general problem of flight flutter testing and the relative merits of various schemes for detecting the nearness to a critical speed are discussed in Chapter 13.

On high-performance airplanes, like medium and heavy bombers, whose structures are designed to withstand relatively small normal accelerations, it is impossible to distinguish clearly between aeroelastic modes and those oscillations which are associated with dynamic stability. The first bending frequency of the wing may be approximately equal to that of the short-period longitudinal mode of the airplane at five or six hundred mph. As seen by the stability analyst, this means that his derivatives must be corrected for aeroelastic effects; the airplane's dynamic characteristics change with flight speed because of both compressibility and structural deformation. To the aeroelastician, it is a case where important flutter modes involve the entire airplane. He must include so-called rigid-body degrees of freedom in his calculations. Occasionally the critical flutter mode may be one which is continuous, not with first bending or first torsion, but with a still-air mode having zero natural frequency.

In such circumstances there is a very good argument for carrying out the complete dynamic analysis of the flexible airplane on a unified, cooperative basis. The accuracy and realism of the results benefit from the contrasting but complementary viewpoints of the engineers thus brought together. A valuable adjunct to theoretical work nearly always proves to be a scaled aeroelastic model of the entire structure. Its usefulness for clarifying the flutter problem has been fully confirmed. What can be accomplished additionally in the way of determining the dynamic behavior of the unrestrained machine depends on the thoroughness and skill of those making the wind-tunnel tests. The problems of supporting a model in neutrally stable motions at low frequencies have not yet been completely solved.

A majority of aircraft do not require the all-inclusive study just described. With regard to flutter, the advent of rather small lifting surfaces attached to massive bodies is leading to a situation where rolling is the only rigid-body degree of freedom which will usually have to be given serious attention.

9–5 Flutter analysis by assumed-mode methods. The commonly accepted way of deriving the flutter equations for a structure is the Rayleigh-Ritz method, in which the motion is assumed to consist of superposition of a finite number of preassigned mode shapes. As it applies to forced motion of restrained and unrestrained aircraft in general, we have described it in Chapter 3. There we regard the structure as nearly flat, lying as close as possible to the xy-plane and very stiff as far as deformation in that plane is concerned. The complete state of strain and stress can be calculated once we know the function $w(x,y,t)$, which gives the normal displacement of the structure (positive upward, as seen by the pilot, in the direction of z) from its initial position. There are two ways in which assumed modes are introduced that form the bases of nearly all practical flutter analyses. The more rigorous and classical of these uses the normal modes of vibration, but the one which is often more efficient aerodynamically is based on artificially prescribed shapes or "primitive modes."

If the orthogonal normal modes of the structure are employed, we write

$$w(x,y,t) = \sum_{i=1}^{\infty} \phi_i(x,y)\xi_i(t). \tag{3–145}$$

Equation (3–145) leads to

$$M_j\ddot{\xi}_j + M_j\omega_j{}^2\xi_j = \Xi_j, \qquad (j = 1, 2, \cdots, \infty), \tag{3–147a}$$

where $\phi_j(x,y)$ and ω_j are the shape and natural frequency of the jth mode. The jth generalized mass is

$$M_j = \iint_S \phi_j{}^2\rho(x,y)dxdy, \tag{3–147b}$$

and the generalized force is

$$\Xi_j = \iint_S F_z(x,y,t)\phi_j(x,y)dxdy, \tag{3–147c}$$

where $F_z(x,y,t)$ is the external force, per unit xy-area, acting in the z-direction. The integrals are taken over the projection of the whole aircraft. In this case, F_z represents the distribution of pressure difference between the lower and upper surfaces of the structure caused by motion of the air. In terms of our aerodynamic symbols,

$$F_z(x,y,t) = (p_L - p_U) = -\Delta p_a(x,y,t). \tag{9-55}$$

When we are dealing with the unrestrained aircraft, the first three modes are rigid-body plunging, pitching, and rolling. They have zero natural frequencies and

$$\phi_1(x,y) = 1, \tag{3-189}$$

$$\phi_2(x,y) = -x, \qquad \phi_3(x,y) = y. \tag{3-203}$$

It follows that their generalized masses are the total mass, the moment of inertia about the y-axis, and the moment of inertia about the x-axis, respectively. The generalized forces are the lift, pitching moment, and rolling moment.

If artificially selected deformation shapes are employed to build up the flutter mode, we start with a finite number of approximate functions $\gamma_i(x,y)$ and assume

$$w(x,y,t) = \sum_{i=1}^{n} \gamma_i(x,y)q_i(t). \tag{3-251}$$

The resulting equations of motion are

$$\sum_{j=1}^{n} m_{ij}\ddot{q}_j + \sum_{j=1}^{n} k_{ij}q_j = Q_i, \qquad (i = 1, 2, \cdots, n). \tag{3-256}$$

Here the generalized forces, masses, and stiffnesses are defined by

$$Q_i = \iint_S F_z(x,y,t)\gamma_i(x,y)dxdy = -\iint_S \Delta p_a(x,y,t)\gamma_i(x,y)dxdy, \tag{3-258}$$

$$m_{ij} = \iint_S \gamma_i(x,y)\gamma_j(x,y)\rho(x,y)dxdy, \tag{3-253b}$$

$$k_{ij} = \iint_S \gamma_i(x,y) \iint_S k(x,y;\xi,\eta)\gamma_j(\xi,\eta)d\xi d\eta dxdy, \tag{3-255}$$

where $k(x,y;\xi,\eta)$ denotes the stiffness influence function. The kinetic and potential energy expressions which enter Lagrange's equations during the derivation of Eqs. (3-256) are

$$T = \tfrac{1}{2}\sum_{i=1}^{n}\sum_{j=1}^{n} m_{ij}\dot{q}_i\dot{q}_j, \tag{3-253a}$$

$$U_E = \tfrac{1}{2}\sum_{i=1}^{n}\sum_{j=1}^{n} k_{ij}q_iq_j. \tag{3-254}$$

We can, of course, replace one or more γ_i with rigid-body degrees of freedom in the manner of Eqs. (3-189) and (3-203); the stiffness coefficients k_{ij} associated with them vanish.

Whether Eq. (3-147a) or Eq. (3-256) is utilized in the flutter calculation, we specify simple harmonic motion,

$$\xi_j(t) = \bar{\xi}_j e^{i\omega t}, \tag{9-56}$$

or

$$q_j(t) = \bar{q}_j e^{i\omega t}, \tag{9-57}$$

with $\bar{\xi}_j$ and $\bar{q}_j$ complex constants. Since the problem is linear, the external force becomes

$$F_z(x,y,t) = \bar{F}_z(x,y)e^{i\omega t} = -\Delta\bar{p}_a(x,y)e^{i\omega t}, \tag{9-58}$$

and the time variable can be eliminated from all formulas. $\Delta\bar{p}_a(x,y)$ is a linear function of the amplitudes of the various modes, so Ξ_j and Q_j are similarly related to these amplitudes. The equations of motion are therefore homogeneous in a finite set of $\bar{\xi}_j$, or in the n $\bar{q}_j$'s, and we are led to an algebraic eigenvalue problem. The requirement for nonzero solutions is that the determinant of the coefficients must vanish, this being the so-called flutter determinant. Its unusual feature is that its elements are complex numbers, introduced by the phase differences among the motions and aerodynamic loads. As in the simpler systems discussed in Sections 9-2 and 9-3, all portions of the structure do not move in phase with one another. The actual flutter mode shape

$$\text{Re}\left\{\sum_{j=1}^{n} \gamma_j(x,y)\bar{q}_j\right\}$$

may be a very complicated affair, quite difficult to measure experimentally.

When the structure whose flutter is being analyzed deforms like a plate, the normal-mode approach is probably the most efficient to use. This is because there is a distinct economy in the vanishing of the inertial and elastic coupling terms m_{ij} and k_{ij} ($i \neq j$). Equal difficulties are encountered in computing generalized forces, whether in the Ξ_j or in the Q_j forms.

On the other hand, it has often been shown more economical to choose artificial modes when the lifting surfaces have rigid chordwise sections. This method is the one most familiar to practicing aeroelasticians, for a majority of wings fall in its special category. Common instances where it is used almost exclusively are bending-torsion and bending-torsion-control-surface flutter of straight wings and tails of large aspect ratio. Propeller blades and cantilever rotor and stator blades in such rotating machinery as turbojet engines can also be treated in this way. In the latter cases, care must be taken to increase the effective bending stiffness when there is a field of centrifugal forces. Also, the aerodynamic loads in compressors and turbines may have to be adjusted to correct for the presence of adjacent blades and blade rows.

We concentrate on the artificial mode scheme in the simple examples and numerical calculations which make up the remainder of the present section.

(a) *Bending-torsion flutter analysis with artificial modes: structural damping.* Consider a wing, tail, etc. with a straight elastic axis normal to the flight direction. The root is assumed cantilever at $y = 0$ for simplicity, although such effects as rigid-body degrees of freedom or tail motions due to fuselage bending and torsion* are easily added. The bending and torsional deflections are just those illustrated in Fig. 9–6, but we no longer specify uniform properties. Each type of deformation is regarded as the superposition of a series of artificial uncoupled modes.

$$w_B(y,t) = \sum_{i=1}^{r} f_{wi}(y)w_i(t), \tag{9-59}$$

$$\theta_T(y,t) = \sum_{i=1}^{(n-r)} f_{\theta i}(y)\theta_i(t). \tag{9-60}$$

The shapes f_{wi} and associated frequencies could be computed, for instance, by placing all the mass along the elastic axis and finding the r lowest modes of the resulting beam. The $f_{\theta i}$ could be found by inserting a rigid pin along the axis to prevent any bending.

The surface deformations, according to Eq. (3–251), are

$$w(x,y,t) = \sum_{i=1}^{r} f_{wi}(y)w_i(t) - \sum_{i=1}^{(n-r)} xf_{\theta i}(y)\theta_i(t), \tag{9-61}$$

with x measured from the local elastic axis, so that

$$\gamma_i(x,y) = f_{wi}(y), \qquad i = 1, 2, \cdots, r, \tag{9-62a}$$

$$\gamma_{(r+i)}(x,y) = -xf_{\theta i}(y), \qquad i = 1, 2, \cdots, (n-r), \tag{9-62b}$$

and the $q_i(t)$ are just the $w_i(t)$ and $\theta_i(t)$. To find the inertial coefficients m_{ij}, we observe that for the half-span between 0 and l

$$T = \tfrac{1}{2}\int_0^l \int_{\text{chord}} [\dot{w}(x,y,t)]^2 \rho(x,y)\,dx\,dy$$

$$= \tfrac{1}{2}\int_0^l \int_{\text{chord}} [\dot{w}_B(y,t) - x\dot{\theta}_T(y,t)]^2 \rho(x,y)\,dx\,dy$$

$$= \int_0^l [\tfrac{1}{2}m\dot{w}_B^2 - S_y\dot{w}_B\dot{\theta}_T + \tfrac{1}{2}I_y\dot{\theta}_T^2]\,dy, \tag{9-63}$$

---

* It should be evident to the reader that the vertical tail is an exception to the assumption that the entire structure lies in the xy-plane. We place the fin and rudder in the xz-plane and assume, for aerodynamic purposes, that its only important displacements are in the y-direction. It is set in motion relative to the wing by fuselage torsion and side bending. The latter represents the other type of deformation within the xy-plane which may be important to the aeroelastician.

where $m(y)$, $S_y(y)$, and $I_y(y)$ are the mass, static unbalance about the elastic axis (positive for aft center of gravity), and moment of inertia about the elastic axis, all per unit spanwise distance. By substituting Eqs. (9–59) and (9–60) into Eq. (9–63), we obtain

$$T = \tfrac{1}{2}\sum_{i=1}^{r}\sum_{j=1}^{r} \dot{w}_i\dot{w}_j \int_0^l mf_{w_i}f_{w_j}dy - \sum_{i=1}^{r}\sum_{j=1}^{(n-r)} \dot{w}_i\dot{\theta}_j \int_0^l S_y f_{w_i}f_{\theta_j}dy$$

$$+ \tfrac{1}{2}\sum_{i=1}^{(n-r)}\sum_{j=1}^{(n-r)} \dot{\theta}_i\dot{\theta}_j \int_0^l I_y f_{\theta_i}f_{\theta_j}dy, \tag{9–64}$$

from which the expressions for m_{ij} are evident by comparison with Eq. (3–253b). The k_{ij} are found by examining the total potential energy of elastic deformation.

As is frequently done, we omit all but a single fundamental bending mode $f_w(y)$ and a single torsion mode $f_\theta(y)$ while completing the present illustration. No essential features of the analysis are lost by this step. We write the kinetic energy from Eq. (9–64):

$$T = \tfrac{1}{2}\dot{w}_R^2 \int_0^l mf_w^2 dy - \dot{w}_R\dot{\theta}_R \int_0^l S_y f_w f_\theta dy + \tfrac{1}{2}\dot{\theta}_R^2 \int_0^l I_y f_\theta^2 dy. \tag{9–65}$$

Subscript R on w_R and θ_R here indicates that they are the actual bending and twist at some conveniently chosen reference station. Since the structure is a simple beam, we can employ Eq. (3–259), in place of the influence functions, to describe the internal strain energy:

$$U_E = \tfrac{1}{2}\int_0^l EI\left[\frac{\partial^2 w_B}{\partial y^2}\right]^2 dy + \tfrac{1}{2}\int_0^l GJ\left[\frac{\partial \theta_T}{\partial y}\right]^2 dy$$

$$= \tfrac{1}{2}w_R^2 \int_0^l EI\left[\frac{d^2 f_w}{dy^2}\right]^2 dy + \tfrac{1}{2}\theta_R^2 \int_0^l GJ\left[\frac{df_\theta}{dy}\right]^2 dy. \tag{9–66}$$

To avoid the need for taking derivatives of the approximations f_w and f_θ, we define free-vibration frequencies ω_w and ω_θ for these modes in such a way that

$$U_E = \tfrac{1}{2}\omega_w^2 w_R^2 \int_0^l mf_w^2 dy + \tfrac{1}{2}\omega_\theta^2 \theta_R^2 \int_0^l I_y f_\theta^2 dy. \tag{9–67}$$

These frequencies are exactly what we would get by artificially decoupling the two types of deformation. For example, if we distribute the mass $m(y)$ along the elastic axis and assume an oscillation

$$w_B(y,t) = f_w(y)\cos \omega_w t, \tag{9–68}$$

we obtain for the kinetic and potential energies during free vibration

$$T_{\text{F.V.}} = \tfrac{1}{2}\int_0^l m\dot{w}_B^2 dy = \tfrac{1}{2}\omega_w^2 \sin^2 \omega_w t \int_0^l mf_w^2 dy, \tag{9–69}$$

$$U_{\text{F.V.}} = \tfrac{1}{2} \int_0^l EI \left[\frac{\partial^2 w_B}{\partial y^2} \right]^2 dy = \tfrac{1}{2} \cos^2 \omega_w t \int_0^l EI \left[\frac{d^2 f_w}{dy^2} \right]^2 dy. \quad (9\text{--}70)$$

The requirement that the maximum values of $T_{\text{F.V.}}$ and $U_{\text{F.V.}}$ are equal reads

$$\tfrac{1}{2} \int_0^l EI \left[\frac{d^2 f_w}{dy^2} \right]^2 dy = \tfrac{1}{2} \omega_w{}^2 \int_0^l m f_w{}^2 dy. \quad (9\text{--}71)$$

If we multiply both sides of Eq. (9–71) by $w_R{}^2$, we discover the identity between the first terms in Eq. (9–66) and (9–67) under this definition of ω_w.

Since $F_z(x,y,t)$ is the negative of the pressure difference $\Delta p_a(x,y,t)$, the generalized forces from Eq. (3–258) are recognized as weighted integrals of familiar aerodynamic quantities.

$$Q_w = -\int_0^l \int_{\text{chord}} \Delta p_a(x,y,t) f_w(y)\,dx\,dy = \int_0^l L(y,t) f_w(y)\,dy, \quad (9\text{--}72)$$

$$Q_\theta = -\int_0^l \int_{\text{chord}} \Delta p_a(x,y,t) [-x f_\theta(y)]\,dx\,dy = \int_0^l M_y(y,t) f_\theta(y)\,dy, \quad (9\text{--}73)$$

where $L(y,t)$ is the lift per unit span (positive upward), and $M_y(y,t)$ is the running pitching moment about the elastic axis (positive to rotate the nose upward).

Lagrange's equations can now be used, in conjunction with (9–65) and (9–67), to derive the forms of Eqs. (3–256) which govern the system. While carrying out this step, we introduce in an approximate fashion the effects of structural friction. As pointed out by Theodorsen and Garrick (Ref. 9–3), this phenomenon has a much stronger controlling influence on a stability problem like flutter than it has, say, on forced motion. The mechanism is always one of energy dissipation, mostly due to rubbing between adjacent parts at their joints, and its presence decreases the extent of any range of instability.

Observations show that the energy per cycle removed from a vibration by structural friction is roughly proportional to the square of the amplitude but independent of the frequency. This fact demonstrates a significant difference from the simpler case of viscous friction. Thus a simple harmonic motion

$$x = x_0 \sin \omega t \quad (9\text{--}74)$$

might be opposed by a viscous damping force

$$F_{\text{D.V.}} = -f \frac{dx}{dt} = -f \omega x_0 \cos \omega t = \mp f \omega \sqrt{x_0{}^2 - x^2}, \quad (9\text{--}75)$$

where the sign is determined from the fact that $F_{\text{D.V.}}$ is always opposite to the instantaneous velocity. The work per cycle

$$W_{\text{D.V.}} = \int_{t=0}^{t=2\pi/\omega} F_{\text{D.V.}} dx \qquad (9\text{--}76)$$

done by $F_{\text{D.V.}}$ on the oscillator is the area inside the ellipse found by plotting $F_{\text{D.V.}}$ vs. x. Since this figure has semiaxes $f\omega x_0$ and x_0, the work per cycle is proportional both to x_0^2 and to ω.

By contrast, a damping force which has the same effect as structural friction would be

$$F_{\text{D.S.}} = -gx_0 \cos \omega t = \mp g \sqrt{x_0^2 - x^2}. \qquad (9\text{--}77)$$

The work is again the area inside an ellipse, but here the semiaxes are gx_0 and x_0. When encountered in cyclic tests of materials, such a force is called hysteresis. If we use the complex representation of simple harmonic motion, $F_{\text{D.S.}}$ lags the position vector by $90°$ and is therefore given by $-igx$.

The precise mechanism of structural friction in aircraft is difficult to describe mathematically. So long as its magnitude is very small, however, we can adequately approximate it with some source of energy dissipation. We might, for instance, imagine the various elastic deformations to be opposed by systems of external forces and moments obtained by multiplying the local elastic restoring force (which is always opposite to the displacement) by i and a small structural damping coefficient g_j. This coefficient differs in magnitude from one type of deformation to another, but it can be estimated experimentally from the rate of decay of free vibrations by observing from the equations of motion that it is proportional to this rate. Since all approximate representations of structural friction are quite artificial, it is important not to try to attach any detailed physical significance to any one of them. We must realize that all they accomplish is to provide a means of energy removal that is correctly dependent on the amplitude and frequency.

In the case of bending and torsion which we are discussing here, coefficients g_w and g_θ may be defined as follows. The elastic restoring force and torque acting on an element dy of the beam, when it is displaced by $w_B(y,t)$ and $\theta_T(y,t)$ in simple harmonic motion are

$$dF = -\frac{\partial^2}{\partial y^2}\left[EI \frac{\partial^2 w_B}{\partial y^2} \right] dy \qquad (9\text{--}78\text{a})$$

and

$$dt = \frac{\partial}{\partial y}\left[GJ \frac{\partial \theta_T}{\partial y} \right] dy. \qquad (9\text{--}78\text{b})$$

With each of these we associate a structural friction

$$dF_{\text{D.S.}} = -ig_w \frac{\partial^2}{\partial y^2}\left[EI \frac{\partial^2 w_B}{\partial y^2} \right] dy, \qquad (9\text{--}79\text{a})$$

$$dt_{\text{D.S.}} = ig_\theta \frac{\partial}{\partial y}\left[GJ\frac{\partial\theta_T}{\partial y}\right]dy. \tag{9-79b}$$

The values of constants g_w and g_θ range around 0.01–0.03 in typical metal aircraft structures, rarely exceeding 0.05. In accordance with Eq. (3–258), specialized for the effect of running load and torque along the elastic axis, $dF_{\text{D.S.}}$ and $dt_{\text{D.S.}}$ make the following contributions to the generalized forces in the bending and torsion equations of motion.

$$(Q_w)_{\text{D.S.}} = \int_0^l \frac{dF_{\text{D.S.}}}{dy} f_w(y)dy = -ig_w w_R \int_0^l \frac{d^2}{dy^2}\left[EI\frac{d^2f_w}{dy^2}\right]f_w dy, \tag{9-80a}$$

$$(Q_\theta)_{\text{D.S.}} = \int_0^l \frac{dt_{\text{D.S.}}}{dy} f_\theta(y)dy = ig_\theta\theta_R \int_0^l \frac{d}{dy}\left[GJ\frac{df_\theta}{dy}\right]f_\theta dy. \tag{9-80b}$$

We simplify Eq. (9–80a) by integrating by parts twice and recognizing that the integrated terms vanish because of the structural boundary conditions. We then employ Eq. (9–71) and get

$$(Q_w)_{\text{D.S.}} = -ig_w w_R \int_0^l EI\left[\frac{d^2f_w}{dy^2}\right]^2 dy = -ig_w\omega_w{}^2 w_R \int_0^l mf_w{}^2 dy. \tag{9-81}$$

A similar operation on Eq. (9–80b), involving a single partial integration, yields

$$(Q_\theta)_{\text{D.S.}} = -ig_\theta\theta_R \int_0^l GJ\left[\frac{df_\theta}{dy}\right]^2 dy = -ig_\theta\omega_\theta{}^2\theta_R \int_0^l I_y f_\theta{}^2 dy. \tag{9-82}$$

To obtain the equations of flutter motion, we substitute Eqs. (9–65), (9–67), (9–72), (9–73), (9–81), and (9–82) into Lagrange's equation

$$\frac{d}{dt}\left(\frac{\partial T}{\partial \dot{q}}\right) + \frac{\partial U_E}{\partial q} = Q_q \tag{9-83}$$

for each of the two degrees of freedom $q = w_R$ and $q = \theta_R$. The results are

$$\ddot{w}_R \int_0^l mf_w{}^2 dy - \ddot{\theta}_R \int_0^l S_y f_w f_\theta dy + w_R\omega_w{}^2[1+ig_w]\int_0^l mf_w{}^2 dy = \int_0^l Lf_w dy, \tag{9-84}$$

$$\ddot{\theta}_R \int_0^l I_y f_\theta{}^2 dy - \ddot{w}_R \int_0^l S_y f_w f_\theta dy + \theta_R\omega_\theta{}^2[1+ig_\theta]\int_0^l I_y f_\theta{}^2 dy = \int_0^l M_y f_\theta dy. \tag{9-85}$$

At this point we must adopt some explicit form for the aerodynamic loads. In accordance with common practice, in the preliminary flutter analysis of straight wings with reasonably large aspect ratios, we assume that two-dimensional strip theory is valid. We know that the lifts and moments due to simple harmonic motion in both incompressible and compressible flow are extensively tabulated in the notation [cf. Eqs. (5–350) and (5–351)]:

$$L(y,t) = \pi\rho_\infty b^3 \omega^2 \left\{ \frac{w_R}{b} f_w(y) L_h - \theta_R f_\theta(y)[L_\alpha - L_h(\tfrac{1}{2} + a)] \right\}, \quad (9\text{–}86)$$

$$M_y(y,t) = \pi\rho_\infty b^4 \omega^2 \left\{ -\frac{w_R}{b} f_w(y)[M_h - L_h(\tfrac{1}{2} + a)] \right.$$

$$\left. + \theta_R f_\theta(y)[M_\alpha - (M_h + L_\alpha)(\tfrac{1}{2} + a) + L_h(\tfrac{1}{2} + a)^2] \right\}.$$

$$(9\text{–}87)$$

Here $b(y)$ is the local wing semichord and $a(y)$ is the distance the elastic axis lies aft of the midchord line, measured in semichords. The dimensionless aerodynamic coefficients L_h, L_α, M_h and M_α are functions of reduced frequency $k(y)$ and Mach number M. We specify simple harmonic motion by means of

$$w_R(t) = \bar{w}_R e^{i\omega t}, \qquad \theta_R(t) = \bar{\theta}_R e^{i\omega t}. \quad (9\text{–}88)$$

Finally, we insert Eqs. (9–86) through (9–88) into Eqs. (9–84) and (9–85), collect terms, and divide by $\pi\rho_\infty b_R^3 \omega^2 l e^{i\omega t}$ and $\pi\rho_\infty b_R^4 \omega^2 l e^{i\omega t}$ respectively, to make the results dimensionless.

$$\frac{\bar{w}_R}{b_R} \left\{ \frac{1}{\pi\rho_\infty b_R^2} \left[1 - (1 + i g_w)\left(\frac{\omega_w}{\omega_\theta}\right)^2 \left(\frac{\omega_\theta}{\omega}\right)^2 \right] \int_0^1 m f_w^2 dy^* + \int_0^1 \left(\frac{b}{b_R}\right)^2 L_h f_w^2 dy^* \right\}$$

$$- \bar{\theta}_R \left\{ \frac{1}{\pi\rho_\infty b_R^3} \int_0^1 S_y f_w f_\theta dy^* + \int_0^1 \left(\frac{b}{b_R}\right)^3 [L_\alpha - L_h(\tfrac{1}{2} + a)] f_w f_\theta dy^* \right\} = 0,$$

$$(9\text{–}89)$$

$$- \frac{\bar{w}_R}{b_R} \left\{ \frac{1}{\pi\rho_\infty b_R^3} \int_0^1 S_y f_w f_\theta dy^* + \int_0^1 \left(\frac{b}{b_R}\right)^3 [M_h - L_h(\tfrac{1}{2} + a)] f_w f_\theta dy^* \right\}$$

$$+ \bar{\theta}_R \left\{ \frac{1}{\pi\rho_\infty b_R^4} \left[1 - (1 + i g_\theta)\left(\frac{\omega_\theta}{\omega}\right)^2 \right] \int_0^1 I_y f_\theta^2 dy^* \right.$$

$$\left. + \int_0^1 \left(\frac{b}{b_R}\right)^4 [M_\alpha - (M_h + L_\alpha)(\tfrac{1}{2} + a) + L_h(\tfrac{1}{2} + a)^2] f_\theta^2 dy^* \right\} = 0. \quad (9\text{–}90)$$

Here all factors in the various integrands are functions of the reduced spanwise variable

$$y^* = \frac{y}{l}. \quad (9\text{–}91)$$

The four coefficients of $\bar{w}_R/b_R$ and $\bar{\theta}_R$ in Eqs. (9–89) and (9–90) form the flutter determinant, which always has an order equal to the number of degrees of freedom remaining in the system after the assumed modes are chosen. As mentioned in foregoing sections, the roots of the determinant cannot be calculated by a direct process to find the flutter eigenvalues U_F

and ω_F (or U_F, ω_F, and M_F). Consequently, a host of indirect schemes has been proposed for its solution, many paralleling those discussed in connection with the simple airfoil in Section 9–2. We can state in a general way that most used in practice start out by assuming one or more values of

$$k_R = \frac{\omega b_R}{U} \tag{9-92}$$

(or M and k_R), since these parameters permit the complex aerodynamic integrals to be evaluated numerically. Then the artificial circumstances are determined under which flutter would occur at the conditions prescribed.

We first enumerate three procedures which suggest themselves after a cursory examination of the flutter determinant. In each case, once the integrals in Eqs. (9–89) and (9–90) are calculated, two real quantities must be left as unknowns to be found by equating the real and imaginary parts of the determinant to zero. One of these should logically be the ratio ω_θ/ω. The other might be chosen in one of the following ways:

(1) *Density variation.* This is the scheme to use when a curve of flutter speed vs. altitude is desired. We make ρ_∞ the second unknown, introducing it in terms of the dimensionless parameter $m_R/\pi\rho_\infty b_R^2$, where m_R is the running mass at the reference station. It is obviously possible to factor this parameter out of each term containing the density, leaving only combinations of known quantities. Simultaneous solution of the flutter determinant for $(\omega_\theta/\omega)^2$ and $m_R/\pi\rho_\infty b_R^2$ is facilitated by the fact that both enter only linearly into the imaginary equation.

(2) *Frequency variation.* The influence of frequency ratio ω_w/ω_θ on flutter, with other dimensionless properties of the system fixed, is computed by choosing $(\omega_\theta/\omega)^2$ and $(\omega_w/\omega_\theta)^2$ as the unknowns. The latter appears as a linear factor in both the real and imaginary equations.

(3) *Mass-balance variation.* The effect of varying the location of the center of gravity relative to the elastic axis can be ascertained by designating $S_{yR}/\pi\rho_\infty b_R^3$ or

$$x_{\alpha R} = \frac{S_{yR}}{m_R b_R^2} \tag{9-93}$$

as the companion unknown to $(\omega_\theta/\omega)^2$. Because $a(y)$ is given, the elastic-axis position is fixed, as are the mass and moment of inertia distributions. Flutter speed is peculiarly sensitive to the magnitude of S_y, as we might surmise from the curves of Fig. 9–5. As in (1), the mass-balance parameter enters only linearly into the imaginary equation obtained by expanding the determinant.

A less obvious but more commonly employed scheme for solving the flutter determinant is based on the use of structural damping coefficients as unknowns. This we illustrate by Example 1 below.

In flutter analyses where compressibility is included, we point out that the need for assuming M in advance introduces an added artificiality into the solution. When ω_F, U_F, and M_F are known, the latter two in combination fix the speed of sound, which, in turn, yields the gas temperature where flutter would occur. This temperature and the density ρ_∞ may not correspond to any state actually encountered during flight in the atmosphere. Such problems are taken up further in Section 9–6.

EXAMPLE 9–1. To compute the bending-torsion flutter speed of a cantilever, tapered wing in incompressible flow at sea level, by the U-g method. For this purpose, we pick the same jet transport wing whose normal modes of free vibration are found in Chapter 4. The planform, mass, static unbalance, and moment of inertia distributions are presented in Figs. 2–28, 4–20, and 4–22, and the Table of Example 4–7. The elastic axis lies along the 35% chord line. The bending and torsional rigidity distributions appear in Fig. 2–29.

Solution: We select as a reference station the one at $y = 3l/4$, which is often regarded as the typical section for flutter prediction. Therefore

$$b_R = 5.468 \text{ ft.} \tag{a}$$

Uncoupled bending and torsion modes $f_w(y)$ and $f_\theta(y)$, suitable for the flutter equations, are calculated in Examples 4–5 and 4–6. They are illustrated in Figs. 4–21(a) and 4–23; the frequencies ω_w and ω_θ are 12.80 and 22.36 rad/sec, respectively.

With all these data assembled, we can set up numerically the flutter equations (9–89) and (9–90). Two modifications are made in them. First we adopt the scheme of Smilg and Wasserman (Ref. 9–1) for solving the flutter determinant by assuming

$$g_w \cong g_\theta = g. \tag{b}$$

This g is regarded as one of the unknowns, an especially convenient selection because it always joins with the other unknown ω_θ/ω in the single complex combination

$$Z = \left(\frac{\omega_\theta}{\omega}\right)^2 [1 + ig]. \tag{c}$$

For a particular value of k_R, the determinant expands into one complex polynomial in Z, rather than two real polynomials. Since the present example has only two degrees of freedom, this polynomial can be factored by the quadratic formula.

Our second modification also follows Ref. 9–1 and greatly simplifies the computation of several terms in Eqs. (9–89) and (9–90). Strictly speaking, the aerodynamic integrals on a tapered wing have integrands like L_h which vary along the span for a particular choice of k_R. Thus, for each k_R, the L_h, etc., must be interpolated from tables for a series of values of k, and the integrals are then evaluated by a trapezoidal rule. However, it turns out that all these aerodynamic coefficients in incompressible flow can be closely approximated by polynomials in $1/k$. Using the notation of Scanlan and Rosenbaum (Ref. 9–11),

$$L_h \cong 1 + \frac{k_R}{k} K_2(L_h) = 1 + \frac{b_R}{b} K_2(L_h), \tag{d}$$

$$L_\alpha \cong 0.5 + \frac{b_R}{b} K_2(L_\alpha) + \left(\frac{b_R}{b}\right)^2 K_3(L_\alpha), \tag{e}$$

$$M_h = 0.5, \tag{f}$$

$$M_\alpha \cong 0.375 + \frac{b_R}{b} K_2(M_\alpha). \tag{g}$$

The functions $K_2(L_h)$, $K_2(L_\alpha)$, etc., depend only on the reduced frequency at the reference station; they are tabulated on page 209 of Ref. 9–1 and page 412 of Ref. 9–11.

With these various substitutions, the literal equations of motion become

$$\frac{\overline{w}_R}{b_R}\left\{\left[1-\left(\frac{\omega_w}{\omega_\theta}\right)^2 Z\right]\int_0^1\left[\frac{m}{\pi\rho b_R{}^2}\right]f_w{}^2 dy^* + \int_0^1\left(\frac{b}{b_R}\right)^2 f_w{}^2 dy^*\right.$$
$$+K_2(L_h)\int_0^1 \frac{b}{b_R}f_w{}^2 dy^*\right\}-\overline{\theta}_R\left\{\int_0^1\left[\frac{S_y}{\pi\rho b_R{}^3}\right]f_w f_\theta dy^*\right.$$
$$-\int_0^1\left(\frac{b}{b_R}\right)^3 af_w f_\theta dy^* + K_2(L_\alpha)\int_0^1\left(\frac{b}{b_R}\right)^2 f_w f_\theta dy^*$$
$$\left. +K_3(L_\alpha)\int_0^1\left(\frac{b}{b_R}\right)f_w f_\theta dy^* - K_2(L_h)\int_0^1[\tfrac{1}{2}+a]\left(\frac{b}{b_R}\right)^2 f_w f_\theta dy^*\right\} = 0, \tag{h}$$

$$-\frac{\overline{w}_R}{b_R}\left\{\int_0^1\left[\frac{S_y}{\pi\rho b_R{}^3}\right]f_w f_\theta dy^* - \int_0^1\left(\frac{b}{b_R}\right)^3 af_w f_\theta dy^* - K_2(L_h)\int_0^1[\tfrac{1}{2}+a]\left(\frac{b}{b_R}\right)^2 f_w f_\theta dy^*\right\}$$
$$+\overline{\theta}_R\left\{[1-Z]\int_0^1\left[\frac{I_y}{\pi\rho b_R{}^4}\right]f_\theta{}^2 dy^* + \int_0^1[\tfrac{1}{8}+a^2]\left(\frac{b}{b_R}\right)^4 f_\theta{}^2 dy^*\right.$$
$$+K_2(M_\alpha)\int_0^1\left(\frac{b}{b_R}\right)^3 f_\theta{}^2 dy^* + K_2(L_h)\int_0^1[\tfrac{1}{2}+a]^2\left(\frac{b}{b_R}\right)^3 f_\theta{}^2 dy^*$$
$$\left. -K_2(L_\alpha)\int_0^1[\tfrac{1}{2}+a]\left(\frac{b}{b_R}\right)^3 f_\theta{}^2 dy^* - K_3(L_\alpha)\int_0^1[\tfrac{1}{2}+a]\left(\frac{b}{b_R}\right)^2 f_\theta{}^2 dy^*\right\} = 0. \tag{i}$$

All integrals in Eqs. (h) and (i) can be carried out numerically, using $a = -0.3$ and the other data referred to above. We are thus led to the following flutter determinant:

$$\begin{vmatrix} A & B \\ C & D \end{vmatrix} = \begin{vmatrix} \{8.2156[1-0.3277Z] & -\{-0.3047+0.3545K_2(L_\alpha) \\ +0.2073+0.2215K_2(L_h)\} & +0.3489K_3(L_\alpha)-0.0709K_2(L_h)\} \\ & \\ \{0.3047 & \{66.1117[1-Z]+0.2807 \\ +0.0709K_2(L_h)\} & +1.0569K_2(M_\alpha)+0.0423K_2(L_h) \\ & -0.2114K_2(L_\alpha)-0.1772K_3(L_\alpha)\} \end{vmatrix} = 0. \tag{j}$$

For several choices of k_R, the coefficients in Eq. (j) were found from the tables, and the two complex roots Z computed by the quadratic formula. These roots, with the corresponding values of speed, frequency, and artificial structural damping g, are listed in Table 9–1.

Figure 9–9 shows a plot of g vs. U for the two modes described by roots Z_1 and Z_2. If we make the conservative assumption that g_w, g_θ, and g are nearly

TABLE 9–1

k_R	Z_1	U_1 mph	ω_1 $\dfrac{\text{rad}}{\text{sec}}$	g_1	Z_2	U_2 mph	ω_2 $\dfrac{\text{rad}}{\text{sec}}$	g_2
0.06	$3.414-2.816i$	748	12.1	-0.825	$1.668+0.021i$	1075	17.3	0.012
0.08	$3.078-2.009i$	590	12.7	-0.653	$1.456+0.0005i$	865	18.6	0.000
0.10	$2.973-1.517i$	484	13.0	-0.510	$1.331-0.030i$	723	19.4	-0.023
0.12	$2.946-1.196i$	403	13.1	-0.406	$1.247-0.049i$	621	20.0	-0.039
0.16	$2.959-0.821i$	303	13.0	-0.277	$1.146-0.058i$	485	20.8	-0.051
0.20	$2.988-0.615i$	240	12.9	-0.206	$1.094-0.054i$	398	21.4	-0.049
0.30	$3.035-0.368i$	159	12.8	-0.121	$1.042-0.038i$	272	21.9	-0.036
0.40	$3.064-0.258i$	119	12.8	-0.084	$1.024-0.029i$	206	22.1	-0.028

zero in the actual structure, we find the following flutter condition where the Z_2-curve crosses the U-axis:

$$U_F = 865 \text{ mph}, \qquad \omega_F = 18.6 \text{ rad/sec}, \qquad k_F = 0.08. \qquad \text{(k)}$$

As in the divergence and reversal calculations of Chapter 8, U_F is far too large to justify the assumption of incompressible flow. It does establish, however, that the jet transport is safe from bending-torsion flutter throughout the operating range of flight speeds. The example loses no value as an illustration of flutter computation procedure, and we might use the 865 mph as the starting point of a more elaborate determination of U_F as a function of Mach number.

We label the particular U-g curve which leads to flutter the torsion branch, since it merges with the first coupled mode of torsional vibration as the speed is brought down to zero (cf. Section 9–4). Points on the two curves where g assumes artificial values represent physically unrealizable situations, so we must be careful when attempting to interpret them. Here we can safely assert, however,

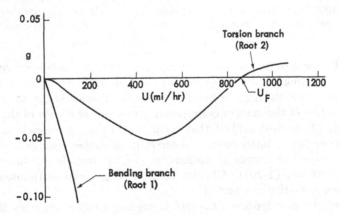

Fig. 9–9. Plot of artificial structural damping g vs. airspeed U from the bending-torsion flutter analysis of a jet transport wing.

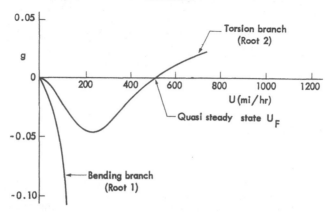

Fig. 9–10. Plot of g vs. U for the jet transport wing with quasi steady-state aerodynamic loads employed in the flutter calculation.

that when g is negative, which means the structure is somehow feeding energy into the vibration, the actual aeroelastic modes are stable. Conversely, small positive values of g describe situations where the dissipation must be augmented to prevent the simple harmonic motion from blowing up and diverging. These are certainly ranges of instability. The slope of the curve of g vs. U where it passes through U_F is a qualitative measure of how violently the oscillations would set in during accelerated flight. As in a minority of cases of bending-torsion flutter, the present one appears only moderately dangerous.

Despite the rather low reduced frequency, unsteady aerodynamic effects are significant in determining U_F. To demonstrate this, we have carried out a parallel analysis using the quasi steady-state aerodynamic theory described in Section 5–6. The force and moment coefficients are the same as the unsteady ones, except with $C(k)$ everywhere replaced by unity. This approximation produces the g vs. U curves pictured in Fig. 9–10. The new flutter speed of 515 mph is 40% lower than its more exact counterpart. Here it appears that quasi-steady aerodynamics yields a highly conservative answer, in contrast to the example of Section 9–1, where it predicted no flutter at all.

(b) *Flutter involving rigid-body degrees of freedom or fuselage deformation.* When predicting the flutter of unrestrained aircraft, we often encounter situations where the mass or moment of inertia of some lifting surface is a large fraction of the mass or corresponding moment of inertia of the entire machine. It is then evident that certain rigid-body motions contribute appreciably to the flutter mode. In deriving the flutter equations, whether by the method of normal or assumed modes, we include one or more of Eqs. (3–189) and (3–203). Chapter 3 outlines in general how the associated equations of motion are formed.

From the aerodynamic standpoint, we can readily compute the lift and moment distributions due to any rigid-body oscillation, since each one is analogous to a simple mode of bending or torsion. Vertical translation of the whole airplane is the same as bending in which the shape function $f_w(y)$ is constant all the way across the wing or tail. Pitching resembles torsion of the wing with $f_\theta(y)$ constant and the axis of rotation coincident with the pitch axis, which usually passes through the center of gravity of the airplane. Vertical translation and pitching do not affect the vertical tail, but pitching is equivalent to rotation of the horizontal tail about an axis far ahead of its leading edge. Rolling is equivalent to an antisymmetrical bending mode of all the aerodynamic surfaces, in which the shape function f_w varies linearly with distance from the fuselage center line.

As a practical illustration of how these facts are used, we assume that the bending-torsion flutter described under Subsection (a) is antisymmetrical and add to it the roll degree of freedom. Because the wing contributes so heavily to the airplane moment of inertia I_x, rolling is likely to be the rigid-body mode considered first in most applications. For simplicity we omit here the aerodynamic effects of the tail and control surfaces.

We denote by

$$\Phi(t) = \bar{\Phi}e^{i\omega t} \tag{9-94}$$

the angular displacement of the undeformed airplane about the x-axis through the center of gravity, which axis we assume fixed and nearly parallel to the flight direction. The second part of Eq. (3–203) shows that $\Phi(t)$ causes a displacement.

$$\Phi(t)\phi_3(x,y) = \Phi(t)y \tag{9-95}$$

of points along the wingspan, so that the total absolute normal displacement and twist are, from Eqs. (9–59) and (9–60),

$$w_B(y,t) = w_R(t)f_w(y) + \Phi(t)y, \tag{9-96}$$

$$\theta_T(y,t) = \theta_R(t)f_\theta(y). \tag{9-97}$$

We use Lagrange's equation to construct the equations of motion, noting that Eq. (9–67) still gives the potential energy U_E, since no elastic deformation is connected with Φ. According to Eq. (9–65), the kinetic energy of the entire airplane can be derived as follows, if we regard f_w and f_θ as odd functions of y which vanish throughout the fuselage. We refer to this as $2T$, since we ultimately plan to work with one-half the airplane:

$$2T = \tfrac{1}{2}\dot{w}_R^2 \int_{-l}^{l} mf_w^2\,dy + \tfrac{1}{2}\dot{\Phi}^2 \int_{\text{airplane}} my^2\,dy + \dot{w}_R\dot{\Phi} \int_{-l}^{l} myf_w\,dy$$

$$- \dot{w}_R\dot{\theta}_R \int_{-l}^{l} S_yf_wf_\theta\,dy - \dot{\Phi}\dot{\theta}_R \int_{-l}^{l} S_yyf_\theta\,dy + \tfrac{1}{2}\dot{\theta}_R^2 \int_{-l}^{l} I_yf_\theta^2\,dy, \tag{9-98a}$$

$$T = \tfrac{1}{2}\dot{\Phi}^2 \frac{I_x}{2} + \tfrac{1}{2}\dot{w}_R^2 \int_{0}^{l} mf_w^2\,dy + \dot{w}_R\dot{\Phi} \int_{0}^{l} myf_w\,dy - \dot{w}_R\dot{\theta}_R \int_{0}^{l} S_yf_wf_\theta\,dy$$

$$- \dot{\Phi}\dot{\theta}_R \int_{0}^{l} S_yyf_\theta\,dy + \tfrac{1}{2}\dot{\theta}_R^2 \int_{0}^{l} I_yf_\theta^2\,dy. \tag{9-98b}$$

The generalized forces Q_w and Q_θ given by Eqs. (9–72) and (9–73) are unchanged except to the extent that Φ contributes to $L(y,t)$ and $M_y(y,t)$. The generalized force associated with rolling is

$$Q_\Phi = \int_0^l L(y,t)y\,dy = \tfrac{1}{2}L_R(t), \qquad (9\text{–}99)$$

one-half the rolling moment applied by the wing to the airplane.

Since the only structural damping forces are those expressed in Eqs. (9–81) and (9–82), we are ready to apply Lagrange's equation (9–83) with q replaced successively by w_R, θ_R, and Φ. The resulting equations of motion read

$$\ddot{w}_R \int_0^l mf_w{}^2dy + \ddot{\Phi}\int_0^l myf_w dy - \ddot{\theta}_R \int_0^l S_y f_w f_\theta dy + w_R \omega_w{}^2[1+ig_w]\int_0^l mf_w{}^2dy$$

$$= \int_0^l Lf_w dy, \quad (9\text{–}100)$$

$$\ddot{\theta}_R \int_0^l I_y f_\theta{}^2 dy - \ddot{\Phi}\int_0^l S_{yy}f_\theta dy - \ddot{w}_R \int_0^l S_y f_w f_\theta dy + \theta_R \omega_\theta{}^2[1+ig_\theta]\int_0^l I_y f_\theta{}^2 dy$$

$$= \int_0^l M_y f_\theta dy, \quad (9\text{–}101)$$

$$\ddot{\Phi}\frac{I_x}{2} + \ddot{w}_R \int_0^l myf_w dy - \ddot{\theta}_R \int_0^l S_{yy}f_\theta dy = \tfrac{1}{2}L_R(t). \qquad (9\text{–}102)$$

No elastic forces appear in the balance equation (9–102), since rolling is coupled with wing bending and torsion only through inertial effects and the antisymmetrical distribution of lift. The w_R and θ_R degrees of freedom are also uncoupled elastically, because of the existence of an elastic axis.

In terms of the aerodynamic coefficients of Ref. 9–1, the three degrees of freedom give rise to the following strip-theory lift and moment per unit span:

$$L(y,t) = \pi\rho_\infty b^3 \omega^2 \left\{ \left[\frac{w_R}{b}f_w(y) + \Phi\frac{y}{b}\right] L_h - \theta_R f_\theta(y)[L_\alpha - L_h(\tfrac{1}{2}+a)] \right\}, \quad (9\text{–}103)$$

$$M_y(y,t) = \pi\rho_\infty b^4 \omega^2 \left\{ -\left[\frac{w_R}{b}f_w(y) + \Phi\frac{y}{b}\right][M_h - L_h(\tfrac{1}{2}+a)] \right.$$

$$\left. + \theta_R f_\theta(y)[M_\alpha - (M_h + L_\alpha)(\tfrac{1}{2}+a) + L_h(\tfrac{1}{2}+a)^2] \right\}. \quad (9\text{–}104)$$

The steps of substituting Eqs. (9–103) and (9–104) into the equations of motion and constructing a flutter determinant composed of dimensionless elements do not differ significantly from those in the previous example. We do not repeat them here. It is worth while to note, however, that the absence of elastic forces from Eq. (9–102) leads to a determinant in which the unknown ω_θ/ω appears only in the first two terms on the principal diagonal. This simplifies the solution appreciably; the quadratic formula can be used in places where an extra elastic degree of freedom would require the factorization of cubic polynomials.

Another situation where degrees of freedom other than lifting-surface deformations may play an important part occurs in flutter analysis of the tail. The fuselage nearly always has appreciable flexibility between its intersections with the wing and tail. Vertical bending of the fuselage causes a symmetrical rotation and angle-of-attack change of the stabilizer relative to the wing. Side bending and torsion produce antisymmetrical displacements, the former changing the fin angle of attack and the latter rocking both the fin and stabilizer.

When we are able to neglect the wing's influence and study flutter of the tail alone, we often treat the fuselage as a nonuniform cantilever beam built in at some experimentally determined station (Fig. 9–11). The tail on the end of its flexible outrigger can then be studied for symmetrical and antisymmetrical flutter, with appropriate fuselage modes and (if necessary) control-surface rotations included in each set of equations of motion. The

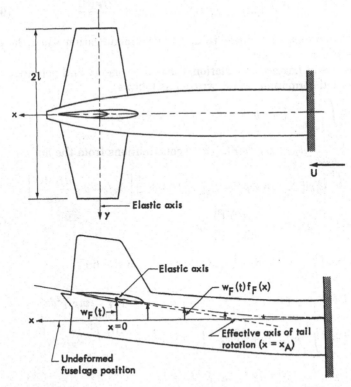

Fig. 9–11. The influence of fuselage bending on the motion of the horizontal tail. The fuselage is assumed built in at some point near the airplane center of gravity.

essential aerodynamic and structural features of all tail surfaces resemble those of the wing, so the novel aspects of the problem come from the fuselage deflections.

To demonstrate how this situation is handled, we repeat the bending-torsion analysis of Subsection (a), now regarding it as symmetrical flutter of the stabilizer and adding fuselage vertical bending to the system. Figure 9-11 pictures the influence of this bending on the tail, along with the single assumed fuselage mode

$$w_{BF}(x,t) = w_F(t)f_F(x). \qquad (9-105)$$

The shape function $f_F(x)$ is so normalized that $w_F(t)$ gives the motion at the point of intersection with the stabilizer's elastic axis, which point projects onto the origin of coordinates. If the elastic axis is perpendicular to the fuselage center line, its absolute normal displacement and the rotation about it are

$$w_B(y,t) = w_R(t)f_w(y) + w_F(t), \qquad (9-106)$$

$$\theta_T(y,t) = \theta_R(t)f_\theta(y) - w_F(t)\frac{df_F(0)}{dx}. \qquad (9-107)$$

These equations are subject to an obvious modification when the elastic axis is swept or crooked.

For use in Lagrange's equation, the kinetic energy and potential energy of elastic deformation can be written as follows:

$$T = \frac{1}{2}\int_{-l}^{l}\int_{\text{chord}}[\dot{w}_B(y,t) - x\dot{\theta}_T(y,t)]^2\rho(x,y)dxdy$$

$$+\frac{1}{2}\int_{\text{fuse.}}[\dot{w}_{BF}(x,t)]^2 m_F(x)dx + [\text{contributions from the fin}]$$

$$=\int_{-l}^{l}\left\{\frac{1}{2}m[\dot{w}_Rf_w+\dot{w}_F]^2 - S_y[\dot{w}_Rf_w+\dot{w}_F]\left[\dot{\theta}_Rf_\theta - \dot{w}_F\frac{df_F(0)}{dx}\right]\right.$$

$$\left.+\frac{1}{2}I_y\left[\dot{\theta}_Rf_\theta - \dot{w}_F\frac{df_F(0)}{dx}\right]^2\right\}dy$$

$$+\frac{1}{2}\dot{w}_F^2\int_{\text{fuse.}}m_Ff_F^2dx + [\text{contributions from the fin}]$$

$$=\frac{1}{2}\dot{w}_F^2\left\{m_T + 2S_{yT}\frac{df_F(0)}{dx} + I_{yT}\left[\frac{df_F(0)}{dx}\right]^2 + \int_{\text{fuse.}}m_Ff_F^2dx\right\}$$

$$+\frac{1}{2}\dot{w}_R^2\int_{-l}^{l}mf_w^2dy + \dot{w}_R\dot{w}_F\left\{\int_{-l}^{l}mf_wdy + \frac{df_F(0)}{dx}\int_{-l}^{l}S_yf_wdy\right\}$$

$$-\dot{w}_R\dot{\theta}_R\int_{-l}^{l}S_yf_wf_\theta dy + \frac{1}{2}\dot{\theta}_R^2\int_{-l}^{l}I_yf_\theta^2dy$$

$$+\dot{\theta}_R\dot{w}_F\left\{-\int_{-l}^{l}S_yf_\theta dy - \frac{df_F(0)}{dx}\int_{-l}^{l}I_yf_\theta dy\right\}. \qquad (9-108)$$

$$U_E = \tfrac{1}{2}w_R{}^2 \int_{-l}^{l} EI \left[\frac{d^2f_w}{dy^2}\right]^2 dy + \tfrac{1}{2}\theta_R{}^2 \int_{-l}^{l} GJ \left[\frac{df_\theta}{dy}\right]^2 dy$$

$$+ \tfrac{1}{2}w_F{}^2 \int_{\text{fuse.}} EI_F \left[\frac{d^2f_F}{dx^2}\right]^2 dx = \tfrac{1}{2}\omega_w{}^2 w_R{}^2 \int_{-l}^{l} mf_w{}^2 dy$$

$$+ \tfrac{1}{2}\omega_\theta{}^2 \theta_R{}^2 \int_{-l}^{l} I_y f_\theta{}^2 dy + \tfrac{1}{2}\omega_F{}^2 w_F{}^2 \int_{\text{fuse.}} m_F f_F{}^2 dx. \qquad (9\text{–}109)$$

In all cases, subscript F here denotes a property of the fuselage. m_T, S_{yT}, and I_{yT} are the mass, static unbalance, and moment of inertia of the entire tail; they include the fin as well as the stabilizer, since it participates in the motion produced by the fuselage.

The generalized forces Q_w and Q_θ in Eqs. (9–72) and (9–73) are unchanged by the new degree of freedom. We neglect aerodynamic loads on the fuselage itself. However, when the coordinate w_F is subjected to a virtual displacement δw_F, both the tail lift and moment do work on the system, in the amount

$$\delta W = Q_F \delta w_F = \delta w_F \int_{-l}^{l} L(y,t)dy - \delta w_F \frac{df_F(0)}{dx} \int_{-l}^{l} M_y(y,t)dy. \qquad (9\text{–}110)$$

Therefore the generalized force is

$$Q_F = L_T(t) - M_T(t) \frac{df_F(0)}{dx}, \qquad (9\text{–}111)$$

where L_T, and M_T are the total instantaneous force on the stabilizer and the moment about the (straight) elastic axis.

In addition to the structural damping effects $(Q_w)_{\text{D.S.}}$ and $(Q_\theta)_{\text{D.S.}}$ [Eqs. (9–81) and (9–82)], we anticipate that some energy is dissipated during the fuselage vibration. In analogy with the others, this effect can be represented by the generalized force

$$(Q_F)_{\text{D.S.}} = -ig_F \omega_F{}^2 w_F \int_{\text{fuse.}} m_F f_F{}^2 dx. \qquad (9\text{–}112)$$

We substitute Eqs. (9–108), (9–109), (9–111), (9–112), and other results from Subsection (a) into Lagrange's equation (9–83) for the degrees of freedom w_R, θ_R, and w_F. Thus we find the equations of motion:

$$\ddot{w}_R \int_{-l}^{l} mf_w{}^2 dy - \ddot{\theta}_R \int_{-l}^{l} S_y f_w f_\theta dy + \ddot{w}_F \left[\int_{-l}^{l} mf_w dy + \frac{df_F(0)}{dx} \int_{-l}^{l} S_y f_w dy\right]$$

$$+ w_R \omega_w{}^2 [1 + ig_w] \int_{-l}^{l} mf_w{}^2 dy = \int_{-l}^{l} Lf_w dy. \qquad (9\text{–}113)$$

$$\theta_R \int_{-l}^{l} I_y f_\theta^2 dy - \ddot{w}_R \int_{-l}^{l} S_y f_w f_\theta dy - \ddot{w}_F \left[\int_{-l}^{l} S_y f_\theta dy + \frac{df_F(0)}{dx} \int_{-l}^{l} I_y f_\theta dy \right]$$

$$+ \theta_R \omega_\theta^2 [1 + ig_\theta] \int_{-l}^{l} I_y f_\theta^2 dy = \int_{-l}^{l} M_y f_\theta dy. \qquad (9\text{-}114)$$

$$\ddot{w}_F \left\{ m_T + 2S_{yT} \frac{df_F(0)}{dx} + I_{yT} \left[\frac{df_F(0)}{dx} \right]^2 + \int_{\text{fuse.}} m_F f_F^2 dx \right\}$$

$$+ \ddot{w}_R \left[\int_{-l}^{l} m f_w dy + \frac{df_F(0)}{dx} \int_{-l}^{l} S_y f_w dy \right]$$

$$- \theta_R \left[\int_{-l}^{l} S_y f_\theta dy + \frac{df_F(0)}{dx} \int_{-l}^{l} I_y f_\theta dy \right]$$

$$+ w_F \omega_F^2 [1 + ig_F] \int_{\text{fuse.}} m_F f_F^2 dx$$

$$= L_T(t) - M_T(t) \frac{df_F(0)}{dx}. \qquad (9\text{-}115)$$

Some care must be exercised in deriving the running lift and moment expressions to be substituted into Eqs. (9–113)–(9–115). This is because the stabilizer is simultaneously subjected to two rotations, one about the elastic axis due to torsion, the other about the line $x = x_A$ due to fuselage bending. Our various aerodynamic theories are arranged to give pitching moment about the same axis which is used in defining the rotations. Hence the easiest stratagem here is to treat the effect of the fuselage as a combined displacement of and rotation about the elastic axis, as described by Eqs. (9–106) and (9–107). The transfer of moment arms is accomplished automatically by our equations of motion. For example, the right side of Eq. (9–115) is the nose-down pitching moment about the axis at $x = x_A$, divided by the magnitude of x_A.

The strip-theory aerodynamic loads due to the displacements $w_B(y,t)$ and $\theta_T(y,t)$ are

$$L(y,t) = \pi \rho_\infty b^3 \omega^2 \left\{ \left[\frac{w_R}{b} f_w(y) + \frac{w_F}{b} \right] L_h \right.$$

$$\left. - \left[\theta_R f_\theta(y) - w_F \frac{df_F(0)}{dx} \right] [L_\alpha - L_h(\tfrac{1}{2} + a)] \right\}, \quad (9\text{-}116)$$

$$M_y(y,t) = \pi \rho_\infty b^4 \omega^2 \left\{ - \left[\frac{w_R}{b} f_w(y) + \frac{w_F}{b} \right] [M_h - L_h(\tfrac{1}{2} + a)] \right.$$

$$\left. + \left[\theta_R f_\theta(y) - w_F \frac{df_F(0)}{dx} \right] [M_\alpha - (M_h + L_\alpha)(\tfrac{1}{2} + a) + L_h(\tfrac{1}{2} + a)^2] \right\}.$$

$$(9\text{-}117)$$

Here $b(y)a(y)$ is the distance from stabilizer midchord to the elastic axis position at any station, positive aft. As in the first example of this subsection, we do not elaborate the details of substituting Eqs. (9–116) and (9–117) into Eqs. (9–113)–(9–115) and setting up the dimensionless flutter determinant.

(c) *Flutter involving a primary control surface.* The vast majority of all cases of flutter encountered on aircraft in flight involve flap-type control surfaces. Although such flutter does not usually have a catastrophic outcome, it is most unpleasant and undesirable. Efforts should be made to anticipate it in advance, and design procedures must be available for increasing critical speeds that fall within the operating range. This is an area in which design experience and rules of thumb play a more important part than bending-torsion flutter. We can state, however, that theoretical calculations of the sort described in this chapter are quite successful, with a few exceptions. One of these is flutter in which motion of a tab and elastic deformation of the control surface are major degrees of freedom. Another is the transonic instability, known as aileron buzz, that is associated with shocks oscillating on the upper and lower wing surfaces and interacting with the boundary layer in an unsteady fashion.

Empirical criteria furnish the only reliable means of avoiding aileron buzz. It is best prevented by making the control system so irreversible that negligibly small rotation occurs in response to the hinge moments experienced in transonic flight. With regard to tab flutter, the classical theory may be used as a qualitative guide, provided the mode shapes used in the analysis are carefully selected. Ground vibration tests of the airplane can be very helpful. For an unusually clear, elementary discussion of this topic and of control-surface balancing in general, by an experienced aeroelastician, the reader is referred to Part III of Broadbent's series of articles in *Aircraft Engineering* (Ref. 9–12).

To demonstrate how a control surface degree of freedom is introduced into the assumed-mode flutter equations, we consider a cantilever wing or tail surface which carries a trailing-edge flap between spanwise stations $y = l_1$ and $y = l_2$. The flap has no aerodynamic balance, so that its leading edge and hinge line practically coincide along a line at a distance be behind midchord; in general, both b and the dimensionless e are functions of y. The presence of aerodynamic balance changes only the airloads in the equations of motion, the effects being fully discussed, for example, in Refs. 9–1 and 9–11. We omit it here for two reasons. First, experimental evidence shows that theoretical estimates of hinge moments on aerodynamically balanced surfaces are inaccurate, even to the extent of having the wrong sign. Hence we believe that the predicted aerodynamic coefficients must be modified to agree with what is known about their measured counterparts; a full discussion of this process is beyond the scope of this chapter. Second, there is a trend in modern aircraft toward power-operated irreversible controls without aerodynamic balance.

Three degrees of freedom are included in our analysis: wing bending and torsion, with the same artificial modes $f_w(y)$ and $f_\theta(y)$ defined in Subsection (a), and rotation of the flap through an angle $\beta(t)$ (positive for trailing edge down) relative to the wing chord. The flap itself is regarded as a rigid body. Therefore β is actually an average displacement over the flap span, since the change in wing twist causes a small amount of variation of the flap angle of attack with y. This approximation has no undesirable effects in practice. The flap is assumed restrained toward $\beta = 0$ by a torsional spring with constant K_β ft·lb/rad. For power-boosted controls or symmetrical flutter of ailerons rigged to a control stick, K_β represents the stiffness of the actuator or the rigidity of the linkage and cables. In cases of antisymmetrical aileron flutter, and elevator or rudder flutter where there is a direct connection with the stick, K_β is difficult to estimate accurately. It lies somewhere between zero and the maximum stiffness of the linkages with the stick firmly fixed. Since this variation usually covers a wide range, it is customary to carry out the flutter calculation with the natural frequency ω_β of the restrained flap as a variable parameter.

Once more we use Lagrange's equation for deriving the equations of motion of the wing. The potential energy is obtained by adding to Eq. (9–67) the energy stored in the deflected control system,

$$U_{\text{C.S.}} = \tfrac{1}{2}K_\beta\beta^2 = \tfrac{1}{2}\omega_\beta^2\beta^2 I_{\beta\,\text{total}}, \qquad (9\text{–}118)$$

where

$$\omega_\beta = \sqrt{\frac{K_\beta}{I_{\beta\,\text{total}}}} \qquad (9\text{–}119)$$

is the uncoupled frequency of free vibration mentioned above. Thus we obtain

$$U_E = \tfrac{1}{2}\omega_w^2 w_R^2 \int_0^l m f_w^2 dy + \tfrac{1}{2}\omega_\theta^2\theta_R^2 \int_0^l I_y f_\theta^2 dy + \tfrac{1}{2}\omega_\beta^2\beta^2 I_{\beta\,\text{total}}. \qquad (9\text{–}120)$$

The upward vertical displacement of points over the wing is

$$w(x,y,t) = \begin{cases} [w_B(y,t) - x\theta_T(y,t)]; & \text{for } 0 \leq y < l_1 \text{ and } l_2 < y \leq l, \\ \{w_B(y,t) - x\theta_T(y,t) - [x - x_F]\beta(t)\}; & \text{for } l_1 \leq y \leq l_2, \end{cases} \qquad (9\text{–}121)$$

where $[x - x_F]$ measures the distance from the hinge line to any point on the flap. By analogy with Eqs. (9–63) and (9–65), the kinetic energy is

$$T = \frac{1}{2}\int_0^l \int_{\text{chord}} [\dot{w}(x,y,t)]^2 \rho(x,y)dxdy$$

$$= \frac{1}{2}\left[\int_0^{l_1} + \int_{l_2}^l\right]\int_{\text{chord}} [\dot{w}_B(y,t) - x\dot{\theta}_T(y,t)]^2\rho(x,y)dxdy$$

$$+ \frac{1}{2}\int_{l_1}^{l_2}\int_{x_l}^{x_F} [\dot{w}_B(y,t) - x\dot{\theta}_T(y,t)]^2\rho(x,y)dxdy$$

$$+ \frac{1}{2}\int_{l_1}^{l_2}\int_{x_F}^{x_t} [\dot{w}_B(y,t) - x\dot{\theta}_T(y,t) - (x - x_F)\dot{\beta}(t)]^2\rho(x,y)dxdy$$

$$= \int_0^l [\frac{1}{2}m\dot{w}_B{}^2 - S_y\dot{w}_B\dot{\theta}_T + \frac{1}{2}I_y\dot{\theta}_T{}^2]dy$$

$$+ \int_{l_1}^{l_2} \{-S_{y\beta}\dot{w}_B\dot{\beta} + [I_{y\beta} + b(e - a)S_{y\beta}]\dot{\theta}_T\dot{\beta} + \frac{1}{2}I_{y\beta}\dot{\beta}^2\}dy, \qquad (9\text{--}122)$$

where x_l, x_F, and x_t are the x-coordinates of the leading edge, hinge line, and trailing edge, respectively. Here $S_{y\beta}$ and $I_{y\beta}$ are the static unbalance and moment of inertia, per unit span, of the flap about its hinge, given by

$$S_{y\beta}(y) = \int_{x_F}^{x_t} [x - x_F]\rho(x,y)dx, \qquad (9\text{--}123)$$

$$I_{y\beta}(y) = \int_{x_F}^{x_t} [x - x_F]^2\rho(x,y)dx. \qquad (9\text{--}124)$$

As in Subsection (a), we have temporarily defined the elastic axis to be at $x = 0$, so the integral representing inertial coupling between flap rotation and torsion is

$$\int_{x_F}^{x_t} x[x - x_F]\rho(x,y)dx = \int_{x_F}^{x_t} [x - x_F][x - x_F]\rho(x,y)dx$$

$$+ x_F\int_{x_F}^{x_t} [x - x_F]\rho(x,y)dx = I_{y\beta}(y) + b(e - a)S_{y\beta}(y). \qquad (9\text{--}125)$$

It is worth pointing out that this coupling does not vanish, as does the coupling with bending, when the flap is completely statically balanced.

Substitution of the assumed bending and torsion deformation modes into Eq. (9–122) leads to

$$T = \frac{1}{2}\dot{w}_R{}^2\int_0^l mf_w{}^2dy - \dot{w}_R\dot{\theta}_R\int_0^l S_yf_wf_\theta dy$$

$$+ \frac{1}{2}\dot{\theta}_R{}^2\int_0^l I_yf_\theta{}^2dy - \dot{w}_R\dot{\beta}\int_{l_1}^{l_2} S_{y\beta}f_wdy$$

$$+ \dot{\theta}_R\dot{\beta}\int_l^{l_2} [I_{y\beta} + b(e - a)S_{y\beta}]f_\theta dy + \frac{1}{2}\dot{\beta}^2 I_{\beta \text{ total}}. \qquad (9\text{--}126)$$

The generalized aerodynamic forces on the three degrees of freedom are

$$Q_w = \int_0^l L(y,t)f_w(y)dy, \tag{9-72}$$

$$Q_\theta = \int_0^l M_y(y,t)f_\theta(y)dy, \tag{9-73}$$

$$Q_\beta = -\int_{l_1}^{l_2}\int_{x_F}^{x_t} \Delta p_a(x,y,t)[-(x - x_F)]dxdy = \int_{l_1}^{l_2} M_\beta(y,t)dy, \tag{9-127}$$

the latter being found from Eq. (5–291) and the fact that the mode shape associated with the flap degree of freedom (over the flap surface only) is

$$\gamma_\beta(x,y) = -(x - x_F). \tag{9-128}$$

Our artificial representation of structural friction has been shown to give rise to additional forces

$$(Q_w)_{\text{D.S.}} = -ig_w\omega_w^2 w_R \int_0^l mf_w^2 dy, \tag{9-81}$$

$$(Q_\theta)_{\text{D.S.}} = -ig_\theta\omega_\theta^2 \theta_R \int_0^l I_y f_\theta^2 dy. \tag{9-82}$$

We also expect friction to be present in the control system, probably to a greater degree than in the primary structure. It may consist of viscous damping built in intentionally to improve stability, or sliding friction in bearings, fair-leads, etc., or a combination of the two. Whatever its mechanism, we can approximate it as a source of energy dissipation in simple harmonic motion by a torque on the flap hinge line which leads the restoring torque by 90°:

$$\delta\beta(Q_\beta)_{\text{D.S.}} = -ig_\beta K_{\beta\beta}\delta\beta = -ig_\beta\omega_\beta^2 I_{\beta\text{ total}}\beta\delta\beta. \tag{9-129}$$

Since $\delta\beta$ here is a small virtual displacement of the flap, $(Q_\beta)_{\text{D.S.}}$ is the generalized force due to the friction.

Equations (9–120), (9–126), (9–72), (9–73), (9–127), (9–81), (9–82), and (9–129) are all that we need to set up Lagrange's equation of motion for the system. Applying Eq. (9–83) to the generalized coordinates w_R, θ_R, and β, successively, we obtain

$$\ddot{w}_R \int_0^l mf_w^2 dy - \ddot{\theta}_R \int_0^l S_y f_w f_\theta dy - \ddot{\beta}\int_{l_1}^{l_2} S_{y\beta}f_w dy$$

$$+ w_R\omega_w^2[1 + ig_w]\int_0^l mf_w^2 dy = \int_0^l Lf_w dy, \tag{9-130}$$

$$\ddot{\theta}_R \int_0^l I_y f_\theta^2 dy - \ddot{w}_R \int_0^l S_y f_w f_\theta dy + \ddot{\beta}\int_{l_1}^{l_2} [I_{y\beta} + b(e - a)S_{y\beta}]f_\theta dy$$

$$+ \theta_R\omega_\theta^2[1 + ig_\theta]\int_0^l I_y f_\theta^2 dy = \int_0^l M_v f_\theta dy, \tag{9-131}$$

$$\ddot{\beta}I_{\beta\,\text{total}} - \ddot{w}_R \int_{l_1}^{l_2} S_{y\beta}f_w dy + \ddot{\theta}_R \int_{l_1}^{l_2} [I_{y\beta} + b(e-a)S_{y\beta}]f_\theta dy$$

$$+ \beta\omega_\beta^2[1 + ig_\beta]I_{\beta\,\text{total}} = \int_{l_1}^{l_2} M_\beta(y,t)dy. \qquad (9\text{--}132)$$

Strip-theory aerodynamic loads for insertion into Eqs. (9–130)–(9–132) can be taken from Eqs. (5–350)–(5–352).* As in previous examples, we must replace h and α, in terms of the structural notation, by $-w_R(t)f_w(y)$ and $\theta_R(t)f_\theta(y)$.

$$L(y,t) = \pi\rho_\infty b^3\omega^2\left\{\frac{w_R}{b}f_w(y)L_h - \theta_R f_\theta(y)[L_\alpha - L_h(\tfrac{1}{2}+a)] - \beta L_\beta\right\}, \quad (9\text{--}133)$$

$$M_y(y,t) = \pi\rho_\infty b^4\omega^2\left\{-\frac{w_R}{b}f_w(y)[M_h - L_h(\tfrac{1}{2}+a)]\right.$$

$$+ \theta_R f_\theta(y)[M_\alpha - (M_h + L_\alpha)(\tfrac{1}{2}+a) + L_h(\tfrac{1}{2}+a)^2]$$

$$\left. + \beta[M_\beta - L_\beta(\tfrac{1}{2}+a)]\right\}, \qquad (9\text{--}134)$$

$$M_\beta(y,t) = \pi\rho_\infty b^4\omega^2\left\{-\frac{w_R}{b}f_w(y)T_h + \theta_R f_\theta(y)[T_\alpha - T_h(\tfrac{1}{2}+a)] + \beta T_\beta\right\}.$$

$$(9\text{--}135)$$

These formulas are limited to simple harmonic motion, so they imply Eqs. (9–88) and

$$\beta(t) = \bar{\beta}e^{i\omega t} \qquad (9\text{--}136)$$

We insert Eqs. (9–133)–(9–135) into Eqs. (9–130)–(9–132), collect terms, and divide out factors which leave the results dimensionless.

$$\frac{\bar{w}_R}{b_R}\left\{\left[1 - (1 + ig_w)\left(\frac{\omega_w}{\omega_\theta}\right)^2\left(\frac{\omega_\theta}{\omega}\right)^2\right]\int_0^1\left[\frac{m}{\pi\rho_\infty b_R^2}\right]f_w^2 dy^*\right.$$

$$+ \int_0^1\left(\frac{b}{b_R}\right)^2 L_h f_w^2 dy^*\right\} - \bar{\theta}_R\left\{\int_0^1\left[\frac{S_y}{\pi\rho_\infty b_R^3}\right]f_w f_\theta dy^*\right.$$

$$+ \int_0^1\left(\frac{b}{b_R}\right)^3[L_\alpha - L_h(\tfrac{1}{2}+a)]f_w f_\theta dy^*\right\} - \bar{\beta}\left\{\int_{l_1^*}^{l_2^*}\left[\frac{S_{y\beta}}{\pi\rho_\infty b_R^3}\right]f_w dy^*\right.$$

$$\left. + \int_{l_1^*}^{l_2^*}\left(\frac{b}{b_R}\right)^3 L_\beta f_w dy^*\right\} = 0. \qquad (9\text{--}137)$$

---

* The reader is reminded to distinguish between $M_\beta(y,t)$, which represents the aerodynamic hinge moment on the control surface, and the dimensionless M_β in Eq. (9–134). The latter is a function of hinge location and reduced frequency employed in constructing the tables of Ref. 9–1.

$$-\frac{\bar{w}_R}{b_R}\left\{\int_0^1\left[\frac{S_y}{\pi\rho_\infty b_R{}^3}\right]f_w f_\theta dy^* + \int_0^1\left(\frac{b}{b_R}\right)^3[M_h - L_h(\tfrac{1}{2}+a)]f_w f_\theta dy^*\right\}$$

$$+\bar{\theta}_R\left\{\left[1 - (1+ig_\theta)\left(\frac{\omega_\theta}{\omega}\right)^2\right]\int_0^1\left[\frac{I_y}{\pi\rho_\infty b_R{}^4}\right]f_\theta{}^2 dy^*\right.$$

$$+\int_0^1\left(\frac{b}{b_R}\right)^4[M_\alpha - (M_h+L_\alpha)(\tfrac{1}{2}+a) + L_h(\tfrac{1}{2}+a)^2]f_\theta{}^2 dy^*\right\}$$

$$+\bar{\beta}\left\{\int_{l_1*}^{l_2*}\frac{I_{y\beta} + b(e-a)S_{y\beta}}{\pi\rho_\infty b_R{}^4}f_\theta dy^*\right.$$

$$+\int_{l_1*}^{l_2*}\left(\frac{b}{b_R}\right)^4[M_\beta - L_\beta(\tfrac{1}{2}+a)]f_\theta dy^*\right\} = 0. \tag{9-138}$$

$$-\frac{\bar{w}_R}{b_R}\left\{\int_{l_1*}^{l_2*}\left[\frac{S_{y\beta}}{\pi\rho_\infty b_R{}^3}\right]f_w dy^* + \int_{l_1*}^{l_2*}\left(\frac{b}{b_R}\right)^3 T_h f_w dy^*\right\}$$

$$+\bar{\theta}_R\left\{\int_{l_1*}^{l_2*}\frac{I_{y\beta} + b(e-a)S_{y\beta}}{\pi\rho_\infty b_R{}^4}f_\theta dy^* + \int_{l_1*}^{l_2*}\left(\frac{b}{b_R}\right)^4[T_\alpha - T_h(\tfrac{1}{2}+a)]f_\theta dy^*\right\}$$

$$+\bar{\beta}\left\{\int_{l_1*}^{l_2*}\left(\frac{b}{b_R}\right)^4 T_\beta dy^* + \left[1 - (1+ig_\beta)\left(\frac{\omega_\beta}{\omega_\theta}\right)^2\left(\frac{\omega_\theta}{\omega}\right)^2\right]\frac{I_{y\beta\,\text{total}}}{\pi\rho_\infty b_R{}^4 l}\right\} = 0.$$
$$\tag{9-139}$$

Here $l_1{}^*$ and $l_2{}^*$ are the coordinates of the inner and outer extremes of the flap, divided by semispan l. The nine coefficients of $\bar{w}_R/b_R$, $\bar{\theta}_R$, and $\bar{\beta}$ form the flutter determinant, which is of third order and produces cubic characteristic equations.

EXAMPLE 9-2. To compute the influence of aileron frequency on the bending-torsion-aileron flutter speed of a cantilever, tapered wing in incompressible flow at sea-level density. We once more choose the jet transport wing, which has been analyzed extensively in Chapters 2, 4, and 8. The inertial and elastic properties and assumed modes for the main wing structure are discussed in Example 9-1, Section 9-5(a). The aileron lies between stations 30.83 ft and 40.58 ft out from midspan, so that

$$l_1{}^* = 0.740, \qquad l_2{}^* = 0.974. \tag{a}$$

The hinge line lies 75% chord behind the leading edge, so that

$$e = 0.5. \tag{b}$$

There is no aerodynamic balance, but the static mass balance is complete ($S_{y\beta} = 0$). The total moment of inertia about the hinge line is

$$I_{y\beta\,\text{total}} = 3.0154 \text{ slug} \cdot \text{ft}^2 \tag{c}$$

distributed along the span in accordance with the following table:

Station (from midspan)	Running moment of inertia $I_{y\beta}$
370 inches	0.4041 slug · ft²/ft
428.5	0.3083
487	0.2202

Solution. We again choose our reference station at $y = 3l/4$. In setting up the flutter equations, we neglect the small variation of k over the aileron span and approximate L_h, L_α, and M_α as shown in Eqs. (d), (e), and (g) of Example 9–1. The structural damping coefficients are all set equal to zero. Thus we get the literal formulas

$$\frac{\bar{w}_R}{b_R} \left\{ \left[1 - \left(\frac{\omega_w}{\omega_\theta}\right)^2 \left(\frac{\omega_\theta}{\omega}\right)^2 \right] \int_0^1 \left[\frac{m}{\pi \rho b_R{}^2} \right] f_w{}^2 dy^* + \int_0^1 \left(\frac{b}{b_R}\right)^2 f_w{}^2 dy^* \right.$$

$$+ K_2(L_h) \int_0^1 \frac{b}{b_R} f_w{}^2 dy^* \right\} - \bar{\theta}_R \left\{ \int_0^1 \left[\frac{S_y}{\pi \rho b_R{}^3} \right] f_w f_\theta dy^* \right.$$

$$- \int_0^1 \left(\frac{b}{b_R}\right)^3 a f_w f_\theta dy^* + K_2(L_\alpha) \int_0^1 \left(\frac{b}{b_R}\right)^2 f_w f_\theta dy^*$$

$$+ K_3(L_\alpha) \int_0^1 \frac{b}{b_R} f_w f_\theta dy^* - K_2(L_h) \int_0^1 [\tfrac{1}{2} + a] \left(\frac{b}{b_R}\right)^2 f_w f_\theta dy^* \right\}$$

$$+ \bar{\beta} \left\{ \int_{l_1{}^*}^{l_2{}^*} \left[\frac{S_{y\beta}}{\pi \rho b_R{}^3} \right] f_w dy^* + L_\beta \int_{l_1{}^*}^{l_2{}^*} \left(\frac{b}{b_R}\right)^3 f_w dy^* \right\} = 0, \qquad (d)$$

$$- \frac{\bar{w}_R}{b_R} \left\{ \int_0^1 \left[\frac{S_y}{\pi \rho b_R{}^3} \right] f_w f_\theta dy^* - \int_0^1 \left(\frac{b}{b_R}\right)^3 a f_w f_\theta dy^* \right.$$

$$- K_2(L_h) \int_0^1 [\tfrac{1}{2} + a] \left(\frac{b}{b_R}\right)^2 f_w f_\theta dy^* \right\} + \bar{\theta}_R \left\{ \left[1 - \left(\frac{\omega_\theta}{\omega}\right)^2 \right] \int_0^1 \left[\frac{I_y}{\pi \rho b_R{}^4} \right] f_\theta{}^2 dy^* \right.$$

$$+ \int_0^1 [\tfrac{1}{8} + a^2] \left(\frac{b}{b_R}\right)^4 f_\theta{}^2 dy^* + K_2(M_\alpha) \int_0^1 \left(\frac{b}{b_R}\right)^3 f_\theta{}^2 dy^*$$

$$+ K_2(L_h) \int_0^1 [\tfrac{1}{2} + a]^2 \left(\frac{b}{b_R}\right)^3 f_\theta{}^2 dy^* - K_2(L_\alpha) \int_0^1 [\tfrac{1}{2} + a] \left(\frac{b}{b_R}\right)^3 f_\theta{}^2 dy^*$$

$$- K_3(L_\alpha) \int_0^1 [\tfrac{1}{2} + a] \left(\frac{b}{b_R}\right)^2 f_\theta{}^2 dy^* \right\} + \bar{\beta} \left\{ \int_{l_1{}^*}^{l_2{}^*} \frac{I_{y\beta} + b(e - a)S_{y\beta}}{\pi \rho b_R{}^4} f_\theta dy^* \right.$$

$$+ M_\beta \int_{l_1{}^*}^{l_2{}^*} \left(\frac{b}{b_R}\right)^4 f_\theta dy^* - L_\beta \int_{l_1{}^*}^{l_2{}^*} [\tfrac{1}{2} + a] \left(\frac{b}{b_R}\right)^4 f_\theta dy^* \right\} = 0, \qquad (e)$$

$$-\frac{\overline{w}_R}{b_R}\left\{\int_{l_1^*}^{l_2^*}\left[\frac{S_{y\beta}}{\pi\rho b_R^3}\right]f_w dy^* + T_h\int_{l_1^*}^{l_2^*}\left(\frac{b}{b_R}\right)^3 f_w dy^*\right\}$$

$$+\bar{\theta}_R\left\{\int_{l_1^*}^{l_2^*}\frac{I_{y\beta}+b(e-a)S_{y\beta}}{\pi\rho b_R^4}f_\theta dy^* + T_\alpha\int_{l_1^*}^{l_2^*}\left(\frac{b}{b_R}\right)^4 f_\theta dy^*\right.$$

$$\left.-T_h\int_{l_1^*}^{l_2^*}[\tfrac{1}{2}+a]\left(\frac{b}{b_R}\right)^4 f_\theta dy^*\right\} + \bar{\beta}\left\{\left[1-\left(\frac{\omega_\beta}{\omega_\theta}\right)^2\left(\frac{\omega_\theta}{\omega}\right)^2\right]\frac{I_{y\beta\,\text{total}}}{\pi\rho b_R^4 l}\right.$$

$$\left.+ T_\beta\int_{l_1^*}^{l_2^*}\left(\frac{b}{b_R}\right)^4 dy^*\right\} = 0. \qquad (f)$$

Here all the aerodynamic coefficients are functions of

$$k_R = \frac{\omega b_R}{U}, \qquad (g)$$

and L_β, M_β, T_α, T_h, and T_β are found in the tables of Ref. 9–1 under the appropriate value of parameter e. After carrying out numerically all the integrations in Eqs. (d), (e), and (f), we are led to the following flutter determinant:

$$\begin{vmatrix} \begin{aligned}&\left\{8.2156\left[1-0.3277\left(\frac{\omega_\theta}{\omega}\right)^2\right]\right.\\ &\left.+0.2073+0.2215K_2(L_h)\right\}\end{aligned} & \begin{aligned}&\{0.3047-0.3545K_2(L_\alpha)\\&+0.3489K_3(L_\alpha)\\&-0.0709K_2(L_h)\}\end{aligned} & \{-0.1325L_\beta\} \\[4ex] \begin{aligned}&\{0.3047\\&+0.0709K_2(L_h)\}\end{aligned} & \begin{aligned}&\left\{66.1117\left[1-\left(\frac{\omega_\theta}{\omega}\right)^2\right]\right.\\&+0.2807\\&+1.0569K_2(M_\alpha)\\&+0.0423K_2(L_h)\\&-0.2114K_2(L_\alpha)\\&\left.-0.1772K_3(L_\alpha)\right\}\end{aligned} & \begin{aligned}&\{0.0107\\&+0.1572M_\beta\\&-0.0314L_\beta\}\end{aligned} \\[4ex] \{-0.1325T_h\} & \begin{aligned}&\{0.0107+0.1572T_\alpha\\&-0.0314T_h\}\end{aligned} & \begin{aligned}&\left\{0.1593T_\beta\right.\\&+0.01083\left[1-\left(\frac{\omega_\beta}{\omega_\theta}\right)^2\right.\\&\left.\left.\times\left(\frac{\omega_\theta}{\omega}\right)^2\right]\right\}\end{aligned} \end{vmatrix}$$

$$=0. \qquad (h)$$

For several values of k_R, the coefficients in Eq. (h) were found from the tables of Ref. 9–1, and the determinant was expanded into real and imaginary equations. These were solved for ω_θ/ω and $\omega_\beta/\omega_\theta$ by algebraic methods. To each root there correspond values of critical speed and frequency. Although three real roots might be found from a single value of k_R, only one of these proved to yield positive $(\omega_\beta/\omega_\theta)^2$, hence an actual flutter condition, in the present example. The physically meaningful solutions are listed in Table 9–2.

Figure 9–12 shows a plot of the tabulated data in the dimensionless form of $U_F/b_R\omega_\theta$ vs. $\omega_\beta/\omega_\theta$, with corresponding values of reduced frequency appended to some of the points. For purposes of reference, the dashed curve shows the variation of torsion-aileron divergence speed $U_D/b_R\omega_\theta$ (discussed in Chapter 8), which can be found from the flutter determinant by multiplying through by ω^2, setting $\omega = 0$, and calculating the eigenvalues. Divergence is seen to be very

TABLE 9–2

k_R	ω_θ/ω	$\omega_\beta/\omega_\theta$	$U_F/b_R\omega_\theta$	U_F mph	ω_F rad/sec
0.037	1.486	0	18.190	1516	15.05
0.040	1.464	11.288	17.074	1423	15.27
0.045	1.409	23.709	15.769	1315	15.87
0.051	1.389	∞	14.111	1176	16.09
0.080	1.202	∞	10.397	865	18.60
0.085	1.193	18.913	9.858	822	18.73
0.090	1.183	9.847	9.395	783	18.91
0.095	1.174	5.345	8.967	747	19.04
0.100	1.167	0	8.568	714	19.15

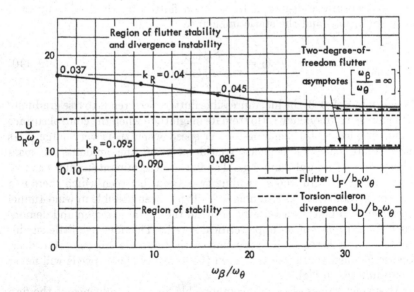

Fig. 9–12. Plot of dimensionless bending-torsion-aileron flutter speed $U_F/b_R\omega_\theta$ and torsion-aileron divergence speed $U_D/b_R\omega_\theta$ vs. frequency ratio $\omega_\beta/\omega_\theta$ obtained from strip-theory analysis of cantilever jet transport wing.

insensitive to changes of aileron frequency, since it is controlled almost entirely by the torsional stiffness of the wing.

Throughout the entire range of frequency ratios, Fig. 9–12 displays a region of flutter instability, bounded by curves of critical speed both above and below. Many three-degrees-of-freedom analyses result in this sort of behavior, although the "flutter region" is often limited to a portion of the ω_β-spectrum, with greatest instability in the vicinity of $\omega_\beta/\omega_\theta = 1$. The lower curve of $U_F/b_R\omega_\theta$, which is the interesting one from the practical standpoint, shows a very gradual rise of critical speed with frequency ratio, going from a minimum of 714 mph when the aileron is completely loose to the asymptotic value of 865 mph (bending-torsion flutter) when the control is locked to the wing.

Further information concerning flutter of a straight wing or tail with a control surface can be obtained from parameter studies such as the one by Theodorsen and Garrick (Ref. 9–3) or the more recent work of Woolston and Huckel (Ref. 9–13). The latter concentrates on the effects of compressibility, and it is difficult to draw general conclusions which are valid for all Mach numbers. The calculations in these two references involve a simplified sectional wing of the type analyzed in Section 9–2; in most cases only two degrees of freedom, either bending-flap or torsion-flap, are employed. These limitations do not impair their value as a qualitative source of parametric data on actual wings.

The dimensionless quantities which govern most strongly the influence of a control-surface degree of freedom on flutter are the frequency ratio $\omega_\beta/\omega_\theta$ (or ω_β/ω_w) and the static unbalance

$$x_\beta = \frac{S_{y\beta\,\text{total}}}{m_R b_R{}^2[l_2 - l_1]}.$$
(9–140)

As shown in Ref. 9–3, the incompressible flutter speed tends to rise gradually with increasing frequency ratio from its magnitude when the control surface is unrestrained to the bending-torsion speed when ω_β is very large. Its intermediate behavior depends considerably on the amounts of mass balance and structural damping. The flutter determinant usually has two physically meaningful roots, yielding two speeds between which there is a range of instability, as in Example 9–2. It has been possible in wind-tunnel experiments to take a model wing through the unstable region and demonstrate the existence of an upper critical speed. This fact has little significance for the aeroelastician dealing with full-scale aircraft, however, because he must guarantee that even the lowest of these speeds will never be encountered in flight.

At smaller values of ω_β in incompressible flow, the influence of the flap can usually be ascertained by studying two-degree-of-freedom flutter. Reference 9–3 points out that if the wing mass ratio $m/\pi\rho b^2$ is low enough, safety from torsion-flap flutter is ensured once the bending-flap type has been controlled. Figure 3 of Ref. 9–13 presents almost the only generally available systematic data on torsion-flap flutter at high $m/\pi\rho b^2$ and suggests quite a different conclusion: mass overbalance can then lead to a reduction of the critical speed. It shows the optimum region to lie near $x_\beta = 0$ when $\omega_\beta = 0$.

In the bending-flap case, according to Ref. 9–3, reducing x_β and increasing ω_β/ω_w are both favorable. Flutter is eliminated entirely either if the frequency ratio is carried slightly past unity or if complete mass balance ($x_\beta = 0$) is achieved. For a given positive x_β, there is a minimum critical frequency ratio below which U_F assumes a low, nearly constant value. In a similar way, x_β's above some limit produce dangerously small U_F's for a given ω_β/ω_w. Friction has a universally beneficial effect, especially in improving the aforementioned critical magnitudes of mass balance and frequency ratio. The comparisons carried out by Goland and Dengler in Ref. 9–14 indicate satisfactory point agreement between theory and experiment for certain control surface configurations of the type we discuss here.

In summary, we emphasize that two alternatives are available to the designer by which he can avoid trouble with most flutter involving primary control surfaces. The first is to achieve as closely as possible a fully mass-balanced condition, thus removing severe coupling with vibrations of the main wing or tail. This measure is often taken for other reasons anyway on controls which are not irreversible. The second is to build so much stiffness into the actuators and back-up structure that the frequency ratio is high enough to ensure no serious influence of the control surface on whatever bending-torsion flutter speed the primary structure already has.

(d) *The selection of degrees of freedom and mode shapes.* When it is planned to study the flutter of a particular aircraft or portion thereof by the Rayleigh-Ritz method, the first problem is to decide which degrees of freedom are to be included and to associate a shape function with each of them. Individual experience is usually a major factor, and it is especially hard to lay down general rules in view of the wide variety of structural and aerodynamic configurations.

A significant feature of mode selection is the saving of effort and computational time that can be gained by judiciously choosing a small number which can combine into a very close approximation of the true flutter mode. This process has perhaps been overemphasized as a result of its unusual success with bending-torsion flutter of bare, straight wings. We must remember that this particular case is one where the critical speed is not sensitive even to large changes of the assumed functions f_w and f_θ. As in the prediction of fundamental vibration modes of simple structures by Rayleigh's method, a rather poor choice may still lead to accurate eigenvalues. The same is not true of many other flutter phenomena. It is therefore fortunate that large, fast computing facilities are eliminating the need for keeping practical flutter calculations down to a very few degrees of freedom.

Certain principles have proved valuable in the past for selecting which motions will probably make major contributions to a given type of flutter mode. The most obvious of these stems from the fact that almost every aircraft has a central plane of symmetry. It follows that all structural and rigid-body oscillations can be separated into those which are symmetrical and those which are antisymmetrical with respect to this plane. Within the framework of linearized mechanics, no member of one set can in any way be coupled with a member of the other, as is proved by a brief examination of equations of motion in which both are included. Symmetry is not the panacea which reduces flutter analysis to a manageable problem, however. If we include rigid-body motions and primary control surfaces, a conventional airplane has some 9 symmetrical degrees of freedom of the type discussed in this section, and 12 antisymmetrical ones. Flexibly mounted external stores can easily run these numbers up to inordinate sizes.

A second principle is that the degrees of freedom which compose a flutter mode usually couple strongly with one another. In numerical terms, this statement means that the coefficients of the various generalized coordinates are of the same order of magnitude in all the equations of motion. Actually, there is enough coupling to be worth examining if a particular coordinate has large coefficients in only two equations of motion. This rule suggests, for example, that tail motions need not be included in a wing flutter calculation on an airplane with such a massive fuselage that wing vibrations do not excite the fin or stabilizer very much.

A third principle concerns natural frequencies of vibration, stating that a degree of freedom with a frequency large compared with the expected flutter frequency does not have to be considered. This might, for example, rule a locked control surface, with stiff linkages, out of bending-torsion flutter. The weakness of the frequency test lies in the difficulty we often encounter in estimating flutter frequency. On low aspect-ratio wings, for instance, flutter has occurred at frequencies higher than the fourth normal mode of vibration.

With regard to the estimation of shape functions to go with particular degrees of freedom, this is a problem only when the artificial-mode method is used. As a general guide, normal modes should be employed whenever the structure or mass distribution is sufficiently novel that we have no assurance, on the basis of previous experience, of the validity of a simpler technique. Naturally, the shape functions can only be roughly determined if one does not have complete information on the system. If not too complicated, an elastic or dynamic-elastic model often provides a useful tool during early design stages.

In a majority of cases, artificial modes should represent as closely as possible the properties of the structure they approximate. We have mentioned, for example, how the inertial coupling can be hypothetically removed from a beam with an elastic axis to facilitate calculating primitive bending and torsion shapes. Very occasionally, flutter eigenvalues are so insensitive to mode shape that simple mathematical functions, chosen to satisfy some or all of the boundary conditions, suffice for assumed modes. This process is so uncertain and requires so much experience that we cannot recommend it for practical applications. The use of normal modes or carefully estimated artificial modes is preferable and often reduces the total computational labor by minimizing the number of degrees of freedom that must be included.

As we have stated before, the only real advantage of artificial over normal modes is on wings with rigid chordwise sections. We demonstrate this by deriving once more a set of equations of motion for the straight, cantilever wing. We take n normal coordinates $\xi_j(t)$, and each contributes to both bending and torsional deformation. The jth shape function $\phi_j(x,y)$ can be put into the form

$$\phi_j(x,y) = b_R f_{wj}(y) - x f_{\theta j}(y), \qquad (9\text{-}141)$$

where x is measured aft from the elastic axis (or arbitrarily designated structural reference line on a wing with rigid cross sections which does not have a clearly defined elastic axis). The total displacement of the wing surface due to $\xi_j(t)$ is

$$w_j(x,y,t) = \xi_j(t)[b_R f_{wj}(y) - x f_{\theta j}(y)]. \qquad (9\text{-}142)$$

Alternatively, we can separate the bending and torsion contributions of the n modes:

$$\frac{w_B(y,t)}{b_R} = \sum_{i=1}^{n} \xi_i(t) f_{wi}(y), \qquad (9\text{-}143)$$

$$\theta_T(y,t) = \sum_{i=1}^{n} \xi_i(t) f_{\theta i}(y). \qquad (9\text{-}144)$$

The equations of motion are

$$M_j \ddot{\xi}_j + M_j \omega_j^2 \xi_j = \Xi_j; \qquad \text{for } j = 1, 2, \cdots, n, \qquad (3\text{-}147a)$$

where M_j is defined by Eq. (3-147b). Structural friction can be introduced by the approximation of replacing ω_j^2 by $\omega_j^2[1 + ig_j]$. The disadvantage relative to the artificial-mode scheme becomes evident when we work out the generalized aerodynamic forces. According to Eqs. (3-147c) and (9-55),

$$\Xi_j = -\iint_S \Delta p_a(x,y,t) \phi_j(x,y) dx dy$$

$$= b_R \int_0^l L(y,t) f_{wj}(y) dy + \int_0^l M_y(y,t) f_{\theta j}(y) dy. \qquad (9\text{-}145)$$

In the notation of Ref. 9–1,

$$L(y,t) = \pi\rho_\infty b^3\omega^2 \left\{ L_h \frac{b_R}{b} \sum_{i=1}^n \xi_i(t)f_{wi}(y) \right.$$

$$\left. - [L_\alpha - L_h(\tfrac{1}{2} + a)] \sum_{i=1}^n \xi_i(t)f_{\theta i}(y) \right\}, \qquad (9\text{–}146)$$

$$M_y(y,t) = \pi\rho_\infty b^4\omega^2 \left\{ -[M_h - L_h(\tfrac{1}{2}+a)] \frac{b_R}{b} \sum_{i=1}^n \xi_i(t)f_{wi}(y) \right.$$

$$\left. + [M_\alpha - (M_h+L_\alpha)(\tfrac{1}{2}+a) + L_h(\tfrac{1}{2}+a)^2] \sum_{i=1}^n \xi_i(t)f_{\theta i}(y) \right\}. \quad (9\text{–}147)$$

Equations (9–146) and (9–147) show what can be gained by locating the structural reference axis at the quarter-chord line, if possible, so that the factor $(\tfrac{1}{2} + a)$ vanishes. After substituting Eqs. (9–146) and (9–147) into Eq. (9–145), and collecting terms, we obtain

$$\frac{\Xi_j}{\pi\rho_\infty\omega^2 b_R{}^4 l e^{i\omega t}}$$

$$= \sum_{i=1}^n \xi_i \int_0^1 \left\langle \left(\frac{b}{b_R}\right)^2 L_h f_{wi} f_{wj} \right.$$

$$- \left(\frac{b}{b_R}\right)^3 \{[L_\alpha - L_h(\tfrac{1}{2}+a)]f_{wj}f_{\theta i} + [M_h - L_h(\tfrac{1}{2}+a)]f_{wi}f_{\theta j}\}$$

$$\left. + \left(\frac{b}{b_R}\right)^4 [M_\alpha - (M_h + L_\alpha)(\tfrac{1}{2}+a) + L_h(\tfrac{1}{2}+a)^2]f_{\theta i}f_{\theta j} \right\rangle dy^*. \quad (9\text{–}148)$$

Here $\bar\xi_i$ is the complex amplitude of ξ_i, defined by Eq. (9–56).

If, finally, we substitute Eq. (9–148) into Eq. (3–147a), we are led to a set of n equations which we can express in the following matrix form:

$$\left(\left[1 - \left(\frac{\omega_j}{\omega}\right)^2 (1 + ig_j) \right] \left[\frac{M_j}{\pi\rho_\infty b_R{}^4 l} \right] + [A] \right) \{\bar\xi_j\} = 0. \qquad (9\text{–}149)$$

Here the aerodynamic matrix $[A]$ has for its elements n^2 integrals similar to the one on the right side of Eq. (9–148). The product of the diagonal unit matrix by the diagonal matrix whose elements are the dimensionless generalized masses

$$\frac{M_j}{\pi\rho_\infty b_R{}^4 l} = \int_0^1 \int_{\text{chord}} \frac{\phi_j{}^2(x,y)}{b_R{}^2} \left[\frac{\rho(x,y)}{\pi\rho_\infty b_R} \right] d\left(\frac{x}{b_R}\right) dy^* \qquad (9\text{–}150)$$

is often called the inertial matrix, because it originates from the acceleration terms in the equations of motion. Similarly, the product of the diagonal matrix of $(1 + ig_j)\omega_j{}^2/\omega^2$ by the inertial matrix is called the stiffness or structural matrix. The flutter determinant is the determinant of the sum of the inertial, stiffness, and aerodynamic matrices.

In matrix notation, the equations of motion for n artificially decoupled modes, including r of bending and $(n - r)$ of torsion [cf. Eqs. (9–59) and (9–60)] would read

$$\left([M] - \left[1 - \left(\frac{\omega_j}{\omega}\right)^2 (1 + ig_j)\right][S] + [A]\right)\begin{Bmatrix} \dfrac{\bar{w}_R}{b_R} \\ \cdots \\ \bar{\theta}_R \end{Bmatrix} = 0. \quad (9\text{--}151)$$

We can see from Eqs. (9–89) and (9–90) the forms of the elements of the various matrices. The inertial matrix $[M]$ is composed of integrals like

$$\int_0^1 \left[\frac{m}{\pi \rho_\infty b_R{}^2}\right] f_{wi} f_{wj} dy^*, \qquad \int_0^1 \left[\frac{S_y}{\pi \rho_\infty b_R{}^3}\right] f_{wi} f_{\theta j} dy^*,$$

and

$$\int_0^1 \left[\frac{I_y}{\pi \rho_\infty b_R{}^4}\right] f_{\theta i} f_{\theta j} dy^*.$$

Since we have assumed an elastic axis and there is no static coupling between the degrees of freedom, the stiffness matrix is diagonal and comprises two types of terms:

$$\int_0^1 \left[\frac{m}{\pi \rho_\infty b_R{}^2}\right] f_{wi}{}^2 dy^* \qquad \text{and} \qquad \int_0^1 \left[\frac{I_y}{\pi \rho_\infty b_R{}^4}\right] f_{\theta i}{}^2 dy^*.$$

Three typical elements of the aerodynamic matrix are

$$\int_0^1 \left(\frac{b}{b_R}\right)^2 L_h f_{wi} f_{wj} dy^*,$$

$$\int_0^1 \left(\frac{b}{b_R}\right)^3 [L_\alpha - L_h(\tfrac{1}{2} + a)] f_{wi} f_{\theta j} dy^*,$$

and

$$\int_0^1 \left(\frac{b}{b_R}\right)^4 [M_\alpha - (M_h + L_\alpha)(\tfrac{1}{2} + a) + L_h(\tfrac{1}{2} + a)^2] f_{\theta i} f_{\theta j} dy^*.$$

The column matrix of unknown amplitudes includes a total of r values of $\bar{w}_i/b_R$ and $(n - r)$ values of $\bar{\theta}_i$.

When we compare Eqs. (9–149) and (9–151), we see no substantial difference between the amounts of labor required to compute the inertial and stiffness matrices, except for a few extra inertial coupling terms between the artificial modes. On the other hand, each element of the aerodynamic matrix in Eq. (9–149) calls for roughly four times the computational work needed for its counterpart in Eq. (9–151). Since most of the effort in setting up any flutter problem goes into the aerodynamic terms, the advantage of the artificial-mode scheme should be evident for this type of wing.

Woolston and Runyan (Ref. 9–16) compare the results of flutter analyses using normal and artificial modes with experimental measurements on a straight, uniform, cantilever wing. To provide an unusually severe test of artificial-mode methods, a large mass was placed near the leading edge at a succession of spanwise stations. Its effects are included in the equations of motion as concentrated contributions to the integrals containing m, S_y, and I_y. The conclusions of Ref. 9–16 are generally favorable to the selection of artificial modes. Normal modes show no better agreement with experiments, either on the bare wing or in tests with the mass added. When the mass is located near the tip, both approaches give poor results if only two degrees of freedom are employed. With additional degrees of freedom put into the analysis, the artificial mode scheme shows more systematic improvement of accuracy than does the other.

For concentrated masses placed ahead of the elastic axis, both schemes have a tendency to be unconservative in predicting flutter speed. In the paper which originally reported the experiments on this particular cantilever model (Ref. 9–9), Runyan and Watkins prove that the differential-equation solution of Goland (Ref. 9–6) is by far the most satisfactory way of dealing with such cases.

With regard to the problem of setting up and solving the flutter determinants of systems with many degrees of freedom, we are prevented by space limitations from doing more than to point to the obvious generalizations of techniques already described in this section. In Ref. 9–17, Williams gives an excellent general discussion of this subject, together with an extensive list of publications on various direct and indirect ways of handling large complex determinants. Reference 9–17 is also one of the best available guides to British literature on flutter of all types. Another valuable treatment appears in van de Vooren's doctoral thesis (Ref. 9–19).

9–6 Inclusion of finite span effects in flutter calculations. In systems of flutter equations such as (9–84)–(9–85) and (9–130)–(9–132), finite span and compressibility effects can, in principle, be easily incorporated. It is necessary merely to replace the running lift and moments with whatever expressions are yielded by the particular theory being employed. The only general specification we must make is that nearly every aerodynamic method is based on advanced knowledge of the mode of simple harmonic motion. So-called aerodynamic influence coefficients may remove this limitation, but time will be required before they become well adapted to routine flutter computation.

We have discussed in several sections of Chapter 7 the difficult problems of organization and systematization associated with most theories for finite wings. Once degrees of freedom and mode shapes are assumed, computation of the inertial and stiffness portions of the flutter equations is straightforward. Well over 90% of the thought and labor is usually devoted to the generalized aerodynamic forces. Considering the troubles which can arise even from strip theory, we are not surprised that three-dimensional effects are so rarely included in practice.

For all but the lowest aspect-ratio wings, it is a good general rule to perform a strip-theory flutter analysis in advance. This procedure has two advantages. First, because of the generally conservative feature of strip calculations, it can eliminate from consideration all but a very few types of instability. Often it shows an airplane to be essentially flutter-free or suggests minor modifications that will make it so. Second, if a more exact analysis is planned, the equations of motion already derived for strip theory need only have certain aerodynamic elements changed. This statement is especially meaningful for methods like Reissner's (see Section 7-2), where the generalized forces are obtained by applying additive corrections to those found from two-dimensional strip theory.

We illustrate how conveniently induction effects can be introduced for straight wings in incompressible fluid by another example based on the jet transport.

EXAMPLE 9-3. To compute the bending-torsion flutter speed of the cantilever wing in Example 9-1, including the influence of finite span by means of Reissner's theory.

Solution. As in Example 9-1, we adopt the U-g method and employ two artificial uncoupled modes $f_w(y)$ and $f_\theta(y)$. We do not repeat here the derivations of the inertial, elastic, and strip-theory elements in the flutter determinant, but concentrate on the manner of adjusting the generalized forces.

It is shown in Section 7-2 and Ref. 9-18 that three-dimensional flow superimposes the following terms on the lift and moment per unit span which appear on the right sides of Eqs. (9-84) and (9-85):

$$\Delta L(y,t) = -2\pi\rho U^2 b_R[ik\sigma_w(y)]\frac{w_R(t)}{b_R}f_w(y)$$

$$+ 2\pi\rho U^2 b[1 + ik(\tfrac{1}{2} - a)]\sigma_\theta(y)\theta_R(t)f_\theta(y), \tag{a}$$

$$\Delta M_y(y,t) = -2\pi\rho U^2 bb_R[(\tfrac{1}{2} + a)\sigma_w(y)]\frac{w_R(t)}{b_R}f_w(y)$$

$$+ 2\pi\rho U^2 b^2[(\tfrac{1}{2} + a) + ik(\tfrac{1}{4} - a^2)]\sigma_\theta(y)\theta_R(t)f_\theta(y). \tag{b}$$

These give rise to additional generalized forces

$$\Delta Q_w = -2\pi\rho U^2 b_R \frac{\overline{w}_R}{b_R} e^{i\omega t}\int_0^l ik\sigma_w(y)f_w{}^2 dy$$

$$+ 2\pi\rho U^2 \overline{\theta}_R e^{i\omega t}\int_0^l b[1 + ik(\tfrac{1}{2} - a)]\sigma_w(y)f_w f_\theta dy, \tag{c}$$

$$\Delta Q_\theta = -2\pi\rho U^2 b_R \frac{\overline{w}_R}{b_R} e^{i\omega t} \int_0^l b[(\tfrac{1}{2} + a)ik]\sigma_w(y)f_w f_\theta dy$$

$$+ 2\pi\rho U^2 \overline{\theta}_R e^{i\omega t} \int_0^l b^2[(\tfrac{1}{2} + a) + ik(\tfrac{1}{4} - a^2)]\sigma_\theta(y)f_\theta{}^2 dy. \qquad (d)$$

To go from Eqs. (9–84)–(9–85) to Eqs. (9–89)–(9–90) (for incompressible flow, with ρ_∞ replaced by ρ), Eqs. (c) and (d) must be divided by $\pi\rho b_R{}^3\omega^2 l e^{i\omega t}$ and $\pi\rho b_R{}^4\omega^2 l e^{i\omega t}$, respectively. They then make the following four contributions to the coefficients of $\overline{w}_R/b_R$ and $\overline{\theta}_R$ in Eqs. (9–89) and (9–90); these coefficients are corrections to the four elements A, B, C, D of the strip-theory flutter determinant:

$$\Delta A = -\frac{2}{k_R{}^2}\int_0^1 ik\sigma_w(y^*)f_w{}^2 dy^*, \qquad (e)$$

$$\Delta B = \frac{2}{k_R{}^2}\int_0^1 \left(\frac{b}{b_R}\right)[1 + ik(\tfrac{1}{2} - a)]\sigma_\theta(y^*)f_w f_\theta dy^*, \qquad (f)$$

$$\Delta C = -\frac{2}{k_R{}^2}\int_0^1 \left(\frac{b}{b_R}\right)[(\tfrac{1}{2} + a)ik]\sigma_w(y^*)f_w f_\theta dy^*, \qquad (g)$$

$$\Delta D = \frac{2}{k_R{}^2}\int_0^1 \left(\frac{b}{b_R}\right)^2 [(\tfrac{1}{2} + a) + ik(\tfrac{1}{4} - a^2)]\sigma_\theta(y^*)f_\theta{}^2 dy^*. \qquad (h)$$

With these adjustments, the numerical form of the flutter determinant [Eq. (j) of Example 9–1] becomes, with $a = -0.3$,

$$\begin{vmatrix} \{A + \Delta A\} & \{B + \Delta B\} \\ \{C + \Delta C\} & \{D + \Delta D\} \end{vmatrix}$$

$$= \begin{vmatrix} \begin{Bmatrix} 8.2156[1 - 0.3277Z] + 0.2073 \\ + 0.2215K_2(L_h) \\ -\dfrac{2}{k_R{}^2}\displaystyle\int_0^1 ik\sigma_w(y^*)f_w{}^2 dy^* \end{Bmatrix} & \begin{Bmatrix} 0.3047 - 0.3545K_2(L_\alpha) \\ - 0.3489K_3(L_\alpha) \\ + 0.0709K_2(L_h) \\ + \dfrac{2}{k_R{}^2}\displaystyle\int_0^1 \left(\dfrac{b}{b_R}\right)[1 + 0.8ik]\sigma_\theta(y^*)f_\theta f_w dy^* \end{Bmatrix} \\[2em] \begin{Bmatrix} 0.3047 + 0.0709K_2(L_h) \\ -\dfrac{0.4}{k_R{}^2}\displaystyle\int_0^1 \left(\dfrac{b}{b_R}\right)ik\sigma_w(y^*)f_w f_\theta dy^* \end{Bmatrix} & \begin{Bmatrix} 66.1117[1 - Z] + 0.2807 \\ + 1.0569K_2(M_\alpha) \\ + 0.0423K_2(L_h) \\ - 0.2114K_2(L_\alpha) \\ - 0.1772K_3(L_\alpha) \\ + \dfrac{2}{k_R{}^2}\displaystyle\int_0^1 \left(\dfrac{b}{b_R}\right)^2 [0.2 + 0.16ik]\sigma_\theta(y^*)f_\theta{}^2 dy^* \end{Bmatrix} \end{vmatrix}$$

$$= 0. \qquad (i)$$

For every selection of k_R, the determinantal equation can be solved for two complex values of Z after the integrals (e), (f), (g), and (h) are evaluated numerically. The last is the most time-consuming step, since it includes solution of the integrodifferential equation (7–25) for $\bar{\Omega}_w(y^*)$ and $\bar{\Omega}_\theta(y^*)$, from which $\sigma_w(y^*)$ and $\sigma_\theta(y^*)$ are found by Eq. (7–29). Assuming symmetrical wing motions, this is solved by taking four terms each in the Fourier series (Eq. 7–36):

$$\bar{\Omega}_w(y^*) = \sum_{n=1,3,5,7} K_{nw} \frac{\sin n\phi}{n} \tag{j}$$

and

$$\bar{\Omega}_\theta(y^*) = \sum_{n=1,3,5,7} K_{n\theta} \frac{\sin n\phi}{n}. \tag{k}$$

With

$$\bar{\Omega}_w^{(2)}(y^*) = -\frac{4iC(k)}{kH_1^{(2)}(k)} \left(\frac{b_R}{b_0}\right) \frac{\overline{w}_R}{b_R} f_w(y^*) \tag{l}$$

and

$$\bar{\Omega}_\theta^{(2)}(y^*) = \frac{4iC(k)}{k_0 H_1^{(2)}(k)} [1 + 0.8ik]\bar{\theta}_R f_\theta(y^*), \tag{m}$$

the systems of simultaneous equations (7–43) for the two degrees of freedom are

$$\sum_{n=1,3,5,7} K_{nw} \left\{ \frac{\sin n\phi}{n} + 0.7069 \frac{b}{b_0} \mu(k) S_n(4.44k_0,\phi) \right\}$$
$$= -2.333 \frac{iC(k)}{kH_1^{(2)}(k)} \frac{\overline{w}_R}{b_R} f_w(4.44 \cos \phi), \quad \text{(n)}$$

$$\sum_{n=1,3,5,7} K_{n\theta} \left\{ \frac{\sin n\phi}{n} + 0.7069 \frac{b}{b_0} \mu(k) S_n(4.44k_0,\phi) \right\}$$
$$= \frac{4iC(k)}{k_0 H_1^{(2)}(k)} [1 + 0.8ik]\bar{\theta}_R f_\theta(4.44 \cos \phi). \quad \text{(o)}$$

Also

$$k_R = 0.583k_0. \tag{p}$$

Equations (n) and (o) were satisfied at the stations $\cos \phi = 0.2, 0.4, 0.6, 0.8$, which correspond with those employed in the tables of Ref. 9–18. Other tables in that report furnish values of $\mu(k)$, $\sin n\phi/n$, $[iC(k)/kH_1^{(2)}(k)]$, and additional functions which facilitate the solution. As a numerical example, Eqs. (n) for $k_R = 0.1$ turn out to be

$$[1.1463-0.1233i]K_{1w}+[-0.4505+0.0887i]K_{3w}+[0.2306-0.0501i]K_{5w}$$
$$+[-0.1060+0.0253i]K_{7w} = -[0.000479+0.002507i]\frac{\overline{w}_R}{b_R}, \quad \text{(q)}$$

$$[1.0700-0.1049i]K_{1w}+[-0.1802+0.0296i]K_{3w}+[-0.1989+0.0418i]K_{5w}$$
$$+[0.3632-0.0782i]K_{7w} = -[0.00132+0.00727i]\frac{\overline{w}_R}{b_R}, \quad \text{(r)}$$

$$[0.9495-0.0749i]K_{1w}+[0.1922-0.0343i]K_{3w}+[-0.4279+0.0835i]K_{5w}$$
$$+[0.0785-0.0152i]K_{7w} = -[0.00285+0.01704i]\frac{\overline{w}_R}{b_R}, \quad \text{(s)}$$

$$[0.7409 - 0.0571i]K_{1w} + [0.5535 - 0.0879i]K_{3w} + [0.0343 + 0.0089i]K_{5w}$$

$$+ [-0.4101 + 0.0734i]K_{7w} = -[0.00376 + 0.02462i]\frac{\bar{w}_R}{b_R}. \quad \text{(t)}$$

These are solved by

$$\frac{K_{1w}}{\bar{w}_R/b_R} = [0.0895 + 1.211i] \times 10^{-2}, \quad \text{(u)}$$

$$\frac{K_{3w}}{\bar{w}_R/b_R} = [-0.0527 + 2.456i] \times 10^{-2}. \quad \text{(v)}$$

$$\frac{K_{5w}}{\bar{w}_R/b_R} = -[0.0470 + 0.2990i] \times 10^{-2}, \quad \text{(w)}$$

$$\frac{K_{7w}}{\bar{w}_R/b_R} = -[0.0327 + 0.4846i] \times 10^{-2}. \quad \text{(x)}$$

The function σ_w is given by Eq. (7–29) as

$$\sigma_w(y^*) = \left[\frac{\bar{\Omega}_w(y^*)}{\bar{\Omega}_w{}^{(2)}(y^*)} - 1\right]\left[C(k) + \frac{iJ_1(k)}{J_0(k) - iJ_1(k)}\right]$$

$$= \left\{\frac{0.429ikH_1{}^{(2)}(k)}{C(k)f_w(y^*)} \sum_{n=1,3,5,7} \frac{K_{nw}}{\bar{w}_R/b_R} \frac{\sin n\phi}{n} - 1\right\}$$

$$\times \left[C(k) + \frac{iJ_1(k)}{J_0(k) - iJ_1(k)}\right]. \quad \text{(y)}$$

A similar expression yields the $\sigma_\theta(y^*)$.

For each k_R, $\sigma_w(y^*)$ and $\sigma_\theta(y^*)$ were computed at six spanwise stations, and the trapezoidal rule was used to evaluate integrals (e), (f), (g), and (h). The roots of matrix (i) were found from the quadratic formula for three values of k_R, using as guides those close to flutter in Example 9–1. These roots, with the corresponding values of speed, frequency, and artificial structural damping g, are listed in Table 9–3.

TABLE 9–3

k_R	Z_1	U_1 mph	ω_1 rad/sec	g_1	Z_2	U_2 mph	ω_2 rad/sec	g_2
0.06	3.392 $-1.390i$	755	12.14	-0.410	1.702 $+0.0562i$	1062	17.14	0.033
0.1	3.094 $-0.719i$	475	12.72	-0.232	1.314 $-0.0651i$	727	19.53	-0.050
0.2	3.084 $-0.289i$	237	12.74	-0.094	1.080 $-0.0564i$	401	21.52	-0.052

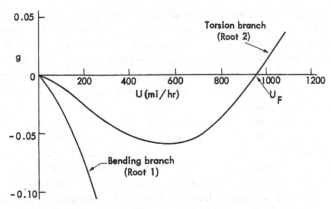

Fig. 9-13. Plot of artificial structural damping g vs. airspeed U from bending-torsion flutter calculation, including finite-span effects by Reissner's theory, on a cantilever jet transport wing.

Figure 9-13 shows a plot of g vs. U for the two artificial aeroelastic modes. If we assume g_w and g_θ to be zero, we find the following flutter condition where the Z_2-curve crosses the U-axis:

$$U_F = 955 \text{ mph}, \qquad \omega_F = 17.9 \text{ rad/sec}, \qquad k_F = 0.070. \qquad (z)$$

This critical speed is 10.4% higher than the 865 mph obtained by strip theory. Experience with the application of Reissner's method in flutter analysis of straight wings indicates that it usually falls closer to experimentally measured results than two-dimensional calculations. Therefore the example we present here confirms the normal conservative tendency of strip theory.

9-7 The effect of compressibility on flutter. Chapter 6 and Section 7-5 list the important sources of aerodynamic coefficients for oscillating airfoils and finite wings in subsonic, sonic, and supersonic flow. As is the case with three-dimensional incompressible theory, only the *aerodynamic* terms in systems of flutter equations like (9-84)-(9-85) and (9-130)-(9-132) have to be modified to account for compressibility. In fact, for bending-torsion flutter analyses using strip theory, Ref. 9-20 furnishes data in the notation of Ref. 9-1, covering all Mach numbers except unity. This permits us to go to dimensionless equations like (9-89) and (9-90) without even changing the symbols. For illustration, we do this on the jet transport wing in Example 9-5 at the end of the present section.

Chapter 6 shows that the functional dependence of compressible airloads on Mach number is just as complicated as that on reduced frequency. We therefore resort to an indirect procedure, based on advance selection of both k and M, in nearly all practical flutter calculations. Two of these stand out as being especially convenient. The first is commonly used in the aircraft industry. It is well suited either to parametric investigations or to the analysis of a particular airplane at a particular altitude. For each of a series of preassigned Mach numbers, the critical speed is determined in one of the standard ways already described for incompressible flow; the U-g method is the most widely accepted in the United States. Once this is done, a plot of U_F vs. M can be drawn and points picked off where the speed and Mach number correspond to the temperature at the assumed altitude in the atmosphere.

The second procedure is an extension of the density-variation idea outlined in Subsection 9–5(a). At several pairs of M and k the flutter determinant is solved for the frequency ω and density ρ_∞. Each ω so obtained, together with k, yields a speed U_F. The speed and Mach number fix the sound velocity a_∞ and therefore the temperature T_∞, from which the pressure is computed by the perfect gas law. The resulting thermodynamic state may not represent a possible set of atmospheric conditions, but the process can be repeated until a flutter surface of U_F vs. p_∞ and ρ_∞ is sketched out. There is a certain line on this surface, found from its intersection with the vertical cylinder describing the standard atmosphere, which is just a curve of flutter speed vs. altitude. The surface can also be employed to find when a wing is predicted to flutter under other ambient states, such as might be encountered in high-speed wind tunnels.

Among the few available published calculations by the first scheme are the parameter studies on simplified two-dimensional models carried out by Garrick and Rubinow (Ref. 9–21), and by Woolston and Huckel (Ref. 9–13). The latter deal with control surface flutter and, as mentioned in Subsection 9–5(c), it is not easy to draw general conclusions from them. The former, however, include valuable extensions of data on the simple bending-torsion case (cf. Section 9–2) and are amenable to some interesting interpretation.

In Figs. 9–14 and 9–15, we reproduce two of the most significant plots from Ref. 9–21 which illustrate how the dimensionless speed $U_F/b\omega_\alpha$ is affected by compressibility. (These figures are actually taken from Ref. 9–22, where points for $M = 1$ have been added.) The thing which first strikes us in both figures is the relative insensitivity of U_F to Mach number for straight wings, of the type exemplified here, in subsonic flight. Only at the highest value of the density ratio is there more than a 10% change below $M = 0.8$. Clearly, the erstwhile compressibility correction based on the steady-flow Prandtl-Glauert rule, which multiplies the incompressible U_F by $\sqrt[4]{1 - M^2}$, is likely to be much too conservative in such cases. We must warn that the quantitative accuracy of these curves through the transonic range is in doubt, because of both the nonpotential flow below $M = 1$ and the erratic behavior of the theoretical aerodynamic coefficients as M approaches unity from above. But this does not invalidate them as indications of trends, and they certainly improve on any law which is strictly limited to two-dimensional, steady fluid motion.

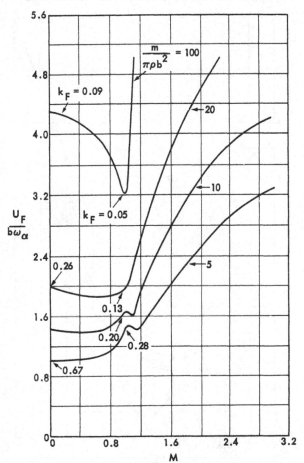

Fig. 9–14. Dimensionless bending-torsion flutter speed $U_F/b\omega_\alpha$ vs. Mach number M for several values of $m/\pi\rho b^2$. Other parameters are $\omega_h/\omega_\alpha = 0$, $x_\alpha = 0.2$, and $a = 0$. Certain points are labeled with the flutter reduced frequency k_F.

When we wish to compute from one of the curves in Fig. 9–14 the critical speed of a model with given geometrical, inertial, and stiffness properties, we use the fact that flight at a particular altitude in the atmosphere is characterized by a straight line through the origin. Thus, if a_∞ is the speed of sound at the known gas state, we have

$$M = \frac{U}{a_\infty}, \tag{6–2}$$

from which

$$\frac{U}{b\omega_\alpha} = \frac{a_\infty}{b\omega_\alpha} M. \tag{9–152}$$

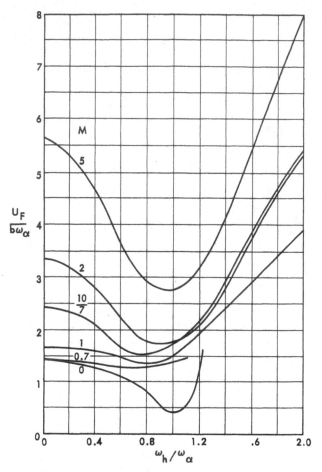

Fig. 9–15. Dimensionless bending-torsion flutter speed $U_F/b\omega_\alpha$ vs. frequency ratio ω_h/ω_α for several values of Mach number M. Other parameters are $m/\pi\rho b^2 = 10$, $x_\alpha = 0.2$, and $a = 0$.

The slope of the curve defined by Eq. (9–152) is lower the higher the altitude, until it becomes constant in the stratosphere. However, only one atmosphere line is rightfully associated with each flutter curve, since the latter implies a choice of the parameter $m/\pi\rho_\infty b^2$ and therefore a fixed altitude. We picture three typical relationships between $U_F/b\omega_\alpha$ plots and an atmosphere line in Fig. 9–16.

Curve 1 shows a tangency at some transonic Mach number, with no flutter at supersonic or subsonic speeds. Any curve lying above 1 (or 1 itself, if the theory can be safely regarded as conservative) ensures a flutter-free system at this altitude. What might happen at other altitudes is not

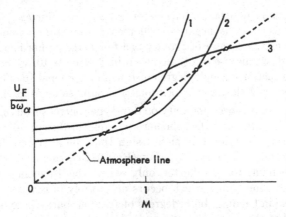

Fig. 9–16. Three possible relationships between curves of flutter speed vs. Mach number and a straight line representing flight at a particular altitude.

firmly established. If the tangency occurs at sea level, however, conditions appear favorable, because increasing altitude swings the atmosphere line downward and normally moves the $U_F/b\omega_\alpha$ points up, through the increase in density ratio.

Curve 2 is representative of many practical straight wings. It shows a range of instability at high subsonic and low supersonic speeds. Safe operation might be accomplished above the upper critical limit, but any excessive deceleration would lead to disaster. These results have special significance for experimental research in wind tunnels with variable supersonic Mach number. To obtain meaningful data, a test must be started at high speed and M reduced in search of the flutter speed.

Curve 3 indicates another theoretical possibility, where instability sets in during accelerated supersonic flight. Figure 9–14 suggests that this might be encountered on relatively light wings at low altitudes.

Much additional information on the effects of various parameters can be found in Ref. 9–21. For instance, as we see from Fig. 9–15, the same reduction of flutter speed near frequency ratio unity occurs at all Mach numbers, the phenomenon being least pronounced around $M = 0.7$ and 1. Garrick and Rubinow point out the important relations between aerodynamic center, elastic axis, and center-of-gravity location at both subsonic and supersonic speeds. The aerodynamic center moves from the quarter-chord point, below the critical Mach number, until it is near midchord, when supersonic flow is fully established. For small values of the frequency ratio ω_h/ω_α, the same controlling influence of the position of center of gravity relative to aerodynamic center is observed as in incompressible flow. At supersonic M, no flutter exists when the center of gravity is ahead of midchord. A sharp drop of $U_F/b\omega_\alpha$ to a relatively low and constant value appears as it passes behind the midchord. When ω_h/ω_α is larger, the elastic axis unbalance parameter x_α governs flutter. There is no instability when x_α is negative, whereas a minimum value of $U_F/b\omega_\alpha$ shows up when the center of gravity is a few percent chordlengths behind.

Although we discuss the topic again in Section 9–10, we call attention here to the fact that single-degree-of-freedom torsional instability is predicted for normal elastic-axis locations and supersonic Mach numbers less than $\sqrt{2.5}$. It is also a possibility on flap-type control surfaces, resulting in both cases from the fact that the aerodynamic coefficient 90° out of phase with pitching oscillation has such a sign that the airstream feeds energy into the motion. As shown by Flax (Ref. 9–23), this instability is influential in bending-torsion flutter only when the frequency ratio ω_h/ω_α exceeds unity, something which occurs only rarely in practice. For wings more flexible in bending, both degrees of freedom participate in the flutter mode, so that the energy input from torsion alone is counteracted. Predictions based on artificial suppression of the bending are then relatively meaningless.

Much of the theoretical research on oscillating supersonic wings in three-dimensional flow has been devoted to examining this pitching instability; general reductions are found in the ranges of Mach number and axis location for which it exists in two-dimensional flow. Care must be exercised, however, when we try to assess the meaning of such information for cases of bending-torsion flutter.

The density-variation method of compressible flutter analysis has not been applied to practical wings in the published literature. Hence we present here the results of a calculation on a simple cantilever wing.

EXAMPLE 9–4. To compute the variation of bending-torsion flutter speed with thermodynamic state for flight in a compressible atmosphere by a uniform model wing having the following properties:

aspect ratio = 4,
center of gravity = 49.1% chord behind leading edge,
elastic axis = 41.3% chord behind leading edge.

$$r_\alpha^2 = \frac{I_y}{mb^2} = 0.28, \quad \frac{\omega_w}{\omega_\theta} = 0.48, \quad b = b_R = 0.1263 \text{ ft}, \quad \omega_\theta = 1747 \text{ rad/sec.} \quad \text{(a)}$$

The model was observed to flutter experimentally at such a density that

$$\frac{m}{\pi\rho_\infty b^2} = 64.9. \tag{b}$$

Solution. A single mode of bending and one of torsion are assumed, the shape functions $f_w(y)$ and $f_\theta(y)$ being those of the uniform cantilever. The flutter equations are Eqs. (9–89) and (9–90), with appropriate numerical values inserted for the dimensionless parameters. We set

$$g_w = g_\theta = 0, \quad a = -0.174, \quad x_\alpha = \frac{S_y}{mb} = 0.172.$$

The flutter determinant is solved for $m/\pi\rho_\infty b^2$ and ω_θ^2/ω^2 at a series of values of k and supersonic M. Aerodynamic coefficients are taken from Ref. 9–20.

We do not elaborate the details of the calculation here, since they closely parallel those by the density-variation method in incompressible flow. Each root of the determinant yields a combination of U_F, p_∞, and ρ_∞, as discussed above. The flutter surface is plotted in Fig. 9–17.

Figure 9–17 shows the trace of the standard atmosphere in the p_∞-ρ_∞ plane, along with the curve on the U_F surface which gives critical speed vs. altitude. Also included is a line indicating how speed might vary with pressure and density in a typical intermittent wind tunnel, with variable Mach number, exhausting into a vacuum tank with rising back pressure. The point of intersection of this

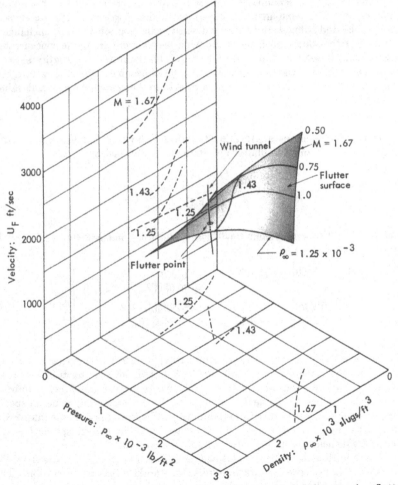

Fig. 9–17. Surface of U_F vs. p_∞ and ρ_∞ for supersonic bending-torsion flutter of a uniform model wing.

line with the surface is where the present model would flutter in the tunnel. Some conclusions about the approach to flutter and the violence with which it would start can be drawn from the angle of intersection between the two. However, there is no easy way of telling from these results which side of the surface represents stability; this can be done by artificially adding structural friction and observing which way U_F goes.

EXAMPLE 9–5. To compute the variation of critical speed with Mach number for bending-torsion flutter of a wing resembling the cantilever jet transport wing in Example 9–1. Strip aerodynamic theory and sea-level atmospheric conditions are to be used.

Solution. As in Example 9–1, we adopt the U-g method for each assumed value of M. The same artificial uncoupled modes $f_w(y)$ and $f_\theta(y)$ are chosen. The inertial and stiffness properties and elastic axis location are kept unchanged, except that the static unbalance of the wing section containing the concentrated engine mass is set equal to -91.44 slug·ft. To facilitate the lengthy aerodynamic calculations, we replace the tapered planform by a straight, rectangular one of the same area and span. The chord is therefore assigned a constant value, corresponding to $y = l/2$,

$$2b = 2b_R = 13.54 \text{ ft.} \tag{a}$$

The flutter equations are Eqs. (9–89) and (9–90) with b/b_R put equal to unity. The numerical form of the flutter determinant is obtained by setting

$$g_w = g_\theta = g, \tag{b}$$

$$Z = \left(\frac{\omega_\theta}{\omega}\right)^2 [1 + ig], \tag{c}$$

and factoring the aerodynamic coefficients, which are independent of y^*, outside of their integrals.

$$\begin{vmatrix} \{5.3589 - 1.7561Z & \{0.1126 \\ + 0.2431L_h\} & + 0.3555[L_\alpha - 0.2L_h]\} \\ \{0.1126 & \{28.1285[1 - Z] \\ + 0.3555[M_h - 0.2L_h]\} & + 0.6941[M_\alpha + 0.04L_h \\ & - 0.2(M_h + L_\alpha)]\} \end{vmatrix} = 0. \tag{d}$$

The coefficients in Eq. (d) were found from Ref. 9–20 for several choices of k at $M = 0$, 0.5, 0.6, 0.7, 10/9, 5/4, 10/7. The two complex roots Z were then calculated by the quadratic formula. For two representative Mach numbers, these roots are listed in Table 9–4, along with the corresponding values of speed, frequency, and artificial structural damping g. We do not present the intermediate graphs of g vs. U for each M, but tabulate in Table 9–5 the speeds U_F and flutter frequencies ω_F at the intersections with $g = 0$.

Figure 9–18 is the summary plot of U_F vs. Mach number. We also show the atmosphere line of speed vs. Mach number for standard sea-level condition. The equation of this line is

$$U = 760.9M \text{ mph.} \tag{e}$$

TABLE 9-4

k	Z_1	U_1 mph	ω_1 rad/sec	g_1	Z_2	U_2 mph	ω_2 rad/sec	g_2
				$M = 0.6$				
0.10	2.79820 −3.19608i	617	13.36	−1.14	1.53036 +0.09607i	835	18.07	0.06
0.15	2.69476 −1.89248i	419	13.62	−0.70	1.40533 +0.010625i	581	18.87	0.01
0.20	2.75570 −1.18989i	311	13.47	−0.69	1.23016 −0.08282i	466	20.16	−0.07
				$M = 10/9$				
0.21	2.51182 −0.86300i	312	14.11	−0.34	0.98377 +0.05134i	498	22.54	0.05
0.24	2.60995 −0.68500i	269	13.84	−0.26	1.01827 −0.02522i	431	22.16	−0.02
0.25,	2.65412 −0.62642i	243	13.72	−0.27	1.01810 −0.04948i	409	22.16	−0.05

TABLE 9-5

M	U_F mph	ω_F rad/sec
0	562	19.8
0.5	549	19.3
0.6	559	19.00
0.7	583	18.67
$\frac{10}{9}$	450	22.23
$\frac{5}{4}$	$\begin{cases}368 \\ 624\end{cases}$	23.02 25.85
$\frac{10}{7}$	stable	

It crosses the flutter curve twice, giving lower and upper critical conditions:*

$$U_{FL} = 582 \text{ mph}, \quad \omega_{FL} = 18.60 \text{ rad/sec}, \quad k_{FL} = 0.148. \quad (f)$$

$$U_{FU} = 800 \text{ mph}, \quad \omega_{FU} = 28.60 \text{ rad/sec}, \quad k_{FU} = 0.165. \quad (g)$$

With all dimensionless properties of the wing fixed, we know that U_F is directly proportional to torsional frequency ω_θ. By translating the entire U_F vs. M curve upward until it becomes tangent to the atmosphere line, we discover that a 162.4% increase in ω_θ would make the wing flutter-free at all Mach numbers. As discussed above, this would probably eliminate bending-torsion flutter

* The upper critical speed is approximate, since it is based on a short extrapolation of the curve of U_F vs. M.

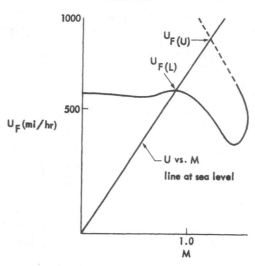

Fig. 9–18. Plot of flutter speed U_F vs. Mach number M for bending-torsion flutter of modified cantilever jet transport wing in compressible flow.

not only at sea level, but throughout the atmosphere. Such a rise in torsional frequency would be brought about by 588% increases in EI and GJ, if they could be achieved without appreciable additions to the wing weight, something which seems highly unlikely.

9–8 Flutter of swept wings. Successful prediction of the flutter of swept wings by assumed-mode methods depends, to an even greater degree than in the absence of sweep, on accurate knowledge of the structural and aerodynamic properties of the system. The state of the art, particularly on the aerodynamic side, is not so far advanced as for straight wings. A planform with appreciable sweep back or sweep forward is characterized by two aspect ratios: a structural one, proportional to the ratio of root-to-tip length l and chord $\bar{c}$ measured normal to the swept span; and a smaller, aerodynamic one. The customary definition of the latter is, of course,

$$AR = \frac{(2l)^2}{S},\qquad (9\text{–}153)$$

where $2l$ is the tip-to-tip distance and S is the plan area.

On a constant-chord wing of sweep angle Λ, the two aspect ratios are related through the equation

$$\frac{(2l)^2}{S} = \frac{2l}{c} \cong \frac{2\bar{l}}{\bar{c}} \cos^2 \Lambda,\qquad (9\text{–}154)$$

c being the chord taken parallel to the flight direction. There is no assurance that AR is the true measure of finite-span effects. Diederich (Ref.

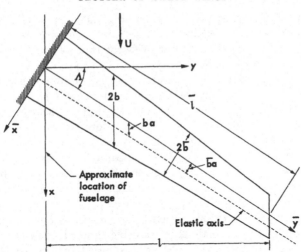

Fig. 9–19. Swept-wing structure with an elastic axis and a cantilever root built in approximately normal to that axis.

9–24) points out how the intermediate parameter

$$\frac{AR}{\cos \Lambda} = \frac{2l}{\bar{c}} \cos \Lambda \tag{9-155}$$

is more important for determining such things as steady-state lift-curve slope, whereas Barmby, Cunningham, and Garrick (Ref. 9–25) conclude that the structural aspect ratio itself may be the relevant one in the case of flutter.

One advantage to the aeroelastician of having larger $l/\bar{c}$ for a given value of AR is that a swept-wing structure is more likely to behave like a simple beam with a true elastic axis. On a great many practical aircraft the wing is observed to act as a cantilever, built in at some section which is normal to that axis and near the junction with the fuselage. This is the type of structure on which we concentrate in the present section. More complicated wings can be treated using strip-theory airloads and superposition of normal (rather than artificial) modes, so long as certain cross sections do not deform during bending and torsion. Wings whose deflections resemble those of a plate must be analyzed by low aspect-ratio methods, which we discuss briefly in Section 9–9. The main difficulty in the latter case is the unavailability of systematic aerodynamic theory for the swept wing in three-dimensional flow.

Figure 9–19 illustrates the configuration for which we derive the flutter equations. The sweep angles Λ of the nearly straight elastic axis and the straight midchord line are almost equal. We omit consideration of rigid-body degrees of freedom, but they can easily be introduced as described in

Subsection 9–5(b). It appears that pitching and vertical translation, for symmetrical modes, and rolling for antisymmetrical modes may have a more important influence on flutter of swept than of unswept wings (cf. remarks in Refs. 9–15 and 9–23).

As proved in Section 7–3, there are two ways of applying aerodynamic theory, one focusing on sections normal to the elastic axis, the other on sections in the flight direction. There correspond two schemes for setting up the flutter determinant. The former is more commonly employed in practice and seems to be most rational when the wing ribs are oriented perpendicular to the span. The latter is suitable for ribs parallel to the fuselage center line; it also does a better job of representing aerodynamic conditions at the root and at the wingtip, which is generally cut off streamwise.

Turning first to flutter calculation by the normal-section approach, we assume bending of the elastic axis (positive upward) and torsion about that axis (positive nose-up) to be approximated by the artificially uncoupled modes

$$\bar{w}_B(\bar{y},t) = \bar{w}_R(t)\bar{f}_w(\bar{y}), \qquad (9\text{--}156)$$

$$\bar{\theta}_T(\bar{y},t) = \bar{\theta}_R(t)\bar{f}_\theta(\bar{y}). \qquad (9\text{--}157)$$

The superscript bar is used throughout to denote coordinates and properties associated with normal sections. If $\bar{l}$ is the root-to-tip distance (Fig. 9–19), etc., Eqs. (9–84) and (9–85) can be adapted to this system by simply barring all the symbols in them:

$$\ddot{\bar{w}}_R \int_0^{\bar{l}} \bar{m}\bar{f}_w^2 d\bar{y} - \ddot{\bar{\theta}}_R \int_0^{\bar{l}} \bar{S}_y\bar{f}_w\bar{f}_\theta d\bar{y} + \bar{w}_R\bar{\omega}_w^2[1 + i\bar{g}_w]\int_0^{\bar{l}} \bar{m}\bar{f}_w^2 d\bar{y}$$

$$= \int_0^{\bar{l}} \bar{L}(\bar{y},t)\bar{f}_w(\bar{y})d\bar{y}, \qquad (9\text{--}158)$$

$$\ddot{\bar{\theta}}_R \int_0^{\bar{l}} \bar{I}_y\bar{f}_\theta^2 d\bar{y} - \ddot{\bar{w}}_R \int_0^{\bar{l}} \bar{S}_y\bar{f}_w\bar{f}_\theta d\bar{y} + \bar{\theta}_R\bar{\omega}_\theta^2[1 + i\bar{g}_\theta]\int_0^{\bar{l}} \bar{I}_y\bar{f}_\theta^2 d\bar{y}$$

$$= \int_0^{\bar{l}} \bar{M}_y(\bar{y},t)\bar{f}_\theta(\bar{y})d\bar{y}. \qquad (9\text{--}159)$$

Here $\bar{m}$, $\bar{S}_y$, $\bar{I}_y$, $\bar{L}$, and $\bar{M}_y$ are all per unit distance in the $\bar{y}$-direction.

The aerodynamic loads are given by Eqs. (7–69) and (7–70), with the dependent variables describing the motion transformed as follows:

$$\bar{h} = -\bar{w}_R\bar{f}_w, \qquad (9\text{--}160a)$$

$$\bar{\alpha} = \bar{\theta}_R\bar{f}_\theta, \qquad (9\text{--}160b)$$

$$\sigma = \frac{\partial h}{\partial \bar{y}} = -\bar{w}_R\frac{d\bar{f}_w}{d\bar{y}}, \qquad (9\text{--}160c)$$

$$\tau = \frac{\partial \bar{\alpha}}{\partial \bar{y}} = \bar{\theta}_R \frac{d\bar{f}_\theta}{d\bar{y}}. \tag{9-160d}$$

For convenience, we employ the simplified coefficients defined in Eqs. (7–71), (7–72), and (7–73), which permit us to write

$$\bar{L}(\bar{y},t) = \pi\rho\omega^2\bar{b}^3 \left\{ \frac{\bar{w}_R}{\bar{b}} \bar{f}_w(\bar{y})L_{hh} + \bar{w}_R \frac{d\bar{f}_w}{d\bar{y}} L_{hh'} \right.$$

$$\left. - \bar{\theta}_R \bar{f}_\theta(\bar{y})L_{h\alpha} - \bar{b}\bar{\theta}_R \frac{d\bar{f}_\theta}{d\bar{y}} L_{h\alpha'} \right\}, \tag{9-161}$$

$$\bar{M}_y(\bar{y},t) = \pi\rho\omega^2\bar{b}^4 \left\{ -\frac{\bar{w}_R}{\bar{b}} \bar{f}_w(\bar{y})M_{\alpha h} - \bar{w}_R \frac{d\bar{f}_w}{d\bar{y}} M_{\alpha h'} \right.$$

$$\left. + \bar{\theta}_R \bar{f}_\theta(\bar{y})M_{\alpha\alpha} + \bar{b}\bar{\theta}_R \frac{d\bar{f}_\theta}{d\bar{y}} M_{\alpha\alpha'} \right\}. \tag{9-162}$$

Finally, we insert Eqs. (9–161) and (9–162) into Eqs. (9–158) and (9–159), collect terms, and divide by $\pi\rho\omega^2\bar{b}_R{}^3\bar{l}$ and $\pi\rho\omega^2\bar{b}_R{}^4\bar{l}$, respectively, to make the results dimensionless. $\bar{b}_R$ is an arbitrarily selected reference semichord. We also utilize the fact of simple harmonic motion to replace $\ddot{\bar{\theta}}_R$ and $\ddot{\bar{w}}_R$ by $-\omega^2\bar{\theta}_R$ and $-\omega^2\bar{w}_R$.

$$\frac{\bar{w}_R}{\bar{b}_R} \left\{ \left[1 - (1 + i\bar{g}_w)\left(\frac{\bar{\omega}_w}{\omega}\right)^2 \right] \int_0^1 \left[\frac{\bar{m}}{\pi\rho\bar{b}_R{}^2} \right] \bar{f}_w{}^2 d\bar{y}^* \right.$$

$$\left. + \int_0^1 \left(\frac{\bar{b}}{\bar{b}_R}\right)^2 L_{hh}\bar{f}_w{}^2 d\bar{y}^* + \frac{\bar{b}_R}{\bar{l}} \int_0^1 \left(\frac{\bar{b}}{\bar{b}_R}\right)^3 L_{hh'}\bar{f}_w \frac{d\bar{f}_w}{d\bar{y}^*} d\bar{y}^* \right\}$$

$$- \bar{\theta}_R \left\{ \int_0^1 \left[\frac{\bar{S}_y}{\pi\rho\bar{b}_R{}^3} \right] \bar{f}_w\bar{f}_\theta d\bar{y}^* + \int_0^1 \left(\frac{\bar{b}}{\bar{b}_R}\right)^3 L_{h\alpha}\bar{f}_w\bar{f}_\theta d\bar{y}^* \right.$$

$$\left. + \frac{\bar{b}_R}{\bar{l}} \int_0^1 \left(\frac{\bar{b}}{\bar{b}_R}\right)^4 L_{h\alpha'}\bar{f}_w \frac{d\bar{f}_\theta}{d\bar{y}^*} d\bar{y}^* \right\} = 0. \tag{9-163}$$

$$-\frac{\bar{w}_R}{\bar{b}_R} \left\{ \int_0^1 \left[\frac{\bar{S}_y}{\pi\rho\bar{b}_R{}^3} \right] \bar{f}_w\bar{f}_\theta d\bar{y}^* + \int_0^1 \left(\frac{\bar{b}}{\bar{b}_R}\right)^3 M_{\alpha h}\bar{f}_w\bar{f}_\theta d\bar{y}^* \right.$$

$$\left. + \frac{\bar{b}_R}{\bar{l}} \int_0^1 \left(\frac{\bar{b}}{\bar{b}_R}\right)^4 M_{\alpha h'}\bar{f}_\theta \frac{d\bar{f}_w}{d\bar{y}^*} d\bar{y}^* \right\}$$

$$+ \bar{\theta}_R \left\{ \left[1 - (1 + i\bar{g}_\theta)\left(\frac{\bar{\omega}_\theta}{\omega}\right)^2 \right] \int_0^1 \left[\frac{\bar{I}_y}{\pi\rho\bar{b}_R{}^4} \right] \bar{f}_\theta{}^2 d\bar{y}^* \right.$$

$$\left. + \int_0^1 \left(\frac{\bar{b}}{\bar{b}_R}\right)^4 M_{\alpha\alpha}\bar{f}_\theta{}^2 d\bar{y}^* + \frac{\bar{b}_R}{\bar{l}} \int_0^1 \left(\frac{\bar{b}}{\bar{b}_R}\right)^5 M_{\alpha\alpha'}\bar{f}_\theta \frac{d\bar{f}_\theta}{d\bar{y}^*} d\bar{y}^* \right\} = 0. \tag{9-164}$$

All factors in the integrands are functions of the reduced spanwise variable

$$\bar{y}^* = \frac{\bar{y}}{l} \,. \tag{9-165}$$

Equations (7-73) show that the aerodynamic coefficients also depend on the sweep angle Λ and the dimensionless location of the elastic axis.

The four brackets mutiplying $\bar{w}_R/\bar{b}_R$ and $\bar{\theta}_R$ in Eqs. (9-163) and (9-164) form the flutter determinant. It can be solved by any of the procedures outlined in Subsection 9-5(a). If compressibility is to be accounted for, ρ is replaced by ρ_∞, and the aerodynamic quantities should be looked up in tables corresponding to the cross-flow Mach number

$$\frac{U \cos \Lambda}{a_\infty} = M \cos \Lambda. \tag{7-61}$$

The flutter equations associated with streamwise wing sections have not received such general acceptance as Eqs. (9-163) and (9-164), probably because the modal deformations used in them seem unnatural for swept wings. We derive them here, however, because they are known to lead to satisfactory results, and there are situations where they are definitely preferable. For instance, rigid-body degrees of freedom like roll fit into them naturally, and aeroelastic models built according to the sectional-construction scheme (e.g., Ref. 9-26) are very well described by the physical approximations underlying them.

When analyzing bending-torsion flutter with two assumed modes, we choose a bending deflection (positive upward) of the elastic axis,

$$w_B(y,t) = w_R(t)f_w(y), \tag{9-166}$$

which is no different from Eq. (9-156), except that f_w is regarded as a function of the coordinate normal to the flight direction (Fig. 9-19). Cross sections parallel to U are presumed not to bend, so we replace their (positive nose-up) twist distribution, *about axes normal to U but intersecting the elastic axis at each station y,* by

$$\theta_T(y,t) = \theta_R(t)f_\theta(y). \tag{9-167}$$

Except on a sectional aeroelastic model, it is hard to see how such a rotation might be produced. Nevertheless, it is physically equivalent to a twisting

$$\bar{\theta}_T = \frac{\theta_T}{\cos \Lambda} \tag{9-168}$$

about the spanwise elastic axis, plus a small amount of camber deformation in perpendicular cross sections.

If we calculate the kinetic and potential energies due to simple harmonic motion in the modes (9-166) and (9-167) and invent an artificial frequency for each of them, we find that T and U_E are identical to Eqs. (9-65) and

(9–67), with the specification that m, S_y, and I_y are inertias per unit distance in the y-direction. $S_y dy$ and $I_y dy$ are the static unbalance and moment of inertia of material contained within a section of width dy, taken about an axis normal to the flight direction and passing through the intersection of the elastic axis with this elementary section. It is easy to confirm formulas like

$$\bar{m} = m \cos \Lambda, \qquad \bar{S}_y = S_y \cos^2 \Lambda, \qquad \bar{I}_y = I_y \cos^3 \Lambda. \qquad (9\text{–}169)$$

There is no alteration in the generalized force expressions coming from structural damping. By examining virtual displacements δw_R and $\delta \theta_R$ of the assumed degrees of freedom, we obtain for the generalized aerodynamic forces

$$Q_w = \int_0^l L(y,t) f_w(y) dy, \qquad (9\text{–}170)$$

$$Q_\theta = \int_0^l M_y(y,t) f_\theta(y) dy, \qquad (9\text{–}171)$$

where L and M_y are the lift and pitching moment per unit y-distance; M_y is also about the axis normal to U through the local elastic axis position.

The equations of motion derived in the manner we have discussed do not differ from Eqs. (9–84) and (9–85), except for the implicit changes in the meanings of terms they contain.

$$\ddot{w}_R \int_0^l m f_w{}^2 dy - \ddot{\theta}_R \int_0^l S_y f_w f_\theta dy$$

$$+ w_R \omega_w{}^2 [1 + ig_w] \int_0^l m f_w{}^2 dy = \int_0^l L f_w dy, \qquad (9\text{–}172)$$

$$\ddot{\theta}_R \int_0^l I_y f_\theta{}^2 dy - \ddot{w}_R \int_0^l S_y f_w f_\theta dy$$

$$+ \theta_R \omega_\theta{}^2 [1 + ig_\theta] \int_0^l I_y f_\theta{}^2 dy = \int_0^l M_y f_\theta dy. \qquad (9\text{–}173)$$

The airloads are expressed by Eqs. (7–58) and (7–59), which fact allows us to adopt the Ref. 9–1 notation with no further change than an over-all factor $\cos \Lambda$.

$$L(y,t) = \pi \rho_\infty b^3 \omega^2 \cos \Lambda \left\{ \frac{w_R}{b} f_w(y) L_h - \theta_R f_\theta(y) [L_\alpha - L_h(\tfrac{1}{2} + a)] \right\}, \qquad (9\text{–}174)$$

$$M_y(y,t) = \pi \rho_\infty b^4 \omega^2 \cos \Lambda \left\{ - \frac{w_R}{b} f_w(y) [M_h - L_h(\tfrac{1}{2} + a)] \right.$$

$$\left. + \theta_R f_\theta(y) [M_\alpha - (M_h + L_\alpha)(\tfrac{1}{2} + a) + L_h(\tfrac{1}{2} + a)^2] \right\}. \qquad (9\text{–}175)$$

Accordingly, the dimensionless flutter equations read

$$\frac{w_R}{b_R}\left\{\left[1-(1+ig_w)\left(\frac{\omega_w}{\omega}\right)^2\right]\int_0^1\left[\frac{m}{\pi\rho_\infty b_R^2}\right]f_w^2 dy^* + \cos\Lambda\int_0^1\left(\frac{b}{b_R}\right)^2 L_h f_w^2 dy^*\right\}$$

$$-\theta_R\left\{\int_0^1\left[\frac{S_y}{\pi\rho_\infty b_R^3}\right]f_w f_\theta dy^* + \cos\Lambda\int_0^1\left(\frac{b}{b_R}\right)^3[L_\alpha-L_h(\tfrac{1}{2}+a)]f_w f_\theta dy^*\right\}$$

$$=0, \quad (9\text{--}176)$$

$$-\frac{w_R}{b_R}\left\{\int_0^1\left[\frac{S_y}{\pi\rho_\infty b_R^3}\right]f_w f_\theta dy^* + \cos\Lambda\int_0^1\left(\frac{b}{b_R}\right)^3[M_h-L_h(\tfrac{1}{2}+a)]f_w f_\theta dy^*\right\}$$

$$+\theta_R\left\{\left[1-(1+ig_\theta)\left(\frac{\omega_\theta}{\omega}\right)^2\right]\int_0^1\left[\frac{I_y}{\pi\rho_\infty b_R^4}\right]f_\theta^2 dy^*\right.$$

$$\left.+\cos\Lambda\int_0^1\left(\frac{b}{b_R}\right)^4[M_\alpha-(M_h+L_\alpha)(\tfrac{1}{2}+a)+L_h(\tfrac{1}{2}+a)^2]f_\theta^2 dy^*\right\}=0.$$

$$(9\text{--}177)$$

EXAMPLE 9-6.* To use the method of streamwise sections to predict the incompressible, bending-torsion flutter of three swept-wing models with constant aspect ratio and varying sweep angle. These models are the ones designated 13, 14, and 15 in Ref. 9-25. The sweep angles are 30°, 45°, and 60°, respectively. Each aspect ratio is 4, as defined in Eq. (9-153), the chord being constant and equal to $\frac{2}{3}$ ft. These were actually tested as half-models from a cantilever support in the wind-tunnel wall.

The model properties entering Eqs. (9-176) and (9-177) are as listed in Table 9-6.

TABLE 9-6

Model No.	13	14	15
Λ (degrees)	30	45	60
m (slugs/ft)	0.0161	0.0138	0.0120
$m/\pi\rho_\infty b_R^2$	6.19	5.50	4.55
I_y/mb_R^2	0.23	0.23	0.23
S_y/mb_R	−0.004	−0.224	−0.324
$b=b_R$(ft)	0.333	0.333	0.333
$\bar{l}$ (ft)	1.540	1.886	2.667
l (ft)	1.333	1.333	1.333
ω_w (rad/sec)	66π	44π	24π
ω_θ (rad/sec)	186π	184π	186π
g_w	0	0	0
g_θ	0	0	0
a	−0.02	0.20	0.30

* This example is taken from an S.M. thesis at the Massachusetts Institute of Technology by John R. Martuccelli.

Solution. A comparison for the 45° model shows that the flutter conditions are insensitive to choice of the modes $f_{\bar{w}}$ and f_θ. We therefore select the easily integrated functions

$$f_w(y^*) = (y^*)^2, \tag{a}$$

$$f_\theta(y^*) = y^*. \tag{b}$$

The U-g graph is employed to solve the determinant of Eqs. (9–176) and (9–177). Before all parameters associated with the individual uniform wings are inserted, this determinant reads

$$
\begin{vmatrix}
0.200 \left\{ \dfrac{m}{\pi \rho_\infty b_R{}^2} \left[1 - \left(\dfrac{\omega_w}{\omega_\theta} \right)^2 Z \right] \right. & -0.250 \left\{ \dfrac{S_y}{\pi \rho_\infty b_R{}^3} \right. \\
\left. + L_h \cos \Lambda \right\} & \left. + [L_\alpha - L_h(\tfrac{1}{2} + a)] \cos \Lambda \right\} \\[2ex]
-0.250 \left\{ \dfrac{S_y}{\pi \rho_\infty b_R{}^3} \right. & 0.333 \left\{ \dfrac{I_y}{\pi \rho_\infty b_R{}^4} [1 - Z] \right. \\
\left. + [M_h - L_h(\tfrac{1}{2} + a)] \cos \Lambda \right\} & + [M_\alpha - (M_h + L_\alpha)(\tfrac{1}{2} + a) \\
& \left. + L_h(\tfrac{1}{2} + a)^2] \cos \Lambda \right\}
\end{vmatrix} = 0, \tag{c}
$$

where

$$Z = \left(\frac{\omega_\theta}{\omega} \right)^2 [1 + ig]. \tag{d}$$

Numerical details of the solution do not differ substantially from those in Example 9–1, so we do not reproduce them. In Table 9–7, we compare the experimental flutter speeds U_e and frequencies ω_e with theoretical values obtained in three different ways. Subscript 1 denotes results obtained in Ref. 9–25 by the normal section type of aerodynamic strip theory suggested there; subscript 2 denotes calculations based on determinant (c) above; and subscript 3 denotes quantities corrected for the influence of finite span by the swept-wing theory of Ref. 9–27. A complete description of the last calculation is beyond the scope of this section.

TABLE 9–7

Model No.	13	14	15
Λ (degrees)	30	45	60
U_e (ft/sec)	297	288	263
U_{F_1} (ft/sec)	234	244	270
U_{F_2} (ft/sec)	277	270	252
U_{F_3} (ft/sec)	343	323	282
ω_e (rad/sec)	384	337	233
ω_{F_1} (rad/sec)	408	383	364
ω_{F_2} (rad/sec)	377	322	272
ω_{F_3} (rad/sec)	325	295	233

The data of Table 9–7 show that for the particular models considered, the streamwise section theory is more accurate than either of the other methods. It is slightly conservative with regard to speed but is very consistent in predicting the trends of both U_F and ω_F with sweep angle. Only the frequency for the 60° model is in error by more than 7%. By comparison, the three-dimensional theory seems to be seriously unconservative, especially at the lower sweep angles. Since the experiments were carried out in a wind tunnel containing Freon-12, a maximum flutter Mach number of 0.62 is represented. Reference 9–25 demonstrates relatively little influence of compressibility on flutter in this range, however.

In Ref. 9–25, an extensive series of tests on uniform cantilever models of various sweep angles and aspect ratios is reported and interpreted to discover the effects of varying different significant parameters. As mentioned above, compressibility appears to have almost no importance for U_F up to a free-stream Mach number of 0.8. It must be remembered that the maximum effective cross-flow Mach number is much lower than this on the more highly swept planforms. Since most of the models have rather low frequency ratios, changing the elastic axis location within reasonable bounds also has only a small influence. The chordwise position of the center of gravity is more critical, however, with U_F increasing as it moves forward on wings of any degree of sweep back. Section 9–2 shows that both these trends are also observed on straight wings. Another tendency which is present at all values of Λ, including zero, is a gradual decrease of U_F with increasing frequency ratio, but this is not examined experimentally in Ref. 9–25.

There are two ways of studying the effects of varying sweep angle on flutter. One is to rotate the wing about its root, keeping the structural aspect ratio constant. The other is to shear it back in such a way that AR in Eq. (9–153) is unchanged. Both of these are tried in Ref. 9–25. Shearing a wing while keeping the running inertial and elastic properties unaltered lowers both ω_w and ω_θ, and a corresponding decrease in U_F results. Both theory and experiment, such as those presented in Example 9–6 above, indicate that U_F falls off slightly more slowly than $\sqrt{\cos \Lambda}$ when the root is clamped.

Rotating a wing about its root cannot alter the mass and stiffness distributions, nor are the fundamental frequencies modified unless the structural aspect ratio $\bar{l}/\bar{c}$ is less than about 3. The elastic axis moves aft but stays relatively straight. We anticipate that a larger U_F will result, and this is what experiments show. On a purely two-dimensional basis, only the cross-flow component produces airloads, and U_F would go up in proportion to $1/\cos \Lambda$. On a finite wing, there appears to be an unfavorable alteration in the coupling between the degrees of freedom, and this causes U_F to rise at a rate somewhere between $1/\cos \Lambda$ and $1/\sqrt{\cos \Lambda}$, but closer to the latter.

In general, the success of the theory for estimating incompressible bending-torsion flutter of swept wings with large enough aspect ratios is satisfactory. It is not as fully reliable as on straight planforms, although better than some of the comparisons in Example 9–6 might suggest. The limited experience we have with strip theory for high subsonic and supersonic flows is not so gratifying, particularly on wings in the most commonly used range of aspect ratios. There appears to be an inordinately high degree of conservatism, which is connected with the unstable moments in pitch experienced by high-speed airfoils when they oscillate about axes ahead of midchord. Some rational correction for the influence of finite span will offer the best hope for escaping the unrealistic answers we obtain in this area.

In closing the section on sweep, we call attention to the related problem of single-degree-of-freedom bending flutter of swept-back wings. As proved by Cunningham (Ref. 9–28), this phenomenon can occur when both the density ratio and the parameter

$$\frac{\tan \Lambda}{l/b_R}$$

are sufficiently large. We have more to say on this subject, along with other similar instabilities, in Section 9–10.

9–9 Wings of low aspect ratio. Although there are undoubtedly many low aspect-ratio wings and tails that are safe from the flutter standpoint because of their great rigidity, this certainly cannot be said of all of them. There is a growing demand for theoretical methods of analyzing these structures, which is augmented by the high costs of model construction, rocket tests, and other experimental substitutes. Such calculations are peculiarly difficult for two reasons that we have discussed in foregoing sections. Since chordwise and spanwise stiffnesses are of the same order of magnitude, platelike deformations must be accounted for. Three-dimensional aerodynamic theory is also needed, and the nearly equal width and breadth of the planform make it irrational to simplify this theory by assuming the form of the pressure distribution in either direction.

Certain promising steps have been taken toward development of the necessary aerodynamics, as we have briefly outlined in Sections 7–2, 7–4, and 7–5. However, few actual flutter computations have been carried to completion or compared with experimental results.

If the Rayleigh-Ritz method is used, it appears most accurate and efficient to adopt normal modes and equations of motion. Just the determination of the shape functions $\phi_j(x,y)$ and frequencies ω_j may require extensive numerical work, difficult vibration tests on the actual structure, or construction of a dynamic model. As an alternative to modal superposition, we describe at the end of this section an interesting scheme due to Voss which applies structural influence coefficients directly to solution of the flutter problem.

The steps in a low aspect-ratio analysis are best described by an illustration. We take the case of a uniform circular wing or simplified "flying saucer," as treated in Ref. 9–29. This has the advantage that the normal modes of vibration of a circular plate are well known. The airload distribution on an oscillating circular wing in incompressible flow can be found from the theories of Kochin (Ref. 9–30) or Krienes and Schade (Ref. 9–31). No tabulated data are given by Kochin, and there is uncertainty about the convergence of series employed in Ref. 9–31; however, their tables are believed accurate enough for flutter purposes at lower reduced frequencies.

EXAMPLE 9–7. To compute the incompressible flutter speed of a uniform, circular wing (Fig. 9–20). The model on which this example is based was built up over a duralumin plate with

$$\text{radius } c = 21 \text{ inches,}$$
$$\text{thickness } t_0 = 0.040 \text{ inch,}$$
$$\text{mass per unit area } \rho(x,y) = 0.1252 \text{ slug/ft}^2, \tag{a}$$

where $\rho(x,y)$ includes the mass of superimposed balsa wood, which was employed to give a smooth airfoil section but did not contribute to the stiffness.

Solution. Since the "flying saucer" is in unrestrained, free flight, its symmetrical flutter motion is assumed to consist of rigid-body vertical translation and pitching, plus the first two normal modes of the circular plate.

$$w(x,y,t) = \sum_{r=1}^{4} \phi_r(x,y)\xi_r(t)$$
$$= c\{A(t) + x^*B(t) + \phi_1(x^*,y^*)D(t) + \phi_2(x^*,y^*)F(t)\}. \tag{b}$$

Here x^* and y^* have been made dimensionless by division with the radius c. From Ref. 9–32, for example, the two vibration modes are

$$\phi_1(x^*,y^*) = \{J_2(2.29\sqrt{x^{*2} + y^{*2}}) + 0.333091I_2(2.29\sqrt{x^{*2} + y^{*2}})\} \cos 2\theta$$
$$\cong 0.8x^{*2} - 0.8y^{*2}, \tag{c}$$

$$\phi_2(x^*,y^*) = J_0(3.01\sqrt{x^{*2} + y^{*2}}) - 0.084058I_0(3.01\sqrt{x^{*2} + y^{*2}})$$
$$\cong 0.915942 - 1.980x^{*2} - 1.980y^{*2}, \tag{d}$$

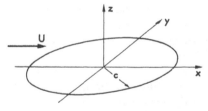

Fig. 9–20. Wing of circular planform in incompressible flow.

where
$$\tan \theta = \frac{y^*}{x^*}. \tag{e}$$

The power series approximations are chosen to fit the deformation shapes for which the tables of Ref. 9–31 are constructed, as are the symbols A, B, D, F.

The kinetic energy corresponding to Eq. (b) is found to be

$$T = \tfrac{1}{2}\pi c^4 \rho(x^*, y^*)\{\dot{A}^2 + \tfrac{1}{4}\dot{B}^2 + 0.099099\dot{D}^2 + 0.207022\dot{F}^2\}. \tag{f}$$

Using the general expression for the elastic energy of a plate (Eq. 2–162), we obtain the strain energy

$$U_E = \frac{\pi E t_0^3}{24(1 - \nu^2)} [4.832216 D^2 + 17.055164 F^2], \tag{g}$$

where E and ν are Young's modulus and Poisson's ratio for the structural material.

In the notation of Krienes and Schade, the generalized forces [Eqs. (3–147c) and (9–55)] assume forms like

$$\Xi_A = 4\rho c^3 U^2 \{l_A A + l_B B + 0.8(l_D - l_F)D + [0.915942 - 1.980(l_D + l_F)]F\}. \tag{h}$$

Here the l's are related to coefficients, tabulated in Ref. 9–31, which are functions only of

$$k = \frac{\omega c}{U}. \tag{i}$$

Structural friction is neglected.

If we substitute Eqs. (f), (g), and the generalized forces into Lagrange's equation (9–83), and assume simple harmonic motion, we are led to the following system:

$$[l_A + k^2 \gamma_0]\bar{A} + l_B \bar{B} + 0.8[l_D - l_F]\bar{D}$$
$$+ [0.915942 l_A - 1.980(l_D + l_F)]\bar{F} = 0, \tag{j}$$

$$m_A \bar{A} + [m_B + \tfrac{1}{4}k^2 \gamma_0]\bar{B} + 0.8[m_D - m_F]\bar{D}$$
$$+ [0.915942 m_A - 1.980(m_D + m_F)]\bar{F} = 0, \tag{k}$$

$$0.8[i_A - j_A]\bar{A} + 0.8[i_B - j_B]\bar{B}$$
$$+ [0.09909 k^2 \gamma_0 + 0.64(i_D - j_D - i_F + j_F) - \alpha_D \delta]\bar{D}$$
$$+ 0.8[0.915942(i_A - j_A) - 1.980(i_D - j_D - i_F + j_F)]\bar{F} = 0, \tag{l}$$

$$[0.915942 l_A - 1.980(i_A + j_A)]\bar{A} + [0.915942 l_B - 1.980(i_B + j_B)]\bar{B}$$
$$+ 0.8[0.915942(l_D - l_F) - 1.980(i_D + j_D - i_F - j_F)]\bar{D}$$
$$+ [0.207002 k^2 \gamma_0 - \alpha_F \delta + 0.838950 l_A - 1.813565(l_D + l_F + i_A + j_A)$$
$$+ 3.9204(i_D + j_D + i_F + j_F)]\bar{F} = 0. \tag{m}$$

The i, j, and m are tabulated functions of k similar to the l's.

$$\gamma_0 = \frac{\pi \rho(x^*, y^*)}{4\rho c}, \tag{n}$$

is a modified density ratio, and

$$\delta = \frac{E t_0^3}{48 \rho U^2 c^2 (1 - \nu^2)}. \tag{o}$$

The flutter determinant consists of the bracketed coefficients in Eqs. (j), (k), (l), and (m). When a value of k is assumed, we obtain a quadratic equation in the remaining unknown δ, which is inversely proportional to dynamic pressure. Flutter occurs at that k which makes the imaginary part of δ vanish.

The particular wing specified in this example turns out to have the theoretical flutter condition

$$U_F = 44.9 \text{ mph}, \qquad \omega_F = 11.3 \text{ rad/sec}, \qquad k_F = 0.30. \qquad \text{(p)}$$

Difficulties with the linearity of the model structure make it impossible to furnish comparable experimental data.

As a substitute for the assumed-mode approach to flutter analysis in general, Voss (Ref. 9–32) has proposed an ingenious application of the reverse-flow theorem (Section 7–5). His method is well suited to systematic computation on a high-speed digital machine and appears to be especially useful for low aspect-ratio wings. By analogy with Eq. (2–55), we write the equation of motion of the oscillating, restrained airplane:

$$\bar{w}(x,y) = \omega^2 \iint_S C^{zz}(x,y;\xi,\eta)\rho(\xi,\eta)\bar{w}(\xi,\eta)d\xi d\eta$$

$$- \iint_S C^{zz}(x,y;\xi,\eta)\Delta\bar{p}_a(\xi,\eta)d\xi d\eta. \qquad (9\text{–}178)$$

Only aerodynamic loads are included here, and $\bar{w}(x,y)$ represents the amplitude distribution of vertical bending

$$w(x,y,t) = \bar{w}(x,y)e^{i\omega t}. \qquad (9\text{–}179)$$

A similar equation of motion describes the unrestrained airplane.

Concentrating on the last term in Eq. (9–178), we multiply by U and divide by $C^{zz}(x,y;x,y)$. Thus we obtain, multiplying $\Delta\bar{p}_a(\xi,\eta)$, the quantity

$$\bar{w}_0 = U\,\frac{C^{zz}(x,y;\xi,\eta)}{C^{zz}(x,y;x,y)}\,. \qquad (9\text{–}180)$$

For a particular choice of (x,y), $\bar{w}_0(\xi,\eta)$ can be interpreted as a distribution of fluid vertical-velocity amplitude over the planform. According to the reverse-flow theorem,

$$U \iint_S \frac{C^{zz}(x,y;\xi,\eta)}{C^{zz}(x,y;x,y)}\,\Delta\bar{p}_a(\xi,\eta)d\xi d\eta$$

$$= \iint_S \left[i\omega\bar{w}(\xi,\eta) + U\,\frac{\partial\bar{w}(\xi,\eta)}{\partial\xi} \right] \bar{r}(x,y;\xi,\eta)d\xi d\eta, \qquad (9\text{–}181)$$

where $\bar{r}(x,y;\xi,\eta)$ is amplitude distribution of pressure difference, over the same planform in reversed flow with velocity U, due to the vertical velocity given by Eq. (9–180). Obviously,

$$\left[i\omega\bar{w}(\xi,\eta) + U\,\frac{\partial\bar{w}(\xi,\eta)}{\partial\xi} \right]$$

in Eq. (9–181) is the vertical velocity distribution which produces $\Delta \bar{p}_a$. We can regard $\bar{r}(x,y;\xi,\eta)$ as a known function, providing we know enough about the bending influence function $C^{zz}(x,y;\xi,\eta)$.

Equation (9–181) permits us to rewrite Eq. (9–178):

$$\bar{w}(x,y) = \omega^2 \iint_S C^{zz}(x,y;\xi,\eta)\rho(\xi,\eta)\bar{w}(\xi,\eta)d\xi d\eta$$

$$- \iint_S \bar{V}(x,y;\xi,\eta)\left[i\frac{\omega}{U}\bar{w}(\xi,\eta) + \frac{\partial \bar{w}(\xi,\eta)}{\partial \xi} \right]d\xi d\eta, \quad (9\text{–}182)$$

where

$$\bar{V}(x,y;\xi,\eta) = C^{zz}(x,y;x,y)\bar{r}(x,y;\xi,\eta). \quad (9\text{–}183)$$

All the aerodynamic information in the problem is now contained in the single function $\bar{V}$, which is independent of the motion $\bar{w}$ but can be calculated for a series of (x,y) from point-by-point data on the influence function. Assuming one of the methods described in Chapter 7 has been employed to obtain $\bar{V}$, Eq. (9–182) can be replaced with a matrix equation by numerical integration of the two integrals on the right. The determinant of the coefficients of the unknown values of $\bar{w}(x,y)$ at a number of chordwise and spanwise stations is the flutter determinant. It can be solved by any of the procedures in the foregoing sections. Alternatively, a sort of matrix iteration may be used on the equations, resembling but more difficult than the one derived for free vibrations in Chapter 4.

Voss' formulation of the flutter problem can be shown to reduce to certain other solutions we have described above when various simplifications of the structural properties are permissible. The reader is referred to Ref. 9–32 for additional details.

9–10 Single-degree-of-freedom flutter. Although most flutter encountered in practice involves the interaction of two or more "degrees of freedom" or distinct types of structural deformation, it has been recognized for twenty-five years that this is not a necessary condition for aeroelastic instability. The majority of examples of so-called single-degrees-of-freedom flutter possess mainly academic interest. They are instructive to study, however, for the light that they throw on an ordinarily much more complicated phenomenon and on the nature of the unsteady aerodynamic forces which govern it. Also, certain cases relating to control surfaces at high subsonic and low supersonic speeds are definite hazards to be considered by the designer.

A comprehensive review and bibliography on single-degree-of-freedom flutter are given by Runyan, Cunningham, and Watkins (Ref. 9–33). All but one of the types they list consist of rotational oscillations of rigid wings and flap-type control surfaces. We re-order these types below, using the nature of the motion as a criterion of separation, and mention a few of the important features of each.

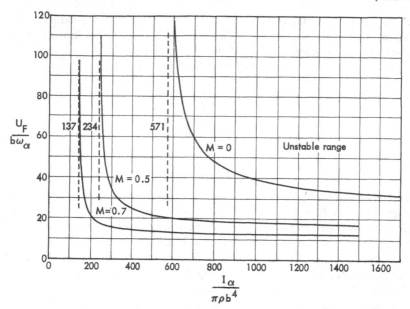

Fig. 9–21. Stability boundaries on a plot of dimensionless flutter speed $U_F/b\omega_\alpha$ vs. inertia ratio $I_\alpha/\pi\rho b^4$ for pitching rotation of an airfoil about its leading edge in subsonic flow.

(a) *Rotation of entire wings in pitch.* In Section 9–1 we discussed flutter of a rigid airfoil oscillating about a spanwise axis in two-dimensional incompressible liquid. Neglecting structural friction, we discovered that the criterion for neutral stability is that the out-of-phase component of the aerodynamic moment about the axis of rotation must vanish (Eq. 9–7b). At speeds below the critical U_F, the moment removes energy from the pitching oscillation and causes it to die out.

For effectively incompressible flow, this instability is exhibited only by unreasonably massive wings and has little practical significance. It turns out, on the other hand, that the effects of compressibility in the higher subsonic and low supersonic flight ranges are strongly destabilizing, although the axis must still lie ahead of the quarter-chord line for $M < 1$. Figure 9–21, which is reproduced from Runyan's study of the subsonic case (Ref. 9–34), illustrates this effect for an airfoil with the rotational axis located at the leading edge ($a = -1.0$). A plot is given of dimensionless flutter speed $U_F/b\omega_\alpha$ vs. the inertial parameter $I_\alpha/\pi\rho_\infty b^4$, which is the chief controlling factor, for Mach numbers of 0, 0.5, and 0.7. There exists in each case a minimum value of the inertial parameter below which instability is impossible; above the minimum the flutter speed rapidly falls off in a manner determined by M.

Small amounts of structural friction produce rapid rises in $U_F/b\omega_\alpha$ when the inertial parameter is large but do not influence the minimum $I_\alpha/\pi\rho_\infty b^4$.

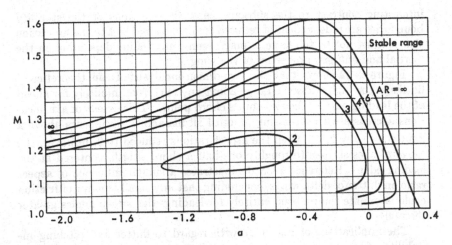

Fig. 9–22. Stability boundaries on a plot of Mach number vs. axis location a for pitching rotation of rectangular wings of various aspect ratios in supersonic flow.

No such friction is present, however, in the case of rigid-body rotation of the fuselage and vertical or horizontal tail about axes through the center of gravity of a complete airplane,[*] yet this is the situation where sufficiently high inertial parameters are encountered to give practical meaning to the curves of Fig. 9–21. We note that most tail surfaces have relatively low aspect ratios and produce highly three-dimensional flow. It is therefore encouraging that the influence of finite span causes marked reductions in the ranges of parameters within which unstable subsonic pitching moments occur. This fact can be deduced for incompressible flow by examining the theoretical and experimental curves of Ashley, Zartarian, and Neilson (Ref. 9–27).

The pitching instability of a two-dimensional airfoil has been demonstrated in a low-speed wind tunnel by Runyan (Ref. 9–34). His work suggests that the theory is quite unconservative for predicting flutter at lower values of $I_\alpha/\pi\rho_\infty b^4$ but improves in accuracy as the inertial parameter increases.

Single-degree-of-freedom pitching flutter can also occur on supersonic airfoils and wings, and again the reduction of aspect ratio is strongly stabilizing. The nature of the aerodynamic theory is such that it is convenient to calculate the axis location for neutral stability as a function of Mach number. Figure 9–22 reproduces a set of these curves from Ref. 9–33. The results are accurate up to the first power in reduced frequency, being based on the work of Watkins (Ref. 9–35) for rectangular planforms. In

---

[*] Negative values of a between -3 and -10 are typical of this problem, and Ref. 9–34 points out that instability is a definite possibility at higher subsonic Mach numbers throughout most of this range of axis locations.

interesting contrast to the subsonic case, the locations of the stability boundaries are independent of reduced frequency. Subject to the limitation that it must not be too large, k may assume any value, depending on the inertial and stiffness properties of the wing.

Figure 9–22 shows that the two-dimensional airfoil can theoretically flutter in the high transonic range about any axis from infinitely far ahead of the leading edge to the two-thirds chord point. This extent of axis locations decreases with increasing M until all instability disappears above a limit close to $\sqrt{2.5}$. When the aspect ratio is reduced to 2, however, the range of axes shrinks to roughly one-half a chord length, centered at the leading edge. Reference 9–33 demonstrates a similar influence of aspect ratio on flutter of delta wings, concluding that any small amount of friction could eliminate the problem entirely for leading-edge sweep angles greater than 60°.

The implications of Fig. 9–22 with regard to flutter and pitching instability of supersonic aircraft are not so alarming as they might seem at first glance. It appears that even a small amount of coupling with other degrees of freedom is very favorable. Thus plunging (vertical translation) motion stabilizes the rigid-body dynamics of a complete airplane or missile, and the possibility of bending deformation leads to bending-torsion flutter properties of elastic wings and tails which can be controlled just as in the subsonic case. The latter fact is apparent from a comparison between Fig. 9–22 and the results of Example 9–5, which were calculated using two-dimensional aerodynamic theory.

(b) *Rotation of trailing-edge control surfaces.* By analogy with Eq. (9–7b), the criterion for neutral stability of a flap-type control surface without hinge friction is that the out-of-phase component of the oscillatory aerodynamic hinge moment must vanish. That is,

$$\text{Im}\{M_\beta\} = 0. \tag{9–184}$$

This equation can be satisfied by control-surface configurations with and without aerodynamic balance in subsonic and low supersonic flow.

As shown by Runyan (Ref. 9–36), subsonic flutter of this type exhibits many of the same characteristics as pitching oscillation of the entire airfoil. The governing parameter is the inertial ratio $I_\beta/\pi\rho_\infty b^4$, I_β being the mass moment of inertia per unit span of the control about its hinge line. Corresponding to each hinge location and Mach number, there exists a minimum value of this parameter below which the system is completely stable. These minima are unreasonably high in incompressible flow but drop sharply with increasing Mach number. Thus, for an unbalanced surface with flap-chord-to-airfoil-chord ratio of 0.15, the minimum $I_\beta/\pi\rho_\infty b^4$ falls from 15 to 5.57 as M rises from 0 to 0.7.

The farther aft the hinge line is along the chord, the lower the minimum inertial parameter for flutter. This does not mean that small control surfaces are necessarily more critical, however, since I_β itself varies in proportion to the second or a higher power of the control-surface chord. For a given chord ratio, increasing the aerodynamic balance has a favorable effect, but this conclusion suffers from the uncertainty of all potential flow theory for balanced controls. A more reliable result is that hinge friction increases the flutter speed but does not alter the minimum value of the inertial parameter for instability.

Equation (9–184) is satisfied by unbalanced supersonic control surfaces below about $M = \sqrt{2}$. The location of the stability boundary on a plot of Mach number vs. hinge-line location is independent of k, as in the case of airfoil pitching, when aerodynamic terms are retained only up to the first power in k. Aerodynamic balance in compressible flow has not been studied because of the inadequacy of existing theory.

It is a safe surmise that the transition from two-dimensional flow to that over control surfaces of finite span will lead to greater stability. No quantitative calculations of this effect have as yet been published. In supersonic flight, however, a wing at zero angle of attack cannot affect the linearized flow over an oscillating control surface behind it. It is accordingly possible to interpret the curves of Fig. 9–22 as applying directly to the present problem. The values of M for neutral stability at $a = -1.0$ are the same as those for full-span flaps of various aspect ratios hinged at the leading edges.

(c) *Aileron buzz.* As mentioned in Section 9–5(c), problems of control-surface flutter are complicated at low transonic flight speeds by a curious instability which is usually referred to as aileron buzz. It appears impossible to predict this phenomenon quantitatively by classical aerodynamic theory, although the aileron's motion is observed to be simple harmonic and closely resembles that described in the preceding subsection. The best physical explanation of buzz associates it with the presence of normal shocks on the surface of the wing ahead of the hinge line. This concept also forms the basis of whatever empirical criteria are available for preventing it.

In subsonic flight above the critical Mach number of the airfoil section, shocks form on the upper and lower wing surfaces in positions shown schematically by Fig. 9–23(a). Since lift is being developed, the flow speeds are higher on top, the extent of local supersonic motion is greater, and the shock is stronger. The essential fact in accounting for buzz is that when the control surface is deflected downward [Fig. 9–23(b)], the flow accelerates further and the upper shock is intensified. This gives rise to lower pressure on top of the control surface, because the strong shock either separates the boundary layer behind itself or augments the existing separation. A restoring hinge moment is thus generated. Similarly, an upward-deflected control experiences a downward hinge moment due to low pressure caused by the intensified lower shock.

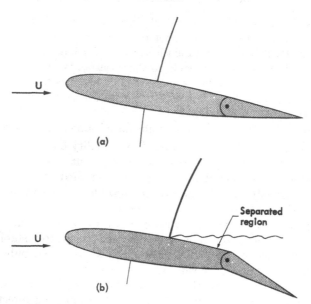

Fig. 9–23. The pattern of shocks on a wing in supercritical flight (a) with no control surface deflection, and (b) with downward control surface deflection.

If the control surface is moved back and forth slowly, this moment stays 180° out of phase with the position. The restoring tendency will lag a more rapid oscillation, however, since it takes time for signals set up by the deflection to be propagated forward to the shock through the nearly sonic flow. Moreover, a further lag arises from the delay of boundary layer separation and reattachment in response to the suddenly applied pressure changes. The combination of these effects can produce quite a large lagging phase angle in the restoring moment, so that it continues to act in the direction of rotation long after the control surface has passed the neutral position. Thus it may feed energy into the oscillation. Such a mechanism might easily lead to single-degree-of-freedom instability. The airloads probably do not depend linearly on the motion, so there may also be limit cycles and attendant variations of stability with amplitude.

The foregoing explanation of aileron buzz is supported by the evidence that it is never observed except when shocks are present both on the top and bottom of the wing, and that its onset is critically dependent on Mach number rather than on altitude or true airspeed.

One suggestion for avoiding buzz is to adjust the aileron's natural frequency so that the phase lag causing instability goes beyond 180°. Another is to build artificial friction into the control system and absorb the energy fed in. Neither of these schemes appears as practical or as attractive to the designer as providing the aileron with the maximum

possible irreversibility.   The ideal of complete irreversibility, which is
never achieved, would, of course, prevent external hinge moments from
disturbing the control system in any way.   Experience has shown, however,
that the amount which can be provided in practice apparently keeps the
amplitude of vibration so low that the boundary layers over the wing
remain unaffected and the lag mechanism is not set in motion.   Another
consideration which will be important in the future is that many aircraft
will be accelerating rapidly through transonic speeds.   The unsteadiness
of the associated flow may not give buzz sufficient time to develop in the
Mach number range where it is severe.   At least, it seems unlikely that
the instability will build up to dangerously large amplitudes before a more
stable supersonic flow pattern is established and suppresses it.

(d) *Single-degree-of-freedom bending flutter of swept-back wings.*   Although pure bending vibration actually involves an infinity of degrees of
freedom, it is logical to classify the instability of swept-back wings mentioned at the end of Section 9–8 as a single-degree-of-freedom flutter.
This is because it can still be predicted even when the structure is constrained to one generalized coordinate by assuming a single mode shape for
bending deformation.   If we adopt aerodynamic strip theory based on
sections normal to a spanwise reference axis of sweep angle Λ, the equation
of motion can be derived from Eq. (9–163) by assuming infinite torsional
stiffness ($\bar{\theta}_R = 0$):

$$\frac{\bar{w}_R}{\bar{b}_R}\left\{\left[1 - (1 + i\bar{g}_w)\left(\frac{\bar{\omega}_w}{\omega}\right)^2\right]\int_0^1\left(\frac{\bar{m}}{\pi\rho\bar{b}_R{}^2}\right)\bar{f}_w{}^2 d\bar{y}^*\right.$$

$$\left. + \int_0^1\left(\frac{\bar{b}}{\bar{b}_R}\right)^2 L_{hh}\bar{f}_w{}^2 d\bar{y}^* + \frac{\bar{b}_R}{l}\int_0^1\left(\frac{\bar{b}}{\bar{b}_R}\right)^3 L_{hh'}\bar{f}_w\frac{d\bar{f}_w}{d\bar{y}^*} d\bar{y}^*\right\} = 0. \quad (9\text{–}185)$$

As before, the superscript bar distinguishes wing properties as observed in
coordinates perpendicular and parallel to the swept spanwise direction.

We use Eqs. (7–73a) and (7–73b) to replace the aerodynamic coefficients
L_{hh} and $L_{hh'}$ by the more familiar quantity L_h, tabulated in Ref. 9–1.   If
L_h is separated into real and imaginary parts

$$L_h = L_{h\,\text{Re}} + iL_{h\,\text{Im}}, \quad (9\text{–}186)$$

we can then identify the real and imaginary components of the bracket in
Eq. (9–185).   These must be equated individually to zero in order to
obtain equations for the flutter condition.

$$\left[1 - \left(\frac{\bar{\omega}_w}{\omega}\right)^2\right]\int_0^1\left[\frac{\bar{m}}{\pi\rho\bar{b}_R{}^2}\right]\bar{f}_w{}^2 d\bar{y}^* + \int_0^1\left(\frac{\bar{b}}{\bar{b}_R}\right)^2 L_{h\,\text{Re}}\bar{f}_w{}^2 d\bar{y}^*$$

$$+ \frac{\tan\Lambda}{l/\bar{b}_R}\int_0^1\left(\frac{\bar{b}}{\bar{b}_R}\right)^3\frac{L_{h\,\text{Im}}}{k}\bar{f}_w\frac{d\bar{f}_w}{d\bar{y}^*} d\bar{y}^* = 0, \quad (9\text{–}187)$$

$$-\bar{g}_w \left(\frac{\bar{\omega}_w}{\omega}\right)^2 \int_0^1 \left[\frac{\bar{m}}{\pi\rho\bar{b}_R^2}\right] \bar{f}_w^2 d\bar{y}^* + \int_0^1 \left(\frac{\bar{b}}{\bar{b}_R}\right)^2 L_{h\,\mathrm{Im}}\bar{f}_w^2 d\bar{y}^*$$

$$- \frac{\tan\Lambda}{\bar{l}/\bar{b}_R} \int_0^1 \left(\frac{\bar{b}}{\bar{b}_R}\right)^3 \frac{L_{h\,\mathrm{Re}}}{k} \bar{f}_w \frac{d\bar{f}_w}{d\bar{y}^*}\, d\bar{y}^* = 0. \tag{9-188}$$

It is evident from Eqs. (9–187) and (9–188) that, in addition to the parameters ordinarily associated with the bending degree of freedom in flutter, the ratio

$$\gamma = \frac{\tan\Lambda}{\bar{l}/\bar{b}_R} \tag{9-189}$$

will have an important influence on the problem. This is especially true in view of the way γ enters Eq. (9–188), which, because the term containing $\bar{g}_w$ is very small, is essentially an aerodynamic relation fixing the value of reduced frequency k_R at which flutter occurs. Equation (9–187) is simply a compatibility condition; once k_R is known, it shows whether there is a real value of the frequency ω at which flutter can occur on a wing with a given mass distribution. If we omit $\bar{g}_w$ and study the last two terms in Eq. (9–188), we must recall that $L_{h\,\mathrm{Im}}$ is always negative, while $L_{h\,\mathrm{Re}}$ changes sign from positive to negative as k drops below about 0.35. This equation can therefore never be satisfied by a straight wing ($\Lambda = 0$), so that single-degree-of-freedom bending flutter is ruled out. On a swept-back wing ($\Lambda > 0$) Eq. (9–188) can be satisfied by some relatively low k_R for each value of γ. The larger γ, the larger the corresponding k_R, and therefore the lower the probable flutter speed (if any) for a given set of wing properties.

Cunningham (Ref. 9–28) has carried out extensive calculations* on uniform wings of constant chord. It turns out that for each γ and $\bar{g}_w = 0$ there is a positive value of the combination

$$\lambda = \frac{\bar{m}}{\pi\rho\bar{b}_R^2}\left[1 - \left(\frac{\bar{\omega}_w}{\omega}\right)^2\right] \tag{9-190}$$

which satisfies Eq. (9–188). Equation (9–190) also establishes the minimum value of density ratio $\bar{m}/\pi\rho\bar{b}_R^2$ for which flutter can occur, since this is equal to the value of λ when the flutter speed and frequency are so high that

$$\left(\frac{\bar{\omega}_w}{\omega}\right)^2 \ll 1.$$

---

* The aerodynamic strip theory used by Cunningham is taken from Ref. 9–25 and differs in minor respects from the one employed here (see Section 7–3). This discrepancy produces no change in the qualitative interpretation of the flutter calculations. Employing a still different type of strip theory, Broding (Ref. 9–37) has extended some of Cunningham's results to sonic and supersonic cross-flow Mach numbers.

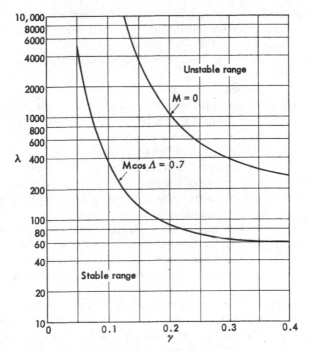

Fig. 9–24. Stability boundaries on a plot of inertial parameter λ vs. sweep-aspect-ratio parameter γ for single-degree-of-freedom bending flutter of uniform swept-back wings at two subsonic Mach numbers.

In Fig. 9–24 we reproduce Cunningham's curves of the stability boundary on a plot of λ vs. γ for cross-flow Mach numbers of 0 and 0.7. As in other examples, the influence of compressibility is strongly destabilizing. A practical swept-back wing would probably never have sufficiently high density ratio to make incompressible flutter a problem, whereas an unstable combination of parameters like $\lambda = 80$, $\gamma = 0.25$ might well be encountered on an aircraft in the transonic range. The actual flutter speed is easily computed in the dimensionless form $U_F/\bar{b}_R\bar{\omega}_w$. As might be expected, this quantity drops off quite rapidly with increasing sweep angle for a given density ratio and structural aspect ratio $\bar{l}/\bar{b}_R$. It approaches a vertical asymptote at some minimum Λ, below which bending flutter is impossible. Structural damping can produce large increases in U_F for higher values of density ratio.

The significance of the influence of parameter γ on bending flutter is that the problem becomes more acute the greater the sweepback and the *smaller* the structural aspect ratio. The latter condition is somewhat surprising, since lowering the aspect ratio of a straight wing improves flutter behavior; no good experimental confirmation is available. The theory shows that decreasing $\bar{l}/\bar{b}_R$ at constant sweep angle actually reduces $U_F/\bar{b}_R\bar{\omega}_w$. This does not mean that U_F itslf goes down, since $\bar{\omega}_w$ is roughly proportional to $1/\bar{l}^2$, but the opposite trend, that is, $U_F/\bar{b}_R\bar{\omega}_w$ increasing with decreasing $\bar{l}/\bar{b}_R$, would be characteristic of most bending-torsion flutter.

All these conclusions with regard to high-speed flutter of swept wings, especially those of moderate and low aspect ratios, are subject to the qualifying remarks made in the next to the last paragraph of Section 9–8. There is reason to suspect an excessive degree of conservatism in the theory, caused by neglecting finite-span effects in compressible flow.

One interesting interpretation of single-degree-of-freedom bending flutter is as a degenerate case of bending-torsion flutter when the frequency ratio $\bar{\omega}_w/\bar{\omega}_\theta$ is forced to vanish by $\bar{\omega}_\theta$ going to infinity. On straight wings, and also below the minimum value of γ for a swept wing with a given mass ratio, the dimensionless combination $U_F/\bar{b}_R\bar{\omega}_\theta$ remains finite during the limiting process, so that U_F becomes indefinitely large. Above the minimum γ, however, $U_F/\bar{b}_R\bar{\omega}_\theta$ approaches zero in such a way that $U_F/\bar{b}_R\bar{\omega}_w$ and U_F possess finite limits of the sort described in this section.

9–11 Certain other interesting types of flutter. (a) *Stall flutter and buffeting.* Stall flutter can be defined as any flutter of a lifting surface in which the airfoil sections are in stalled flow during at least part of each cycle of oscillation. Since aircraft rarely come close to the stall when flying at the maximum velocities and dynamic pressures for which they are designed, the problem is not a serious one on wings and tails. High-speed stalling of propellers and helicopter rotors is much more common, and the majority of past examples of stall flutter have been encountered on them. The subject has recently acquired added interest because of the occurrence of what may be stall flutter of compressor and turbine blades in turbojet engines operating off their design points.

It is important to distinguish between stall flutter and the buffeting of a wing due to forced excitation from its own wake. The former involves a nearly pure simple harmonic motion and is a true instability, in that the fluctuating flow and aerodynamic forces are generated by the oscillation and maintain it by supplying energy. In buffeting, the forces are little affected by the motion and would be present even though the structure were infinitely rigid; at normal flight Reynolds numbers these forces have a random frequency distribution. If the wing response has a predominant frequency, this is because the structure acts as a filter and singles out the excitation components in resonance with one or more of its natural aero-elastic modes. A similar forced excitation is encountered on poorly stream-lined structures, such as cylindrical smokestacks, at relatively low Reynolds numbers. This is a resonance with the natural frequency of the wake composed of alternating shed vortices (Kármán "vortex street"). It can usually be identified by the fact that the driving force has a reduced frequency, based on the cylinder's diameter or the body's thickness normal to the flow, of about 0.4π (Ref. 9–38).

The earliest systematic investigations of stall flutter on wind-tunnel models were those of Rauscher (Ref. 9–39) and Studer (Ref. 9–40). Their work showed that a substantial drop of critical speed can occur when the wing's mean angle of attack is increased from zero up to the neighborhood of maximum steady-state lift. This drop is especially marked when the reduced frequency of flutter is low, and Studer finds reductions in U_F as large as 60%. The motion was principally torsional oscillation in nearly all cases, even when potential-flow flutter of the same model involved a great deal of bending. Stall flutter speed appeared relatively insensitive to the amount of static unbalance of the model. A great many other parameters have a stronger influence on the phenomenon, notably mean angle of attack, Reynolds number, reduced frequency, location of the elastic axis, airfoil shape, and even amplitude of oscillation. The last two of these suggest mathematical nonlinearity, which complicates the matter of theoretical prediction immensely.

Light has been thrown on the problem of estimating stall flutter speed by experiments reported by Victory (Ref. 9–41) and Halfman, Johnson, and Haley (Ref. 9–42). The authors present time histories of forces and moments on rigid wings performing pitching and translational oscillations of known amplitude and frequency at high mean angles of attack. They also calculate the work per cycle done by these loads on the motion producing them and employ these data in approximate ways to determine U_F for flexible wings. The method proposed by Ref. 9–42 seems particularly accurate for this purpose. It is also shown in this report that very small amounts of bending motion can substantially alter the energy balance, so that flutter predictions based on the assumption of pure torsion will not always be successful.

Recent theoretical and experimental studies of stall flutter in cascades by Schnittger (Ref. 9–43) and Sisto (Ref. 9–51) point the way toward better understanding of aeroelastic instability in compressors and turbines. Schnittger makes the observation that the density ratios of solid blades are so high that the critical frequency and mode shape can often be taken to coincide with one of the normal modes of free vibration in torsion or bending. By employing an approximate, step-function representation of the nonlinear aerodynamic loads, he is able to predict flutter as a limit-cycle type of oscillation, a phenomenon which was also demonstrated theoretically and experimentally by Sisto's earlier work.

(b) *Supersonic panel flutter.* Since the time when sustained, and occasionally destructive, vibrations were observed on certain skin sections of the German V-2 rocket, interest has grown in the subject of panel flutter. At subsonic speeds, the only aeroelastic instability that might be exhibited by an elastic plate with a uniform stream on one side and air at rest on the other is a steady-state divergence or self-induced buckling. Apparently the skin thicknesses required before this can happen are so small that it is not a practical problem. In supersonic flight, however, a true flutter is possible. This was first studied theoretically by Miles (Ref. 9–44). A more recent paper of Shen (Ref. 9–45) gives a rigorous treatment of the two-dimensional case with a single panel supported at its front and rear edges.

The system that Shen analyzes consists of a simply supported, uniform plate of length L embedded in an otherwise plane surface. Air streams past one side normal to the supported edges, while the plate executes small lateral oscillations. The normal modes of vibration of the plate are sine waves. Since the upper and lower surfaces of a two-dimensional supersonic surface are independent, the pressure distribution is easily found from the theory of Garrick and Rubinow (Section 6–6 and Ref. 9–21). With the rth mode in the form

$$\xi_r \phi_r(x) = \xi_r \sin \frac{r \pi x}{L}, \qquad (9\text{–}190)$$

the amplitude distribution of vertical velocity to be substituted into the aerodynamic theory (Eq. 6–241) is

$$\bar{w}_a(x) e^{i \omega t} = \xi_r \left[i \omega \sin \frac{r \pi x}{L} + \frac{r \pi U}{L} \cos \frac{r \pi x}{L} \right]. \qquad (9\text{–}191)$$

The corresponding velocity potential and pressure are calculated by straightforward integrations.

The equations of simple harmonic motion read

$$[\omega_r{}^2 - \omega^2] M_r \xi_r = \sum_{s=1}^{\infty} Q_{rs} \xi_s, \qquad (9\text{–}192)$$

where M_r is a generalized mass proportional to the mass per unit span M_{panel} of the plate. Each generalized force is given by

$$Q_{rs} = - \int_0^L \frac{\Delta \bar{p}_r}{\xi_r} \sin \frac{s \pi x}{L} \, dx, \qquad (9\text{–}193)$$

where $\Delta \bar{p}_r$ is the amplitude of pressure difference $p - p_\infty$ over the plate produced by simple harmonic vibration in the rth mode. This implies that the pressure behind the plate is equal to atmospheric ambient. Shen finds that the Q_{rs} depend on tabulated functions similar to the f_0 of Ref. 9–21.

In Ref. 9–45 the system of Eqs. (9–192) is solved using only the first two normal modes. The resulting determinant depends in a complicated fashion on Mach number and reduced frequency, as with compressible-flow flutter of wings. For a given Mach number, however, the results can be expressed as a stability boundary on a plot of air-to-plate density ratio

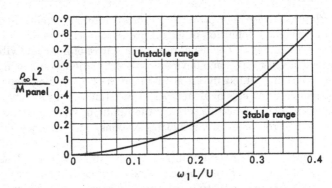

Fig. 9–25. Panel flutter stability boundary for $M = 2$ on a plot of density ratio $\rho_\infty L^2 / M_{panel}$ vs. reduced natural frequency $\omega_1 L/U$. The uniform panel is simply supported in two-dimensional flow.

$\rho_\infty L^2 / M_{panel}$ vs. the reduced frequency of the fundamental mode $\omega_1 L/U$. Figure 9–25 shows such a curve for $M = 2$.

As might be expected, the panel is stable when its mass is large enough relative to the air and it is sufficiently stiff. Curiously, for Mach number $\sqrt{2}$ Shen discovers a second region of instability corresponding to very high natural frequencies and low density ratios. There is reason to believe that inclusion of additional vibration modes in the analysis might eliminate this region, or at least that the stability within it is so nearly marginal that any very small amount of structural damping would prevent flutter. An interesting preliminary study in support of the latter statement is presented by Hedgepeth and Budiansky (Ref. 9–46). They have discovered that a long panel on a large number of equally spaced supports is amenable to especially simple mathematical treatment, and they examine its instabilities in considerable detail. Although there is some question about the quantitative value of the calculations of Ref. 9–46, they do show a behavior closely resembling those of Shen for higher Mach numbers.

There is a close relationship between static buckling and supersonic panel flutter, as proved, for example, in a paper by Fung (Ref. 9–47). Plates buckled in certain modes seem to be peculiarly susceptible to this sort of instability. By contrast, initial tension has a strong stabilizing influence. An appreciable pressure difference between the two sides can produce such tension, and this is believed to explain why panel flutter is so rarely encountered in practice on supersonic aircraft and missiles. The nature of the external flow gives rise to a pressure difference one way or the other on all but a very small part of the total skin area, and cabin or cockpit pressurization often produces tension in many of the largest unstiffened panels. Reduced skin stiffnesses caused by the high temperatures of flight faster than about $M = 2.5$ will evidently have an adverse effect on both the buckling and panel flutter problems.

(c) *Flutter of other than aircraft structures.* Aerodynamic instabilities which fall within our definition of flutter are encountered on many familiar objects besides aircraft. The usual result is a source of noise and vibration, or inconvenience because some device does not function as it was intended to. The problem is cured by an adjustment such as adding damping or stiffness, reducing an angle of attack below the stall, relocating a hinge line, etc. Common incidents of this sort are the luffing of sails, flutter of flags,* vibration of control vanes in ventilating systems, and wavering of poorly designed weathervanes.

One example of catastrophic flutter stands out, however. This is the unstable oscillation of very flexible suspension bridges, which was brought dramatically to public attention by the self-destruction of the Tacoma Narrows bridge in a 42-mph crosswind on November 7, 1940. Actually there has been a long history of similar undesirable oscillations and failures; the earliest modern records describe damage to the Menai Straights Bridge in 1826 and the Brighton Chain Pier in 1836. Most of the bridges involved had roadbeds that were abnormally weak in torsion. The dangerous vibration mode was first antisymmetrical torsion, which has a node at midspan and is restrained only poorly by the supporting cables unless special stays are installed to remedy the condition. Improved behavior has generally been achieved by increasing the torsional stiffness and modifying the cross-sectional shape to reduce the instability of the airloads which are developed.

The side trusses and roadbed of the average suspension bridge have many sharp corners and are anything but aerodynamically streamlined, so that the flow over them involves separation and the creation of complicated vortex patterns. Potential theory has little value for predicting the unsteady pressure distribution, and practically all research on flutter has been conducted with various sorts of wind-tunnel models. Two particularly extensive series of tests are reported by Farquharson *et al.* (Ref. 9–48) at the University of Washington, and by Frazer and Scruton (Ref. 9–49) in England. The former was initiated as a consequence of the Tacoma Narrows disaster; the authors describe experiments with 1/50-scale models of the original bridge and its replacement, along with many auxiliary investigations needed to achieve a satisfactory redesign. Reference 9–48 also reprints the results of tests made at California Institute of Technology by von Kármán and Dunn.

British experiments were principally related to the design of an aeroelastically satisfactory bridge over the River Severn estuary. Reference 9–49 contains a number of important fundamental contributions to the art of bridge modeling and testing. It is established that the influence of Reynolds number on such tests is not serious down to linear scale reductions of 1/100. Moreover, it is shown that most of the important information

---

* In this connection, see Thoma, Ref. 9–52.

about a given bridge can be obtained by measurements on properly supported typical-section models of large aspect ratio, so that it is not always necessary to reproduce the entire shape and structure to ensure safety from flutter.

No review of research on suspension-bridge instability would be complete without reference to the work of Steinman, who recounts his own extensive contributions on the problem in Ref. 9–50. This paper contains a list of empirical criteria which can quickly be applied to a proposed design as tests for freedom from flutter.

One conclusion of all studies on bridges is that the critical parameter is the combination U/Nb, where U is the wind speed, N the oscillation frequency of the mode under examination, and b some typical width such as the cable spacing. This quantity is proportional to the inverse of the reduced frequency familiar to flutter analysts. As is the case with compressor blades, suspension bridge structures appear to have such high relative densities that the frequencies of their aeroelastic modes do not change appreciably as the speed is brought up from zero. This means that N in the reduced velocity U/Nb can be replaced by its free-vibration counterpart, and each normal mode of the structure can be assigned a critical value of the parameter. This fact makes the prediction of the effects of stiffness or mass changes much easier than on typical airplane wings, and facilitates the use of sectional models, etc., in studying new and complicated designs.

CHAPTER 10

DYNAMIC RESPONSE PHENOMENA

10-1 Introduction. Static aeroelastic phenomena treated in Chapter 8 are characterized by the absence of the independent variable, time, whereas the flutter problem in Chapter 9 involves essentially a harmonic variation of displacement with time. In this chapter we turn our attention to a class of problems in which the motion of the system may vary in any arbitrary manner with time, depending upon the initial conditions, the character of the applied forces, and the response properties of the system.

In computing the dynamic response of an airplane in flight, it is often assumed that the structure is perfectly rigid. On the basis of this assumption, externally applied forces, such as air or ground loads, are put into equilibrium with aerodynamic and inertial forces which arise as a result of translational and rotational motion of the airplane as a rigid body, and these forces are considered to comprise the total force system acting on the airplane. This procedure introduces error in two ways. Deformation of the structure may induce additional aerodynamic forces which affect the over-all response of the airplane. Moreover, if the dynamic response is the direct result of rapidly applied external forces, the airplane is not only caused to translate and rotate, but structural vibrations are induced as well. The latter have a significant effect upon the internal stress distribution in the structure, and a somewhat lesser influence on the over-all response of the airplane.

Our interests in this chapter will be directed toward two ends. The first is that of studying the effect of aeroelastic phenomena on the stresses in the structure when the airplane has been subjected to rapidly applied external forces or to an arbitrarily specified set of initial conditions. The second is that of studying the effect of aeroelastic phenomena on the over-all response of the airplane when it is subjected to similar conditions.

Rapidly applied external forces may arise from many sources during the life of a modern airplane. One of the most important disturbing forces is that produced by gusts. In civil aircraft, gusts are the result of atmospheric turbulence, and in military aircraft the gust may be due not only to atmospheric turbulence but also to strong blast-induced shock waves. Impact forces during landing are a source of large rapidly and locally applied external forces, of interest primarily in connection with the internal stress aspects of dynamic response problems. In addition, rapidly varying external forces may be associated with catapulting, dropping of bombs, rapid maneuvering resulting from abrupt deflection of control surfaces, and turbulent wake behind wings, nacelles, or other components of the airplane.

The first part of this chapter contains a discussion of the equations of motion common to dynamic response problems of aircraft in flight and their solution. Subsequent sections contain application of these equations and their solutions to particular problems.

10–2 Equations of disturbed motion of an elastic airplane. A completely general formulation of the equations of motion of an elastic airplane in flight involves consideration of a three-dimensional elastic body with six translational and rotational degrees of freedom in addition to its many elastic degrees of freedom. Because of the complexity of such a system, we hypothesized for purposes of discussion in Chapter 3, a somewhat simpler mathematical model in which the structure is compressed into an elastic plate in the xy-plane. We adopt this model as a basis for our discussion in the present chapter, and although some generality is lost, this approach will allow us to demonstrate the basic principles in a concise manner.

Consider an unrestrained elastic airplane compressed into an elastic plate in the xy-plane, as illustrated by Fig. 10–1. Assume that the airplane is permitted freedom in vertical translation, pitch, and roll. A disturbing force per unit area, $F^D(x,y,t)$, with arbitrary spatial and time dependence, is assumed applied to the surface of the airplane. An initial steady-state condition of equilibrium is assumed, and the problem is one of calculating the time histories of deformations and displacements from this condition. As a result of the disturbed motion, aerodynamic pressures $\Delta p_a^M(x,y,t)$ (for example, aerodynamic damping forces) are brought into play. The equations of motion which result can be derived by applying the procedures explained in Section 3–7(b). For equilibrium of total forces along the z-axis:

$$\iint_S \ddot{w}(x,y,t)\rho\,dx\,dy = \iint_S (F^D + \Delta p_a^M)\,dx\,dy. \qquad (10\text{–}1)$$

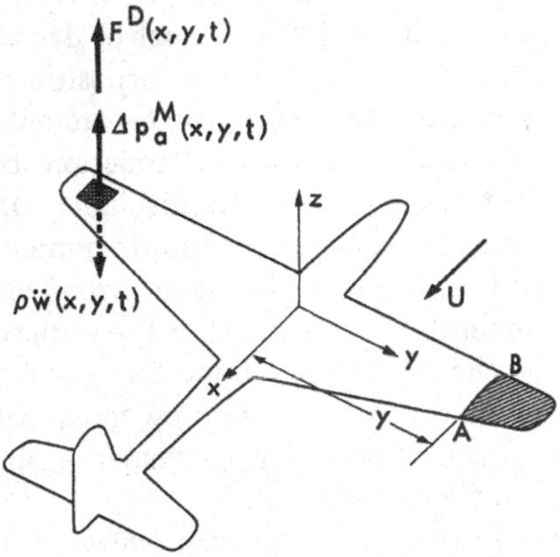

Fig. 10–1. Elastic airplane subjected to disturbing force.

For equilibrium of total moments about the x-axis,

$$\iint_S \ddot{w}(x,y,t)\rho y\,dx\,dy = \iint_S (F^D + \Delta p_a{}^M)y\,dx\,dy. \tag{10-2}$$

For equilibrium of total moments about the y-axis,

$$\iint_S \ddot{w}(x,y,t)\rho x\,dx\,dy = \iint_S (F^D + \Delta p_a{}^M)x\,dx\,dy. \tag{10-3}$$

For equilibrium of an element, we obtain from Eq. (2–55),

$$w(x,y,t) - w(0,0,t) - x\,\frac{\partial w(0,0,t)}{\partial x} - y\,\frac{\partial w(0,0,t)}{\partial y}$$

$$= \iint_S C(x,y;\xi,\eta)[F^D(\xi,\eta,t) + \Delta p_a{}^M(\xi,\eta,t) - \rho(\xi,\eta)\ddot{w}(\xi,\eta,t)]\,d\xi\,d\eta. \tag{10-4}$$

In this system of equations, the unknown quantity is $w(x,y,t)$, which represents the disturbed displacement of the elastic airplane from its original equilibrium configuration. The disturbance $F^D(x,y,t)$ is assumed to be given explicitly with respect to both its space and time variation. The aerodynamic pressures $\Delta p_a{}^M$ may, in general, depend upon the instantaneous values of the displacement, velocity, and acceleration of the airplane, as well as the past history of the motion, as explained in Chapter 5.

We have seen in Chapter 3 that solutions to the equations of motion in a practical application are invariably commenced by assuming that the space configuration of the deformed structure, which is actually an infinite degree-of-freedom system, can be approximated by an equivalent system with a finite number of degrees of freedom. Once this initial step is taken, the equations of motion of the continuous system are reduced to simultaneous ordinary differential equations with time the independent variable. Three methods of approach were described in Chapter 3 for reducing a continuous system to one with a finite number of degrees of freedom. In the first, the deformation is assumed as a superposition of a finite number of natural modes of the structure. These modes are obtained, as described in Chapter 4, from the eigenvalues and eigenfunctions of the homogeneous equations of motion of the unrestrained airplane. In the second, the deformation is taken as a superposition of a finite number of assumed mode shapes, and in the third, the deformation is described by the deflections at a number of discrete points on the surface of the structure. In discussing transient solutions of the equations of motion we shall apply the first method, in which we assume that the natural modes of the unrestrained structure are known. This is merely a convenience, and the reader may extend the principles to the other methods.

When the solution is expressed in terms of natural modes, we introduce into Eqs. (10–1), (10–2), (10–3), and (10–4) the following series:

$$w(x,y,t) = \sum_{i=1}^{n} \phi_i(x,y)\xi_i(t), \tag{10-5}$$

where the $\phi_i(x,y)$ are normalized natural mode shapes of the unrestrained airplane including rigid body modes and the $\xi_i(t)$ are normal coordinates. It has been shown in Section 3–7(b) that when the solution is expressed in terms of normal coordinates, the equations of motion, Eqs. (10–1), (10–2), (10–3), and (10–4), reduce to

$$M_i\ddot{\xi}_i + M_i\omega_i^2\xi_i = \Xi_i, \qquad (i = 1, 2, \cdots, n; \omega_1 = \omega_2 = \omega_3 = 0), \tag{10-6}*$$

with initial conditions

$$\xi_i(0) = \dot{\xi}_i(0) = 0, \tag{10-7}$$

and where

$$M_i = \iint_S \phi_i^2(x,y)\rho(x,y)dxdy, \tag{10-8}$$

$$\Xi_i = \iint_S [F^D(x,y,t) + \Delta p_a^M(x,y,t)]\phi_i(x,y)dxdy. \tag{10-9}$$

The generalized force Ξ_i is composed of a disturbance component, which is an explicit function of space and time, and a component which is dependent upon the motion of the system. The latter is, in general, a function of all the normal coordinates and their first and second time derivatives. It will be convenient to rewrite Eqs. (10–6) in the following form:

$$M_i\ddot{\xi}_i + M_i\omega_i^2\xi_i = \Xi_i^M + \Xi_i^D, \qquad (i=1, 2, \cdots, n; \omega_1=\omega_2=\omega_3=0), \tag{10-10}$$

where Ξ_i^M is the component of the generalized force resulting from the disturbed motion,

$$\Xi_i^M(\xi_1, \cdots, \xi_n; \dot{\xi}_1, \cdots, \dot{\xi}_n; \ddot{\xi}_1, \cdots, \ddot{\xi}_n) = \iint_S \Delta p_a^M(x,y,t)\phi_i(x,y)dxdy, \tag{10-11}$$

and Ξ_i^D is the component of the generalized force resulting from the disturbing force,

$$\Xi_i^D(t) = \iint_S F^D(x,y,t)\phi_i(x,y)dxdy. \tag{10-12}$$

10-3 Systems with prescribed time-dependent external forces. The method to be employed in the solution of the ordinary simultaneous differential equations of motion, Eqs. (10–10), depends on the nature of the generalized force. Let us first consider systems in which this force derives

---

* Equations (10–6), if desired, could be supplemented by additional equations of rigid-body motion representing yawing, fore and aft, and lateral motions. Since the airplane is assumed compressed into a thin plate in the xy-plane, these additional degrees of freedom do not affect the unrestrained modes of the elastic plate.

entirely from external forces of known time history, so that it is independent of the motion of the system, that is, $\Xi_i{}^M = 0$. The generalized force thus becomes simply a function of time, as follows:

$$\Xi_i = \Xi_i{}^D(t). \tag{10-13}$$

Since the choice of normal coordinates has already eliminated the elastic and inertial coupling in Eqs. (10–10), the equations of motion are independent and can be solved separately.

(a) *Solutions of the equations of motion.* Solutions to Eqs. (10–10) for $i = 1, 2, 3$ can be obtained by double integration. Since the system is assumed at rest prior to applying the disturbing force, we have simply

$$\xi_i(t) = \frac{1}{M_i} \int_0^t \int_0^t \Xi_i{}^D(t)dt dt, \qquad (i = 1, 2, 3). \tag{10-14}$$

Considering next solutions to Eqs. (10–10) for $i = 4, \cdots, n$, their form is identical to that of a simple undamped single-degree-of-freedom system consisting of a mass on a spring, and methods of solution of this simple system can be applied (Ref. 10–1). Let us assume first that the disturbing generalized force has the simple functional form shown by Fig. 10–2. This function is known as a unit step function (Ref. 10–1). It jumps instantaneously from zero to unity at $t = 0$, and remains indefinitely at the value unity. The unit step function is designated by the symbol $1(t)$. Substituting a unit step function for the disturbing force in Eq. (10–10) gives

$$M_i\ddot{\xi}_i + M_i\omega_i{}^2\xi_i = 1(t), \qquad (i = 4, 5, \cdots, n). \tag{10-15}$$

The general solution is

$$\xi_i = A_i \sin \omega_i t + B_i \cos \omega_i t + \frac{1}{M_i\omega_i{}^2}, \tag{10-16}$$

and applying initial conditions $\xi_i(0) = \dot{\xi}_i(0) = 0$, we obtain

$$A_i(t) = \frac{1}{M_i\omega_i{}^2} (1 - \cos \omega_i t), \qquad (i = 4, 5, \cdots, n). \tag{10-17}$$

The function designated by $A_i(t)$, representing the response to a unit step function $1(t)$, is called the *indicial admittance* (Ref. 10–1). Figure 10–3 illustrates graphically the form of the indicial admittance. It is apparent from Fig. 10–3 that the effect of applying a unit step force to a simple undamped single-degree-of-freedom system is to produce a response twice as large as that produced when the force is applied statically. The unit step is thus said to produce a *dynamic magnification* of two. When the force is applied in infinite time (static case) the dynamic magnification is one. For times of application between these extremes, the dynamic magnification lies between one and two.

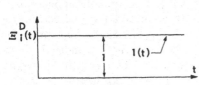

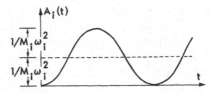

Fig. 10–2. Unit step function.

Fig. 10–3. Indicial admittance of undamped single-degree-of-freedom system.

It is also of interest to calculate the solution to Eqs. (10–10) for $n \geq 4$, for a disturbance force of the form of the Dirac delta-function introduced in Section 3–2(b), Eq. (3–79). This type of disturbance is referred to as a unit impulse function. In this case, Eq. (10–10) has the form

$$M_i \ddot{\xi}_i + M_i \omega_i^2 \xi_i = \delta(t), \qquad (i = 4, 5, \cdots, n), \qquad (10\text{–}18)$$

where $\delta(t)$ has the properties specified by Eqs. (3–80) and (3–81). Special remarks are necessary concerning the initial conditions to be assigned to Eq. (10–18). Since it is assumed that the system is at rest prior to application of the unit impulse, the displacement $\xi_i(0)$ is obviously zero at $t = 0$; however, the velocity condition requires a more searching analysis. If we integrate Eq. (10–18) with respect to time as follows:

$$\int_0^{\Delta t} d\dot{\xi}_i + \omega_i^2 \int_0^{\Delta t} \xi_i dt = \frac{1}{M_i} \int_0^{\Delta t} \delta(t) dt, \qquad (10\text{–}19)$$

we observe that when Eq. (3–80) is applied, and when Δt becomes vanishingly small, we obtain

$$\dot{\xi}_i(\Delta t \to 0) = \frac{1}{M_i}. \qquad (10\text{–}20)$$

Thus we conclude that the initial velocity condition is $\dot{\xi}_i(0) = 1/M_i$. Applying these initial conditions to Eq. (10–18), we obtain the following expression:

$$h_i(t) = \frac{1}{M_i \omega_i} \sin \omega_i t, \qquad (i = 4, 5, \cdots, n), \qquad (10\text{–}21)$$

where the symbol $h(t)$ is employed to designate response to a unit impulse.

By comparing Eqs. (10–17) and (10–21), we see that the response to a unit impulse can be obtained by taking the time derivative of the indicial admittance. That is,

$$h_i(t) = \frac{dA_i(t)}{dt}. \qquad (10\text{–}22)$$

This useful result is not only true for the simple system under discussion but holds also for any linear system, as demonstrated in Ref. 10–2.

Since the special forms $1(t)$ and $\delta(t)$ are not common in applied problems, let us consider the response of the ith mode to a generalized force of arbitrary time history designated by $\Xi_i{}^D(t)$. It is possible to express the response to an arbitrary applied force in terms of either the indicial admittance $A(t)$, or the response to a unit impulse, $h(t)$. This is accomplished by means of Duhamel's integral, derived in Appendix C, which states that when a system is initially at rest, the response can be expressed in terms of the indicial admittance by

$$\xi_i(t) = \int_0^t \Xi_i{}^D(\tau)A_i{}'(t-\tau)d\tau, \tag{10-23}$$

where

$$A_i{}'(t-\tau) = \frac{dA_i(t-\tau)}{d(t-\tau)} \tag{10-24}$$

or, alternatively, in terms of the response to a unit impulse by

$$\xi_i(t) = \int_0^t \Xi_i{}^D(\tau)h(t-\tau)d\tau. \tag{10-25}$$

Substituting Eq. (10–17) in Eq. (10–23) or Eq. (10–21) in Eq. (10–25) both lead to the following result for the response to an arbitrary force:

$$\xi_i(t) = \frac{1}{M_i\omega_i}\int_0^t \Xi_i{}^D(\tau)\sin\omega_i(t-\tau)d\tau, \qquad (i = 4, 5, \cdots, n). \tag{10-26}$$

We can now construct a result for the complete response of the unrestrained elastic airplane by substituting Eqs. (10–14) and (10–26) into Eq. (10–5), as follows:

$$w(x,y,t) = \sum_{i=1}^{3} \frac{\phi_i}{M_i}\int_0^t\int_0^t \Xi_i{}^D(\tau)dtdt + \sum_{i=4}^{n} \frac{\phi_i}{M_i\omega_i}\int_0^t \Xi_i{}^D(t)\sin\omega_i(t-\tau)d\tau. \tag{10-27}$$

In many cases, the disturbing forces are not known in analytical form, but are given simply as a plotted curve. In these circumstances, numerical or graphical techniques are required to evaluate Duhamel's integral. Numerous approximate methods have been devised for this purpose; for example, a useful graphical construction is described by Jones in Ref. 10–3.

EXAMPLE 10–1. To calculate the response of an unrestrained uniform wing attached to a central fuselage mass under the influence of a uniform force $f(t)$ per unit length, as illustrated by Fig. 10–4. Assume that the disturbing force has a time history of the form of a half sine wave.

Solution. The lateral response of a slender beam can be represented by a superposition of natural mode shapes of the freely vibrating beam, as follows:

$$w(\xi,t) = \sum_{i=1}^{n} \phi_i(\xi)\xi_i(t), \tag{a}$$

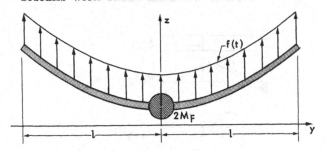

Fig. 10–4. Uniform wing attached to fuselage mass subjected to external forces.

where ξ is a dimensionless spanwise variable defined by $\xi = y/l$. In the present example, since the wing is loaded symmetrically, we need consider only symmetrical natural modes of the wing-fuselage mass system. These modes and their corresponding frequencies have already been computed in Example 3–2. Since the system is free, Eq. (a) contains a single-rigid body mode of zero frequency. Introducing the mode shapes obtained in Example 3–2, Eq. (a) becomes

$$w(\xi,t) = \xi_1(t) + \sum_{i=2}^{n} \phi_i(\xi)\xi_i(t),$$ (b)

where

$$\phi_i(\xi) = B_i[e^{-\Omega_i\xi} + K_1^{(i)} \cos \Omega_i\xi + K_2^{(i)} \sinh \Omega_i\xi + (1 - K_2^{(i)}) \sin \Omega_i\xi],$$
$$(i = 2, 3, \cdots, \infty).$$ (c)

The normalizing constant B_i and the constants $K_1^{(i)}$, $K_2^{(i)}$, and $K_3^{(i)}$ are defined in Example 3–2. The functions of time, $\xi_1(t)$, $\xi_2(t)$, $\cdots$, $\xi_n(t)$, are obtained as solutions to differential equations of the form of Eqs. (10–10). In each of these equations, the generalized mass M_i is equal to the half mass of the total system, since the mode shapes in Example 3–2 were normalized in this way.

$$\int_0^l \phi_1^2 dm = \int_0^l \phi_2^2 dm = \cdots = \int_0^l \phi_n^2 dm = M = M_F + ml,$$ (d)

where $2M_F$ is the fuselage mass and m is the mass per unit length of the wing. Thus, the equations of motion defining the normal coordinates are

$$M\ddot{\xi}_1(t) = \Xi_1(t),$$ (e)

$$M\ddot{\xi}_i(t) + M\omega_i^2\xi_i(t) = \Xi_i(t), \qquad (i = 2, 3, \cdots, n).$$ (f)

The generalized forces have the values

$$\Xi_i(t) = C_i f(t), \qquad (i = 1, 2, 3, \cdots, n),$$ (g)

where

$$C_1 = l,$$ (h)

$$C_i = \frac{A_i l}{\Omega_i}[K_1^{(i)} \sin \Omega_i + K_2^{(i)} \cosh \Omega_i - (1 - K_2^{(i)}) \cos \Omega_i - e^{-\Omega_i} + 2(1 - K_2^{(i)})],$$
$$(i = 2, 3, \cdots, n).$$ (i)

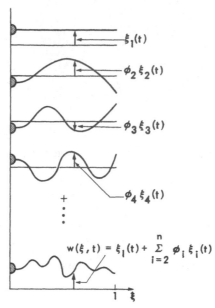

Fig. 10–5. Illustration of superposition of natural modes.

If the system is initially at rest prior to application of $f(t)$, it follows from Eq. (10–14) that the solution to Eq. (e) is given by

$$\xi_1(t) = \frac{C_1}{M} \int_0^t \int_0^t f(t)\,dt\,dt. \tag{j}$$

Similarly, it can be seen by Eq. (10–26) that the solution to Eq. (f) is

$$\xi_i(t) = \frac{C_i}{M\omega_i} \int_0^t f(\tau) \sin \omega_i(t - \tau)\,d\tau, \qquad (i = 2, 3, \cdots, n). \tag{k}$$

Thus the final result is given by

$$w(\xi,t) = \frac{C_1}{M} \int_0^t \int_0^t f(t)\,dt\,dt + \frac{1}{M} \sum_{i=2}^n \frac{C_i\phi_i(\xi)}{\omega_i} \int_0^t f(\tau) \sin \omega_i(t - \tau)\,d\tau, \tag{l}$$

where $\phi_i(\xi)$, C_1, and C_i are defined by Eqs. (c), (h), and (i) respectively.

The displacement of the rigid-body mode $\xi_1(t)$ can be interpreted physically as the displacement of the center of gravity of the vibrating system. It is apparent that Eq. (e) does indeed represent merely an application of Newton's second law to the system as a whole. The displacements $\xi_2(t), \cdots, \xi_n(t)$ represent physically the displacements of the deformation modes of the system with respect to its rigid-body displacement. This is illustrated schematically by Fig. 10–5, where we have represented the total displacement of the system as a displacement of a horizontal line through the center of gravity of the system plus a superposition of displacements of the system in its various vibration modes with respect to the line through the center of gravity. Let us now assume that the disturbing force $f(t)$ has the form of a half sine wave, as follows:

$$f(t) = \sin \Omega t, \qquad (0 \le t \le \pi/\Omega),$$

$$f(t) = 0, \qquad (t > \pi/\Omega). \tag{m}$$

Substituting these values into Eq. (l), we obtain, after integration,

$$w(\xi,t) = \frac{C_1}{M\Omega}\left(t - \frac{1}{\Omega}\sin \Omega t\right) + \frac{1}{M}\sum_{i=2}^{n}\frac{C_i\phi_i(\xi)}{(\omega_i^2 - \Omega^2)}\left(\sin \Omega t - \frac{\Omega}{\omega_i}\sin \omega_i t\right),$$
$$(0 \le t \le \pi/\Omega), \tag{n}$$

$$w(\xi,t) = \frac{C_1\pi}{M\Omega^2} + \frac{1}{M}\sum_{i=2}^{n}\frac{C_i\phi_i(\xi)}{(\omega_i^2 - \Omega^2)}\left\{\sin \Omega t + \sin \Omega\left(t - \frac{\pi}{\Omega}\right)\right.$$
$$\left. - \frac{\Omega}{\omega_i}\left[\sin \omega_i t + \sin \omega_i\left(t - \frac{\pi}{\Omega}\right)\right]\right\}, \qquad (t \ge \pi/\Omega). \tag{o}$$

(b) *Transient stresses in systems with prescribed time-dependent external forces.* It is always possible, in principle, to determine the stress distribution corresponding to a known deformation pattern. When the elastic characteristics of the structure are available in the form of influence coefficients or influence functions, the loads required to hold the structure in the given deformation configuration can be obtained. Using these loads, a stress analysis of the structure may be carried out in the conventional manner. It is also, of course, possible in principle to determine the stresses from the strains associated with the space curvature of the deformed structure.

Although straightforward in principle, these methods may not be the most direct, particularly if the deformation configuration has been computed in terms of displacements of normal modes. Two alternative methods for the calculation of transient stresses are now considered.

The mode displacement method. In the first it is observed that the principle of superposition applies, and the total stress pattern for any deformation of the structure is a superposition of the stress patterns corresponding to displacements of each of the normal modes. Therefore, the stress at a particular point p in the structure may be expressed by

$$\sigma_p = \sum_{i=4}^{n} A_p^{(i)}\xi_i, \tag{10-28}$$

where the quantities $A_p^{(i)}$ are constants that represent the stress at the point p due to a unit displacement in the ith normal mode.

The stress corresponding to a unit displacement in a normal mode is the same as the stress produced by the inertial loading associated with a free vibration of unit amplitude in that mode. This inertial loading may be expressed in terms of the mode shape and frequency of the mode, and the constant $A_p^{(i)}$ may be determined by a conventional stress analysis.

Let us consider in some detail, for example, the calculation of the shear and bending moment at the section A–B of the wing of the airplane illustrated in Fig. 10–1. It is assumed that the time histories of the normal coordinates $\xi_4(t), \cdots, \xi_n(t)$ are known. The inertial loading associated with vibration in the ith normal mode or, in other words, the external loading which must be applied to the structure to cause it to deform in the ith mode shape, with amplitude ξ_i, is simply

$$Z_i(x,y) = \phi_i(x,y)\rho(x,y)\omega_i^2\xi_i. \tag{10-29}$$

By superposition, the total external loading necessary to deform the structure into a shape which can be regarded as a superposition of n normal mode shapes is obtained by merely summing Eq. (10–29), as follows:

$$Z(x,y) = \sum_{i=4}^{n} \phi_i(x,y)\rho(x,y)\omega_i^2\xi_i. \tag{10-30}$$

The quantity $Z(x,y)$ can be regarded as an external loading, and a stress analysis of the structure may be carried out in a conventional manner. For example, the shear at section $A–B$, located at distance y from the airplane center line, as shown by Fig. 10–1, is obtained by integrating $Z(x,y)$ over the wing area A, outboard of section $A–B$.

$$S_{AB} = \iint_A Z(\xi,\eta)d\xi d\eta. \tag{10-31}$$

Inserting Eq. (10–30) results in

$$S_{AB} = \sum_{i=4}^{n} S_{AB}^{(i)}\xi_i, \tag{10-32}$$

where

$$S_{AB}^{(i)} = \omega_i^2 \iint_A \phi_i(\xi,\eta)\rho(\xi,\eta)d\xi d\eta. \tag{10-33}$$

Similarly, the bending moment at section $A–B$ is

$$M_{AB} = \iint_A Z(\xi,\eta)(\eta - y)d\xi d\eta. \tag{10-34}$$

Inserting Eq. (10–30),

$$M_{AB} = \sum_{i=4}^{n} M_{AB}^{(i)}\xi_i, \tag{10-35}$$

where

$$M_{AB}^{(i)} = \omega_i^2 \iint_A \phi_i(\xi,\eta)\rho(\xi,\eta)(\eta - y)d\xi d\eta. \tag{10-36}$$

Thus, it can be seen by Eqs. (10–32) and (10–35) that the shear and moment are linear functions of the *displacements* of the normal modes in the same form as expressed by Eq. (10–28); hence the origin of the term "mode displacement method."

The mode acceleration method. In the second method of approach, the stresses that would result if the system were restrained against vibrating are first calculated, and the additional stresses resulting from the vibratory motion are added. The stresses computed in the first step, assuming no vibration, are those ordinarily computed by the airplane designer when he assumes that the structure is rigid; they will be referred to as static stresses. The additional stresses resulting from the vibratory motion represent the dynamic overstress. This method (which is particularly suited, but not restricted, to undamped systems subjected to prescribed external forces) was first suggested by Williams (Ref. 10–4).

In the vibratory equations of motion of the normal modes, the vibratory inertial effects may be eliminated by setting $\ddot{\xi}_i = 0$. This yields the static mode displacements

$$\xi_{i\,(\text{static})} = \Xi_i{}^D(t)/M_i\omega_i^2, \qquad (i = 4, 5, \cdots, n). \qquad (10\text{–}37)$$

Substituting Eq. (10–37) into Eq. (10–28), the static stress at the point p becomes

$$\sigma_{p\,(\text{static})} = \sum_{i=4}^{n} A_p{}^{(i)} \left[\frac{\Xi_i{}^D(t)}{M_i\omega_i^2} \right]. \qquad (10\text{–}38)$$

It is apparent that the stresses given by Eq. (10–38), if all of the normal modes of vibration are taken into account, i.e., $n = \infty$, are the same as the stresses that are normally computed by the airplane designer when he makes the assumption of structural rigidity.

The total displacement of the ith vibration mode can be written as

$$\xi_i = \frac{\Xi_i{}^D(t)}{M_i\omega_i^2} - \frac{\ddot{\xi}_i}{\omega_i^2}, \qquad (i = 4, 5, \cdots, n), \qquad (10\text{–}39)$$

or when Eq. (10–37) is introduced,

$$\xi_i = \xi_{i\,(\text{static})} - \frac{\ddot{\xi}_i}{\omega_i^2}, \qquad (i = 4, 5, \cdots, n). \qquad (10\text{–}40)$$

Substituting Eq. (10–40) into Eq. (10–28) and introducing Eq. (10–38), the total stress at the point p becomes

$$\sigma_p = \sigma_{p\,(\text{static})} - \sum_{i=4}^{n} \frac{A_p{}^{(i)}}{\omega_i^2} \ddot{\xi}_i. \qquad (10\text{–}41)$$

This result indicates that the stress in the structure at any instant is equal to the static stress at that instant minus an additional stress that is a linear function of the accelerations of the normal modes; hence the origin of the term "mode acceleration method." It should be observed that the so-called "static stress" is really a pseudostatic stress, since it includes the effect of the inertial forces associated with the rigid-body acceleration. Equation (10–41) is especially useful, since $\sigma_{p\,(\text{static})}$ is the stress that the engineer ordinarily computes when the airplane structure is assumed rigid. Thus, the terms proportional to acceleration are merely correction terms which account for the dynamic overstress.

The accelerations of the generalized coordinates can be obtained by transposing Eq. (10–6):

$$\ddot{\xi}_i = \frac{1}{M_i} \left[\Xi_i{}^D - M_i\omega_i^2\xi_i \right] \qquad (10\text{–}42a)$$

and introducing Eq. (10–26):

$$\ddot{\xi}_i = \frac{1}{M_i}\left[\Xi_i^{D}(t) - \omega_i \int_0^t \Xi_i^{D}(\tau) \sin \omega_i(t - \tau)d\tau\right]. \qquad (10\text{–}42\text{b})$$

Equation (10–42b) can be reduced by integration by parts to

$$\ddot{\xi}_i = \frac{1}{M_i}\left[\Xi_i^{D}(0) \cos \omega_i t + \int_0^t \frac{d}{d\tau}\left(\Xi_i^{D}(\tau)\right) \cos \omega_i(t - \tau)d\tau\right]. \qquad (10\text{–}42\text{c})$$

Both the mode displacement and mode acceleration methods give equivalent results if sufficient vibration modes are considered. However, for practical purposes, only a relatively few modes can be taken. It is desirable, therefore, to use, whenever possible, the method giving the more rapid convergence. In Eq. (10–41), the components corresponding to the static application of the loads are all contained implicitly in the first term, while in Eq. (10–28) they are contained in each term of the series. This suggests a more rapid convergence of the series in Eq. (10–41). In fact, in most practical problems involving aircraft structures, significant vibration is excited in only the lowest two or three modes of the structure, and if Eq. (10–41) is employed, only two or three terms of the series need be included. On the other hand, if Eq. (10–28) is used, enough terms must be taken to ensure that all significant components of the static deformation have been included.

Example 10–2.   To calculate the transient distribution of shear and bending moment in the unrestrained uniform wing attached to a central fuselage mass analyzed in Example 10–1. Apply the mode displacement and mode acceleration methods and compare the results.

Solution. (a) *Mode displacement method.* According to the mode displacement, method, the shear * distribution is given by

$$S(\xi,t) = \sum_{i=2}^{n} S^{(i)}(\xi)\xi_i(t), \qquad (a)$$

where $S^{(i)}(\xi)$ is the distribution of shear corresponding to a unit displacement of the ith normal mode. Since the load distribution which produces a unit displacement of a vibration mode must be the same as the distribution of inertial forces associated with a free vibration of unit amplitude in that mode, we have

$$S^{(i)}(\xi) = ml\omega_i^2 \int_{\xi}^{1} \phi_i(\lambda)d\lambda, \qquad (i = 2, 3, \cdots, n). \qquad (b)$$

Similarly, the bending moment * is given by

$$M(\xi,t) = \sum_{i=2}^{n} M^{(i)}(\xi)\xi_i(t), \qquad (c)$$

---

* Upward shear is positive and bending moments which produce tension in the bottom fiber are positive.

where $M^{(i)}(\xi)$ is the distribution of bending moment corresponding to a unit displacement in the ith normal mode of vibration.

$$M^{(i)}(\xi) = l \int_\xi^1 S^{(i)}(\lambda)d\lambda, \qquad (i = 2, 3, \cdots, n). \tag{d}$$

The mode shapes of the uniform wing attached to the central fuselage mass are given in Example 3–2, as follows:

$$\phi_i(\xi) = B_i[e^{-\Omega_i\xi} + K_1^{(i)} \cos \Omega_i\xi + K_2^{(i)} \sinh \Omega_i\xi + (1 - K_2^{(i)}) \sin \Omega_i\xi],$$
$$(i = 2, 3, \cdots, n), \quad \text{(e)}$$

where the normalizing constant B_i and the constants $K_1^{(i)}$, $K_2^{(i)}$, and $K_3^{(i)}$ are defined in Example 3–2.

Substituting Eq. (e) into Eq. (b), $S^{(i)}(\xi)$ reduces to

$$S^{(i)}(\xi) = ml\omega_i^2[D_1^{(i)} + D_2^{(i)}(\xi)], \qquad (i = 2, 3, \cdots, n), \tag{f}$$

where

$$D_1^{(i)} = \frac{B_i}{\Omega_i}[K_1^{(i)} \sin \Omega_i + K_2^{(i)} \cosh \Omega_i - (1 - K_2^{(i)}) \cos \Omega_i - e^{-\Omega_i}],$$

$$D_2^{(i)} = -\frac{B_i}{\Omega_i}[K_1^{(i)} \sin \Omega_i\xi + K_2^{(i)} \cosh \Omega_i\xi - (1 - K_2^{(i)}) \cos \Omega_i\xi - e^{-\Omega_i\xi}].$$

Similarly $M^{(i)}(\xi)$ can be written as

$$M^{(i)}(\xi) = ml^2\omega_i^2[D_3^{(i)} + D_4^{(i)}(\xi)], \qquad (i = 2, 3, \cdots, n), \tag{g}$$

where

$$D_3^{(i)} = D_1^{(i)} - \frac{B_i}{\Omega_i^2}[K_2^{(i)} \sinh \Omega_i - K_1^{(i)} \cos \Omega_i - (1 - K_2^{(i)}) \sin \Omega_i + e^{-\Omega_i}],$$

$$D_4^{(i)} = -D_1^{(i)}\xi - \frac{B_i}{\Omega_i^2}[K_1^{(i)} \cos \Omega_i\xi + (1 - K_2^{(i)}) \sin \Omega_i\xi - K_2^{(i)} \sinh \Omega_i\xi - e^{-\Omega_i\xi}].$$

The spanwise distributions of shear and bending moment due to a disturbing force $f(t)$ in the form of a half sine wave (Expression (m), Example 10–1) are given, according to the mode displacement method, by

$$S(\xi,t) = ml \sum_{i=2}^n \omega_i^2[D_1^{(i)} + D_2^{(i)}(\xi)]\xi_i(t) \tag{h}$$

and

$$M(\xi,t) = ml^2 \sum_{i=2}^n \omega_i^2[D_3^{(i)} + D_4^{(i)}(\xi)]\xi_i(t), \tag{i}$$

where $D_1^{(i)}, \cdots, D_4^{(i)}$ are defined above and the $\xi_i(t)$ are given by

$$\xi_i(t) = \frac{C_i}{M(\omega_i^2 - \Omega^2)}\left[\sin \Omega t - \frac{\Omega}{\omega_i} \sin \omega_i t\right], \qquad (i = 2, 3, \cdots, n; 0 \le t \le \pi/\Omega),$$

$$\xi_i(t) = \frac{C_i}{M(\omega_i^2 - \Omega^2)}\left\{\sin \Omega t + \sin \Omega\left(t - \frac{\pi}{\Omega}\right) - \frac{\Omega}{\omega_i}\left[\sin \omega_i t + \sin \omega_i\left(t - \frac{\pi}{\Omega}\right)\right]\right\},$$
$$(i = 2, 3, \cdots, n; t \ge \pi/\Omega). \quad \text{(j)}$$

(b) *Mode acceleration method.* In the mode acceleration method we find the inertial loads on the beam under the assumption of rigidity and then add the inertial loads associated with the vibration modes. The expression for shear can thus be written, according to Eq. (10–41), as

$$S(\xi,t) = S_{\text{static}}(\xi,t) - \sum_{i=2}^{n} \frac{S^{(i)}(\xi)}{\omega_i^2} \ddot{\xi}_i(t), \qquad (k)$$

where $S_{\text{static}}(\xi,t)$ is the pseudostatic shear obtained when the beam is assumed rigid. It can easily be seen that

$$S_{\text{static}}(\xi,t) = (M - ml)(1 - \xi)\ddot{\xi}_1(t). \qquad (l)$$

where

$$M = M_F + ml.$$

The corresponding expressions for the bending moment are

$$M(\xi,t) = M_{\text{static}}(\xi,t) - \sum_{i=2}^{n} \frac{M^{(i)}(\xi)}{\omega_i^2} \ddot{\xi}_i(t) \qquad (m)$$

and

$$M_{\text{static}}(\xi,t) = l(M - ml) \frac{(1 - \xi)^2}{2} \ddot{\xi}_1(t). \qquad (n)$$

The mode acceleration for the first mode is simply

$$\ddot{\xi}_1(t) = \frac{C_1}{M} \sin \Omega t, \qquad (0 \le t \le \pi/\Omega),$$

$$\ddot{\xi}_1(t) = 0, \qquad (t \ge \pi/\Omega). \qquad (o)$$

The mode accelerations for modes above the first are obtained by differentiating Eqs. (j) twice with respect to time:

$$\ddot{\xi}_i(t) = -\frac{C_i\Omega^2}{M(\omega_i^2 - \Omega^2)} \left[\sin \Omega t - \frac{\omega_i}{\Omega} \sin \omega_i t \right], \qquad (i = 2, 3, \cdots, n; 0 \le t \le \pi/\Omega),$$

$$\ddot{\xi}_i(t) = -\frac{C_i\Omega^2}{M(\omega_i^2 - \Omega^2)} \left\{ \sin \Omega t + \sin \Omega \left(t - \frac{\pi}{\Omega} \right) \right.$$
$$\left. - \frac{\omega_i}{\Omega} \left[\sin \omega_i t + \sin \omega_i \left(t - \frac{\pi}{\Omega} \right) \right] \right\}, \qquad (i = 2, 3, \cdots, n; t \ge \pi/\Omega). \qquad (p)$$

Thus, the distributions of shear and bending moment, according to the mode acceleration method, are, respectively,

$$S(\xi,t) = (M - ml)(1 - \xi)\ddot{\xi}_1(t) - ml \sum_{i=2}^{n} [D_1^{(i)} + D_2^{(i)}(\xi)]\ddot{\xi}_i(t), \qquad (q)$$

$$M(\xi,t) = l(M - ml) \frac{(1 - \xi)^2}{2} \ddot{\xi}_1(t) - ml^2 \sum_{i=2}^{n} [D_3^{(i)} + D_4^{(i)}(\xi)]\ddot{\xi}_i(t), \qquad (r)$$

where the constants $D_1^{(i)}, \cdots, D_4^{(i)}$ are defined by Eqs. (f) and (g) and the accelerations $\ddot{\xi}_1(t)$ and $\ddot{\xi}_i(t)$ are given by Eqs. (o) and (p).

(c) *Numerical calculations and comparison of the methods.* The first step in a numerical application is to compute the constants $S^{(i)}(\xi)$ and $M^{(i)}(\xi)$, defining the shear and bending moment distributions, respectively, in the ith natural mode.

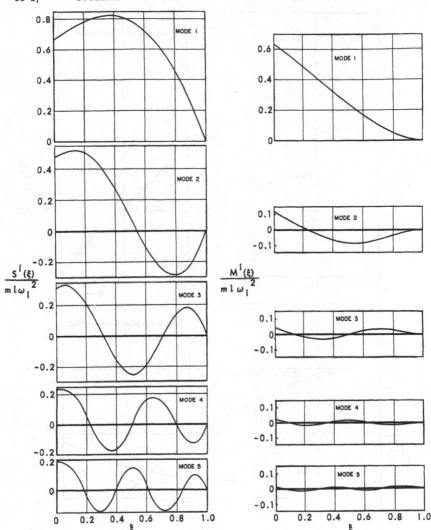

Fig. 10–6. Shear and bending moment distributions for unit displacements of the vibratory modes ($R = 1$).

Results for these constants calculated from Eqs. (f) and (g) for the first five symmetrical modes are given in Fig. 10–6 for the case where the fuselage and wing masses are equal ($R = 1$).

The results of calculations of maximum shear and maximum bending moment along the span by the two methods are shown in Figs. 10–7 and 10–8. These calculations have been made for the case of $T_I/T_1 = \frac{1}{2}$ where T_I is the time to peak value of the forcing function and T_1 is the period of the fundamental natural mode.

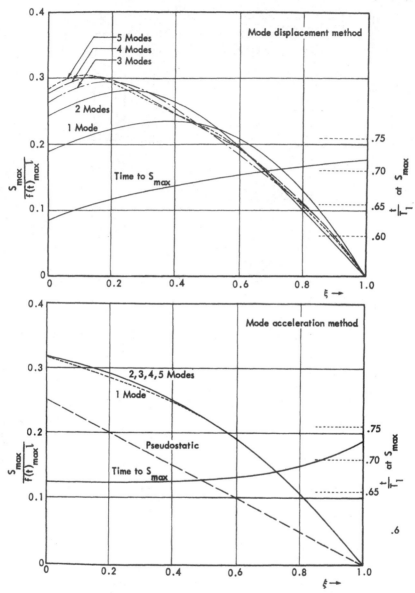

Fig. 10–7. Maximum shear distribution ($T_I/T_1 = \frac{1}{2}$, $R = 1$).

Figure 10–7 shows the spanwise variation of maximum shear, and the instants of time that the maximum shears occur, as determined by the two methods. It is apparent that the mode acceleration method gives much better convergence than the mode displacement method, particularly in the vicinity of the root, where the higher modes contribute very substantially to the shear when the mode

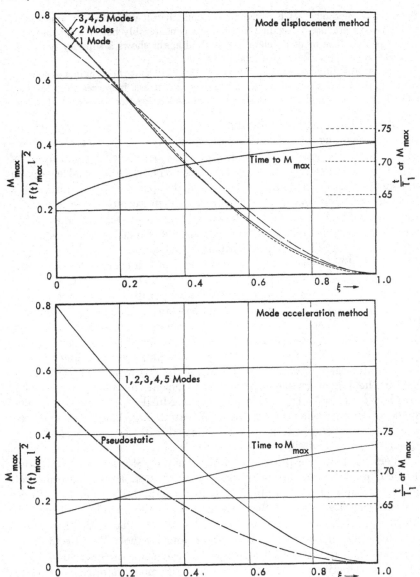

Fig. 10–8. Maximum bending moment distribution ($T_I/T_1 = \frac{1}{2}$, $R = 1$).

displacement method is used. In fact, it can be seen that even with five modes, satisfactory convergence of the shear is not obtained by the mode displacement method.

Figure 10–8 shows the spanwise variation of maximum bending moment, and the instants of time that the maximum bending moments occur, as determined by the two methods. It is apparent by comparison with the previous results that the convergence of bending moments is considerably better than the shear convergence. The mode acceleration method again shows a superiority in convergence over the mode displacement method.

It is interesting to observe in Figs. 10–7 and 10–8 that the dynamic overstress, which is the ratio of actual stress to pseudostatic stress, increases from the root to the tip. This illustrative example is discussed in greater detail in Ref. 10–5.

10–4 Transient stresses during landing. During landing, the vertical velocity or sinking speed of an airplane is quickly reduced to zero when the wheels strike the ground. This process is accomplished by a transfer of energy from kinetic energy of the sinking airplane to internal energy in the shock absorption system, where it is dissipated. If the airplane is completely "rigid" the shock struts must transform into potential energy and dissipate as heat all the kinetic energy of the sinking airplane. However, if the airplane is flexible, some of the kinetic energy is transformed into strain energy in the airplane structure. The fraction of kinetic energy thus diverted from the shock absorption system is a function of the vibration response characteristics of the structure and the properties of the shock struts. The properties of airplane shock struts, for practical reasons, are such that the vertical velocity of the airplane is brought to zero within a fraction of a second, and hence the forces applied to the structure through the shock strut change from zero to a maximum, also in a fraction of a second. This rapid change in velocity, or equally rapid application of force, excites the lower vibration modes of the structure. The excitation may, in the case of flexible airplanes, be sufficient to produce stresses of a destructive nature. Serious dynamic landing load problems were not encountered until recent years, during which time large landplanes and seaplanes have experienced structural failures due to dynamic stresses not considered in their design. Useful methods of analysis were available to designers by 1946 due primarily to the work of Stowell, Schwartz, and Houbolt, published in several reports, of which Ref. 10–6 is an example; Biot and Bisplinghoff, Ref. 10–7; Ramberg and McPherson, Ref. 10–8; and Williams, Ref. 10–4. Other investigators which include, for example, Wasserman, Ref. 10–9; Kroll and Levy, Ref. 10–10; Pian and Flomenhoft, Ref. 10–11; O'Brien and Pian, Ref. 10–12; and Houbolt, Ref. 10–13, have since extended the methods of analysis and provided experimental verification. The majority of the methods employ the modal approach, with the exception of those described by Kroll and Levy, Ref. 10–10, and Houbolt, Ref. 10–13. In their approach, the deformation of the structure is described by the displacements of discrete points on the structure, thereby circumventing the use of modes.

The problem of determining transient landing stresses may be approached in two different ways. For the landplane, the structure and the landing gear may be considered as a single system, and the actual force-displacement characteristics of the landing gear may be introduced into the theory. In this procedure, the dynamic stresses result from the sudden application of moving constraints imposed on the airplane during landing. Similarly, for the seaplane the elastic structure and the water surrounding the hull may be considered as interacting bodies. This method of approach involves inherent complexities, such as those resulting from the nonlinear properties of the landing gear of the landplane and the variable-mass effects of the water surrounding the hull of a seaplane. As a result, although the method is not precluded in the investigation of specific cases for research purposes, it is laborious in practical application.

In the second method of approach, in which stresses in the structure are considered to be caused by a landing impact force applied directly to the structure, it is assumed that the time history of the impact force may be investigated independently of the properties of the structure. In this way, the problem can be treated in two separate phases—a study of the landing forces and a study of the transient response of the structure under such forces. This procedure involves the assumption that a landing impact force may be defined in such a way that its time history is for all practical purposes independent of the response characteristics of the structure. It has been verified by analytical and model studies, such as, for example, those conducted by McPherson, Evans, and Levy (Ref. 10–14), that this procedure gives results sufficiently precise for engineering purposes in the case of landplanes. However, in the case of seaplanes, the hydrodynamic forces may be appreciably altered by structural response, especially in the case of tip float impact, and the first method of approach may be required.

It is usually assumed in treating the landing problem that in first approximation the damping due to aerodynamic forces may be neglected. The validity of this assumption has been investigated in model tests by Pian and Flomenhoft (Ref. 10–11). Wind-tunnel tests conducted on a simple drop test model indicate that although aerodynamic forces tend to damp out oscillations rapidly, they have little effect on the maximum transient stress that occurs at the first peak. Thus, when the result given by the first peak value of stress is of primary importance, neglect of aerodynamic damping is justified; however, if subsequent oscillations are also important, neglect of aerodynamic damping may introduce appreciable error.

We discuss the landing problem by applying the assumptions that the landing impact force is independent of structural response and that aerodynamic damping can be neglected. The problem thus reduces to one of computing the response of an undamped elastic structure subjected to external loads of prescribed time history. We assume that the airplane is making a landing approach with constant sinking speed, so that prior to landing it is in a state of static equilibrium under gravity and aerodynamic forces. During impact, the landing forces, assumed to be predetermined,

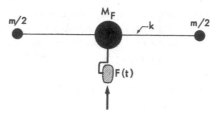

Fig. 10–9. Simplified airplane.

are applied directly to the structure. The resulting time histories of transient stresses are superimposed upon the stresses corresponding to the initial steady conditions.

We shall discuss first the case of a simplified airplane consisting of a fuselage and two tip tanks, with a single shock strut at its center line, as illustrated by Fig. 10–9. The wing is assumed weightless, with a stiffness k. Let us assume that upon contact with the ground, a force

$$F(t) = N_{\max}Mgf(t) \qquad (10\text{–}43)$$

is developed in the shock strut, where $N_{\max}$ is the landing load factor used in the conventional stress analysis and M is the total mass:

$$M = M_F + m, \qquad (10\text{–}44)$$

where M_F is the fuselage mass and $m/2$ is the mass of each tip tank. It can be easily seen that the simple model of Fig. 10–9 has a single rigid-body mode and a single deformation mode with the following properties:

Rigid-body mode $\begin{cases}\phi^{(1)}(0) = 1, \\ \phi^{(1)}(l) = 1,\end{cases}$ $(\omega_1 = 0)$. $\qquad (10\text{–}45)$

Deformation mode $\begin{cases}\phi^{(2)}(0) = -\dfrac{m}{M_F}, \\ \phi^{(2)}(l) = 1,\end{cases}$ $(\omega_2 = \sqrt{kM/M_Fm})$. $\quad (10\text{–}46)$

If we wish, for example, to compute the maximum shear in the wing just inboard of the tip tank, we can apply the mode acceleration method* (Eq. 10–41), as follows:

$$S(t) = S_{\text{static}} - \frac{S^{(2)}}{\omega_2{}^2}\ddot{\xi}_2(t), \qquad (10\text{–}47)$$

which reduces to

$$S(t) = -\frac{m}{2}(\ddot{\xi}_1(t) + \ddot{\xi}_2(t)). \qquad (10\text{–}48)$$

---

* The mode acceleration and mode displacement methods give identical results in this example. This is true of any system if all the modes of vibration are included.

Positive shear is assumed upward in Eqs. (10–47) and (10–48). The accelerations of the generalized coordinates are obtained from

$$\ddot{\xi}_1 = N_{\max} g f(t), \tag{10–49}$$

$$\ddot{\xi}_2 + \omega_2{}^2 \xi_2 = -N_{\max} g f(t). \tag{10–50}$$

Introducing solutions to Eqs. (10–49) and (10–50) for $\ddot{\xi}_1$ and $\ddot{\xi}_2$, respectively, into Eq. (10–48), we obtain

$$S(t) = \frac{mg}{2} N_{\max} \left[-f(t) + f(0) \cos \omega_2 t + \int_0^t \frac{df(\tau)}{d\tau} \cos \omega_2(t-\tau) d\tau \right]. \tag{10–51}$$

Assuming that the force in the shock strut has the following time variation:

$$\begin{aligned} f(t) &= \sin \Omega t, &(0 \le t \le \pi/\Omega), \\ f(t) &= 0, &(t \ge \pi/\Omega), \end{aligned} \tag{10–52}$$

we obtain an explicit result for the wing shear:

$$S(t) = \frac{mg}{2} N_{\max} \left[-\sin \Omega t + \frac{\Omega}{\omega_2} \frac{\sin \omega_2 t - \dfrac{\Omega}{\omega_2} \sin \Omega t}{1 - \left(\dfrac{\Omega}{\omega_2}\right)^2} \right],$$
$$(0 \le t \le \pi/\Omega). \tag{10–53}$$

Since, in a practical case, the maximum shear is usually reached prior to $t = \pi/\Omega$, Eq. (10–53) suffices to give the desired result. Figure 10–10 illustrates the nature of Eq. (10–53). It compares the shear force on the wing of a rigid structure with that obtained from Eq. (10–53). The total shear force is obtained by superimposing a $1g$ shear load on Eq. (10–53). The effect of elasticity shown in Fig. 10–10 is to amplify the peak value of the shear force and to delay its time of occurrence.

Let us consider the more general case of an airplane executing a two-wheel symmetrical landing, as illustrated by Fig. 10–11. The airplane, as a whole, is assumed free to translate vertically and the wing is permitted both bending and twisting action. The problem is one of computing the time histories and spanwise distributions of bending moment, torsional moment, and shear along the wing due to vertical landing reactions $\frac{1}{2}F(t)$ at each landing gear. For simplicity, the drag forces are not included here. We assume that the elastic axis of the wing is a straight line perpendicular to the plane of symmetry, and that the wing is rigid in the chordwise direction.

Applying the mode acceleration method, the time history of the stress in the wing can be written in the following form:

$$\sigma = B_1 f(t) + \sum_{i=2}^{n} B_i K_i(t), \tag{10–54}$$

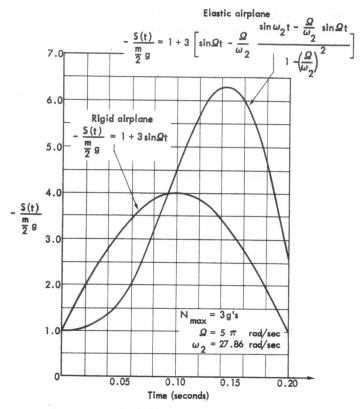

Elastic airplane

$$-\frac{S(t)}{\frac{m}{2}g} = 1 + 3\left[\sin\Omega t - \frac{\Omega}{\omega_2} \frac{\sin\omega_2 t - \frac{\Omega}{\omega_2}\sin\Omega t}{1 - \left(\frac{\Omega}{\omega_2}\right)^2}\right]$$

Rigid airplane

$$-\frac{S(t)}{\frac{m}{2}g} = 1 + 3\sin\Omega t$$

$N_{max} = 3\text{g's}$

$\Omega = 5\pi$ rad/sec

$\omega_2 = 27.86$ rad/sec

$-\dfrac{S(t)}{\frac{m}{2}g}$

Time (seconds)

Fig. 10–10. Wing shear during landing.

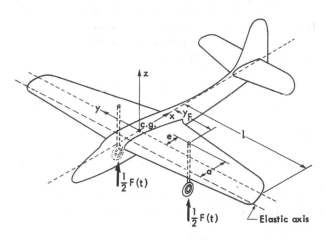

Fig. 10–11. Two-wheel symmetrical landing condition.

where
$$F(t) = N_{max}Mgf(t), \qquad (10\text{–}55)$$

$$K_i(t) = \int_0^t \cos \omega_i(t - \tau) \frac{df(\tau)}{d\tau} d\tau, \qquad (i = 2, 3, \cdots, n). \quad (10\text{–}56)$$

N_{max} is the landing load factor used in the conventional stress analysis, and M is the total mass of the airplane. The coefficients B_1 and B_i are appropriately determined depending upon the characteristics of the airplane. For example, for the airplane under consideration, we have the following values of the coefficients B_1 and B_i:

Shear distribution, $\sigma = S(y,t)$; *positive upward:*

$$B_1 = -N_{max}g \int_y^l \left\{ m + \frac{eM}{I_y} [S_\alpha - m(a-e)] \right\} dy, \qquad (y > y_F),$$

$$B_1 = -N_{max}g \int_y^l \left\{ m + \frac{eM}{I_y} [S_\alpha - m(a-e)] \right\} dy + \tfrac{1}{2}MgN_{max}, \qquad (y < y_F),$$

$$B_i = -N_{max}g\phi_F{}^{(i)} \int_y^l [mh^{(i)} + S_\alpha \alpha^{(i)}] dy, \qquad (i = 2, 3, \cdots, n). \quad (10\text{–}57)$$

Bending moment distribution, $\sigma = M(y,t)$; *positive when bottom surface is in tension:*

$$B_1 = -N_{max}g \int_l^y \int_l^y \left\{ m + \frac{eM}{I_y} [S_\alpha - m(a-e)] \right\} dy dy, \qquad (y > y_F),$$

$$B_1 = \tfrac{1}{2}(y_F - y)MN_{max}g - N_{max}g \int_l^y \int_l^y \left\{ m + \frac{eM}{I_y} [S_\alpha - m(a-e)] \right\} dy dy,$$
$$(y < y_F),$$

$$B_i = -N_{max}g\phi_F{}^{(i)} \int_l^y \int_l^y (mh^{(i)} + S_\alpha \alpha^{(i)}) dy dy, \qquad (i = 2, 3, \cdots, n).$$
$$(10\text{–}58)$$

Torsional moment distribution, $\sigma = T(y,t)$; *positive when nose up:*

$$B_1 = N_{max}g \int_y^l \left\{ \frac{eM}{I_y} [I_{ea} - S_\alpha(a-e)] + S_\alpha \right\} dy, \qquad (y > y_F),$$

$$B_1 = N_{max}g \int_y^l \left\{ \frac{eM}{I_y} [I_{ea} - S_\alpha(a-e)] + S_\alpha \right\} dy - \tfrac{1}{2}aMN_{max}g,$$
$$(y < y_F),$$

$$B_i = N_{max}g\phi_F{}^{(i)} \int_y^l (I_{ea}\alpha^{(i)} + S_\alpha h^{(i)}) dy, \qquad (i = 2, 3, \cdots, n). \quad (10\text{–}59)$$

Fig. 10–12.   Photograph of drop test model.

In the above formulas:

e = Moment arm of the vertical reaction about the c.g. of the air-plane; positive when the load is behind the c.g.

a = Moment arm of the vertical reaction about the elastic axis of the wing; positive when the load is behind the elastic axis.

I_y = Pitching moment of inertia of airplane about the y-axis.

y_F = Distance of the landing gear from the airplane center line.

$h^{(i)}(y)$ = Function describing the mode shape of bending of the elastic axis for the ith normalized symmetrical mode.

$\alpha^{(i)}(y)$ = Function describing the mode shape of twisting about the elastic axis for the ith normalized symmetrical mode.

$\phi_F^{(i)}$ = Amplitude of the ith normalized symmetrical mode at the point in the wing structure where the landing gear is located.

$I_{ea}(y)$ = Moment of inertia about the elastic axis per unit wing span.

$S_\alpha(y)$ = Static unbalance about the elastic axis per unit wing span; positive when the c.g. is behind the elastic axis.

$m(y)$ = Mass of wing per unit span.

The mode shapes are assumed normalized such that

$$M = \iint_S [\phi]^2 \rho \, dx \, dy = \int_{-l}^{+l} (mh^2 + I_{ea}\alpha^2 + 2S_\alpha h\alpha) \, dy. \quad (10\text{–}60)$$

When the constants B_1 and B_i have been obtained, the procedure reduces to an evaluation of Eq. (10–54).   This entails a calculation of the time

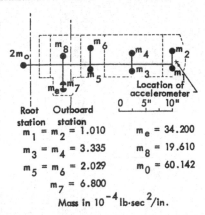

Fig. 10-13. Dynamic model of the model wing.

histories defined by the integral $K_i(t)$ of Eq. (10-56). The latter computation may be somewhat tedious if $f(t)$ is not known in analytical form, and numerical or graphical methods are required.

In order to confirm experimentally the method described above, it has been applied to the model illustrated by the photograph of Fig. 10-12. The computations were carried out for the conditions resulting from a $2\frac{1}{2}$-inch drop of the model at zero forward speed, giving a maximum impact force of 45.5 lb.

For simplicity in analysis, the wing is replaced by the dynamic model shown in Fig. 10-13. The wing is divided into five sections. The mass of each section is considered concentrated in the spanwise direction at its center of gravity, except for the center section mass, which is included with that of the fuselage. In the chordwise direction, each section is replaced by a pair of concentrated masses. The masses in each pair are equal in the three outboard sections and are located at equal distances from the chordwise center of gravity location. In the engine section, the masses are distributed so that the location of the front mass coincides with the center of gravity of the lead weight used for simulating the overhanging engine. Since this lead weight is flexibly supported, the front mass of the engine section is divided into two elastically connected masses.

Employing three deformation modes, the transient stress at an arbitrary point in the model is given by

$$\sigma(t) = B_1 f(t) + B_2 K_2(t) + B_3 K_3(t) + B_4 K_4(t). \qquad (10\text{-}61)$$

In order to obtain a better form for comparison of experimental and analytical results, the stresses are computed in terms of nondimensional quantities. The calculated maximum stresses corresponding to the condition that the structure is restrained against vibratory motion are chosen as the

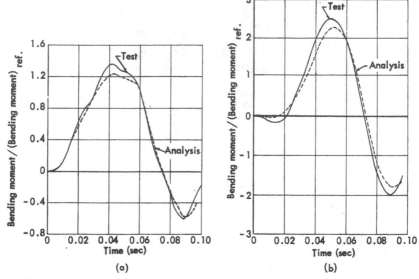

Fig. 10–14. (a) Time history of wing root bending moment. (b) Time history of outboard bending moment.

reference quantities. When the results are presented in terms of nondimensional quantities, the following equation applies:

$$\frac{\sigma(t)}{\sigma_{ref.}} \cdot \frac{\sigma(t)}{B_1} = f(t) + \frac{B_2}{B_1} K_2(t) + \frac{B_3}{B_1} K_3(t) + \frac{B_4}{B_1} K_4(t). \quad (10\text{–}62)$$

The impact force-time curve used in the analysis was taken from an experimentally determined record of the time history of the shock strut load obtained when the strut was dropped 2.5 inches, with the elastic wing replaced by a rigid mass.

Numerical results have been obtained by applying Eq. (10–62) to the calculation of bending moments at the wing root station, and at a station just outboard of the engine station (cf. Fig. 10–13). The constants B_1, B_2, B_3, and B_4 are computed from Eqs. (10–58) and (10–59). Calculated time histories of the nondimensional bending moments at the two stations are compared with experimental results in Figs. 10–14(a) and 10–14(b). It can be observed that the use of three modes yields an acceptable result for bending moments. It is interesting to notice that, whereas the dynamic overstress at the root is only approximately 20%, it is over 100% at the outboard station. Wing-tip acceleration is often of interest in practical calculations. Equation (10–62) can be applied to the calculation of the acceleration at a specified point p on the wing by using the following constants:

$$B_1 = N_{max}g, \qquad B_i = N_{max}g\phi_F{}^{(i)}\phi_p{}^{(i)}, \qquad (10\text{–}63)$$

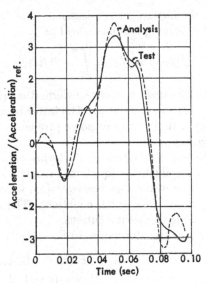

Fig. 10-15.   Time history of wing-tip acceleration.

where $\phi_F{}^{(i)}$ and $\phi_p{}^{(i)}$ are the amplitudes of the ith normalized symmetrical mode at the landing gear location and at the point p, respectively.   Comparison of theory with experimental wing-tip acceleration is illustrated by Fig. 10-15.   Figure 10-13 shows the position of the accelerometer on the model.   Although the agreement here is not as close as the bending moment agreement, the trend of the experiment is satisfactorily predicted.   Further details of this comparison between experiment and theory may be found in Ref. 10-15.

10-5 Systems with external forces depending upon the motion.   We turn our attention next to the processes required for transient solutions when the forces due to motion are included.   In aeroelastic problems, these forces arise primarily from aerodynamic sources.   The problem is one of solving for transient solutions of n simultaneous ordinary differential equations.   If we assume that the disturbance generalized forces can be expressed as

$$\Xi_i{}^D(t) = C_i f(t), \tag{10-64}$$

where $f(t)$ is a nondimensional function representing the time variation of the disturbing force, the differential equations are

$$M_i \ddot{\xi}_i + M_i \omega_i^2 \xi_i = \Xi_i{}^M(\xi_1, \cdots, \xi_n; \dot{\xi}_1, \cdots, \dot{\xi}_n; \ddot{\xi}_1, \cdots, \ddot{\xi}_n) + C_i f(t),$$
$$(i = 1, 2, \cdots, n), \tag{10-65}$$

with initial conditions $\xi_i(0) = \dot{\xi}_i(0) = 0$.

(a) *Solution by the Laplace transformation.* The Laplace transformation of a function $f(t)$ is an operation defined as

$$\mathcal{L}\{f(t)\} = \bar{f}(p) = \int_0^\infty e^{-pt}f(t)dt \qquad (10\text{-}66)$$

where $\mathcal{L}\{\quad\}$ denotes the operation of transformation (Ref. 10–16). This yields a function $\bar{f}(p)$ of the variable p. The latter is called the Laplacian operator. In dealing with dynamic response problems, the function $f(t)$ is usually zero for negative values of t.

By applying this transformation to all terms of a linear differential equation, we obtain an algebraic equation. Applying an inverse transformation to the solution of the algebraic equation, we obtain in final form the solution of the original differential equation. The method is particularly useful in calculating transient solutions.

Applying the Laplace transformation to Eq. (10–65), we obtain

$$M_i p^2 \bar{\xi}_i(p) + M_i \omega_i^2 \bar{\xi}_i(p) = \mathcal{L}\{\Xi_i{}^M(\xi_1,\cdots,\xi_n;\dot{\xi}_1,\cdots,\dot{\xi}_n;\ddot{\xi}_1,\cdots,\ddot{\xi}_n) + C_i f(t)\},$$

$$(i = 1, 2, \cdots, n), \quad (10\text{-}67)$$

where

$$\bar{\xi}_i(p) = \mathcal{L}\{\xi_i(t)\}.$$

As a consequence of the assumption of linearity of the applied forces, Eq. (10–67) can be put in the form

$$\{(p^2 + \omega_i{}^2)M_i - \Xi_{ii}{}^M(p)\}\bar{\xi}_i(p) - \sum_{j \neq i}^n \Xi_{ij}{}^M(p)\bar{\xi}_j(p) = C_i\bar{f}(p),$$

$$(i = 1, 2, \cdots, n), \quad (10\text{-}68)$$

where $\Xi_{ii}{}^M\bar{\xi}_i(p)$ is the transformed component of the generalized force $\Xi_i{}^M$ due to the motion in the coordinate ξ_i; $\Xi_{ij}{}^M(p)\bar{\xi}_j(p)$, $(j \neq i)$, are the transformed coordinates of the generalized force $\Xi_i{}^M$ due to the motion in coordinates other than ξ_i; and $\bar{f}(p)$ is the transform of $f(t)$.

Equation (10–68) represents a set of n simultaneous linear algebraic equations in the unknown functions $\bar{\xi}_1(p), \cdots, \bar{\xi}_n(p)$.

$$C_{11}\bar{\xi}_1 + C_{12}\bar{\xi}_2 + \cdots + C_{1n}\bar{\xi}_n = C_1\bar{f},$$
$$C_{21}\bar{\xi}_1 + C_{22}\bar{\xi}_2 + \cdots + C_{2n}\bar{\xi}_n = C_2\bar{f},$$
$$\cdot$$
$$\cdot \qquad (10\text{-}69)$$
$$\cdot$$
$$C_{n1}\bar{\xi}_1 + C_{n2}\bar{\xi}_2 + \cdots + C_{nn}\bar{\xi}_n = C_n\bar{f}.$$

Solving Eqs. (10–69) for $\bar{\xi}_i(p)$, we obtain

$$\bar{\xi}_i(p) = \sum_{j=1}^n \frac{a_{ij}(p)}{N(p)} C_j\bar{f}(p), \qquad (i = 1, 2, \cdots, n), \qquad (10\text{-}70)$$

where $N(p)$ is the determinant of the matrix of coefficients of $\xi_1(p), \cdots, \xi_n(p)$, and $a_{ij}(p)$ is the cofactor of the term C_{ij}. Equation (10–70) can be written as

$$\xi_i(p) = \frac{M_i(p)}{N(p)} \bar{f}(p), \qquad (i = 1, 2, \cdots, n), \qquad (10\text{–}71)$$

where

$$M_i(p) = \sum_{j=1}^{n} a_{ij}(p)C_j.$$

The process of computing the response $\xi_i(t)$ requires that the inverse of Eq. (10–71) be taken. This is accomplished by applying the convolution integral (Ref. 10–16)

$$\mathcal{L}^{-1}\{\bar{g}(p)\bar{h}(p)\} = \int_0^t g(\tau)h(t - \tau)d\tau \qquad (10\text{–}72)$$

to Eq. (10–71), which gives

$$\xi_i(t) = \mathcal{L}^{-1}\left\{\frac{M_i(p)}{N(p)} \bar{f}(p)\right\} = \int_0^t \mathcal{L}^{-1}\left\{\frac{M_i(p)}{N(p)}\right\}f(t - \tau)d\tau,$$

$$(i = 1, 2, \cdots, n). \quad (10\text{–}73)$$

Since the elements C_{ij} in the determinant $N(p)$ are algebraic functions of p, the determinant can be expanded into a polynomial which may be factored to

$$N(p) = \prod_{i}^{m} (p - p_i), \qquad (10\text{–}74)$$

where m is the number of roots of the polynomial.

The inverse transformation of $M_i(p)/N(p)$ is carried out by Heaviside's partial fraction expansion (Ref. 10–16):

$$\mathcal{L}^{-1}\left\{\frac{M_i(p)}{N(p)}\right\} = \sum_{k=1}^{m} \frac{M_i(p_k)}{N'(p_k)} e^{p_k\tau}, \qquad (i = 1, 2, \cdots, n), \quad (10\text{–}75)$$

where

$$N'(p_k) = \left[\frac{d}{dp} N(p)\right]_{p=p_k}.$$

Substituting Eq. (10–75) into Eq. (10–73), we obtain the final solution for the ith normal coordinate:

$$\xi_i(t) = \sum_{k=1}^{m} \frac{M_i(p_k)}{N'(p_k)} \int_0^t f(t - \tau)e^{p_k\tau}d\tau, \qquad (i = 1, 2, \cdots, n). \quad (10\text{–}76)$$

In obtaining this result, it has been necessary to factor the denominator polynomial $N(p)$. This feature of the Laplace transform solution is one of its most unfortunate aspects, since the factoring of a polynomial of high degree with a number of complex roots is an exceptionally difficult task.

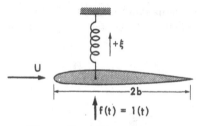

Fig. 10–16. Elastically suspended wing in an airstream.

EXAMPLE 10–3. To calculate, by application of the Laplace transform, the transient response of a thin rigid airfoil suspended by a spring in an incompressible airstream of velocity U, as illustrated by Fig. 10–16. Assume that the system is disturbed by a unit step force $1(t)$. The airfoil is free to translate vertically.

Solution. The equation of motion of the single-degree-of-freedom system is

$$M\ddot{\xi} + M\omega^2\xi = \Xi^M(\xi) + 1(t),\tag{a}$$

where M is the mass of the airfoil and ω is its natural frequency in the absence of air forces. Applying quasi-steady aerodynamic theory,

$$\Xi^M = -\pi\rho S U\dot{\xi},\tag{b}$$

where S is the area of the rigid airfoil. Substituting (b) into Eq. (a) gives

$$M\ddot{\xi} + \pi\rho S U\dot{\xi} + M\omega^2\xi = 1(t).\tag{c}$$

It will be convenient to transform from the time variable t to a dimensionless independent variable

$$s = \frac{Ut}{b},\tag{d}$$

where b is the half-chord of the airfoil. The variable s represents the distance traveled by particles of the moving air measured in half-chord lengths. The equation of motion in terms of the independent variable s is

$$\lambda\xi''(s) + \xi'(s) + \lambda\Omega^2\xi = \frac{b1(s)}{\pi\rho S U^2},\tag{e}$$

where $\lambda = M/\pi\rho Sb$ is a dimensionless quantity defined as the mass parameter of the wing and $\Omega = \omega b/U$ is a dimensionless reduced frequency of the system.

Let us next transform Eq. (e) from the dependent variable $\xi(s)$ to $\zeta(s)$, where the latter is defined as the ratio of the actual deflection to the static deflection

$$\zeta(s) = \frac{\xi(s)}{\xi_{(static)}} = \frac{\xi(s)}{1/M\omega^2}.\tag{f}$$

Equation (e) reduces to

$$\zeta''(s) + \frac{1}{\lambda}\zeta'(s) + \Omega^2\zeta(s) = \Omega^2 1(s).\tag{g}$$

Applying the Laplace transformation to Eq. (g),

$$p^2\bar{\zeta}(p) + \frac{1}{\lambda}\,p\bar{\zeta}(p) + \Omega^2\bar{\zeta}(p) = \frac{\Omega^2}{p}, \tag{h}$$

where $\mathcal{L}\{1(s)\} = 1/p$. Solving for $\bar{\zeta}(p)$ yields

$$\bar{\zeta}(p) = \frac{\Omega^2}{p[p^2 + (p/\lambda) + \Omega^2]}. \tag{i}$$

The final result is obtained by taking the inverse transformation of Eq. (i). The denominator can be factored as follows:

$$p[p^2 + (p/\lambda) + \Omega^2] = (p - 0)(p - p_1)(p - p_2), \tag{j}$$

where

$$p_1 = -\frac{1}{2\lambda} + \sqrt{\frac{1}{4\lambda^2} - \Omega^2}, \qquad p_2 = -\frac{1}{2\lambda} - \sqrt{\frac{1}{4\lambda^2} - \Omega^2}. \tag{k}$$

Applying Heaviside's partial fraction expansion (Eq. 10–75),

$$\zeta(s) = \Omega^2\left(\frac{1}{\Omega^2} + C_1 e^{p_1 s} + C_2 e^{p_2 s}\right), \tag{l}$$

where

$$C_1 = \frac{1}{p_1{}^2 - \Omega^2}, \qquad C_2 = \frac{1}{p_2{}^2 - \Omega^2}.$$

If we assume that $\Omega^2 > 1/4\lambda^2$, p_1 and p_2 are complex conjugate quantities and Eq. (l) can be written in the form

$$\zeta(s) = 1 + \Omega^2 e^{-s/2\lambda}$$

$$\times\left[(C_1 + C_2)\cos\left(\sqrt{\Omega^2 - \frac{1}{4\lambda^2}}\,s\right) + (C_1 - C_2)i\sin\left(\sqrt{\Omega^2 - \frac{1}{4\lambda^2}}\,s\right)\right]. \tag{m}$$

The response is a function of the two dimensionless parameters, λ and Ω. If we assume, for example, that $\lambda = \frac{1}{4}$ and $\Omega = 20$, $\zeta(s)$ is obtained from Eq. (m) as follows:

$$\zeta(s) = 1 - e^{-2s}[\cos(19.899s) + 0.1005\sin(19.899s)]. \tag{n}$$

(b) *Solution by the mechanical admittance or frequency response method.* The term *admittance*, as defined in electrical engineering, is the ratio of the amplitude of the current in a circuit to the amplitude of an applied voltage which varies sinusoidally with time. In an analogous manner, we can define a *mechanical admittance* as the ratio of the amplitude of a displacement of a mechanical system to the amplitude of a sinusoidally varying force that causes the displacement. The mechanical admittance is, in general, a function of the frequency of the applied force.

In the case of conservative or undamped systems, the response is always either in phase or 180° out of phase with the applied force. In the case of systems in which damping forces are present, the response is generally out of phase with the applied force. In order to properly relate the response to the forcing function, the mechanical admittance should therefore consist of two quantities: the ratio of the amplitude and the phase lag of the response with respect to the forcing function.

A comprehensive presentation of the concept of mechanical admittance and its application in the solution of vibration problems is given by Duncan (Ref. 10–17). The term mechanical admittance as used here is not restricted in its application to a forcing function and a response in the form of a force and displacement, respectively, but is considered to apply to any kind of response. For example, the forcing function may be a sinusoidal force applied at a gun mount, a sinusoidal gust velocity pattern, or a sinusoidal control surface movement; and the response may be the displacement of a normal mode or the stress at a point in the wing. The concept of mechanical admittance as used here is identical to the *frequency response* method widely used in instrumentation work (Ref. 10–18).

Let us suppose that the nondimensional function $f(t)$, representing the time variation of the disturbing force, has the form

$$f(t) = e^{i\omega t}. \tag{10–77}$$

Equation (10–65) becomes

$$M_r \ddot{\xi}_r + M_r \omega_r{}^2 \xi_r = \Xi_r{}^M(\xi_1,\cdots,\xi_n; \dot{\xi}_1,\cdots,\dot{\xi}_n; \ddot{\xi}_1,\cdots,\ddot{\xi}_n) + C_r e^{i\omega t},$$

$$(r = 1, 2, \cdots, n). \tag{10–78}$$

Solutions to Eq. (10–78) are given by

$$\xi_r(t) = H_r(i\omega)e^{i\omega t}, \qquad (r = 1, 2, \cdots, n), \tag{10–79}$$

where the $H_r(i\omega)$ are complex quantities called the *mechanical admittances* of the system. Their reciprocals are the complex impedances $Z_r(i\omega)$. That is,

$$Z_r(i\omega) = \frac{1}{H_r(i\omega)}. \tag{10–80}$$

Substituting Eq. (10–79) into Eq. (10–78) yields

$$[(\omega_r{}^2 - \omega^2)M_r - \Xi_{rr}{}^M(i\omega)]H_r(i\omega) + \sum_{s \neq r} \Xi_{rs}{}^M(i\omega)H_s(i\omega) = C_r,$$

$$(r = 1, 2, \cdots, n). \tag{10–81}$$

The quantities $\Xi_{rr}{}^M(i\omega)$ and $\Xi_{rs}{}^M(i\omega)$ are analogous to the similarly represented quantities in the Laplace transform solution (Eq. 10–68), except that they are now functions of the forcing frequency ω. The mechanical admittances are obtained by solution of the n simultaneous equations (10–81):

$$H_r(i\omega) = \sum_{s=1}^{n} \frac{a_{rs}(i\omega)}{\Delta(i\omega)} C_s, \qquad (r = 1, 2, \cdots, n), \tag{10–82}$$

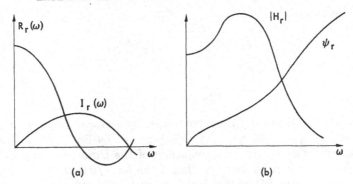

Fig. 10–17. Graphical representations of the admittance function.

where $\Delta(i\omega)$ is the determinant of the coefficients of $H_1(i\omega), \cdots, H_n(i\omega)$, and $a_{rs}(i\omega)$ is the cofactor of the term of $\Delta(i\omega)$ in the rth row and sth column. It is convenient to write Eq. (10–82) in the form

$$H_r(i\omega) = \frac{A_r(i\omega)}{\Delta(i\omega)}, \qquad (r = 1, 2, \cdots, n), \qquad (10\text{–}83)$$

where

$$A_r(i\omega) = \sum_{s=1}^{n} a_{rs}(i\omega)C_s.$$

In practical aeroelastic problems, it is usually not possible to derive $\Xi_{rr}{}^{M}(i\omega)$ and $\Xi_{rs}{}^{M}(i\omega)$ in explicit functional form. However, if the forcing frequency is given a particular numerical value ω_k, Eqs. (10–81) become a set of simultaneous linear algebraic equations with complex coefficients, and the mechanical admittances $H_1(i\omega)_k$, $H_2(i\omega_k), \cdots,$ $H_n(i\omega_k)$ for that frequency can be obtained by numerical solution. By solving the simultaneous equations for different values of ω, curves such as those of Fig. 10–17 can be plotted. In Fig. 10–17 we have represented the admittance function graphically in two ways, where the following notation is employed:

$$H_r(i\omega) = R_r(\omega) + iI_r(\omega) = |H_r|e^{i\psi_r}. \qquad (10\text{–}84)$$

The real and imaginary parts of $H_r(i\omega)$, which are designated $R_r(\omega)$ and $I_r(\omega)$, respectively, may be plotted versus ω as shown by Fig. 10–17(a); or the amplitude and phase, designated by $|H_r|$ and ψ_r, respectively, may be plotted as shown by Fig. 10–17(b). If the system has n degrees of freedom, there are n curves of the type shown by Fig. 10–17.

Let us consider next the process required to obtain the response to an arbitrary disturbance $f(t)$ when the mechanical admittances are known. In principle, we can apply the complex form of the Fourier integral derived in Appendix C:

$$f(t) = \int_{\omega=-\infty}^{\omega=+\infty} G(i\omega)e^{i\omega t}d\omega, \qquad (10\text{--}85)$$

$$G(i\omega) = \frac{1}{2\pi} \int_{\lambda=-\infty}^{\lambda=+\infty} f(\lambda)e^{-i\omega\lambda}d\lambda. \qquad (10\text{--}86)$$

In Eq. (10–85) the force $f(t)$ is synthesized by a superposition of harmonic functions of time. The quantity $G(i\omega)$ defined by Eq. (10–86) describes the spectral distribution with frequency of the harmonic components of $f(t)$. It is shown in Appendix C that if the force $f(t)$ can be expressed in the form of Eq. (10–85), that is, if the integral of Eq. (10–86) is convergent, the response to $f(t)$ can be expressed as

$$\xi_r(t) = \int_{\omega=-\infty}^{\omega=+\infty} H_r(i\omega)G(i\omega)e^{i\omega t}d\omega. \qquad (10\text{--}87)$$

The infinite integral in Eq. (10–87) provides an explicit result for the response to an arbitrary forcing function $f(t)$, providing the integral in Eq. (10–86) is convergent.

Evaluation of Eq. (10–87) is usually not practical when the admittance cannot be expressed as a simple analytical function of the frequency. In such cases, it has been found convenient (Ref. 10–18) to derive an approximate expression for calculating the indicial admittances in terms of the admittance functions, and then to apply Duhamel's integral to calculate the response to an arbitrary forcing function. Bromwich's integral formula, derived in Appendix C (Eq. 44), provides an explicit formula for calculating the indicial admittance when the admittance function is given:

$$A_r(t) = \frac{1}{2\pi} \int_{\omega=-\infty}^{\omega=+\infty} \frac{H_r(i\omega)}{i\omega} e^{i\omega t}d\omega, \qquad (10\text{--}88)$$

where the path of integration makes an infinitesimal loop below the origin. Care must be exercised in evaluation of the integral since, in general, the integrand has a pole at the origin. That is, when $\omega = 0$, the integrand becomes infinite, and unless $H_r(0) = 0$, the integral will not converge. This difficulty can be avoided by rewriting Eq. (10–88) in the form

$$A_r(t) = \frac{H_r(0)}{2\pi} \int_{\omega=-\infty}^{\omega=+\infty} \frac{e^{i\omega t}}{i\omega} d\omega + \frac{1}{2\pi} \int_{\omega=-\infty}^{\omega=+\infty} \frac{H_r(i\omega)-H_r(0)}{i\omega} e^{i\omega t}d\omega, \qquad (10\text{--}89)$$

where $H_r(0)$ is the value of the admittance function $H_r(i\omega)$, when $\omega = 0$. Since it can be shown by applying the residue theorem (Ref. 10–1 and Appendix C) that

$$\frac{1}{2\pi} \int_{\omega=-\infty}^{\omega=+\infty} \frac{e^{i\omega t}}{i\omega} d\omega = 1(t), \qquad (10\text{--}90)$$

Eq. (10–89) reduces to

$$A_r(t) = H_r(0)1(t) + \frac{1}{2\pi} \int_{\omega=-\infty}^{\omega=+\infty} \frac{H_r(i\omega) - H_r(0)}{i\omega} e^{i\omega t} d\omega. \quad (10\text{–}91)$$

Equation (10–91) can be transformed to real form by substituting

$$e^{i\omega t} = \cos \omega t + i \sin \omega t, \quad (10\text{–}92)$$

$$H_r(i\omega) = R_r(\omega) + iI_r(\omega), \quad (10\text{–}93)$$

which gives

$$A_r(t) = R_r(0)1(t) + \frac{1}{2\pi} \int_{\omega=-\infty}^{\omega=+\infty} \left(\frac{R_r(\omega) - R_r(0)}{\omega} \sin \omega t + \frac{I_r(\omega)}{\omega} \cos \omega t \right) d\omega$$

$$- \frac{i}{2\pi} \int_{\omega=-\infty}^{\omega=+\infty} \left(\frac{R_r(\omega) - R_r(0)}{\omega} \cos \omega t - \frac{I_r(\omega)}{\omega} \sin \omega t \right) d\omega,$$

$$(10\text{–}94)$$

where we have assumed $I_r(0) = 0$. The latter assumption, which requires merely that when the forcing function is statically applied the lag is zero, is valid in aeroelastic systems.

Since $A_r(t)$ is real, the last term in Eq. (10–94) must vanish, and since (Ref. 10–2)

$$\int_{\omega=-\infty}^{\omega=+\infty} \frac{\sin \omega t}{\omega} d\omega = \begin{cases} \pi \text{ for } t > 0, \\ -\pi \text{ for } t < 0, \end{cases} \quad (10\text{–}95)$$

we obtain finally

$$A_r(t) = \tfrac{1}{2}R_r(0) + \frac{1}{2\pi} \int_{\omega=-\infty}^{\omega=+\infty} \left(\frac{R_r(\omega)}{\omega} \sin \omega t + \frac{I_r(\omega)}{\omega} \cos \omega t \right) d\omega. \quad (10\text{–}96)$$

Expression (10–96) provides an explicit formula for computing the indicial admittance $A_r(t)$ if the real and imaginary parts of the mechanical admittance functions are known. It can be reduced to a somewhat simpler form by introducing the conditions that $R_r(\omega)$ is an even function, $R_r(-\omega) = R_r(\omega)$, and $I_r(\omega)$ is an odd function, $I_r(-\omega) = -I_r(\omega)$. These conditions are generally satisfied in aeroelastic systems.

$$A_r(t) = \tfrac{1}{2}R_r(0) + \frac{1}{\pi} \int_{\omega=0}^{\omega=\infty} \left(\frac{R_r(\omega)}{\omega} \sin \omega t + \frac{I_r(\omega)}{\omega} \cos \omega t \right) d\omega. \quad (10\text{–}97)$$

Since $H_r(\omega)$ and $I_r(\omega)$ are not often available in explicit function form but are usually plotted curves, as illustrated by Fig. (10–17), the summation form of Eq. (10–97) is more useful for practical application:

$$A_r(t) = \tfrac{1}{2}R_r(0) + \frac{2}{\pi} \sum_{n=1,3,5,\cdots}^{\infty} \frac{1}{n} [R_r(n\omega_0) \sin n\omega_0 t + I_r(n\omega_0) \cos n\omega_0 t].$$

$$(10\text{–}98)$$

The frequency ω_0 is arbitrarily selected so as to obtain satisfactory convergence. The criterion for convergence is that $R_r(n\omega_0)$ and $I_r(n\omega_0)$ approach zero for large values of n. This requires a damped system and, for example, for lightly damped systems a small value of ω_0 must be selected. This numerical technique is not satisfactory near the flutter speed. An identical result is obtained by computing the response of the system to a square wave having a period $T = 2\pi/\omega_0$ (Ref. 10–15). It is obvious that the response to a square wave can represent an indicial admittance only when the system is damped. When the admittance has been obtained, the response to an arbitrary disturbance can be computed by application of Duhamel's integral.

(c) *Solution by finite differences.* The calculus of finite differences provides a useful numerical tool for transient solutions of the differential equations of motion of aeroelastic systems. This approach is especially useful in transient motion problems where the first peak value is desired. Such problems are common in aeroelasticity, the principal one being the gust problem which is discussed in the next section. Techniques of applying finite differences vary with the type of problem and usually some ingenuity is required to obtain a satisfactory solution. In the present subsection we shall illustrate a simple problem rather than attempt to outline a general scheme applicable to all transient aeroelastic problems. Let us consider, for simplicity, the special case where the modal coupling produced by the forces due to the motion can be neglected and where the forces due to the motion are simply viscous damping forces. That is,

$$\Xi_r{}^M = -\beta\xi_r. \tag{10-99}$$

The equation of motion of the rth normal mode thus becomes

$$M\ddot{\xi} + \beta\dot{\xi} + M\omega^2\xi = \Xi^D(t), \tag{10-100}$$

where the subscript r has been dropped. Equation (10–100) is identical to the equation of motion of a simple mass on a spring with viscous damping subjected to a disturbance force of arbitrary time history.

Let us consider first how Eq. (10–100) can be transformed from a differential equation to a difference equation. The approach is to replace the time derivatives in the differential equation by appropriate difference expressions and solve the resulting algebraic equation for the response in successive steps. Let us assume that the function $\xi(t)$ is to be defined for equidistant values of t, as illustrated by Fig. 10–18. We designate Δ as the interval between successive values of t. In applications of difference solutions to engineering problems it is often assumed that $\xi(t)$ can be approximated by a parabola passing through three successive ordinates. If we suppose that the ordinates ξ_{n-1}, ξ_n, and ξ_{n+1} are connected by the arcs of a parabola as illustrated by Fig. 10–18, then it can be easily shown

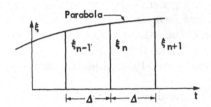

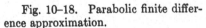

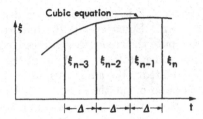

Fig. 10–18. Parabolic finite difference approximation.

Fig. 10–19. Cubic finite difference
approximation.

(see, for example, Ref. 10–19) that the first and second time derivatives
at the ordinate ξ_n are given by

$$\dot{\xi}_n = \frac{\xi_{n+1} - \xi_{n-1}}{2\Delta}, \tag{10–101}$$

$$\ddot{\xi}_n = \frac{\xi_{n+1} - 2\xi_n + \xi_{n-1}}{\Delta^2}. \tag{10–102}$$

These expressions, which give derivatives at the nth ordinate in terms of
the $(n-1)$st and $(n+1)$st ordinates, are satisfactory for solution of
many engineering problems.   However, for aeroelastic problems in general,
and for the gust problem in particular, it has been shown by Houbolt
(Ref. 10–19) that difference expressions are needed which give the derivatives at the end ordinate of several successive ordinates.   If the end ordinate
is used, however, it is desirable to obtain the result in terms of four successive ordinates rather than three in order to preserve accuracy equivalent
to that of Eqs. (10–101) and (10–102).   Consider next Fig. 10–19, in which
four successive ordinates are connected by a cubic equation.   Houbolt has
shown (Ref. 10–19) that the first and second time derivatives at the
ordinate ξ_n can be expressed in terms of the three preceding ordinates by

$$\dot{\xi}_n = \frac{11\xi_n - 18\xi_{n-1} + 9\xi_{n-2} - 2\xi_{n-3}}{6\Delta}, \tag{10–103}$$

$$\ddot{\xi}_n = \frac{2\xi_n - 5\xi_{n-1} + 4\xi_{n-2} - \xi_{n-3}}{\Delta^2}. \tag{10–104}$$

If we now replace the derivatives in Eq. (10–100) by their difference expressions given by Eqs. (10–103) and (10–104), we obtain

$$\left(2 + \frac{11}{6}\frac{\beta\Delta}{M} + \omega^2\Delta^2\right)\xi_n = \left(5 + \frac{3\beta\Delta}{M}\right)\xi_{n-1} - \left(4 + \frac{3\beta\Delta}{2M}\right)\xi_{n-2}$$

$$+ \left(1 + \frac{\beta\Delta}{3M}\right)\xi_{n-3} + \frac{\Delta^2\Xi_n{}^D}{M}, \tag{10–105}$$

where $\Xi_n{}^D$ is the value of the disturbance generalized force at the time $t = n\Delta$. Equation (10–105) is a recurrence formula which can be used to compute the response at successive values of n. If the values of ξ at three preceding stations are known, the value of ξ at the next station can be calculated. We are thus able to carry out a step-by-step numerical computation of the time history of the response.

Some special remarks are necessary concerning the early stages of the process at the outset of the motion. Houbolt has shown (Ref. 10–19) that the recurrence process can be commenced by applying the initial conditions together with the difference expressions for the first and second derivatives at the third ordinate of four successive ordinates. These expressions are

$$\dot{\xi}_n = \frac{1}{6\Delta} (2\xi_{n+1} + 3\xi_n - 6\xi_{n-1} + \xi_{n-2}), \qquad (10\text{–}106)$$

$$\ddot{\xi}_n = \frac{1}{\Delta^2} (\xi_{n+1} - 2\xi_n + \xi_{n-1}). \qquad (10\text{–}107)$$

Applying these expressions at $n = 0$, we obtain

$$\dot{\xi}_0 = \frac{1}{6\Delta} (2\xi_1 + 3\xi_0 - 6\xi_{-1} + \xi_{-2}), \qquad (10\text{–}108)$$

$$\ddot{\xi}_0 = \frac{1}{\Delta^2} (\xi_1 - 2\xi_0 + \xi_{-1}). \qquad (10\text{–}109)$$

Let us assume that the initial velocity and displacement are zero. We will also need a secondary initial condition giving the acceleration at $t = 0$. If we assume that $\Xi^D(0)$ is finite, we can see from Eq. (10–100) that the acceleration at $t = 0$ is $\Xi^D(0)/M$. The two initial conditions and the secondary initial condition are thus

$$\xi(0) = 0, \qquad (10\text{–}110)$$

$$\dot{\xi}(0) = 0, \qquad (10\text{–}111)$$

$$\ddot{\xi}(0) = \frac{\Xi^D(0)}{M}. \qquad (10\text{–}112)$$

Substituting the initial conditions into Eqs. (10–108) and (10–109) yields

$$2\xi_1 - 6\xi_{-1} + \xi_{-2} = 0, \qquad (10\text{–}113\text{a})$$

$$\xi_1 + \xi_{-1} = \frac{\Delta^2 \Xi^D(0)}{M}. \qquad (10\text{–}113\text{b})$$

Setting $n = 1$ in Eq. (10–105) gives

$$\left(2 + \frac{11}{6} \frac{\beta\Delta}{M} + \omega^2\Delta^2\right)\xi_1 = -\left(4 + \frac{3\beta\Delta}{2M}\right)\xi_{-1} + \left(1 + \frac{\beta\Delta}{3M}\right)\xi_{-2} + \frac{\Delta^2 \Xi_1{}^D}{M}.$$

$$(10\text{–}114)$$

We now have three equations which can be solved simultaneously for ξ_1, ξ_{-1}, and ξ_{-2}. The values of ξ_1 and ξ_{-1}, together with $\xi_0 = 0$, can be used to compute ξ_2 by Eq. (10–105) and thus start the recurrence process. No physical meaning can be attached to ξ_{-1} and ξ_{-2}, since the actual system is at rest until $t = 0$. The assumption of finite values of deflection prior to the start of the motion is required to start the recurrence process, and is merely an artifice for expressing the initial conditions in finite difference form.

EXAMPLE 10–4. To obtain the solution of the problem stated in Example 10–3 by the finite difference method.* Assume $\lambda = \frac{1}{4}$ and $\Omega = 20$.

Solution. The differential equation of motion derived in Example 10–3 is

$$\zeta''(s) + \frac{1}{\lambda}\zeta'(s) + \Omega^2\zeta(s) = \Omega^2 1(s). \tag{a}$$

Introducing the numerical values of the parameters λ and Ω, Eq. (a) becomes

$$\zeta''(s) + 4\zeta'(s) + 400\zeta(s) = 400 \times 1(s). \tag{b}$$

Adapting Eq. (10–105) to the present example and taking an interval $\Delta = 0.01$, we obtain the following recurrence formula:

$$[2 + (\tfrac{11}{6})(4)(0.01) + 400(0.01)^2]\zeta_n$$
$$= [5 + (3)(4)(0.01)]\zeta_{n-1} - [4 + (\tfrac{3}{2})(4)(0.01)]\zeta_{n-2}$$
$$+ [1 + (\tfrac{1}{3})(4)(0.01)]\zeta_{n-3} + (0.01)^2(400), \tag{c}$$

which reduces to

$$\zeta_n = 0.018927 + 2.42272\zeta_{n-1} - 1.92114\zeta_{n-2} + 0.47949\zeta_{n-3}. \tag{d}$$

In order to commence the recurrence process, we must express the initial conditions in finite difference form. The initial conditions are $\zeta(0) = \zeta'(0) = 0$. A secondary initial condition is $\zeta''(0) = 400$. Adapting Eqs. (10–108) and (10–109) to the notation of the present example,

$$\zeta_0' = \frac{1}{(6)(0.01)}(2\zeta_1 + 3\zeta_0 - 6\zeta_{-1} + \zeta_{-2}), \tag{e}$$

$$\zeta_0'' = \frac{1}{(0.01)^2}(\zeta_1 - 2\zeta_0 + \zeta_{-1}). \tag{f}$$

Applying the initial conditions and the secondary initial condition, we can write, with the aid of Eqs. (e) and (f),

$$\zeta_0 = 0, \quad \zeta_{-2} = 6\zeta_{-1} - 2\zeta_1, \quad \zeta_{-1} = 0.04 - \zeta_1. \tag{g}$$

Equations (g) may also be expressed as

$$\zeta_0 = 0, \quad \zeta_{-1} = 0.04 - \zeta_1, \quad \zeta_{-2} = 0.24 - 8\zeta_1. \tag{h}$$

---

* This example is due to Houbolt, who gives a detailed solution in Ref. 10–19.

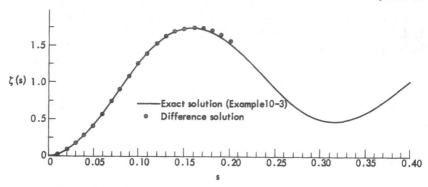

Fig. 10–20. Comparison of difference solution with exact solution.

Setting $n = 1$ in Eq. (d) and substituting the values of ζ_0, ζ_{-1}, and ζ_{-2} given above, an explicit numerical value of ζ_1 can be computed:

$$\zeta_1 = 0.019610. \tag{i}$$

Having obtained ζ_1, ζ_2 can be calculated by putting $n = 2$ into Eq. (d) together with known values of ζ_{-1}, ζ_0, and ζ_1. With the initial values thus established the process continues step by step by successive applications of Eq. (d). The results of the numerical calculation are plotted in Fig. 10–20, where they are compared with the exact result obtained by the Laplace transform method (Example 10–3).

(d) *Transient stresses in systems with external forces depending upon the motion.* The mode displacement method described in Subsection 10–3(b) is a simple and direct method of computing transient stresses in systems with external forces depending upon the motion. The mode acceleration method, which is simple and rapidly convergent when applied to systems with prescribed time-dependent external forces, has certain inherent disadvantages when $\Xi_r{}^M \neq 0$. Let us consider briefly some of the features of the latter method. The total displacement of the ith vibration mode can be derived from Eq. (10–11) in the following form:

$$\xi_i = \xi_{i\,(\text{static})} + \frac{\Xi_i{}^M}{M_i \omega_i{}^2} - \frac{\ddot{\xi}_i}{\omega_i{}^2}, \qquad (i = 4, 5, \cdots, n), \tag{10–115}$$

where

$$\xi_{i\,(\text{static})} = \frac{\Xi_i{}^D}{M_i \omega_i{}^2}.$$

Substituting Eq. (10–115) into Eq. (10–28), we obtain

$$\sigma_p = \sigma_{p\,(\text{static})} + \sum_{i=4}^{n} A_p{}^{(i)} \frac{\Xi_i{}^M}{M_i \omega_i{}^2} - \sum_{i=4}^{n} A_p{}^{(i)} \frac{\ddot{\xi}_i}{\omega_i{}^2}, \tag{10–116}$$

where

$$\sigma_{p\,(\text{static})} = \sum_{i=4}^{n=\infty} A_p{}^{(i)} \xi_{i\,(\text{static})}.$$

The first term in Eq. (10–116) represents the stress that would be obtained if vibration were suppressed; the second term represents the stress due to forces, other than inertial forces, resulting from the motion of the system; and the last term is the stress due to vibratory inertial forces. Since $\Xi_i{}^M$ is, in general, a function of all the coordinates and their first and second time derivatives, the second term in Eq. (10–116) may be exceedingly complicated. A third approach can be applied in which the stresses due to the various external forces are merely superimposed. This method, which can be termed the *force summation method*, is applied to the gust problem in Sections 10–6 and 10–7.

10–6 Dynamic response to a discrete gust. From the viewpoint of the structural designer, the response of an airplane in flight due to atmospheric gusts is perhaps one of the most important dynamic response problems. Loads due to gusts often control the wing design of large aircraft. Aeroelastic effects may have a significant influence on the magnitude and distribution of wing loads due to gusts.

The calculation of gust loads on an airplane in flight is a twofold problem involving consideration of both the character of the gust and the response of the airplane. Structural design for gust loading has been largely based upon a single discrete gust and structural strength provided to meet the resulting stresses. Fundamental to any method of predicting dynamic response to a discrete gust is a rational representation of the gust configuration. Such parameters as gust intensity, gradient, and profile, as well as spanwise distribution of gust velocity, have important effects upon the airplane response and the resulting stresses. This book is devoted for the most part to the response aspect of the gust problem, and matters relating to properties of the atmosphere are beyond its scope. The reader is referred, for example, to Ref. 10–20 for further information on this important aspect of gust analyses.

Development of methods of analysis of the response of elastic aircraft during gust encounter was undertaken as early as 1932. Perhaps the earliest contributor was Küssner (Ref. 10–21), who concluded that the stress in an elastic wing due to a gust may be considerably higher than in a rigid wing. In 1936, Bryant and Jones (Ref. 10–22) investigated the effect of wing elasticity on gust loads by assuming that the wing consists of a rigid center section and rigid outer panels connected by elastic joints. They concluded that the aerodynamic forces on an elastic wing are practically the same as those on a rigid wing subject to the same gust, so that the forces calculated for a rigid wing could be applied to the elastic wing to find its bending motion. They also concluded that stresses in flexible wings are not likely to exceed by more than 10% those in rigid wings. Williams and Hanson (Ref. 10–23) in 1937 considered the effects of both longitudinal stability and wing elasticity. They took into account elasticity in bending only and assumed a deflection mode. Their conclusions tend to bear out those of Bryant and Jones. This early work is mentioned because of its comprehensive nature. The most recent of the pre-World War II investigations was that of Sears and Sparks in 1941 (Ref. 10–24). They took wing damping into account and showed that at high flying speeds damping has an important effect on the magnitude of the dynamic response. The earliest postwar research was conducted by Pierce in 1947 (Ref. 10–25). Pierce developed a semiempirical method based partly on test results from a model in a gust tunnel. Only the fundamental bending mode of the wing is taken into account, and the effect of unsteady flow is included approximately in the form of an equivalent viscous damping coefficient. The results indicate that a dynamic overstress of about 12% may be expected for gusts of short gradient distance. Putnam (Ref. 10–26), using the same assumptions, improved on Pierce's methods by taking damping effects into account in a somewhat more satisfactory manner.

The work of Goland, Luke, and Kahn in 1947 (Ref. 10–27) represented an important step forward. Their analysis takes account of unsteady flow effects in a more rational manner than any of the previous investigations and includes the effect of torsion as well as that of bending. A similar method, with some refinements and similar limitations, is that due to Jenkins and Pancu (Ref. 10–28). The latter makes use of the operational calculus, while the former uses a numerical integration process in the solution of the equations of motion. Subsequent work published in 1949 by Bisplinghoff, Isakson, Pian, Flomenhoft, and O'Brien (Ref. 10–15) applied mechanical admittance or frequency response methods to the problem and took into account factors which had heretofore been neglected, such as airplane pitching, spanwise air load effects, unsteady downwash on the tail, and a larger number of wing vibratory modes. This work showed that magnifications in root bending moment of 15% to 20% can be expected in multi-engine straight-wing airplanes.

All of the early work outlined above applied to straight wings, and the purposes of the investigations were primarily to evaluate orders of magnitude of the various effects.

Subsequent analytical research has been directed somewhat more towards developing systematic procedures of analysis for use by designers. The work of Houbolt in Ref. 10–19 and Houbolt and Kordes (Refs. 10–29 and 10–30) has proven to be extremely useful for analyses of complex structures. The primary contribution has been the solution of the time-dependent equations of motion in terms of finite difference equations and matrix notation. Subsequent work by Codik, Lin, and Pian (Ref. 10–31) applies the methods of Houbolt and Kordes to the swept-wing problem.

(a) *Rigid wing free to translate vertically.* Prior to taking up the more general case in which wing elasticity is included, it is instructive to consider

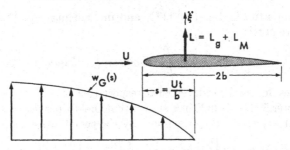

Fig. 10–21. Rigid wing encountering a gust.

first a simpler problem of gust encounter of a uniform rigid wing with a single translatory degree of freedom. Referring to Fig. 10–21, we assume that a thin wing of chord $2b$ and area S encounters a discrete gust of arbitrary distribution of velocity w_G in the flight direction, and uniform velocity distribution in the spanwise direction. The wing is permitted freedom to translate vertically. Wing thickness and amplitude of vertical translation are assumed small compared with the chord. The forward flight speed U is assumed unchanged as the wing progresses through the gust. The response is governed by the single differential equation

$$M\ddot{\xi} = \Xi^M + \Xi^D,\tag{10–117}$$

where M is the wing mass and ξ is its vertical displacement, positive upward. In computing the generalized forces, we assume that they derive entirely from unsteady aerodynamic forces on the wing, since the gravity force and the steady aerodynamic forces acting in the initial steady-flight condition are in equilibrium and therefore cancel. Moreover, the unsteady lift per unit span, positive upward, is equal to the sum of the aerodynamic forces due to the gust and the disturbed motion:

$$L = L_G + L_M,\tag{10–118}$$

where the subscript G refers to the gust and the subscript M to the disturbed motion. The generalized forces are therefore

$$\Xi^D = \int_{\text{span}} L_G dy,\tag{10–119}$$

$$\Xi^M = \int_{\text{span}} L_M dy.\tag{10–120}$$

It will be convenient to change from the time variable t to the dimensionless variable s defined by

$$s = \frac{Ut}{b},\tag{10–121}$$

which represents the distance traveled in half-chords of the wing. Transforming from s to t in Eq. (10–117), and introducing Eqs. (10–119) and (10–120), we obtain

$$\frac{U^2}{b^2} M\xi''(s) = \int_{\text{span}} L_G dy + \int_{\text{span}} L_M dy, \qquad (10\text{–}122)$$

where primes denote derivatives with respect to s.

In Chapter 5 (Eq. 5–382) it is shown that the lift per unit span due to a gust of arbitrary profile $w_G(s)$ is given by the two-dimensional result

$$L_G(s) = 2\pi\rho Ub\left[w_G(0)\psi(s) + \int_0^s \frac{dw_G(\sigma)}{d\sigma} \psi(s-\sigma)d\sigma \right], \quad (5\text{–}382)$$

where $\psi(s)$ is the Küssner function which describes the growth of lift on a thin airfoil during penetration of a sharp-edged gust. Integration of Eq. (5–382) by parts gives a form more convenient for the present application, as follows:

$$L_G(s) = 2\pi\rho Ub \int_0^s w_G(\sigma)\psi'(s-\sigma)d\sigma. \qquad (10\text{–}123)$$

In calculating the aerodynamic forces due to translatory motion of the airfoil, the formula given by Eq. (5–370), taking account of unsteady aerodynamic effects, can be adapted to the present application:

$$L_M(s) = -\pi\rho U^2 \left[\xi''(s) + 2\int_0^s \xi''(\sigma)\phi(s-\sigma)d\sigma \right], \quad (10\text{–}124)$$

where $\phi(s)$ is the Wagner function which describes the growth of lift on an airfoil following a sudden change in angle of attack.

Introducing Eqs. (10–123) and (10–124) into Eq. (10–122) and assuming that strip theory applies, we obtain the differential equation of motion

$$\frac{U^2}{b^2} M\xi''(s) = \pi\rho U^2 S\left[\int_0^s \frac{w_G(\sigma)}{U} \psi'(s-\sigma)d\sigma \right.$$

$$\left. - \frac{\xi''(s)}{2b} - \frac{1}{b} \int_0^s \xi''(\sigma)\phi(s-\sigma)d\sigma \right]. \qquad (10\text{–}125)$$

Applying the Laplace transformation to Eq. (10–125), this time with respect to the variable s rather than t, yields

$$\frac{U^2}{b^2} Mp^2\bar{\xi}(p) = \pi\rho U^2 S\left[\frac{\bar{w}_G(p)}{U} p\bar{\psi}(p) - \frac{p^2}{2b} \bar{\xi}(p) - \frac{p^2}{b} \bar{\xi}(p)\bar{\phi}(p) \right],$$

$$(10\text{–}126)$$

where the initial conditions are $\xi(0) = \xi'(0) = 0$, and where we have applied the following Laplace transform formulas (Ref. 10–16):

$$\mathcal{L}\left\{ \int_0^s w_G(\sigma)\psi'(s-\sigma)d\sigma \right\} = \bar{w}_G(p)p\bar{\psi}(p), \qquad (10\text{–}127)$$

$$\mathcal{L}\left\{\int_0^s \xi''(\sigma)\phi(s - \sigma)d\sigma\right\} = p^2\bar{\xi}(p)\bar{\phi}(p). \tag{10–128}$$

Solving Eq. (10–126) for $\bar{\xi}(p)$ yields

$$\bar{\xi}(p) = \frac{b}{2} \frac{(\bar{w}_G(p)/U)\bar{\psi}(p)}{p(\lambda + \frac{1}{4} + \frac{1}{2}\bar{\phi}(p))}, \tag{10–129}$$

where $\lambda = M/\pi\rho Sb$ is a dimensionless quantity defined as the mass parameter of the wing.

We have seen in Chapter 5 that the exact functions $\psi(s)$ and $\phi(s)$ cannot be expressed in simple analytical form. Satisfactory approximations are, however, given by the following expressions:

$$\psi(s) = 1 - 0.500e^{-0.130s} - 0.500e^{-s}, \tag{5–383}$$

$$\phi(s) = 1 - 0.165e^{-0.0455s} - 0.335e^{-0.300s} \tag{5–371}$$

The Laplace transforms of these functions are

$$\bar{\psi}(p) = \frac{1}{p} - \frac{0.500}{p + 0.130} - \frac{0.500}{p + 1}, \tag{10–130}$$

$$\bar{\phi}(p) = \frac{1}{p} - \frac{0.165}{p + 0.0455} - \frac{0.335}{p + 0.300}. \tag{10–131}$$

Introducing these expressions into Eq. (10–129), we obtain, after reduction,

$$\bar{\xi}(p) = \frac{0.2824b}{\lambda + 0.250} \times \frac{p^3 + 0.5756p^2 + 0.09315p + 0.003141}{p(p + 0.130)(p + 1)(p^3 + a_1p^2 + a_2p + a_3)} \times \frac{\bar{w}_G(p)}{U}, \tag{10–132}$$

where

$$a_1 = \frac{0.3455\lambda + 0.3364}{\lambda + 0.250}, \quad a_2 = \frac{0.01365\lambda + 0.1438}{\lambda + 0.250}, \quad a_3 = \frac{0.006825}{\lambda + 0.250}.$$

Let us specialize Eq. (10–132) to the case of a sharp-edged gust by putting

$$w_G(s) = w_0 1(s), \tag{10–133}$$

which has the Laplace transform (Ref. 10–16)

$$\mathcal{L}\{w_0 1(s)\} = \frac{w_0}{p}. \tag{10–134}$$

Noting that $\ddot{\xi} = (U^2/b^2)\xi''(s)$ and that the initial velocity and acceleration are zero, we obtain the following expression for the vertical acceleration due to a sharp-edged gust:

$$\ddot{\xi}(s) = \frac{U^2}{b^2} \mathcal{L}^{-1}\{p^2\bar{\xi}(p)\} = \frac{0.2824}{\lambda + 0.250} \frac{Uw_0}{b} \mathcal{L}^{-1}\left\{\frac{M(p)}{N(p)}\right\}, \tag{10–135}$$

where $M(p) = p^3 + 0.5756p^2 + 0.09315p + 0.003141,$ (10-136)

$N(p) = (p + 0.130)(p + 1)(p^3 + a_1p^2 + a_2p + a_3).$ (10-137)

The inverse transformation can be carried out by application of Heaviside's partial fractions expansion to yield a sum of exponential terms with real or complex exponents. Thus

$$\ddot{\xi}(s) = \frac{0.2824}{\lambda + 0.250} \frac{U w_0}{b}$$

$$\times (A_1 e^{-0.130s} + A_2 e^{-s} + B_1 e^{\gamma_1 s} + B_2 e^{\gamma_2 s} + B_3 e^{\gamma_3 s}),   (10-138)$$

where γ_1, γ_2, and γ_3 are roots of the third degree polynomial

$$(p^3 + a_1p^2 + a_2p + a_3) = (p - \gamma_1)(p - \gamma_2)(p - \gamma_3)   (10-139)$$

and

$$A_1 = \frac{M(-0.130)}{N'(-0.130)},   A_2 = \frac{M(-1)}{N'(-1)},   B_k = \frac{M(\gamma_k)}{N'(\gamma_k)},   (k = 1, 2, 3).$$

The acceleration is defined explicitly by Eq. (10-138) when a numerical value of the mass parameter λ is specified.

Let us introduce a hypothetical acceleration due to a sharp-edged gust computed by assuming that the aerodynamic forces due to the motion are neglected and that the forces due to the gust are obtained from steady-state aerodynamic theory, as follows:

$$\ddot{\xi}_{(s.s.)} = \frac{w_0 U}{2 \lambda b} .   (10-140)$$

Expression (10-140) is called the sharp-edged gust formula (Ref. 10-20).

The nondimensional ratio of the actual acceleration to the hypothetical value of Eq. (10-140) is called the acceleration ratio:

$$\frac{\ddot{\xi}}{\ddot{\xi}_{(s.s.)}} = \text{acceleration ratio}$$

$$= \frac{0.5648\lambda}{\lambda + 0.25r} (A_1 e^{-0.130s} + A_2 e^{-s} + B_1 e^{\gamma_1 s} + B_2 e^{\gamma_2 s} + B_3 e^{\gamma_3 s}).   (10-141)$$

The maximum value of the acceleration ratio is the alleviation factor. When unsteady flow is considered there is an alleviation effect even for a sharp-edged gust, and the alleviation factor is a function of the mass parameter λ. As an illustration of the nature of the results obtained from Eq. (10-141), numerical calculations for four values of the mass parameter are plotted in Fig. 10-22. These curves illustrate that the influences of unsteady aerodynamic effects and forces due to motion are large for small values of the mass parameter and small for large values of the mass parameter. Further details on this particular problem may be found in Ref. 10-32.

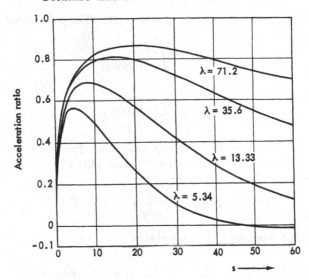

Fig. 10–22. Nondimensional acceleration responses using unsteady aerodynamic theory.

Finally, the response of the rigid wing to an arbitrary gust profile can be obtained by applying Duhamel's integral to the sharp-edged gust response calculated above.

(b) *Airplane free to translate vertically with wing bending.* We take up next the case of an airplane free in vertical translation with a wing elastic in bending but restrained against twisting. The wing is assumed slender and straight with arbitrary mass, stiffness, and chord variations along the span. The response is given by solution of the differential equations

$$M_i\ddot{\xi}_i + M_i\omega_i^2\xi_i = \Xi_i^M + \Xi_i^D, \qquad (i = 1,2,\cdots,n; \omega_1 = 0), \quad (10\text{–}142)$$

where ξ_1 is the response of the rigid-body mode of vertical translation and $\xi_2, \xi_3, \cdots, \xi_n$ are the normal coordinates which represent responses of the vibratory bending modes of the wing. Positive directions are as illustrated by Fig. 10–23. Here M is the total mass of the airplane and the ω_i are the bending frequencies of the wing. The mode shapes are represented by

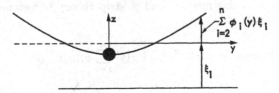

Fig. 10–23. Airplane free to plunge with wing bending.

$\phi_i(y)$, normalized such that $2\int_0^l \phi_i{}^2 m\,dy = M$. The generalized forces are defined by

$$\Xi_i{}^D = \int_{-l}^{+l} L_G \phi_i \, dy, \tag{10-143}$$

$$\Xi_i{}^M = \int_{-l}^{+l} L_M \phi_i \, dy, \tag{10-144}$$

where L_G and L_M are the forces per unit span due to the gust and the motion, respectively. Transforming Eq. (10-142) to the nondimensional independent variable

$$s = \frac{Ut}{b_R}, \tag{10-145}$$

where $2b_R$ is a reference chord, we obtain

$$\frac{U^2}{b_R{}^2} M\xi_i{}''(s) + M\omega_i{}^2 \xi_i(s) = \Xi_i{}^M + \Xi_i{}^D, \quad (i = 1,2,\cdots,n\,; \omega_1 = 0). \tag{10-146}$$

We have seen that in the determination of the unsteady aerodynamic forces per unit span, the significant variable is the distance traveled in half-chords of the airfoil. The chord is now, however, a variable with respect to the spanwise coordinate. In order to retain a single independent variable, we relate the chord at any station to the reference chord by $b = a(y)b_R$, where the quantity $a(y)$ is a variable with respect to the spanwise coordinate y.

Let us assume that the airplane encounters a discrete gust of arbitrary distribution $w_G(s)$ in the flight direction, and uniform spanwise distribution. If we assume in addition that the wing is sufficiently slender so that strip theory can be applied, the force per unit span due to the gust, L_G, is obtained from Eq. (5-382) as follows:

$$L_G(y,s) = 2\pi\rho U a(y) b_R \int_0^s w_G(\sigma)\psi'(s - \sigma)\,d\sigma. \tag{10-147}$$

The quantity L_M in Eq. (10-144) represents the force per unit length due to translatory motion of the wing. Both the plunging and vibratory bending contribute to this motion, and if strip theory is assumed, we obtain from Eq. (5-370)

$$L_M(y,s) = -\pi\rho U^2 a(y) \left[a(y) \sum_{j=1}^n \phi_j(y)\xi_j{}''(s) \right.$$
$$\left. + 2\int_0^s \left(\sum_{j=1}^n \phi_j(y)\xi_j{}''(\sigma) \right) \phi(s - \sigma)\,d\sigma \right]. \tag{10-148}$$

Substituting $L_G(y,s)$ and $L_M(y,s)$ from Eqs. (10–147) and (10–148) into the expressions for the generalized forces given by Eqs. (10–143) and (10–144), and putting the results into Eq. (10–146), the differential equations of motion are obtained:

$$\lambda \xi_i''(s) + \lambda \Omega_i^2 \xi_i(s) + \sum_{j=1}^{n} A_{ij} \xi_j''(s) + 2 \sum_{j=1}^{n} B_{ij} \int_0^s \xi_j''(\sigma) \phi(s - \sigma) d\sigma$$

$$= 2 b_R B_{1i} \int_0^s \frac{w_G(\sigma)}{U} \psi'(s - \sigma) d\sigma, \qquad (i = 1, 2, \cdots, n; \, \omega_1 = 0), \quad (10\text{–}149)$$

where
$$\lambda = \frac{M}{\pi \rho S b_R}, \qquad \Omega_i = \frac{\omega_i b_R}{U},$$

$$A_{ij} = \frac{b_R}{S} \int_{-l}^{+l} a^2 \phi_i \phi_j \, dy, \qquad B_{ij} = \frac{b_R}{S} \int_{-l}^{+l} a \phi_i \phi_j \, dy.$$

Solutions to Eqs. (10–149) for the normal coordinates $\xi_i(s)$ may be obtained by the Laplace transform, the admittance, or the finite difference methods described in Subsections 10–5(a), (b), and (c), respectively. Once the solutions are known, the motion of the airplane and the internal stresses in the wing can be computed. For example, the shear in the wing at a distance y from the airplane center line is obtained by merely summing all of the aerodynamic loads and inertial loads outboard of y, as follows:[*]

$$S(y,s) = \int_y^l \left[L_G + L_M - \frac{U^2}{b_R^2} \left(\sum_{i=1}^{n} \phi_i(\eta) \xi_i''(s) \right) m(\eta) \right] d\eta. \quad (10\text{–}150)$$

The bending moment in the wing at a distance y from the center line is obtained in a similar manner:

$$M(y,s) = \int_y^l \left[L_G + L_M - \frac{U^2}{b_R^2} \left(\sum_{i=1}^{n} \phi_i(\eta) \xi_i''(s) \right) m(\eta) \right] (\eta - y) d\eta, \quad (1\text{–}151)$$

where L_G and L_M are defined by Eqs. (10–147) and (10–148), respectively. The first two terms in the integrand represent the contribution of aerodynamic loads, and the last term the contribution of inertial loads.

The equations of motion of the gust problem are formulated in a similar manner when other rigid-body and vibratory modes are taken into consideration. The above derivation of a simple case serves only to illustrate the procedure and does not necessarily represent a result which can be applied to all aircraft. The inclusion of wing torsion, as well as wing bending, is often necessary. In the case of very flexible airplanes it is desirable to use normal modes of the entire airplane. An example of the use of normal modes in formulating the gust problem for a large airplane is given in Ref. 10–15.

---

[*] This is an application of the force summation method mentioned in Section 10–5(d).

Many other refinements may be necessary in particular cases. For example, in the case of swept wings, the pitching degree of freedom plays a relatively more important role than in the case of straight wings. In general, the inclusion of pitching tends to increase the loads, and hence neglect of the pitching degree of freedom is a nonconservative practice. The pitching degree of freedom brings into the problem complicating tail effects such as the influence of wing downwash. In general, it can be said that satisfactory results can be obtained by a simple approximation to the lag in downwash. This approximation involves taking the downwash on the tail at any second as that downwash which leaves the wing at a time L_T/U seconds earlier, where L_T is the tail length and U is the forward velocity. This simple correction accounts for the lag in time required for the particles of air to travel from the wing to the tail.

In treating high-speed airplanes with low aspect-ratio swept planforms, it becomes desirable to employ a three-dimensional aerodynamic theory taking account of spanwise induction effects in computing gust loads. No such rational theory exists which can be easily applied to arbitrary swept planforms, and the only practical means of handling the gust build-up problem is by means of a strip-theory or modified strip-theory approach.

EXAMPLE 10–5. To calculate the dynamic bending moment at the wing root of the jet transport airplane at a speed of $U = 475$ mph due to a sharp-edged gust of value w_0. Take into consideration plunging of the airplane and the fundamental bending mode of the wing. Assume that the aerodynamic forces due to the motion are defined satisfactorily by quasi-steady aerodynamic theory.[*]

Solution. The equations of motion are obtained from Eqs. (10–149) by putting $i = 1, 2$ and assuming quasi-steady aerodynamic forces due to motion.

$$\lambda \xi_1''(s) + 2B_{11}\xi_1'(s) + 2B_{12}\xi_2'(s) = 2b_R B_{11} \int_0^s \frac{w_G(\sigma)}{U} \psi'(s - \sigma)d\sigma, \quad \text{(a)}$$

$$\lambda \xi_2''(s) + \lambda \Omega_2{}^2 \xi_2(s) + 2B_{21}\xi_1'(s) + 2B_{22}\xi_2'(s)$$

$$= 2b_R B_{12} \int_0^s \frac{w_G(\sigma)}{U} \psi'(s - \sigma)d\sigma, \quad \text{(b)}$$

where

$$\lambda = \frac{M}{\pi \rho S b_R}, \qquad \Omega_2 = \frac{\omega_2 b_R}{U}, \qquad b_R = \frac{S}{4l},$$

$$B_{12} = B_{21} = \frac{b_R}{S} \int_{-l}^{+l} a(y)\phi_2(y)dy, \qquad B_{11} = \frac{b_R}{S} \int_{-l}^{+l} a(y)dy,$$

$$B_{22} = \frac{b_R}{S} \int_{-l}^{+l} a(y)\phi_2{}^2(y)dy.$$

---

[*] In the present example a negligible error is incurred by this assumption, since a relatively low frequency mode is involved.

It will be convenient to evaluate first the response to a hypothetical gust input of $(w_G(s)/U)\psi(s) = 1(s)$ and then to apply Duhamel's integral to obtain the final result. Putting $(w_G(s)/U)\psi(s) = 1(s)$ in Eqs. (a) and (b) and taking the Laplace transform of both sides gives

$$(\lambda p^2 + 2B_{11}p)\bar{\xi}_1(p) + 2B_{12}p\bar{\xi}_2(p) = 2b_R B_{11}\frac{1}{p}, \tag{c}$$

$$(2B_{21}p)\bar{\xi}_1(p) + (\lambda p^2 + \lambda\Omega_2{}^2 + 2B_{22}p)\bar{\xi}_2(p) = 2b_R B_{12}\frac{1}{p}. \tag{d}$$

The following data are required to evaluate the constants:

$$M = 2606.1 \text{ slugs}, \qquad l = 41.666 \text{ ft}, \qquad b_R = 6.771 \text{ ft},$$

$$S = 1128.47 \text{ ft}^2, \qquad \omega_2 = 15.316 \text{ rad/sec}.$$

The wing planform of the jet transport is illustrated in Fig. 8–18a. The mode shape of the first unrestrained bending mode, $\phi_2(y)$, given in Example 4–5(b), is normalized such that $2\int_0^l \phi_2{}^2(y)m\,dy = M$. Substituting these data into the formulas for the appropriate constants results in

$$\lambda = 45.6560, \qquad B_{12} = B_{21} = 0.3963,$$

$$\lambda\Omega_2{}^2 = 1.0122, \qquad 2b_R B_{12} = 5.3665,$$

$$B_{11} = 0.5000, \qquad 2b_R B_{11} = 6.7708.$$

$$B_{22} = 1.4460,$$

Substituting the constants into Eqs. (a) and (b) and solving for $\bar{\xi}_1(p)$ and $\bar{\xi}_2(p)$ yields

$$\bar{\xi}_1(p) = \frac{309.1276[p^2 + .049584p + .022170]}{2084.470p^2[p + .0222179][p^2 + .0630284p + .0218558]}, \tag{e}$$

$$\bar{\xi}_2(p) = \frac{245.0129}{2084.470[p + .0222179][p^2 + .0630284p + .0218558]}. \tag{f}$$

The inverse transformation is obtained by applying Heaviside's expansion theorem. This results in

$$\xi_1(s) = -309.1276 + 6.77079s + 309.22e^{-.0222179s} \tag{g}$$
$$+ e^{-.0315142s}[-.087901\cos(.144439s) + .66852\sin(.144439s)],$$

$$\xi_2(s) = 5.6108e^{-.0222179s}$$
$$+ e^{-.0315142s}[-5.6108\cos(.144439s) - .36112\sin(.144439s)]. \tag{h}$$

The bending moment at the wing root $(y = 0)$ is obtained from Eq. (10–151) as

$$M(0,s) = \int_0^l \left[L_G + L_M - \frac{U^2}{b_R{}^2}(\xi_1''(s) + \phi_2(\eta)\xi_2''(s))m(\eta) \right]\eta\,d\eta, \tag{i}$$

where, for the hypothetical gust input $(w_G(s)/U)\psi(s) = 1(s)$, the terms L_G

and L_M have the forms

$$L_G = 2\pi\rho U^2 a(y) b_R,$$

$$L_M = -2\pi\rho U^2 a(y)[\xi_1'(s) + \phi_2(y)\xi_2'(s)]. \tag{j}$$

Substituting expressions (g) and (h) into Eqs. (j) and putting the results obtained thereby into Eq. (i) yields

$$M(0,s) = 2\pi\rho U^2 b_R (750.250) - 2\pi\rho U^2 (750.250)\xi_1'(s)$$

$$-2\pi\rho U^2 (1309.435)\xi_2'(s) - \frac{U^2}{b_R^2}(13,286.058)\xi_1''(s)$$

$$-\frac{U^2}{b_R^2}(12,657.658)\xi_2''(s). \tag{k}$$

Expression (k) represents the response to a hypothetical gust input $(w_G(s)/U)$ $\psi(s) = 1(s)$. The response due to a gust input $(w_G(s)/U)\psi(s)$ is obtained by applying Duhamel's integral:

$$M(0,s) = \int_0^s \left[\frac{w_G(\sigma)}{U}\psi(\sigma)\right]' M(s-\sigma)\,d\sigma, \tag{l}$$

where $\psi'(\sigma)$ is the first derivative of the Küssner function (Eq. 5–383):

$$\psi'(\sigma) = \frac{d}{d\sigma}(1 - 0.500e^{-0.130\sigma} - 0.500e^{-\sigma}) = 0.065e^{-0.130\sigma} + 0.500e^{-\sigma}. \tag{m}$$

Putting, for example, $w_G(s) = w_0$,* inserting expressions (k) and (m) into Eq. (l), integrating, and evaluating gives

$$\frac{M(0,s)}{w_0/U} = 10^7\{-.622068e^{-.130s} + 1.820448e^{-.022179s} + .0200273e^{-s}$$

$$- e^{-.0315142s}[1.218407 \cos(.144439s) + .624602 \sin(.144439s)]\}. \tag{n}$$

Figure 10–24 illustrates a plot of Eq. (n) showing the history of oscillations of the root bending moment. For comparison purposes, Fig. 10–24 shows also the result obtained if the airplane is assumed rigid. The reader should keep in mind that the results portrayed by Fig. 10–24 should be superimposed upon steady level flight stresses.

---

* A more realistic assumption for atmospheric gusts would be a gust velocity distribution of the form

$$w_G(s) = \tfrac{1}{2}w_0\left(1 - \cos\frac{\pi s}{s_G}\right)$$

for $0 \le s \le 2s_G$ and $w_G(s) = 0$ for $s \ge 2s_G$. The quantity w_0 in this formula represents the peak gust velocity and the quantity s_G the gust gradient expressed in half-chord length units. A matter of some interest to military aircraft designers is that of blast-induced gusts where the gust velocity distribution has the general form $w_G(s) = w_0(1 - k_1 s)e^{-k_2 s}$. It should be mentioned that in this case it is more nearly correct to employ the Wagner function in Eq. (l) instead of the Küssner function.

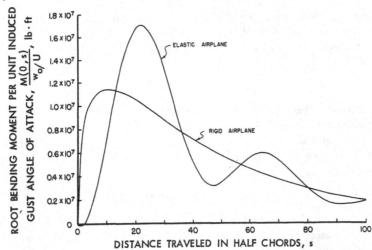

Fig. 10–24. Root bending moment per unit induced gust angle of attack, $\dfrac{M(0,s)}{w_0/U}$.

10–7 Dynamic response to continuous atmospheric turbulence. The artificiality of the discrete gust assumption as a basis for airplane structural design has long been recognized. Atmospheric turbulence is in reality a continuous phenomenon in which the airplane is subjected to repeated gustiness. Trends in airplane design to widely different configurations and applications of high-strength alloys, which require a more careful examination of fatigue properties, have tended to magnify the irrational character of the discrete gust viewpoint. To account more fully for the continuous properties of atmospheric turbulence, it is apparent that some sort of statistical approach is required. Statistical concepts applied to studying the response of linear systems to random input have been developed extensively in connection with communication theory by Wiener (Ref. 10–33) and others. These concepts, which are discussed in general terms in Appendix C, have also been systematically applied in the statistical theory of turbulence. Clementson (Ref. 10–34) obtained experimentally the power spectrum of the response of an airplane as it was flown through turbulent air. Subsequently, Liepmann (Ref. 10–35) and Press and Mazelsky (Ref. 10–36) have called attention to the use of statistical concepts in aeronautics and demonstrated their application to aeroelastic problems.

(a) *Rigid wing free to translate vertically.* We consider first the vertical response of a rigid wing of constant chord $2b$ moving through turbulent air at a forward velocity U. The airfoil thickness and amplitude of vertical translation are assumed small compared with the chord. Components of the fluctuating turbulence velocities u, v, and w are assumed small compared with the forward velocity U. Since the wing is free only in vertical translation, it is assumed that the effects of the velocity components u and v can be neglected. Thus the airfoil is subjected to a fluctuating angle of attack

$$\alpha = \frac{w}{U}. \tag{10-152}$$

It is assumed also that the fluctuating angle of attack does not vary in the spanwise direction and that the turbulence pattern does not change during the time required for a given particle of air to pass the airfoil; i.e., during a time $2b/U$. As a consequence of the latter assumption, the turbulence w is expressible in the form

$$w = w\left(t - \frac{x}{U}\right), \tag{10-153}$$

where the coordinate axes are assumed fixed to the wing. Expression (10-153) merely expresses the fact that the turbulence pattern moves past the airfoil with a velocity U.

We now assume that the turbulence w is not susceptible to Fourier analysis, but is a stationary random function, as defined in Appendix C. Thus we will be unable to calculate an explicit time history of response of the wing but must be content with certain mean values. In particular, it is shown in Appendix C that if the mean square velocity fluctuation $\overline{w^2}$ is known, the mean square response of the rigid wing (for example, vertical acceleration $\overline{\ddot{z}^2}$) can be calculated. This is accomplished by use of the power spectral density functions $\Phi(\omega)$ and $\Psi(\omega)$, of the input and output, respectively, as defined in Appendix C. In the present application, the function $\Phi(\omega)$ satisfies

$$\overline{w^2} = \int_0^\infty \Phi(\omega)d\omega \tag{10-154}$$

and $\Psi(\omega)$ satisfies

$$\overline{\ddot{z}^2} = \int_0^\infty \Psi(\omega)d\omega. \tag{10-155}$$

In addition, it is shown in Appendix C that the following relation holds:

$$\Psi(\omega) = |H(i\omega)|^2 \Phi(\omega), \tag{10-156}$$

where $H(i\omega)$ is the mechanical admittance of the linear system. Thus the calculation of the response to a random input entails a knowledge of the power spectral density of the forcing function and the mechanical admittance of the system. Then the power spectral density of the response is obtained from Eq. (10-156), and the mean square of the response from expression (10-155).

For the rigid wing, the mechanical admittance of the system can be calculated by subjecting the wing to a sinusoidal gust velocity described by the real part of

$$w_G = \overline{w}_G e^{ik(s-x^*)}, \tag{5-374}$$

as given in Chapter 5. The response of the wing to this gust velocity is obtained by solution of the equation of motion given by Eq. (10–122):

$$\frac{U^2}{b^2} M\xi''(s) = \int_{\text{span}} L_G dy + \int_{\text{span}} L_M dy, \qquad (10\text{–}122)$$

where, according to Eq. (5–376), the lift due to the sinusoidal gust is

$$L_G = 2\pi\rho U b \overline{w}_G \{C(k)[J_0(k) - iJ_1(k)] + iJ_1(k)\}e^{iks} \qquad (10\text{–}157)$$

and, according to Eq. (5–311), the lift due to the motion is

$$L_M = -\pi\rho U^2 \xi''(s) - 2\pi\rho U^2 C(k)\xi'(s). \qquad (10\text{–}158)$$

The Theodorsen function $C(k)$ is defined explicitly by Eq. (5–309). Substituting expressions (10–157) and (10–158) into Eq. (10–122) and applying strip theory, we obtain

$$(2\lambda + 1)\xi'' + 2C(k)\xi' = 2b\frac{\overline{w}_G}{U} K(k)e^{iks}, \qquad (10\text{–}159)$$

where $\lambda = M/\pi\rho S b$ and

$$K(k) = C(k)[J_0(k) - iJ_1(k)] + iJ_1(k).$$

Putting as a solution $\xi = \bar{\xi}e^{iks}$, we obtain

$$\frac{\bar{\xi}}{\overline{w}_G} = \frac{b}{U} \frac{2K(k)}{k[2iC(k) - (2\lambda + 1)k]}. \qquad (10\text{–}160)$$

This quantity is the admittance function with respect to the wing displacement. If we put

$$\ddot{\xi} = \ddot{\bar{\xi}}e^{iks} = -\frac{U^2}{b^2}k^2\bar{\xi}e^{iks}, \qquad (10\text{–}161)$$

the admittance function with respect to the acceleration is obtained as follows:

$$\frac{\ddot{\bar{\xi}}}{\overline{w}_G} = -\frac{U}{b} \frac{2kK(k)}{2iC(k) - (2\lambda + 1)k}. \qquad (10\text{–}162)$$

The mean square of the acceleration is

$$\overline{\ddot{\xi}^2} = \int_0^\infty \left| \frac{\ddot{\bar{\xi}}}{\overline{w}_G} \right|^2 \Phi(\omega)d\omega. \qquad (10\text{–}163)$$

If we assume, for example, that the turbulence in the gust field is isotropic, it is shown by Liepmann (Ref. 10–35) that the input power spectrum is given by

$$\Phi(\omega) = \overline{w}_G{}^2 \frac{L}{\pi U} \frac{1 + 3(\omega L/U)^2}{[1 + (\omega L/U)^2]^2}, \qquad (10\text{–}164)$$

where $\overline{w}_G{}^2$ is the mean square of the gust velocity and L is the integral scale of turbulence. The latter is defined by the area under the correlation curve (Ref. 10–35) and can be considered as a measure of the average eddy size in the turbulence.

Substituting expressions (10–162) and (10–164) into Eq. (10–163) yields

$$\overline{\xi^2} = \frac{4U^2}{b^2} \frac{\overline{w_G^2}}{\pi} \int_0^\infty \frac{k^2 |K(k)|^2}{|2iC(k) - (2\lambda + 1)k|^2} \frac{1 + 3\zeta^2}{(1 + \zeta^2)^2} d\zeta, \quad (10\text{–}165)$$

where $\zeta = \omega L / U$. Liepmann (Ref. 10–35) has shown that a simple approximation to $|K(k)|^2$ is

$$|K(k)|^2 = \frac{1}{1 + 2\pi k}. \quad (10\text{–}166)$$

If, in addition, we assume that a reasonable approximation is given by

$$|2iC(k) - (2\lambda + 1)k|^2 = 4 + k^2(2\lambda + 1)^2, \quad (10\text{–}167)$$

the formula for the mean square of the acceleration reduces to

$$\overline{\xi^2} = \frac{4U^2}{b^2} \frac{\overline{w_G^2}}{\pi} \frac{1}{(2\lambda + 1)^2} \int_0^\infty \frac{\zeta^2}{a^2 + \zeta^2} \frac{1}{1 + 2\pi r\zeta} \frac{1 + 3\zeta^2}{(1 + \zeta^2)^2} d\zeta, \quad (10\text{–}168)$$

where $r = b/L$ and $a = 2/r(2\lambda + 1)$. Fung (Ref. 10–37) has shown that Eq. (10–168) can be integrated explicitly to yield

$$\overline{\xi^2} = \overline{w_G^2} \left[\frac{4U^2}{\pi b^2 (2\lambda+1)^2} \right] \left[\frac{a^2(3a^2 - 1)}{(a^2 - 1)^2} I_1(a, r) - \frac{5a^2 - 3}{(a^2 - 1)^2} I_1(l, r) + \frac{2}{a^2 - 1} I_2(a, r) \right],$$

$$(10\text{–}169)$$

where

$$I_1 = \frac{2\pi r}{1 + 4\pi^2 r^2 a^2} \left(\ln 2\pi r + \ln a + \frac{1}{4ar} \right),$$

$$I_2 = \frac{8\pi^3 r^3}{(1 + 4\pi^2 r^2 a^2)^2} \left(\ln 2\pi r + \ln a + \frac{1}{4ar} \right) + \frac{\pi r}{a^2(1 + 4\pi^2 r^2 a^2)} \left(\frac{1}{4ra} - 1 \right).$$

Examination of this result indicates that the mean square of the acceleration is critically dependent upon the ratio of wing semichord to turbulence scale, $r = b/L$. In particular, when this ratio tends to zero, that is, when the turbulence scale is large compared with the semichord, the mean square of the acceleration tends to zero. In this case, the rigid wing is, in effect, progressing through gusts with very large gradients which produce small accelerations. When the ratio of semichord to turbulence scale is large and tends to infinity, the mean square of the acceleration also tends to zero. This result is plausible, since the small eddies distributed in the chordwise direction tend to cancel one another and the resulting rigid-wing acceleration is thereby small. Between these two extremes there is a critical value of r at which peak accelerations would be experienced.

Although the mean square acceleration computed above has a certain quantitative significance, especially in comparing one aircraft with another, more information is needed in order to apply these methods of analysis to structural design. Additional relations between the power spectrum of response and the actual time history of the response are needed. For example, the probable number of times a given acceleration will be exceeded while flying a given distance through air of a given mean square turbulence intensity is of practical importance in fatigue studies. Such results can be predicted statistically, and when more is known about the statistical aspects of atmospheric turbulence they may eventually form a practical basis for structural design. Space does not permit discussion of these important statistical concepts, and the reader is referred to Refs. 10–35 and 10–36 for further elaboration.

(b) *Airplane free to translate vertically with wing bending.* Consider the response to turbulent air of an airplane free in vertical translation and having a wing elastic in bending but restrained against twisting. The wing is assumed slender and straight with arbitrary mass, stiffness, and chord variations along the span. In Section 10–6(b) we have seen that the equations of motion are given by Eqs. (10–146). These equations can be adapted to the prediction of the response to a sinusoidal gust of the form of expression (5–374) by substituting the solution

$$\xi_r = \bar{\xi}_r e^{iks}, \tag{10–170}$$

where $k = \omega b_R / U$ and $s = Ut/b_R$. This substitution yields

$$-\left(\frac{U}{b_R}\right)^2 M k^2 \bar{\xi}_r e^{iks} + M \omega_r^2 \bar{\xi}_r e^{iks} = \Xi_r^M + \Xi_r^D,$$

$$(r = 1, 2, \cdots, n; \omega_1 = 0). \tag{10–171}$$

The right side is evaluated by substituting the following into Eqs. (10–143) and (10–144):

$$L_G = 2\pi\rho U^2 b_R \frac{\bar{w}_G}{U} a(y) K(k) e^{iks}, \tag{10–172}$$

where $K(k)$ is defined by Eq. (10–159), and

$$L_M = \pi\rho U^2 [a^2(y)k^2 - 2a(y)ikC(k)] \left[\sum_{j=1}^{n} \phi_j \bar{\xi}_j \right] e^{iks}, \tag{10–173}$$

where $C(k)$ is defined by Eq. (5–309). Thus when strip theory is applied, Eq. (10–171) reduces to*

$$-\lambda k^2 \bar{\xi}_r + \lambda \Omega_r^2 \bar{\xi}_r - k^2 \sum_{j=1}^{n} A_{rj} \bar{\xi}_j + 2ikC(k) \sum_{j=1}^{n} B_{rj} \bar{\xi}_j = 2b_R B_{1r} \frac{\bar{w}_G}{U} K(k),$$

$$(r = 1, 2, \cdots, n; \omega_1 = 0). \tag{10–174}$$

---

* The assumption that the turbulence is two-dimensional and constant along the spanwise axis introduces error when dealing with scales of turbulence such that $2l/L$ is of the order of one or greater.

where λ, Ω_r, A_{rj}, and B_{rj} are defined by Eq. (10–149). Equations (10–174) are a set of simultaneous linear equations with complex coefficients and complex unknown quantities $\xi_1, \cdots, \xi_n$. In general, solutions can be obtained for the complex quantities

$$\frac{\xi_1}{b_R(\bar{w}_G/U)}, \quad \frac{\xi_2}{b_R(\bar{w}_G/U)}, \quad \cdots, \quad \frac{\xi_n}{b_R(\bar{w}_G/U)}. \quad (10\text{–}175)$$

These quantities may be regarded as the admittance functions with respect to displacements of the normal modes. Let us suppose, for example, that we are interested in studying the effect of turbulence on wing bending moment. The admittance function with respect to wing bending moment, $\bar{M}/\bar{w}_G$, can be computed from Eq. (10–151) by putting $M(y,s) = \bar{M}e^{iks}$:

$$\bar{M}e^{iks} = \int_y^l \left[L_G + L_M - \frac{U^2}{b_R{}^2}\left(\sum_{r=1}^n \phi_r(\eta)\xi_r{}''(s) \right) m(\eta) \right](\eta - y)d\eta. \quad (10\text{–}176)$$

When ξ_r, L_G, and L_M are introduced from Eqs. (10–170), (10–172), and (10–173), respectively, the admittance function with respect to the bending moment becomes

$$\frac{\bar{M}}{\bar{w}_G} = \frac{MU}{\lambda b_R} \left\{ 2K(k)\frac{b_R}{S}\int_y^l a(\eta)(\eta - y)d\eta \right.$$

$$+ k^2 \sum_{r=1}^n \frac{\xi_r}{b_R(\bar{w}_G/U)}\frac{b_R}{S}\int_y^l a^2(\eta)\phi_r(\eta)(\eta - y)d\eta$$

$$- 2ikC(k) \sum_{r=1}^n \frac{\xi_r}{b_R(\bar{w}_G/U)}\frac{b_R}{S}\int_y^l a(\eta)\phi_r(\eta)(\eta - y)d\eta$$

$$\left. + \lambda k^2 \sum_{r=1}^n \frac{\xi_r}{b_R(\bar{w}_G/U)}\frac{1}{M}\int_y^l \phi_r(\eta)m(\eta)(\eta - y)d\eta \right\}. \quad (10\text{–}177)$$

When the quantities given by (10–175) are introduced into Eq. (10–177) the admittance function is defined explicitly. The mean square of the bending moment is

$$\overline{M^2} = \int_0^\infty \left| \frac{\bar{M}}{\bar{w}_G} \right|^2 \Phi(\omega)d\omega. \quad (10\text{–}178)$$

Equations of motion taking account of other degrees of rigid-body and elastic freedom can be derived in a similar manner. Perhaps the most significant omission in the above illustrative derivation is the pitching degree of freedom. Experimental and theoretical studies of aircraft response in continuous rough air have indicated that the pitching degree of freedom and the associated symmetrical stability properties are important parameters.

It is important to observe the similarity of the form of the equations of motion (10–174) and those employed in flutter analyses. The same aerodynamic force coefficients are involved on the left side, and advantage may thereby be taken of tables of aerodynamic coefficients at various Mach numbers which have been prepared for flutter work.

EXAMPLE 10–6. To calculate the power spectrum of root bending moment of the jet transport airplane at a speed of $U = 475$ mph. Take into consideration plunging of the airplane and the fundamental bending mode of the wing. Assume that the power spectrum of the atmospheric turbulence is defined by the empirical expression* (cf. Fig. 10–25)

$$\Phi\left(\frac{\omega}{U}\right) = \frac{0.060}{0.000004 + (\omega/U)^2}. \tag{a}$$

Compare the results with those obtained for a rigid airplane and deduce a factor of root-mean-square dynamic overstress.

  Solution. The equations of motion derive from Eqs. (10–174) by putting $n = 2$, as follows:

$$\bar{a}_{11}\left(\frac{\xi_1}{b_R(\overline{w}_G/U)}\right) + \bar{a}_{12}\left(\frac{\xi_2}{b_R(\overline{w}_G/U)}\right) = \bar{f},$$

$$\bar{a}_{21}\left(\frac{\xi_1}{b_R(\overline{w}_G/U)}\right) + \bar{a}_{22}\left(\frac{\xi_2}{b_R(\overline{w}_G/U)}\right) = \frac{B_{12}}{B_{11}}\bar{f},$$

(b)

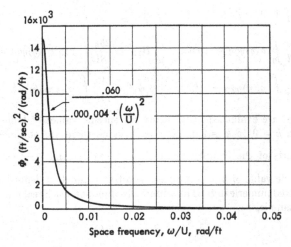

Fig. 10–25. Power spectrum of atmospheric turbulence.

---

*This expression is an approximation of the curve given in Ref. 10–36, Fig. 2. Experimental and theoretical evidence indicates that Φ is proportional to $1/(\omega/U)^2$ for large values of ω/U. The constant is added to the denominator to make Φ finite at $\omega/U = 0$.

where
$$\bar{a}_{11}(k) = -\lambda k^2 - k^2 A_{11} + 2ikC(k)B_{11},$$
$$\bar{a}_{12}(k) = \bar{a}_{21}(k) = -k^2 A_{12} + 2ikC(k)B_{12},$$
$$\bar{a}_{22}(k) = -\lambda k^2 + \lambda \Omega_2{}^2 - k^2 A_{22} + 2ikC(k)B_{22},$$
$$\tilde{f}(k) = 2B_{11}K(k).$$

Evaluations of the constants λ, Ω_2, B_{11}, B_{12}, B_{21}, and B_{22} have been carried out in Example 10–5. In the present example we require, in addition, the constants A_{11}, A_{12}, and A_{22}. The latter are defined by Eqs. (10–149), and their numerical values are

$$A_{11} = \frac{b_R}{S} \int_{-l}^{+l} a^2(y)dy = 0.5246, \qquad A_{12} = \frac{b_R}{S} \int_{-l}^{+l} a^2(y)\phi_2(y)dy = 0.2610,$$

$$A_{22} = \frac{b_R}{S} \int_{-l}^{+l} a^2(y)\phi_2{}^2(y)dy = 1.1078.$$

The admittance function with respect to the root bending moment derives from Eq. (10–177) by putting $n = 2$ and $y = 0$:

$$\frac{\overline{M}}{\overline{w}_G} = \frac{MUl}{\lambda b_R} \left\{ B_{11}'K(k) + \tfrac{1}{2}[A_{11}'k^2 - 2B_{11}'ikC(k) + 2\lambda k^2 \gamma_1'] \frac{\xi_1}{b_R(\overline{w}_G/U)} \right.$$
$$\left. + \tfrac{1}{2}[A_{12}'k^2 - 2B_{12}'ikC(k) + 2\lambda k^2 \gamma_2'] \frac{\xi_2}{b_R(\overline{w}_G/U)} \right\}, \quad \text{(c)}$$

where

$$A_{11}' = \frac{2b_R}{Sl} \int_0^l a^2(\eta)\eta d\eta = .1965, \qquad A_{12}' = \frac{2b_R}{Sl} \int_0^l a^2(\eta)\phi_2(\eta)\eta d\eta = .2841,$$

$$B_{11}' = \frac{2b_R}{Sl} \int_0^l a(\eta)\eta d\eta = .2163, \qquad B_{12}' = \frac{2b_R}{Sl} \int_0^l a(\eta)\phi_2(\eta)\eta d\eta = .3779,$$

$$\gamma_1' = \frac{1}{Ml} \int_0^l m(\eta)\eta d\eta = .1224, \qquad \gamma_2' = \frac{1}{Ml} \int_0^l m(\eta)\phi_2(\eta)\eta d\eta = .1166.$$

Equations (b) are evaluated numerically for specific values of k chosen in the interval from $k = 0$ to $k = 2.0$. For each value there are obtained real and imaginary parts of the complex quantities $\dfrac{\xi_1}{b_R(\overline{w}_G/U)}$ and $\dfrac{\xi_2}{b_R(\overline{w}_G/U)}$. Thus for each specific value of k, the complex admittance function defined by Eq. (c) can be evaluated numerically. The admittance function has a real and imaginary part. For example,

$$\frac{\overline{M}(k)}{\overline{w}_G} = R(k) + iI(k) \tag{d}$$

where R and I are the real and imaginary parts, respectively. The square of the absolute value of the admittance function, which is a real number, is computed by simply summing the square of the real part and the square of the imaginary part.

$$\left| \frac{\overline{M}(k)}{\overline{w}_G} \right|^2 = R^2(k) + I^2(k). \tag{e}$$

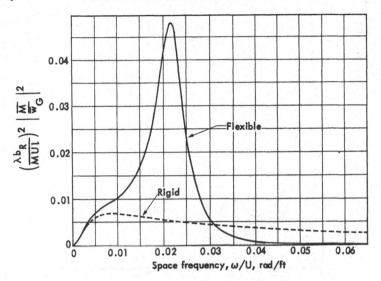

Fig. 10–26. Admittance function of root bending moment.

In the present problem, the square of the absolute value of the admittance function is evaluated numerically for the same originally selected values of k. The results are plotted in Fig. 10–26. In this figure, the ordinate scale is chosen as $(\lambda b_R/MUl)^2|\overline{M}(\omega/U)/\overline{w}_G|^2$ and the abscissa scale as ω/U for convenience. It should be observed that Fig. 10–26 illustrates also the result for a rigid airplane, as shown by the dashed curve. This result is obtained by using the same equations given above with ξ_2 put equal to zero. The power spectrum of the root bending moment is obtained by applying Eq. (10–156):

$$\Psi\left(\frac{\omega}{U}\right) = \left|\frac{\overline{M}(\omega/U)}{\overline{w}_G}\right|^2 \Phi\left(\frac{\omega}{U}\right). \tag{f}$$

When corresponding ordinates of Figs. 10–25 and 10–26 are multiplied there are obtained ordinates to a curve of $(\lambda b_R/MUl)^2\Psi$. These values, which are plotted vs. ω/U in Fig. 10–27, can be transformed into the power spectrum of the root bending moment by multiplying by $(\lambda b_R/MUl)^2$.

The curves in Fig. 10–27 illustrate clearly the different response characteristics which result from various frequency components in the atmosphere. The first peak at $\omega/U = .003$ is due to vertical translation, and wing flexibility has a small effect on the peak value. A second peak in the solid curve is associated with the fundamental wing bending degree of freedom; and a similar peak does not exist in the dashed curve of the rigid airplane. Thus the fundamental bending degree of freedom of the elastic airplane seeks out or is "tuned" to the atmospheric frequency components ω/U of approximately .0215.

The mean square of the gust input is obtained by taking the area under the curve of Fig. 10–25.

$$\overline{w_G^2} = \int_0^\infty \frac{0.060}{0.000004 + (\omega/U)^2}\, d\left(\frac{\omega}{U}\right) = 47.12 \text{ (ft/sec)}^2. \tag{g}$$

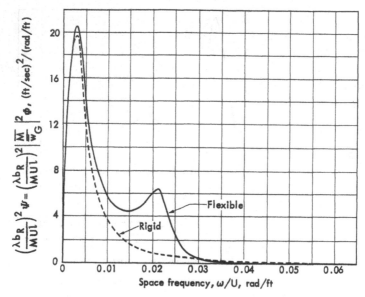

Fig. 10–27. Power spectrum of root bending moment.

The mean squares of the bending moments follow from the areas under the curves of Fig. 10–27 by application of Eq. (10–178).

$$\overline{M^2}_{(elastic)} = 0.2023 \left(\frac{MUl}{\lambda b_R}\right)^2 = 1.2115 \times 10^{10} (\text{lb·ft})^2,$$

$$\overline{M^2}_{(rigid)} = 0.1400 \left(\frac{MUl}{\lambda b_R}\right)^2 = 0.8384 \times 10^{10} (\text{lb·ft})^2. \tag{h}$$

Finally, the root-mean-square dynamic overstress factor, of some significance in appraising the influence of wing flexibility on stresses, is given by

$$\sqrt{\frac{\overline{M^2}_{(elastic)}}{\overline{M^2}_{(rigid)}}} = 1.202. \tag{i}$$

CHAPTER 11

AEROELASTIC MODEL THEORY

11-1 Introduction. Experimental investigations in the field of aeroelasticity have served two major purposes. They have been the guiding influence necessary to the development of useful theory and they have produced solutions to immediate practical problems in the large areas where existing theory is not yet dependable. Particularly in dealing with flutter, the testing of wind-tunnel models with properly scaled mass and stiffness properties has often been more rewarding than equivalent efforts using analytical techniques or even full-scale airplanes. In the course of this experimentation many new and valuable methods and techniques have been developed and have been transmitted only by word of mouth and by published paper. In order to provide a more adequate source of information, the remaining chapters are devoted to a presentation of the basic concepts of experimental aeroelasticity.

An intelligent approach to the design and use of models requires a thorough comprehension of model theory, which is the subject of this chapter. Chapter 12 deals with the problems of model design and construction and Chapter 13 with testing techniques.

11-2 Dimensional concepts. Model theory, whether applied to airplanes or chemical processes, must be based on a clear understanding of the concept of dimensions and a familiarity with the principles of dimensional analysis. For this reason we shall start with a review of basic properties of scientific systems of measurement. Fundamentally, the need for a "dimension" arises when it is desired to measure some physical quantity. Thus the measurement of a distance requires the concept of length. Similarly, the measurement of velocity requires the concept of time as well as length. Of course, measurement of a distance requires an explicit rule by which numbers are associated with distances. A moment's reflection will reveal that the rule in common use states that the measure of a given distance is the number of times a unit length can be fitted into that distance. In a similar manner, the measure of a period of time requires the use of a unit time interval. As a consequence, the measure of a distance depends on the size of the unit length which is employed, in the manner that the measure is inversely proportional to unit size. This characteristic of the usual systems of measurement leads to the important result that the ratio of the measures of two distances is quite independent of the size of the length unit which is used to measure them. For instance, the aspect ratio of a lifting surface does not depend on whether inches or meters are used to measure the span and chord. All systems of measurement in scientific use have this characteristic, which is fundamental to the concepts of dimensional analysis.

In the process of measuring physical quantities it soon becomes evident that there are two types. The first and simplest type is made up of quantities each of which can easily be measured directly in units of its own kind. The second type is made up of quantities which are derived from quantities of the first type according to some rule of combination. In a problem involving a certain set of physical quantities all of the quantities can be derived from a small number of quantities of the first type, which are then called *primary* quantities. The remaining quantities are called *secondary* quantities. For example, in the problems of static structures all of the secondary quantities such as bending moment and modulus of elasticity can be derived from the primary quantities, force and length. In mechanics, mass, length, and time are conveniently chosen as the primary quantities, whereas in compressible aerodynamics and thermodynamics, temperature must often be added as a fourth primary quantity. It should be noted that there is no fundamental reason why certain quantities must always be chosen as primary quantities, but rather it is the system of measurement which has been adopted which makes it convenient to call some quantities primary and others secondary.

An important restriction which must be embodied in the rules of combination for the derivation of secondary from primary quantities results from the previously mentioned characteristic of the usual scientific systems of measurement, that the ratio of the measures in two instances of a physical quantity, whether primary or secondary, is independent of the sizes of the units of the primary quantities. Thus the ratio of the measures of two velocities is the same whether the units are consistently miles and hours or feet and seconds. Bridgman (Ref. 11–1) shows that this restriction limits the manner in which the measures of a secondary quantity can be derived from measures $P_1, P_2, \cdots, P_m$ of the appropriate set of m primary quantities to the form

$$S = C P_1{}^{d_1} P_2{}^{d_2} P_3{}^{d_3} \cdots P_m{}^{d_m}, \tag{11–1}$$

where C is a pure number. Thus every secondary quantity is expressible as a number multiplied by positive or negative powers of the measures of the primary quantities. The exponent d_k is called the "dimension" of the secondary quantity in the primary quantity P_k and is an integer for the usual sets of primary quantities.

To formulate some examples in the familiar field of mechanics, we arbitrarily (and customarily) choose mass, length, and time as the primary quantities and give their measures the symbols M, L, and T, respectively. Then the measure of a force is expressible as $CM^1L^1T^{-2}$. Similarly, the measure of a density is CM^1L^{-3}, of pressure is $CM^1L^{-1}T^{-2}$, and of velocity is CL^1T^{-1}. If the units in the force expression are pounds, slugs, feet, and seconds, or dynes, grams, centimeters, and seconds, the constant C has a value of unity. The other physical quantities of density, pressure, and velocity, unlike force, do not have special names for their commonly used units but use instead the names of other quantities. Thus the unit of density is written in terms of the primary quantities, slugs and feet, that is, slugs/feet^3. Pressure is usually written in terms of a secondary and a primary quantity, pounds/feet^2. In both cases the constant C has, of course, the value unity. In the case of velocity there is a less common unit, the knot, for which the constant C is not unity unless length is measured in nautical miles and time in hours.

It is important to realize that the form of the expression for the measure of a force,

$$F = CM^1L^1T^{-2}, \tag{11–2}$$

is in general use primarily because of the practical importance of Newton's Second Law, which states that the mass of a body times its acceleration is proportional to the applied force. There is no fundamental reason why Newton's Law of Gravitation could not have determined the dimensions of force. This law states that the force of attraction between two masses is proportional to the product of the masses divided by the square of the distance between them. Thus the measure of the force could be written

$$F = CM^2L^{-2}. \tag{11–3}$$

Clearly, these two formulations of the secondary quantity, force, are incompatible and must not be used simultaneously. Consistent use of the first formulation requires that the Law of Gravitation be written as

$$F = \gamma \frac{m_1 m_2}{r^2}, \tag{11–4}$$

where γ has the dimensional form $M^{-1}L^3T^{-2}$ and is the universal gravitational constant. A consistent (but less customary) use of the second formulation would require that the Second Law be written as

$$F = \beta m \frac{d^2 s}{dt^2}, \tag{11–5}$$

where β might be called the "universal inertial constant" and would have the dimensional form $M^1L^{-3}T^2$. In this text, the first and more usual formulation based on Newton's Second Law will be employed. Nevertheless, it is important to recognize the existence of dimensional constants, such as the gravitational constant, the speed of light, and the gas constant, and not to confuse them with the pure number C, which depends only on the unit sizes.

11-3 Equations of motion. In order to describe a physical situation in terms of a mathematical equation, one or more physical principles must be applied. Typical of these principles in both mechanics and aerodynamics are Newton's Laws and the principle of conservation of matter. Application of appropriate principles to a given situation results in one or more equations containing the pertinent physical quantities as aruguments, and these may be primary quantities, secondary quantities, or dimensional constants (which have the same form as secondary quantities). An equation derived in the foregoing manner, if it includes the necessary dimensional constants, will be essentially independent of changes in size of the units of the primary quantities. Such a change in primary unit size will simply change each term in the equation by the same factor because each term has the same dimensions. In other words, since the physical situation is not altered by changes in size of the primary units, an equation which completely describes the physical situation should likewise be independent of the primary unit size. Such an equation is called a "complete equation" and is the only kind considered in this book.

The only instance in which changing the size of a primary unit does not alter the coefficient of each term in a complete equation is when the terms are each dimensionless in that particular primary quantity. For example, the equation for the unaccelerated motion of a particle,

$$x = x_0 + v_0 t, \tag{11-6}$$

is not altered if the primary quantity, time, is measured in hours rather than seconds. In fact, dividing the equation by x_0 conveniently makes each term dimensionless in both primary quantities, time and length. Thus the dimensionless form,

$$\frac{x}{x_0} = 1 + \frac{v_0 t}{x_0}, \tag{11-7}$$

is completely independent of unit size. Furthermore, the apparent complexity of the equation is seen to be reduced in the dimensionless form in that only two dimensionless variables, (x/x_0) and $(v_0 t/x_0)$, need be considered, rather than the four original physical quantities x, x_0, v_0, and t.

This idea of writing complete equations in dimensionless form with fewer arguments has been generalized by Buckingham (Ref. 11-2) and explicitly demonstrated by Bridgman (Ref. 11-1). It is summarized in the "Π Theorem," which states that if a physical situation can be represented by means of a complete equation,

$$\psi(S_1, S_2, S_3, S_4, \cdots, S_n) = 0 \tag{11-8}$$

(where the n arguments include all the primary quantities, secondary quantities, and dimensional constants which must be considered in the problem), the equation can be written in the form

$$\phi(\Pi_1, \Pi_2, \Pi_3, \Pi_4, \cdots, \Pi_{n-m}) = 0, \tag{11-9}$$

in which Π_1, Π_2, $\cdots$, Π_{n-m} are the $(n-m)$ independent products of the arguments S_1, $\cdots$, S_n, which are dimensionless in the m primary quantities. The form of these dimensionless Π's can be found by a formal procedure (Ref. 11–1) but can usually be constructed by inspection and the application of common sense. Typical Π's in general use are aspect ratio, reduced frequency, Mach number, and Reynolds number.

In model theory the advantages of using dimensionless variables are twofold. First, the problem is conveniently expressed in terms of the minimum number of variables. Secondly, since the dimensionless equation of motion is completely unaffected by scale effects (that is, changes in the size of the primary units) the values of the dimensionless variables must be the same for both the original problem and its model. This idea will be applied in detail in the following sections of this chapter.

11–4 Vibration model similarity laws. As a first example of model theory, consider the problem of determining by the use of a conveniently small model the natural frequencies and mode shapes of a *cantilever tapered wing with a concentrated mass at its tip*. As shown in Fig. 11–1, the wing is of semispan l with tip weight M_T. It fits into the class of structures adequately described by simple beam theory and has a running mass $m(y)$, a varying area moment of inertia $I(y)$, and a deflection $w(y,t)$.

The problem of designing a suitable model of this wing will be attacked in the fashion suggested by the previous section. Thus, the governing equation of motion will be studied to ascertain the parameters which must enter the problem. These parameters will be formed into a suitable set of dimensionless parameters so that the equation of motion can be written in a dimensionless form involving the minimum number of variables. Since this equation is not affected in any way by changes in the size of the primary units, it applies equally well to the full-scale wing and to any model which is designed to simulate the full-scale wing. It shows directly the dimensionless parameters which must be built into the model so that the model will produce the same dimensionless answers as would be obtained from tests on the full-scale wing.

The actual scale factors to be used in the design of the model are found by considering the practical limitations imposed on the model. These may

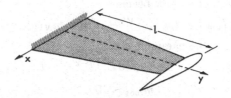

Fig. 11–1. Cantilever tapered wing with tip mass.

be allowable model size, methods of model fabrication, or perhaps available instrumentation and apparatus. If all of the dimensionless parameters appear independently in the equation of motion, the number of separate design limitations which can be imposed on the model is equal to the number of scale factors to be determined, that is, the number of primary quantities involved in the problem. In the simpler problems, one or more pairs of dimensionless parameters may appear only in a fixed combination. In these cases, additional design limitations may be imposed.

Following this plan of attack, then, the first step is to write the applicable equation of motion,

$$[EI(y)w''(y,t)]'' + m(y)\ddot{w}(y,t) = 0 \qquad (3\text{-}7)$$

with appropriate boundary conditions at $y = 0$ and at $y = l$ to account for the cantilever root mounting and the effect of the concentrated tip mass M_T. As shown in Chapter 3, this equation can be separated into two total differential equations [Eqs. (3-10) and (3-11)]. The mode shapes $W(y)$ and frequencies ω which we seek are solutions of Eq. (3-11):

$$[EI(y)W''(y)]'' - \omega^2 m(y)W(y) = 0. \qquad (3\text{-}11)$$

The dimensions of each of the parameters in this equation can be expressed in terms of the set of primary quantities mass, length, and time (M, L, and T). Before doing this, however, it should be noted that some of the parameters involve two concepts; they involve a variation with y as well as the idea of dimensional magnitude. The use of a notation which separates these concepts makes the coming steps more readily understandable and will be accomplished in the following manner. Let

$$m(y) = m_{\text{root}} m_N(y), \qquad (11\text{-}10)$$

where m_{root} is the dimensional magnitude of the running mass at the root station and $m_N(y)$ is the variation of running mass along the wing normalized to have a value of unity at the root station. Thus $m_N(y)$ represents the mass distribution independent of the actual magnitude of the running mass at any station. It is, of course, a dimensionless quantity. Similarly, let

$$W(y) = W_{\text{tip}} \phi(y), \qquad (11\text{-}11)$$

$$I(y) = I_{\text{root}} I_N(y). \qquad (11\text{-}12)$$

The equation of motion now becomes

$$[EI_{\text{root}} I_N(y)\phi''(y)]'' - \omega^2 m_{\text{root}} m_N(y)\phi(y) = 0. \qquad (11\text{-}13)$$

The dimensions of the parameters are

$$y, l \sim L, \qquad E \sim ML^{-1}T^{-2}, \qquad I_{\text{root}} \sim L^4,$$

$$\omega \sim T^{-1}, \qquad m_{\text{root}} \sim ML^{-1}, \qquad M_T \sim M, \qquad (11\text{-}14)$$

$$I_N(y), \ \phi(y), \ m_N(y) \sim \text{dimensionless}.$$

These can be combined into the dimensionless set

$$\frac{y}{l}, \quad \frac{I_{root}}{l^4}, \quad \frac{M_T}{m_{root}l}, \quad \frac{\omega^2 m_{root}}{E}, \quad I_N(y), \quad \phi(y), \quad m_N(y). \quad (11\text{-}15)$$

As predicted by the Π Theorem, the dimensionless parameters (11–15) are three less in number than the dimensional parameters (11–14) and form a complete set because any other dimensionless parameters formed from the same dimensional set (11–14) can also be constructed from combinations of the dimensionless parameters in the complete set (11–15).

The dimensionless form of the equation of motion can be now written as

$$\left[I_N\left(\frac{y}{l}\right)\phi''\left(\frac{y}{l}\right)\right]'' - \left[\frac{l^4\omega^2 m_{root}}{EI_{root}}\right]m_N\left(\frac{y}{l}\right)\phi\left(\frac{y}{l}\right) = 0 \quad (11\text{-}16)$$

with appropriate boundary conditions at $y/l = 0$ and at $y/l = 1$ which involve M_T/m_{root} as well as parameters in the above equation. Here the primes denote differentiation with respect to y/l rather than y. It is apparent from the form of the equation that the parameters $\omega^2 m_{root}/E$ and I_{root}/l^4 do not appear independently in this problem but only in the combination $\omega^2 l^4 m_{root}/EI_{root}$.

The desired mode shapes and natural frequencies, in their dimensionless form $\phi(y/l)$ and $\omega^2 l^4 m_{root}/EI_{root}$ apparently can be obtained from any model with the correct area moment of inertia distribution $I_N(y/l)$, with the correct running mass distribution $m_N(y/l)$, and with the correct proportion of mass concentrated at the tip.

To obtain the dimensional properties to be designed into the model, the independent limitations which determine the three scale factors must be examined. Assuming that the model size is limited for practical and economic reasons to $\frac{1}{10}$ of the wing size, the length scale factor can be written

$$\frac{L_m}{L_w} = \frac{1}{10}. \quad (11\text{-}17)$$

Assuming also that the measuring equipment which is available is best suited to the frequency range above that expected of the wing vibrations (full-scale), the model should be designed to have frequencies perhaps three times those of the wing; that is,

$$\frac{T_m}{T_w} = \frac{1}{3}. \quad (11\text{-}18)$$

The third design limitation may be that the model must have a metal structure so that the internal damping is of the same order as for the full-scale wing. Thus the ratio of moduli of elasticity is determined by choice of material. If steel is used in the model, and aluminum alloy in the wing,

$$\frac{E_m}{E_w} = \frac{3}{1} = \frac{(ML^{-1}T^{-2})_m}{(ML^{-1}T^{-2})_w}. \quad (11\text{-}19)$$

Solving for the scale factor in the third primary quantity,

$$\frac{M_m}{M_w} = \frac{3}{1} \frac{L_m}{L_w} \left(\frac{T_m}{T_w}\right)^2 = \frac{1}{30} . \tag{11-20}$$

From these three scale factors in the primary quantities mass, length, and time, the model size of all the necessary dimensional quantities which appear in the equation of motion (11–16) can be determined. For instance, the tip weight on the model can be found from

$$(M_T)_m = \tfrac{1}{30} (M_T)_w, \tag{11-21}$$

and the reference value of the running mass becomes

$$(m_{\text{root}})_m = \tfrac{1}{30}(\tfrac{1}{10})^{-1} (m_{\text{root}})_w = \tfrac{1}{3}(m_{\text{root}})_w. \tag{11-22}$$

The area moment of inertia becomes

$$(I_{\text{root}})_m = \frac{1}{10^4} (I_{\text{root}})_w. \tag{11-23}$$

When the sizes of all these model properties have been computed they may be used to determine all other model properties which do not specifically appear in Eq. (11–16). It is not ordinarily correct to calculate these latter quantities by scaling them down according to the scale factors for mass, length, and time.

For example, we cannot conclude that the maximum stresses in the model for a scaled deflection in the first natural mode can be found by applying the appropriate scale factors to the full-scale stresses. The maximum stress at a given cross section depends not only on the local value of the bending moment and area moment of inertia but also on the distance between the neutral axis and the outermost fiber. Although the first two quantities will be properly scaled, the last quantity depends, for instance, on the cross-sectional shape chosen for the model structure. There is no reason why this shape must be exactly the same as on the full-scale structure; the model cross section need only have the correct area moment of inertia to produce the desired frequencies and mode shapes. On most aeroelastic models the structural properties are actually obtained with the use of somewhat inefficient but easily fabricated structures which bear but slight resemblance to the full-scale structure.

Thus the model may be constructed of steel with $\tfrac{1}{10}$ the semispan, $\tfrac{1}{30}$ the total mass, and with the same distributions of mass and stiffness as the full-scale wing. The model frequencies will then be three times those of the wing in the corresponding modes of vibration, and the model mode shapes $\phi(y/l)$ will be identical with those of the wing. Of course, if the limitations on the model design were different from those assumed above, the scale factors would have been different.* The measured frequencies, when converted to full scale, would have been exactly the same.

As a second example of model theory, we shall consider a more generalized vibration problem written in terms of an integral equation rather than the differential equation used above. Suppose that we desire to build a vibration model of a *complete unrestrained airplane* in order to determine its many mode shapes and frequencies of vibration.

A complete equation describing the vibration of the airplane of Fig. 3-21 is derived in Chapter 3 as

$$\omega^2 \iint_S G(x,y;\xi,\eta)W(\xi,\eta)\rho(\xi,\eta)d\xi d\eta = W(x,y), \tag{3-172}$$

where the $W(x,y)$ are the natural mode shapes, the ω's are the natural frequencies, $\rho(x,y)$ is the two-dimensional mass distribution, and

$$G(x,y;\xi,\eta) = C(x,y;\xi,\eta) - \iint_S C(r,s;\xi,\eta)\left[\frac{1}{M}+\frac{ys}{I_x}+\frac{xr}{I_y}\right]\rho(r,s)drds. \tag{3-173}$$

In Eq. (3-173), $C(x,y;\xi,\eta)$ is the two-dimensional flexibility influence function and M, I_y, and I_x are the mass properties of the rigid airplane. The area of integration S can be characterized by the semispan l. As in the previous problem, some of the variables involve both the concept of a distribution over the plan of the airplane and a dimensional magnitude. To separate these concepts, let

$$\rho(x,y) = \rho_0\rho_N(x,y), \tag{11-24}$$

where $\rho_N(x,y)$ is normalized to have the value unity at the station $x = 0$, $y = 0$. Similarly, let

$$C(x,y;\xi,\eta) = C_{\text{tip}}C_N(x,y;\xi,\eta), \tag{11-25}$$

where C_{tip} is the value of C at $x = \xi = 0$, $y = \eta = l$. The mode shape is expressed in terms of a normalized mode shape by

$$W(x,y) = W_{\text{tip}}\phi(x,y). \tag{11-26}$$

In order to cast the integral equation into dimensionless form, the dimensions of each parameter must first be written in terms of the three

---

* In this example, as mentioned earlier, the two dimensionless ratios I_{root}/l^4 and $\omega^2 m_{\text{root}}/E$ do not appear independently but only in the combination $\omega^2 l^4 m_{\text{root}}/EI_{\text{root}}$. As a result of this unusual situation, only the combined parameter must be simulated; a fourth independent limitation may be applied. For instance, if the scale factor for I_{root} is chosen different from 10^4, and if the scale factor for m_{root} differs in the same ratio from 3, the value of $\omega^2 l^4 m_{\text{root}}/EI_{\text{root}}$ will be unchanged. This modification also requires a change in the scale factor for M_T in order to leave unchanged $M_T/m_{\text{root}}l$.

primary quantities mass, length, and time. Thus,

$$\omega \sim T^{-1}, \qquad \rho_0 \sim ML^{-2}, \qquad \left.\begin{matrix} x,y \\ \xi,\eta \\ r,s,b \end{matrix}\right\} \sim L, \qquad (11\text{–}27)$$

$$C_{\text{tip}} \sim M^{-1}T^2, \qquad W_{\text{tip}} \sim L,$$

$$M \sim M, \qquad I_y, I_x \sim ML^2,$$

$$\rho_N(x,y), \; C_N(x,y\,;\xi,\eta), \; \phi(x,y) \sim \text{dimensionless.}$$

A convenient set of dimensionless ratios can easily be constructed from these parameters. According to the Π Theorem, there should be three less independent dimensionless ratios than there are parameters listed above, since there are three primary quantities in this problem. The dimensionless parameters are

$$\omega^2 M C_{\text{tip}}, \qquad \frac{\rho_0 l^2}{M}, \qquad \frac{I_x}{Ml^2}, \qquad \frac{I_y}{Ml^2}, \qquad \frac{W_{\text{tip}}}{l}, \qquad \frac{x}{l}, \qquad \frac{y}{l},$$

$$\frac{\xi}{l}, \qquad \frac{\eta}{l}, \qquad \frac{r}{l}, \qquad \frac{s}{l}, \qquad \rho_N, \qquad C_N, \qquad \phi. \qquad (11\text{–}28)$$

The integral equation can be written in the form

$$(\omega^2 M C_{\text{tip}}) \left(\frac{\rho_0 l^2}{M}\right) \iint_S \frac{G\left(\dfrac{x}{l}, \dfrac{y}{l}; \dfrac{\xi}{l}, \dfrac{\eta}{l}\right)}{C_{\text{tip}}} \, \phi\left(\frac{\xi}{l}, \frac{\eta}{l}\right) \rho_N\left(\frac{\xi}{l}, \frac{\eta}{l}\right) d\frac{\xi}{l} d\frac{\eta}{l}$$

$$= \phi\left(\frac{x}{l}, \frac{y}{l}\right), \qquad (11\text{–}29)$$

where

$$\frac{G\left(\dfrac{x}{l}, \dfrac{y}{l}; \dfrac{\xi}{l}, \dfrac{\eta}{l}\right)}{C_{\text{tip}}} = C_N\left(\frac{x}{l}, \frac{y}{l}; \frac{\xi}{l}, \frac{\eta}{l}\right) - \iint_S C_N\left(\frac{r}{l}, \frac{s}{l}; \frac{\xi}{l}, \frac{\eta}{l}\right)$$

$$\times \left[1 + \left(\frac{Ml^2}{I_x}\right)\frac{y}{l}\frac{r}{l} + \left(\frac{Ml^2}{I_y}\right)\frac{x}{l}\frac{s}{l}\right]\left(\frac{\rho_0 l^2}{M}\right) \rho_N\left(\frac{r}{l}, \frac{s}{l}\right) d\frac{r}{l} d\frac{s}{l}. \qquad (11\text{–}30)$$

Thus any model which has the same mass distribution ρ_N (and thus the same values of $\rho_0 l^2/M$, I_y/Ml^2, I_x/Ml^2) and the same flexibility distribution C_N over the same geometry (dimensionless area of integration) will produce the same mode shapes ϕ and dimensionless frequencies $\omega^2 M C_{\text{tip}}$ as the full-scale airplane. The three scale factors which determine for the model the dimensional magnitudes of the mass and flexibility distributions $(\rho_0)_m$ and $(C_{\text{tip}})_m$ as well as model size are determined, as in the previous problem, by three independent design limitations imposed on the model. They may be expressed, for example, as a restriction on model size, a desire to simulate the flexibility of various components using a single metal spar of reasonable dimensions, or in numerous other ways.

11-5 Similarity laws for systems under steady airloads. As a further example of model theory, we consider the problem of studying by means of a model the static elastic deformations of a lifting surface under its airload in incompressible flow. In Chapter 8 analytical methods have been presented for predicting such deformations and the accompanying airload distribution for the case of a simple two-dimensional wing and for the case of swept and tapered surfaces.

The angle of twist of a *two-dimensional elastically supported* wing in an airstream of velocity U is obtained in Chapter 8 as

$$\theta = \frac{C^{\theta\theta}[(\partial C_L/\partial\alpha)e\alpha^r + C_{MAC}c]qS}{1 - C^{\theta\theta}(\partial C_L/\partial\alpha)qSe}, \qquad (8\text{--}4b)$$

where the symbols are defined in Section 8–2(a).

Since this is a problem in statics, we have only two primary quantities which may be chosen as force (F) and length (L). It is evident that the dimensions of the quantities in Eq. (8–4b) are

$$C^{\theta\theta} \sim F^{-1}L^{-1}, \qquad S \sim L^2, \qquad e \sim L,$$

$$q \sim FL^{-2}, \qquad c \sim L, \qquad (11\text{--}31)$$

$$\theta, \ \alpha^r, \ \frac{\partial C_L}{\partial\alpha}, \ C_{MAC} \sim \text{dimensionless}.$$

The independent dimensionless ratios which can be formed from these quantities are, by inspection,

$$C^{\theta\theta}qSc, \quad \frac{e}{c}, \quad \frac{c^2}{S}, \quad \theta, \quad \alpha^r, \quad \frac{\partial C_L}{\partial\alpha}, \quad C_{MAC} \quad (11\text{--}32)$$

or two less in number, as predicted by the Π Theorem. In terms of dimensionless parameters, Eq. (8–4b) becomes

$$\theta = \frac{[(\partial C_L/\partial\alpha)(e/c)\alpha^r + C_{MAC}](C^{\theta\theta}qSc)}{1 - (\partial C_L/\partial\alpha)(C^{\theta\theta}qSc)(e/c)}. \qquad (11\text{--}33)$$

We can see from this dimensionless equation that any model of this semirigid wing must present the same airfoil shape to the airstream (C_{MAC} and $\partial C_L/\partial\alpha$),* must have the same dimensionless location of the elastic axis (e/c), the same initial angle of attack (α^r), and the same ratio of aerodynamic to elastic forces ($C^{\theta\theta}qSc$). Since the parameter c^2/S does not actually appear in Eq. (11–33), its simulation is not required for this prob-

---

* It should be mentioned that to obtain the same kinds of aerodynamic reactions on the model as on the wing the Reynolds number $\rho Uc/\mu$ should be the same. On a small model in air this would require a high airstream velocity which might be impractical. Since Reynolds number effects are often of secondary importance, this restriction is usually modified to restrict the Reynolds number from being less than a certain minimum value.

lem. The two scale factors are determined by two independent design limitations.

Proceeding now from this relatively simple example to the case of the airload distribution on a *slender swept and tapered flexible lifting surface*, it is assumed that the airplane is in a maneuver which imposes a given steady acceleration on the lifting surface. The applicable integral equation can be formed from Eqs. (8–170) and (8–173) as

$$\mathcal{Q}[cc_l{}^e] = q \int_0^l \bar{C}(y,\eta)cc_l{}^e d\eta + \tilde{f}(y), \tag{11–34}$$

where

$$\tilde{f}(y) = q \int_0^l \bar{C}(y,\eta)cc_l{}^r d\eta + q \int_0^l C^{\theta\theta}(y,\eta)c_{mAC}c^2 d\eta$$

$$- g \int_0^l [C^{\theta z}(y,\eta) + C^{\theta\theta}(y,\eta)d(\eta)]mN d\eta, \tag{8–171}$$

$$\bar{C}(y,\eta) = C^{\theta z}(y,\eta) + e(\eta)C^{\theta\theta}(y,\eta). \tag{8–172}$$

The operator $\mathcal{Q}$ is the functional relation between incidence distribution and lift distribution, and its form depends on the type of aerodynamic theory which is used. The product mNg represents the running weight multiplied by the load factor in g's. $d(y)$ defines the position of the center of gravity at any section.

In order to separate the concepts of distribution and dimensional magnitude, it is helpful to define

$$C^{\theta z}(y,\eta) = C_{tip}{}^{\theta z}C_N{}^{\theta z}(y,\eta), \qquad c(y) = c_{tip}c_N(y),$$
$$C^{\theta\theta}(y,\eta) = C_{tip}{}^{\theta\theta}C_N{}^{\theta\theta}(y,\eta), \qquad e(y) = e_{tip}e_N(y),$$
$$m(y) = m_{tip}m_N(y), \qquad d(y) = d_{tip}d_N(y). \tag{11–35}$$

The dimensions of the parameters appearing in Eqs. (11–34) and (11–35) are

$$C_{tip}{}^{\theta z} \sim F^{-1}, \qquad C_{tip}{}^{\theta\theta} \sim F^{-1}L^{-1}, \qquad \left. \begin{array}{l} l, c_{tip} \\ \\ y, \eta \\ \\ d_{tip}, e_{tip} \end{array} \right\} \sim L, \tag{11–36}$$
$$q \sim FL^{-2}, \qquad (gm_{tip}) \sim FL^{-1},$$
$$\mathcal{Q} \sim L^{-1},$$

$$\left. \begin{array}{l} c_l{}^e, c_l{}^r, c_{mAC}, N, C_N{}^{\theta z}, c_N, \\ m_N, C_N{}^{\theta\theta}, e_N, d_N \end{array} \right\} \sim \text{dimensionless.}$$

These can be combined into the dimensionless set:

$$qc_{tip}lC_{tip}{}^{\theta z}, \qquad \frac{c_{tip}C_{tip}{}^{\theta\theta}}{C_{tip}{}^{\theta z}}, \qquad \frac{gm_{tip}}{qc_{tip}}, \qquad \frac{d_{tip}}{c_{tip}},$$
$$\frac{e_{tip}}{c_{tip}}, \qquad \mathcal{Q}c_{tip}, \qquad \frac{y}{l}, \qquad \frac{\eta}{l}, \qquad \frac{c_{tip}}{l}, \tag{11–37}$$

to which must be added the dimensionless quantities already listed. The equation thus becomes

$$c_{\text{tip}}\alpha[c_N c_l{}^e] = (qc_{\text{tip}}lC_{\text{tip}}{}^{\theta z})\int_0^1 \frac{\bar{C}}{C_{\text{tip}}{}^{\theta z}}\, c_N c_l{}^e d\left(\frac{\eta}{l}\right) + \tilde{f}, \quad (11\text{–}38a)$$

where

$$\tilde{f} = (qc_{\text{tip}}lC_{\text{tip}}{}^{\theta z})\left\{\int_0^1 \frac{\bar{C}}{C_{\text{tip}}{}^{\theta z}}\, c_N c_l{}^r d\left(\frac{\eta}{l}\right) + \left(\frac{c_{\text{tip}}C_{\text{tip}}{}^{\theta\theta}}{C_{\text{tip}}{}^{\theta z}}\right)\int_0^1 C_N{}^{\theta\theta} c_{mA}c c_N{}^2 d\left(\frac{\eta}{l}\right)\right.$$

$$\left. -\left(\frac{gm_{\text{tip}}}{qc_{\text{tip}}}\right) N\int_0^1 \left[C_N{}^{\theta z} + \left(\frac{d_{\text{tip}}}{c_{\text{tip}}}\right)\left(\frac{c_{\text{tip}}C_{\text{tip}}{}^{\theta\theta}}{C_{\text{tip}}{}^{\theta z}}\right) C_N{}^{\theta\theta} d_N\right] m_N d\left(\frac{\eta}{l}\right)\right\}, \quad (11\text{–}38b)$$

$$\frac{\bar{C}}{C_{\text{tip}}{}^{\theta z}} = C_N{}^{\theta z} + \left(\frac{e_{\text{tip}}}{c_{\text{tip}}}\right)\left(\frac{c_{\text{tip}}C_{\text{tip}}{}^{\theta\theta}}{C_{\text{tip}}{}^{\theta z}}\right) C_N{}^{\theta\theta} e_N. \quad (11\text{–}38c)$$

Since N always appears in a product with $gm_{\text{tip}}/qc_{\text{tip}}$, these two parameters are not independent in this problem and can be combined to form a single parameter $Ngm_{\text{tip}}/qc_{\text{tip}}$.

In order to find the load distribution $c_N c_l{}^e$ for the full-scale wing, it is necessary to construct a model which has:

(a) The same aerodynamic shape as the full-scale wing. That is, the planform, airfoil section, and initial "rigid" twist must be the same to produce same aerodynamic center distribution, moment coefficient distribution, and aerodynamic operator $c_{\text{tip}}\alpha$ (of course, parameters such as Mach number and Reynolds number may also be important).

(b) The same ratio of aerodynamic to elastic stiffness $qc_{\text{tip}}lC_{\text{tip}}{}^{\theta z}$.

(c) The same elastic stiffness distributions $C_N{}^{\theta z}$, $C_N{}^{\theta\theta}$, and e_N.

(d) The same ratio of inertial forces to aerodynamic forces $Ngm_{\text{tip}}/qc_{\text{tip}}$.

(e) The same mass distribution m_N and d_N.

(f) The same ratio of bending to torsional stiffness $c_{\text{tip}}C_{\text{tip}}{}^{\theta\theta}/C_{\text{tip}}{}^{\theta z}$.

(g) The same location of elastic axis and center of gravity $e_{\text{tip}}/c_{\text{tip}}$ and $d_{\text{tip}}/c_{\text{tip}}$.

This model will then have exactly the same load distribution as well as elastic deformation shape as the full-scale wing.

The scale factors of force and length are again determined by limitations on model design. Since the model is probably to be built for a particular wind tunnel, the model size and thus length scale are determined by the test section size. The dynamic pressure for the model is determined by the wind-tunnel capabilities or perhaps by the type of model construction. In any case, only two independent limitations can be imposed, since all of the dimensionless parameters are independent.

11–6 Flutter model similarity laws. As a first example of flutter model theory, we shall consider a *uniform straight cantilever wing of moderate aspect ratio* whose behavior can be adequately described by Theodorsen's typical section equations [Eqs. (9–13) through (9–19)]. After the assumption of simple harmonic motion at the circular frequency ω in both vertical motion and pitching, the equations have the form

$$-m\omega^2 h_0 + K_h h_0 - S_\alpha \omega^2 \alpha_0 e^{i\phi} + \pi \rho b^2 (iU\omega\alpha_0 e^{i\phi} - \omega^2 h_0 + ba\omega^2 \alpha_0 e^{i\phi})$$
$$+ 2\pi\rho UbC(k)[U\alpha_0 e^{i\phi} + i\omega h_0 + ib(\tfrac{1}{2} - a)\omega\alpha_0 e^{i\phi}] = 0, \quad (11\text{–}39a)$$

$$-S_\alpha \omega^2 h_0 - I_\alpha \omega^2 \alpha_0 e^{i\phi} + K_\alpha \alpha_0 e^{i\phi}$$
$$+ \pi\rho b^3 [i(\tfrac{1}{2} - a)U\omega\alpha_0 e^{i\phi} - \omega^2 b(\tfrac{1}{8} + a^2)\alpha_0 e^{i\phi} + a\omega^2 h_0]$$
$$- 2\pi\rho b^2 U(\tfrac{1}{2} + a)C(k)[U\alpha_0 e^{i\phi} + i\omega h_0 + ib(\tfrac{1}{2} - a)\omega\alpha_0 e^{i\phi}] = 0. \quad (11\text{–}39b)$$

The parameters which appear in the two equations of motion number thirteen and, in terms of the primary quantities mass, length, and time, have the dimensions

$$h_0, b \sim L, \qquad K_h \sim ML^{-1}T^{-2}, \qquad K_\alpha \sim MLT^{-2},$$
$$\omega \sim T^{-1}, \qquad S_\alpha \sim M, \qquad \rho \sim ML^{-3}, \qquad (11\text{–}40)$$
$$m \sim ML^{-1}, \qquad I_\alpha \sim ML, \qquad U \sim LT^{-1},$$
$$\alpha_0, \ \phi, \ a \sim \text{dimensionless.}$$

The reduced frequency k is not listed because it can be formed from ω, b, and U. A convenient set of dimensionless parameters can be formed as

$$\frac{h_0}{b}, \qquad \frac{\omega b}{U} (=k), \qquad \frac{m}{\rho b^2}, \qquad \sqrt{\frac{K_h/m}{K_\alpha/I_\alpha}} \left(= \frac{\omega_h}{\omega_\alpha} \right), \qquad \frac{\omega}{\sqrt{K_\alpha/I_\alpha}} \left(= \frac{\omega}{\omega_\alpha} \right),$$
$$\frac{S_\alpha}{mb} (= x_\alpha), \qquad \sqrt{\frac{I_\alpha}{mb^2}} (= r_\alpha), \qquad \alpha_0, \qquad \phi, \qquad a, \qquad (11\text{–}41)$$

and are ten in number, as predicted by the Π Theorem. Rewriting Eq. (11–39) in terms of these dimensionless parameters, we obtain

$$-\frac{h_0}{b} + \left(\frac{\omega_h}{\omega_\alpha}\right)^2 \left(\frac{\omega_\alpha}{\omega}\right)^2 \frac{h_0}{b} - x_\alpha \alpha_0 e^{i\phi} + \frac{\pi\rho b^2}{m}\left(\frac{i}{k}\alpha_0 e^{i\phi} - \frac{h_0}{b} + a\alpha_0 e^{i\phi}\right)$$
$$+ \frac{2\pi\rho b^2}{m}\frac{1}{k}C(k)\left[\frac{1}{k}\alpha_0 e^{i\phi} + i\frac{h_0}{b} + i(\tfrac{1}{2} - a)\alpha_0 e^{i\phi}\right] = 0, \quad (11\text{–}42a)$$

$$-x_\alpha \frac{h_0}{b} - r_\alpha^2 \alpha_0 e^{i\phi} + r_\alpha^2 \left(\frac{\omega_\alpha}{\omega}\right)^2 \alpha_0 e^{i\phi}$$
$$+ \frac{\pi\rho b^2}{m}\left[\frac{i}{k}(\tfrac{1}{2} - a)\alpha_0 e^{i\phi} - (\tfrac{1}{8} + a^2)\alpha_0 e^{i\phi} + a\frac{h_0}{b}\right]$$
$$- \frac{2\pi\rho b^2}{m}\frac{1}{k}(\tfrac{1}{2} + a)C(k)\left[\frac{1}{k}\alpha_0 e^{i\phi} + i\frac{h_0}{b} + i(\tfrac{1}{2} - a)\alpha_0 e^{i\phi}\right] = 0. \quad (11\text{–}42b)$$

Since the equations are homogeneous in h_0/b and α_0, the number of dimensionless ratios can be further reduced to nine by dividing through by α_0 to form an amplitude-ratio parameter $h_0/b\alpha_0$. Of course, the effect of the viscosity μ of the air was not included in the equations, nor were the speed of sound a_∞ and the ratio of specific heats γ. If these had been included, the number of original variables becomes 16 and the number of dimensionless variables 13, where the additional ones are Reynolds number $\rho U b/\mu$, Mach number U/a_∞, and γ.

Now that the equations for the wing motion are in a form which is independent of primary unit size, all other wings which have the same dimensionless properties under the same dimensionless flight conditions will give the same dimensionless answers. Thus any one of these wings can be considered to be a flutter model of the original wing.

The incompressible case. First we shall consider the simpler case of a wind-tunnel model where compressibility (and thus Mach number and γ) is presumed to have no effect. The quantities we seek are $\omega_F b/U_F$, ω_F/ω_α, $h_0/b\alpha_0$, and ϕ, where the subscript F refers to values at flutter. To obtain these quantities, the model must have the correct frequency ratio ω_h/ω_α, mass coupling x_α, elastic axis location a, radius of gyration r_α, mass ratio $m/\rho b^2$, external shape, and Reynolds number.

The number of independent limitations to which the model can be designed cannot exceed three because all of the dimensionless parameters are independent and there are only three primary quantities in the problem whose scale can be changed. The length scale factor between model and wing, L_m/L_w, is usually determined by the ratio of the allowable model span in the wind tunnel to the actual span. This length scale factor, through the dimensionless parameters x_α, a, and r_α, determines the dimensional chordwise location in the model of the center of gravity and elastic axis, as well as determining the model radius of gyration. In the incompressible testing being considered, the operating speed range of the tunnel or perhaps the type of model construction sets an upper limit on the airspeed range in which flutter must occur. If, for instance, the desired model flutter speed is in the neighborhood of 70 mph and the estimated flutter speed of the actual wing is 350 mph, a velocity scale factor U_m/U_w of $\frac{1}{5}$ is required. From this velocity scale factor and the previously determined length scale factor the scale factor of a second primary quantity, time, can be determined:

$$\frac{T_m}{T_w} = \frac{U_w}{U_m} \frac{L_m}{L_w}. \tag{11–43}$$

The scale factor of the final primary quantity, mass, usually is set by consideration of the air density to be employed in the wind tunnel. Thus, since the mass ratio $m/\pi\rho b^2$ must be the same for both model and wing,

$$\frac{m_m}{m_w} = \frac{(\pi\rho b^2)_m}{(\pi\rho b^2)_w} = \frac{\rho_m b_m^2}{\rho_w b_w^2}. \tag{11-44}$$

Since m is defined as wing mass per foot of span, the mass scale factor can be written from Eq. (11–44) as

$$\frac{M_m}{M_w} = \left(\frac{L_m}{L_w}\right)^3 \frac{\rho_m}{\rho_w}, \tag{11-45}$$

where the length scale factor and the air density ratio are known. The quantity ρ_m refers to the density of the air in the wind-tunnel test section, whereas ρ_w refers to the density of the air at the flight altitude of the wing.

It is important to note that the Reynolds number of the model test in the wind tunnel will usually be much lower than that of the wing at flight altitude and there is little that can be done about it. The three limitations already imposed on the model test, those of size, speed, and air density, together with the known viscosity of the air in the test section, completely determine the model Reynolds number. Fortunately, the effect of changes of Reynolds number on the oscillatory air forces is relatively small, provided the Reynolds number is above about 4×10^5, and the values, particularly of flutter speed and frequency, are relatively insensitive to Reynolds number variations. The test conditions are customarily chosen to make the Reynolds number as high as conveniently possible but the full-scale value is seldom approached.

When the three basic scale ratios have been obtained, it is possible to determine all of the dimensional properties of the model. For example, choose an airplane which weighs 10,000 lb, has a 50-ft span, and an estimated 500 mph flutter speed. If the maximum-span model which can be accommodated in the test section is 5 ft,

$$\frac{L_m}{L_w} = \frac{1}{10}. \tag{11-46}$$

If model design limits testing to speeds of 70 mph, an appropriate velocity ratio is

$$\frac{U_m}{U_w} = \frac{1}{8}, \tag{11-47}$$

and from Eq. (11–43),

$$\frac{T_m}{T_w} = \frac{8}{10}. \tag{11-48}$$

Thus the time scale of the model is almost the same as that of the airplane, and the corresponding natural frequencies will be almost the same. If the wind-tunnel density has a sea-level value, whereas flight altitude, full scale, is such that

$$\frac{\rho_w}{\rho_m} = .5, \tag{11-49}$$

then Eq. (11–45) determines the third and final basic scale factor as

$$\frac{M_m}{M_w} = \frac{1}{500}. \tag{11-50}$$

The weight of the model (the gravitational constant g is not scaled) is then 10,000/500 or 20 lb. Similarly,

$$\frac{m_m}{m_w} = \frac{M_m}{M_w}\frac{L_w}{L_m} = \frac{1}{50}, \tag{11-51}$$

$$\frac{I_{\alpha m}}{I_{\alpha w}} = \frac{M_m}{M_w}\frac{L_m}{L_w} = \frac{1}{5000}. \tag{11-52}$$

To find the bending stiffness on the model which will give the correct effective K_h at the typical section, remembering that the wing properties in this example do not vary along the span,

$$\frac{(EI)_m}{(EI)_w} = \frac{M_m}{M_w}\left(\frac{L_m}{L_w}\right)^3\left(\frac{T_w}{T_m}\right)^2 = \frac{1}{320,000}. \tag{11-53}$$

As was pointed out in Section 11–4, other model properties which do not appear in the equations of motion should be computed from those that do rather than by direct application of the scale factors.

When the results of the flutter tests on the model are known, Eqs. (11–47) and (11–48) can be used for converting to the dimensional full-scale flutter speed and frequency.

The compressible case. If the assumption cannot be made that the fluid is incompressible, that is, that the velocity of sound is very high compared with velocities over the wing, then the two additional parameters, Mach number and γ, must be considered. To make a true flutter model of a high-speed lifting surface, it is necessary that these dimensionless quantities as well as those of the incompressible case must have the same values in the model tests as in full-scale flight. To have γ the same implies that model tests must also be in ordinary air rather than some other gas such as Freon-12. (However, if it is simply desired to compare experimental flutter points with those predicted by calculation from a theoretical basis, there is no reason why tests cannot be run in gases such as Freon-12. The effect of changes in γ, such as from air to Freon-12, will be of secondary importance, as illustrated by the absence of γ from the linearized supersonic unsteady air forces.)

Just as in the incompressible case, no more than three independent limitations can be imposed on the model, since all of the dimensionless parameters are independent and there are only three scale factors to be determined, M, L, and T. Once again, the model size is definitely limited by the test equipment to be used, whether the testing method involves rockets, bomb drops, or wind tunnels. Thus the length scale factor is quickly established, at least within close limits. A second stringent limitation involves the attainment of the desired Mach number range, which usually also determines the model flight velocity and air density.

In rocket tests or bomb drops, flight velocity is directly related to Mach number as a function of altitude (or, more correctly, temperature, which is itself a function of altitude). Thus at any altitude a given desired Mach number fixes the model velocity, and from it and the previously determined length scale factor, the time scale factor can be found. The chosen altitude fixes the air density and thus the mass scale factor because the dimensionless parameter $m/\rho b^2$ must be kept constant [Eq. (11–45)].

In wind-tunnel tests, tunnel design considerations usually determine the air velocity and density which go with a given desired Mach number. These relationships may vary considerably from one tunnel to another. For example, a blow-down intermittent tunnel will have a higher test-section density than a suck-down intermittent tunnel because the stagnation pressure is considerably higher. The test-section speed of sound and thus velocity for a given Mach number is a function of stagnation or stilling chamber temperature, which, of course, varies with different tunnels.

In the compressible case, it is usually impossible to obtain full-scale Reynolds numbers, and it is unfortunate that we do not yet know for unsteady compressible flow the Reynolds number range in which we can test with confidence.

11–7 The unrestrained flutter model. As a second example of flutter model theory, we consider the difficult case of a flutter model of a complete airplane simulating unrestrained straight level flight. This lack of restraint of the "rigid-body freedoms" is desirable in the many cases when the flutter modes involve considerable "rigid-body motion" and when the flutter and rigid airplane dynamic stability frequencies are close together.

The equations of motion of an airplane unrestrained in pitch, roll, and vertical translation are derived in Chapter 3, in terms of normal modes. If the axis directions coincide with the airplane's principal axes, these equations can be extended to include all six rigid-body freedoms as well as the normal modes of vibration. Thus Eqs. (3–204) become, in the more general case,

$$M_1\ddot{\xi}_1 = \Xi_1$$

$$\cdot$$
$$\cdot$$
$$\cdot$$

$$M_6\ddot{\xi}_6 = \Xi_6,$$

$$M_7\ddot{\xi}_7 + M_7\omega_7{}^2\xi_7 = \Xi_7 \tag{11–54}$$

$$\cdot$$
$$\cdot$$
$$\cdot$$

$$M_\infty\ddot{\xi}_\infty + M_\infty\omega_\infty{}^2\xi_\infty = \Xi_\infty$$

where $\xi_1, \cdots, \xi_6$ are the perturbation amplitudes for the rigid-body modes, $M_1, \cdots, M_6$ are the total mass and principal moments of inertia of the airplane, and the normalized deformation modes ϕ and frequencies ω are solutions to Eq. (3–172).

The coefficients of the left sides of Eqs. (11–54) involve the magnitude and distribution parameters describing the mass, stiffness, and geometric properties of the airplane. These are analogous to the parameters m, S_α, I_α, K_h, K_α, b, and a in the previous example. The generalized forces Ξ on the right sides of Eqs. (11–54) are calculated as indicated in Chapter 3 and involve the external aerodynamic, gravitational, and propulsive system loadings. The coupling between these equations is quite extensive, in that each generalized force contains terms in many of the coordinates. If the coordinates $\xi_7, \cdots, \xi_\infty$ are set equal to zero in the rigid-body equations, they reduce identically to the familiar rigid airplane dynamic stability equations except for the use of the unconventional axis system.

If we examine the parameters which appear as coefficients in the generalized forces, we find the ordinary dynamic stability coefficients such as C_L, C_D, $C_{L\alpha}$, $C_{M\alpha}$, C_{Mq}, etc., which are purely aerodynamic in origin and are functions of airplane shape and attitude. We also find unsteady aerodynamic terms similar to those in the previous problem which are functions of airplane geometry, vibration mode shape, and reduced frequency. If compressibility is important, the coefficients will involve Mach number and they are always to some extent functions of Reynolds number.

If care is used in writing those equations which involve balances between equilibrium forces and moments, we will find terms involving the airplane weight which we can write in terms of the total mass and the gravitational constant g, and terms involving the equilibrium propulsive force as well as its variations with altitude, airspeed, etc. We will find a moment balance for longitudinal trim between wing, tail, and fuselage contributions. These equilibrium or steady-state balance relations should not be canceled out of the airplane equations unless they can be similarly canceled out of the model equations.

Consider, for example, the terms involving airplane weight. These terms are balanced by lift terms which are a function of airplane attitude. In fact, it is the airplane weight under given flight conditions which determines the equilibrium attitude. The model must have the same attitude in order to have the same C_L and the other coefficients which depend on it (e.g., roll due to sideslip, x force due to speed change). Thus, for the model to fly at the correct equilibrium attitude, the gravitational force terms must be scaled in the same manner as the aerodynamic, inertial, and elastic force terms. Let us see if this can be done when the scale factors are determined by typical model test restrictions.

As in the previous example, we are dealing with a problem in dynamics and will use the primary quantities mass, length, and time. Since all of the dimensionless parameters appear independently in the equations, only three independent model design limitations can be applied. As before, the length scale factor is determined by the ratio of allowable model size to airplane size. The time scale factor is determined by combining the length scale with the ratio of allowable test speed to estimated flutter speed in the incompressible case. For compressibility simulation, the test and full-scale Mach numbers must be the same. In conjunction with the temperatures, this condition determines the velocity and thus the time scale for the compressible case. In both cases the mass scale is determined by the condition that model and airplane density ratio be the same.

The dimensionless parameter involving weight will appear as a ratio of weight to aerodynamic force, $mbg/\rho U^2 b^2$, if the usual procedure is used. (If it appears as a ratio of weight to inertial force or elastic force, it can be reduced to a ratio of weight to aerodynamic force by proper multiplication by other parameters such as $m/\rho b^2$ and $\omega b/U$.) To see if the weight, or gravitational force, is correctly scaled, note that the ratio of gravitational to aerodynamic force can be written as

$$\frac{mbg}{\rho U^2 b^2} = \frac{m}{\rho b^2} \cdot \frac{g}{U^2/b}. \tag{11–55}$$

The mass ratio $m/\rho b^2$ is already the same on both model and airplane; however, the ratio* $g/(U^2/b)$ may be entirely different. The gravitational constant g must be the same, but U^2/b is generally not the same on the model and the full-scale airplane. That is, the acceleration scale factor is not normally unity for a flutter model. To illustrate the range over which it can vary, Table 11–1 has been prepared, in which typical scale factors

TABLE 11–1

	L_w/L_m	U_w/U_m	Acceleration$_w$/Acceleration$_m$
Fighter— Low-speed model	10	10	10
High-speed model	10	1	$\frac{1}{10}$
Transport—Low-speed model	25	5	1
High-speed model	25	1	$\frac{1}{25}$

have been assumed for a small high-speed fighter and a large jet transport for both low-speed and high-speed flutter models. The last column can also be interpreted as the ratio of model weight to model lift at the attitude which produced equilibrium on the full-scale airplane. It is apparent that the low-speed fighter model is probably incapable of supporting its own weight at any attitude, whereas the high-speed version flies at a very small value of the lift coefficient. It is fortunate and useful that the models of

---

* This ratio is the inverse square of the Froude number, so familiar to naval architects in their studies of wave drag.

large transports and bombers fly at the correct attitudes at least in the case where compressibility effects are not simulated, because the flutter of these large flexible airplanes is usually inextricably related to freedom in the rigid-body motions.

The fact that gravitational force is not properly scaled for most flutter models is one of the major difficulties in testing models with rigid-body freedom. Either we must accept results obtained with a model flying at lift coefficients which are too high or too low, or the model support must provide a suitable supplementary vertical force. The provision of such a force is discussed in Chapter 13 along with other problems involved in the simulation of rigid-body freedom.

The disproportionately high weights characteristic of many low-speed flutter models give rise to another difficulty whether or not body freedoms are simulated. If the model is supported at the fuselage, the wings will exhibit an inordinate amount of droop unless the wing structure is initially designed with the correct amount of "negative droop."

11-8 The dynamic stability model. Because of the increasing importance of the effect of structural flexibility on airplane dynamic stability and automatic control, we will look at the possibility of designing aeroelastic models to aid in the solution of this problem. The governing equations are Eqs. (11-54), which we wrote down for the unrestrained flutter model. The difference is simply one of emphasis. In the flutter model we are interested in a neutral dynamic stability condition in which the elastic modes are dominant. In the dynamic stability model we are interested in a subcritical dynamic behavior which is dominated by the rigid-body modes of motion. In the flutter model we often accept the simplification of suppressing those rigid-body modes which are unimportant to flutter. Similarly, the dynamic stability model can have a reduced degree of simulation of the less important elastic deflection modes without compromising the subcritical behavior under investigation.

Nevertheless, the model design problem in both cases is basically the same. There are only three scale factors and these are fixed by considerations of model size, speed, and relative mass. It is necessary to build into the model the correct magnitudes and distributions of mass, stiffness, and shape. There is the redundant restriction on model attitude or weight which is difficult to meet for low-speed models and impossible when compressibility is simulated, unless the model support mechanism is capable of providing a supplementary vertical force (cf. Section 13-6).

The designer of the dynamic stability model cannot escape this redundancy even in the case where vibration frequencies are high compared with dynamic stability frequencies, because mass and stiffness simulation are required to obtain correct vibration frequencies. The correct mass distribution is also needed to simulate inertial loadings on the structure due to essentially rigid-body motions, and the correct stiffness distribution is needed to ensure proper deflection shapes under the air and inertial loadings arising from the rigid-body motions.

Even in the extreme case of designing a completely rigid dynamic stability model the four redundant restrictions must be satisfied. This is a result of the necessity of scaling the model so that the scale factor is the same for all four types of forces which are important—aerodynamic, elastic, inertial, and gravitational.

There is another more promising approach to the use of elastic models in the dynamic stability problem. Up to this point we have discussed the use of an entire unrestrained airplane model in an airstream as an analog computer which indicates over-all degree of stability. It solves the entire set of equations. The other approach uses the model to determine the effect of elasticity on the individual coefficients of the basic rigid-body motion equations. This procedure is primarily applicable to airplanes for which the stability and elastic frequencies are well separated and is somewhat analogous to the use of rigid sinusoidally-forced models to check the aerodynamic coefficients of the flutter equations (cf. Sections 5–6, 7–2, and 13–6).

For the "steady-state" stability coefficients such as lift, side force, pitching moment, and yawing moment due to the angular displacement rather than its time derivatives, the model design problem is essentially that of Section 11–5. For the coefficients associated with time derivatives of angular displacements the model design problem is still that of Section 11–5 or this section without the weight restriction. The major difficulties are usually related to model support and testing techniques and are discussed in Chapter 13 (cf. Sections 13–4 and 13–6).

CHAPTER 12

MODEL DESIGN AND CONSTRUCTION

12-1 Introduction. The three fundamental airplane properties which appear in examples of aeroelastic model theory are
  (1) the distribution of structural stiffness,
  (2) the external shape presented to the airstream, and
  (3) the distribution of mass.
It is the problem of the model designer to reproduce these airplane properties in a manner which is adequate for the execution of the particular tests which are planned. For instance, in the case of a model designed for control effectiveness measurements both (1) and (2) are of paramount importance, whereas for a vibration model (2) need not be considered. A flutter model requires the reproduction of all three airplane properties and is often quite difficult to design. Because the flutter model does constitute a relatively complex design job it serves as an instructive illustration of the general approach to model design. Of the two types of flutter models—the series of generally representative wings to determine parameter trends and the correctly scaled model of an actual full-scale airplane—the latter is the more difficult to design and will be the focus of our attention.

A common first approach to flutter model design is to consider building to the desired scale an exact replica of the actual airplane structure. Thus the scaled spars, ribs, skin, etc., are supposed to give the correct stiffness distribution as well as the basic mass distribution and external shape. Unfortunately, this procedure, when it can be used at all, must be modified to a considerable extent. For the usual low-speed flutter model the skin thicknesses which are required, if the same material is employed as on the airplane, are prohibitively small. Even on large, high-speed models the required skin thicknesses may be less than a hundredth of an inch. In order to obtain workable thicknesses, a softer material must be used. On the high-speed model illustrated in Fig. 12-1 magnesium sheet replaced the aluminum alloy used on the airplane, but on low-speed models it is usually necessary to use a plastic.

This use of thicker components tends to reduce the accuracy of the stiffness reproduction and limits the structural detail which can be obtained. Often such items as rib spacing on the model must be modified in order to control skin buckling, and a variety of gluing and cementing techniques must be developed because of the impracticality of riveting many parts of the model. Major disadvantages of the use of plastics are the temperature sensitivity of the stiffness properties, the tendency to creep, and the high value of internal damping.

Normally the use of a simplified structure which is easily designed to meet the model stiffness requirements is preferable to the use of directly scaled structural members. If the full-scale airplane has such an unconventional structure that the stiffness properties cannot be accurately calculated, a separate structural model may be necessary. It is usually unreasonable to try to build a flutter model which is at the same time a structural model, since the requirements of the two types of models are often incompatible.

Thus the flutter model designer most often finds that the best plan of attack is to first lay out a simplified structure, taking care that it does not use up too large a share of the available mass and that it fits into the available space. He next designs the shell which encloses the structure and forms the external surface over which the air flows. Lastly, he calculates the amount of mass which must still be added to each section and arranges to distribute it in the required manner. Of course, in some instances the shell may contribute a portion of the stiffness, as in the case of rubber-covered low-speed flutter models, while in other instances part of the structure may also form an external surface.

The type of construction used both for the structure and the external shell depends upon the size of the model, the speed range in which it is to be tested, and the ratio of air density at airplane flight altitude to air density in the model test. This last parameter determines whether the designer must employ a very efficient structure and save weight at every opportunity or whether he can use convenient and simple, although perhaps less efficient, means to obtain the correct stiffness and shape distributions. Representative flutter models and components are shown in Figs. 12–1 through 12–10.

12–2 Structural simulation. Stiffness distribution in the various major components of an airplane structure may generally be most simply describable in one of two ways. The first and simplest way can be used only for airplane components whose elastic properties are primarily functions of only one variable. Thus a high aspect-ratio lifting surface which has essentially completely rigid ribs can be described by using the concepts of an elastic axis and bending and torsional stiffness distributions along the span. High fineness-ratio fuselages fall into this same class. If the lifting surface is of low aspect ratio or if the ribs are comparatively flexible, the second method of description is necessary. The elastic properties in this type of structure are functions of two variables; that is, the surface can deform in the manner of a plate and is not limited to simple bending and torsion about an elastic axis. For these structures the stiffness distribution is best described by a matrix of influence coefficients. The problem of structural simulation is quite different in these two cases and they will be treated separately.

Fig. 12–1.　Early high-subsonic flutter model (Grumman F6F) in NACA-Langley wind tunnel.

Fig. 12–2.　Internal construction of F6F high-subsonic model before covering with magnesium skin.

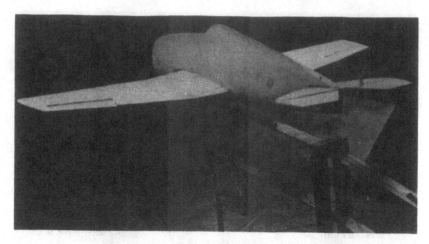

Fig. 12–3. Small low-speed flutter model of Grumman F6F wing using two-spar torque-tube rubber-covered construction.

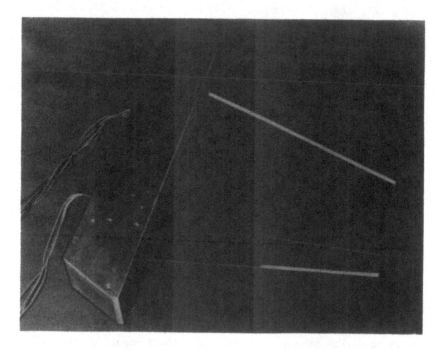

Fig. 12–4. Supersonic cantilever flutter model employing metal-spar solid-balsa construction.

Fig. 12–5.   Low-speed two-spar torque-tube flutter model with built-up balsa sections.

Fig. 12–6.   Internal construction of large high-subsonic swept wing flutter model.   (Courtesy of Glenn L. Martin Company)

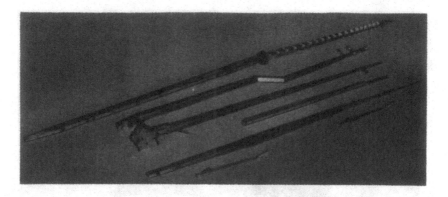

Fig. 12–7. Typical wing and fuselage spars employed in single-spar model construction. (Courtesy of Boeing Airplane Company)

Fig. 12–8. Internal construction of a model wing with flexibly mounted masses. (Courtesy of Boeing Airplane Company)

Fig. 12–9. Low-speed delta-wing flutter model employing a beam-network structure and sectional balsa covering.

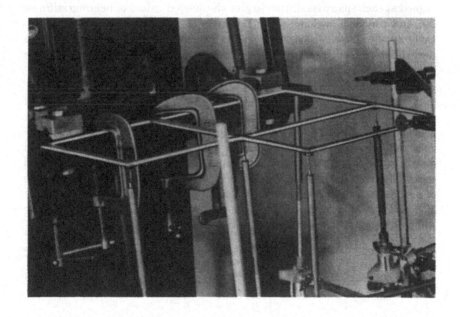

Fig. 12–10. Static test of a beam network similar to the one employed in the model of Fig. 12–9.

12-3 Elastic properties as functions of one variable. The wing of the jet transport introduced in Example 2-1 is typical of the structures which fall into the first class mentioned above, and its stiffness properties are described in Fig. 12–11. The elastic axis is represented as the locus of the shear centers of each section, and the spanwise distribution of stiffness is given in terms of the bending stiffness EI and the torsional stiffness GJ at each station. It is assumed that the fore and aft bending stiffness need not be simulated for a flutter model of this wing and that the chordwise ribs are so stiff that an arbitrary deflection at a station can be represented by a simple combination of bending and twisting; that is, the position of each rigid rib can be described by the linear deflection of the rib at the elastic axis and its rotation about the elastic axis.

The types of structural arrangement which can be used for this model can be separated into two categories: in the first, the internal structure is surrounded by a shell which does not contribute markedly to the bending and torsional stiffness, and in the second, the contribution of the shell is highly important.

The simplest of the structural arrangements which fall into the first category consists of a single spar whose axis coincides with the desired location of the elastic axis of the model. The cross-sectional shape of the spar is designed at each spanwise station to give the desired values of bending stiffness EI_x and torsional stiffness GJ. The ratio of stiffnesses is determined entirely by the shape of the cross section, whereas the magnitude of the stiffness is determined by its size (Ref. 12–1). A round rod has a J which is twice I_x, so that its torsion-to-bending stiffness ratio is

$$\frac{GJ}{EI_x} = \frac{2G}{E} \sim \frac{2}{3}. \tag{12-1}$$

Thus a stiffness ratio different from $\frac{2}{3}$ cannot be obtained with a solid circular rod or even a round tube. A spar with a rectangular cross section can meet a range of stiffness ratio requirements because its width can be varied separately from its height. As shown in Fig. 12–12, the torsional stiffness of such a spar can be defined with the aid of a form factor β. The variation of β and $(a^2/b^2)\beta$ with cross-section shape is given in Fig. 12–13. The stiffness ratio for the rectangular cross section can be written as

$$\frac{GJ}{EI_x} = \begin{cases} \dfrac{12G}{E}\,\beta, & \dfrac{a}{b} \geq 1, \\[2mm] \dfrac{12G}{E}\,\dfrac{a^2\beta}{b^2}, & \dfrac{a}{b} \leq 1. \end{cases} \tag{12-2}$$

Since the value of the form factor varies from 0 to $\frac{1}{3}$, the rectangular cross section can simulate desired torsion-to-bending stiffness ratios up to a maximum of about $\frac{4}{3}$.

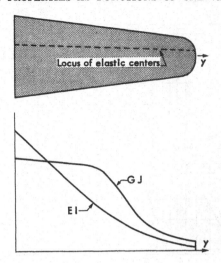

Fig. 12–11. Stiffness properties of jet transport wing.

In some cases it may be necessary to simulate the fore-and-aft bending stiffness EI_z as well as EI_x and GJ. This is particularly true in the case of high aspect-ratio wings with underslung masses and also for high fineness-ratio fuselages when both side and vertical bending must be taken into account. Since the two parameters a and b of a spar of rectangular cross

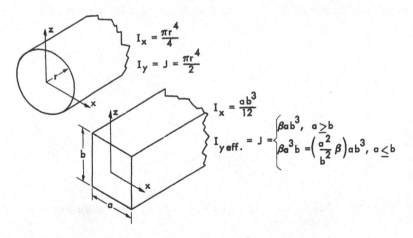

Fig. 12–12. Area moment of inertia properties for spars of circular and rectangular cross section.

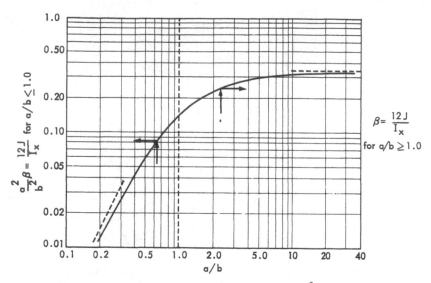

Fig. 12-13. Torsional stiffness form factors β and $\dfrac{a^2}{b^2}\beta$ for the rectangular spar of Fig. 12-12.

section can meet no more than two conditions, this third requirement necessitates the use of a more complicated cross section.*

The shape shown in Fig. 12–14 is based on the assumption that the rectangular center portion is designed to obtain the desired I_x and J. The flanges are added to increase I_z to the desired value. (If I_z is smaller than I_x, the center portion is designed for I_z and J, and the flanges for I_x.) If $b/t \geq 10$, the influence of the flanges on I_x and J will be less than 5%. If

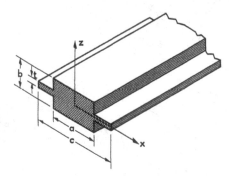

Fig. 12-14. Spar cross section which permits matching of torsional stiffness and both bending stiffnesses.

---

* Both this technique and the following one for higher stiffness ratios are the developments of the Boeing Airplane Company.

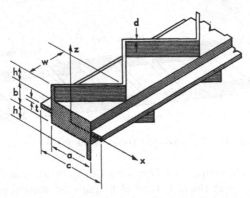

Fig. 12–15. Unusual spar cross section designed for high ratios of torsional to bending stiffness.

$b/t < 10$, a design iteration is needed to correct the original values of a and b for the effect of the flanges. Since the four disposable quantities a, b, c, and t must meet only three design conditions, I_x, J, and I_z, the flange thickness t may be chosen to be less than 10% of b subject only to limits of machinability.

To permit the use of single-spar structures when the desired ratio of torsional stiffness to the smaller bending stiffness is greater than $\frac{4}{3}$, a lattice section spar has been developed (Fig. 12–15). This consists of a rectangular, flanged center section with zigzag ribs on the under sides. At a section where the upper rib is angling forward, the lower rib is angling backward, and vice versa. The ribs twist when the beam bends about x, they bend when the beam twists about y, and offer no appreciable restraint when the beam bends about z.

For the particular case of aluminum alloy, with 90° rib intersections and $d = 0.06w$, so that $a = 1.085w$, it has been determined empirically that

$$J = ab^3\beta + 0.31w^{1.1}h^{1.92}b,$$

$$I_x = \frac{ab^3}{12} + 0.018w^{1.25}h^{1.43}b^{1.32}, \tag{12-3}$$

$$I_z = \frac{ba^3}{12} + \frac{t(c^3 - a^3)}{12}.$$

Another type of structural arrangement in which the spar structure provides the entire stiffness requirements is the two-spar torque-tube structure shown in plan view in Fig. 12–16 and also in Figs. 12–5 and 12–33. The two spars provide the bending stiffness EI_x, whereas the torque tubes increase the torsional stiffness GJ to the desired value. This structure cannot simply simulate a fore-and-aft bending stiffness distribution. It does, however, provide a differential-bending type of stiffness, as will be

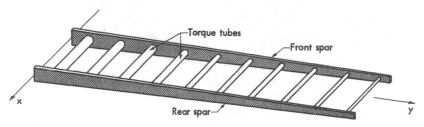

Fig. 12–16. Two-spar torque-tube model structure.

shown later. The torque tubes are very stiff in bending and ensure that the general deflection of the structure at a spanwise station can be described as a deflection at the elastic axis plus a rotation about it.

Since the torque tubes are perpendicular to the spars and do not contribute to the bending stiffness EI_x, the sum of the bending stiffnesses of the front and rear spars must at each station equal the desired stiffness $(EI_x)_d$:

$$(EI_x)_f + (EI_x)_r = (EI_x)_d. \tag{12-4}$$

Also, the relative sizes and locations of the spars must be such as to ensure the correct location of the local elastic centers. With the concept that the elastic center at a given section is the point at which a load would have to be applied to produce pure bending if the local conditions extended over the whole span, it can be shown that the moment of the spar stiffness EI_x about the elastic center must be zero. To demonstrate this, assume that a load is applied at the elastic axis at the tip of the uniform structure which has everywhere the local properties. The resulting pure bending of the structure requires that both spars have exactly the same deflection shape. If w_f and w_r are the vertical deflections (positive upward) of the front and rear spars, respectively, if positive bending moment M induces compression in the upper fibers of the spar, and if positive shear S tends to cause the outboard section to slide upward with respect to the inboard section,

$$S_f = -\frac{d}{dy} M_f = -\frac{d}{dy}\left[(EI_x)_f \frac{d^2 w_f}{dy^2} \right] = -(EI_x)_f \frac{d^3 w_f}{dy^3}. \tag{12-5}$$

However,

$$\frac{d^3 w_f}{dy^3} = \frac{d^3 w_r}{dy^3}, \tag{12-6}$$

so that

$$\frac{S_f}{(EI_x)_f} = \frac{S_r}{(EI_x)_r}. \tag{12-7}$$

For equilibrium of the outboard section,

$$S_f l_f + S_r l_r = 0 \tag{12-8}$$

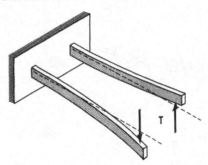

Fig. 12–17. Differential bending type of resistance offered by two spars to an applied torque.

when l_f and l_r are the distances from the elastic center to the front and rear spars, respectively, positive forward. Substituting Eq. (12–7), we obtain the desired result:

$$(EI_x)_f l_f + (EI_x)_r l_r = 0. \tag{12-9}$$

Thus Eqs. (12–4) and (12–9) govern the spacing and bending stiffnesses of the spars.

To obtain the desired torsional stiffnesses for the model it is necessary to supplement the torsional stiffnesses of the spars by installing suitable torque tubes. In most models of this type the major portion of the torsional stiffness comes from the torque tubes. It is important to recognize, however, that on multispar models, as well as on multispar wings, the differential bending of the spars also contributes to the torsional stiffness, although in a complicated fashion. Even in the absence of connections between the spars they will offer some restraint to an applied torque T, as illustrated in Fig. 12–17. In order to evaluate this effect at a station along the span, we observe that the two spars will resist the applied couple such that

$$S_f = -S_r, \qquad M_f = -M_r, \qquad T = S_f D, \tag{12-10}$$

where D is the distance between the parallel spars. Thus

$$T = -\frac{d}{dy}[DM_f] = -\frac{d}{dy}[l_f M_f + l_r M_r]$$

$$= -\frac{d}{dy}\left[l_f (EI_x)_f \frac{d^2 w_f}{dy^2} + l_r (EI_x)_r \frac{d^2 w_r}{dy^2}\right] = -\frac{d}{dy}\left[l_f (EI_x)_f \frac{d^2(w_f - w_r)}{dy^2}\right]$$

$$= -\frac{d}{dy}\left[l_f (EI_x)_f D \frac{d^2\theta}{dy^2}\right], \tag{12-11}$$

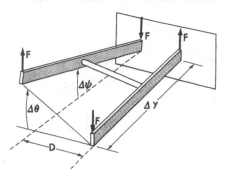

Fig. 12–18. Distortion of one bay of a two-spar torque-tube structure under the action of an applied torque.

where the angle of twist is

$$\theta = \frac{w_f - w_r}{D} . \qquad (12\text{–}12)$$

Rearranging,

$$T = -\frac{d}{dy}\left\{[l_f^2(EI_x)_f + l_r^2(EI_x)_r]\frac{d^2\theta}{dy^2}\right\} . \qquad (12\text{–}13a)$$

Thus the torsional restraint set up by differential bending of the spars is a function of the second moment of bending stiffness about the elastic center and the third derivative of the twist θ. This second moment of the bending stiffness, σ, can be written in several forms:

$$\sigma = l_f^2(EI_x)_f + l_r^2(EI_x)_r = D^2\,\frac{(EI_x)_f(EI_x)_r}{(EI_x)_f + (EI_x)_r} . \qquad (12\text{–}13b)$$

In order to evaluate accurately enough for design purposes the effectiveness of the torque tubes in providing torsional stiffness, consider the distorted spar-torque-tube assembly in Fig. 12–18 and assume for the moment that the spars themselves are untwisted. If the length of spars which may properly be considered as acting with this torque tube rather than its neighbors is Δy, and if the assembly is acted on by the torque $T = FD$, the moment absorbed by the torque tube is $F\Delta y$. If the torsional stiffness of the torque tube about its own axis is $(GJ)_{TT}$,

$$F\Delta y = (GJ)_{TT}\,\frac{\Delta\psi}{D} = (GJ)_{TT}\,\frac{\Delta\theta}{\Delta y} , \qquad (12\text{–}14)$$

if we make the approximation that this spar length is relatively undistorted. Then,

$$T = FD = \left[(GJ)_{TT}\,\frac{D}{\Delta y}\right]\frac{\Delta\theta}{\Delta y} = (GJ)_{TT(\text{eff})}\,\frac{\Delta\theta}{\Delta y} . \qquad (12\text{–}15)$$

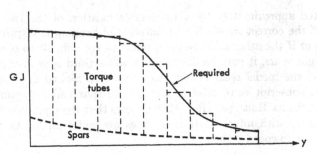

Fig. 12–19. Contributions of spars and torque tubes to the required torsional stiffness.

Thus the effective torsional stiffness of the torque tube about a spanwise axis is $(GJ)_{TT}(D/\Delta y)$. Remembering that the torque tubes are fairly closely spaced and are used to represent a continuous stiffness distribution, it is often convenient to write

$$T = (GJ)_{TT(\text{eff})}\frac{d\theta}{dy}. \tag{12–16}$$

If we now take into account the fact that when the wing twists the spars twist also, it is apparent that each spar offers a torsional restraint about the spanwise axis proportional to its own GJ. This spar stiffness is in addition to the differential bending stiffness and the contribution of the torque tubes, so that the actual twist of the entire two-spar torque-tube structure can be written as

$$T = \left[(GJ)_{\text{spars}} + (GJ)_{TT(\text{eff})}\right]\frac{d\theta}{dy}$$

$$- \frac{d}{dy}\left\{\left[l_f{}^2(EI_x)_f + l_r{}^2(EI_x)_r\right]\frac{d^2\theta}{dy^2}\right\}. \tag{12–17}$$

Equation (12–17) is only an approximate representation of the behavior of this lattice truss. A more exact analysis can be carried out by the method of minimum potential energy in Chapter 2 to obtain a set of torsional influence coefficients, but the design of models is usually based on Eq. (12–17) as well as Eqs. (12–4) and (12–9). Thus the required bending stiffness, elastic axis location, and differential bending stiffness are obtained to the desired degree of accuracy by suitable choice of the sizes and locations of the spars. The required torsional stiffness is supplied by the spars and torque tubes, as shown in Fig. 12–19.

The three requirements of total bending stiffness and its first and second moments about the elastic axis generally cannot be satisfied by the use of two parallel spars. If the simulation of differential bending at each spanwise station is of secondary importance, its over-all effect can be represented approximately by a reasonable location of the two straight spars. If the correct simulation of differential bending is of primary importance or if the other stiffness requirements are difficult to obtain with two straight spars, it may be desirable to add a third spar over at least a portion of the model span rather than resort to kinked or offset spars. Using the subscript m to refer to the middle spar and σ for differential bending stiffness, Rauscher (Ref. 12–2) shows that the equations for three parallel spars with interspar torque tubes are quite analogous to those derived for two spars. Thus,

$$(EI_x)_f + (EI_x)_m + (EI_x)_r = (EI_x)_d,$$

$$l_f(EI_x)_f + l_m(EI_x)_m + l_r(EI_x)_r = 0,$$

$$l_f^2(EI_x)_f + l_m^2(EI_x)_m + l_r^2(EI_x)_r = \sigma_d,$$

$$\sum(GJ)_{\text{spars}} + \sum(GJ)_{TT(\text{eff})} = (GJ)_d. \qquad (12\text{--}18)$$

It should be remembered that in this structure the interspar members are very stiff in bending, so that chordwise sections rotate and translate without deforming when the wing twists and bends.

In order to visualize more effectively the effects of differential bending on the torsional behavior of a parallel-spar, torque-tube model, consider the special case of a wing with a straight elastic axis and constant values of EI_x, GJ, and σ across the span. For a cantilever wing loaded with a concentrated moment T at the tip $(y = l)$, the differential equation becomes

$$\sigma \frac{d^3\theta}{dy^3} - GJ \frac{d\theta}{dy} = -T. \qquad (12\text{--}19)$$

The cantilever root condition for each spar leads through Eq. (12–12) to the root boundary conditions

$$\theta = \frac{d\theta}{dy} = 0. \qquad (12\text{--}20)$$

Since there is no bending moment in the spars at the tip, we have the tip boundary condition

$$\left(\frac{d^2w_f}{dy^2} - \frac{d^2w_r}{dy^2}\right) = D \frac{d^2\theta}{dy^2} = 0. \qquad (12\text{--}21)$$

The solution to Eq. (12–19) subject to boundary conditions (12–20) and (12–21) is

$$\theta = \frac{T}{GJ}y - \frac{T}{GJ}\sqrt{\frac{\sigma}{GJ}}\left\{\tanh\sqrt{\frac{GJ}{\sigma}}\,l - \frac{\sinh\sqrt{GJ/\sigma}\,(l-y)}{\cosh\sqrt{GJ/\sigma}\,l}\right\}, \qquad (12\text{--}22)$$

which for typical flutter model properties is shown by the cantilever results of Fig. 12–20. Also shown is the variation of angle of attack under a concentrated moment applied at the midspan station of the same model.

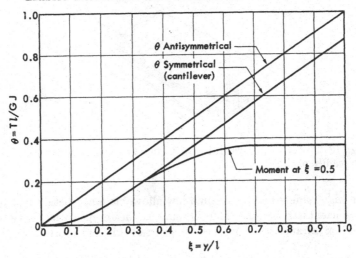

Fig. 12–20. Angular deflections under applied moments of a typical uniform two-spar torque-tube wing.

Consider next the same uniform wing unrestrained at the root but subjected to equal and opposite moments at the tips. This loading may be thought of as an antisymmetrical loading, whereas the tip loading with cantilever root corresponds to a symmetrical loading with T at both tips opposed by a moment $2T$ at the root. In the antisymmetrical case, the boundary conditions at the root become

$$\theta = \frac{d^2\theta}{dy^2} = 0, \tag{12-23}$$

since the deflection mode shape must be antisymmetrical. This leads, for the uniform wing, to the simple solution

$$\theta = \frac{T}{GJ} y, \tag{12-24}$$

which is plotted as the antisymmetrical case in Fig. 12–20. The deflections under antisymmetrical loading are thus seen to be larger than for the corresponding symmetrical loading. At the tip the deflection ratio is

$$\frac{\theta_{\text{sym.}}}{\theta_{\text{anti.}}} = 1 - \sqrt{\frac{\sigma}{GJ}} \tanh \sqrt{\frac{GJ}{\sigma}} l. \tag{12-25}$$

For this uniform wing the effects of differential bending disappear under antisymmetrical loading.

Although the single-spar and the two-spar torque-tube types of structure have been most often used in flutter models, sometimes the combina-

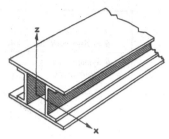

Fig. 12–21. A simple welded-steel torque-box structure whose stiffnesses can be readily modified.

tion of high required stiffnesses and low allowable mass dictates the use of more efficient structures. Thus it may be desirable to use a two-spar torque-box arrangement or even a stressed-skin structure (Figs. 12–1 and 12–6).

In Fig. 12–21 a simple welded steel torque-box structure is illustrated. If designed slightly overstiff, this structure can easily be brought to the correct stiffnesses at any section by first thinning the top and bottom cover plates to obtain the correct torsional stiffness and then filing the cover plate edges to reduce the bending stiffness without modifying appreciably the torsional stiffness.

Flutter models which simulate flight at high altitudes and which are to be tested at near-sea-level densities have a relatively large allowable mass and need not have efficient structures. This situation often occurs with the modern high-speed aircraft, and results in the use of solid models. They may be solid wood, metal, or a combination of both, and may or may not require lead weights to obtain the desired mass distribution. A combination which has proved useful consists of an aluminum spar embedded in solid balsa which forms the desired external shape (Figs. 12–4 and 12–22). Although the stiffness properties of the balsa vary considerably and are strongly affected by grain direction, simple static tests on the slabs from which the models are to be built provide the designer with adequate information. This is particularly true for the typical model in which the spar contributes the major share of both bending and torsional stiffness.

The design procedure for the aluminum-spar balsa model is quite analogous to that used for the two-spar torque-tube structure, particularly

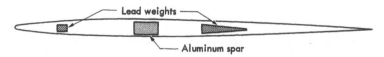

Fig. 12–22. Cross section of a supersonic model wing with weights and spar embedded in solid balsa.

insofar as bending stiffness is concerned.   As before, the designer must match at each spanwise station the desired bending stiffness, the desired torsional stiffness, and the desired elastic axis position.   He can conveniently consider the model as made up of a solid balsa beam and a spar which has a Young's modulus and a shear modulus equal to the difference between the corresponding moduli of the aluminum and the balsa.   This enables him to insert the spar anywhere in the balsa beam without having to make subsequent corrections for the changes in the balsa-beam properties caused by hollowing a cavity for the spar.   The bending and torsional stiffnesses of the solid balsa beam at a given spanwise station are fixed by the airfoil section * and can be subtracted from the desired values at that section to leave the stiffnesses which must be supplied by the "reduced-modulus" spar.

The same reasoning which led to Eq. (12–9) in the case of the two-spar torque-tube structure leads in this case to the placing of the spar such that the "center of gravity" of the bending stiffnesses of the solid balsa beam and the reduced-modulus spar lies at the desired elastic axis location.   This reduced-modulus spar must then be proportioned according to Eq. (12–2) to provide the remaining requirements of bending and torsional stiffness.

12–4 Elastic properties as functions of two variables.   The problem of structural stiffness simulation for flutter model design is far more difficult when elastic properties of the structure are not basically functions of only one variable but require the use of two variables.   This problem has been emphasized by the increased use of low aspect-ratio lifting surfaces which do not bend and twist like their higher aspect-ratio counterparts.   Rather, the flexibility in the chordwise direction is of the same order as in the spanwise direction and the deflection modes have the pronounced double curvature characteristic of plate deformations (see Fig. 12–23).

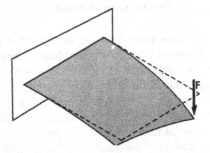

Fig. 12–23.   Plate-like deformation under load characteristic of thin low aspect-ratio wings.

* The effective GJ for a typical thin solid airfoil section is given approximately by the formula $\frac{1}{6}Gct^3$, where c is the chord and t is the maximum thickness of the airfoil.

These modes cannot be expressed in terms of bending and torsion about a specified axis but are actually complicated deflection surfaces in terms of x and y.

In order to describe in a manageable form the elastic characteristics of such a structure, a matrix of influence coefficients is defined as described in Chapter 2. This matrix of influence coefficients is usually known to the model designer as a result of calculations or experimental data resulting from wing tests. His problem is to build a structure which has the correctly scaled influence coefficients at a sufficient number of points. The structure he uses may not resemble the full-scale structure except possibly in the case of large, high-speed models because of the impracticality of building scaled replicas, as outlined in the introduction to this chapter. Whatever type of "simplified" structure he chooses, it cannot be designed point by point as were the single-variable structures, because a configuration which gives the correct deflections for a load at one point will not, in general, give the correct deflections for loads at other points.

Since the low aspect-ratio structure deforms somewhat like a plate, the first approach which suggests itself is to use a plate as the simplified structure. It would be a convenient surface upon which to mount forms for producing the desired external shape, if it could be designed to have the correct influence coefficients. Since a constant-thickness plate will not normally suffice, the designer must use a built-up plate (Fig. 12–24), a plate with suitable cutouts (Fig. 12–25), a plate with varying thickness (Fig. 12–26), or perhaps some combination of the three. Unfortunately, two important factors prevent this approach from working well. First, the designer would have a difficult task in computing the influence coefficients of a given built-up, cutout, or tapered plate. The inverse problem of designing a plate to a given set of influence coefficients is much more difficult. Alternatively, the use of an experimental iterative process for such a design seems prohibitively tedious except for the most elementary sets of influence coefficients.

The second and perhaps more basic factor is related to the small plate thicknesses usually required for correct simulation, particularly for low-speed models. Plate deformations are linear only if they are of the order of the plate thickness, and if the plate is essentially flat. For relatively flexible flutter models neither of these conditions can normally be met even by the use of a somewhat thicker plastic plate rather than the thin metal plate. Of course, the use of plastic introduces problems of creep, damping, thermal dependence, and in some cases a difference between static and dynamic stiffness properties.

If the thin plate is discarded as the basic simplified structure, a new arrangement must be devised which is not seriously limited by the scale of permissible deflections and which is more amenable to analysis. Of the plate configurations mentioned above, the plate with cutouts is least

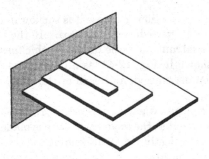

Fig. 12–24. Built-up plate structure for low aspect-ratio models.

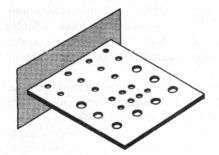

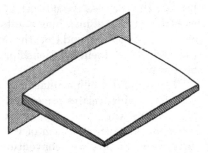

Fig. 12–25. Plate structure with suitable cutouts for simulating low aspect-ratio wings.

Fig. 12–26. Contoured plate for simulating the stiffness distribution of low aspect-ratio wings.

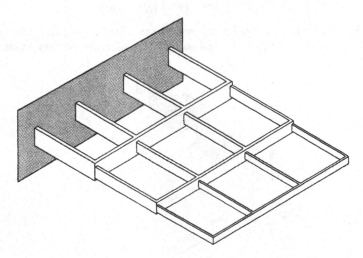

Fig. 12–27. Beam-network structure designed to match influence coefficients of a low aspect-ratio wing.

restrictive to scale of deflections because it is thicker in order to develop the
necessary stiffness with its reduced plan area. If the cutouts are enlarged
the remaining material must increase further in thickness, with the network
of simple beams shown in Fig. 12–27 as a limiting case. Here the "plate
thickness" has been maximized, and each of the elements can be analyzed
by familiar means.

For the purpose of investigating the practicality of this arrangement it
is convenient to assume that the model structure is made up of a rectangular
array of simple beams effective in both bending and torsion. It is assumed
also that the reference locations for the influence coefficients to be matched
coincide in each case with beam intersections. In general, not all of the
available wing influence coefficients will be used in designing the model
because the complexity and cost of the matching process increases rapidly
with the number of matching points. To obtain some idea of the difficulty
of the design process and thus the degree of compromise which is necessary,
the various pertinent relationships involving influence coefficients should
be examined.

The set of flexibility influence coefficients $[C^{zz}]$, which is usually measured for a wing, relates forces and linear deflections perpendicular to the
plane of the wing. This set is part of a more general set relating external
forces and moments at each of the reference points to linear and angular
deflections. Using the conventions indicated in Fig. 12–28, the more
general relationship is

$$\begin{bmatrix} C^{zz} & C^{z\alpha} & C^{z\theta} \\ C^{\alpha z} & C^{\alpha\alpha} & C^{\alpha\theta} \\ C^{\theta z} & C^{\theta\alpha} & C^{\theta\theta} \end{bmatrix} \begin{Bmatrix} P \\ M \\ N \end{Bmatrix} = \begin{Bmatrix} z \\ \alpha \\ \theta \end{Bmatrix}. \tag{12–26}$$

Even though the coefficients which are usually measured are the set
$[C^{zz}]$, it seems probable that the set $[C^{z\alpha}]$ would also be very valuable to

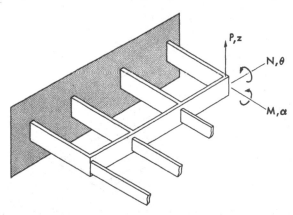

Fig. 12–28. Sign conventions used in the analysis of the beam network.

the model designer because he is attempting to build a structure whose deformations influence the surrounding air in the same manner as the structure of the full-scale wing. This influence is more directly a function of the slope than of the deflection of the wing surface at a point. It is conceivable that fewer force-slope coefficients would have to be matched than force-deflection coefficients in order to achieve the same degree of simulation in a model.

In anticipation of the fact that calculated influence coefficients may be available to the model designer in stiffness rather than flexibility form, because of the ease with which the stiffness coefficients can be written down for each of the many elementary parts of a structure, the stiffness coefficient approach should also be examined. The corresponding basic relationship is

$$
\begin{bmatrix} K^{zz} & K^{z\alpha} & K^{z\theta} \\ K^{\alpha z} & K^{\alpha\alpha} & K^{\alpha\theta} \\ K^{\theta z} & K^{\theta\alpha} & K^{\theta\theta} \end{bmatrix} \begin{Bmatrix} z \\ \alpha \\ \theta \end{Bmatrix} = \begin{Bmatrix} P \\ M \\ N \end{Bmatrix}. \tag{12-27}
$$

If the external torques M and N are set equal to zero, α and θ can be eliminated algebraically, so that Eq. (12–27) becomes

$$
[\bar{K}^{zz}]\{z\} = \{P\}, \tag{12-28}
$$

where $[\bar{K}^{zz}]$ is the inverse of the usual flexibility matrix $[C^{zz}]$ and is given by

$$
[\bar{K}^{zz}] = [C^{zz}]^{-1} = [K^{zz}] - [K^{z\alpha}K^{z\theta}] \begin{bmatrix} K^{\alpha\alpha} & K^{\alpha\theta} \\ K^{\theta\alpha} & K^{\theta\theta} \end{bmatrix}^{-1} \begin{bmatrix} K^{\alpha z} \\ K^{\theta z} \end{bmatrix}. \tag{12-29}
$$

For the rectangular array of Fig. 12–27 the terms in the stiffness matrix of Eq. (12–27) are simple linear sums of the individual beam bending and torsional stiffnesses modified by factors related to the network geometry. If the corresponding wing stiffness matrix is available to the model designer the procedure of equating corresponding terms in the two matrices produces a set of simultaneous algebraic equations which can be solved for the unknown beam stiffnesses.

Unfortunately, the known quantity is usually $[C^{zz}]$ in numerical form, and this cannot be linearly related to the unknown beam stiffnesses. Even though for the rectangular array the submatrices $[K^{\theta\alpha}]$ and $[K^{\alpha\theta}]$ are identically zero, the simplest relationship which can be deduced from Eq. (12–27) is

$$
[C^{zz}]^{-1} = [K^{zz}] - [K^{z\alpha}][K^{\alpha\alpha}]^{-1}[K^{\alpha z}] - [K^{z\theta}][K^{\theta\theta}]^{-1}[K^{\theta z}]. \tag{12-30}
$$

The inversion and multiplication of the submatrices produces a highly complicated nonlinear function of the unknown stiffnesses.

A relationship similar to Eq. (12–29) but entirely in terms of flexibility influence coefficients is derived in Chapter 2 by the Principle of Minimum Strain Energy [Eqs. (2–133) to (2–147)]. When the external loads are just

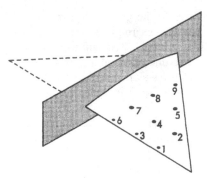

Fig. 12–29. Assumed influence-coefficient stations on a delta wing.

the forces, Eq. (2–147) can be written as

$$[C^{zz}] = [H_{QQ}] - [H_{QP}][H_{PP}]^{-1}[H_{PQ}], \qquad (12\text{--}31)$$

where the submatrices of $[H]$ are linear in the individual beam flexibilities rather than stiffnesses.

The relationship between unknown beam properties and given data is, however, no less complicated and nonlinear than with Eq. (12–29) or (12–30). It should be pointed out that if external moments M and N are included in Eqs. (2–133) through (2–147), $[C]$ becomes the larger flexibility coefficient matrix of Eq. (12–26). It is apparent that even if this larger matrix were available as basic wing data to the model designer, no linear relationship of the unknown beam properties exists in terms of flexibility influence coefficients such as is the case for the corresponding stiffness influence coefficient formulation.

To illustrate the magnitude of the complexity of the matching process for a delta wing whose flexibility influence coefficients $[C^{zz}]$ are known at the points indicated in Fig. 12–29, assume a model structure made up of the simple 8-beam network illustrated in Fig. 12–30 (solid lines). For this case there are 16 unknown beam properties which can be designed to give no more than 16 independent influence coefficients. Since there are 15 such influence coefficients in a 5-point array and 21 in a 6-point array, only the coefficients associated with points 1 through 5 can be matched using the 8-beam network. If the 4 dotted beams in Fig. 12–30 are added to the network a maximum of only 30 influence coefficients out of the given 45 can be matched, because each beam is assumed to possess two unknown properties.

The number of matching points and the number of beams which are employed in a given model design may be limited by the paucity of available influence coefficient data. It is far more likely, however, that the limiting factor is the capacity of the available computing facilities.

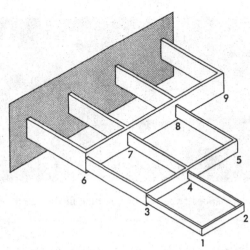

Fig. 12–30. Beam-network structure designed to simulate the influence coefficients of the delta wing of Fig. 12–29.

Even the 8-beam network involves the handling of a matrix of 24 × 24 elements. Since the relationships between the beam properties and the given flexibility matrix $[C^{zz}]$ are nonlinear, we must resort to trial and error or at best a cumbersome iterative process. In either case, the number of successive mathematical operations is so large that the accumulation of round-off errors requires the carrying of a very large number of significant figures.

Obviously, the design of this type of beam-network model is feasible only with the help of the rapidly developing art of high-speed machine computation. Although we do not yet fully understand the ramifications of the computational problem associated with such cumbersome manipulations, the use of beam networks for simulation of low aspect-ratio structures appears promising for both low-speed and high-speed flutter models.

12–5 Shape simulation. In many aeroelastic model tests it is necessary that the model present the correct external shape to the airstream in order to ensure the correct simulation of the steady or unsteady airloads. Since the model structure deflects and twists under these airloads, the external shell must be capable of presenting a suitably distorted shape to the airstream. Also, since in most instances it is very desirable for the model stiffness properties to be supplied by an internal structure, the external shell must be capable of supporting the airloads while distorted without modifying seriously the stiffness properties of the model. The types of external shell used by the model designer thus depend markedly on the manner in which the shell must be able to distort, on the relative level of

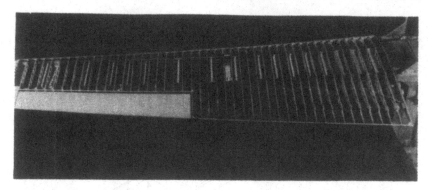

Fig. 12–31. Low-speed two-spar torque-tube flutter model with thin rubber cover removed.

stiffness of the model structure, and, of course, on the mass available for the shell.

One basic type of distortion which must be considered is that associated with the structures whose elastic properties are functions of one variable such as high fineness-ratio fuselages and high aspect-ratio lifting surfaces. Low-speed models of this class which require a relatively weak structure and thus an external shell which adds almost no stiffness usually use either a sectional construction or ribs covered with a thin continuous sheet of rubber (see Figs. 12–5, 12–8, 12–31, 12–32). In the sectional construction, the shell is built up from a convenient number of separate rigid sections or pods, each attached independently to the model spar structure. Enough clearance is left between sections to prevent contact when the wing deforms, although, from an aerodynamic point of view, the cracks may be covered with strips of thin rubber sheet if necessary. The individual sections may

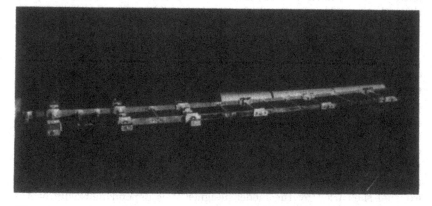

Fig. 12–32. Sectional structure with same properties as model in Fig. 12–31.

Fig. 12–33. Balsa sections mounted on the structure illustrated in Fig. 12–32.

be fabricated from balsa stiffened, if necessary, by a coating of silkspan or fiberglass. If enough mass is available, it may even be desirable to use built-up metal sections. If very light sections are required, the mass of the glue which is used may form a considerable portion of the shell section mass and should be carefully controlled. Particularly in swept wings, the surface discontinuities which are created between sections when the wing twists are undesirable and can be minimized by the use of many sections if the higher running mass of the narrow sections can be tolerated. A construction employing a thin rubber skin stretched over numerous ribs avoids surface discontinuities but is susceptible to distortion under the airloads. Above a velocity of sixty miles per hour the ballooning of the rubber skin is very difficult to control. On many of the low-speed, relatively flexible models the static deflections or droop due to gravity loading are excessive because the strength of the gravity field cannot be scaled

Fig. 12–34. High-subsonic flutter model of a flying wing (also shown in Fig. 12–6). (Courtesy of Glenn L. Martin Company)

down. This effect can be minimized by building "negative droop" into the model structure such that no sagging is apparent on the final model.

High-speed models whose elastic properties are functions of only one variable are usually much stiffer than those discussed above, so that it is often possible to use a continuous balsa covering (see Figs. 12–4 and 12–35). In fact, it may be feasible, if the mass allowance is high, to enclose the spar structure with solid balsa which has the correct external shape. If the mass allowance is low on a high-stiffness model, the more efficient main structure which must be used for stiffness simulation may form part or all of the desired external shell (Figs. 12–1, 12–2, 12–6).

A second basic type of distortion is that associated with structures whose elastic properties are functions of two variables. Here it is necessary to build up an external shape which is doubly flexible and which must often contribute very little to the model's stiffness. For models with relatively low stiffness it is again possible to build up the shell in sections such that the shell is sectional in two directions. That is, the surface is divided in both directions into a sufficient number of shell sections so that they form reasonably smooth surfaces when the structure to which they are attached distorts. Here again the cracks between sections can be covered with rubber sheet to prevent disruption of the airflow. For models with relatively high stiffness a solid balsa covering can again be used for those portions of the surface which are not already formed by some part of the structure.

Fig. 12–35. Lower half of swept supersonic flutter model showing instrumented spar and numerous lead weights.

12–6 Inertial simulation. An aeroelastic model which has been designed to have the desired stiffness distribution and external shape will normally require some adjustment of its initial properties in order to have the required mass distribution. The model designer should always be careful to use internal structure and external shell designs which are light enough to leave a workable margin of disposable mass for all sections of the model. It is simpler to add mass in the correct amounts than it is to lighten a section already constructed.

For models whose structural properties are properly functions of only one variable it is usually convenient to think of the inertial properties as functions of the same variable. Thus the relevant characteristics of the entire mass distribution of an ordinary relatively high aspect-ratio surface can be expressed in curves such as those shown in Fig. 12–36. Since each chord section is essentially rigid, its inertial effect will be reproduced in a model whose corresponding section has the same chordwise position of the center of gravity, the same running mass, and the same moment of inertia. The fact that these properties may be obtained by the use of a different chordwise distribution of mass is unimportant.

To obtain the desired inertial properties at each station of the model wing it is first necessary to calculate or measure the inertial properties of all of the material at each station which makes up the structure, shell, instrumentation, etc. The remainder of the required mass at each station can then be added as lead in a manner which produces the desired section center of gravity and moment of inertia. As an example, consider a section of the wing of convenient width. This width may be determined by the rib spacing in a rubber-covered model or a section width in a sectional model. The inertial properties of the complete but unweighted wing section can be computed by filling out a table such as is shown in Table 12–1. Here m represents the mass of an element, $\bar{x}$ the distance of the element's center

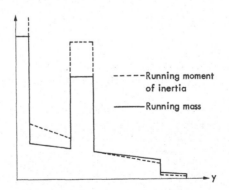

Fig. 12–36. Inertial properties as a function of the spanwise coordinate for the jet transport wing.

TABLE 12-1

Element	m	x	r	mx	mx^2	mr^2	I
Front spar Rear spar Torque tube							
$\sum m =$			$\sum mx =$			$\sum I =$	

of gravity from a convenient reference line such as the leading edge, r the element's own radius of gyration, and I the element's moment of inertia about the reference line. From the three totals shown, the unweighted section mass m_0, center of gravity position x_0, and moment of inertia I_0 about the center of gravity can be computed.

If the desired section properties are m_d, x_d, and I_d, the necessary additional mass is

$$m_\Delta = m_d - m_0. \qquad (12\text{-}32)$$

Lc x_Δ be the location of the center of gravity of the additional mass m_Δ and let

$$d_\Delta = x_d - x_\Delta, \qquad d_0 = x_0 - x_d, \qquad (12\text{-}33)$$

as shown in Fig. 12-37.

The center of gravity of the additional mass must be so located that the section with its desired mass has the desired center of gravity. That is,

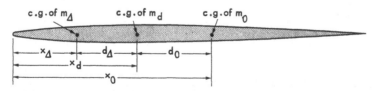

Fig. 12-37. Locations of the centers of gravity of the unweighted section m_0, the added mass m_Δ, and the properly weighted section m_d.

$$d_\Delta = \frac{m_0}{m_\Delta} d_0 \cdot \qquad (12\text{–}34)$$

This additional mass must also provide enough moment of inertia I_Δ about its own center of gravity to ensure that

$$I_\Delta + m_\Delta d_\Delta{}^2 + I_0 + m_0 d_0{}^2 = I_d. \qquad (12\text{–}35)$$

To obtain this moment of inertia I_Δ it is usually convenient to split the additional mass into two (or sometimes more) pieces m_r and m_f. Whatever their relative sizes,

$$m_r + m_f = m_\Delta \qquad (12\text{–}36)$$

and

$$m_f d_f = m_r d_r, \qquad (12\text{–}37)$$

where d_f is the distance to the front mass m_f and d_r is the distance to the rear mass m_r from the center of gravity of m_Δ (Fig. 12–38). The two masses and their distances from the center of gravity of m_Δ must be such that

$$m_f d_f{}^2 + m_r d_r{}^2 = I_\Delta \qquad (12\text{–}38)$$

provided their own radii of gyration are negligibly small. They must also satisfy Eqs. (12–36) and (12–37). These three conditions still leave some choice in determining the four unknowns, so that the model designer may try a number of combinations to see which is most satisfactory. To systematize this trial procedure, note that

$$I_\Delta = r_\Delta{}^2 (m_f + m_r), \qquad (12\text{–}39)$$

where r_Δ is the radius of gyration of m_Δ. Combining Eqs. (12–37), (12–38), and (12–39), it can be shown that

$$d_f = \frac{r_\Delta{}^2}{d_r}, \qquad (12\text{–}40)$$

which suggests the following scheme. As shown in Fig. 12–38, lay off the known distance r_Δ perpendicular to the chordline of the section at the

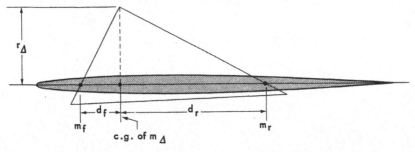

Fig. 12–38. The use of a draftsman's triangle for locating the added masses to obtain simultaneously the desired center of gravity and moment of inertia.

center of gravity of the mass m_Δ to be added.   If a right triangle is laid on the figure as shown, any pair of distances defined by the intersections of the sides and the chordline will be in the relation prescribed by Eq. (12–40). It is a simple matter to rotate the triangle about its apex until the most satisfactory pair of distances is discovered.   The two pieces of m_Δ can then be found from the expressions

$$m_f = m_\Delta \frac{d_r}{d_r + d_f}, \qquad m_r = m_\Delta \frac{d_f}{d_r + d_f}. \qquad (12\text{–}41)$$

For doubly flexible surfaces the inertial properties of the model cannot be described as in Fig. 12–36 but must be known as functions of both span and chord coordinates.   If the surface is divided into small sections for the purposes of structural simulation, it may be convenient to think of the distributed mass of each section as lumped at its center, so that it is merely necessary to add lead to bring each small section up to its desired mass. If the sections are not small, it is necessary to simulate to some extent the distribution of mass within the section.

CHAPTER 13

TESTING TECHNIQUES

13–1 Introduction. In the two preceding chapters we have examined the principles of design and construction of aeroelastic models for various kinds of tests. It is the purpose of this chapter to describe the associated testing techniques and the principles on which they are based. Since most of the tests are also often made on full-scale airplanes, we will differentiate where necessary between model and full-scale testing techniques.

Three general areas make up the field of aeroelastic testing. The first includes experiments in which no airstream is present, e.g., static testing to find stiffness distributions and vibration testing to find natural frequencies and mode shapes. Tests in the second and third areas both require the presence of an airstream; tests in the second area involve only "steady-state" aeroelastic phenomena such as control effectiveness, whereas those in the third area include only "unsteady" phenomena such as flutter and dynamic stability.

13–2 Measurement of structural flexibility. Accompanying the growth of the importance of aeroelastic effects has been a trend toward experimental determination of airplane stiffness as well as strength characteristics. This measurement of stiffness properties has been particularly valuable and necessary for the difficult-to-analyze thin low aspect-ratio wings. For flutter models no other types of static tests are ordinarily employed. Low-speed flutter models in particular are usually more than strong enough to withstand any steady loads which can be applied in the wind tunnel, despite their relatively inefficient structures, and the model designer performs static tests only to see how well he has reproduced the desired magnitude and distribution of structural stiffness.

The type of stiffness or flexibility measurements which are made in a particular case depends on the form of the data which is to be checked as well as on the type of structure under test. In the case of flutter models, the tests should be designed not only to check the degree of simulation but also to indicate as clearly as possible the modifications of the model structure which may be required to obtain the desired degree of simulation.

If we look first at structures whose properties can be considered functions of only one variable (Section 12–3), the basic data to be checked are in the form of bending and torsional stiffness distributions along an elastic axis. If the elastic axis is straight, a comparison between measured and calculated deflection curves for simple loads and torques applied at the tip is sufficient to prove the existence of adequate simulation. If modifications to the structure are required, however, these will take the form of changes in bending and torsional stiffnesses at certain sections of the structure. For bending, these corrections are related directly to errors in curvature at the sections in question and these errors are not easily obtained from experimental deflection curves. In principle, it is merely necessary to differentiate the deflection curves twice and divide the result by the applied bending moment in order to obtain a plot of the inverse of the bending stiffness. For torsion, only one differentiation is required. However, the small inaccuracies and errors always present in experimentally determined functions are tremendously accentuated by the differentiation process and lead to prohibitively large errors, particularly in the second derivative.

In cases where bending stiffness corrections must be deduced from the static tests the measurement of slopes rather than linear deflections may provide a better indication of the size and location of the necessary corrections. The interpretation of torsion tests is no simpler in the cases where a considerable amount of differential bending stiffness is present. In many cases, the effect of differential bending can be isolated by the comparison of torsional deflections or slopes obtained by symmetrical and antisymmetrical torque loadings (see Section 12–3 and Fig. 12–20). For structures which have significant discontinuities in stiffness and elastic axis or which are difficult to modify, the original model design must be good enough so that the static tests are needed only to confirm the adequacy of the design. In these cases, a comparison of measured and calculated influence coefficients may be the most satisfactory process.

For structures whose properties are functions of two variables (Section 12–4), the stiffness information is available to the model designer in the form of a set of influence coefficients which can be compared directly with a corresponding set measured on the model. If modifications to the structure are necessary, they can be estimated by the same sort of iteration procedure as was used in the original model design.

The performance of static tests involves techniques of deformation measurement, application of loads, and support of structure during testing. If we look first at the requirements of deformation measurement techniques we find that they are basically the same for both model and full-scale. The devices which are used must not only have adequate accuracy over appropriate ranges and be capable of rapid and simple interpretation, but must not distort the results by introducing extraneous and indeterminate loadings.

Although this last requirement may seem obvious, it is the most difficult to meet in the testing of flutter models. Ordinary deflection measuring devices such as dial gauges incorporate more friction and stiffness than can be tolerated on most models, even though they may have the necessary range and accuracy. The loads applied to the model structures by the

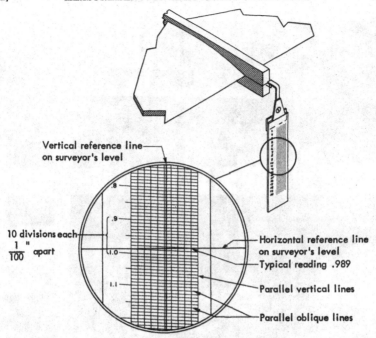

Vertical reference line
on surveyor's level

10 divisions each
$\frac{1}{100}$ " apart

Horizontal reference line
on surveyor's level
Typical reading .989
Parallel vertical lines
Parallel oblique lines

Fig. 13-1.   Deflection measurements using glass scales viewed through a surveyor's level.

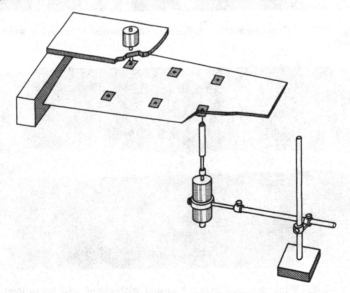

Fig. 13-2.   Deflection measurements using the signal generated in an energized coil by the displacement of its core.

Fig. 13-3. Measurement of influence coefficients of a delta wing using Schaevitz coils.

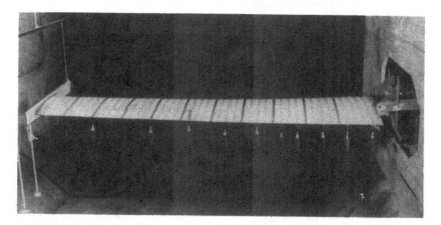

Fig. 13-4. Wing mounted in wind tunnel with string and beam loading rig at tip. Deflections of the suspended scales are observed through a surveyor's level.

measuring devices must, in most cases, vary no more than a few grams during the test cycle. The deflection amplitudes to be measured are of the order of a half inch or less, and accuracy to within a few thousandths or even a few ten-thousandths of an inch is required. For these reasons, beams of light and optical magnification are often employed. Precise measurements of slope are possible by observing the position on a fixed scale of a spot of light reflected from a small mirror attached to the model structure. Slope measurements can also be made by mounting a very sensitive accelerometer in a vertical position on the structure and noting that the accelerometer output changes with angular deflection ("$2g$" for a 180° rotation). Linear deflections can be measured to within a few thousandths of an inch by observing calibrated glass scales through a sighting device such as a surveyor's level (Fig. 13-1). The deflection of the scale (Fig. 13-2) with respect to the crosshairs in the level can be read directly in hundredths of an inch and the next place can be estimated by noting the intersection of the horizontal crosshair with the slant line vernier-type markings on the glass scale. Although the method is tedious and leads to eyestrain, it can be used at many stations on a structure by hanging a scale at each station and focusing the level on each one in succession. The loading imposed on the wing by the lightweight scales is small and does not vary during a test.

A somewhat more accurate scheme (Ref. 13-1) for measuring deflections, which is particularly suited to rapid determination of influence coefficients on small, stiff, high-speed models, is shown in Fig. 13-3. Metal cores suspended from the wing move vertically in energized coils held rigidly below the wing when it deflects. By the use of linear elements carefully matched to a master element which is motor driven through a null-balance network, the deflection of any core can be read on the master dial gauge by throwing a switch.

The loading and support of model structures for static test is a relatively simple problem (Fig. 13-4) because of the small scale. Conversely, the loading and support of full-scale structures can involve costly mounting frameworks, tension pad connectors, and high-capacity hydraulic jacks.

13-3 Measurement of natural frequencies and mode shapes. The test programs for the prototypes of all but the smallest and simplest of aircraft and missiles include "shake testing," to determine the normal modes of vibration. The primary use of this information is to check the estimated mass and stiffness properties used in the flutter and dynamic-loads calculations by comparing the experimental frequencies and mode shapes with the ones that are calculated. Sometimes the experimental data serve as the basis for a new set of calculations. In the case of dynamic models, such as flutter models, vibration tests are always performed, in order to evaluate the accuracy with which the full-scale structure has been simulated. The measured modes may be compared either with corresponding measurements made on the full-scale structure or with calculated modes.

The basic techniques of exciting the structure and interpreting the data are the same regardless of the scale of the structure, although, of course, the scale of the exciting and measuring equipment must be appropriate. We have seen in Chapter 3 that the forced motion of the structures with which we are concerned can be expressed directly by superposition of the normal modes we are trying to measure. Thus the deflection $w(x,y,t)$ of a point in the structure can be written in terms of the normal modes $\phi(x,y)$ and the corresponding normal coordinates $\xi(t)$ as

$$w(x,y,t) = \sum_{i}^{\infty} \phi_i(x,y)\xi_i(t). \tag{3-145}$$

The equations governing the response to external excitation are

$$M_i\ddot{\xi}_i + M_i\omega_i^2\xi_i = \Xi_i, \qquad (i = 1, 2, 3, \cdots), \tag{3-147a}$$

where

$$M_i = \iint_S \phi_i^2(x,y)\rho(x,y)dxdy \tag{3-147b}$$

is the generalized mass, ω_i is the frequency, and

$$\Xi_i = \iint_S F_z(x,y,t)\phi_i(x,y)dxdy \tag{3-147c}$$

is the generalized force associated with the ith normal mode.

We can see from these equations that the response to an arbitrary shaking force F_z is, in general, a superposition of all of the normal modes of the structure. To measure the separate modes and frequencies, we shall certainly have to use ingenuity in the choice and application of the shaking force. It is usually advantageous to use a sinusoidal excitation rather than the pulse type often used in measurements on linear systems, because the sinusoidal excitation often permits a concentration on one mode after another so that the desired answers can be obtained, and to some extent evaluated, during the tests rather than after an elaborate record analysis. (Although sinusoidal excitations are used in this section, other excitations are discussed in Section 13-5.) Assuming that sinusoidal excitation is used, we must choose the points of excitation, the phases and amplitudes of the applied forces, and the forcing frequency to accentuate most clearly the mode under investigation. The degree to which we can succeed is, of course, related to the limitations of the available equipment but is most of all dependent on our skill in employing our equipment in an intelligent and purposeful fashion.

(a) *The response to single-point excitation.* In order to derive the principles which govern the general application of a sinusoidal excitation to a structure, let us first study the response of a structure with only one

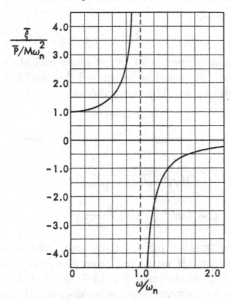

Fig. 13-5. The amplitude of the dimensionless response of an undamped single-degree-of-freedom system to a sinusoidal generalized force.

natural mode excited by a single sinusoidal force of amplitude $\bar{F}$ and frequency ω. The response is sinusoidal and is given by

$$\bar{\xi} = \frac{\bar{P}}{M\omega_n{}^2} \frac{1}{1 - (\omega/\omega_n)^2}, \tag{13-1}$$

where $\bar{\xi}$ is the amplitude of the response, $\bar{P}$ is the amplitude of the generalized force, and ω_n is the structure's natural frequency. This response is shown in dimensionless form in Fig. 13-5. The size of the generalized force compared with the applied force depends on the point of application. It is a maximum when applied at the point which has the largest amplitude and is zero when applied at a nodal point.

A measured response would differ from that shown in Fig. 13-5 because of the presence of some damping in the actual structure. This damping is usually called "structural damping" (see Section 9-5) and exhibits the property that the amount of damping is proportional to the amplitude of the motion.[*] For sinusoidal motion, damping can best be represented by an additional term in Eq. (3-147a). Using complex notation and a cosine-

---

[*] Structural damping differs from the more familiar damping proportional to velocity, which would give the equation

$$M\ddot{\xi} + f\dot{\xi} + M\omega_n{}^2\xi = \Xi.$$

type generalized force,

$$M\ddot{\xi} + M\omega_n^2(1 + ig)\xi = \bar{P}e^{i\omega t}. \tag{13-2}$$

In this way the damping term is 90° out of phase with the displacement but is proportional to it. It should be recognized that g seldom exceeds a value of 0.05. If we substitute a solution of the form

$$\xi = \bar{\xi}e^{i(\omega\tau+\psi)}, \tag{13-3}$$

we obtain the response (exclusive of the transient terms)

$$\bar{\xi} = \frac{\bar{P}}{M\omega_n^2} \frac{1}{\sqrt{[1 - (\omega/\omega_n)^2]^2 + g^2}},$$

$$\psi = \tan^{-1}\left(\frac{-g}{1 - (\omega/\omega_n)^2}\right). \tag{13-4}$$

This is plotted in Fig. 13–6 for $g = 0.03$, along with a comparative plot for the undamped case. If we make a polar plot of $\bar{\xi}e^{i\psi}$, we find it to be a nearly complete circle, as shown in Fig. 13–7 (see Ref. 13–2). It is important to note that the maximum response occurs when the forcing frequency equals the natural frequency and the displacement lags 90° behind the exciting force.

Let us now assume that our structure has two normal modes but is still excited by a single force. For example, consider the first unrestrained antisymmetrical bending mode ($\omega_b = 28.85$ rad/sec) and the first torsion mode ($\omega_t = 22.36$ rad/sec) calculated for the jet transport wing in Section 4–6. We shall assume that the torsion mode is symmetrical and that $g = 0.03$ for both modes. Under the action of a sinusoidal force applied to the right wing $\sqrt{10}$ feet ahead of the elastic axis at station 368 (see Fig. 2–28) the response in each normal mode can be calculated from Eq. (13–4) if the generalized forces are first obtained according to Eq. (3–147c). In terms of the subscripts b and t for the bending and torsion modes, respectively, the responses can be written as

$$\bar{\xi}_b = \frac{3.351 \times 10^{-6}\bar{P}}{\sqrt{\left[1 - \left(\frac{\omega}{28.85}\right)^2\right]^2 + .0009}}, \quad \psi_b = \tan^{-1}\left(\frac{-.03}{1 - \left(\frac{\omega}{28.85}\right)^2}\right),$$

$$\bar{\xi}_t = \frac{1.658 \times 10^{-7}b_{\text{tip}}\bar{P}}{\sqrt{\left[1 - \left(\frac{\omega}{22.36}\right)^2\right]^2 + .0009}}, \quad \psi_t = \tan^{-1}\left(\frac{-.03}{1 - \left(\frac{\omega}{22.36}\right)^2}\right). \tag{13-5}$$

where the bending mode is normalized at the tip station of the elastic axis and the torsion mode at the point half of a tip chord ahead of the elastic axis at the tip station.

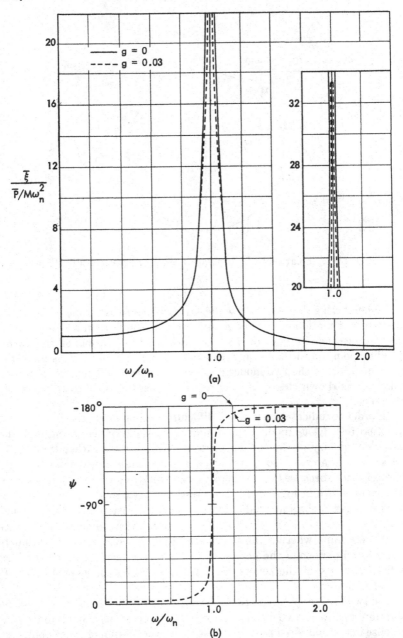

Fig. 13–6. The amplitude (a) and phase (b) of the dimensionless response to a sinusoidal generalized force of a single-degree-of-freedom system with structural damping ($g = 0.03$).

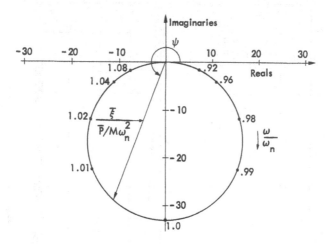

Fig. 13–7. Polar plot of the damped response shown in Fig. 13–6.

Remembering that we observe the wing's response by noting [cf. (3–145)] the resultant motion of specific points, we plot the motion of the point of force application in Fig. 13–8 as curves A. The motion of the corresponding point on the left wing is shown as curves B. In each instance, the resonant peaks of the two modes are unmistakable, and symmetry or antisymmetry is shown clearly by the relative phases of the two responses at resonance.

In order to indicate the kinds of difficulties which may arise in an actual vibration test, let us modify our simplified example by assuming that the natural frequency of the torsion mode is 27.36 rad/sec rather than 22.36 rad/sec. It is now not so easy to locate the resonance peaks under single free excitation, because the model responses overlap considerably to produce the curves shown in Fig. 13–9. The presence of two modes is indicated by the two peaks in the amplitude plot of A, but the existence of the torsional mode is far less evident in B. Further confusion is introduced by attempting to determine whether the somewhat hidden torsion mode is symmetrical. A comparison of the phases of the points on the right and left wings at the frequency of the torsion peak for A indicates an inconclusive 90° difference in phase.

The two basic approaches to the problem of separating a difficult pair of vibration modes are to place the exciting force so as to minimize the response in the undesired mode and to observe the motion of those points on the wing which are least influenced by the undesired mode. In our example, a pickup placed on the nodal line of the torsion mode at station 368 would produce the pure bending mode response shown as curve C in Fig. 13–10.

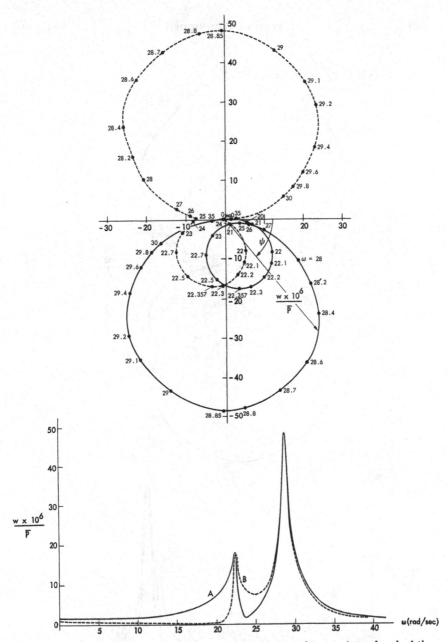

Fig. 13–8. Amplitude and phase of the response of two points ahead of the elastic axis of the jet transport wing, considering only first antisymmetrical bending ($w_b = 28.85$ rad/sec) and first torsion ($w_t = 22.36$ rad/sec). Curves A refer to the point of sinusoidal force application and curves B to the corresponding point on the opposite wing. In both modes $g = 0.03$.

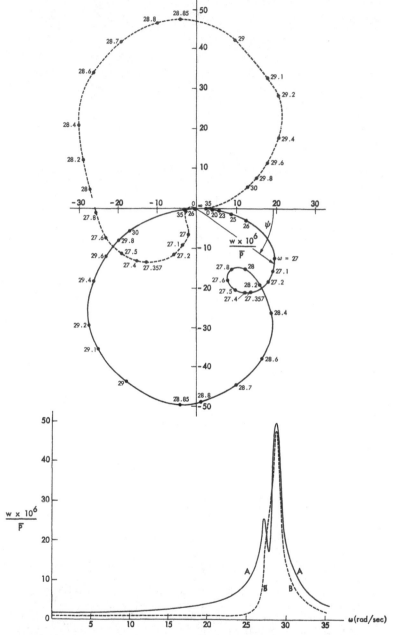

Fig. 13–9. Same responses as in the preceding figure except that the torsion frequency is assumed as 27.36 rad/sec.

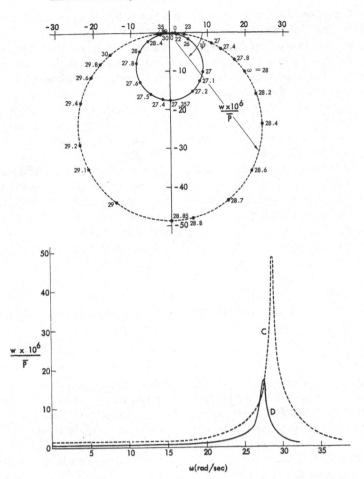

Fig. 13–10. Responses in bending (curve C) and torsion (curve D) as separated by proper location of either force application or response measurement.

The same result could have been observed by moving the shaking force to the nodal point and not changing the point of observation.

(b) *Multipoint excitation.* All of the above responses were produced by an exciting force located on one wing. If two forces were employed symmetrically on the wing the response in the antisymmetrical mode would be eliminated because the corresponding generalized force would be zero and only the torsion mode would remain (curve D in Fig. 13–10). If these symmetrically located forces were applied with a 180° phase difference the symmetrical mode would be eliminated for the same reason, and the bending mode would be emphasized (curve C in Fig. 13–10).

If we remember that our airplane actually has a large number of normal modes and that many of them may be grouped in quite a small frequency range, we can see that just separating the symmetrical from the antisymmetrical modes may still leave a complicated response to decipher wherever we place our pickups. Adding another pair of symmetrically placed exciting forces which are driven synchronously with the first pair allows a shifting of the point of application of the resultant applied force by adjusting the ratio of the applied forces on each wing. If the point of application can be put on a node line for one of the interfering modes, that mode can be eliminated from the response (Ref. 13–2).

In principle, the addition of more pairs of shaking forces will allow the response in the target mode to be accentuated while decreasing the responses in the interfering modes. The limiting case would involve an exciting force distributed such that the generalized forces were zero in all but the target mode. The orthogonality condition for the normal modes [cf. Eq. (3–182)],

$$\iint_S \phi_i\phi_j\rho dx dy = 0, \qquad i \neq j, \tag{13–6}$$

when compared with Eq. (3–147a) for the generalized force, indicates that the necessary force distribution for the ith mode is

$$\bar{F}_z(x,y) \sim \phi_i(x,y)\rho(x,y). \tag{13–7}$$

That is, the amplitude of the exciting force at every point on the structure must be proportional to the product of the amplitude of the normalized target mode at that point and the airplane density at that point. With this exciting force distribution the response is entirely in the ith mode, regardless of the forcing frequency. The amplitude and phase of the response at any point (except a nodal point) behave as indicated in Figs. 13–6 and 13–7.

Another line of reasoning (Ref. 13–3) which leads to the same conclusion starts with the assumption that the complex structure is forced to oscillate with the mode shape and frequency of the target mode. At every instant the distributed inertial loading produces shears and moments which are exactly balanced by the shears and moments in the elastically deformed structure. If no damping forces are present the oscillation will continue unabated without requiring external excitation. If damping is present in the structure, then the oscillation can be maintained undisturbed if the damping shears and moments are just canceled by the external forces. Since the basic concept of structural damping assumes that each elastic shear and moment is accompanied by a proportional damping shear and moment and since the elastic shears and moments are balanced by the inertial loading, it can be reasoned that the damping shears and moments can be balanced by external forces distributed according to the inertial loading and 90° out of phase with it. Since the inertial loading at a point is proportional to the product of the oscillation amplitude at that point and the density at that point, this line of reasoning also recommends the force distribution described by Eq. (13-7).

(c) *Identification and measurement of normal modes.* Up to this point in this section we have looked at the responses to external excitation of structures whose vibration modes were known. However, the problem usually facing us concerns the scanning of the responses to judiciously applied shaking forces of a structure whose vibration modes we are attempting to measure. It is this problem which we shall consider next.

The amount of skill and physical equipment required to "shake test" a structure depends very markedly on the characteristics of the structure in the frequency range of interest. If the various normal modes are well separated and the structural damping is small, only one shaking force need be used, although for structures with a plane of symmetry it is usually quite advantageous to use a pair of symmetrically located shaking forces. (This permits the independent investigation of symmetrical and antisymmetrical modes, as discussed in an earlier paragraph.) If the shaking frequency is varied slowly while the magnitude of the shaking force is maintained even approximately constant, a response-sensitive pickup will give a dramatic indication of the approach to a natural frequency. A matched pair of pickups symmetrically located on the structure indicates directly whether the mode is symmetrical. The mode shape can be determined by maintaining a steady oscillation at the frequency for peak response and measuring the output of a pickup as it is placed at successive stations along the structure. Of course, if there is any suspicion that a mode may have been missed because the shaking force was applied at a node, responses should also be viewed for an alternate point of force application.

If two or more modes are grouped together in a small frequency range, the problem of determining their separate properties is far more difficult. This is particularly true if the structural damping is relatively high. In this case a major problem is the identification of all of the modes, to say nothing of measuring their properties once they are identified. As can be seen from the calculated responses of the bending and torsion modes of the jet transport, as shown in Fig. 13-9, the approach to the frequency of a normal mode does not necessarily show up as a distinct resonance peak. The peaks which are apparent may be displaced slightly from the normal mode frequencies, and often the mode shapes measured at a resonance peak are so influenced by the responses in neighboring modes that they are very seriously distorted. As discussed in Ref. 13-2, the natural frequencies in this situation may often be more easily seen and more accurately obtained by looking at the resonance peaks on curves of the derivative of response amplitude with respect to frequency versus frequency. Also, some experi-

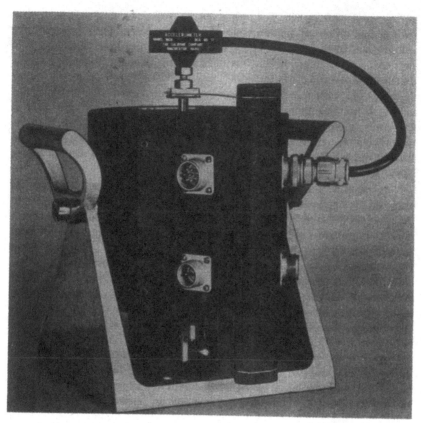

Fig. 13–11. Electromagnetic shaker of 25 pound-maximum-force rating with accelerometer attached to the "voice coil."   (Courtesy of the Calidyne Company)

ence in interpreting polar-type response plots permits the identification of troublesome modes.

When only one or two pairs of shaking forces are used an experienced operator can separate the interfering modes by locating and adjusting the forces by a reasoned trial-and-error procedure in the cases which are not too difficult.   However, more and more frequently it is necessary to find the modes of increasingly complex structures, e.g., large flexible aircraft with many flexibly supported engines and tanks.   In these cases, the operator feels the need for many more shaking forces in order to locate and accentuate adequately the target mode, because even the most painstaking analysis of the responses to a few carefully controlled shaking forces may not identify all of the modes in a difficult group and certainly cannot give accurate knowledge of the mode shapes.   The simple addition of more shaking forces does not by itself solve the problem, since the operator is then faced with the necessity of making the correct adjustments of a large number of forces. The number of possible adjustments is extremely large, and some systematic iterative procedure must be coupled with the operator's experience to permit isolating the target mode with an acceptable amount of effort.

One such systematic procedure has been effectively developed by Lewis and Wrisley (Ref. 13–3), together with the equipment needed for its implementation. The equipment has a capacity of twenty-four independently controllable electromagnetic shakers, each with an attached accelerometer (Fig. 13–11). All the shakers are driven either in phase or 180° out of phase with each other by a single variable frequency source. The iterative procedure which was devised is based on attempting to obtain for exciting the target mode the unique force distribution which was derived above,

$$\bar{F}_z(x,y) \sim \phi_i(x,y)\rho(x,y). \tag{13–7}$$

The continuous distribution of exciting force is replaced by twenty-four concentrated forces, and the total mass is considered to be prorated among the shakers according to the amount of structure which can reasonably be assigned to each shaker. These data are used to adjust the twenty-four mass-ratio potentiometers, which then attenuate the signals indicating shaker force. Indications of the amplitudes of the responses at the shakers are given by the accelerometers, whose outputs are integrated to make them proportional to velocity. By the use of switches, the response velocity and modified shaking force for any point can be displayed on the vertical and horizontal axes of a cathode-ray oscilloscope. The slope of the resulting elliptical figure is indicative of the ratio of the signals, and the width of the ellipse is a measure of their phase difference. (Figure 13–12 shows the control console.)

The procedure for identifying and measuring a target mode which is in close proximity to other modes is normally along the following lines. The operator (who should still be experienced and ingenious in order to handle this previously insoluble task) first glimpses the target mode as he makes frequency sweeps through the range of interest using various combinations of shakers and observing resonances and phase shifts at suitable stations. With the forcing frequency adjusted to produce the best indication of resonance, he selects the forcing station whose response best indicates the resonance. He calls this station the master station and adjusts the oscilloscope gains until the ellipse leans at a 45° angle. He then switches to each remaining station in turn and adjusts the amplitude of each force to produce a 45° ellipse with its response. Of course, when additional shaking forces are introduced the response at the master station is influenced and the ratio of force to amplitude there is changed. This difficulty is effectively met by using an automatic normalizer driven by the master station

Fig. 13–12. Control console for a multiple shaker system with some of the 24 associated shakers. (Courtesy of the Calidyne Company)

signals which appropriately modifies each velocity signal before it is presented on the scope. In this manner each shaking force is adjusted to have the same ratio to response velocity as the master station. The later adjustments affect the ratios at some of the stations previously adjusted, but a few iterations of this adjustment procedure produce a rapid convergence on pure excitation of the target mode. This is indicated by the fact that as the adjustments proceed the velocities become more and more nearly in phase with the exciting forces. If the excitation is suddenly removed the responses at all stations decay smoothly without beats, which is another excellent indication that a pure mode was excited. In between iterations the operator may have found it advisable to adjust the driving frequency nearer to resonance. When the pure mode is finally satisfactorily excited he records the responses of all of the pickups simultaneously on the accompanying recording oscillograph. It is interesting to note that the operator usually finds it unnecessary to look at the vibrating structure itself while locating and measuring a mode.

(d) *Shakers and pickups.* In the previous discussion we have assumed that shaking forces can be applied to structures with little thought as to how the forces could be produced and where they may actually be applied to a practical structure. Let us consider next the requirements of the ideal shaking device and classify existing types according to their capabilities.

The most fundamental requirement for a shaker is that it apply the desired force to the structure without significantly influencing the mass and stiffness properties of the structure. To put it another way, the presence of the exciting and measuring apparatus must not change and distort the very quantities we seek to measure. Thus we must be careful to use equipment of a scale appropriate to the structure to be tested, which may be a small model, a light aircraft, or a heavy bomber. In any particular case we can estimate the amount of additional mass which can be safely added to a point in the structure without modifying its vibration modes by estimating the mass segment of the structure which has essentially the same motion as the point in question. Of course, the relative size of the distortions in the neighboring structure generally increases as the mode frequency increases, so that the estimate of the size of the locally rigid mass should be made using modes typical of the highest to be measured. If the additional mass is less than a few percent of the locally rigid mass its effect can be neglected.

Similarly, if the shaking device adds stiffness to the wing that is not part of the measured exciting force, this added stiffness must produce force increments which are small compared with the internal elastic shears. Since the elastic shears are balanced by inertial shears in vibration at a natural mode it is sufficient to say that the added stiffness should be small compared with the product of the locally rigid mass and the square of the frequency. This criterion may be most restrictive at the lower-frequency modes.

Another requirement for the ideal shaker is that it be capable of synchronization with other shakers of the same variety. In this multiple shaker operation the various shakers must have not only the same frequency but also the same phase. Further, each shaker should have adequate and controllable force output over the necessary frequency range, should be readily attachable to the structure, and should not require an unreasonable amount of auxiliary equipment and facilities.

Looking first at small-scale shakers such as are used on typical low-speed flutter models, we find that they are usually userbuilt and operate on the following principles:

(1) Electromagnetic shaking, in which the very light "voice" coil is mounted on the structure and the field coil and iron flux path are externally

Fig. 13–13.   Elaborate flutter-model vibration-test arrangement using electromagnetic excitation.   (Courtesy of Boeing Airplane Company)

Fig. 13-14. Variable-frequency pulsating air jets can provide adequate excitation for many model vibration tests.

supported (Fig. 13-13). A sinusoidally varying current in the voice coil induces a corresponding force on the structure. This type of shaker is quite versatile and is particularly adaptable to the synchronized use of multiple units. For very light models the mass of the voice coils may become excessive. The usable frequency range can easily extend from a few cycles per second to a few hundred cycles per second.

(2) Air-jet shaking, in which the structure deflects pulsing streams of air directed against opposite sides (Fig. 13-14). Rotating valves in the air-lines modulate the flow to each side to produce a net sinusoidal reaction on the structure. This device has a frequency range of 1 to 200 cps and can be used in multiple sets. Its chief advantage is that no direct connection to the structure is needed and the chief disadvantage is that the exact phase and amplitude of the applied force cannot be measured.

(3) Rotating unbalance and reaction-jet shakers are sometimes used for vibration testing of models, particularly when the shaking force must also be provided in the presence of an airstream. These methods are discussed in the section on flight flutter testing.

Turning now to large-scale shakers such as are employed on aircraft, we find the following types:

(1) Electromagnetic shakers similar to the one described above but

Fig. 13–15. Electromagnetic shakers can be practical devices over a large range of sizes. This one has a maximum force rating of 12,500 lb. (Courtesy of the Calidyne Company)

built to almost any desired scale (Fig. 13–15). These shakers are very versatile and are widely used, although it is difficult to provide the shakers with adequate power at the lowest airplane frequencies on large-scale installations.

(2) Rotating unbalance excitation in which an off-balance flywheel driven through a flexible shaft is attached to the structure. This device is used primarily for fairly rough determination of natural frequencies when better equipment is not available. It is not readily adaptable to multiple use, and it has relatively high mass, particularly at low excitation frequencies. The force amplitude and phase are not easily adjusted.

(3) Spring-connected eccentric in which one end of a spring is attached to the structure and the other end is displaced sinusoidally. The force amplitude and phase depend on the relative displacement of the structure and driven end of the spring and thus are not readily controllable. The device can have a fairly low mass but always modifies the stiffness properties of the structure. It cannot be used easily in multiple sets.

The measurement of a normal mode of a structure, once it is excited, involves the measurement of the relative amplitudes of vibration at numerous stations either during the steady-state excitation at the resonant frequency or during the decay transient following the sudden removal of the excitation. The usual instrument is some variation of a seismic pickup, such as an accelerometer, which produces an electrical signal proportional to the amplitude of the structure to which it is attached. Such a pickup is usually small enough so that its mass becomes a troublesome factor only on small, light structures. If it is desired to measure only natural frequencies and elementary mode-shape characteristics, one or two pickups or even a single resistance-wire strain gage is sufficient. These can be used to indicate resonance and the experimeter's eye or finger can locate nodal lines. It is almost always worth while, however, in full-scale testing to employ two matched pickups which can be easily moved from point to point on the structure to indicate relative phase between any pair of points. This provides a simple, direct, and accurate means of establishing the symmetry or lack of symmetry of a mode of vibration and helps in tracking down elusive nodal lines. An excellent pictorial record of nodal line positions in model testing can often be made by sprinkling salt or some similar substance on the vibrating surface. The particles will congregate on the nodal lines if the surface is approximately horizontal, and can be photographed as shown in Fig. 13–16. In the frequency range between about 10 and 100 cps stroboscopic lighting devices can be used to "slow down" the vibration so that its detailed shape is apparent to the eye of the observer.

(e) *Support of structures for vibration testing.* The problem of correctly supporting a structure during its vibration tests is often quite difficult. To obtain a desired set of normal modes, not only must the excitation be suitably applied but the boundary conditions must be accurately simulated.

The most usual set of boundary conditions (and one of the most difficult) which must be provided corresponds to the aircraft structure in a free-flight configuration. In this case the weight of the structure must be supported without introducing any external constraints or distortions which will affect the vibration modes to be measured. Ideally, devices are needed which can apply constant vertical forces to the various parts of the structure without adding any mass or stiffness. In practice, the mass and stiffness which are added at a point of the structure must be small enough to produce a negligible effect on the measured vibration modes, and the criterion is the same as described above for the attachment of shaking devices.

Thus the added mass must be small compared with the mass of the neighboring structure, which has essentially the same motion as the point of attachment. The added stiffness must produce force increments which are small compared with the local elastic shears or, equivalently, the inertial force of the locally rigid mass.

For example, let us assume that we want to find the normal modes in a flight condition of a small fighter-type aircraft. Its total weight is 15,000 lb, the engine is in the fuselage, and it is convenient to hang the aircraft

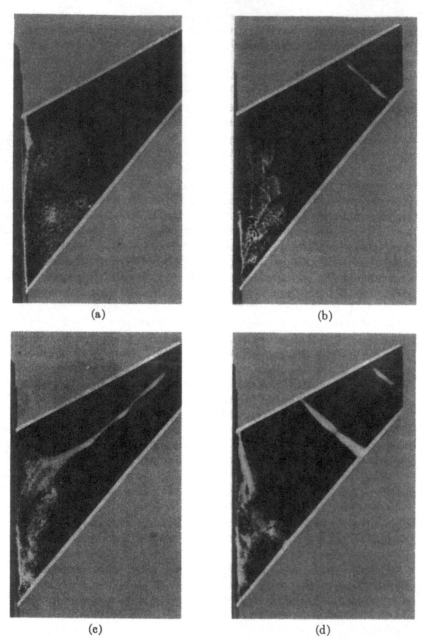

(a)   (b)

(c)   (d)

Fig. 13–16.   Nodal lines for the four lowest modes of a swept supersonic flutter model as indicated by the congregation of salt crystals sprinkled on the surface during steady-state excitation at each of the natural frequencies.

from a hoisting point in back of the cockpit. We must first estimate the mass of the fuselage section adjacent to the hoisting point, which remains essentially undistorted during the vibrations. A reasonable guess might be 3000 lb. If it is anticipated that the lowest frequency of interest is about 3 cps, the stiffness of the support must be quite small compared with $(2\pi \times 3)^2(3000/32)$ or 36,000 lb/ft. Choosing a stiffness of 3000 lb/ft, we find that the static deflection under the aircraft's weight is 5 ft. Thus a spring must be obtained with this stiffness which can carry a load of 15,000 lb. In addition, the effective mass of the spring must be small compared with 3000 lb, if only the first few modes are to be measured. Otherwise it must be small compared with the locally rigid mass at the worst case.

The natural frequency of the aircraft on this support is about $\frac{1}{3}$ of a cycle per second, which is small compared with the lowest frequency of interest (assumed to be 3 cps). However, it is not sufficient, in general, to say that a supporting spring is weak enough if it provides a support frequency which is small compared with the lowest frequency of interest. If, for instance, this fighter had been suspended by two springs, one near each wing tip, so as to give this same support frequency, it is quite probable that the support stiffness would be large enough to alter significantly some of the lower modes because of the much-reduced size of the locally rigid mass.

If it is possible to support the structure only at nodal points, rigid supports can be used, but usually we do not know beforehand where the nodal lines of a given mode will be. An exception is the case where a single support which does not restrain rotation can be used at the plane of symmetry in the measurement of antisymmetrical modes.

Although the simulation of free-flight boundary conditions is normally quite difficult, the simulation of other boundary conditions can at times tax the engineer's ingenuity. As an illustration, consider the problem of exciting the cantilever modes of the very light, stiff, small model shown in Fig. 12–4. Available shaking devices could not be attached to the wing without seriously modifying its characteristics, and the air-jet shaker, which requires no attachment, was not capable of excitation at the very high frequencies required. The solution which was finally worked out is shown in Fig. 13–17. The mass and moments of inertia of the wing's supporting base were increased until they were orders of magnitude larger than corresponding properties of the wing. This massive base was then supported on very soft springs and attached to a pair of large aircraft-type electromagnetic shakers. The excitation was applied to the base and through it to the wing. At a wing natural frequency the amplitude of base motion was almost imperceptible, whereas the wing amplitudes were considerable. For all practical purposes, the wing was vibrating on a fixed base and thus in its cantilever modes.

Fig. 13–17. Electromagnetic excitation as applied to light stiff supersonic flutter models with a simulated cantilever root restraint.

(f) *The effect of changes in the structure and its boundary conditions.* A knowledge of the qualitative effect of modifications to the "shake test" setup is of considerable value when the desired conditions cannot be achieved completely. Thus the attempted simulation of free-flight boundary conditions often means that some mass and some stiffness must be added to the test specimen in order to even approximate the boundary conditions. How does the added mass, even though small, affect the natural frequencies and mode shapes? What is the effect of the added stiffness? Rayleigh (Ref. 13–4), among others, has examined this kind of problem in considerable detail and the discussion below follows his general line of reasoning.

For an elastic system which can be considered to have a finite number of degrees of freedom and which is performing small oscillations about its equilibrium position the following general statements can be made:

(1) The addition of mass to the system will decrease all the natural

frequencies of the system or, at worst, leave unchanged those for which the added mass is at a node.

(2) The addition of stiffness to the system will increase or leave unchanged the natural frequencies of the system.

(3) The natural frequencies of the modified systems in either (1) or (2) above will, in general, alternate with or separate the frequencies of the original system.

(4) The suppression of a degree of freedom reduces by one the number of natural frequencies, such that the $(n - 1)$ remaining frequencies alternate with the n original frequencies.

(5) The addition of a degree of freedom increases by one the number of natural frequencies, such that the n original frequencies alternate with the $(n + 1)$ frequencies of the augmented system.

(6) If a spring-mass combination (such as a flexibly mounted engine) is added to a system, the frequencies originally higher than the basic spring-mass frequency $(\sqrt{K/m})$ are increased, those originally lower are decreased, and a new frequency appears between the original pair of frequencies nearest the spring-mass frequency.

(7) If a spring is added to attach the system to a fixed reference and mass is also added to the system at the point of spring attachment, the frequencies of the original system are all shifted toward the basic spring-mass frequency $\sqrt{K/m}$. (This might correspond to the soft spring suspension described earlier, and is Rayleigh's so-called "vibrator.")

In applying the foregoing statements, it is helpful to consider the essentially continuous structure of an airplane or model to be represented by a lumped-parameter system (see Section 4–6) with a large but finite number of degrees of freedom. The accuracy of this representation is sufficient for all practical purposes.

To demonstrate the validity of these statements, consider a system with n degrees of freedom. We can express the kinetic and potential energies in terms of the normal coordinates ξ, the generalized masses M, and the natural frequencies ω:

$$T = \tfrac{1}{2}M_1\dot{\xi}_1{}^2 + \tfrac{1}{2}M_2\dot{\xi}_2{}^2 + \cdots + \tfrac{1}{2}M_n\dot{\xi}_n{}^2, \tag{13–8}$$

$$U = \tfrac{1}{2}M_1\omega_1{}^2\xi_1{}^2 + \tfrac{1}{2}M_2\omega_2{}^2\xi_2{}^2 + \cdots + \tfrac{1}{2}M_n\omega_n{}^2\xi_n{}^2. \tag{13–9}$$

Suppose that constraints are put on the system such that only one degree of freedom is left. Since any motion of this system can be represented by a superposition of its normal modes, these constraints can be represented by relating the normal coordinates to the single variable θ. Thus

$$\xi_1 = A_1\theta, \qquad \xi_2 = A_2\theta, \qquad \cdots, \qquad \xi_n = A_n\theta, \tag{13–10}$$

where the A's are given for a given constraint. To find the frequency ω at which the system will oscillate sinusoidally under the given constraint, set $\theta = \cos \omega t$ and substitute Eq. (13–10) into Eqs. (13–8) and (13–9), so that

$$T = \{\tfrac{1}{2}M_1A_1{}^2 + \tfrac{1}{2}M_2A_2{}^2 + \cdots \}\omega^2 \sin^2 \omega t, \qquad (13\text{–}11)$$

$$U = \{\tfrac{1}{2}M_1\omega_1{}^2A_1{}^2 + \tfrac{1}{2}M_2\omega_2{}^2A_2{}^2 + \cdots \} \cos^2 \omega t. \qquad (13\text{–}12)$$

Since in the oscillation of a conservative system the amplitudes of variation of the kinetic and potential energies are equal, the square of the frequency of the constrained oscillation is given by

$$\omega^2 = \frac{M_1\omega_1{}^2A_1{}^2 + M_2\omega_2{}^2A_2{}^2 + \cdots + M_n\omega_n{}^2A_n{}^2}{M_1A_1{}^2 + M_2A_2{}^2 + \cdots + M_nA_n{}^2}. \qquad (13\text{–}13)$$

It can be seen by inspection or differentiation that the frequency is stationary with respect to the constraint A_m $(\partial\omega/\partial A_m = 0)$ when all the other A's are zero, that is, when the system is vibrating in the normal mode ϕ_m at the frequency ω_m. To say it another way, if a system is constrained such that it vibrates very nearly in a normal mode, the difference between the frequency of the constrained system and the natural frequency is of second order compared with the difference in mode shape.

Let us use this idea to see what happens when a very small amount of mass is added to the system. Although the normal modes will be altered slightly, the system can be made to vibrate in any one of the original normal modes by the imposition of suitable constraints. The added mass will decrease the frequency of any one of these original modes in which it is not placed at a point of zero displacement. If the constraints are now relaxed the concurrent shift of frequency as the mode shape changes is only of second order and the general decrease of natural frequencies is still apparent. An integration of this effect leads to the first of the statements listed above. A similar argument concerning an increase in stiffness or potential energy leads to the second statement above.

To demonstrate the third statement concerning the relative positions of the natural frequencies before and after the addition of mass and stiffness, suppose we add a mass m at the point A whose displacement, expressed in terms of normal coordinates, is

$$q = \phi_1(A)\xi_1 + \phi_2(A)\xi_2 + \cdots + \phi_n(A)\xi_n. \qquad (13\text{–}14)$$

The expression for kinetic energy becomes

$$T = \tfrac{1}{2}M_1\xi_1{}^2 + \cdots + \tfrac{1}{2}M_n\xi_n{}^2 + \tfrac{1}{2}m(\phi_1\xi_1 + \phi_2\xi_2 + \cdots + \phi_n\xi_n)^2, \qquad (13\text{–}15)$$

whereas the expression for potential energy is unchanged. If we consider simultaneously the case of adding stiffness K between two points A and B in the system, and use the notation that $\phi(A) - \phi(B) = \phi'$, the expression for kinetic energy is unchanged but the potential energy becomes

$$U = \tfrac{1}{2}M_1\omega_1{}^2\xi_1{}^2 + \cdots + \tfrac{1}{2}M_n\omega_n\xi_n{}^2 + \tfrac{1}{2}K(\phi_1'\xi_1 + \phi_2'\xi_2 + \cdots + \phi_n'\xi_n)^2.$$
$$(13\text{--}16)$$

Substitution into Lagrange's equation of free vibration for an undamped system,

$$\frac{d}{dt}\left(\frac{\partial T}{\partial \dot{\xi}}\right) + \frac{\partial U}{\partial \xi} = 0, \tag{13--17}$$

produces the equations of motion for the modified system:

$$M_1\ddot{\xi}_1 + m\phi_1(\phi_1\ddot{\xi}_1 + \phi_2\ddot{\xi}_2 + \cdots + \phi_n\ddot{\xi}_n) + M_1\omega_1{}^2\xi_1 + K\phi_1'(\phi_1'\xi_1 + \cdots + \phi_n'\xi_n) = 0,$$

$$M_2\ddot{\xi}_2 + m\phi_2(\phi_1\ddot{\xi}_1 + \phi_2\ddot{\xi}_2 + \cdots + \phi_n\ddot{\xi}_n) + M_2\omega_2{}^2\xi_2 + K\phi_2'(\phi_1'\xi_1 + \cdots + \phi_n'\xi_n) = 0,$$

$$\vdots$$

$$M_n\ddot{\xi}_n + m\phi_n(\phi_1\ddot{\xi}_1 + \phi_2\ddot{\xi}_2 + \cdots + \phi_n\ddot{\xi}_n) + M_n\omega_n{}^2\xi_n + K\phi_n'(\phi_1'\xi_1 + \cdots + \phi_n'\xi_n) = 0.$$
$$(13\text{--}18)$$

With the assumption of a sinusoidal solution of frequency ω, the equations of motion become

$$M_1(\omega_1{}^2 - \omega^2)\xi_1 - m\omega^2\phi_1(\phi_1\xi_1 + \cdots + \phi_n\xi_n) + K\phi_1'(\phi_1'\xi_1 + \cdots + \phi_n'\xi_n) = 0,$$

$$M_2(\omega_2{}^2 - \omega^2)\xi_2 - m\omega^2\phi_2(\phi_1\xi_1 + \cdots + \phi_n\xi_n) + K\phi_2'(\phi_1'\xi_1 + \cdots + \phi_n'\xi_n) = 0,$$

$$\vdots$$

$$M_n(\omega_n{}^2 - \omega^2)\xi_n - m\omega^2\phi_n(\phi_1\xi_1 + \cdots + \phi_n\xi_n) + K\phi_n'(\phi_1'\xi_1 + \cdots + \phi_n'\xi_n) = 0.$$
$$(13\text{--}19)$$

For the moment, let the added stiffness K be zero, multiply the first equation by $\phi_1/M_1(\omega_1{}^2 - \omega^2)$, the second by $\phi_2/M_2(\omega_2{}^2 - \omega^2)$, and so on, and add the equations. The common factor $(\phi_1\xi_1 + \cdots + \phi_n\xi_n)$ cancels out, leaving

$$1 - \frac{m\omega^2\phi_1{}^2}{M_1(\omega_1{}^2 - \omega^2)} + \frac{m\omega^2\phi_2{}^2}{M_2(\omega_2{}^2 - \omega^2)} + \cdots + \frac{m\omega^2\phi_n{}^2}{M_n(\omega_n{}^2 - \omega^2)} = 0. \quad (13\text{--}20)$$

Rewriting, this characteristic equation for the modified system becomes

$$M_1 M_2 M_3 \cdots M_n (\omega_1{}^2 - \omega^2)(\omega_2{}^2 - \omega^2) \cdots (\omega_n{}^2 - \omega^2)$$
$$- m\omega^2\phi_1{}^2 M_2 M_3 \cdots M_n (\omega_2{}^2 - \omega^2)(\omega_3{}^2 - \omega^2) \cdots (\omega_n{}^2 - \omega^2)$$
$$- M_1 m\omega^2\phi_2{}^2 M_3 \cdots M_n (\omega_1{}^2 - \omega^2)(\omega_3{}^2 - \omega^2) \cdots (\omega_n{}^2 - \omega^2)$$
$$- \cdots = 0. \quad (13\text{--}21)$$

The first line is the characteristic equation of the unmodified system and vanishes when ω equals an original natural frequency, as do all but one of the remaining terms. The sign of this term determines the sign of the expression. Notice that when $\omega = 0$ the expression is positive, when $\omega = \omega_1$ it is negative, when $\omega = \omega_2$ it is positive, etc. Thus one root of Eq. (13–21) and therefore one natural frequency of the system with added mass lies below ω_1, the next lies between ω_1 and ω_2, and, in general, the new frequencies are lower than but alternate with the original frequencies.

As an example, suppose a tip tank is added to a cantilever wing.* The new natural frequencies will all be lower than for the bare wing but will still alternate with them no matter how large the tip tank mass is. This can easily be checked in the limit by comparing the natural frequencies of a cantilever uniform beam and those of the same beam pinned at the free end.

Returning to Eq. (13–18), let the added mass be zero and investigate the effect of adding stiffness. The characteristic equation can be obtained by an analogous process of eliminating the term $(\phi_1'\xi_1 + \phi_2'\xi_2 + \cdots + \phi_n'\xi_n)$ such that

$$M_1 M_2 M_3 \cdots M_n (\omega_1{}^2 - \omega^2)(\omega_2{}^2 - \omega^2) \cdots (\omega_n{}^2 - \omega^2)$$
$$+ K(\phi_1')^2 M_2 M_3 \cdots M_n (\omega_2{}^2 - \omega^2)(\omega_3{}^2 - \omega^2) \cdots (\omega_n{}^2 - \omega^2)$$
$$+ M_1 K(\phi_2')^2 M_3 \cdots M_n (\omega_1{}^2 - \omega^2)(\omega_3{}^2 - \omega^2) \cdots (\omega_n{}^2 - \omega^2)$$
$$+ \cdots = 0. \quad (13\text{–}22)$$

In this case the expression is positive when $\omega = 0$ and is still positive when $\omega = \omega_1$. The sign alternates for higher frequencies, indicating that all the new natural frequencies are higher than those of the unmodified system but alternate with them. It is interesting to note that the transition from the natural frequencies of the free-free beam to those of the beam with clamped ends can also be thought of as being produced by the addition of infinitely stiff springs at the ends. In this case the frequencies, including the zero frequency, of the free-free beam go up, whereas analogous results were obtained when we added mass and watched them go down.

Statements (6) and (7) can be verified in a similar manner after making the following additions to the energy expressions. For statement (7), in which a spring is attached between a fixed reference and the point on the structure where the mass is added, $\phi = \phi'$ and the energy expressions can be written

$$T = \tfrac{1}{2}M_1\xi_1{}^2 + \cdots + \tfrac{1}{2}M_n\xi_n{}^2 + \tfrac{1}{2}m(\phi_1\xi_1 + \cdots \phi_n\xi_n)^2,$$
$$U = \tfrac{1}{2}M_1\omega_1{}^2\xi_1{}^2 + \cdots + \tfrac{1}{2}M_n\omega_n{}^2\xi_n{}^2 + \tfrac{1}{2}K(\phi_1\xi_1 + \cdots + \phi_n\xi_n)^2. \quad (13\text{–}23)$$

For statement (6) another coordinate x is needed to describe the motion of

---

* Although the example here is a continuous system which ideally has an infinite number of degrees of freedom, we can apply the results of this section for any finite number of lower modes, because the continuous system can be approximated to any desired degree of accuracy by a system with a finite number of degrees of freedom.

the mass relative to the point of attachment. Since the kinetic energy of the mass must be referred to a fixed reference, the energy expressions can be written

$$T = \tfrac{1}{2}M_1\dot{\xi}_1{}^2 + \cdots + \tfrac{1}{2}M_n\dot{\xi}_n{}^2 + \tfrac{1}{2}m(\phi_1\dot{\xi}_1 + \cdots \phi_n\dot{\xi}_n + \dot{x})^2, \quad (13\text{-}24)$$

$$U = \tfrac{1}{2}M_1\omega_1{}^2\xi_1{}^2 + \cdots + \tfrac{1}{2}M_n\omega_n{}^2\xi_n{}^2 + \tfrac{1}{2}Kx^2. \quad (13\text{-}25)$$

The proof of statement (4) reduces to that of statement (7) above if we start with a system of $(n + 1)$ degrees of freedom and represent the suppression of a degree of freedom as an algebraic relationship between the normal coordinates. Using this expression, ξ_{n+1} can be eliminated to give

$$T = \tfrac{1}{2}M_1\dot{\xi}_1{}^2 + \cdots + \tfrac{1}{2}M_n\dot{\xi}_n{}^2 + \tfrac{1}{2}M_{n+1}(A_1\dot{\xi}_1 + \cdots + A_n\dot{\xi}_n)^2, \quad (13\text{-}26)$$

$$U = \tfrac{1}{2}M_1\omega_1{}^2\xi_1{}^2 + \cdots + \tfrac{1}{2}M_n\omega_n{}^2\xi_n{}^2 + \tfrac{1}{2}M_{n+1}\omega_{n+1}{}^2(A_1\xi_1 + \cdots + A_n\xi_n)^2. \quad (13\text{-}27)$$

The proof of statement (5) follows from statement (4) by taking the augmented system and showing what happens when the added degree of freedom is suppressed.

13-4 Steady-state aeroelastic testing. In this section we shall consider testing in which an airstream is present and which is essentially steady state or nonvibratory in character. Included in this category are tests for surface load distribution, control effectiveness, roll rate, divergence, and stability derivatives. We shall limit our discussion almost entirely to model techniques, since full-scale techniques in this field are well known. The model techniques are growing in complexity and importance following the general trend of aeroelastic effects in these days of high-speed flight.

All of the aeroelastic problems mentioned above have the common basis that they are concerned with elastic deformations under airloads, with the consequent modification of the airloads until a state of equilibrium is reached. The model must simulate the airplane's aerodynamic shape, stiffness distribution, and ratio of elastic to aerodynamic stiffness. It must be placed in an airstream of as high a Reynolds number as possible and of the same Mach number if compressibility effects are to be simulated. Mass and weight distributions need not be simulated but should be such as to preclude flutter in the test range. The length scale is fixed by the ratio of desired model size to airplane size. The force scale is fixed by the length scale and the ratio of full-scale dynamic pressure to the dynamic pressure which can be obtained in the model tests.

If compressibility effects are not simulated, the models are usually tested at convenient low speeds and thus must be correspondingly flexible. If the model tests are made at full-scale Mach number, the dynamic pressure is high and the model must be stiff. The strength of the low-speed models is usually relatively greater than that of the airplane, whereas for

Fig. 13–18. Single-spar sectional wing with deflected ailerons free to achieve a steady rolling velocity.

high-speed models the reverse is true. In both cases, relatively inefficient but simple structures are commonly used (Section 12–2).

The two most common means of testing these models are to mount them in a suitable wind tunnel or to use rocket-propelled test vehicles in free flight. If we consider first the low-speed model in the wind tunnel, we find that the standard model support system must be modified. If we want to use the wind-tunnel balance system to measure over-all forces and moments, it must be connected to the model in such a way as to permit the model to deform elastically. In order to evaluate aeroelastic effects, tests are usually run over as large a range of dynamic pressure as possible. This range is often seriously limited on the lower end by the inability of the balance system to measure accurately low levels of force and moment (Ref. 13–5). Measurement of quantities such as spanwise force and moment distribution can be made simply on sectional single-spar models by spacing wire strain gages along the spar; with this construction, all of the airload on a given wing station is transmitted directly to the spar (Ref. 13–6).

An interesting sidelight on the use of flexible models in wind tunnels is the consternation of many wind tunnel personnel on the occasion of the first test. The elastic model illustrates quite graphically the many imperfections of the airstream in the test section. Similarly, the model which is allowed freedom to roll in roll rate tests (Fig. 13–18) is a discouragingly sensitive indicator of airstream twist.

The sting mount and internal balance system of a typical high-speed wind tunnel are well adapted to the use of flexible models. The most serious difficulty is the relative weakness of the models. They have trouble surviving the starting roughness of supersonic tunnels and are seriously limited by their inability to withstand the airloads at large angles of attack.

The use of the rocket vehicle for testing elastic models is particularly advantageous in the transonic range. The absence of wind-tunnel walls and the smoothness and uniformity of the airflow outweigh the disadvantages of trying to make measurements on a small, distant, fast-traveling system.

13–5 Dynamic aeroelastic testing—full scale. Let us now turn our attention to dynamic aeroelastic testing in which we consider first tests carried out on the full-scale airplane. The more important of these tests are designed to investigate possible flutter conditions, evaluate dynamic stability and gust response, and establish buffet boundaries. Broadly speaking, these tests may be divided into two categories, according to their primary purposes. In the first category are flutter and dynamic stability tests, which are concerned with stability characteristics of the airplane. In the second category are gust response and buffeting tests, in which airplane strength is the primary consideration.

(a) *Flight flutter testing.* Looking first at the problems involved in flight flutter testing (Ref. 13–7) it must be recognized that we are usually dealing with a prototype airplane whose testing must be completed as quickly as possible. However, such airplanes represent tremendous investments both in time and money and must not be placed in unduly dangerous situations if they can reasonably be avoided. That flight flutter testing can be very dangerous even when approached with caution is borne out by the number of fatal accidents which have occurred during such tests. In many cases the tests must be carried out near limit design speeds being attained by the aircraft for the first time, and an actual occurrence of flutter can be spectacularly destructive. In some cases it may be advantageous to modify the test airplane to lessen the possibility of encountering flutter. This new stabilizing influence can be removed analytically from the test results to find the behavior of the unmodified airplane. Of course, the probability of encountering flutter except in a

minor tab or control surface mode is not large, because of the many preventive measures taken by the designer as a result of analytical and model
investigations. Nevertheless, the possibility of such an occurrence cannot
be overlooked. Thus the test procedure must embody the following
principles:

(1) The tests must be carried out as expeditiously as possible without
introducing too large an element of danger.

(2) The approach to a critical flutter condition must be recognized by
observation of subcritical airplane response.

(3) Means of excitation must be used which will permit adequate observation of subcritical response.

(4) The added mass of the test equipment should not be so located as to
increase the possibility of a dangerous flutter.

The means of excitation which have been used and proposed vary from
the one extreme of asking the pilot to shake or kick the controls to the other
extreme of installing complex and costly forcing equipment weighing many
hundreds of pounds. The particular scheme appropriate to a given situation depends on the probability and expected seriousness of a flutter condition, the part of the airplane which it involves, the frequency range which is
important, and the time and money available for flight tests and parallel
analytical or model investigations. The more important techniques are
outlined in the following paragraphs, along with their advantages and short-
comings.

Excitation through existing control systems. This approach is adopted in
the many instances where only a cursory investigation is needed to prove
the absence of flutter. It can assume a variety of forms. As mentioned
above, in the case of a manually powered system the pilot can "pulse,"
kick, or jerk the controls and excite modes with frequencies up to about 6
or 7 cps. If there is an autopilot installed, simple signal sources can be connected to it to produce pulse-type or sinusoidal excitation through about
the same frequency range. For airplanes with power-operated control
surfaces, similar low-power-level signal generators can be used successfully
up to about 10 or 12 cps, depending on the capabilities of the servos. Refinements such as the provision for symmetrical as well as antisymmetrical
aileron forcing are easily incorporated in this type of system. It is seldom
possible to excite to a reasonable degree modes above 15 cps through existing or even specially modified control systems.

Pulse excitation. The simplest type of excitation equipment to install
in an airplane is usually the pulse generator. This device may take the
form of a modified gun or cannon which fires projectiles in the direction
opposite to that in which the applied pulse is desired. The projectiles
can be fired singly or in bursts and several guns can be synchronized to
obtain different force and moment patterns. Another form of pulse exciter
consists of one or more firing chambers in which carefully arranged charges

can be fired. By suitable choice of powder and shaping of grains a large variety of pulse shapes and durations can be obtained. Control over pulse duration is particularly advantageous because it permits considerable selective emphasis of mode response throughout the frequency range of interest. A smooth distribution of force over a pulse interval equal to about one-half of the period of a given mode maximizes the energy fed into the mode.

Some advantages of pulse-type shaking are that it requires only short testing times during which the airplane must maintain the near-limit speeds peculiar to flight flutter testing, and that the necessary equipment can be made quite light, is not bulky, and consequently can be installed in many possible locations. The major disadvantage of the technique is the relative difficulty with which sufficient information can be extracted from the recorded data.

Sinusoidal excitation. An alternative to applying short pulses of energy to the airplane and observing its transient responses is the technique of steady sinusoidal forcing throughout the frequency range of importance, and observation of amplitude and phase of the forced response. The favorite excitation device consists basically of a rotating unbalanced wheel which imparts a rotating force vector to the structure to which it is attached. A basic difficulty lies in the fact that the excitation amplitude is proportional to the amount of unbalance and the square of the frequency. Thus a considerable excess of shaking force is available at the high frequencies when only the barest minimum is available at the low frequencies. This inherent characteristic leads to such refinements as variable unbalance settings which may even be automatically controlled as a function of frequency. The use of matched counter-rotating unbalances produces a simple sinusoidal shaking force instead of a rotating force vector.

A major difficulty in the use of this kind of shaker is the need for a light, powerful motor with very positive frequency control characteristics. Not only are the power requirements continuous and high in sinusoidal shaking, but the control of forcing frequency in the neighborhood of resonance peaks is very difficult. Also synchronization of multiple units some distance apart is quite difficult, particularly in the low-frequency high-power range. Rotating unbalances and the associated driving equipment tend to be heavy and bulky and are difficult to bury inside lifting surfaces without requiring large protuberances. Housing within a special external store may often be the best solution for wing installation. For symmetrical tail excitation a location in the aft fuselage is often used.

Flight test procedure with sinusoidal shaking requires comparatively long periods of steady flight near limit speed while the series of constant frequency runs is being made. The programing of a slow sweep through the frequency range may shorten the testing time at the expense of complicating the already formidable array of equipment.

A sinusoidal exciter often discussed but seldom used is that of a mass which is forced to oscillate linearly with respect to the airplane by either mechanical or electromagnetic means. In the latter arrangement it is often proposed to use a heavy field coil with its iron case as the oscillating mass and attach the voice coil to the airplane structure. Since the use of the necessary mass by this scheme is less efficient in producing shaking force than with rotating unbalances the latter technique is preferred.

A means of excitation, which can produce sinusoidal torques in the lower frequency range quite efficiently with a limiting magnitude proportional to forcing frequency, also uses rotating flywheels, but in a different way. In this arrangement the flywheel is spun at a very high constant rate about an axis perpendicular to the plane of the wing in the case of a wing installation. It is mounted in a gimbal hinged about a fore-and-aft axis so that torques which tend to twist the wing can be transmitted from the flywheel to the wing. These torques are produced by using a linear actuator to apply a sinusoidal force to the gimbal, tending to rotate it about the fore-and-aft axis. This torque is transformed by the spinning flywheel into a wing twisting torque. The output torque amplitude is limited by the angular oscillation amplitude of the flywheel about the fore-and-aft axis permitted by the available mounting space. It is also a function of the spin rate, the flywheel inertia, and the input torque. The upper limit on forcing frequency is primarily determined by the characteristics of the device used to produce the input torque to the flywheel.

Random excitation. The bane of all the techniques mentioned so far is the unavoidable presence of random "hash" on the records, which obscures the desired information. Much of this hash comes from airplane response to air turbulence and for this reason tests are often carried out in the "turbulence-free" early morning hours. With the growth of engineering application of statistical analysis techniques, particularly in the fields of turbulence, buffeting, and gust response, it appears increasingly advantageous to use atmospheric turbulence itself as the source of excitation for flight flutter testing. Instead of fighting to minimize its effects, we might use them to our own ends.

By this procedure a frequency response curve for the airplane similar to that obtained by sinusoidal excitation techniques can be calculated if certain statistical properties of the input turbulence and output wing motion are known (see Chapter 10 and Ref. 13–8). From input and output time histories which are essentially random, autocorrelation functions can be obtained. These can be transformed into a curve of power spectral density versus frequency for both input and output. The ratio of the output curve to the input is equal to the square of the airplane frequency response curve. If simultaneous time histories of input and output are measured, cross-correlation and cross-spectral density curves can be found which permit the determination of the phase as well as the amplitude characteristics of the airplane frequency response.

A basic difficulty with the technique is the lack of control over the amplitude and frequency characteristics of the input turbulence. There is some indication that most free-air turbulence has approximately the same power spectral density, and it is probably not feasible to try to search out a kind of turbulence which accentuates the frequency range of interest in a particular set of tests. Nevertheless, the possibility of flight flutter testing without the necessity of providing special excitation equipment is quite attractive.

Another difficulty is the measurement of the characteristics of the turbulence as it approaches the airplane. The vibrating and bouncing airplane, whose motions distort the flow field (in the subsonic case) some distance ahead, does not serve as a very adequate reference system. However, the present attacks on the problem by those interested in the effect of turbulence and gusts on wing stresses, missile and airplane guidance systems, and fatigue promise to be quite fruitful.

Observation and interpretation. The entire approach to the problem of instrumentation and data interpretation in flight flutter testing is governed by the short analysis times available to the engineer. He must deduce quickly from a set of data at a given velocity whether or not it appears safe to proceed with another test at a higher velocity. In a simple situation where no serious flutter condition is expected and the "stick-shaking" excitation technique is adequate the engineer may use his own senses as pickup instruments while flying in the airplane and rely on his experience and intuition in deciding on the potential danger of the situation. However, if a flutter condition is anticipated a more sophisticated technique is required.

In the case of pulse excitation the airplane response is interpreted in terms of the frequencies and damping rates of the various modes which make up the response. In principle, an applied pulse excites all of the modes and thus a complete description of airplane stability is available from just one response record. This is not true in practice for a variety of reasons. The location, duration, and type of pulse all serve to emphasize some modes at the expense of others. The amplitude and frequency characteristics of the "hash," as well as the location and frequency characteristics of the pickups, tend to suppress and give apparent distortion to the records. These effects can be minimized by suitable choice of pulse lengths and careful spacing of pickups, as well as by employing analytical and model techniques to indicate *in advance* the subcritical behavior which can be expected. It is difficult to overemphasize the usefulness during a hectic flight flutter test program of a large body of information on the probable subcritical response of the airplane to the particular forcing means which are employed. Every means must be used to minimize the time-consuming process of record analysis and speed the time when engineering interpretations and decisions can be made.

In the case of sinusoidal excitation, a similar analysis in terms of frequencies and damping ratios can be quickly made for a given airspeed if decay records are produced in the following manner. The excitation frequency is varied until the response indicates a resonance condition. The excitation is suddenly removed and the decay of the vibration is observed. This process is repeated at all of the resonance peaks of interest. In many instances, this technique may provide better data than pulse excitation but it is seldom used, for two reasons. First, it requires the use of considerable engineering judgment under very difficult flight conditions. Second, if sinusoidal excitation equipment is installed, it can be better used to obtain, with about the same in-flight testing time, the complete frequency response curves at the given airspeed.

These frequency response curves relate the phase and amplitude of the sinusoidal response of a carefully chosen point in the airplane to the input excitation as a function of frequency. The interpretation of the data can be aided by a variety of plotting techniques. Plots of response amplitude per unit forcing amplitude versus frequency and response phase versus frequency at each of the test airspeeds permit a rapid estimation of trends toward flutter, particularly if they can be compared with similar curves predicted by analytical or model techniques. A somewhat more powerful plotting technique for this purpose is the polar plot of forcing amplitude per unit response amplitude at constant airspeed (Ref. 13–9). In this case, the approach of the higher airspeed curves to the origin indicates decreasing stability. In some instances the use of logarithmic scales is advantageous (Ref. 13–10).

(b) *Dynamic stability, gust response, and buffeting.* Since the effects of aeroelasticity are not of major importance in full-scale dynamic stability investigations, we shall consider the subject only briefly. The basic difference between dynamic stability and flight flutter testing (the former is concerned with subcritical stability, the latter with a controlled approach to a critical instability) is reflected in the techniques which are employed. In dynamic stability testing much of the sense of urgency and danger is missing and test data can be examined at leisure after a set of tests is completed. More cumbersome data analysis techniques may be used, and the tests must cover a large range of Mach numbers, gross weight, and mass distributions.

Such thoroughness has been necessary only since the advent of "systems" concepts in connection with autopilot and automatic control system design. To permit adequate matching of autopilot to airplane the dynamic as well as the steady-state characteristics of the "black box" which represents the airplane must be known over the entire flight range. The test techniques are essentially the same as in flight flutter testing except that it is seldom necessary to install special excitation systems. The frequencies of importance in dynamic stability are well within the capabilities of control system servos.

There are a number of aeroelastic effects which may be of importance. Even though the elastic frequencies may be much higher than dynamic stability frequencies, the aircraft structure bends and twists under the action of air and inertial loads. This effect can strongly influence the placing of pickups and the interpretation of their signals. In the case of large, flexible airplanes the elastic frequencies may be quite low and the dynamic stability motion may look more like subcritical flutter with a good deal of rigid-body motion. Particularly in these latter situations, the automatic control system may further complicate the situation because its frequency capabilities may extend up into the elastic frequency range, and it may serve to accentuate aeroelastic effects (Ref. 13–11).

Full-scale investigations of gust response are still in their infancy for several reasons. First, present methods of analysis for gust response, when coupled with the other design criteria, have been largely successful in preventing structural failure due to flight through turbulent air. Thus extensive flight testing has been unnecessary. Second, satisfactory means of measuring and interpreting the gustiness of the air encountered by an airplane have not yet been developed (Ref. 13–8). The growing need for better analyses and gust response criteria on large, flexible, high-performance aircraft, coupled with a spreading understanding and use of statistical analysis techniques, makes certain the development of adequate flight test techniques in the near future. The close relationship between gust response and buffeting and the importance of buffeting in the transonic range further motivates this development. Another strong influence aiding this development is being exerted by automatic control system designers who must optimize system performance in rough as well as smooth air.

13–6 Dynamic aeroelastic testing—model scale. The testing of dynamic models is a rapidly developing art. The inadequacy of many theoretical techniques for making accurate predictions in the high-speed range and for the ever-growing variety of airplane shapes has led to increasing demands on experimental procedures. In the field of airplane dynamic stability, the accurate description of airplane characteristics demanded by guidance and control engineers is a strong stimulus. In the field of flutter the mounting of large massive objects on only slightly stiffened wings has seriously aggravated an already difficult situation. The use of models for dynamic testing has already provided very useful data and holds considerable promise for the future. It is the purpose of this section to explore the fundamental principles underlying the many techniques which have been developed.

There are two basically different approaches to the use of models in dynamic problems. In one approach the model tests are designed to evaluate coefficients in the differential equation governing the problem. Solutions are then carried out by numerical procedures. In the other approach the model test is designed as an analog of the full-scale problem and is expected to produce solutions directly. The models can simulate parts of airplanes, such as wings or tails, or whole airplanes. They may be mounted in wind tunnels, on rocket vehicles, on bombs dropped from high altitudes, on sleds which skim along level tracks, in local flow regions on test aircraft, or in the blasts of shock tubes. They may even be true flying models. Each scheme has its peculiar advantages and drawbacks as well as the usual problems of excitation, mounting, and measurement.

In order to gain a needed perspective for considering test fundamentals, consider the broad general problem of the dynamic behavior of an elastic airplane in free flight. Think of flutter as emphasizing the elastic high-frequency part of this stability and response problem and of dynamic stability as emphasizing the low-frequency free-body motions. Gust response and landing loads emphasize strength rather than stability. Think of the various testing techniques in terms of their ability to illumine the many facets of this general problem.

(a) *Coefficient evaluation in the wind tunnel.* Perhaps the most important techniques are based on the use of a wind tunnel to move air past the model. However, the wind tunnel test section must provide not only acceptable boundaries to the airstream but also suitable supporting means for the model. If we look first at the tests designed to evaluate coefficients of the governing equations, we usually find that the required data involve measurement of aerodynamic reactions on oscillating rigid surfaces and bodies. These rigid models are either forced in a specified sinusoidal motion or are disturbed in order to observe the decaying oscillations. Tests of this nature always become more difficult as the required oscillation frequencies are increased. These frequencies are dictated by the requirements of maintaining (for airplane work, if not for turbine blades, propellor blades, and some missiles) Reynolds numbers of the order of at least 4×10^5 throughout the desired range of reduced frequency.

In low-speed testing the forcing of light, rigid models can be accomplished by rigid linkages connected to suitable mechanical drive systems (Refs. 13–12, 13–13, 13–14, 13–15). Either total forces and moments or pressure distributions can be measured. Total forces and moments have been determined by measuring with strain-sensitive gages the model's reactions on its supports and by noting changes in the drive power input. The oscillating pressures can be measured by installing a suitable pickup (Ref. 13–16) in the wing. In either case, it is very difficult to eliminate the effects on the measurements of the high accelerations to which model and instrumentation are subjected.

In high-speed testing the required frequencies are very high in order to maintain reduced frequency in the face of large test velocities and smaller models. It becomes very difficult to build rigid models and more difficult to build linkages to drive them. It is helpful to introduce flexibility into the support and drive system so that the natural frequency of the model on its support is in the range of test frequencies. A relatively small shaking force is then sufficient to maintain a reasonable amplitude of oscillation near resonance. The problems of precise frequency and amplitude control under these conditions are difficult and can best be solved by the use of a suitable feedback loop which senses the difference between the desired and the actual model motions (Ref. 13–17). The very high test frequencies also complicate the measurement problem. The sensing of model reactions or changes in power input are less practicable because of the much decreased ratio of aerodynamic to inertial forces, and methods of examining the flow directly must be used if possible. These may include pressure measurement by pickups submerged in the model or in a stationary probe close to the model. Also, in certain cases interferometric techniques are applicable.

The mounting and forcing techniques for high-speed testing described above are most easily employed with wall-mounted models rather than models mounted on the stings commonly present in high-speed wind tunnels. Because of sting flexibility and the large distance between power source and model, it is difficult to force a model in a prescribed motion. The task becomes somewhat easier if the sting flexibility is treated as an asset rather than a liability. Thus if the model is supposed to perform a pure pitching motion about its center of gravity, the sting flexibility can be adjusted so that a natural mode of the sting-model combination has a node at the model center of gravity and a frequency in the center of the testing range.

(b) *Wind-tunnel wall interference.* The models which are used in dynamic wind-tunnel tests must be of such a size as to keep interference effects of the walls to an acceptably small value. In supersonic tests the model must be small enough so that disturbances originating on the model are swept downstream of the model before they can travel to the wall and back. In subsonic flow the interference can never be zero and is a complicated function of tunnel size and shape, model size and shape, Mach number, and reduced frequency. At very low reduced frequency the interference can be predicted on a quasi-steady basis using the ordinary steady-state correction factors. At high frequencies a resonance effect may be expected when disturbances require just one-half the period of oscillation to travel from the model to the wall and back. In the two-dimensional case (Ref. 13–18) with the model midway between plane parallel walls, this

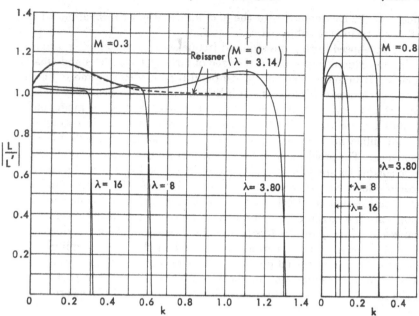

Fig. 13-19. Ratio of lift in pitch about mid-chord in the wind tunnel to its free-air value as a function of reduced frequency for various values of Mach number and height-to-chord ratio (λ). (From I.A.S. Preprint No. 446 by Woolston and Runyan)

resonant frequency ω is given in terms of test section height H, speed of sound a, and Mach number as

$$\frac{\omega H}{a} = \pi\sqrt{1 - M^2}. \tag{13-28}$$

This relationship for predicting resonance can also be written in terms of reduced frequency, Mach number and height-to-chord ratio ($\lambda = H/2b$) as

$$\frac{M}{\sqrt{1 - M^2}} k\lambda = \frac{\pi}{2}. \tag{13-29}$$

At intermediate frequencies the subsonic interference effects in a practical situation can be estimated roughly from two-dimensional calculations such as are shown in Fig. 13-19. Here the ratio of lift in pitch (about midchord) to its free-air value is plotted against reduced frequency for various values of Mach number and height-to-chord ratio. The effect of the walls is more pronounced as the Mach number is increased and may be large even at frequencies well removed from resonance. Increasing the tunnel height reduces the effect of the walls but also lowers the resonant frequency.

(c) *Simulating free flight in the wind tunnel.* Let us now consider the use of the wind tunnel for dynamic tests in which the aim is not the measurement of coefficients but rather the solution of the governing equations of motion. Let us look at the difficult general problem of simulating the dynamic behavior of the flexible airplane in free flight and consider the simpler special cases in later paragraphs. The wind-tunnel test section and associated apparatus must not exert any appreciable influences on the model which would not be present in free flight. Thus the model must be small enough to minimize wall interference effects and allow adequate space for perturbations from straight and level flight. Means must be provided to simulate thrust by applying a force to the model which is either independent of model motion or is a prescribed function of velocity. Since, in general, the weight of the model is incorrectly scaled when the mass is properly scaled (Section 11–7), a vertical force must be applied to the model which is completely independent of model motion in order to have it fly at the attitude and lift coefficient corresponding to level flight. For many low-speed models an upward force is needed, whereas high-speed models require a downward force. Some low-speed models of large bombers and transports need no vertical force. If perturbations about an accelerated flight condition are to be simulated, large vertical forces and sometimes even side forces must be applied which are independent of model motion. In addition to these requirements, the model must be controllable and must have suitable means of excitation and measurement.

The provision of a constant vertical supporting force is not an easy task. Low-speed flutter model designers have grappled with the problem for many years and have not been successful for two reasons. First, their consideration of an adequate support mechanism was always severely limited in time and funds and was of secondary importance. Second, the provision of a force independent of model motion over distances of more than about an inch calls for devices whose moving mass is not small compared with model mass. Since any mass which is effectively attached to the model and moves with it must be included as part of the model's mass, the support designer is forced into an impossible situation, except perhaps in models with large disposable fuel and bomb loads near the point of attachment. Among the devices which have been tried are the loose-fitting air piston, the modified electromagnetic shaker, and various spring arrangements. The combined requirements of large constant forces, large travel, and small moving mass have always forced severe compromises.

It is apparent then that the model cannot properly supply the large forces necessary to move a reasonable support mass and that light constant-force devices are practical only over small amplitudes. If we introduce the new idea that we use an external source of power to move most of the support mass in approximately the correct manner and use a light constant-force device to bridge the small remaining gap, the problem becomes tractable. Since the large-amplitude model motions are essentially rigid-body motions, the drive system of the main support mass need respond only up to dynamic stability frequencies. The smaller motions associated with the higher frequency elastic responses of the model can be absorbed by the constant-force device.

A sketch of a possible support mechanism for simulating free flight in a low-speed wind tunnel is shown in Fig. 13–20. Three separate hydraulic control servos of the standard aircraft variety provide the three components of linear displacement for the main support mass. A constant-force device and sensing elements for the hydraulic actuators are located at the top of the support and are shown in Fig. 13–21. The constant force is applied to the ball mounted in the model at the proper angle to give the desired forward, vertical, and side forces, regardless of the angular orientation of the model. The sensing elements measure the displacement of the constant-force rod and the bending strains caused by the offset of the ball from the rod axis. They send suitable signals to the three hydraulic actuators.

Fig. 13–20. Sketch of a servo-support system for a low-speed tunnel capable of simulating free-flight mounting conditions while providing necessary constant horizontal and vertical forces (see Fig. 13–21).

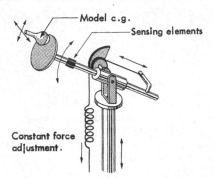

Fig. 13–21. Schematic view of upper support mechanism for applying a constant force to the model's center of gravity despite its motion. The cam-spring arrangement provides a constant but adjustable force to the small disk, which transmits it to the model through the rolling contact.

An arrangement more suited to a transonic tunnel is illustrated in Fig. 13–22. The constant-force device is located in the head of the vertical support member and the force is transmitted through a cable to the center of gravity of the model. Angular and linear displacements of the cable

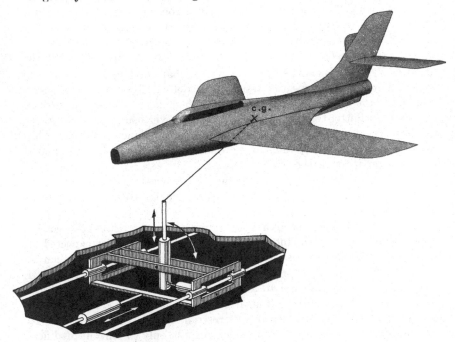

Fig. 13–22. A servo-support for simulating free flight at higher tunnel speeds.

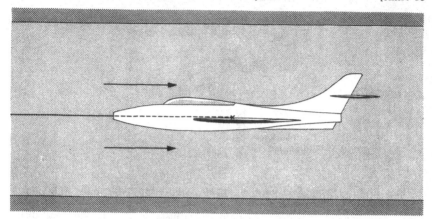

Fig. 13-23. A simple drag-wire support for partial simulation of free flight at high tunnel speeds.

are used to generate signals to the three servos. The frequency characteristics of these servos must be far superior to those used in the low-speed tunnel because the dynamic stability frequencies of high-speed models are much higher than for low-speed models of the same size.

(d) *Flutter models in the wind tunnel.* In the testing of many flutter models it is not necessary to provide the complete simulation of free flight by using supports as in Figs. 13-20 and 13-22. When the model Froude number $(U/\sqrt{gL})$ is about equal to or greater than the full-scale value such as in all but the very low-speed tests, the model will fly at a lift coefficient well below the stall. Since exact simulation of lift coefficient is seldom necessary for flutter testing, no external constant vertical force is necessary for this class of models. It is merely necessary to balance the drag. This can be done with a "drag wire," as shown in Fig. 13-23. If the wire cannot readily enter through the model's nose, a bridle can be used to transmit the force to the desired point in the model. A constant-force device can be mounted at either end of the drag wire to provide freedom in the fore-and-aft direction.

If the modification of dynamic stability characteristics caused by applying restraints to the model's rigid-body freedoms is not expected to affect its flutter characteristics, many other support configurations can be used. In the excellent support system developed by Boeing (Fig. 13-24 and Ref. 13-19) the fore-and-aft and lateral degrees of freedom are subject to a spring restraint arising from bending of the stretched vertical rod. Vertical and pitching motions are not restrained. Extreme care must be exercised in the constraint of model freedoms, not only to prevent modification of flutter characteristics, but to keep from making the model dynamically unstable. It is extremely easy to create an instability by eliminating

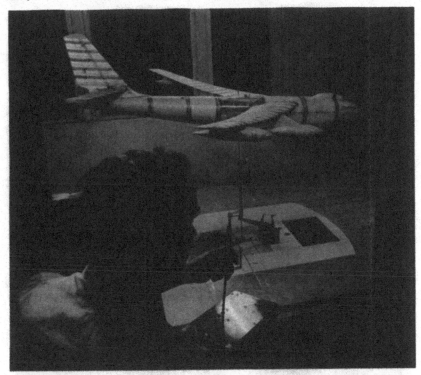

Fig. 13–24.   Early version of vertical rod support for large flutter models at moderate tunnel speeds.   (Courtesy of Boeing Airplane Company)

the fore-and-aft degree of freedom and introducing some spring restraint in vertical motion.   This situation is likely to occur in the testing of low-speed flutter models in which freedom in pitch and vertical translation is necessary to avoid serious modification of the flutter characteristics (Figs. 13–25 and 13–26).   In some cases the direct effect on the flutter character-istics of support stiffness and excess mass can be estimated by noting their effect on the pertinent natural frequencies of the model, according to Section 13–3.

The testing of components rather than entire models is often a justi-fiable procedure, particularly when the tests are designed as a check against calculations rather than as directly applicable to the full-scale air-plane.   Typical tail and wing models are shown in Figs. 13–27 to 13–30. The support of the tail model is designed to simulate fuselage flexibility. The wall mount of the model in Fig. 13–28 provides freedom either in roll or in vertical translation.   This permits separate investigation of sym-metrical and antisymmetrical modes.   It should be noted that the wall-mounted model has an effectively symmetrical aerodynamic boundary

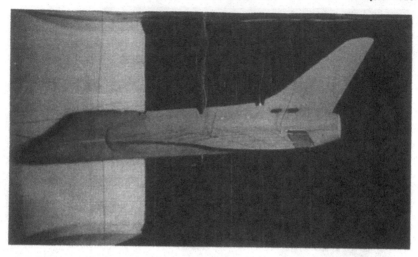

Fig. 13–25. A flutter model mounted on spring-restrained wires in a low-speed wind tunnel to permit "rigid-body motions." The wing structure consists of a built-up plastic plate supporting the sectional external shell. (Courtesy of Douglas Aircraft Company)

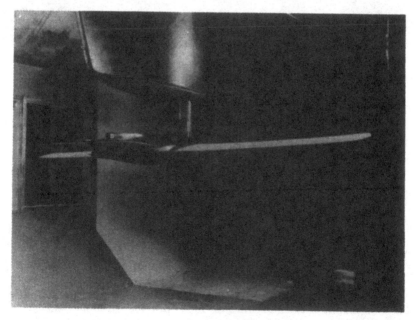

Fig. 13–26. Flutter model of a wing with rigid fuselage and adjustable horizontal tail. The overhead support provides the model with freedom in pitch, roll, vertical translation, and horizontal translation.

Fig. 13–27. Early flutter model of horizontal tail including rod support to simulate fuselage flexibilities.

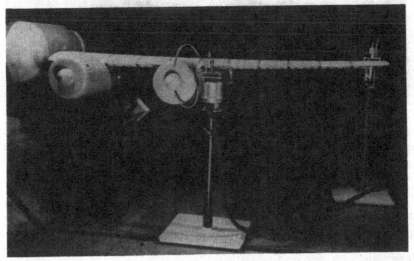

Fig. 13–28. Semispan flutter model restrained either in roll or in vertical translation at the root to simulate either symmetrical or antisymmetrical structural root conditions.

Fig. 13–29. A nonparallel four-bar linkage and a tilted roll axis provided this model with some artificial stability.

Fig. 13–30. This supersonic flutter model is free to roll about an axis canted a few degrees from the stream direction in order to provide a self-centering tendency.

condition at the mounting wall regardless of the freedoms provided at the root. Artificial "rigid-body stability" can be built into component models where needed. The straight wing in Fig. 13–29 is allowed freedom in vertical translation through the use of a parallel bar linkage and freedom in roll by the provision of a roll axis. A slight positional stability is achieved by introducing some nonparallelism in the linkage. Thus an upward motion decreases the angle of attack and a downward motion increases it. A slight positive inclination of the roll axis introduces some stability into the roll motion.

The free-to-roll supersonic flutter model shown in Fig. 13–30 was stabilized in roll by giving the roll axis a few degrees of yaw. It is constructed as a half-model to permit simple injection into an established supersonic flow. It could not withstand the turbulence associated with the starting of a supersonic tunnel. The best type of wind tunnel for supersonic flutter testing has a nozzle which can be easily adjusted while the tunnel is running (Ref. 13–20). This permits the injection of the model in a stable flow condition. The Mach number can then be varied until the model flutters. It is just as desirable to find a supersonic flutter condition by varying stream rather than model parameters as it is in subsonic testing (Ref. 13–1).

(e) *Control and excitation in the wind tunnel.* The control of flutter models during testing may be quite difficult, particularly for the models with numerous rigid-body freedoms. These models must often be flown by means of a fairly complete system of control surfaces and actuators. Pilots situated outside the wind tunnel can manage quite effectively those low-speed models whose time scales are not far different from full scale. High-speed models are likely to be beyond human control and must be managed by gyrostabilizing loops or even autopilots.

The excitation of models to measure degree of stability or to detect the nearness to flutter can be achieved by means similar to those described for flight flutter testing (Section 13–4). The location of the model in the tunnel, however, often permits the use of less sophisticated techniques. For many low-speed models, loose cords attached to the spars near the wingtips can be jerked to provide subcritical excitation and can be pulled taut to suppress flutter and prevent destruction of the model. In a large number of wind tunnels there is sufficient turbulence in the airstream to preclude the need for special excitation equipment. If necessary, controlled gusts may even be introduced into the stream.

To prevent destruction of models during flutter a very careful control of airspeed is required. The amplitude of the self-sustaining flutter oscillation in many cases varies rapidly with airspeed over a very small range just above the critical speed. This speed range characterized by nondestructive amplitudes may be as small as a fraction of one percent of flutter speed, although it is occasionally as large as ten percent. In the case of large, expensive models it may be desirable to build into the model a safety device which can achieve a sudden change in model parameters designed to extend substantially the subcritical range. Thus a solenoid can quickly shift the position of a mass, or the precessional motion of a high-speed flywheel can be suddenly and suitably restrained.

(f) *Testing with rocket and sled.* There are two means other than a wind tunnel for subjecting a model to controlled flow conditions, which have assumed considerable importance particularly in the high-speed range. First, the mounting of aeroelastic models of airplane components on free-flight rocket vehicles offers some advantages over wind-tunnel testing. Especially in the high subsonic and supersonic speed ranges, a carefully controlled rocket flight subjects the model to the same smooth unbounded air flow which is experienced by the full-scale airplane. Furthermore, the tests are usually carried out in the dense air at low altitudes, which fact leads to high model Reynolds numbers and large allowable mass. The time histories of the vehicle flights can be programed within rather wide bounds.

Among the serious disadvantages of rocket-vehicle testing are the general difficulties of observation and measurement on a model speeding through the air over large distances. The time of flight is short. The space and weight allowance for instrumentation is not large. The chance of recovery of model and instrumentation is often small. Acceleration loads are high and there is little margin for error. Once a flight is initiated it cannot be recalled. The difficulties of mounting a model in a wind tunnel are balanced by the need for free-flight stability for the rocket vehicle whatever the model configuration.

The testing of aeroelastic models on rocket-propelled sleds which skim along straight and level tracks is somewhat of a cross between free-flight rocket and wind-tunnel testing. The model can be subjected to smooth, high-speed, relatively unbounded airflow at high density. More elaborate and expensive instrumentation can be installed on the sled than in the rocket, and it is recoverable. Even the models are recoverable if they are not destroyed by flutter. However, sled testing does not have the general ease of observation and control available in the wind tunnel. Sled tests are perhaps most comparable to tests in an intermittent blowdown wind tunnel.

Both the sled and the rocket vehicle can be adapted to either coefficient measurement or the investigation of aeroelastic stability. Many of the excitation and mounting techniques discussed in previous sections can be used with little or no basic modification. The mounting of components in a free-flight rocket vehicle is only a few steps from the testing of a complete rocket-powered aeroelastic scale model. Unfortunately, such a test may be a very expensive gamble. It is beset with the difficulties of finding space for rockets and adequate instrumentation, of providing the correct mass distribution during the proper portion of the flight, of obtaining sufficient data to make the test worth while, and especially of getting everything working at once so that a minor failure does not nullify the whole test. Nevertheless, the idea of testing a completely free aeroelastic model in a high-speed unbounded airstream is quite intriguing.

(g) *Measurement techniques and instrumentation.* To the aeronautical engineer, the most unusual feature of the measurement problem in dynamic aeroelastic testing is the large range of frequencies which must be handled. Large, low-speed models may have frequencies as low as 2 or 3 cps, whereas small low aspect-ratio supersonic models may be in the range of 200 cps. Except for problems associated with the higher frequencies, the engineer can draw heavily on experience in ordinary flight-test measurement techniques. The fundamental tool is a photographic recording system which usually takes the form of a multichannel recording oscillograph. This device, with appropriate galvanometers, can cover the desired frequency range and produce on the same time base the time histories of any measured quantities which can be put into suitable electrical form.

If it is advantageous to retain certain signals in electrically available form for later processing through wave analyzers, phase meters, etc., the high-performance multichannel tape recorder is an excellent device. It cannot easily replace the oscillograph because it does not afford a qualitative over-all picture to the data analyst, but it is unexcelled in its ability to recreate in the useful electrical form at a convenient later time whatever input signals may be needed. Anyone who has carried out a detailed analysis of oscillograph records is impressed with the tedium and expense which is entailed.

The out-of-ordinary quantities which must be measured in a flutter test are the frequency of the flutter oscillation and the flutter mode shape. Resistance wire strain gages mounted on the main structure can be relied upon to give a very accurate indication of flutter frequency. With suitable location and orientation they can easily give a qualitative indication of mode shape. The best device for indicating the flutter mode shape is the high-speed camera. Although accurate quantitative information may not be readily obtainable from the pictures, they do give an invaluable qualitative view of flutter and often indicate clearly unexpected phenomena which might not otherwise have been detected or properly interpreted. Synchronization of the camera and oscillograph records by a timing signal is a valuable aid to proper data interpretation.

MATHEMATICAL TOOLS

APPENDIX A

MATRICES

A-1 Introduction. Appendix A is written primarily to serve as a convenient summary of the matrix operations required in aeroelastic problems. No attempt is made to include rigorous derivations and definitions and the operations are illustrated by simple numerical examples. The reader is referred to the end of the list of references for more complete discussions of the subject.

A-2 Definitions. Matrices are defined as arrays of numbers or other elements arranged in rows and columns and which obey certain rules of matrix algebra (Ref. A-1). Let us consider, for example, a system of linear equations, such as

$$3x + 5y - z = 6,$$
$$-9x - 4y + 5z = 7. \tag{1}$$

It is often more efficient to investigate this system by rewriting it in the following matrix notation:

$$\begin{bmatrix} 3 & 5 & -1 \\ -9 & -4 & 5 \end{bmatrix} \begin{bmatrix} x \\ y \\ z \end{bmatrix} = \begin{bmatrix} 6 \\ 7 \end{bmatrix}, \tag{2}$$

which leads at once to the concept of arrays of numbers. The first array of numbers in Eq. (2), which represents the coefficients of the linear system, has two rows and three columns. The array of numbers on the right side of Eq. (2), which represents the constant terms on the right side of Eq. (1), has two rows and one column. Arrays of numbers such as these are called *matrices*, and if the array has m rows and n columns it is called a matrix of order $m \times n$. For example, some of the basic types of matrices and their shorthand notation are as follows:

Column matrix (3×1):
$$\{A\} = \begin{bmatrix} 4 \\ -5 \\ 3 \end{bmatrix}.$$

Row matrix (1×3):
$$\lfloor A \rfloor = \begin{bmatrix} 5 & -2 & 7 \end{bmatrix}.$$

Square matrix (3×3):
$$[A] = \begin{bmatrix} 1 & 3 & 0 \\ -5 & 4 & 1 \\ 6 & 2 & -8 \end{bmatrix}.$$

Diagonal matrix (3 × 3):
$$[A] = \begin{bmatrix} 1 & 0 & 0 \\ 0 & 6 & 0 \\ 0 & 0 & -2 \end{bmatrix}.$$

Unit matrix (3 × 3):
$$[I] = \begin{bmatrix} 1 & 0 & 0 \\ 0 & 1 & 0 \\ 0 & 0 & 1 \end{bmatrix}.$$

A–3 Matrix algebra. (a) *Equality of matrices.* Equal matrices are matrices of the same order which have equal corresponding elements.

(b) *Addition and subtraction.* Addition and subtraction of two matrices can be performed only on matrices of the same order. Addition is performed by simply adding corresponding elements. For example,

$$\begin{bmatrix} 1 & 2 & 3 \\ 2 & -3 & 4 \end{bmatrix} + \begin{bmatrix} 4 & -5 & 6 \\ 5 & 6 & 7 \end{bmatrix} = \begin{bmatrix} 5 & -3 & 9 \\ 7 & 3 & 11 \end{bmatrix}. \tag{3}$$

Subtraction is performed in a similar manner by subtracting corresponding elements.

(c) *Scalar multipliers.* When a matrix is multiplied by a scalar number, every element in the matrix is multiplied by the number. For example,

$$2 \times \begin{bmatrix} 1 & 2 & 3 \\ 2 & 3 & 4 \end{bmatrix} = \begin{bmatrix} 2 & 4 & 6 \\ 4 & 6 & 8 \end{bmatrix}. \tag{4}$$

(d) *Multiplication of matrices.* Two matrices, $[A]$ and $[B]$ can be multiplied in a given order, as follows:

$$[A][B] = [C], \tag{5}$$

provided that a certain condition is satisfied. If the number of columns in $[A]$ is equal to the number of rows in $[B]$, the matrices are said to be conformable and they can be multiplied in the order $[A]$ times $[B]$. To define the process of multiplication, suppose that $[A]$ and $[B]$ are defined by

$$[A] = \begin{bmatrix} a_{11} & a_{12} & a_{13} \\ a_{21} & a_{22} & a_{23} \end{bmatrix}, \qquad [B] = \begin{bmatrix} b_{11} & b_{12} \\ b_{21} & b_{22} \\ b_{31} & b_{32} \end{bmatrix}.$$

Then the matrix $[C]$ is given by

$$[C] = \begin{bmatrix} c_{11} & c_{12} \\ c_{21} & c_{22} \end{bmatrix},$$

where, for example, the element

$$c_{11} = a_{11}b_{11} + a_{12}b_{21} + a_{13}b_{31} \tag{6}$$

in $[C]$ has been obtained from $[A]$ and $[B]$ by multiplying the elements in the first row of $[A]$ by the corresponding elements in the first column of

$[B]$ and adding the results. In general, the element c_{ik} is defined by the row by column rule, which can be stated in terms of the formula

$$c_{ik} = \sum_{j=1}^{n} a_{ij}b_{jk}, \tag{7}$$

where n is the number of columns in the matrix $[A]$ and the number of rows in $[B]$. In addition, if the order of the matrix $[A]$ is $m \times n$, and the order of the matrix $[B]$ is $n \times r$, the order of the product matrix $[C]$ is $m \times r$.

Matrix multiplication is, in general, associative and distributive but not commutative, that is,

$$[A]([B][C]) = ([A][B])[C],$$
$$[A]([B] + [C]) = [A][B] + [A][C], \tag{8}$$
$$[A][B] \neq [B][A].$$

The formula given by (7) can be applied to compute the product of all types of matrices, as illustrated by the following examples.

Row matrix times a column matrix:

$$\begin{bmatrix} 1 & -3 & 5 \end{bmatrix} \begin{bmatrix} -2 \\ 4 \\ 6 \end{bmatrix} = (1 \times -2) + (-3 \times 4) + (5 \times 6) = 16. \tag{9}$$

Square matrix times a column matrix:

$$\begin{bmatrix} 3 & -1 & 2 \\ 1 & 0 & -2 \\ 2 & 1 & 0 \end{bmatrix} \begin{bmatrix} 1 \\ -2 \\ 1 \end{bmatrix} = \begin{bmatrix} (3\times1)+(-1\times-2)+(2\times1) \\ (1\times1)+(0\times-2)+(-2\times1) \\ (2\times1)+(1\times-2)+(0\times1) \end{bmatrix} = \begin{bmatrix} 7 \\ -1 \\ 0 \end{bmatrix}. \tag{10}$$

Rectangular matrix times a rectangular matrix:

$$\begin{bmatrix} 1 & 2 & -3 \\ 2 & -1 & 3 \end{bmatrix} \begin{bmatrix} 1 & 2 \\ 0 & 1 \\ -1 & 3 \end{bmatrix}$$

$$= \begin{bmatrix} (1 \times 1) + (2 \times 0) + (-3 \times -1) & (1 \times 2) + (2 \times 1) + (-3 \times 3) \\ (2 \times 1) + (-1 \times 0) + (3 \times -1) & (2 \times 2) + (-1 \times 1) + (3 \times 3) \end{bmatrix}$$

$$= \begin{bmatrix} 4 & -5 \\ -1 & 12 \end{bmatrix}. \tag{11}$$

(e) *Partitioned matrices.* A given matrix may be partitioned into smaller submatrices by horizontal and vertical lines, as follows:

$$\begin{bmatrix} 4 & 0 & 1 \\ 1 & -1 & 2 \\ \hline -3 & 2 & -4 \end{bmatrix} = \begin{bmatrix} \{A\} & [B] \\ \hline [C] & [D] \end{bmatrix}, \tag{12}$$

where

$$\{A\} = \begin{bmatrix} 4 \\ 1 \end{bmatrix}, \quad [B] = \begin{bmatrix} 0 & 1 \\ -1 & 2 \end{bmatrix}, \quad [C] = [-3], \quad [D] = [2 \quad -4]$$

are submatrices of the original matrix. In general, partitioned matrices can be added, subtracted, and multiplied as though the submatrices were ordinary matrix elements, and the result is the same as if the same operation is performed on the original unpartitioned matrices. It is, of course, necessary during multiplication that the partitioned matrices satisfy the conformability requirement stated previously for ordinary matrices.

(f) *Transposition.* The transpose of a matrix $[A]$, designated by $[A]'$, is obtained by interchanging the rows and columns of the matrix $[A]$. When a matrix is equal to its transpose it is said to be symmetrical.

(g) *Reciprocation.* The process of division in matrix algebra is not as directly analogous to the corresponding process on numbers as are the other elementary operations. Division is accomplished by reciprocal matrices. The reciprocal $[A]^{-1}$ of a matrix $[A]$ is defined by

$$[A][A]^{-1} = [I]. \tag{13}$$

Let us consider, for example, a set of linear equations in matrix form, as follows:

$$[A]\{x\} = \{a\}, \tag{14}$$

where

$$[A] = \begin{bmatrix} 3 & -2 & 0 & -1 \\ 0 & 2 & 2 & 1 \\ 1 & -2 & -3 & -2 \\ 0 & 1 & 2 & 1 \end{bmatrix}, \quad \{x\} = \begin{bmatrix} x_1 \\ x_2 \\ x_3 \\ x_4 \end{bmatrix}, \quad \{a\} = \begin{bmatrix} 1 \\ 1 \\ 1 \\ 1 \end{bmatrix}.$$

A solution to this set of equations for the unknowns x_i can be obtained by inversion of the matrix $[A]$:

$$\{x\} = [A]^{-1}\{a\} = \begin{bmatrix} -4 \\ 0 \\ 7 \\ -13 \end{bmatrix}, \tag{15}$$

where

$$[A]^{-1} = \begin{bmatrix} 1 & 1 & -2 & -4 \\ 0 & 1 & 0 & -1 \\ -1 & -1 & 3 & 6 \\ 2 & 1 & -6 & -10 \end{bmatrix}$$

is the reciprocal of the matrix of coefficients, $[A]$. It can be easily verified that the product of $[A]$ and $[A]^{-1}$ is the unit matrix. Numerical methods of obtaining reciprocal matrices are discussed in Refs. A–2 and A–3. Crout's method, for example (Ref. A–3), provides a method very convenient for machine computation.

INTEGRATION BY WEIGHTING NUMBERS

B–1 Introduction. In the solution of practical aeroelastic problems it is frequently necessary to evaluate definite integrals such as

$$I = \int_a^b f(y)dy, \tag{1}$$

where the explicit form of $f(y)$ is unknown or is a function such that its definite integral cannot be computed conveniently in terms of other known functions. If ordinates to the $f(y)$ curve, $f_0, f_1, f_2, \cdots, f_n$, are known at $(n + 1)$ points within the interval (a,b), the definite integral may be expressed approximately by the summation

$$\int_a^b f(y)dy \approx \sum_{i=0}^n f_i W_i, \tag{2}$$

where W_i are weighting numbers which depend upon the method of numerical integration that is employed. We summarize in the following sections several methods of obtaining weighting numbers useful in aeroelastic problems.

B–2 Trapezoidal rule. The trapezoidal rule method is based on the assumption that the function $f(y)$ between adjacent sets of ordinates can be replaced by straight lines, as illustrated by Fig. B–1. Thus the definite integral can be approximated by the sum of the areas of the trapezoids formed by the ordinates and the straight lines connecting the ordinates. If we choose $(n + 1)$ equally spaced ordinates, as illustrated by Fig. B–1, the area, according to the trapezoidal rule, is

$$\int_a^b f(y)dy \approx \lambda(\tfrac{1}{2}f_0 + f_1 + f_2 + \cdots + f_{n-1} + \tfrac{1}{2}f_n), \tag{3}$$

where λ is the interval between ordinates.

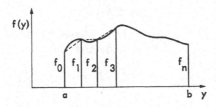

Fig. B–1. Representation of area by a sum of trapezoids.

B-3 Simpson's rule. If the interval (a,b) is divided into an even number n of equal intervals λ, and if the curve $f(y)$ between adjacent ordinates is approximated by parabolic arcs, the area is given by an approximate formula known as Simpson's Rule (Ref. B-1), as follows:

$$\int_a^b f(y)dy \approx \frac{\lambda}{3}(f_0 + 4f_1 + 2f_2 + 4f_3 + 2f_4 + \cdots + 2f_{n-2} + 4f_{n-1} + f_n). \quad (4)$$

Simpson's rule is probably the most generally useful of all of the approximate quadrature formulas. It yields an exact result if the function $f(y)$ which is being approximated is a polynomial of the third degree or less.

B-4 Lagrange's interpolation formula. Another useful method of approximate quadrature is based on expressing the function $f(y)$ within the interval (a,b) in terms of ordinates to the curve according to Lagrange's interpolation formula, as follows:

$$f(y) \approx \sum_{j=0}^{n} L_j(y)f_j, \quad (5)$$

where the f_j are ordinates to $f(y)$ at $(n+1)$ points and the $L_j(y)$ are functions of y called Lagrangian interpolation functions (Ref. B-2). A form such as Eq. (5) is possible only if

$$\begin{aligned} L_j(y_i) &= 0, \qquad (i \neq j), \\ L_j(y_i) &= 1, \qquad (i = j). \end{aligned} \quad (6)$$

A convenient form of the interpolation function $L_j(y)$ is that of a polynomial of degree n. In order that the first condition of Eq. (6) be satisfied, it is evident that the polynomial can be expressed in the factored form

$$L_j(y) = B_k(y - y_0)(y - y_1) \cdots (y - y_{j-1})(y - y_{j+1}) \cdots (y - y_n). \quad (7)$$

The constant B_k can be evaluated using the second condition of Eq. (6):

$$B_k(y_j - y_0)(y_j - y_1) \cdots (y_j - y_{j-1})(y_j - y_{j+1}) \cdots (y_j - y_n) = 1. \quad (8)$$

Combining Eqs. (7) and (8) yields

$$L_j(y) = \frac{(y - y_0)(y - y_1) \cdots (y - y_{j-1})(y - y_{j+1}) \cdots (y - y_n)}{(y_j - y_0)(y_j - y_1) \cdots (y_j - y_{j-1})(y_j - y_{j+1}) \cdots (y_j - y_n)}. \quad (9)$$

The definite integral in Eq. (1) can be evaluated by introducing Lagrange's interpolation formula given by Eqs. (5) and (9):

$$\int_a^b f(y)dy \approx \sum_{j=0}^{n} \left\{ \int_a^b L_j(y)dy \right\} f_j. \quad (10)$$

When the definite integrals in Eq. (10) are evaluated, we obtain explicit numerical values, and Eq. (10) reduces to the form of Eq. (2). In the

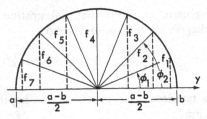

Fig. B-2. Division of interval (a,b) into Multhopp stations $(m = 7)$.

case where the ordinates to the curve are equally spaced, such that

$$y_1 - y_0 = y_2 - y_1 = \cdots = y_n - y_{n-1} = \lambda, \tag{11}$$

the explicit form of Eq. (10) is greatly simplified. For example, for $n = 5$,

$$\int_{y_0}^{y_5} f(y)\,dy \approx \frac{5\lambda}{288} (19f_0 + 75f_1 + 50f_2 + 50f_3 + 75f_4 + 19f_5); \tag{12}$$

and for $n = 10$,

$$\int_{y_0}^{y_{10}} f(y)\,dy \approx \frac{\lambda}{3.7266} (f_0 + 6.61605f_1 - 3.02017f_2 + 16.95401f_3$$
$$- 16.21647f_4 + 26.59910f_5 - 16.21647f_6$$
$$+ 16.95401f_7 - 3.02017f_8 + 6.61605f_9 + f_{10}). \tag{13}$$

Formulas for other values of n are given, for example, in Ref. B-3.

B-5 Multhopp's quadrature formula. A process of approximate quadratures developed by Multhopp in Ref. B-4 is convenient and accurate when applied to functions which arise in lifting-line theory:

$$\int_a^b f(y)\,dy \approx \frac{\pi(b - a)}{m + 1} \sum_{n=1}^m f(\phi_n) \sin \phi_n. \tag{14}$$

In constructing formula (14) the interval between a and b has been divided into m stations, not including the end points, as illustrated by Fig. B-2 for the case of $m = 7$. The division is such that the intervals are symmetric with respect to the midpoint.

Multhopp's method yields exact results when applied to functions which can be represented by the trigonometric series

$$f(\phi) = \sum_{n=1}^{2m} A_n \sin n\phi. \tag{15}$$

Formula (14) is a simple application of the well-known Gauss method of approximate quadratures.

In applying formula (14) to symmetric lift distribution problems, it is convenient to consider only the wing semispan. If the interval (a,b)

represents the wing span, an approximate integration over the semispan, according to Multhopp's method, has the form

$$\int_0^l f(y)dy \approx \frac{\pi l}{m+1} \left(\sum_{n=1}^{(m-1)/2} f(\phi_n) \sin \phi_n + \tfrac{1}{2}f(\phi_{(m+1)/2}) \right). \quad (16)$$

B–6 Weighting matrices. We have seen in Chapter 2 (Section 2–10) that the integral form

$$w(y) = \int_0^l C^{zz}(y,\eta)Z(\eta)d\eta \quad (17)$$

can be expressed approximately in terms of matrices by

$$\{w\} = [C^{zz}][\overline{W}]\{Z\}, \quad (18)$$

where $[\overline{W}]$ is a diagonal matrix of weighting numbers. This matrix can be constructed by applying the methods and formulas summarized above. For example, the weighting matrix obtained by applying Multhopp's formula to an integration over the semispan (Eq. 16) becomes

$$[\overline{W}] = \frac{\pi l}{8} \begin{bmatrix} \sin \pi/8 & & & \\ & \sin \pi/4 & & \\ & & \sin 3\pi/8 & \\ & & & \tfrac{1}{2}\sin \pi/2 \end{bmatrix}. \quad (19)$$

APPENDIX C

LINEAR SYSTEMS

C-1 Definition. A linear system is some clearly defined portion of the physical universe which has the property of linearity. To show what we mean by linearity, we must describe the system quantitatively with one or more measurable variables $y(t)$. These $y(t)$ may be locations or velocities of points in the system, temperatures, voltages, readings of meters, or anything else that characterizes the instantaneous state. They are often termed "outputs." Each y is a function of at least one independent variable t, which represents time in most applications but may also be length along a beam, position on a curve, or some similar vector or scalar dimension in space.

Also associated with the system are one or more "input" variables $x(t)$ which define how it is excited through contact with its surroundings. These may be voltages impressed on an electrical circuit, motions of the control column in an airplane, velocities of gusts striking a wing or tail, etc. In the most general case the spatial and temporal variation of any output will depend in a completely arbitrary manner on a set of inputs.

The portion of the system which generates a particular output $y_r(t)$ in response to inputs $x_1(t)$, $x_2(t)$, $\cdots$, $x_n(t)$ is said to be linear if the effects of different sets of inputs can be superimposed linearly. Thus if $y_r^1(t)$ and $y_r^2(t)$ denote the responses to separately applied inputs $x_1^1(t)$, $x_2^1(t)$, $\cdots$, $x_n^1(t)$ and $x_1^2(t)$, $x_2^2(t)$, $\cdots$, $x_n^2(t)$, then the inputs $[x_1^1(t) + x_1^2(t)]$, $[x_2^1(t) + x_2^2(t)]$, $\cdots$, $[x_n^1(t) + x_n^2(t)]$ will give rise to $[y_r^1(t) + y_r^2(t)]$. Since we can choose the two sets of inputs arbitrarily within the physical limitations of the system, it follows that the effects of any number of them can be linearly superimposed. This possibility permits us to equate all but one of the inputs to zero and concentrate on the linear relationship between any one input and any one output.

The fact of linearity could be established by purely experimental means, but normally the connections between outputs and inputs are expressible accurately enough for engineering purposes by linear mathematical formulas. Four types comprise all those met in theoretical aeroelasticity: ordinary and partial differential equations, integral equations, and integro-differential equations wherein the dependent variable is subjected to both differentiation and integration. In the present appendix we shall focus on the ordinary differential equation having constant coefficients, one output, and one input. It will be evident to the reader, however, that the various techniques of solution which are grounded in the fundamental concept of superposition can be applied to any linear system whatever.

In practice, we must examine the effects of a large number of different mathematical forms of input. When a system is not linear, the outputs have to be laboriously calculated one at a time, and often the analytical difficulties render this process practically impossible. In contrast, linearity enables us to compute the responses to a few simple input functions, and then employ superposition in finding from them any other desired output. Moreover, much can be learned about the stability and efficiency of performance of a linear system by studying its outputs for a few simple inputs, such as the step function and the sinusoidal function of t.

C-2 Duhamel's integral. The most straightforward aeroelastic application of the principle of superposition is the use of Duhamel's integral to calculate the response to a driving force which varies arbitrarily with time. Let us consider the rth normal mode of the linear elastic airplane, whose equation of forced motion for $\Xi_r{}^M = 0$ reads

$$M_r\ddot{\xi}_r + M_r\omega_r{}^2\xi_r = \Xi_r{}^D(t) = f(t). \tag{1}$$

Here ξ_r plays the part of $y(t)$, and $f(t)$ that of $x(t)$. We first assume the mode to be at rest for $t < 0$ and determine by elementary methods its responses to a unit step function

$$f(t) = 1(t) = \begin{cases} 0, & t < 0, \\ 1, & t \geq 0, \end{cases} \tag{2}$$

and to a unit impulse at $t = 0$. This latter input is zero for all values of t except in an infinitesimal interval surrounding the instant when it is applied. If this instant of application is $t = t_1$, the unit impulse $\delta(t - t_1)$ can be defined by the equation

$$\int_0^t \delta(t - t_1)dt = \begin{cases} 0, & \text{if interval 0 to } t \text{ excludes } t_1, \\ 1, & \text{if interval 0 to } t \text{ includes } t_1. \end{cases} \tag{3}$$

It follows from Eq. (3) that

$$\int_0^t f(t)\delta(t - t_1)dt = \begin{cases} 0, & \text{if interval 0 to } t \text{ excludes } t_1, \\ f(t_1), & \text{if interval 0 to } t \text{ includes } t_1. \end{cases} \tag{4}$$

For the system defined by Eq. (1), initially at rest, an impulse $\delta(t)$ at $t_1 = 0$ is physically equivalent to starting the system with initial conditions

$$\xi_r(0) = 0, \tag{5a}$$

$$\dot{\xi}_r(0) = \frac{1}{M_r}, \tag{5b}$$

since the impulse produces unit generalized momentum $M_r\dot{\xi}_r$.

We designate the responses to the unit step and unit impulse by $A_r(t)$ and $h_r(t)$, respectively. $A_r(t)$ is often called the indicial admittance of the

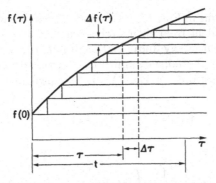

Fig. C-1. Superposition of step functions.

system. If we now consider the response to an arbitrary generalized force $f(t)$, we can superimpose a number of step functions in such a way as to construct the time history $f(t)$. Figure C-1 illustrates this process, with τ an independent variable denoting the position of any particular step.

The response to a step function having an ordinate $\Delta f(\tau)$ and applied at the instant $\tau + \Delta\tau$ is given by

$$\Delta\xi_r(t) = \Delta f(\tau)A_r[t - (\tau + \Delta\tau)]. \tag{6}$$

The response to the entire aggregate of step functions from $\tau = 0$ to $\tau = t$ is found by superposition:

$$\xi_r(t) = f(0)A_r(t) + \sum_{\tau=\Delta\tau}^{\tau=t} \Delta f(\tau)A_r(t - (\tau + \Delta\tau))$$

$$= f(0)A_r(t) + \sum_{\tau=\Delta\tau}^{\tau=t} \frac{\Delta f(\tau)}{\Delta\tau} A_r(t - (\tau + \Delta\tau))\Delta\tau. \tag{7}$$

Letting $\Delta\tau \to 0$ and using the definition of the definite integral,

$$\xi_r(t) = f(0)A_r(t) + \int_0^t \frac{df(\tau)}{d\tau} A_r(t - \tau)d\tau. \tag{8}$$

Integrating by parts, we can reduce Eq. (8) to

$$\xi_r(t) = -\int_0^t f(\tau)\frac{d}{d\tau} A_r(t - \tau)d\tau \tag{9}$$

or to

$$\xi_r(t) = \int_0^t f(\tau)A_r{'}(t - \tau)d\tau, \tag{10}$$

where

$$A_r{'}(t - \tau) = \frac{dA_r(t - \tau)}{d(t - \tau)}. \tag{11}$$

The integral forms in Eqs. (9), (10), and (11) are known as Duhamel's integrals. It is evident from their derivation that they are applicable to any output and input of any linear system, if $\xi_r(t)$ is replaced by $y(t)$, $f(t)$ by $x(t)$, and $A_r(t)$ by the indicial admittance function relating $x(t)$ and $y(t)$. For the system described by Eq. (1), it is shown in Chapter 10 [Eq. (10–17)] that

$$A_r(t) = \frac{1}{M_r\omega_r{}^2} [1 - \cos \omega_r t], \tag{12}$$

so that Eq. (10) reads

$$\xi_r(t) = \frac{1}{M_r\omega_r} \int_0^t f(\tau) \sin \omega_r(t - \tau)d\tau. \tag{13}$$

Just as we constructed the force $f(\tau)$ by superposition of a series of steps, we can regard it as the effect of a succession of impulses

$$\Delta\delta(t - \tau) = f(\tau)\Delta\tau \tag{14}$$

acting at $t/\Delta\tau$ equally spaced instants τ between 0 and t. These would give rise to the cumulative response

$$\xi_r(t) = \sum_{\tau=0}^{\tau=t} f(\tau)\Delta\tau h_r(t - \tau). \tag{15}$$

Letting $\Delta\tau \to 0$ and the number of impulses become infinitely large, we get the integral representation

$$\xi_r(t) = \int_0^t f(\tau)h_r(t - \tau)d\tau. \tag{16}$$

As before, Eq. (16) can be employed for any linear system whatever if $\xi_r(t)$, $f(t)$, and $h_r(t)$ are replaced by the appropriate output, input, and unit impulse response.

C–3 Response to a harmonic input.   Another input of fundamental importance is the sinusoidal force

$$f(t) = a \sin \omega t. \tag{17}$$

If such a force acts on a stable linear system for a sufficiently long time, any starting transients associated with its initial application will die out, and the system will settle down to a sinusoidal response

$$\xi_r(t) = A \sin \omega t + B \cos \omega t. \tag{18a}$$

The relationship between input amplitude a and the constants A and B, which determine the amplitude and relative phase of the output, is a function of ω that can be calculated from the equations of motion or measured experimentally. For example, the rth normal mode of the elastic airplane has*

$$\frac{A}{a} = \frac{1}{M_r(\omega_r^2 - \omega^2)}, \qquad \frac{B}{a} = 0. \tag{18b}$$

Because of the efficiency of the complex representation of simple harmonic motion, we often replace Eq. (17) by

$$f(t) = ce^{i\omega t}, \tag{19}$$

where c is a complex number. The physical interpretation of Eq. (19) is that its real or imaginary part defines the actual driving force. But the operation of taking the real part is linear and can be interchanged with all the other operations required in computing the response of a linear system, so that we can find the output in complex form and then say that its real part describes the actual motion. The complex output due to input Eq. (19) can be written

$$\xi_r(t) = H_r(i\omega)ce^{i\omega t}. \tag{20}$$

For example, in the case of the rth mode of the elastic airplane,

$$H_r(i\omega) = \frac{1}{M_r[\omega_r^2 + (i\omega)^2]}. \tag{21}$$

The complex function $H_r(i\omega)$ is called the mechanical admittance of that portion of the system which connects the input and output concerned. The name admittance is chosen because its size is a measure of the amount of energy the system admits to pass through itself or to be generated in response to a given amplitude of sinusoidal input. The reciprocal of the admittance

$$Z_r(i\omega) = \frac{1}{H_r(i\omega)} \tag{22}$$

is, logically, referred to as the impedance, and

$$\xi_r(t) = \frac{1}{Z_r(i\omega)} ce^{i\omega t}. \tag{23}$$

C–4 Fourier integral. In the technique of Fourier series we have a natural tool by which the superposition principle can be used to find the response to a *periodic* input. If the period is

$$T = \frac{2\pi}{\omega_0}, \tag{24}$$

---

* Since no energy dissipation agency is included in Eq. (1), the theoretical transient part of the response would never die out. In practice, however, structural or aerodynamic damping would ultimately destroy all but the permanent, forced motion.

any such input can be written

$$f(t) = \sum_{n=1}^{\infty} a_n \sin n\omega_0 t + \sum_{n=0}^{\infty} b_n \cos n\omega_0 t, \tag{25a}$$

where

$$a_n = \frac{2}{T} \int_0^T f(\tau) \sin (n\omega_0 \tau) d\tau, \tag{25b}$$

$$b_0 = \frac{1}{T} \int_0^T f(\tau) d\tau, \tag{25c}$$

$$b_n = \frac{2}{T} \int_0^T f(\tau) \cos (n\omega_0 \tau) d\tau, \quad \text{for } n \geq 1. \tag{25d}$$

If we introduce complex notation into Eqs. (25) through the Euler formulas

$$\sin \omega_0 t = \frac{e^{i\omega_0 t} - e^{-i\omega_0 t}}{2i}, \tag{26}$$

$$\cos \omega_0 t = \frac{e^{i\omega_0 t} + e^{-i\omega_0 t}}{2}, \tag{27}$$

we obtain the complex Fourier series

$$f(t) = \sum_{n=-\infty}^{\infty} c_n e^{in\omega_0 t}, \tag{28a}$$

where

$$c_n = \frac{1}{T} \int_0^T f(\tau) e^{-in\omega_0 \tau} d\tau. \tag{28b}$$

For a linear system, we see from Eq. (20) that the response to the periodic force in the form (28a) must be

$$\xi_r(t) = \sum_{n=-\infty}^{\infty} H_r(in\omega_0) c_n e^{in\omega_0 t}. \tag{29}$$

In this formula, there is no question of taking real or imaginary parts. The summation yields a real number, because c_n and c_{-n} are complex conjugates [cf. Eq. (28b)], as are

$$H_r(in\omega_0) \quad \text{and} \quad H_r(-in\omega_0).$$

For driving forces that do not have the property of being periodic, the Fourier integral furnishes a generalization of Fourier series. We can make the transition from the latter to the former by considering a function $f(t)$ represented by Eq. (28). Substituting Eq. (28b) into Eq. (28a), we get

$$f(t) = \sum_{n=-\infty}^{\infty} \frac{\omega_0}{2\pi} \int_{-T/2}^{T/2} f(\tau) e^{-in\omega_0 \tau} d\tau e^{in\omega_0 t}. \tag{30}$$

In this series, T may be made as large as we please. The Fourier integral is established if we take the result obtained as $T \to \infty$. During this process ω_0 becomes very small and can be replaced by the small increment $\Delta\omega$. In the limit, $\Delta\omega$ becomes the differential in the definite integral that replaces the sum from $n = -\infty$ to $n = +\infty$. The substitution

$$n\omega_0 = n\Delta\omega = \omega \tag{31}$$

then leads to

$$f(t) = \int_{\omega=-\infty}^{\omega=\infty} \left\{ \frac{1}{2\pi} \int_{\tau=-\infty}^{\tau=\infty} f(\tau)e^{-i\omega\tau}d\tau \right\} e^{i\omega t}d\omega, \tag{32}$$

which is the complex form of the Fourier integral. Equation (32) can be written in the dual form

$$f(t) = \int_{\omega=-\infty}^{\omega=\infty} G(i\omega)e^{i\omega t}d\omega, \tag{33}$$

where*

$$G(i\omega) = \frac{1}{2\pi} \int_{\tau=-\infty}^{\tau=\infty} f(\tau)e^{-i\omega\tau}d\tau \tag{34}$$

and $G(i\omega)$ is called the Fourier transform of $f(t)$, and conversely. In mathematical terms, Eq. (34) may be regarded as the solution or inversion of the integral equation (33).

It is not difficult to show that if $f(t)$ is an even function of t, Eqs. (33) and (34) reduce to the completely symmetrical Fourier cosine transform pair

$$f_E(t) = \sqrt{\frac{2}{\pi}} \int_0^\infty G_E(\omega) \cos \omega t\, d\omega, \tag{35}$$

$$G_E(\omega) = \sqrt{\frac{2}{\pi}} \int_0^\infty f_E(\tau) \cos \omega\tau\, d\tau. \tag{36}$$

Similarly, an odd $f(t)$ is represented by the Fourier sine transform pair

$$f_0(t) = \sqrt{\frac{2}{\pi}} \int_0^\infty G_0(\omega) \sin \omega t\, d\omega, \tag{37}$$

$$G_0(\omega) = \sqrt{\frac{2}{\pi}} \int_0^\infty f_0(\tau) \sin \omega\tau\, d\tau. \tag{38}$$

A serious weakness of the Fourier integral method is that for many inputs of practical interest, the integrals defined by Eqs. (33) and (34) do not converge and are therefore meaningless. In fact, the requirement for

---

*Equation (34) illustrates clearly the property of $G(i\omega)$ that its value for negative ω is the complex conjugate of its value for equal positive ω.

their convergence is that $f(\tau)$ be absolutely integrable in the range $\tau = -\infty$ to $\tau = \infty$:

$$\int_{-\infty}^{\infty} |f(\tau)| d\tau = \text{finite number,} \tag{39}$$

a condition not strictly met, for example, by the step function. If we assume that the integrals do converge, we can find an expression for the output $\xi_r(t)$ corresponding to the input $f(t)$ by superposition. One way of doing this is to observe that the response to the infinitesimal portion

$$df = G(i\omega)e^{i\omega t} d\omega \tag{40}$$

of the frequency spectrum in Eq. (33) between ω and $(\omega + d\omega)$ is given by Eq. (20) as

$$d\xi_r = H_r(i\omega)G(i\omega)e^{i\omega t} d\omega. \tag{41}$$

Hence, the integrated response to the entire frequency spectrum must be

$$\xi_r(t) = \int_{\omega=-\infty}^{\omega=\infty} d\xi_r = \int_{-\infty}^{\infty} H_r(i\omega)G(i\omega)e^{i\omega t} d\omega. \tag{42}$$

The integral in Eq. (42) will converge whenever Eq. (33) does, because $H_r(i\omega)$ usually behaves like $1/(i\omega)$ or $1/(i\omega)^2$ for large values of the argument.

The function $H_r(p)$, with $i\omega$ supplanted by a real or complex independent variable p, is called the admittance function of the linear system. It provides a connection between the Fourier integral and the Laplace transform methods of finding responses. By comparing the steps involved in calculating the steady-state output corresponding to an input $e^{i\omega t}$ with those involved in taking the Laplace transformation of the equations of motion, one can prove, for any system described by linear differential or integral equations with constant coefficients, that $H_r(p)$ is just the denominator polynomial in the expression for the transform of the response to any input involving zero initial conditions at $t = 0$. Moreover, we are led naturally to the idea of the Laplace transform if we attempt to use the Fourier integral in computing the indicial admittance $A_r(t)$. Thus it is known that the step function can be written

$$1(t) = \frac{1}{2\pi} \int_{-\infty}^{\infty} \frac{e^{i\omega t}}{i\omega} d\omega, \tag{43}$$

where convergence is ensured by requiring that the path of integration from $-\infty$ to $+\infty$ make a small semicircular loop around the origin $\omega = 0$. According to Eq. (42), the response to the input defined by Eq. (43) is

$$A_r(t) = \frac{1}{2\pi} \int_{-\infty}^{\infty} \frac{H_r(i\omega)e^{i\omega t}}{i\omega} d\omega. \tag{44}$$

This expression, known as Bromwich's integral formula, can be applied to computation of the indicial admittance when the admittance function is known. Such applications to aeroelastic problems are discussed in Chapter 10, Subsection 10-5(b).

C-5 Response to a random input. In connection with buffeting, gust loads on aircraft flying through turbulent air, fatigue life of structural materials, and similar problems, we encounter inputs which are so-called random functions of time. They vary in a highly irregular fashion and can be defined only in terms of certain average values, and manipulated by the mathematical tools of statistical theory. An extended account of this subject is beyond the scope of our work, but we can reproduce with some degree of rigor a few definitions and relations that play an important role in aeroelasticity.

Let $f(t)$ be a random input. If we measure the values of $f(t)$ over a sufficiently long interval of time $2T$, then in many cases the nth mean value of f:

$$\overline{f^n} = \lim_{T \to \infty} \frac{1}{2T} \int_{-T}^{T} f^n(t)dt, \qquad (45)$$

settles down to a clearly defined limit. A stationary random process is one in which the value of $\overline{f^n}$ is independent of where, throughout the entire history of $f(t)$, the particular interval is chosen which is used to compute the mean. The present discussion is confined to such processes. We often pick our scale of reference for measuring f in such a way that the first mean value or average,

$$\bar{f} = \lim_{T \to \infty} \frac{1}{2T} \int_{-T}^{T} f(t)dt, \qquad (46)$$

is equal to zero. The most significant property of f is then its mean-square value

$$\overline{f^2} = \lim_{T \to \infty} \frac{1}{2T} \int_{-T}^{T} f^2(t)dt, \qquad (47)$$

where $\overline{f^2}$ provides an estimate of the energy or power contained in the input f. For example, if this input were a gust velocity w, then $\frac{3}{2}\overline{w^2}$ would be the mean kinetic energy per unit mass in the turbulent air. If $\overline{w^2}$ and the other statistical properties of the turbulence are independent of position in the sample of air and of the direction of the component w being averaged, we say that the turbulence is statistically homogeneous and isotropic.

The statistical relationship between the values of $f(t)$ at different instants of time is measured by the autocorrelation function

$$\phi(\tau) = \lim_{T \to \infty} \frac{1}{2T} \int_{-T}^{T} f(t)f(t + \tau)dt. \qquad (48)$$

Clearly, $\phi(0)$ equals the mean-square $\overline{f^2}$. We expect ϕ to approach zero for large τ (whenever $\bar{f} = 0$), because the values of the random quantity at times very far apart should be unrelated to one another, so that, for positive $f(t)$, $f(t + \tau)$ would be equally likely to have a positive or a negative value. In stationary random processes, ϕ is unaffected by where the interval $2T$ is picked, so it is a function of τ only. Also it is an even function, since the statistical properties would not be changed by reversing the direction of the flow of time.

Let us imagine for the moment that the random input $f(t)$ is susceptible of Fourier analysis. It is not periodic, so we would have to calculate its spectrum or frequency distribution by means of the Fourier integral. Referring to Eq. (34), we might define a function

$$F(i\omega) = \frac{1}{2\pi} \int_{-T}^{T} f(t)e^{-i\omega t}dt \qquad (49)$$

which, for large enough T, would give us some idea of the relative magnitudes of the various simple harmonic components of $f(t)$. Unfortunately, the limiting value of $F(i\omega)$ as $T \to \infty$ turns out to be mathematically undefined. However, we are led to a meaningful description of the spectrum of $f(t)$ if we examine the limiting behavior of the function

$$|F(i\omega)|^2 = F(i\omega)F(-i\omega). \qquad (50)$$

For a particular finite value of T the inverse Fourier transform [Eq. (33) with t replaced by τ] of this function reads

$$\int_{-\infty}^{\infty} e^{i\omega\tau}F(i\omega)F(-i\omega)d\omega = \int_{-\infty}^{\infty} e^{i\omega\tau}F(-i\omega)\left\{\frac{1}{2\pi}\int_{-T}^{T} f(t)e^{-i\omega t}dt\right\} d\omega. \qquad (51)$$

If we invert the order of integration and make the variable change $\omega' = -\omega$, we get

$$2\pi\int_{-\infty}^{\infty} e^{i\omega\tau}|F(i\omega)|^2 d\omega = \int_{-T}^{T} f(t)\left\{\int_{-\infty}^{\infty} e^{i\omega'(t-\tau)}F(i\omega')d\omega'\right\} dt. \qquad (52)$$

Now the inner integral of Eq. (52) may be regarded as the inverse of the Fourier integral (49), so that*

$$\int_{-\infty}^{\infty} e^{i\omega'(t-\tau)}F(i\omega')d\omega' = f(t - \tau). \qquad (53)$$

Therefore Eq. (52) reads

$$2\pi\int_{-\infty}^{\infty} e^{i\omega\tau}|F(i\omega)|^2 d\omega = \int_{-T}^{T} f(t)f(t - \tau)dt. \qquad (54)$$

---

* This step is made more plausible by imagining that the function $f(t)$ employed in Eq. (49) is not the random function itself, but one which coincides with it in the interval $-T \leq t \leq T$ and has the value zero everywhere else.

The integral on the right side of Eq. (54) itself diverges as $T \to \infty$, but the same integral multiplied by $1/2\,T$ approaches the (even) autocorrelation function $\phi(\tau)$. Hence we are led to replace the function $|F(i\omega)|^2$, which is meaningless in the limit, by another quantity

$$\Phi(\omega) = \frac{2\pi |F(i\omega)|^2}{T}, \tag{55}$$

which reduces Eq. (54) to

$$\tfrac{1}{2} \int_{-\infty}^{\infty} e^{i\omega\tau} \Phi(\omega) d\omega = \lim_{T \to \infty} \frac{1}{2T} \int_{-T}^{T} f(t) f(t - \tau) dt = \phi(-\tau) = \phi(\tau). \tag{56}$$

The quantity $\Phi(\omega)$ is known as the power spectral density of $f(t)$, since it is proportional to the square of the magnitude (the power in certain electrical applications) of the component of $f(t)$ having frequency ω. We could have defined $\Phi(\omega)$ directly as the Fourier transform of the autocorrelation function, but its physical significance is partly obscured by that procedure. Through Eq. (34), the inverse relation between $\Phi(\omega)$ and $\phi(\tau)$ reads

$$\Phi(\omega) = \frac{1}{\pi} \int_{-\infty}^{\infty} \phi(\tau) e^{-i\omega\tau} d\tau = \frac{2}{\pi} \int_{0}^{\infty} \phi(\tau) \cos \omega\tau d\tau, \tag{57}$$

the latter expression resulting from the evenness of the autocorrelation function. Since $\Phi(\omega)$ is also evidently an even function of ω, Eq. (56) shows that

$$\phi(0) = \overline{f^2} = \int_{0}^{\infty} \Phi(\omega) d\omega. \tag{58}$$

Equation (58) is interpreted physically by the statement that the average energy or power of the function $f(t)$ is the integrated sum of the energies associated with all its frequency components.

Having defined some of the statistical properties of a random input to a linear system, we assume that the corresponding output $\xi_r(t)$ is also a random function and attempt to relate its properties to those of $f(t)$. We call the autocorrelation function and power spectral density of $\xi_r(t)$ by the symbols $\psi(\tau)$ and $\Psi(\omega)$. They are related through the definition

$$\psi(\tau) = \tfrac{1}{2} \int_{-\infty}^{\infty} e^{i\omega\tau} \Psi(\omega) d\omega. \tag{59}$$

For any instant of time, $\xi_r(t)$ can be calculated from $f(t)$ through the unit-impulse-response equation [cf. Eq. (16)]

$$\xi_r(t) = \int_{-\infty}^{t} f(\tau) h_r(t - \tau) d\tau, \tag{60}$$

where the lower limit is replaced by $\tau = -\infty$, since the applied force is stationary and has been acting for all time. Equation (60) is left unchanged if we set the upper limit equal to $+\infty$, for h_r is obviously zero when its argument is negative, and the integrand vanishes for $t < \tau < +\infty$. We make this substitution and also introduce the integration variable $\sigma = t - \tau$, thus obtaining

$$\xi_r(t) = \int_{-\infty}^{\infty} f(\tau)h_r(t - \tau)d\tau = \int_{-\infty}^{\infty} f(t - \sigma)h_r(\sigma)d\sigma. \qquad (61)$$

Finally, we compute the autocorrelation function of the output by manipulating Eq. (61):

$$\psi(\tau) = \lim_{T \to \infty} \frac{1}{2T} \int_{-T}^{T} \xi_r(t)\xi_r(t + \tau)dt$$

$$= \lim_{T \to \infty} \frac{1}{2T} \int_{-T}^{T} \int_{-\infty}^{\infty} \int_{-\infty}^{\infty} f(t - \sigma)f(t + \tau - \sigma')h_r(\sigma)h_r(\sigma')d\sigma d\sigma' dt$$

$$= \int_{-\infty}^{\infty} \int_{-\infty}^{\infty} h_r(\sigma)h_r(\sigma') \left\{ \lim_{T \to \infty} \frac{1}{2T} \int_{-T}^{T} f(t - \sigma)f(t + \tau - \sigma')dt \right\} d\sigma d\sigma'$$

$$= \int_{-\infty}^{\infty} \int_{-\infty}^{\infty} h_r(\sigma)h_r(\sigma')\phi(\tau + \sigma - \sigma')d\sigma d\sigma'. \qquad (62)$$

The substitution of the autocorrelation function of the input in the last term of Eq. (62) is justified by the stationary nature of the process. Since the location of the averaging interval $2T$ has no effect on the result, we let $t' = t - \sigma$ and get

$$\lim_{T \to \infty} \frac{1}{2T} \int_{-T}^{T} f(t - \sigma)f(t + \tau - \sigma')dt$$

$$= \lim_{T \to \infty} \frac{1}{2T} \int_{(-T-\sigma)}^{(T-\sigma)} f(t')f(t' + (\sigma + \tau - \sigma'))dt' = \phi(\tau + \sigma - \sigma'). \qquad (63)$$

Equation (62) connects the autocorrelation functions. A simpler and more important relation is the one between the power spectral densities, which we derive by replacing ψ and ϕ through the Fourier integrals (56) and (59):

$$\frac{1}{2} \int_{-\infty}^{\infty} \Psi(\omega)e^{i\omega\tau}d\omega = \frac{1}{2} \int_{-\infty}^{\infty} \int_{-\infty}^{\infty} \int_{-\infty}^{\infty} \Phi(\omega)e^{i\omega\tau}h_r(\sigma)e^{i\omega\sigma}h_r(\sigma')e^{-i\omega\sigma'}d\sigma' d\sigma d\omega$$

$$= \frac{1}{2} \int_{-\infty}^{\infty} \Phi(\omega)e^{i\omega\tau} \left\{ \left[\int_{-\infty}^{\infty} h_r(\sigma)e^{i\omega\sigma}d\sigma \right] \left[\int_{-\infty}^{\infty} h_r(\sigma')e^{-i\omega\sigma'}d\sigma' \right] \right\} d\omega. \qquad (64)$$

The two integrals in the inner brackets are expressions for the response of the linear system to a sinusoidal input of unit amplitude which has acted for a long period of time. They are therefore related to the mechanical admittance $H_r(i\omega)$:

$$\int_{-\infty}^{\infty} h_r(\sigma')e^{-i\omega\sigma'}d\sigma' = \int_{0}^{\infty} h_r(\sigma')e^{-i\omega\sigma'}d\sigma' = -\int_{t}^{-\infty} h_r(t-\tau)e^{i\omega(\tau-t)}d\tau$$

$$= e^{-i\omega t}\int_{-\infty}^{t} e^{i\omega\tau}h_r(t-\tau)d\tau = H_r(i\omega) \tag{65}$$

and

$$\int_{-\infty}^{\infty} h_r(\sigma)e^{i\omega\sigma}d\sigma = e^{i\omega t}\int_{-\infty}^{t} e^{-i\omega\tau}h_r(t-\tau)d\tau = H_r(-i\omega), \tag{66}$$

so that

$$\left[\int_{-\infty}^{\infty} h_r(\sigma)e^{i\omega\sigma}d\sigma\right]\left[\int_{-\infty}^{\infty} h_r(\sigma')e^{-i\omega\sigma'}d\sigma'\right] = H_r(-i\omega)H_r(i\omega) = |H_r(i\omega)|^2. \tag{67}$$

Substituting Eq. (67) into Eq. (64), we get

$$\int_{-\infty}^{\infty} \Psi(\omega)e^{i\omega\tau}d\omega = \int_{-\infty}^{\infty} |H_r(i\omega)|^2\Phi(\omega)e^{i\omega\tau}d\omega. \tag{68}$$

Since Eq. (68) must hold for all values of τ, the integrands are equal:

$$\Psi(\omega) = |H_r(i\omega)|^2\Phi(\omega) = \frac{1}{|Z_r(i\omega)|^2}\Phi(\omega). \tag{69}$$

This shows that the power spectral densities of the input and output differ only by a factor which is the square of the magnitude of the mechanical admittance of the system.

REFERENCES

CHAPTER 1

1-1. COLLAR, A. R., The Expanding Domain of Aeroelasticity, *Journal of the Royal Aeronautical Society*, Vol. L, pp. 613–636, August, 1946.

1-2. HILL, G. T. R., Advances in Aircraft Structural Design, *Proceedings of the Third Anglo-American Aeronautical Conference, Brighton, England*, September, 1951.

1-3. BREWER, G., The Collapse of Monoplane Wings, *Flight*, 1913.

1-4. LANCHESTER, F. W., *Torsional Vibration of the Tail of an Airplane*, Br. A.R.C., R. & M. 276, Part 1, July, 1916.

1-5. BAIRSTOW, L. AND FAGE, A., *Oscillations of the Tailplane and Body of an Aeroplane in Flight*, Br. A.R.C., R. & M. 276, Part 2, July, 1916.

1-6. FOKKER, A. H. G., *The Flying Dutchman*, Henry Holt and Co., Inc., 1931.

1-7. BAUMHAUER, A. F., AND KONING, C., On the Stability of Oscillations of an Airplane Wing, International Air Congress, London, 1922.

1-8. REISSNER, H., Neuere Probleme aus der Flugzeugstatik, *Z. für Flugtechnik und Motorluftschiffahrt*, Vol. 17, No. 7, April, 1926.

1-9. ROXBEE COX, H., AND PUGSLEY, A. G., *Theory of Loss of Lateral Control Due to Wing Twisting*, Br. A.R.C., R. & M. 1506, October, 1932.

1-10. GLAUERT, H., *The Force and Moment of an Oscillating Airfoil*, Br. A.R.C. R. & M. 1216, November, 1928.

1-11. FRAZER, R. A., AND DUNCAN, W. J., *The Flutter of Aeroplane Wings*, Br. A.R.C., R. & M. 1155, August, 1928.

1-12. KÜSSNER, H. G., Schwingungen von Flugzeugflügeln, *Luftfahrtforschung*, Vol. 4, June, 1929.

1-13. THEODORSEN, T., *General Theory of Aerodynamic Instability and the Mechanism of Flutter*, N.A.C.A. Report 496, 1935.

CHAPTER 2

2-1. FRAZER, R. A., DUNCAN, W. J., AND COLLAR, A. R., *Elementary Matrices*, The Macmillan Co., New York, 1946.

2-2. NILES, A. S., AND NEWELL, J. S., *Airplane Structures*, Vol. 1, John Wiley & Sons, Inc., New York, 1945.

2-3. SOKOLNIKOFF, I. S., *Mathematical Theory of Elasticity*, McGraw-Hill Book Company, Inc., New York, 1946.

2-4. LOVITT, W. V., *Linear Integral Equations*, McGraw-Hill Book Company, Inc., New York, 1924.

2-5. BENSCOTER, S. U., AND GOSSARD, M. L., *Matrix Methods for Calculating Cantilever-beam Deflections*, N.A.C.A. T.N. 1827, March 1949.

2-6. *Tables of Lagrangian Interpolation Coefficients*, National Bureau of Standards, Columbia University Press, New York, 1944.

2-7. MILNE, W. E., *Numerical Calculus*, Princeton University Press, Princeton, New Jersey, 1949.

2-8. TIMOSHENKO, S., AND GOODIER, J. N., *Theory of Elasticity*, McGraw-Hill Book Company, Inc., New York, 1951.

2-9. CARSLAW, H. S., AND JAEGER, J. C., *Operational Methods in Applied Mathematics*, Oxford University Press, Oxford, England, 1941.

2-10. PEERY, D. J., *Aircraft Structures*, McGraw-Hill Book Company, Inc., New York, 1950.

2-11. KUHN, P., *Deformation Analysis of Wing Structures*, N.A.C.A. T.N. 1361, July, 1947.

2-12. LEVY, S., Computation of Influence Coefficients for Aircraft Structures with Discontinuities and Sweepback, *Journal of the Aeronautical Sciences*, Vol. 14, No. 10, pp. 547–560, October, 1947.

2-13. LANG, A. L., AND BISPLINGHOFF, R. L., Some Results of Swept-back Wing Structural Studies, *Journal of the Aeronautical Sciences*, Vol. 18, No. 11, pp. 705–717, November, 1951.

2-14. LANGEFORS, B., Analysis of Elastic Structures by Matrix Transformation with Special Regard to Semimonocoque Structures, *Journal of the Aeronautical Sciences*, Vol. 19, No. 7, pp. 451–458, July, 1952.

2-15. WEHLE, L. B., AND LANSING, W., A Method for Reducing the Analysis of Complex Redundant Structures to a Routine Procedure, *Journal of the Aeronautical Sciences*, Vol. 19, No. 10, pp. 677–684, October, 1952.

2-16. SCHUERCH, H. U., *Delta Wing Design Analysis*. Paper presented at the S.A.E. National Aeronautic Meeting, Los Angeles, September 29–October 3, 1953. Preprint No. 141.

2-17. LEVY, S., Structural Analysis and Influence Coefficients for Delta Wings, *Journal of the Aeronautical Sciences*, Vol. 20, No. 7, pp. 449–454, July, 1953.

2-18. TURNER, M. J., CLOUGH, R. W., MARTIN, H. C., AND TOPP, L. J., *Stiffness and Deflection Analysis of Complex Structures*. Paper presented at the January 1954 Annual Meeting of the Institute of Aeronautical Sciences, New York.

2-19. REISSNER, E., AND STEIN, M., *Torsion and Transverse Bending of Cantilever Plates*, N.A.C.A. T.N. 2369, June, 1951.

2-20. FUNG, Y.-C., *Elastostatic and Aeroelastic Problems Relating to Thin Wings of High-Speed Airplanes*, Ph.D. thesis, California Institute of Technology, 1948.

2-21. FUNG, Y.-C., Bending of Thin Elastic Plates of Variable Thickness, *Journal of the Aeronautical Sciences*, Vol. 20, No. 7, pp. 455–468, July, 1953.

2-22. SCHUERCH, H. U., Structural Analysis of Swept, Low Aspect-ratio Multispar Aircraft Wings, *Aeronautical Engineering Review*, Vol. 11, No. 11, pp. 34–41, November, 1952.

CHAPTER 3

3-1. HILDEBRAND, F. B., *Advanced Calculus for Engineers*, Prentice-Hall, Inc., New York, 1949.

3-2. SIDDALL, J. N., AND ISAKSON, G., *Approximate Analytical Methods for Determining Natural Modes and Frequencies of Vibration*. Report of M.I.T. Aeroelastic and Structures Research Laboratory for Office of Naval Research, January 15, 1951.

3–3. PINNEY, E., Vibration Modes of Tapered Beams, *American Mathematical Monthly*, p. 391, August–September, 1947.

3–4. WATSON, G. N., *Treatise on the Theory of Bessel Functions*, p. 107, Cambridge University Press, Cambridge, England, 1932.

3–5. BISPLINGHOFF, R. L, ISAKSON, G., AND O'BRIEN, T. F., *Gust Loads on Rigid and Elastic Airplanes*. Report of M.I.T. Aeroelastic and Structures Research Laboratory for Navy Bureau of Aeronautics, August, 1949.

3–6. CARSLAW, H. S., AND JAEGER, J. C., *Operational Methods in Applied Mathematics*, Oxford University Press, Oxford, England, 1941.

3–7. LOVITT, W. V., *Linear Integral Equations*, McGraw-Hill Book Company, Inc., New York, 1924.

3–8. LEVY, S., Computation of Influence Coefficients for Aircraft Structures with Discontinuities, *Journal of the Aeronautical Sciences*, Vol. 14, No. 10, pp. 547–560, October, 1947.

3–9. WHITTAKER, E. T., *Analytical Dynamics*, Cambridge University Press, Cambridge, England, 1937.

CHAPTER 4

4–1. SIDDALL, J. N., AND ISAKSON, G., *Approximate Analytical Methods for Computing Natural Modes and Frequencies of Vibration*. Report of M.I.T. Aeroelastic and Structures Laboratory for the Office of Naval Research, January, 1951.

4–2. COLLATZ, L., *Eigenwertprobleme*, Chelsea Publishing Co., New York, 1948.

4–3. REISSNER, E., Note on the Method of Complementary Energy, *Journal of Mathematics and Physics*, Vol. 27, pp. 159–160, 1948.

4–4. DUNCAN, W. J., *Mechanical Admittances and Their Application to Oscillation Problems*, Br. A.R.C., R. & M. 2000, 1947.

4–5. WILLIAMS, D., *The Use of the Principle of Minimum Potential Energy in Problems of Static Equilibrium*, Br. A.R.C., R. & M. 1827, 1937.

4–6. RAUSCHER, M., Station Functions and Air Density Variations in Flutter Analysis, *Journal of the Aeronautical Sciences*, Vol. 16, No. 6, pp. 345–353, June, 1949.

4–7. BESKIN, L., AND ROSENBERG, R. M., Higher Modes of Vibration by a Method of Sweeping, *Journal of the Aeronautical Sciences*, Vol. 13, No. 11, pp. 597–604, November, 1946.

4–8. DUNCAN, W. J., *Galerkin's Method in Mechanics and Differential Equations*, Br. A.R.C., R. & M. 1798, 1937.

4–9. HOLZER, H., *Die Berechung der Drehschwingungen*, Springer-Verlag, Berlin, 1921.

4–10. MYKLESTAD, N. O., *Vibration Analysis*, McGraw-Hill Book Company, Inc., New York, 1944.

4–11. WAYLAND, H., Expansion of Determinantal Equations into Polynomial Form, *Quarterly of Applied Mathematics*, Vol. 11, No. 4, pp. 277–306, January, 1945.

4–12. ISAKSON, G., A Method for Accelerating the Convergence of an Iteration Procedure, *Journal of the Aeronautical Sciences*, Vol. 16, p. 443, July, 1949.

4-13. FLOMENHOFT, H. I., A Method for Determining Mode Shapes and Frequencies above the Fundamental by Matrix Iteration, *Journal of Applied Mechanics*, Vol. 17, No. 3, pp. 249–256, September, 1950.

4-14. LEVY, S., Computation of Influence Coefficients for Aircraft Structures with Discontinuities, *Journal of the Aeronautical Sciences*, Vol. 14, No. 10, pp. 547–560, October, 1947.

CHAPTER 5

5-1. RAUSCHER, M., *Introduction to Aeronautical Dynamics*, John Wiley & Sons, Inc., New York, 1953.

5-2. LAMB, H., *Hydrodynamics*, 6th ed., Dover Publications, New York, 1945.

5-3. KEENAN, J. H., *Thermodynamics*, John Wiley & Sons, Inc., New York, 1941.

5-4. PHILLIPS, H. B., *Vector Analysis*, John Wiley & Sons, Inc., New York, 1933.

5-5. LICHTENSTEIN, L., *Grundlagen der Hydromechanik*, Springer-Verlag, Berlin, 1929.

5-6. MUNK, M. M., *The Aerodynamic Forces on Airship Hulls*, N.A.C.A. Report 184, 1924.

5-7. THEODORSEN, T., *Impulse and Momentum in an Infinite Fluid*, Theodore von Kármán Anniversary Volume, California Institute of Technology, pp. 49–58, 1941.

5-8. ALLEN, H. J., *Pressure Distribution and Some Effects of Viscosity on Slender Inclined Bodies of Revolution*, N.A.C.A. T.N. 2044, 1950.

5-9. VON KÁRMÁN, T. AND BURGERS, J. M., General Aerodynamic Theory of Perfect Fluids, Division E, Vol. II of *Aerodynamic Theory* (Editor: W. F. Durand), Durand Reprinting Committee, Pasadena, California, 1943. (Originally published 1934.)

5-10. STRATTON, J. A., *Electromagnetic Theory*, McGraw-Hill Book Company, Inc., New York, 1941.

5-11. MANGLER, K. W., *Improper Integrals in Theoretical Aerodynamics*, British Royal Aircraft Establishment Report Aero 2424, 1951.

5-12. ALLEN, H. J., *General Theory of Airfoil Sections Having Arbitrary Shape or Pressure Distributions*, N.A.C.A. Report 833, 1945.

5-13. SÖHNGEN, H., Die Lösungen der Integralgleichung und deren Anwendung in der Tragflügeltheorie, *Math. Z.*, Band 45, pp. 245–264, 1939.

5-14. ABBOTT, I. A., VON DOENHOFF, A. E., AND STIVERS, L. S., *Summary of Airfoil Data*, N.A.C.A. Report 824, 1945.

5-15. REISSNER, E., *Effect of Finite Span on the Airload Distributions for Oscillating Wings, I — Aerodynamic Theory of Oscillating Wings of Finite Span*, N.A.C.A. T.N. 1194, 1947.

5-16. REISSNER, E., *On the General Theory of Thin Airfoils for Nonuniform Motion*, N.A.C.A. T.N. 946, 1944.

5-17. KINNER, W., *The Principle of the Potential Theory Applied to the Circular Wing*, British Ministry of Aircraft Production, R.T.P. Translation 2345. (Translated from Ingenieur Archiv, Bd. VIII, pp. 47–80, 1937.)

5–18. JONES, W. P., *The Calculation of Aerodynamic Derivative Coefficients for Wings of Any Planform in Nonuniform Motion*, Br. A.R.C., R. & M. 2470, 1952.

5–19. REISSNER, E., Note on the Theory of Lifting Surfaces, *Proceedings of the National Academy of Sciences*, Vol. 35, No. 4, April, 1949.

5–20. SEARS, W. R., AND PAI, S. I., Some Aeroelastic Properties of Swept Wings, *Journal of the Aeronautical Sciences*, Vol. 16, No. 2, February, 1949.

5–21. MULTHOPP, H., *The Calculation of the Lift Distribution of Aerofoils*, British Ministry of Aircraft Production, R.T.P. Translation 2392. (Translated from Luftfahrtforschung, Bd. 15, Nr. 4, June, 1938.)

5–22. DE YOUNG, J., AND HARPER, C. W., *Theoretical Symmetrical Span Loading at Subsonic Speeds for Wings Having Arbitrary Planform*, N.A.C.A. Report 921, 1948.

5–23. WEISSINGER, J., *The Lift Distribution of Swept-back Wings*, N.A.C.A. T.M. 1120, 1947. (Translated from *ZWB Forschungsbericht* Nr. 1553, Berlin-Adlershof, 1942.)

5–24. JONES, R. T., *Properties of Low Aspect-ratio Pointed Wings at Speeds Below and Above the Speed of Sound*, N.A.C.A. Report 835, 1946.

5–25. LAWRENCE, H. R., The Lift Distribution on Low Aspect-ratio Wings at Subsonic Speeds, *Journal of the Aeronautical Sciences*, Vol. 18, No. 10, October, 1951.

5–26. LAIDLAW, W. R., *Theoretical and Experimental Pressure Distributions on Low Aspect-ratio Wings Oscillating in an Incompressible Flow*, Technical Report 51-2 of the M.I.T. Aeroelastic and Structures Research Laboratory for Naval Bureau of Aeronautics, September, 1954.

5–27. DIEDERICH, F. W., *A Planform Parameter for Correlating Certain Aerodynamic Characteristics of Swept Wings*, N.A.C.A. T.N. 2335, 1951.

5–28. DIEDERICH, F. W., AND BUDIANSKY, B., *Divergence of Swept Wings*, N.A.C.A. T.N. 1680, 1948.

5–29. THEODORSEN, T., *General Theory of Aerodynamic Instability and the Mechanism of Flutter*, N.A.C.A. Report 496, 1935.

5–30. GLAUERT, H., *The Force and Moment on an Oscillating Aerofoil*, Br. A.R.C., R. & M. 1242, 1929.

5–31. CICALA, P., Le Axioni aerodinamiche sui profili di ala oscillanti in presenza di corrente uniforme, *Mem. R. Accad. Torino*, Ser. 2, Pt. I, t. 68, 1934–1935.

5–32. ELLENBERGER, G., Bestimmung der Luftkräfte auf einen ebenen Tragflügel mit Querruder, *Z. angew. Math. Mech.*, Bd. 16, Heft 4, August, 1936.

5–33. VON BORBELY, S., Mathematischer Beitrag zur Theorie der Flügelschwingungen, *Z. angew. Math. Mech.*, Bd. 16, Heft 1, February, 1936.

5–34. KASSNER, R., AND FINGADO, H., Das ebene Problem der Flügelschwingung, *Luftfahrtforsch.*, Bd. 13, Nr. 11, November, 1936.

5–35. KÜSSNER, H. G., Zusammenfassender Bericht über den instationären Auftrieb von Flügeln, *Luftfahrtforsch.*, Bd. 13, Nr. 12, December, 1936.

5–36. GRAY, A., MATHEWS, G. B., AND MAC ROBERT, T. M., *A Treatise on Bessel Functions*, 2nd ed., Macmillan and Co., Ltd., London, 1931.

5–37. SCHWARZ, L., Berechnung der Druckverteilung einer harmonisch sich

Verformenden Tragfläche in ebener Strömung, *Luftfahrtforsch.*, Bd. 17, Nr. 11 & 12. December, 1940.

5–38. BISPLINGHOFF, R. L., ISAKSON, G., PIAN, T. H. H., FLOMENHOFT, H. I., AND O'BRIEN, T. F., *An Investigation of Stresses in Aircraft Structures Under Dynamic Loading*, Report of the M.I.T. Aeroelastic and Structures Research Laboratory for Naval Bureau of Aeronautics, 1949.

5–39. THEODORSEN, T., AND GARRICK, I. E., *Nonstationary Flow About a Wing-Aileron-Tab Combination, Including Aerodynamic Balance*, N.A.C.A. Report 736, 1942

5–40. KÜSSNER, H. G., AND SCHWARZ, L., *The Oscillating Wing with Aerodynamically Balanced Elevator*, N.A.C.A. T.M. 991, 1941. (Translated from *Luftfahrtforsch.*, Bd. 17, Nr. 11 & 12, December, 1940.)

5–41. SMILG, B., AND WASSERMAN, L. S., *Application of Three-dimensional Flutter Theory to Aircraft Structures*, Air Force Technical Report 4798, 1942.

5–42. WASSERMAN, L. S., MYKYTOW, W. S., AND SPIELBERG, I. N., *Tab Flutter Theory and Applications*, Air Force Technical Report 5153, 1944.

5–43. HALFMAN, R. L., *Experimental Aerodynamic Derivatives of a Sinusoidally Oscillating Airfoil in Two-dimensional Flow*, N.A.C.A. Report 1108, 1952.

5–44. GREIDANUS, J. H., VAN DE VOOREN, A. I., AND BERGH, H., *Experimental Determination of the Aerodynamic Coefficients of an Oscillating Wing in Incompressible, Two-dimensional Flow*, Parts I–IV, National Luchtvaartlaboratorium, Amsterdam, Reports F-101, 102, 103, and 104, 1952.

5–45. ANDREOPOLIS, T. C., CHEILEK, H. A., AND DONOVAN, A. F., *Measurements of the Aerodynamic Hinge Moments of an Oscillating Flap and Tab*, Air Force Technical Report 5784, 1949.

5–46. SPIELBERG, I. N., *The Two-dimensional Incompressible Aerodynamic Coefficients for Oscillatory Changes in Airfoil Camber*, Wright Air Development Center, U. S. Air Force Technical Note WCNS 52-7, 1952.

5–47. LUKE, Y. L., AND DENGLER, M. A., Tables of the Theodorsen Circulation Function for Generalized Motion, *Journal of the Aeronautical Sciences*, Vol. 8, No. 7, July, 1951.

5–48. SÖHNGEN, H., Bestimmung der Auftriebsverteilung für beliebige instationäre Bewegungen (Ebenes Problem), *Luftfahrtforschung*, Bd. 17, Nr. 11 & 12, December, 1940.

5–49. JONES, W. P. *Aerodynamic Forces on Wings in Nonuniform Motion*, Br. A.R.C., R. & M. No. 2117, 1945.

5–50. FRAEYS DE VEUBEKE, B., *Aérodynamique Instationnaire des Profils Minces Déformables*, Bulletin du Service Technique de l'Aeronautique (Brussels, Belgium), No. 25, November, 1953.

5–51. HILDEBRAND, F. B., *Advanced Calculus for Engineers*, Prentice-Hall, Inc., New York, 1948.

5–52. SNEDDON, I. N., *Fourier Transforms*, McGraw-Hill Book Company, Inc., New York, 1951.

5–53. WAGNER, H., Über die Entstehung des dynamischen Auftriebes von Tragflügeln, *Z. angew. Math. Mech.*, Bd. 5, Heft 1, February, 1925.

5–54. GARRICK, I. E., *On Some Fourier Transforms in the Theory of Nonstationary Flows*. Proceedings of the Fifth International Congress for

Applied Mechanics, pp. 590–593. John Wiley & Sons, Inc., New York, 1939.

5-55. SEARS, W. R., Operational Methods in the Theory of Airfoils in Non-uniform Motion, *Journal of the Franklin Institute*, Vol. 230, pp. 95–111, 1940.

5-56. VON KÁRMÁN, T., AND SEARS, W. R., Airfoil Theory for Nonuniform Motion, *Journal of the Aeronautical Sciences*, Vol. 5, No. 10, August, 1938.

5-57. NEUMARK, S., Pressure Distribution on an Airfoil in Nonuniform Motion, *Journal of the Aeronautical Sciences*, Readers' Forum, Vol. 19, No. 3, March, 1952.

5-58. CHURCHILL, R. V., *Modern Operational Mathematics in Engineering*, McGraw-Hill Book Company, Inc., New York, 1944.

CHAPTER 6

6-1. HILDEBRAND, F. B., *Advanced Calculus for Engineers*, Prentice-Hall, Inc., New York, 1948.

6-2. DE YOUNG, J., AND HARPER, C. W., *Theoretical Symmetric Span Loading at Subsonic Speeds for Wings Having Arbitrary Planform*, N.A.C.A. Report 921, 1948.

6-3. LIEPMANN, H. W., AND PUCKETT, A. E., *Introduction to Aerodynamics of a Compressible Fluid*, John Wiley & Sons, Inc., New York, 1947.

6-4. KATZOFF, S., *Wind-tunnel Wall Corrections* (Appendix A), N.A.C.A. University Conference on Aerodynamics, Langley Field, Virginia, June, 1948.

6-5. HEASLET, M. A., LOMAX, H., AND SPREITER, J. R., *Linearized Compressible-flow Theory for Sonic Flight Speeds*, N.A.C.A. Report 956, 1950.

6-6. DIEDERICH, F. W., *A Planform Parameter for Correlating Certain Aero-dynamic Characteristics of Swept Wings*, N.A.C.A. T.N. 2335, 1951.

6-7. VINCENTI, W., *Comparison Between Theory and Experiment for Wings at Supersonic Speeds*, N.A.C.A. T.N. 2100, 1950.

6-8. ACKERET, J., *Air Forces on Airfoils Moving Faster than Sound*, N.A.C.A. T.M. 317, 1925.

6-9. PUCKETT, A. E., Supersonic Wave Drag of Thin Airfoils, *Journal of the Aeronautical Sciences*, Vol. 13, No. 9, September, 1946.

6-10. EVVARD, J. C., *Use of Source Distributions for Evaluating Theoretical Aero-dynamics of Thin Finite Wings at Supersonic Speeds*, N.A.C.A. Report 951, 1950.

6-11. LAGERSTROM, P. A., *Linearized Supersonic Theory of Conical Wings*, N.A.C.A. T.N. 1685, 1950.

6-12. LIGHTHILL, M. J., *Supersonic Theory of Wings of Finite Span*, Br. A.R.C., R. & M. 2001, 1944.

6-13. HEASLET, M. A., LOMAX, H., AND JONES, A. L., *Volterra's Solution of the Wave Equation as Applied to Three-dimensional Supersonic Airfoil Problems*, N.A.C.A. Report 889, 1947.

6-14. LOMAX, H., HEASLET, M. A., AND FULLER, F. B., *Formulas for Source, Doublet, and Vortex Distributions in Supersonic Wing Theory*, N.A.C.A. T.N. 2252, 1950.

834 REFERENCES

6–15. HEASLET, M. A., AND LOMAX, H., *The Use of Source-Sink and Doublet Distributions Extended to the Solution of Boundary-value Problems in Supersonic Flow*, N.A.C.A. Report 900, 1948.

6–16. LAPIN, E., *Charts for the Computation of Lift and Drag of Finite Wings at Supersonic Speeds*, Douglas Aircraft Co., Report No. SM-13480, 1949.

6–17. POSSIO, C., L'Azione Aerodinamica sul Profilo Oscillante in un Fluido Compressibile a Velocità Iposonora, *L'Aerotecnica*, t. XVIII, fasc. 4, April, 1938. (Also available as British Ministry of Aircraft Production R.T.P. translation 987, and as a Chance Vought Aircraft Co. translation.)

6–18. DIETZE, F., *The Air Forces of the Harmonically Vibrating Wing at Subsonic Velocity (Plane Problem)*, Parts I and II, U. S. Air Force translations F-TS-506-RE and F-TS 948-RE, 1947. (Originally, *Luftfahrtforsch.*, Bd. 16, Lfg. 2, 1939, S. 84–96.)

6–19. FETTIS, H. E., *An Approximate Method for the Calculation of Nonstationary Air Forces at Subsonic Speeds*, U. S. Air Force, Wright Air Development Center Technical Report 52–56, 1952.

6–20. REISSNER, E., AND SHERMAN, S., *Compressibility Effects in Flutter*, Report No. S.B. 240-S-1, Airplane Division, Curtiss-Wright Corp., January, 1944.

6–21. HASKIND, M. D., *Oscillations of a Wing in a Subsonic Gas Flow*, Brown University translation No. A9-T-22, 1948. [Originally *Prikl. Mat. i Mekh.* (Moskow), Vol. XI, No. 1, pp. 129–146, 1947.]

6–22. REISSNER, E., *On the Application of Mathieu Functions in the Theory of Subsonic Compressible Flow Past Oscillating Airfoils*, N.A.C.A. T.N. 2363, 1951.

6–23. TIMMAN, R., AND VAN DE VOOREN, A. I., *The Oscillating Wing with Aerodynamically Balanced Control Surface in a Two-Dimensional Subsonic, Compressible Flow*, Nationaal Luchtvaartlaboratorium, Amsterdam, Report F.54, 1949.

6–24. TIMMAN, R., VAN DE VOOREN, A. I., AND GREIDANUS, J. H., Aerodynamic Coefficients of an Oscillating Airfoil in Two-dimensional Subsonic Flow, *Journal of the Aeronautical Sciences*, Vol. 18, No. 12, December, 1951.

6–25. VAN SPIEGEL, E., AND VAN DE VOOREN, A. I., *On the Theory of the Oscillating Wing in Two-dimensional Subsonic Flow*, Nationaal Luchtvaartlaboratorium, Amsterdam, Report F.142, 1953.

6–26. KARP, S. N., SHU, S. S., AND WEIL, H., *Aerodynamics of the Oscillating Airfoil in Compressible Flow*, U. S. Air Force, Air Materiel Command, Technical Report F-TR-1167-ND, 1947.

6–27. JAHNKE, E., AND EMDE, F., *Tables of Functions with Formulas and Curves*, Dover Publications, Inc., New York, 1945.

6–28. FETTIS, H. E., *Tables of Lift and Moment Coefficients for an Oscillating Wing-Aileron Combination in Two-dimensional Subsonic Flow*, Air Force Technical Report 6688 (with supplementary pages), 1951.

6–29. SMILG, B., AND WASSERMAN, L. S., *Application of Three-dimensional Flutter Theory to Aircraft Structures*, Air Force Technical Report 4798, 1942.

6–30. LUKE, Y. L., *Tables of Coefficients for Compressible Flutter Calculations*, Air Force Technical Report 6200, 1950.

6-31. TURNER, M. J., AND RABINOWITZ, S., *Aerodynamic Coefficients for an Oscillating Airfoil with Hinged Flap, with Tables for a Mach Number of 0.7*, N.A.C.A. T.N. 2213, 1950.

6-32. McLACHLAN, N. W., *Theory and Application of Mathieu Functions*, Oxford University Press, New York, 1947.

6-33. *Tables Relating to Mathieu Functions*, Computation Laboratory, U. S. National Bureau of Standards, Columbia University Press, New York, 1951.

6-34. MAZELSKY, B., *Numerical Determination of Indicial Lift of a Two-dimensional Sinking Airfoil at Subsonic Mach Numbers from Oscillatory Lift Coefficients with Calculations for Mach Number 0.7*, N.A.C.A. T.N. 2562, 1951.

6-35. MAZELSKY, B., *Determination of Indicial Lift and Moment of a Two-dimensional Pitching Airfoil at Subsonic Mach Numbers from Oscillatory Coefficients with Numerical Calculations for a Mach Number of 0.7*, N.A.C.A. T.N. 2613, 1952.

6-36. MAZELSKY, B., AND DRISCHLER, J. A., *Numerical Determination of Indicial Lift and Moment Functions for a Two-dimensional Sinking and Pitching Airfoil at Mach Numbers 0.5 and 0.6*, N.A.C.A. T.N. 2793, 1952.

6-37. LOMAX, H., HEASLET, M. A., FULLER, F. B., AND SLUDER, L., *Two- and Three-dimensional Unsteady Lift Problems in High-speed Flight*, N.A.C.A. Report 1077, 1952.

6-38. GOODMAN, T. R., The Lift Distribution on Conical and Nonconical Flow Regions of Thin Finite Wings in a Supersonic Stream, *Journal of the Aeronautical Sciences*, Vol. 16, No. 6, June, 1949.

6-39. GARRICK, I. E., *On Some Fourier Transforms in the Theory of Nonstationary Flows.* Proceedings of the Fifth International Congress for Applied Mechanics, pp. 590–593, New York, 1939.

6-40. VON BORBELY, S., *Aerodynamic Forces on a Harmonically Oscillating Wing at Supersonic Velocity*, British Ministry of Aircraft Production, R.T.P. translation 2019. (Translated from *Z. angew. Math. Mech.*, Bd. 22, Heft 4, August, 1942.)

6-41. GARRICK, I. E., AND RUBINOW, S. I., *Flutter and Oscillating Air-force Calculations for an Airfoil in Two-dimensional Supersonic Flow*, N.A.C.A. Report 846, 1946.

6-42. SCHWARZ, L., Untersuchung einiger mit den Zylinderfunktionen nullter Ordnung verwandter Funktionen, *Luftfahrtforsch.*, Bd. 20, Lfg. 12, February, 1944.

6-43. HUCKEL, V., AND DURLING, B. J., *Tables for Wing-Aileron Coefficients of Oscillating Air Forces for Two-dimensional Supersonic Flow*, N.A.C.A. T.N. 2055, 1950.

6-44. VAN DYKE, M. D., *Supersonic Flow Past Oscillating Airfoils Including Nonlinear Thickness Effects*, N.A.C.A. T.N. 2982, 1953.

6-45. JONES, W. P., *The Influence of Thickness Chord Ratio on Supersonic Derivatives for Oscillating Aerofoils*, Br. A.R.C., R. & M. 2679, 1947.

6-46. LIGHTHILL, M. J., Oscillating Airfoils at High Mach Number, *Journal of the Aeronautical Sciences*, Vol. 20, No. 6, June, 1953.

6-47. TEMPLE, G., AND JAHN, H. A., *Flutter at Supersonic Speeds: Derivative Coefficients for a Thin Aerofoil at Zero Incidence*, Br. A.R.C., R. & M. 2140, 1945.

6-48. STEWARTSON, K., On the Linearized Potential Theory ot Unsteady Supersonic Motion, *Quarterly Journal of Mechanics and Applied Mathematics*, Vol. III, Part 2, June, 1950.

6-49. GARDNER, M. F., AND BARNES, J. L., *Transients in Linear Systems*, John Wiley & Sons, Inc., New York, 1942.

6-50. BIOT, M. A., *Loads on a Supersonic Wing Striking a Sharp-edged Gust*, Cornell Aeronautical Laboratory Report SA-247-S-7, 1948.

6-51. CHANG, C. C., *Transient Aerodynamic Behavior of an Airfoil due to Different Arbitrary Modes of Nonstationary Motions in a Supersonic Flow*, N.A.C.A. T.N. 2333, 1951.

6-52. ROTT, N., Flügelschwingungsformen in ebener kompressibler Potentialströmung, *Z. angew. Math. Phys.*, Vol. 1, Fasc. 6, pp. 380–410, 1950.

6-53. NELSON, H. C., AND BERMAN, J. H., *Calculations on the Forces and Moments for an Oscillating Wing-Aileron Combination in Two-dimensional Potential Flow at Sonic Speed*, N.A.C.A. T.N. 2590, 1952.

CHAPTER 7

7-1. SCHADE, T., AND KRIENES, K., *The Oscillating Circular Airfoil on the Basis of Potential Theory*, N.A.C.A. T.M. 1098, 1947.

7-2. KÜSSNER, H. G., A General Method for Solving Problems of the Unsteady Lifting Surface Theory in the Subsonic Range, *Journal of the Aeronautical Sciences*, Vol. 21, No. 1, January, 1954.

7-3. CICALA, P., *Comparison of Theory with Experiment in the Phenomenon of Wing Flutter*, N.A.C.A. T.M. 887, 1939.

7-4. REISSNER, E., *Effect of Finite Span on the Airload Distributions for Oscillating Wings, I—Aerodynamic Theory of Oscillating Wings of Finite Span*, N.A.C.A. T.N. 1194, 1947.

7-5. REISSNER, E., AND STEVENS, J. E., *Effect of Finite Span on the Airload Distributions for Oscillating Wings, II—Methods of Calculation and Examples of Application*, N.A.C.A. T.N. 1195, 1947.

7-6. ASHLEY, H., ZARTARIAN, G., AND NEILSON, D. O., *Investigation of Certain Unsteady Aerodynamic Effects in Longitudinal Dynamic Stability*, Air Force Technical Report 5986, 1951.

7-7. BISPLINGHOFF, R. L., ISAKSON, G., PIAN, T. H. H., FLOMENHOFT, H. I., AND O'BRIEN, T. F., *An Investigation of Stresses in Aircraft Structures Under Dynamic Loading*, Report by Aeroelastic and Structures Research Laboratory, M.I.T. for U. S. Navy Bureau of Aeronautics, 1949.

7-8. JONES, R. T., *The Unsteady Lift of a Wing of Finite Aspect Ratio*, N.A.C.A. Report 681, 1940.

7-9. SHEN, S. F., *A New Lifting-Line Theory for the Unsteady Lift of a Swept or Unswept Wing in an Incompressible Fluid*, Part X of Air Force Technical Report 6358 (General Title: Effect of Structural Flexibility on Aircraft Loading), 1953.

7-10. TURNER, M. J., Aerodynamic Theory of Oscillating Swept-back Wings, *Journal of Mathematics and Physics,* Vol. XXVIII, No. 4, January, 1950.

7-11. HILDEBRAND, F. B., AND REISSNER, E., *Studies for an Aerodynamic Theory of Oscillating Swept-back Wings of Finite Span,* Part III, Chance Vought Aircraft Report 7039, 1948.

7-12. DENGLER, M. A., AND GOLAND, M., The Subsonic Calculation of Circulatory Spanwise Loading for Oscillating Airfoils by Lifting-line Techniques, *Journal of the Aeronautical Sciences,* Vol. 19, No. 11, November, 1952.

7-13. JONES, W. P., *The Calculation of Aerodynamic Derivative Coefficients for Wings of Any Planform in Nonuniform Motion,* Br. A.R.C., R. & M. 2470, 1952.

7-14. FETTIS, H. E., *Calculations of the Flutter Characteristics of Swept Wings at Subsonic Speeds,* Air Force Air Materiel Command Memorandum Report TSEAC 5-4595-2-9, 1946.

7-15. SPIELBERG, I. N., FETTIS, H. E., AND TONEY, H. S., *Methods for Calculating the Flutter and Vibration Characteristics of Swept Wings at Subsonic Speeds,* Air Force Air Materiel Command Memorandum Report No. MCREXA 5-4595-8-4, 1948.

7-16. SPIELBERG, I. N., *The Two-dimensional Incompressible Aerodynamic Coefficients for Oscillatory Changes in Airfoil Camber,* U. S. Air Force, Wright Air Development Center, Technical Note WCNS 52-7, 1952.

7-17. BROADBENT, E. G., *Some Considerations of the Flutter Problems of High-speed Aircraft,* Second International Aeronautical Conference. Published by Institute of the Aeronautical Sciences, Inc., New York, 1949.

7-18. BARMBY, J. G., CUNNINGHAM, H. J., AND GARRICK, I. E., *Study of Effects of Sweep on the Flutter of Cantilever Wings,* N.A.C.A. Report 1014, 1951.

7-19. LAWRENCE, H. R., AND GERBER, E. H., The Aerodynamic Forces on Low Aspect-ratio Wings Oscillating in an Incompressible Flow, *Journal of the Aeronautical Sciences,* Vol. 19, No. 11, November, 1952.

7-20. LAIDLAW, W. R., *Theoretical and Experimental Pressure Distributions on Low-Aspect Ratio Wings Oscillating in an Incompressible Flow,* Technical Report 51-2 of the M.I.T. Aeroelastic and Structures Research Laboratory for Naval Bureau of Aeronautics, September, 1954.

7-21. LOMAX, H., AND SLUDER, L., *Chordwise and Compressibility Corrections to Slender-wing Theory,* N.A.C.A. T.N. 2295, 1951.

7-22. VOSS, H. M., AND HASSIG, H. J., *Introductory Study of Flutter of Low Aspect-ratio Wings at Subsonic Speeds,* M.I.T. Aeroelastic and Structures Research Laboratory Report on Contract NOa(s) 51-109-c, 1952.

7-23. MERBT, H., AND LANDAHL, M., *Aerodynamic Forces on Oscillating Low Aspect-ratio Wings in Compressible Flow,* Swedish K.T.H. AERO T.N. 30, Royal Institute of Technology, Stockholm, 1953.

7-24. BRYSON, A. E., Stability Derivatives for a Slender Missile with Application to a Wing-Body-Vertical-Tail Configuration, *Journal of the Aeronautical Sciences,* Vol. 20, No. 5, May, 1953.

7-25. MILES, J. W., On Nonsteady Motion of Slender Bodies, *Aeronautical Quarterly,* Vol. 2, pp. 183-194, November, 1950.

7–26. LAIDLAW, W. R. AND HSU, P. T., A Semi-empirical Method for Determining Delta Wing Pressure Distributions in an Incompressible Flow, *Journal of the Aeronautical Sciences*, Readers' Forum, Vol. 21, No. 12, December, 1954.

7–27. MILES, J. W., On the Compressibility Correction for Subsonic Unsteady Flow, *Journal of the Aeronautical Sciences*, Readers' Forum, Vol. 17, No. 3, March, 1950.

7–28. REISSNER, E., *On the Theory of Oscillating Airfoils of Finite Span in Subsonic Compressible Flow*, N.A.C.A. T.N. 1953, 1949.

7–29. WATKINS, C. E., RUNYAN, H. L., AND WOOLSTON, D. S., *On the Kernel Function of the Integral Equation Relating the Lift and Downwash Distributions of Oscillating Finite Wings in Subsonic Flow*, N.A.C.A. T.N. 3131, January, 1954.

7–30. GARRICK, I. E., AND WATKINS, C. E., *Some Recent Developments in the Aerodynamic Theory of Oscillating Wings*. Preprint of Proceedings of N.A.C.A.–University Conference on Aerodynamics, Construction and Propulsion, Lewis Flight Propulsion Laboratory, Cleveland, Ohio, October, 1954.

7–31. VOSS, H. M., *Low Aspect-ratio Flutter*, Sc.D. Thesis, M.I.T., June, 1954.

7–32. VOSS, H. M., ZARTARIAN, G., AND HSU, P. T., *Application of Numerical Integration Techniques to the Low Aspect-ratio Flutter Problem in Subsonic and Supersonic Flows*, M.I.T., Aeroelastic and Structures Research Laboratory Report on Contract NOa(s) 53-564-c, October, 1954.

7–33. WALSH, J., ZARTARIAN, G., AND VOSS, H. M., *Generalized Aerodynamic Forces on the Delta Wing with Supersonic Leading Edges*, Institute of the Aeronautical Sciences Preprint No. 433, January, 1954.

7–34. GARRICK, I. E., AND RUBINOW, S. I., *Theoretical Study of Air Forces on an Oscillating or Steady Thin Wing in a Supersonic Main Stream*, N.A.C.A. T.N. 1383, 1947.

7–35. NELSON, H. C., *Lift and Moment on Oscillating Triangular and Related Wings with Supersonic Edges*, N.A.C.A. T.N. 2494, 1951.

7–36. MILES, J. W., *On Harmonic Motions of Wide Delta Airfoils at Supersonic Speeds*, U. S. Naval Ordnance Laboratory Report NAVORD 1234, 1950.

7–37. FROEHLICH, J. E., Nonstationary Motion of Purely Supersonic Wings, *Journal of the Aeronautical Sciences*, Vol. 18, No. 5, May, 1951.

7–38. STEWART, H. J., AND LI, T. Y., Source-Superposition Method of Solution of a Periodically Oscillating Wing at Supersonic Speed, *Quarterly of Applied Mathematics*, Vol. IX, No. 1, April, 1951.

7–39. STEWART, H. J., and LI, T. Y., Periodic Motions of a Rectangular Wing Moving at Supersonic Speed, *Journal of the Aeronautical Sciences*, Vol. 17, No. 9, September, 1950.

7–40. LI, T. Y., Purely Rolling Oscillations of a Rectangular Wing in Supersonic Flow, *Journal of the Aeronautical Sciences*, Vol. 18, No. 3, March, 1951.

7–41. CHANG, C. C., *The Aerodynamic Behavior of a Harmonically Oscillating Finite Swept-back Wing in Supersonic Flow*, N.A.C.A. T.N. 2467, 1951.

7–42. WATKINS, C. E., *Effect of Aspect Ratio on the Air Forces and Moments of Harmonically Oscillating Thin Rectangular Wings in Supersonic Potential Flow*, N.A.C.A. Report 1028, 1951.

7–43. NELSON H. C., RAINEY, R. A., AND WATKINS, C. E., *Lift and Moment Coefficients Expanded to the Seventh Power of Frequency for Oscillating Rectangular Wings in Supersonic Flow and Applied to a Specific Flutter Problem*, N.A.C.A. T.N. 3076, 1954.

7–44. WATKINS, C. E., AND BERMAN, J. H., *Air Forces and Moments on Triangular and Related Wings with Subsonic Leading Edges Oscillating in Supersonic Potential Flow*, N.A.C.A. Report 1099, 1952.

7–45. WATKINS, C. E., AND BERMAN, J. H., *Velocity Potential and Air Forces Associated with a Triangular Wing in Supersonic Flow, with Subsonic Leading Edges, and Deforming Harmonically According to a General Quadratic Equation*, N.A.C.A. T.N. 3009, 1953.

7–46. HASKIND, M. D., AND FALKOVICH, S. V., *Vibration of a Wing of Finite Span in Supersonic Flow*, N.A.C.A. T.M. 1257, 1950.

7–47. KRASILSCHIKOVA, E. A., *Disturbed Motion of Air Caused by Vibrations of a Wing Moving at Supersonic Speed*, A.M.C. Technical Report 102-AC49/4-34 (GDAM A9-T-24).

7–48. FRANKL, F. L., AND KARPOVICH, E. A., *Gas Dynamics of Slender Bodies*, Moscow, 1948 (translated from Russian by Interscience Publishers).

7–49. ROTT, N., On the Unsteady Motion of a Rectangular Wing in Supersonic, Flow, *Journal of the Aeronautical Sciences*, Readers' Forum, Vol. 18, No. 11, November, 1951.

7–50. LOMAX H., HEASLET, M. A., FULLER, F. B., AND SLUDER, L., *Two- and Three-dimensional Unsteady Lift Problems in High-speed Flight*, N.A.C.A. Report 1077, 1952.

7–51. LIBAN, E., NEURINGER, J., AND RABINOWITZ, S., *Flutter Analysis of an Elastic Wing with Supersonic Edges*, Republic Aviation Corp. Report E-SAF-1, April, 1953.

7–52. PINES, S., AND DUGUNDJI, J., *Application of Aerodynamic Flutter Derivatives to Flexible Wings with Supersonic and Subsonic Edges*, Republic Aviation Corp. Report E-SAF-2, April, 1954.

7–53. MILES, J. W., *The Oscillating Rectangular Airfoil at Supersonic Speeds*, U. S. Naval Ordnance Laboratory Report NAVORD 1170, 1949.

7–54. MILES, J. W., A General Solution for the Rectangular Airfoil in Supersonic Flow, *Quarterly of Applied Mathematics*, Vol. XI, No. 1, April, 1953.

7–55. STEWARTSON, K., On the Linearized Potential Theory of Unsteady Supersonic Motion, *Quarterly Journal of Mechanics and Applied Mathematics*, Vol. III, pp. 182–199, 1950.

7–56. TEMPLE, G., "Unsteady Motion," Volume I, Chapter IX, pp. 325–374, of *Modern Developments in Fluid Dynamics — High-speed Flow*, Clarendon Press, Oxford, 1953.

7–57. HEASLET, M. A., AND SPREITER, J. R., *Reciprocity Relations in Aerodynamics*, N.A.C.A. T.N. 2700, May, 1952.

7–58. FLAX, A. H., Reverse Flow and Variational Theorems for Lifting Surfaces in Nonstationary Compressible Flow, *Journal of the Aeronautical Sciences*, Vol. 20, No. 2, February, 1953.

7–59. LIN, C. C., REISSNER, E., AND TSIEN, H. S., On Two-dimensional Nonsteady Motion of a Slender Body in a Compressible Fluid, *Journal of Mathematics and Physics*, Vol. 27, No. 3, October, 1948.

7-60. MILES, J. W., *Monograph on Unsteady Flow*, to be published.

7-61. LAMB, SIR HORACE, *Hydrodynamics*, 6th ed., Dover Publications, New York, 1945.

7-62. MILES, J. W., *Unsteady Supersonic Flow Past Slender Pointed Bodies* (unclassified paper). Proceedings of the Naval Bureau of Ordnance Symposium on Aeroballistics, Report NAVORD 1651, pp. 429–480, 1950.

7-63. ZAHM, A. F., *Flow and Force Equations for a Body Revolving in a Fluid*, N.A.C.A. Report 323, 1928.

CHAPTER 8

8-1. THEODORSEN, T., *General Theory of Aerodynamic Instability and the Mechanism of Flutter*, N.A.C.A. Report 496, 1935.

8-2. NILES, A. S., AND NEWELL, J. S., *Airplane Structure*, Vol. I, 3rd ed., John Wiley & Sons, Inc., 1935.

8-3. REISSNER, H., Neuere Probleme aus der Flugzeugstatik, *Zeitschrift für Flugtechnik und Motorluftschiffahrt*, Vol. 17, No. 7, pp. 154–169, 1926.

8-4. COX, H. R., AND PUGSLEY, A. G., *Theory of Loss of Lateral Control Due to Wing Twisting*, Br. A.R.C., R. & M., 1506, October, 1932.

8-5. COX, H. R., AND PUGSLEY, A. G., Stability of Static Equilibrium of Elastic and Aerodynamic Actions on a Wing, Br. A.R.C., R. & M. 1059, April, 1934.

8-6. PUGSLEY, A. G., AND COX, H. R., *The Aileron Power of a Monoplane*, Br. A.R.C., R. & M. 1640, April, 1934.

8-7. DÄTWYLER, G., Der Einfluss der Elastischen Deformationen auf die Luftkräfte am Tragflügel, *Schweizer Aero-Revue*, Vol. 6, No. 20, p. 264, October, 1931.

8-8. DÄTWYLER, G., *Calculations of the Effect of Wing Twist on the Air Forces Acting on a Monoplane Wing*, N.A.C.A. T.N. 520, 1935.

8-9. FLAX, A. H., The Influence of Structural Deformation on Airplane Characteristics, *Journal of the Aeronautical Sciences*, Vol. 12, No. 1, January, 1945.

8-10. HILDEBRAND, F. B., AND REISSNER, E., *The Influence of the Aerodynamic Span Effect on the Magnitude of the Torsional Divergence Velocity and on the Shape of the Corresponding Deflection Mode*, N.A.C.A. T.N. 926, February, 1944.

8-11. LAWRENCE, H. R., AND SEARS, W. R., *Three-dimensional Wing Theory for the Elastic Wing*, Northrop Aircraft, Inc., Report No. A-59, June, 1944.

8-12. PINES, S., A Unit Solution for the Load Distribution of a Nonrigid Wing by Matrix Methods, *Journal of the Aeronautical Sciences*, Vol. 16, No. 8, August, 1949.

8-13. VON KÁRMÁN, T., AND BIOT, M. A., *Mathematical Methods in Engineering*, McGraw-Hill Book Company, Inc., 1940.

8-14. PEARSON, H. A., AND AIKEN, W. S., JR., *Charts for the Determination of Wing Torsional Stiffness Required for Specified Rolling Characteristics of Aileron Reversal Speed*, N.A.C.A. Report 799, 1944.

8-15. GROTH, E., *Evaluation of Methods for Calculating the Rolling Effectiveness and Aileron Reversal Speed of a Straight Wing*, Air Materiel Command

Engr. Division Memorandum Report, Ser. No. MCREXA5-4595-8-6, December, 1948.

8-16. COLLAR, A. R., Aeroelastic Problems at High Speed, *Journal of the Royal Aeronautical Society*, Vol. 51, No. 1, Jan., 1947.

8-17. DIEDERICH, F. W. AND BUDIANSKY, B., *Divergence of Swept Wings*, N.A.C.A. T. N. 1680, August, 1948.

8-18. PAI, S. I., AND SEARS, W. R., Some Aeroelastic Properties of Swept Wings, *Journal of the Aeronautical Sciences*, Vol. 16, No. 2, February, 1949.

8-19. MILES, J. W., A Formulation of the Aeroelastic Problem for a Swept Wing, *Journal of the Aeronautical Sciences*, Vol. 16, No. 8, August, 1949.

8-20. PIAN, T. H. H., AND LIN, H., *Effect of Structural Flexibility on Aircraft Loading, Part II—Spanwise Airload Distribution*, Air Force Technical Report No. 6358, Part II, May, 1951.

8-21. HILDEBRAND, F. B., *Advanced Calculus for Engineers*, Prentice-Hall, Inc., New York, 1949.

8-22. LOVITT, W. V., *Linear Integral Equations*, Dover Publications, New York, 1950.

8-23. DIEDERICH, F. W., AND FOSS, K. A., *Charts and Approximate Formulas for the Estimation of Aeroelastic Effects on the Loading of Swept and Unswept Wings*, N.A.C.A. T.N. 2608, February, 1952.

8-24. DE YOUNG, JOHN, *Theoretical Antisymmetric Span Loading for Wings of Arbitrary Planform at Subsonic Speeds*, N.A.C.A. T.N. 2140, July, 1950.

8-25. COURANT, R. AND HILBERT, D., *Methods of Mathematical Physics*, Interscience Publishers, Inc., New York, 1953.

8-26. FLAX, A. H., *Aeroelastic Problems at Supersonic Speeds*. Proceedings of 2nd International Aeronautical Conference, New York, 1949.

8-27. SEIFERT, GEORGE, A Third Order Boundary Value Problem Arising in Aeroelastic Wing Theory, *Quarterly of Applied Mathematics*, Vol. IX, No. 2, 1951.

8-28. BROWN, R. B., HOLTBY, K. F., AND MARTIN, H. C., A Superposition Method for Calculating the Aeroelastic Behavior of Swept Wings, *Journal of the Aeronautical Sciences*, Vol. 18, No. 8, August, 1951.

8-29. BIOT, M. A., *Aeroelastic Stability of Supersonic Wings, Report No. I, Chordwise Divergence — The Two-dimensional Case*, Cornell Aeronautical Laboratory Report, CAL/CM-427, CAL-1-E-1, December 8, 1947.

8-30. TIMOSHENKO, S., AND GOODIER, J. N., *Theory of Elasticity*, 2nd ed., McGraw-Hill Book Company, Inc., 1951.

8-31. BIOT, M. A., *Aeroelastic Stability of Supersonic Wings, Report No. 2, An Approximate Treatment of Some Simple Three-dimensional Cases*, Cornell Aeronautical Laboratory Report CAL/CM-470, CAL-1-E-1, May 12, 1948.

8-32. FUNG, Y. C., On the Behavior of a Sharp Leading Edge, *Journal of the Aeronautical Sciences*, Readers Forum, Vol. 20, Number 9, September, 1953.

8-33. BIOT, M. A., *Aeroelastic Stability of Supersonic Wings, Report No. 3, General Method for the Two-dimensional Case and its Application to the Chordwise Divergence of a Biconvex Section*, Cornell Aeronautical Laboratory Report, CAL/CM-506, CAL-1-E-1, September 23, 1948.

8–34. LIN, H., AND PIAN, T. H. H., *Effect of Structural Flexibility on Aircraft Loading*, Air Force Technical Report 6358, Part VIII, September, 1953.

CHAPTER 9

9–1. SMILG, B., AND WASSERMAN, L. S., *Application of Three-dimensional Flutter Theory to Aircraft Structures*, Air Force Technical Report, 4798, 1942.

9–2. SMILG, B., The Instability of Pitching Oscillations of an Airfoil in Subsonic Incompressible Potential Flow, *Journal of the Aeronautical Sciences*, Vol. 16, No. 11, November, 1949.

9–3. THEODORSEN, T., AND GARRICK, I. E., *Mechanism of Flutter, a Theoretical and Experimental Investigation of the Flutter Problem*, N.A.C.A. Report 685, 1940.

9–4. THEODORSEN, T., *General Theory of Aerodynamic Instability and the Mechanism of Flutter*, N.A.C.A. Report 496, 1935.

9–5. MYKLESTAD, N., *Vibration Analysis*, McGraw-Hill Book Company, Inc., New York, 1944.

9–6. GOLAND, M., The Flutter of a Uniform Cantilever Wing, *Journal of Applied Mechanics*, Vol. 12, No. 4, December, 1945.

9–7. WOOLSTON, D. S., AND CASTILE, G. E., *Some Effects of Variations in Several Parameters Including Fluid Density on the Flutter Speed of Light, Uniform Cantilever Wings*, N.A.C.A. T.N. 2558, 1951.

9–8. GOLAND, M., AND LUKE, Y. L., The Flutter of a Uniform Wing with Tip Weights, *Journal of Applied Mechanics*, Vol. 15, No. 1, March, 1948.

9–9. RUNYAN, H. L., AND WATKINS, C. E., *Flutter of a Uniform Wing with an Arbitrarily Placed Mass According to a Differential-equation Analysis and a Comparison with Experiment*, N.A.C.A. T.N. 1848, 1949.

9–10. GOLAND, M., AND LUKE, Y. L., A Study of the Bending-Torsion Aeroelastic Modes for Airplane Wings, *Journal of the Aeronautical Sciences*, Vol. 16, No. 7, July, 1949.

9–11. SCANLAN, R. H., AND ROSENBAUM, R., *Introduction to the Study of Aircraft Vibration and Flutter*, The Macmillan Co., New York, 1951.

9–12. BROADBENT, E. G., The Elementary Theory of Aeroelasticity, Part III — Flutter of Control Surfaces and Tabs, *Aircraft Engineering*, Vol. XXVI, No. 303, May, 1954.

9–13. WOOLSTON, D. S., AND HUCKEL, V., *A Calculation Study of Wing-Aileron Flutter in Two Degrees of Freedom for Two-dimensional Supersonic Flow*, N.A.C.A. T.N. 3160, 1954.

9–14. GOLAND, M., AND DENGLER, M. A., *Comparison Between Calculated and Observed Flutter Speeds*, Air Force Technical Report 6184, 1950.

9–15. BROADBENT, E. G., *Some Considerations of the Flutter Problems of High-speed Aircraft*, Second International Aeronautical Conference, published by Institute of the Aeronautical Sciences, Inc., New York, 1949.

9–16. WOOLSTON, D. S., AND RUNYAN, H. L., *On the Use of Coupled Modal Functions in Flutter Analysis*, N.A.C.A. T.N. 2375, 1951.

9–17. WILLIAMS, J., *Aircraft Flutter*, Br. A.R.C., R. & M. 2492, 1951.

9–18. REISSNER, E., AND STEVENS, J. E., *Effect of Finite Span on the Airload Distributions for Oscillating Wings, II—Methods of Calculation and Examples of Application*, N.A.C.A. T.N. 1195, 1947.

9–19. VAN DE VOOREN, A. I., *Theory and Practice of Flutter Calculations for Systems with Many Degrees of Freedom*, Doctoral Thesis, Technical Institute of Delft, published by Eduard Ijdo, Leiden, 1952.

9–20. LUKE, Y. L., *Tables of Coefficients for Compressible Flutter Calculations*, Air Force Technical Report 6200, 1950.

9–21. GARRICK, I. E., AND RUBINOW, S. I., *Flutter and Oscillating Air-force Calculations for an Airfoil in Two-dimensional Supersonic Flow*, N.A.C.A. Report 846, 1946.

9–22. NELSON, H. C., AND BERMAN, J. H., *Calculations on the Forces and Moments for an Oscillating Wing-Aileron Combination in Two-dimensional Potential Flow at Sonic Speed*, N.A.C.A. T.N. 2590, 1952.

9–23. FLAX, A. H., *Aeroelastic Problems at Supersonic Speed*, Second International Aeronautical Conference, published by Institute of the Aeronautical Sciences, Inc., New York, 1949.

9–24. DIEDERICH, F. W., *A Planform Parameter for Correlating Certain Aerodynamic Characteristics of Swept Wings*, N.A.C.A. T.N. 2335, 1951.

9–25. BARMBY, J. G., CUNNINGHAM, H. J., AND GARRICK, I. E., *Study of Effects of Sweep on the Flutter of Cantilever Wings*, N.A.C.A. Report 1014, 1951.

9–26. BECKLEY, L. E., *Report on the Design of Sectional Type Flutter and Dynamic Models of Aircraft Structures*, M.I.T. Aeroelastic and Structures Research Laboratory Report on Contract NOa(s) 7493, 1948.

9–27. ASHLEY, H., ZARTARIAN, G., AND NEILSON, D. O., *Investigation of Certain Unsteady Aerodynamic Effects in Longitudinal Dynamic Stability*, Air Force Technical Report 5986, 1951.

9–28. CUNNINGHAM, H. J., *Analysis of Pure-bending Flutter of a Cantilever Swept Wing and Its Relation to Bending-torsion Flutter*, N.A.C.A. T.N. 2461, 1951.

9–29. VOSS, H. M., HASSIG, H. J., AND ASHLEY, H., *Introductory Study of Flutter of Low Aspect-ratio Wings at Subsonic Speeds*, M.I.T. Aeroelastic and Structures Research Laboratory Report on Contract NOa(s) 51-109-c, 1952.

9–30. KOCHIN, N. E., *Steady Vibrations of Wing of Circular Planform* and *Theory of Wing of Circular Planform*, N.A.C.A. T.M. 1324, 1953. (Translated from Russian.)

9–31. SCHADE, T., AND KRIENES, K., *The Oscillating Circular Airfoil on the Basis of Potential Theory*, N.A.C.A. T.M. 1098, 1947.

9–32. VOSS, H. M., ZARTARIAN, G., AND HSU, P. T., *Application of Numerical Integration Techniques to the Low Aspect-ratio Flutter Problem in Subsonic and Supersonic Flows*, M.I.T., Aeroelastic and Structures Research Laboratory, Report on Contract NOa(s) 53-564c, October, 1954.

9–33. RUNYAN, H. L., CUNNINGHAM, H. J., AND WATKINS, C. E., *Theoretical Investigation of Several Types of Single-degree-of-freedom Flutter*, *Journal of the Aeronautical Sciences*, Vol. 19, No. 2, February, 1952.

9-34. RUNYAN, H. L., *Single-degree-of-freedom Flutter Calculations for a Wing in Subsonic Potential Flow and Comparison with an Experiment*, N.A.C.A. T.N. 2396, 1951.

9-35. WATKINS, C. E., *Effect of Aspect Ratio on the Air Forces and Moments of Harmonically Oscillating Thin Rectangular Wings in Supersonic Potential Flow*, N.A.C.A. T.N. 2064, 1950.

9-36. RUNYAN, H. L., *Effect of Various Parameters Including Mach Number on the Single-degree-of-freedom Flutter of a Control Surface in Potential Flow*, N.A.C.A. T.N. 2551, 1951.

9-37. BRODING, W. C., *Criteria for Single-degree-of-freedom Flutter in Bending of a Cantilever Swept Wing*, Chance Vought Aircraft Report SISM No. 1211, March, 1953.

9-38. KRZYWOBLOCKI, M. Z., Investigation of the Wing-wake Frequency with Application of the Strouhal Number, *Journal of the Aeronautical Sciences*, Vol. 12, No. 1, January, 1945.

9-39. RAUSCHER, M., Model Experiments on Flutter at the Massachusetts Institute of Technology, *Journal of the Aeronautical Sciences*, Vol. 3, No. 2, March, 1936.

9-40. STUDER, H. L., Experimentelle Untersuchungen über Flügelschwingungen, *Mitt. Inst. Aerodyn. Zürich*, No. 4/5, 1936.

9-41. VICTORY, M., *Flutter at High Incidence*, Br. A.R.C., R. & M. 2048, 1943.

9-42. HALFMAN, R. L., JOHNSON, H. C., AND HALEY, S. M., *Evaluation of High-angle-of-attack Aerodynamic-derivative Data and Stall-flutter Prediction Techniques*, N.A.C.A. T.N. 2533, 1951.

9-43. SCHNITTGER, J. R., *The Aerodynamic Mechanism of Vibrating Compressor Blades*, Sc.D. Thesis, M.I.T., 1953 (to be published in Journal of the Aeronautical Sciences).

9-44. MILES, J. W., *Dynamic Chordwise Stability at Supersonic Speeds*, North American Aviation Report AL-1140, 1950.

9-45. SHEN, S. F., *Flutter of a Two-dimensional Simply Supported Uniform Panel in a Supersonic Stream*, M.I.T., Aeroelastic and Structures Research Laboratory Report on Contract N5-ori-07833, 1952.

9-46. HEDGEPETH, J. M., AND BUDIANSKY, B., *Analysis of Flutter in Compressible Flow of a Panel on Many Supports*, Institute of the Aeronautical Sciences Preprint No. 443, January, 1954 (to appear in Journal of the Aeronautical Sciences).

9-47. FUNG, Y. C., The Static Stability of a Two-dimensional Curved Panel in a Supersonic Flow, with an Application to Panel Flutter, *Journal of the Aeronautical Sciences*, Vol. 21, No. 8, August, 1954.

9-48. FARQUHARSON, F. B., et al., *Aerodynamic Stability of Suspension Bridges with Special Reference to the Tacoma Narrows Bridge*, University of Washington Engineering Experiment Station Bulletin No. 116, Parts I through V (various dates up to June, 1954).

9-49. FRAZER, R. A., AND SCRUTON, C., *A Summarized Account of the Severn Bridge Aerodynamic Investigation*, British National Physical Laboratory Report N.P.L./Aero./222, 1952.

9-50. STEINMAN, D. B., *Suspension Bridges: the Aerodynamic Problem and Its Solution*, American Scientist, Vol. 42, No. 3, July, 1954.

9-51. SISTO, F., Stall Flutter in Cascades, *Journal of the Aeronautical Sciences*, Vol. 20, No. 9, September, 1953.

9-52. THOMA, D., *Why Does the Flag Flutter?*, Cornell Aeronautical Laboratory translation, 1949. (Translated from *Mitt. hydraul. Inst. Münch.* No. 9, pp. 30-34, 1939.)

CHAPTER 10

10-1. BUSH, V., *Operational Circuit Analysis*, John Wiley & Sons, Inc., New York, 1936.

10-2. VON KÁRMÁN, T., AND BIOT, M. A., *Mathematical Methods in Engineering*, McGraw-Hill Book Company, Inc., New York, 1940.

10-3. JONES, R. T., Calculation of the Motion of an Airplane Under the Influence of Irregular Disturbances, *Journal of the Aeronautical Sciences*, Vol. 3, No. 12, October, 1936.

10-4. WILLIAMS, D., *Dynamic Loads in Aeroplanes Under Given Impulsive Loads with Particular Reference to Landing and Gust Loads on a Large Flying Boat*, Great Britain Royal Aircraft Establishment Reports SME 3309 and 3316, 1945.

10-5. BISPLINGHOFF, R. L., ISAKSON, G., AND O'BRIEN, T. F., *Gust Loads on Rigid and Elastic Airplanes*, M.I.T., Aeroelastic and Structures Research Laboratory Report for the Navy Bureau of Aeronautics, August 15, 1950.

10-6. STOWELL, E. Z., SCHWARTZ, E. B., AND HOUBOLT, J. C., *Bending and Shear Stresses Developed by the Instantaneous Arrest of a Moving Cantilever Beam*, N.A.C.A. A.R.R. No. L4127, 1944.

10-7. BIOT, M. A. AND BISPLINGHOFF, R. L., *Dynamic Load on Airplane Structures During Landing*, N.A.C.A. A.R.R. No. 4H10, 1944.

10-8. RAMBERG, W., AND MCPHERSON, A. E., *Experimental Verification of Theory of Landing Impact.* Paper presented at the Sixth International Congress of Applied Mechanics, Paris, 1946.

10-9. WASSERMAN, L. S., The Prediction of Dynamic Landing Loads, *Proceedings of the Seventh International Congress of Applied Mechanics*, London, 1948.

10-10. KROLL, WILHEMINA D., AND LEVY, S., *A Step-by-step Method of Determining the Dynamic Response of Aircraft in Landing*, National Bureau of Standards Report No. 6.4/1-181 PR6.

10-11. PIAN, T. H. H., AND FLOMENHOFT, H. I., Analytical and Experimental Studies on Dynamic Loads in Airplane Structures During Landing, *Journal of the Aeronautical Sciences*, Vol. 17, No. 12, December, 1950.

10-12. O'BRIEN, T. F., AND PIAN, T. H. H., *Effect of Structural Flexibility on Aircraft Loading-ground Loads*, Air Force Technical Report 6358, Part 1, July, 1951.

10-13. HOUBOLT, J. C., A Recurrence Matrix Solution for the Dynamic Response of Elastic Aircraft, *Journal of the Aeronautical Sciences*, Vol. 17, No. 9, September, 1950.

10-14. MCPHERSON, A. E., EVANS, J. JR., AND LEVY, S., *Influence of Wing Flexibility on Force-Time Relation in Shock Strut Following Vertical Landing Impact*, N.A.C.A. T.N. 1995, November, 1949.

10–15. Bisplinghoff, R. L., Isakson, G., Pian, T. H. H., Flomenhoft, H. I., and O'Brien, T. F., *An Investigation of Stresses in Aircraft Structures Under Dynamic Loading*, M.I.T., Aeroelastic and Structures Research Laboratory Report for the Navy Bureau of Aeronautics, January 21, 1949.

10–16. Hildebrand, F. G., *Advanced Calculus for Engineers*, Prentice-Hall, Inc., New York, 1949.

10–17. Duncan, W. J., *Mechanical Admittances and their Application to Oscillation Problems*, Br. A.R.C., R. & M. 2000, 1946.

10–18. Seamans, R. C., Bromberg, B. G., and Payne, L. E., Application of the Performance Operator to Aircraft Automatic Control, *Journal of the Aeronautical Sciences*, Vol. 15, No. 9, September, 1948.

10–19. Houbolt, J. C., *A Recurrence Matrix Solution for the Dynamic Response of Aircraft in Gusts*, N.A.C.A. T.N. 2060, March, 1950.

10–20. Donely, Phillip, *Summary of Information Relating to Gust Loads on Airplanes*, N.A.C.A. Report 997, 1950.

10–21. Küssner, H. G., *Stresses Produced in Airplane Wings by Gusts*, N.A.C.A. T.M. 654, 1932.

10–22. Bryant, L. W., and Jones, I. M. W., *Stressing of Aeroplane Wings Due to Symmetrical Gusts*, Br. A.R.C., R. & M. 1690, February, 1936.

10–23. Williams, D., and Hanson, J., *Gust Loads on Wings and Tails*, Br. A.R.C., R. & M. 1823, 1937.

10–24. Sears, W. R., and Sparks, B. O., On the Reaction of an Elastic Wing to Vertical Gusts, *Journal of the Aeronautical Sciences*, Vol. 9, No. 2, December, 1941.

10–25. Pierce, H. B., *Investigation of the Dynamic Response of Airplane Wings to Gusts*, N.A.C.A. T.N. 1320, June, 1947.

10–26. Putnam, A. A., *An Improved Method for Calculating the Dynamic Response of Flexible Airplanes to Gusts*, N.A.C.A. T.N. 1321, May, 1947.

10–27. Goland, M., Luke, Y. L., and Kahn, E. A., *Prediction of Wing Loads Due to Gusts Including Aeroelastic Effects*, Midwest Research Institute Report 1-S36-E-9, 1947.

10–28. Jenkins, E. S., and Pancu, C. D. P., Dynamic Loads on Airplane Structures, *S.A.E. Quarterly Transactions*, Vol. 3, No. 3, July, 1949.

10–29. Houbolt, J. C., and Kordes, E. E., *Gust Response Analysis of an Airplane Including Wing Bending Flexibility*, N.A.C.A. T.N. 2736, September, 1952.

10–30. Kordes, E. E., and Houbolt, J. C., *Evaluation of Gust Response Characteristics of Some Existing Aircraft with Wing Bending Flexibility Included*, N.A.C.A. T.N. 2897, February, 1953.

10–31. Codik, A., Lin, H., and Pian, T. H. H., *The Gust Response of a Sweptback Tapered Wing Including Bending Flexibility*, Air Force Technical Report 6358, Part XII, October, 1953.

10–32. Bisplinghoff, R. L., Isakson, G., and O'Brien, T. F., Gust Loads on Rigid Airplanes with Pitching Neglected, *Journal of the Aeronautical Sciences*, Vol. 18, No. 1, January, 1951.

10–33. Wiener, N., *Extrapolation, Interpolation, and Smoothing of Stationary Time Series*, John Wiley & Sons, Inc., New York, 1950.

10-34. CLEMENTSON, GERHARDT C., *An Investigation of the Power Spectral Density of Atmospheric Turbulence*, Sc.D. Thesis, M.I.T., 1950.

10-35. LIEPMANN, H. W., On the Application of Statistical Concepts to the Buffeting Problem, *Journal of the Aeronautical Sciences*, Vol. 19, No. 12, December, 1952.

10-36. PRESS, H., AND MAZELSKY, B., *A Study of the Application of Power-spectral Methods of Generalized Harmonic Analysis to Gust Loads on Airplanes*, N.A.C.A. T.N. 2853, 1953.

10-37. FUNG, Y. C., Statistical Aspects of Dynamic Loads, *Journal of the Aeronautical Sciences*, Vol. 20, No. 5, May, 1953.

CHAPTER 11

11-1. BRIDGEMAN, P. W., *Dimensional Analysis*, Yale University Press (revised edition), 1931.

11-2. BUCKINGHAM, E., On Physically Similar Systems; Illustrations of the Use of Dimensional Equations, *Phys. Rev.*, Vol. IV, No. 4, 1914.

CHAPTER 12

12-1. TIMOSHENKO, S. AND GOODIER, J. N., *Theory of Elasticity* (Chapter 11—Torsion), McGraw-Hill Book Company, Inc., 1951.

12-2. RAUSCHER, M., *Report on the Suitability of Various Materials and Methods of Construction for Wind Tunnel Models Representing Flutter Characteristics of Actual Airplanes*, M.I.T., Aeroelastic and Structures Research Laboratory Report, June, 1942.

CHAPTER 13

13-1. MCCARTHY, J. F., JR., HALFMAN, R. L., PRIGGE, J. S., JR., AND LEVEY, G. M., *Supersonic Flutter Model Tests, Part 1 — Model Design and Testing Techniques*, Wright Air Development Center TR 54-113, 1955.

13-2. KENNEDY, C. C., AND PANCU, C. D. P., Use of Vectors in Vibration Measurement and Analysis, *Journal of the Aeronautical Sciences*, Vol. 14, No. 11, November, 1947.

13-3. LEWIS, R. C., AND WRISLEY, D. L., A System for the Excitation of Pure Natural Modes of Complex Structures, *Journal of the Aeronautical Sciences*, Vol. 17, No. 11, November, 1950.

13-4. STRUTT, J. W. (LORD RAYLEIGH), *The Theory of Sound*, Vol. 1, 2nd ed., Dover Publications, Inc., New York, 1945.

13-5. PENGELLEY, C. D., AND BENUM, D., Aeroelastic Studies on High-performance Swept-wing Airplane. *Proceedings of the First U. S. National Congress of Applied Mechanics*, June, 1951.

13-6. JOHNSON, H. C., AND FOTIEO, G., *Rolling Effectiveness and Aileron Reversal Characteristics of Straight and Swept-back Wings*, Air Force Technical Report 6198, February, 1951.

13–7. SCHWARTZ, M. D., *Investigation of Flight Flutter Testing Techniques*, M.I.T., Aeroelastic and Structures Research Laboratory Report on Contract NOa(s) 10921, September, 1951.

13–8. SUMMERS, R. A., *A Statistical Description of Large-scale Atmospheric Turbulence*, Sc.D. Thesis, M.I.T., May, 1954.

13–9. PEPPING, R. A., *A Theoretical Investigation of the Oscillating Control Surface Frequency Response Technique of Flight Flutter Testing*, Aircraft Industries Association ATC Report ARTC-6, January, 1953.

13–10. DRAPER, C. S., McKAY, W., AND LEES, S., *Instrument Engineering*, Vol. 2, McGraw-Hill Book Company, Inc., New York, 1953.

13–11. WHITE, R. J., Investigation of Lateral Dynamic Stability in the XB-47 Airplane, *Journal of the Aeronautical Sciences*, Vol. 17, No. 3, March, 1950.

13–12. HALFMAN, R. L., *Experimental Aerodynamic Derivatives of a Sinusoidally Oscillating Airfoil in Two-dimensional Flow*, N.A.C.A. Report 1108, 1952.

13–13. ASHLEY, H., ZARTARIAN, G., AND NEILSON, D. O., *Investigation of Certain Unsteady Aerodynamic Effects in Longitudinal Dynamic Stability*, Air Force Technical Report 5986, December, 1951.

13–14. BRATT, J. B., AND SCRUTON, C., *Measurements of Pitching Moment Derivatives for an Airfoil Oscillating about the Half-chord Axis*, Br. A.R.C., R. & M. 1921, November, 1938.

13–15. GREIDANUS, J. H., VAN DE VOOREN, A. I., AND BERGH, H., *Experimental Determination of the Aerodynamic Coefficients of an Oscillating Wing in Incompressible, Two-dimensional Flow* (in four parts), National Aeronautical Research Institute Reports F. 101-4, Amsterdam, 1952.

13–16. LAIDLAW, W. R., *Theoretical and Experimental Pressure Distributions on Oscillating Low-Aspect Ratio Wings*, M.I.T. Aeroelastic and Structures Research Laboratory Report 51-2 on Contract NOa(s) 52-576c, November, 1954.

13–17. MOLLO-CHRISTENSEN, E. L., *An Experimental and Theoretical Investigation of Unsteady Transonic Flow*, Sc.D. Thesis, M.I.T., May, 1954.

13–18. WOOLSTON, D. S., AND RUNYAN, H. L., *Some Considerations on the Air Forces on a Wing Oscillating Between Two Walls for Subsonic Compressible Flow*, Institute of the Aeronautical Sciences Preprint No. 446, January, 1954.

13–19. KINNAMAN, E. B., Flutter Analysis of Complex Airplanes by Experimental Methods, *Journal of the Aeronautical Sciences*, Vol. 19, No. 9, September, 1952.

13–20. HALFMAN, R. L., McCARTHY, J. F., JR., PRIGGE, J. S., JR., AND WOOD, G. A., JR., *A Variable Mach Number Supersonic Test Section for Flutter Research*, Wright Air Development Center TR 54-114, November, 1954.

APPENDIX A

A–1. PERLIS, SAM, *Theory of Matrices*, Addison-Wesley Publishing Company, Inc., Cambridge, Mass., 1952.

A–2. FRAZER, R. A., DUNCAN, W. J., AND COLLAR, A. R., *Elementary Matrices*, Cambridge University Press, Cambridge, England, 1950.

A-3. CROUT, P. D., A Short Method of Evaluating Determinants and Solving Systems of Linear Equations with Real or Complex Coefficients, *Trans. A.I.E.E.*, Vol. 60, pp. 1235–1240, 1941.

APPENDIX B

B-1. MARGENAU, H., AND MURPHY, G. H., *The Mathematics of Physics and Chemistry*, D. Van Nostrand Co., Inc., New York, 1943.

B-2. MILNE, W. E., *Numerical Calculus*, Princeton University Press, Princeton, New Jersey, 1949.

B-3. *Tables of Lagrangian Interpolation Coefficients*, National Bureau of Standards, Columbia University Press, New York, 1944.

B-4. MULTHOPP, H., Die Anwendung der Tragflügel Theorie auf Fragen der Flugmechanik, *Bericht S2 der Lilienthal-Gesellschaft fur Luftfahrtforschung, Preisausschreiben*, pp. 53–64, 1938–39.

AUTHOR INDEX

851

A CATALOG OF SELECTED
DOVER BOOKS
IN SCIENCE AND MATHEMATICS

Astronomy

CHARIOTS FOR APOLLO: The NASA History of Manned Lunar Spacecraft to 1969, Courtney G. Brooks, James M. Grimwood, and Loyd S. Swenson, Jr. This illustrated history by a trio of experts is the definitive reference on the Apollo spacecraft and lunar modules. It traces the vehicles' design, development, and operation in space. More than 100 photographs and illustrations. 576pp. 6 3/4 x 9 1/4. 0-486-46756-2

EXPLORING THE MOON THROUGH BINOCULARS AND SMALL TELESCOPES, Ernest H. Cherrington, Jr. Informative, profusely illustrated guide to locating and identifying craters, rills, seas, mountains, other lunar features. Newly revised and updated with special section of new photos. Over 100 photos and diagrams. 240pp. 8 1/4 x 11. 0-486-24491-1

WHERE NO MAN HAS GONE BEFORE: A History of NASA's Apollo Lunar Expeditions, William David Compton. Introduction by Paul Dickson. This official NASA history traces behind-the-scenes conflicts and cooperation between scientists and engineers. The first half concerns preparations for the Moon landings, and the second half documents the flights that followed Apollo 11. 1989 edition. 432pp. 7 x 10. 0-486-47888-2

APOLLO EXPEDITIONS TO THE MOON: The NASA History, Edited by Edgar M. Cortright. Official NASA publication marks the 40th anniversary of the first lunar landing and features essays by project participants recalling engineering and administrative challenges. Accessible, jargon-free accounts, highlighted by numerous illustrations. 336pp. 8 3/8 x 10 7/8. 0-486-47175-6

ON MARS: Exploration of the Red Planet, 1958-1978--The NASA History, Edward Clinton Ezell and Linda Neuman Ezell. NASA's official history chronicles the start of our explorations of our planetary neighbor. It recounts cooperation among government, industry, and academia, and it features dozens of photos from Viking cameras. 560pp. 6 3/4 x 9 1/4. 0-486-46757-0

ARISTARCHUS OF SAMOS: The Ancient Copernicus, Sir Thomas Heath. Heath's history of astronomy ranges from Homer and Hesiod to Aristarchus and includes quotes from numerous thinkers, compilers, and scholasticists from Thales and Anaximander through Pythagoras, Plato, Aristotle, and Heraclides. 34 figures. 448pp. 5 3/8 x 8 1/2. 0-486-43886-4

AN INTRODUCTION TO CELESTIAL MECHANICS, Forest Ray Moulton. Classic text still unsurpassed in presentation of fundamental principles. Covers rectilinear motion, central forces, problems of two and three bodies, much more. Includes over 200 problems, some with answers. 437pp. 5 3/8 x 8 1/2. 0-486-64687-4

BEYOND THE ATMOSPHERE: Early Years of Space Science, Homer E. Newell. This exciting survey is the work of a top NASA administrator who chronicles technological advances, the relationship of space science to general science, and the space program's social, political, and economic contexts. 528pp. 6 3/4 x 9 1/4.
0-486-47464-X

STAR LORE: Myths, Legends, and Facts, William Tyler Olcott. Captivating retellings of the origins and histories of ancient star groups include Pegasus, Ursa Major, Pleiades, signs of the zodiac, and other constellations. "Classic." – *Sky & Telescope.* 58 illustrations. 544pp. 5 3/8 x 8 1/2. 0-486-43581-4

A COMPLETE MANUAL OF AMATEUR ASTRONOMY: Tools and Techniques for Astronomical Observations, P. Clay Sherrod with Thomas L. Koed. Concise, highly readable book discusses the selection, set-up, and maintenance of a telescope; amateur studies of the sun; lunar topography and occultations; and more. 124 figures. 26 halftones. 37 tables. 335pp. 6 1/2 x 9 1/4. 0-486-42820-6

Browse over 9,000 books at www.doverpublications.com

Chemistry

MOLECULAR COLLISION THEORY, M. S. Child. This high-level monograph offers an analytical treatment of classical scattering by a central force, quantum scattering by a central force, elastic scattering phase shifts, and semi-classical elastic scattering. 1974 edition. 310pp. 5 3/8 x 8 1/2.   0-486-69437-2

HANDBOOK OF COMPUTATIONAL QUANTUM CHEMISTRY, David B. Cook. This comprehensive text provides upper-level undergraduates and graduate students with an accessible introduction to the implementation of quantum ideas in molecular modeling, exploring practical applications alongside theoretical explanations. 1998 edition. 832pp. 5 3/8 x 8 1/2.   0-486-44307-8

RADIOACTIVE SUBSTANCES, Marie Curie. The celebrated scientist's thesis, which directly preceded her 1903 Nobel Prize, discusses establishing atomic character of radioactivity; extraction from pitchblende of polonium and radium; isolation of pure radium chloride; more. 96pp. 5 3/8 x 8 1/2.   0-486-42550-9

CHEMICAL MAGIC, Leonard A. Ford. Classic guide provides intriguing entertainment while elucidating sound scientific principles, with more than 100 unusual stunts: cold fire, dust explosions, a nylon rope trick, a disappearing beaker, much more. 128pp. 5 3/8 x 8 1/2.   0-486-67628-5

ALCHEMY, E. J. Holmyard. Classic study by noted authority covers 2,000 years of alchemical history: religious, mystical overtones; apparatus; signs, symbols, and secret terms; advent of scientific method, much more. Illustrated. 320pp. 5 3/8 x 8 1/2.
0-486-26298-7

CHEMICAL KINETICS AND REACTION DYNAMICS, Paul L. Houston. This text teaches the principles underlying modern chemical kinetics in a clear, direct fashion, using several examples to enhance basic understanding. Solutions to selected problems. 2001 edition. 352pp. 8 3/8 x 11.   0-486-45334-0

PROBLEMS AND SOLUTIONS IN QUANTUM CHEMISTRY AND PHYSICS, Charles S. Johnson and Lee G. Pedersen. Unusually varied problems, with detailed solutions, cover of quantum mechanics, wave mechanics, angular momentum, molecular spectroscopy, scattering theory, more. 280 problems, plus 139 supplementary exercises. 430pp. 6 1/2 x 9 1/4.   0-486-65236-X

ELEMENTS OF CHEMISTRY, Antoine Lavoisier. Monumental classic by the founder of modern chemistry features first explicit statement of law of conservation of matter in chemical change, and more. Facsimile reprint of original (1790) Kerr translation. 539pp. 5 3/8 x 8 1/2.   0-486-64624-6

MAGNETISM AND TRANSITION METAL COMPLEXES, F. E. Mabbs and D. J. Machin. A detailed view of the calculation methods involved in the magnetic properties of transition metal complexes, this volume offers sufficient background for original work in the field. 1973 edition. 240pp. 5 3/8 x 8 1/2.   0-486-46284-6

GENERAL CHEMISTRY, Linus Pauling. Revised third edition of classic first-year text by Nobel laureate. Atomic and molecular structure, quantum mechanics, statistical mechanics, thermodynamics correlated with descriptive chemistry. Problems. 992pp. 5 3/8 x 8 1/2.   0-486-65622-5

ELECTROLYTE SOLUTIONS: Second Revised Edition, R. A. Robinson and R. H. Stokes. Classic text deals primarily with measurement, interpretation of conductance, chemical potential, and diffusion in electrolyte solutions. Detailed theoretical interpretations, plus extensive tables of thermodynamic and transport properties. 1970 edition. 590pp. 5 3/8 x 8 1/2.   0-486-42225-9

Engineering

FUNDAMENTALS OF ASTRODYNAMICS, Roger R. Bate, Donald D. Mueller, and Jerry E. White. Teaching text developed by U.S. Air Force Academy develops the basic two-body and n-body equations of motion; orbit determination; classical orbital elements, coordinate transformations; differential correction; more. 1971 edition. 455pp. 5 3/8 x 8 1/2.   0-486-60061-0

INTRODUCTION TO CONTINUUM MECHANICS FOR ENGINEERS: Revised Edition, Ray M. Bowen. This self-contained text introduces classical continuum models within a modern framework. Its numerous exercises illustrate the governing principles, linearizations, and other approximations that constitute classical continuum models. 2007 edition. 320pp. 6 1/8 x 9 1/4.   0-486-47460-7

ENGINEERING MECHANICS FOR STRUCTURES, Louis L. Bucciarelli. This text explores the mechanics of solids and statics as well as the strength of materials and elasticity theory. Its many design exercises encourage creative initiative and systems thinking. 2009 edition. 320pp. 6 1/8 x 9 1/4.   0-486-46855-0

FEEDBACK CONTROL THEORY, John C. Doyle, Bruce A. Francis and Allen R. Tannenbaum. This excellent introduction to feedback control system design offers a theoretical approach that captures the essential issues and can be applied to a wide range of practical problems. 1992 edition. 224pp. 6 1/2 x 9 1/4.   0-486-46933-6

THE FORCES OF MATTER, Michael Faraday. These lectures by a famous inventor offer an easy-to-understand introduction to the interactions of the universe's physical forces. Six essays explore gravitation, cohesion, chemical affinity, heat, magnetism, and electricity. 1993 edition. 96pp. 5 3/8 x 8 1/2.   0-486-47482-8

DYNAMICS, Lawrence E. Goodman and William H. Warner. Beginning engineering text introduces calculus of vectors, particle motion, dynamics of particle systems and plane rigid bodies, technical applications in plane motions, and more. Exercises and answers in every chapter. 619pp. 5 3/8 x 8 1/2.   0-486-42006-X

ADAPTIVE FILTERING PREDICTION AND CONTROL, Graham C. Goodwin and Kwai Sang Sin. This unified survey focuses on linear discrete-time systems and explores natural extensions to nonlinear systems. It emphasizes discrete-time systems, summarizing theoretical and practical aspects of a large class of adaptive algorithms. 1984 edition. 560pp. 6 1/2 x 9 1/4.   0-486-46932-8

INDUCTANCE CALCULATIONS, Frederick W. Grover. This authoritative reference enables the design of virtually every type of inductor. It features a single simple formula for each type of inductor, together with tables containing essential numerical factors. 1946 edition. 304pp. 5 3/8 x 8 1/2.   0-486-47440-2

THERMODYNAMICS: Foundations and Applications, Elias P. Gyftopoulos and Gian Paolo Beretta. Designed by two MIT professors, this authoritative text discusses basic concepts and applications in detail, emphasizing generality, definitions, and logical consistency. More than 300 solved problems cover realistic energy systems and processes. 800pp. 6 1/8 x 9 1/4.   0-486-43932-1

THE FINITE ELEMENT METHOD: Linear Static and Dynamic Finite Element Analysis, Thomas J. R. Hughes. Text for students without in-depth mathematical training, this text includes a comprehensive presentation and analysis of algorithms of time-dependent phenomena plus beam, plate, and shell theories. Solution guide available upon request. 672pp. 6 1/2 x 9 1/4.   0-486-41181-8

Browse over 9,000 books at www.doverpublications.com

HELICOPTER THEORY, Wayne Johnson. Monumental engineering text covers vertical flight, forward flight, performance, mathematics of rotating systems, rotary wing dynamics and aerodynamics, aeroelasticity, stability and control, stall, noise, and more. 189 illustrations. 1980 edition. 1089pp. 5 5/8 x 8 1/4.   0-486-68230-7

MATHEMATICAL HANDBOOK FOR SCIENTISTS AND ENGINEERS: Definitions, Theorems, and Formulas for Reference and Review, Granino A. Korn and Theresa M. Korn. Convenient access to information from every area of mathematics: Fourier transforms, Z transforms, linear and nonlinear programming, calculus of variations, random-process theory, special functions, combinatorial analysis, game theory, much more. 1152pp. 5 3/8 x 8 1/2.   0-486-41147-8

A HEAT TRANSFER TEXTBOOK: Fourth Edition, John H. Lienhard V and John H. Lienhard IV. This introduction to heat and mass transfer for engineering students features worked examples and end-of-chapter exercises. Worked examples and end-of-chapter exercises appear throughout the book, along with well-drawn, illuminating figures. 768pp. 7 x 9 1/4.   0-486-47931-5

BASIC ELECTRICITY, U.S. Bureau of Naval Personnel. Originally a training course; best nontechnical coverage. Topics include batteries, circuits, conductors, AC and DC, inductance and capacitance, generators, motors, transformers, amplifiers, etc. Many questions with answers. 349 illustrations. 1969 edition. 448pp. 6 1/2 x 9 1/4.
0-486-20973-3

BASIC ELECTRONICS, U.S. Bureau of Naval Personnel. Clear, well-illustrated introduction to electronic equipment covers numerous essential topics: electron tubes, semiconductors, electronic power supplies, tuned circuits, amplifiers, receivers, ranging and navigation systems, computers, antennas, more. 560 illustrations. 567pp. 6 1/2 x 9 1/4.   0-486-21076-6

BASIC WING AND AIRFOIL THEORY, Alan Pope. This self-contained treatment by a pioneer in the study of wind effects covers flow functions, airfoil construction and pressure distribution, finite and monoplane wings, and many other subjects. 1951 edition. 320pp. 5 3/8 x 8 1/2.   0-486-47188-8

SYNTHETIC FUELS, Ronald F. Probstein and R. Edwin Hicks. This unified presentation examines the methods and processes for converting coal, oil, shale, tar sands, and various forms of biomass into liquid, gaseous, and clean solid fuels. 1982 edition. 512pp. 6 1/8 x 9 1/4.   0-486-44977-7

THEORY OF ELASTIC STABILITY, Stephen P. Timoshenko and James M. Gere. Written by world-renowned authorities on mechanics, this classic ranges from theoretical explanations of 2- and 3-D stress and strain to practical applications such as torsion, bending, and thermal stress. 1961 edition. 560pp. 5 3/8 x 8 1/2.   0-486-47207-8

PRINCIPLES OF DIGITAL COMMUNICATION AND CODING, Andrew J. Viterbi and Jim K. Omura. This classic by two digital communications experts is geared toward students of communications theory and to designers of channels, links, terminals, modems, or networks used to transmit and receive digital messages. 1979 edition. 576pp. 6 1/8 x 9 1/4.   0-486-46901-8

LINEAR SYSTEM THEORY: The State Space Approach, Lotfi A. Zadeh and Charles A. Desoer. Written by two pioneers in the field, this exploration of the state space approach focuses on problems of stability and control, plus connections between this approach and classical techniques. 1963 edition. 656pp. 6 1/8 x 9 1/4.
0-486-46663-9

Mathematics–Bestsellers

HANDBOOK OF MATHEMATICAL FUNCTIONS: with Formulas, Graphs, and Mathematical Tables, Edited by Milton Abramowitz and Irene A. Stegun. A classic resource for working with special functions, standard trig, and exponential logarithmic definitions and extensions, it features 29 sets of tables, some to as high as 20 places. 1046pp. 8 x 10 1/2.   0-486-61272-4

ABSTRACT AND CONCRETE CATEGORIES: The Joy of Cats, Jiri Adamek, Horst Herrlich, and George E. Strecker. This up-to-date introductory treatment employs category theory to explore the theory of structures. Its unique approach stresses concrete categories and presents a systematic view of factorization structures. Numerous examples. 1990 edition, updated 2004. 528pp. 6 1/8 x 9 1/4.   0-486-46934-4

MATHEMATICS: Its Content, Methods and Meaning, A. D. Aleksandrov, A. N. Kolmogorov, and M. A. Lavrent'ev. Major survey offers comprehensive, coherent discussions of analytic geometry, algebra, differential equations, calculus of variations, functions of a complex variable, prime numbers, linear and non-Euclidean geometry, topology, functional analysis, more. 1963 edition. 1120pp. 5 3/8 x 8 1/2.   0-486-40916-3

INTRODUCTION TO VECTORS AND TENSORS: Second Edition--Two Volumes Bound as One, Ray M. Bowen and C.-C. Wang. Convenient single-volume compilation of two texts offers both introduction and in-depth survey. Geared toward engineering and science students rather than mathematicians, it focuses on physics and engineering applications. 1976 edition. 560pp. 6 1/2 x 9 1/4.   0-486-46914-X

AN INTRODUCTION TO ORTHOGONAL POLYNOMIALS, Theodore S. Chihara. Concise introduction covers general elementary theory, including the representation theorem and distribution functions, continued fractions and chain sequences, the recurrence formula, special functions, and some specific systems. 1978 edition. 272pp. 5 3/8 x 8 1/2.
0-486-47929-3

ADVANCED MATHEMATICS FOR ENGINEERS AND SCIENTISTS, Paul DuChateau. This primary text and supplemental reference focuses on linear algebra, calculus, and ordinary differential equations. Additional topics include partial differential equations and approximation methods. Includes solved problems. 1992 edition. 400pp. 7 1/2 x 9 1/4.   0-486-47930-7

PARTIAL DIFFERENTIAL EQUATIONS FOR SCIENTISTS AND ENGINEERS, Stanley J. Farlow. Practical text shows how to formulate and solve partial differential equations. Coverage of diffusion-type problems, hyperbolic-type problems, elliptic-type problems, numerical and approximate methods. Solution guide available upon request. 1982 edition. 414pp. 6 1/8 x 9 1/4.   0-486-67620-X

VARIATIONAL PRINCIPLES AND FREE-BOUNDARY PROBLEMS, Avner Friedman. Advanced graduate-level text examines variational methods in partial differential equations and illustrates their applications to free-boundary problems. Features detailed statements of standard theory of elliptic and parabolic operators. 1982 edition. 720pp. 6 1/8 x 9 1/4.   0-486-47853-X

LINEAR ANALYSIS AND REPRESENTATION THEORY, Steven A. Gaal. Unified treatment covers topics from the theory of operators and operator algebras on Hilbert spaces; integration and representation theory for topological groups; and the theory of Lie algebras, Lie groups, and transform groups. 1973 edition. 704pp. 6 1/8 x 9 1/4.
0-486-47851-3

Browse over 9,000 books at www.doverpublications.com

A SURVEY OF INDUSTRIAL MATHEMATICS, Charles R. MacCluer. Students learn how to solve problems they'll encounter in their professional lives with this concise single-volume treatment. It employs MATLAB and other strategies to explore typical industrial problems. 2000 edition. 384pp. 5 3/8 x 8 1/2.   0-486-47702-9

NUMBER SYSTEMS AND THE FOUNDATIONS OF ANALYSIS, Elliott Mendelson. Geared toward undergraduate and beginning graduate students, this study explores natural numbers, integers, rational numbers, real numbers, and complex numbers. Numerous exercises and appendixes supplement the text. 1973 edition. 368pp. 5 3/8 x 8 1/2.   0-486-45792-3

A FIRST LOOK AT NUMERICAL FUNCTIONAL ANALYSIS, W. W. Sawyer. Text by renowned educator shows how problems in numerical analysis lead to concepts of functional analysis. Topics include Banach and Hilbert spaces, contraction mappings, convergence, differentiation and integration, and Euclidean space. 1978 edition. 208pp. 5 3/8 x 8 1/2.   0-486-47882-3

FRACTALS, CHAOS, POWER LAWS: Minutes from an Infinite Paradise, Manfred Schroeder. A fascinating exploration of the connections between chaos theory, physics, biology, and mathematics, this book abounds in award-winning computer graphics, optical illusions, and games that clarify memorable insights into self-similarity. 1992 edition. 448pp. 6 1/8 x 9 1/4.   0-486-47204-3

SET THEORY AND THE CONTINUUM PROBLEM, Raymond M. Smullyan and Melvin Fitting. A lucid, elegant, and complete survey of set theory, this three-part treatment explores axiomatic set theory, the consistency of the continuum hypothesis, and forcing and independence results. 1996 edition. 336pp. 6 x 9.   0-486-47484-4

DYNAMICAL SYSTEMS, Shlomo Sternberg. A pioneer in the field of dynamical systems discusses one-dimensional dynamics, differential equations, random walks, iterated function systems, symbolic dynamics, and Markov chains. Supplementary materials include PowerPoint slides and MATLAB exercises. 2010 edition. 272pp. 6 1/8 x 9 1/4.   0-486-47705-3

ORDINARY DIFFERENTIAL EQUATIONS, Morris Tenenbaum and Harry Pollard. Skillfully organized introductory text examines origin of differential equations, then defines basic terms and outlines general solution of a differential equation. Explores integrating factors; dilution and accretion problems; Laplace Transforms; Newton's Interpolation Formulas, more. 818pp. 5 3/8 x 8 1/2.   0-486-64940-7

MATROID THEORY, D. J. A. Welsh. Text by a noted expert describes standard examples and investigation results, using elementary proofs to develop basic matroid properties before advancing to a more sophisticated treatment. Includes numerous exercises. 1976 edition. 448pp. 5 3/8 x 8 1/2.   0-486-47439-9

THE CONCEPT OF A RIEMANN SURFACE, Hermann Weyl. This classic on the general history of functions combines function theory and geometry, forming the basis of the modern approach to analysis, geometry, and topology. 1955 edition. 208pp. 5 3/8 x 8 1/2.   0-486-47004-0

THE LAPLACE TRANSFORM, David Vernon Widder. This volume focuses on the Laplace and Stieltjes transforms, offering a highly theoretical treatment. Topics include fundamental formulas, the moment problem, monotonic functions, and Tauberian theorems. 1941 edition. 416pp. 5 3/8 x 8 1/2.   0-486-47755-X